Climate Change and Terrestrial Ecosystem Modeling

Climate models have evolved into Earth system models with representation of the physics, chemistry, and biology of terrestrial ecosystems. This companion book to Gordon Bonan's *Ecological Climatology: Concepts and Applications, Third Edition*, builds on the concepts introduced there and provides the mathematical foundation upon which to develop and understand ecosystem models and their relevance for these Earth system models. The book bridges the disciplinary gap among land surface models developed by atmospheric scientists; biogeochemical models, dynamic global vegetation models, and ecosystem demography models developed by ecologists; and ecohydrology models developed by hydrologists. Review questions, supplemental code, and modeling projects are provided to aid with understanding how the equations are used. The book is an invaluable guide to climate change and terrestrial ecosystem modeling for graduate students and researchers in climate change, climatology, ecology, hydrology, biogeochemistry, meteorology, environmental science, mathematical modeling, and environmental biophysics.

Gordon Bonan is Senior Scientist and Head of the Terrestrial Sciences Section at the National Center for Atmospheric Research in Boulder, Colorado. He studies the interactions of terrestrial ecosystems with climate, using models of Earth's biosphere, atmosphere, hydrosphere, and geosphere. He is the author of *Ecological Climatology: Concepts and Applications* (now in its third edition) and has published 150 peer-reviewed articles in atmospheric science, geoscience, and ecological journals on terrestrial ecosystems, climate, and their coupling. He is a fellow of the American Geophysical Union and the American Meteorological Society and has served on advisory boards for numerous national and international organizations and as an editor for several journals.

"This is a thoroughly comprehensive overview, and must be the definitive text for this field. The breadth of topics and the depth of knowledge displayed here is unparalleled. For students and expert practitioners alike, this is an extraordinarily useful resource, expertly presented using a thoughtful balance of text, figures, data tables and equations."

Mathew Williams, University of Edinburgh

"Terrestrial ecosystem models are a fundamental basis for the science of global change impacts and biospheric climate, and an essential tool to investigate options to understand and mitigate climate change. This comprehensive overview of the history, theoretical underpinnings and application of terrestrial ecosystem models, paired with an in-depth discussion of the mathematical fundaments of the model, and complemented with clear and helpful practical examples, is a must-read for anyone interested in using or developing models of land ecosystems, from the scale of individual ecosystems, to studies of the role of land surface dynamics in comprehensive Earth system models. I wish this book had been available at the beginning of my career."

Sönke Zaehle, Max Planck Institute for Biogeochemistry

"Bonan's excellent text is comprehensive, covering all the components needed to model the land surface and its interactions with the atmosphere and climate system. For each topic Bonan provides not only historical perspective and a comprehensive theoretical overview, but also the critically important discussion of how these theoretical approaches can be put into practice numerically. This book is the new definitive reference for anyone working on understanding and modeling all aspects of the land surface."

Abigail L. S. Swann, University of Washington

"I highly recommend this groundbreaking book to everyone who wants to learn the nuts and bolts of land surface modeling and/or Earth system modeling. For many years I have been teaching *Ecological Climatology*: *Concepts and Applications* to geoscience and engineering students but struggling to supplement them with appropriate materials on modeling blocks and for modeling projects. Now this book arrives as a perfect addition and companion as it provides exactly the kind of materials I wanted to have."

Zong-Liang Yang, University of Texas at Austin

Climate Change and Terrestrial Ecosystem Modeling

Gordon Bonan

National Center for Atmospheric Research, Boulder, Colorado

CAMBRIDGE
UNIVERSITY PRESS

CAMBRIDGE
UNIVERSITY PRESS

University Printing House, Cambridge CB2 8BS, United Kingdom

One Liberty Plaza, 20th Floor, New York, NY 10006, USA

477 Williamstown Road, Port Melbourne, VIC 3207, Australia

314–321, 3rd Floor, Plot 3, Splendor Forum, Jasola District Centre, New Delhi – 110025, India

79 Anson Road, #06–04/06, Singapore 079906

Cambridge University Press is part of the University of Cambridge.

It furthers the University's mission by disseminating knowledge in the pursuit of
education, learning, and research at the highest international levels of excellence.

www.cambridge.org
Information on this title: www.cambridge.org/9781107043787
DOI: 10.1017/9781107339217

First published 2019

Printed in the United Kingdom by TJ International Ltd. Padstow Cornwall

A catalogue record for this publication is available from the British Library.

Library of Congress Cataloging-in-Publication Data
Names: Bonan, Gordon B., author.
Title: Climate change and terrestrial ecosystem modeling / Gordon B. Bonan,
    National Center for Atmospheric Research, Boulder, Colorado.
Description: Cambridge, United Kingdom ; New York, NY : Cambridge University Press, 2019. |
    Includes bibliographical references and index.
Identifiers: LCCN 2018035968| ISBN 9781107043787 (hardback) | ISBN 9781107619074 (paperback)
Subjects: LCSH: Biotic communities–Mathematical models. | Ecology–Mathematical models. |
    Climatic changes. | Global environmental change.
Classification: LCC QH541.15.M3 B66 2019 | DDC 577.8/2–dc23
LC record available at https://lccn.loc.gov/2018035968

ISBN 978-1-107-04378-7 Hardback
ISBN 978-1-107-61907-4 Paperback

Additional resources for this publication at www.cambridge.org/bonanmodeling

To Amie, again

Contents

Preface

Writing a modeling textbook is daunting. There are many equations to introduce and explain, and writing a book that also shows how to solve the equations in a computer model is even more difficult. Why would one undertake such a task? There are several excellent topical textbooks, mostly in the area of environmental biophysics. David Gates's *Biophysical Ecology* is the classic text, but the equations, units, etc. are dated. *Principles of Environmental Physics* (John Monteith and Mike Unsworth), *An Introduction to Environmental Biophysics* (Gaylon Campbell and John Norman), *Evaporation into the Atmosphere* (Wilfried Brutsaert), and *Plants and Microclimate* (Hamlyn Jones) are still definitive introductions to plants and environmental physics. Daniel Hillel's *Environmental Soil Physics* is an essential text for soil physics. While these books provide excellent treatment of radiative transfer, turbulent fluxes, leaf biophysics, or soil physics, they do not integrate these concepts into a complete model of biosphere–atmosphere coupling. Nor do they, or other available textbooks, extend the subject matter to include ecosystems and biogeochemical cycles. This present book covers the fundamentals of environmental biophysics, integrates those principles into a complete model, and additionally broadens the scope to cover ecosystems and biogeochemical cycles. More importantly, the book explains the numerical methods needed to implement and solve the equations in a computer model. There is an enormous leap between seeing a mathematical equation in a research paper and actually using that equation in a model. In this respect, *Soil Physics with Basic* (Gaylon Campbell), *Soil Water Dynamics* (Arthur Warrick), *Numerical Methods in the Hydrological Sciences* (George Hornberger and Patricia Wiberg), and *A Theory of Forest Dynamics* (Hank Shugart) are indispensable reference books in that they also provide computer programs.

Many colleagues have contributed to this book in various ways. The roughness sublayer parameterization in Chapter 6 is an outgrowth of collaborations with Ned Patton and Ian Harman. The surface energy balance bucket model in Chapter 7 was developed in collaboration with Marysa Laguë as part of her simple land model. Martyn Clark introduced me to the Picard iteration used to solve the Richards equation in Chapter 8 and also the probability distributed model of rainfall–runoff in Chapter 9, which is a generalization of the VIC model. Sönke Zaehle provided details of the implementation of the Kull and Kruijt (1998) photosynthesis in his O-CN terrestrial biosphere model. Collaboration with Danica Lombardozzi is seen in her work on photosynthesis and stomatal conductance, as is collaboration with Peter Franks on stomatal conductance. Mat Williams introduced me to his SPA model during a sabbatical visit to NCAR; the optimal stomatal conductance is described in Chapter 12, the plant hydraulics in Chapter 13, and the Norman radiative transfer in Chapter 14. Rosie Fisher also provided many insights to plant hydraulics. Ryan Knox shared details of ED2, its turbulence parameterization, and multilayer two-stream radiation. Ying-Ping Wang shared his CASA-CNP code, and Melannie Hartmann provided thorough documentation of the model; this model is the basis for Chapter 17. In his many visits to NCAR and his collaborations with the Community Land Model, Yiqi Luo shared details of the traceability analysis of biogeochemical models in Chapter 17. Quinn Thomas's and Will Wieder's insights to carbon–nitrogen biogeochemistry are also seen in Chapter 17. Collaborations with Melannie Hartmann and Will Wieder are evident in the DAYCENT soil organic matter model and microbial model, both described in Chapter 18. Rosie Fisher, Ryan Knox, Charlie Koven, and Jackie Shuman provided essential background on ecosystem demography models in Chapter 19.

Review questions and modeling projects are provided with each chapter. The review questions are intended to help students and readers assess whether they understand the concepts that have been introduced. These questions are structured around several essential modeling requirements: understanding units and how to convert among the different units favored in various scientific communities, deriving equations, and knowing what the equations mean and how to use them. Modeling projects are included in each chapter. The projects aid with understanding how the equations are used, bringing various concepts together into a mathematical model, and using models to answer research questions. The projects utilize and build upon sample code provided with each chapter. It is hoped that the sample code will inspire students and instructors to devise their own projects, tailored to their specific interests. MATLAB® is used as the computational framework to illustrate code.[1] The sample code does not take advantage of the matrix capabilities of MATLAB or its full mathematical sophistication. Rather, the code is written to explicitly show in a step-by-step manner how the calculations are performed. In particular, many example codes explicitly loop over a number of calculations, such as for soil layers, rather than using matrix algebra. The same code could be written utilizing matrix capabilities, but then the calculations would be less obvious to non–technical specialists. MATLAB purists and scientific coding experts will likely find the code offensive, but I prefer to err on the side of understandability rather than elegance.

Finally, I am indebted to Matt Lloyd at Cambridge University Press, who has supported this endeavor over the many years it took to complete, and also Zoë Pruce for her patience with this project.

The National Center for Atmospheric Research is sponsored by the National Science Foundation.

---

[1] MATLAB is a registered trademark of The MathWorks, Inc., Natick, Massachusetts, United States.

Mathematical Symbols

The following is a list of major symbols. Duplication of symbols is unavoidable, and the reader should always check the specific usage of a symbol within a particular chapter.

$a(z)$	leaf area density at height z (m^2 m^{-3})
A	area (m^2); *subscript*: A_C crown; A_L leaf; A_S sapwood
A	gross photosynthesis rate (μmol CO$_2$ m^{-2} s^{-1}); *subscript*: A_c Rubisco-limited assimilation; A_j RuBP regeneration-limited assimilation; A_p product-limited assimilation
A_n	net photosynthesis rate (μmol CO$_2$ m^{-2} s^{-1})
c_a, c_c, c_i, c_s	CO$_2$ concentration in air, chloroplast, intercellular space, and leaf surface, respectively (μmol mol^{-1})
c_i	carbon mass per area of biogeochemical pool i (kg C m^{-2})
c_j	mole fraction of constituent j (mol mol^{-1})
c_d	leaf aerodynamic drag coefficient (–)
c_{dry}	specific heat of dry biomass (J kg^{-1} K^{-1})
c_{ice}, c_{wat}	specific heat of ice and water, respectively (J kg^{-1} K^{-1}); Table A.4
c_L	leaf heat capacity (J m^{-2} K^{-1})
c_p	specific heat of moist air at constant pressure (J mol^{-1} K^{-1}); *subscript*: c_{pd} dry air; c_{pw} water vapor; Table A.4
c_v	volumetric heat capacity (J m^{-3} K^{-1}); *subscript*: c_{vf} frozen soil; c_{vu} unfrozen soil; $c_{v,ice}$ ice; $c_{v,sol}$ soil solids; $c_{v,wat}$ water
$C(\psi)$	specific moisture capacity (m^{-1})
C_h, C_m	drag coefficient (–); *subscript*: C_h heat; C_m momentum
C_p	plant hydraulic capacitance (mol H$_2$O m^{-2} Pa^{-1})
d	displacement height (m)
D	stem diameter (cm)
D	vapor pressure deficit (Pa); *subscript*: D_ℓ leaf-to-air, $e_{sat}(T_\ell) - e_a$; D_s leaf surface, $e_{sat}(T_\ell) - e_s$
D_f	drag force from vegetation (mol m^{-2} s^{-2})
D_{ij}	element in the Lagrangian dispersion matrix (s m^{-1})
D_j	molecular diffusivity of constituent j (m^2 s^{-1}); *subscript*: D_c CO$_2$; D_h heat; D_w H$_2$O; Table A.3
$D(\theta)$	hydraulic diffusivity (m^2 s^{-1})
e	vapor pressure (Pa); *subscript*: e_a air; $e_{sat}(T_\ell)$ stomatal pore; e_s leaf surface
$e_{sat}(T)$	saturation vapor pressure at temperature T (Pa); Table 3.3
E	evaporation (mol H$_2$O m^{-2} s^{-1})
f_{sun}	fractional area on a horizontal plane that is sunlit (–)
F_j	flux of constituent j (mol m^{-2} s^{-1})
g	gravitational acceleration (m s^{-2}); Table A.4
g_a	aerodynamic conductance (mol m^{-2} s^{-1}); *subscript*: g_{ac} scalars; g_{am} momentum
g_b	leaf boundary layer conductance (mol m^{-2} s^{-1}); *subscript*: g_{bc} CO$_2$; g_{bh} heat; g_{bj} constituent j; g_{bw} H$_2$O
g_b^*	canopy excess conductance between z_{0m} and z_{0c} (mol m^{-2} s^{-1})
g_c	canopy conductance (mol m^{-2} s^{-1})
g_h	conductance for heat (mol m^{-2} s^{-1})
g_j	conductance of constituent j (mol m^{-2} s^{-1})
g_ℓ	total leaf conductance of stomata and boundary layer (mol m^{-2} s^{-1}); *subscript*: $g_{\ell c}$ CO$_2$; $g_{\ell w}$ H$_2$O
g_m	leaf mesophyll conductance (mol CO$_2$ m^{-2} s^{-1})
g_r	radiative conductance (mol m^{-2} s^{-1})
g_s	leaf stomatal conductance (mol m^{-2} s^{-1}); *subscript*: g_{sc} CO$_2$; g_{sw} H$_2$O
g_w	conductance for water vapor (mol m^{-2} s^{-1})
g_0	minimum stomatal conductance (mol H$_2$O m^{-2} s^{-1})
g_1	slope parameter for stomatal conductance models
G	soil heat flux (W m^{-2})
$G(Z)$	projection of leaf area in the direction of the solar beam (–)
Gr	Grashof number (–)
h	height in canopy (m); *subscript*: h_c top of canopy
h_s	leaf surface humidity (–), $e_s/e_{sat}(T_\ell)$
h_{s1}	relative humidity of pore space in surface soil layer (–)
H	sensible heat flux (W m^{-2}); *subscript*: H_v virtual
H_f	energy flux from soil freezing or thawing (W m^{-2})

ΔH_a	activation energy (J mol^{-1})	K_e	Kersten number (–)
ΔH_d	deactivation energy (J mol^{-1})	K_L	leaf-specific plant hydraulic conductance (mol H$_2$O m^{-2} s^{-1} Pa^{-1})
i_c	infiltration rate (m s^{-1})		
I_c	cumulative infiltration (m)	K_m	Michaelis–Menten constant (μmol mol^{-1})
$I^\downarrow$	solar radiation in the downward direction (μmol photon m^{-2} s^{-1}); subscript: $I_b^\downarrow$, scattered direct beam flux; $I_d^\downarrow$ scattered diffuse flux	K_n	canopy nitrogen decay coefficient (–)
		K_o, K_{o25}	Michaelis–Menten constant for O$_2$ and its value at 25°C (mmol mol^{-1})
		K_p	leaf-specific stem hydraulic conductance (mol H$_2$O m^{-2} s^{-1} Pa^{-1})
$I_{sky}^\downarrow$	solar radiation incident on the top of the canopy (μmol photon m^{-2} s^{-1}); subscript: $I_{sky,b}^\downarrow$ direct beam; $I_{sky,d}^\downarrow$ diffuse	ℓ	characteristic leaf dimension (m)
		l	mixing length (m); subscript: l_c scalar; l_m momentum
$I^\uparrow$	solar radiation in the upward direction (μmol photon m^{-2} s^{-1}); subscript: $I_b^\uparrow$ scattered direct beam flux; $I_d^\uparrow$ scattered diffuse flux	$L^\downarrow$	downward longwave radiation (W m^{-2}); subscript: $L_{sky}^\downarrow$ atmospheric
		$L^\uparrow$	upward longwave radiation (W m^{-2}); subscript: $L_g^\uparrow$ ground
$\vec{I}_c$	solar radiation absorbed by the canopy, per ground area (μmol photon m^{-2} s^{-1}); subscript: $\vec{I}_{cb}$ direct beam; $\vec{I}_{cd}$ diffuse; $\vec{I}_{cSun}$ sunlit canopy; $\vec{I}_{cSha}$ shaded canopy	$\vec{L}$	absorbed longwave radiation, per ground area (W m^{-2}); subscript: $\vec{L}_c$ canopy; $\vec{L}_{cSun}$ sunlit canopy; $\vec{L}_{cSha}$ shaded canopy; $\vec{L}_g$ ground
$\vec{I}_g$	solar radiation absorbed by the ground (μmol photon m^{-2} s^{-1}); subscript: $\vec{I}_{gb}$ direct beam; $\vec{I}_{gd}$ diffuse	$\vec{L}_\ell$	longwave radiation absorbed by foliage, per leaf area (W m^{-2})
$\vec{I}_\ell$	solar radiation absorbed by foliage, per leaf area (μmol photon m^{-2} s^{-1}); subscript: $\vec{I}_{\ell b}$ direct beam; $\vec{I}_{\ell bb}$ unscattered direct beam; $\vec{I}_{\ell bs}$ scattered direct beam; $\vec{I}_{\ell d}$ diffuse	L	leaf area index (m^2 m^{-2}); subscript: L_{sha} shaded leaf area; L_{sun} sunlit leaf area
		ΔL	leaf area index of a canopy layer (m^2 m^{-2}); subscript: ΔL_{sha} shaded leaf area; ΔL_{sun} sunlit leaf area
$\vec{I}_{\ell sha},$ $\vec{I}_{\ell sun}$	solar radiation absorbed by shaded and sunlit leaves, per leaf area (μmol photon m^{-2} s^{-1})	L_c	canopy length scale (m)
		L_f	latent heat of fusion (J kg^{-1}); Table A.4
I_{PSII}	light utilized in electron transport by photosystem PS II (μmol m^{-2} s^{-1})	L_{MO}	Obukhov length scale (m)
		L_r	root length density (m m^{-3})
J	electron transport rate (μmol m^{-2} s^{-1})	m_a	leaf mass per area (kg C m^{-2}, or kg dry mass m^{-2})
$J_{max},$ J_{max25}	maximum electron transport rate and its value at 25°C (μmol m^{-2} s^{-1})	m_i	biomass per area of biogeochemical pool i (kg m^{-2})
k	von Karman constant (–); Table A.4		
k_i	loss rate, or turnover rate, of biogeochemical pool i (per unit time, e.g., d^{-1})	m_j	mass of constituent j (kg); subscript: m_d dry air; m_v water vapor
		M_j	molecular mass of constituent j (kg mol^{-1}); subscript: M_a dry air; M_w water; Table A.4
k_p, k_{p25}	initial slope of CO$_2$ response for C$_4$ photosynthesis and its value at 25°C (mol m^{-2} s^{-1})	M_r	root biomass (kg m^{-2})
		M_i	plant biomass per individual (kg); subscript: M_L leaf; M_H heartwood; M_S sapwood; M_R fine root
K	eddy diffusivity (m^2 s^{-1}); subscript: K_c scalar; K_m momentum		
$K(\theta)$	hydraulic conductivity (m s^{-1}); subscript: K_{sat} saturated conductivity	n_a	leaf nitrogen per unit leaf area (kg N m^{-2})
		n_i	nitrogen per area of biogeochemical pool i (kg N m^{-2})
K_b, K_d	extinction coefficient for direct beam and diffuse radiation, respectively (–)	n_j	number of moles of constituent j
		n_m	leaf nitrogen per unit leaf mass (kg N kg^{-1} C, or kg N kg^{-1} dry mass)
K_c, K_{c25}	Michaelis–Menten constant for CO$_2$ and its value at 25°C (μmol mol^{-1})	Nu	Nusselt number (–)
		o_i	intercellular O$_2$ concentration (mmol mol^{-1})

P	atmospheric pressure (Pa); *subscript*: P_d dry air	T	temperature (K); *subscript*: T_a air; T_e equilibrium; T_f freezing point of water (Table A.4); T_g ground; T_ℓ leaf; T_s surface; T_v virtual; T_A acclimation growth temperature
P	precipitation (kg H_2O m^{-2} s^{-1} = mm H_2O s^{-1}); *subscript*: P_R rain; P_S snow		
P_j	partial pressure of constituent j (Pa); *subscript*: P_d dry air	T_L	Lagrangian time scale (s)
Pr	Prandtl number (–)	T_p, T_{p25}	triose phosphate utilization rate and its value at 25°C (µmol m^{-2} s^{-1})
q	mole fraction (e/P; mol mol^{-1}); *subscript*: q_a air; q_{ref} at some reference height; q_s surface; q_* characteristic scale	u	wind speed, or also specifically the velocity component in the x-directional (m s^{-1}); *subscript*: u_{ref} at some reference height
$q_{sat}(T)$	saturation water vapor mole fraction at temperature T (mol mol^{-1}), $e_{sat}(T)/P$	u_*	friction velocity (m s^{-1})
Q	water flux (m s^{-1}, or mol m^{-2} s^{-1})	U	carbon input from plant production (kg C m^{-2} d^{-1}, or kg C m^{-2} y^{-1})
Q_a	radiative forcing (W m^{-2})		
Q_{10}	Q_{10} temperature parameter (–)	v	velocity component in the y-direction (m s^{-1})
r_c	leaf Nusselt number (heat) or Stanton number (scalar) (–)	v	general symbol for a rate; *subscript*: v_{base} base rate; v_{max} maximum rate; v_{25} rate at 25°C
r_j	resistance of constituent j (s m^2 mol^{-1})		
r_l	specific root length (m kg^{-1})	v_d	dry deposition velocity (m s^{-1})
r_r	fine root radius (m)	V	volume (m^3)
R	runoff (kg H_2O m^{-2} s^{-1} = mm H_2O s^{-1})	$V_{c\,max}$, $V_{c\,max\,25}$	maximum Rubisco carboxylation rate and its value at 25°C (µmol m^{-2} s^{-1})
R	respiration rate (kg C m^{-2} s^{-1}); *subscript*: R_A autotrophic; R_g growth; R_m maintenance; R_H heterotrophic	w	vertical velocity (m s^{-1})
R_A	isotope ratio of a sample A (–); *subscript*: R_a air; R_E evaporation flux; R_l liquid water; R_p photosynthate; R_v water vapor	w	mole fraction of water vapor (mol mol^{-1}), e/P; *subscript*: w_ℓ leaf-to-air, $[e_{sat}(T_\ell) - e_a]/P$; w_s leaf surface , $[e_{sat}(T_\ell) - e_s]/P$
R_d, R_{d25}	leaf mitochondrial respiration in light, or day respiration, and its value at 25°C (µmol CO_2 m^{-2} s^{-1})	w_c	Rubisco-limited carboxylation rate (µmol m^{-2} s^{-1})
		w_i	light harvesting-limited carboxylation rate (µmol m^{-2} s^{-1})
R_n	net radiation (W m^{-2})	w_j	RuBP regeneration-limited carboxylation rate (µmol m^{-2} s^{-1})
Re	Reynolds number (–)		
Ri_B	bulk Richardson number (–)	w_p	product-limited carboxylation rate (µmol m^{-2} s^{-1})
$\Re$	gas constant (J K^{-1} mol^{-1}); Table A.4		
s	temperature derivative of q_{sat} at temperature T (mol mol^{-1} K^{-1}), dq_{sat}/dT	W	soil water (kg H_2O m^{-2} = mm H_2O); *subscript*: W_{ice} ice; W_{liq} liquid; W_{snow} snow
S	storage heat flux (W m^{-2}); *subscript*: S_h sensible heat; S_q latent heat; S_v biomass	z	vertical distance (m)
$S^\downarrow$	downward solar radiation (W m^{-2}); *subscript*: $S_\Lambda^\downarrow$ at specified wavelength	z_0	roughness length (m); *subscript*: z_{0c} scalars; z_{0m} momentum
$S^\uparrow$	upward solar radiation (W m^{-2}); *subscript*: $S_\Lambda^\uparrow$ at specified wavelength	α_{A-B}	isotopic fractionation factor for samples A and B (–); *subscript*: α_k kinetic fractionation; α_{kb} kinetic fractionation for leaf boundary layer; α_{ks} kinetic fractionation for stomata; α_{l-v} liquid–vapor transition; α_p photosynthate
S_c	scalar source or sink flux (mol m^{-3} s^{-1})		
$S_{c/o}$	relative specificity of Rubisco (–)		
S_e	relative soil wetness (–), θ/θ_{sat} or $(\theta - \theta_{res})/(\theta_{sat} - \theta_{res})$	α_ℓ	leaf absorptance (–)
		α_Λ	absorptance at a specified wavelength (–)
Sc	Schmidt number (–)	A	azimuth angle; *subscript*: A_ℓ leaf
Sh	Sherwood number (–)	β	ratio of friction velocity to wind speed at the canopy height (–); *subscript*: β_N neutral value of β
ΔS	entropy term (J K^{-1} mol^{-1})		
t	time (s, seconds; h, hours; y, years)		

β, β_0	two-stream upscatter parameters for diffuse and direct beam radiation, respectively (–)	$\xi(z)$	cumulative leaf drag area (–)
		ρ, ρ_d, ρ_v	density of moist air ($\rho = \rho_d + \rho_v$), its dry air component, and water vapor, respectively (kg m^{-3})
β_w	soil wetness factor (–)		
B	solar elevation angle	ρ_a	density of dry air at pressure P and temperature T (kg m^{-3})
γ	psychrometric constant (Pa K^{-1})		
γ_ℓ	incidence angle of the solar beam on a leaf	ρ_b	soil bulk density (kg m^{-3})
Γ	CO_2 compensation point (µmol mol^{-1})	ρ_{ice}, ρ_{wat}	density of ice and water (kg m^{-3}); Table A.4
Γ_*, Γ_{*25}	CO_2 compensation point in the absence of non-photorespiratory respiration, and its value at 25°C (µmol mol^{-1})	ρ_j	mass concentration of constituent j (kg m^{-3})
		ρ_m	molar density (mol m^{-3})
		ρ_r	root tissue density (kg m^{-3})
δ_A	isotope ratio of sample A relative to a standard (–)	ρ_s	soil particle density (kg m^{-3})
		ρ_{snow}	density of snow (kg m^{-3})
Δ_{A-B}	isotopic fractionation between samples A and B (–)	ρ_w	density of wood (kg m^{-3})
		ρ	reflectance (–); *subscript*: ρ_g ground; ρ_ℓ leaf; ρ_Λ at a specified wavelength
ε	emissivity (–); *subscript*: ε_g ground; ε_ℓ leaf; ε_Λ at a specified wavelength		
		ρ_c	canopy albedo (–); *subscript*: ρ_{cb} direct beam; ρ_{cd} diffuse
ε_{A-B}	isotopic enrichment factor for samples A and B (–)		
		ρ_g	ground albedo (–); *subscript*: ρ_{gb} direct beam; ρ_{gd} diffuse
ε_j	emission factor for chemical compound j		
E	quantum yield (mol CO_2 mol^{-1} photon, or mol CO_2 J^{-1})	σ	Stefan–Boltzmann constant (W m^{-2} K^{-4}); Table A.4
ζ	Monin–Obukhov stability parameter (–)	σ_w	standard deviation of vertical wind velocity (m s^{-1})
Z	solar zenith angle		
θ	potential temperature (K); *subscript*: θ_{ref} air at some reference height; θ_s surface; θ_v virtual; θ_* characteristic temperature scale	τ	transmittance (–); *subscript*: τ_b direct beam; τ_d diffuse radiation; τ_ℓ leaf; τ_Λ at a specified wavelength
		τ	momentum flux (mol m^{-1} s^{-2}); *subscript*: τ_x zonal; τ_y meridional
θ	volumetric soil water content (m^3 m^{-3}); *subscript*: θ_{ice} ice; θ_{liq} liquid; θ_{res} residual water; θ_{sat} saturation (porosity)		
		τ_L	leaf temperature time constant (s)
		$T_{b,i}$	transmittance of direct beam radiation through the cumulative leaf area above canopy layer i (–)
Θ	curvature factor for co-limitation (–); *subscript*: Θ_A photosynthesis; Θ_J electron transport		
		ϕ	ratio of Rubisco oxygenation to carboxylation rates (–)
Θ_ℓ	leaf inclination angle		
ι	marginal carbon gain of water loss (µmol CO_2 mol^{-1} H_2O)	ϕ_1, ϕ_2	terms in the Ross–Goudriaan $G(B)$ function
κ	thermal conductivity (W m^{-1} K^{-1}); *subscript*: κ_{air} air (Table A.4); κ_{dry} dry soil; κ_f frozen soil; κ_{ice} ice (Table A.4); κ_o other soil minerals; κ_q quartz; κ_{sat} saturated soil; κ_{snow} snow; κ_{sol} soil solids; κ_u unfrozen soil; κ_{wat} water (Table A.4)	ϕ_c, ϕ_m	Monin–Obukhov similarity theory flux–gradient relationships for scalars and momentum, respectively (–)
		$\hat{\phi}_c, \hat{\phi}_m$	roughness sublayer functions for ϕ_c and ϕ_m, respectively (–)
		Φ_c, Φ_m	roughness sublayer-modified flux–gradient relationships for scalars and momentum, respectively (–)
λ	latent heat of vaporization (J mol^{-1}); Table 3.6		
Λ	wavelength (m)	Φ_{PSII}	quantum yield of photosystem II (mol mol^{-1})
μ	cosine solar zenith angle		
$\bar{\mu}$	average inverse of the optical depth of diffuse radiation per unit leaf area	χ	mass mixing ratio (kg kg^{-1}); *subscript*: χ_v water vapor
υ	molecular diffusivity for momentum, or kinematic viscosity (m^2 s^{-1}); Table A.3	χ_ℓ	departure of leaf angle distribution from spherical (–)

ψ soil water potential (m or Pa); *subscript*: ψ_w at the wetting front; ψ_{sat} at saturation

ψ_ℓ leaf water potential (Pa); *subscript*: $\psi_{\ell\,min}$ minimum leaf water potential

ψ_c, ψ_m integrated form of Monin–Obukhov functions for scalars and momentum, respectively (–)

$\hat{\psi}_c, \hat{\psi}_m$ integrated form of the roughness sublayer functions for scalars and momentum, respectively (–)

ω scattering coefficient (–); *subscript*: ω_ℓ leaf; ω_Λ at a specified wavelength

Ω foliage clumping factor for light transmission (–)

Terrestrial Biosphere Models

Chapter Overview

Earth system models simulate climate as the outcome of interrelated physical, chemical, and biological processes. With these models, it is recognized that the biosphere not only responds to climate change, but also influences the direction and magnitude of change. Earth system models contain component atmosphere, land, ocean, and sea ice models. The land component model simulates the world's terrestrial ecosystems and their physical, chemical, and biological functioning at climatically relevant spatial and temporal scales. These models are part of a continuum of terrestrial ecosystem models, from models with emphasis on biogeochemical pools and fluxes, dynamic vegetation models with focus on individual plants or size cohorts, canopy models with focus on coupling leaf physiological processes with canopy physics, and global models of the land surface for climate simulation. This latter class of models incorporates many features found in other classes of ecosystem models but additionally includes physical meteorological processes necessary for climate simulation. This book describes these models and refers to them as terrestrial biosphere models.

1.1 | Introduction

The global nature of environmental problems has transformed our scientific understanding and study of the biosphere. Global change can be broadly taken to mean the interactive physical, chemical, and biological processes that regulate Earth as a system, maintain planetary habitability, and sustain life, and the changes that are occurring in the Earth system, both natural and anthropogenic. Terrestrial ecosystems are central to solving the environmental and socioeconomic threats posed by changes in climate, atmospheric composition, and air quality; land use and land-cover change; habitat loss, species extinction, and invasive species; appropriation of freshwater, net primary production, and other ecosystems goods and services for human uses; and anthropogenic addition of reactive nitrogen. Devising suitable solutions to these global change challenges require not only strong empirically and experimentally based research at the local scale to understand how ecosystems are structured and how they function, but also sound theoretical foundations to generalize this understanding to regional, continental, and global scales and to make projections of the future. Computer models of terrestrial ecosystems are essential to this generalization.

Models of terrestrial ecosystems take many different forms depending on scientific disciplines. Ecologists develop models of community composition and biogeochemical cycles to study ecosystem response to climate change. Hydrologists develop models of watersheds to study freshwater availability and stormflow and must represent leaves, stomata, and plant canopies in some manner to calculate evapotranspiration loss. Atmospheric chemists must include reactive gas exchanges between ecosystems and the atmosphere in their chemistry models. Likewise, atmospheric scientists have been developing models of atmospheric general circulation and planetary climate since the 1960s. These models require a

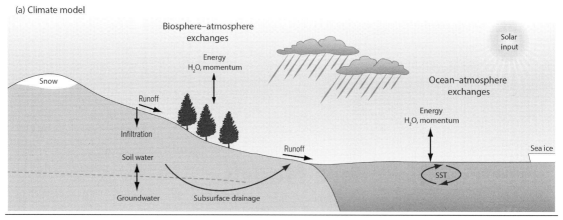

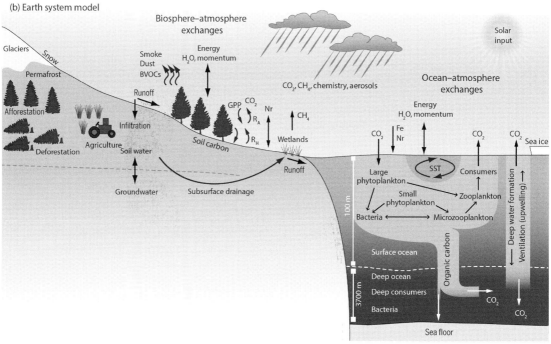

Figure 1.1 Scientific scope of (a) climate models and (b) Earth system models. Climate models simulate biogeophysical fluxes of energy, water, and momentum on land and also the hydrologic cycle. Terrestrial and marine biogeochemical cycles are new processes in Earth system models. The terrestrial carbon cycle includes carbon uptake through gross primary production (GPP) and carbon loss from autotrophic respiration R_A, heterotrophic respiration R_H, and wildfire. Many models also include the nitrogen cycle. Anthropogenic land use and land-cover change are additional processes. The fluxes of CO_2, CH_4, Nr, aerosols, biogenic volatile organic compounds (BVOCs), and wildfire chemical emissions are passed to the atmosphere to simulate atmospheric chemistry and composition. Nitrogen is carried in freshwater runoff to the ocean. Adapted from Bonan and Doney (2018)

mathematical formulation of the exchanges of energy, water, and momentum between land and atmosphere to solve the equations of atmospheric physics and dynamics. These fluxes are mediated by plants, and so models of Earth's climate and its land surface require depictions of terrestrial ecosystems. In climate models, vegetation is represented by plant

canopies with a focus on physical flux exchanges with the atmosphere, soil moisture hydrology, and snow. With the broadening of the science from the physical climate system to the Earth system, the models have expanded to include biogeochemical cycles, biogeography, and dynamic vegetation – typically the realm of ecosystem models (Figure 1.1).

Many types of models consider terrestrial ecosystems, and the particular way in which terrestrial ecosystems are depicted varies among disciplines (Table 1.1). The next three sections examine how ecosystems are represented in ecological models, atmospheric models, and hydrologic models. For ecologists, the focus may be biogeochemical cycling and the distribution of carbon and nitrogen within an ecosystem. For an atmospheric scientist, this may be the manner in which terrestrial ecosystems affect weather, climate, and atmospheric composition through energy, water, and chemical flux exchanges with the atmosphere. Both disciplines use mathematics to describe and model terrestrial ecosystems, their functioning, and their response to environmental changes but with very different meaning to ecologists and climate scientists.

The distinction between ecological and atmospheric depictions of ecosystems has become blurred over the past few decades, and these two viewpoints of terrestrial ecosystems are merging into a common depiction of the global terrestrial biosphere, particularly as atmospheric scientists have embraced a broad Earth system perspective to understand planetary climate (Figure 1.2). This book describes this type of model, hereafter referred to as a terrestrial biosphere model. In this sense, terrestrial biosphere models is used broadly to represent the intersection among the atmosphere, hydrosphere, geosphere, and biosphere. Such models have become an essential, albeit imperfect, research tool to study global change. Bonan (2016) reviewed the influence of terrestrial ecosystems on climate and more broadly their role in the Earth system, and also provided an introduction to terrestrial biosphere models and how they are used to study climate. The present book is concerned with how to model the terrestrial biosphere.

Table 1.1	Classes of terrestrial ecosystem models	
Type of model	Description	Example
Biogeochemical	Ecosystem models with emphasis on biogeochemical pools and fluxes (e.g., C, N, P) using prescribed biogeography	TEM, CASA, BIOME-BGC, CENTURY, CASA-CNP
Forest gap models	Individual trees, population dynamics, demography, community composition	JABOWA, FORET
Ecosystem demography	As in gap models, but cohort based	ED
Dynamic global vegetation models	Biogeochemistry, community composition, global biogeography	IBIS, LPJ, SDGVM, LPJ-GUESS, SEIB-DGVM
Land surface models	Global models of the land surface for weather and climate simulation with an emphasis on hydrometeorological processes and biogeophysical coupling with the atmosphere; the models now additionally include biogeochemical cycles and vegetation dynamics	See Table 1.2
Plant canopy	Multilayer canopy-scale models with focus on coupling leaf physiological processes and canopy physics	CUPID, CANOAK
Canopy–chemistry	Plant canopy models that additionally include chemical transport and reactions in the canopy airspace	CACHE, CAFE, ACCESS, FORCAsT
Ecohydrology	Similar to land surface models (without coupling to atmospheric models), but spatially distributed within a watershed and with lateral flow connectivity	RHESSys

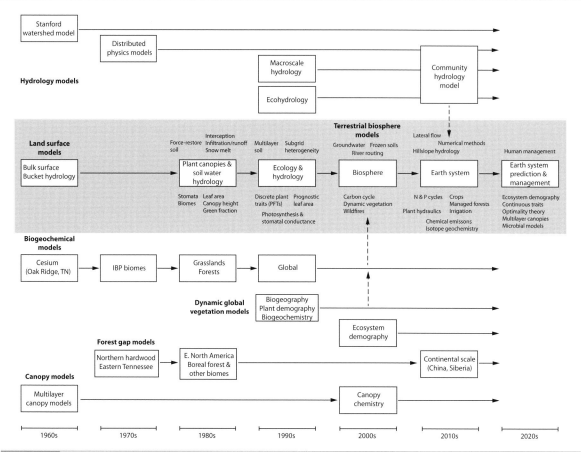

Figure 1.2 Timeline of model development. Shown are the broad classes of hydrology, biogeochemical, dynamic global vegetation, individual-based forest gap models, canopy, and land surface models. For each class of models, boxes denote major model developments and approximate timelines. The shading highlights land surface models, and the text around these boxes describes model capabilities. Vertical dashed lines show when the various classes of models merged with land surface models. The most significant is the incorporation of biogeochemical and dynamic global vegetation models in the 2000s to form terrestrial biosphere models. A community hydrology model has been discussed as a means to more authentically represent hydrologic processes in land surface models.

1.2 | The Ecological Ecosystem

The archetypal ecological view of an ecosystem emphasizes material flows. The structure of an ecosystem is measured by the amount of carbon, nitrogen, or other materials in various compartments. The functioning of an ecosystem is measured by the cycling of materials among these compartments. Odum's (1957) study of Silver Springs, an aquatic spring ecosystem in Florida, is a classic example of this type of ecosystem analysis (Figure 1.3). Odum abstracted the ecosystem into five trophic groups of producers, herbivores, carnivores, top carnivores, and decomposers and described energy transfers (in

the sense of caloric value of biomass) among these groups. Energy flows into the ecosystem via photosynthesis, cycles among the various trophic groups, is lost as respiration, or accumulates as biomass in the ecosystem. Odum later formalized this view of energy flows as analogous to an electrical circuit using Ohm's law (Odum 1960).

This conceptualization of an ecosystem lends itself to a system of first-order, linear differential equations to describe material flows among various compartments. This type of model, known as a box or compartment model, came to dominate ecosystem modeling in the late 1960s and early 1970s with the advent of systems ecology. One of the first examples was a model of differential equations to simulate

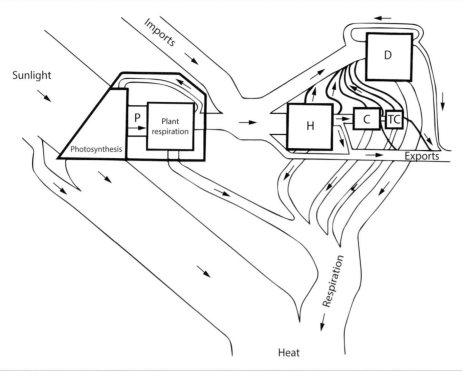

Figure 1.3 Energy flow in the Silver Springs ecosystem (Odum 1957). The five tropic levels are producers (P), herbivores (H), carnivores (C), top carnivores (TC), and decomposers (D). Redrawn from Odum (1960)

cesium in a forest (Olson 1965), and box models gained prominence during the International Biological Program (IBP). The IBP was a coordinated research program in the late 1960s and early 1970s that studied various biomes with an emphasis on biomass, nutrient, and water flows and the development of computer models to simulate these flows (Patten 1975). The models were extraordinary in their mathematical representation of ecosystems but did not achieve their goals or fulfill their potential; they were seen as too large and unnecessarily mathematically complex but biologically simple (Golley 1993; Kwa 1993, 2005). The grassland biome model, for example, consisted of submodels of plant productivity, mammalian and insect consumption, decomposition, nitrogen and phosphorus cycles, and temperature and water in the plant canopy and soil (Innis 1975, 1978). It used 120 state variables and more than 1000 parameters (Kwa 1993). The coniferous forest biome model employed 29 state variables connected by 65 flows to model water and carbon dynamics (Sollins et al. 1979) and was also considered unsuccessful (Long 2005).

Despite the shortcomings of the IBP models, compartment models remain in use today, known now more generally as biogeochemical models, and are commonly used to simulate the terrestrial carbon cycle in Earth system models. Biogeochemical models simulate the carbon balance of terrestrial ecosystems given a specified geographic distribution of biomes as input to the model. The models represent an ecosystem by aggregate pools of foliage, stem, and root biomass without regard to individual plants or species and use additional pools to represent litter and soil carbon (Figure 1.4). Flows among the pools are described in terms of net primary production, allocation, and other plant physiological and microbial processes specific to the different biomes. Concurrent with carbon flows are transfers of nitrogen and other nutrients. A typical model time step is daily to monthly. Early such models include: CENTURY for grasslands (Parton et al. 1987, 1988, 1993), which was subsequently modified from a monthly to daily time step with DAYCENT (Parton et al. 1998; Del Grosso et al. 2005b, 2009; Hartman et al. 2011); FOREST-BGC for forests (Running and Coughlan 1988; Running and Gower 1991) and subsequently the more generalized BIOME-BGC model (Running and Hunt 1993; Thornton et al. 2002); the Terrestrial

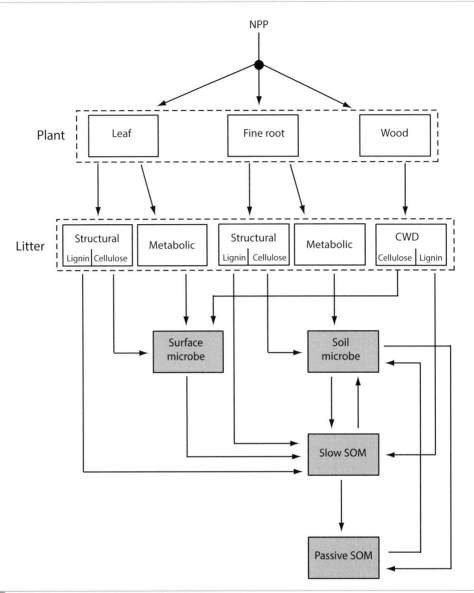

Figure 1.4 Carbon pools and associated transfers in CASA. Shown are the allocation of net primary production (NPP) to leaf, fine root, and wood plant compartments; turnover and litterfall to structural and metabolic litter pools for foliage and root and to coarse woody debris (CWD) for wood; and soil carbon pools consisting of microbes and soil organic matter (SOM). The litter and soil organic matter submodel is the same as in CENTURY. Redrawn from Randerson et al. (1996)

Ecosystem Model (TEM; Raich et al. 1991; McGuire et al. 1992; Melillo et al. 1993); and the Carnegie–Ames–Stanford Approach (CASA; Potter et al. 1993; Randerson et al. 1996). A more recent example of such models is CASA-CNP (Wang et al. 2010), which builds upon the CASA framework for carbon to include nitrogen and phosphorus. Many biogeochemical models focus on carbon and nutrient flows and represent the physical environment (evapotranspiration, soil water, temperature, plant canopies) in a simplified manner. BIOME-BGC is a notable exception and uses the concept of a big-leaf canopy to simulate the physical environment similar to the land surface models used in atmospheric models, albeit with a daily time step.

An alternative depiction of terrestrial ecosystems considers the behavior of an ecosystem as the

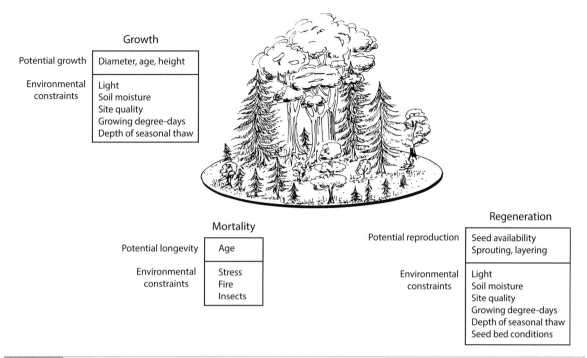

Figure 1.5 Depiction of a boreal forest gap model. The growth of an individual tree depends on its diameter, age, and height as modified by environmental constraints. Mortality depends on the age of the tree as modified by stress, wildfire, and insects. Regeneration depends on seed availability, the ability to sprout or layer, and site conditions. Redrawn from Bonan (1989)

outcome of individual plants competing for light and space (Figure 1.5). Such individual-based models were a response to the IBP box-and-arrow biogeochemical representation of ecosystems, and, indeed, the first individual-based models of forest dynamics were developed at about the same time as the IBP models. In contrast with biogeochemical models, however, individual-based models have their roots in population dynamics and the natural history and life cycle of species. A broad class of individual tree forest dynamics models called gap models simulate the dynamics of trees in an area of approximately 0.1 ha, which corresponds to the size of a gap in the canopy (Shugart 1984; Botkin 1993; Bugmann 2001). These models simulate size structure and community composition by directly representing demographic processes such as establishment, competition, and mortality. Ecosystem properties such as carbon storage and its distribution among foliage, stem, root, litter, and soil are the aggregate outcome of the interactions among and demography of individual plants. The first such model JABOWA (Botkin et al. 1972) was designed for northern

hardwood forests in northeastern United States, followed by FORET (Shugart and West 1977) for the forests of eastern Tennessee. Forest dynamics, community composition, and size structure in these early models were primarily driven by light availability in the canopy. Later models added soil water and nutrient availability and were generalized to eastern North America (Pastor and Post 1986, 1996) or for boreal forests (Bonan 1989, 1990a,b). Forest gap models have been developed for numerous locations throughout the world (Shugart and Woodward 2011). They are still being developed and used, for example, in the forests of China (Yan and Shugart 2005) and Russia (Shuman et al. 2013, 2014, 2015). A newer class of model termed ecosystem demography (ED; Hurtt et al. 1998; Moorcroft et al. 2001; Medvigy et al. 2009; Fisher et al. 2010b, 2015, 2018) reduces the individual trees to cohorts of similar age and size so as to reduce the computational demands.

Another type of vegetation dynamics model is known as dynamic global vegetation models (DGVM; Prentice et al. 2007). These models also simulate changes in community composition, biomass, productivity, and nutrient cycling. Because the models are applied globally, they do not recognize individual species. Rather, they employ plant functional types, typically distinguished by woody or herbaceous biomass, broadleaves or needleleaves, and evergreen or deciduous leaf longevity. Most models do not formally simulate individual plants as in gap models. Instead, they represent cohorts of individuals based on similar size distribution, or the model may represent separately an average individual plant and the density of plants. One such model is the Lund–Potsdam–Jena (LPJ) model (Sitch et al. 2003). This model characterizes vegetation as patches of plant functional types within a model grid cell. Each plant functional type is represented by an individual plant with the average biomass, crown area, height, and stem diameter of its population, by the number of individuals in the population, and by the fractional cover in the grid cell. Vegetation is updated in response to resource competition, allocation, mortality, biomass turnover,

litterfall, establishment, and fire. There are many such models, as described by Fisher et al. (2014). Other examples include the Sheffield DGVM (SDGVM; Woodward et al. 1995; Woodward and Lomas 2004), LPJ-GUESS (Smith et al. 2001, 2014), and the Spatially Explicit Individual-Based DGVM (SEIB-DGVM; Sato et al. 2007).

1.3 | The Atmospheric Ecosystem

Deardorff (1978) outlined the basic equations needed to represent energy and water fluxes from vegetation and soil, and this approach was adopted for use in climate models with the Biosphere–Atmosphere Transfer Scheme (BATS; Dickinson et al. 1986, 1993) and the Simple Biosphere model (SiB; Sellers et al. 1986). Figure 1.6 illustrates this framework for SiB. The canopy is treated as a single exchange surface without any vertical structure (known as a big-leaf

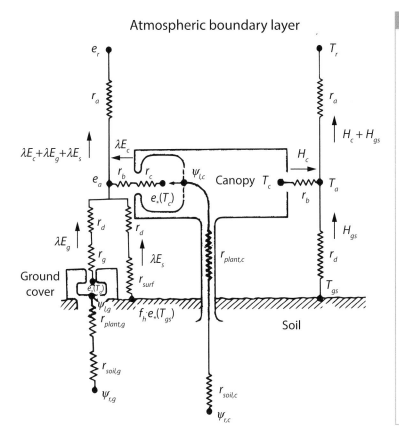

Figure 1.6 Latent heat fluxes (left) and sensible heat fluxes (right) in SiB. These fluxes are modeled as a network of resistances. The sensible heat flux consists of a ground flux H_{gs} that depends on the ground T_{gs} and canopy air T_a temperatures and a within-canopy aerodynamic resistance r_d, a vegetation flux H_c from the canopy with temperature T_c and bulk boundary layer resistance r_b, and the total flux to the reference height with temperature T_r and aerodynamic resistance r_a. Plant water uptake occurs from the soil with matric potential $\psi_{r,c}$ to the canopy with leaf water potential $\psi_{l,c}$ in relation to soil $r_{soil,c}$ and plant $r_{plant,c}$ resistances acting in series. Transpiration is from the stomatal cavity with vapor pressure $e_*(T_c)$ to the canopy air with vapor pressure e_a in relation to the bulk stomatal r_c and boundary layer r_b resistances acting in series. Similar fluxes occur in ground cover. Soil evaporation is a separate water flux and depends on the vapor pressure of the surface soil layer $f_h e_*(T_{gs})$ and a soil surface resistance r_{surf}. Additional processes include radiative transfer and intercepted water, but they are omitted for clarity. Redrawn from Sellers et al. (1986)

canopy), but with separate fluxes for soil. Processes in the models include radiative transfer in the plant canopy; momentum transfer arising from vegetated canopies, including turbulence within the canopy; sensible heat exchange from foliage and soil; latent heat exchange from evaporation of intercepted water, soil evaporation, and transpiration; the control of transpiration by stomata; and heat transfer in soil. Associated with the energy fluxes is the hydrologic cycle consisting of interception, throughfall, stemflow, infiltration, runoff, snowmelt, soil water, evaporation, and transpiration. BATS and SiB were pioneering in showing that vegetation is essential to model climate. The second version of SiB was particularly groundbreaking in its linkage of canopy physiology (including leaf photosynthesis), canopy fluxes, and remote sensing (Sellers et al. 1996a). The computational framework was quickly adopted by the major climate modeling centers and remains the standard in the current generation of models (Table 1.2). The same approach is used in numerical weather prediction models and air quality models (Chen and Dudhia 2001; Ek et al. 2003; Niu et al. 2011; Pleim and Ran 2011; Ran et al. 2017).

With the advent of global biogeochemical models and DGVMs, climate modelers readily adapted their models of the land surface to allow for simulation of the carbon cycle and biogeography and their feedbacks with climate change. The coupling with these models was enabled by the ability of land surface models to simulate photosynthesis in conjunction with stomatal conductance. This carbon input to the ecosystem is then tracked in ecological submodels. In contrast with their ecological counterparts, the models are specifically designed for coupling with atmospheric models and link biogeophysical, hydrologic, physiological, demographic, and biogeochemical processes into a unified representation of surface energy fluxes, hydrology, photosynthesis, respiration, allocation, and the cycling of carbon and nutrients within ecosystems. One of the first such models to include a DGVM was the Integrated Biosphere Simulator (IBIS; Foley et al. 1996; Kucharik et al. 2000), and most climate modeling centers now include either a biogeochemical or DGVM submodel (Table 1.2).

Land surface models with a DGVM simulate ecosystem processes at multiple timescales (Figure 1.7). The land and atmosphere exchange energy, water, momentum, and CO_2 over periods of minutes to hours through short timescale biogeophysical, biogeochemical, physiological, and hydrologic processes. Leaf phenology includes the timing of budburst, senescence, and leaf abscission in response to temperature and soil water over periods of days to weeks. Changes in community composition, vegetation structure, and soil carbon occur over periods of years or longer in relation to gross primary production and respiration; allocation of net primary production to grow foliage, stem, and root biomass; and mortality as a result of low growth rate or fire. The growth and success of particular plant functional types are dependent on life history patterns such as evergreen and deciduous phenology, needleleaf and broadleaf foliage, C_3 and C_4 photosynthetic pathway, and temperature and precipitation preferences for biogeography. Plant growth is linked to soil biogeochemistry through litterfall, decomposition, and nitrogen availability. Research frontiers include representing managed croplands, pastures, and forests; chemistry–climate interactions from biogenic volatile organic compounds (BVOCs), reactive nitrogen, methane, ozone, secondary organic aerosols, dust, and wildfires; and nitrogen and phosphorus biogeochemistry.

Although the big-leaf concept with its vertically unstructured canopy is the prevailing paradigm in land surface models, multilayer models that vertically resolve physiological and microclimatic gradients within the canopy have been developed for many years. An early example is the model of Waggoner and Reifsnyder (1968) and Waggoner et al. (1969), which considered radiative transfer and leaf energy fluxes in a multilayer canopy. Goudriaan (1977) extended this type of model to include leaf physiology and gas exchange (photosynthesis, stomatal conductance). The current generation of such models provides a comprehensive depiction of the soil–plant–atmosphere system. The models simulate canopy processes by linking radiative transfer, mechanistic parameterizations of leaf energy fluxes, photosynthesis, and stomatal conductance, and parameterization of turbulent processes within and above the plant canopy. They depict a canopy based on the physiology of leaves, leaf gas exchange, and canopy architecture (the vertical distribution of leaf and stem area, leaf angle distribution, foliage clumping). They account for vertical structure but, similar to big-leaf models, treat the canopy as spatially homogenous layers of phytoelements

Table 1.2 Models of the land surface and biogeochemistry or vegetation dynamics used in Earth system models

Center	Land	Ecosystem	Reference
Beijing Climate Center	BCC-AVIM	same	Ji (1995); Wu et al. (2013)
Canadian Center for Climate Modelling and Analysis	CLASS	CTEM	Verseghy (1991); Verseghy et al. (1993); Arora et al. (2009)
Centre National de Recherches Météorologiques (France)	ISBA	same	Noilhan and Planton (1989); Séférian et al. (2016); Boone et al. (2017)
Commonwealth Scientific and Industrial Research Organization (Australia)	CABLE	CASA-CNP	Kowalczyk et al. (2006, 2013); Wang et al. (2010, 2011)
European Center for Medium-Range Weather Forecasts	H-TESSEL	–	van den Hurk et al. (2000); Balsamo et al. (2015)
Geophysical Fluid Dynamics Laboratory (USA)	LM2	–	Milly and Shmakin (2002); Anderson et al. (2004)
	LM3	same	Shevliakova et al. (2009); Milly et al. (2014)
Goddard Institute for Space Studies (USA)	ModelE2	–	Rosenzweig and Abramopoulos (1997); Friend and Kiang (2005); Schmidt et al. (2014)
Hadley Center (UK)	MOSES	TRIFFID	Cox et al. (1999); Cox (2001); Essery et al. (2001)
	JULES	TRIFFID	Best et al. (2011); Clark et al. (2011a)
Institut Pierre Simon Laplace (France)	SECHIBA	ORCHIDEE	Ducoudré et al. (1993); Krinner et al. (2005)
Japan Agency for Marine-Earth Science and Technology	MATSIRO	SEIB-DGVM	Takata et al. (2003); Sato et al. (2007)
Max Planck Institute for Meteorology (Germany)	JSBACH	same	Raddatz et al. (2007); Reick et al. (2013)
National Center for Atmospheric Research (USA)	BATS	same	Dickinson et al. (1986, 1993, 1998, 2002)
	LSX	IBIS	Pollard and Thompson (1995); Foley et al. (1996)
	NCAR LSM	DGVM	Bonan (1996); Bonan et al. (2003)
	NCAR LSM	CASA'	Fung et al. (2005)
	CLM3	DGVM	Oleson et al. (2004); Levis et al. (2004)
	CLM4	CN	Oleson et al. (2010b); Thornton et al. (2009)
	CLM4.5	BGC	Oleson et al. (2013); Koven et al. (2013)
	CLM4.5	ED	Fisher et al. (2015)
	CLM5	FATES	Lawrence et al. (2018)

without regard to individual plants or mixtures of species. They have mostly been applied at the local scale to simulate a single stand of vegetation. Examples include CUPID (Norman 1979, 1982, 1989; Norman and Campbell 1983; Kustas et al. 2007) and CANOAK (Baldocchi and Harley 1995; Baldocchi and Meyers 1998; Baldocchi and Wilson 2001; Baldocchi et al. 2002). The latter model was

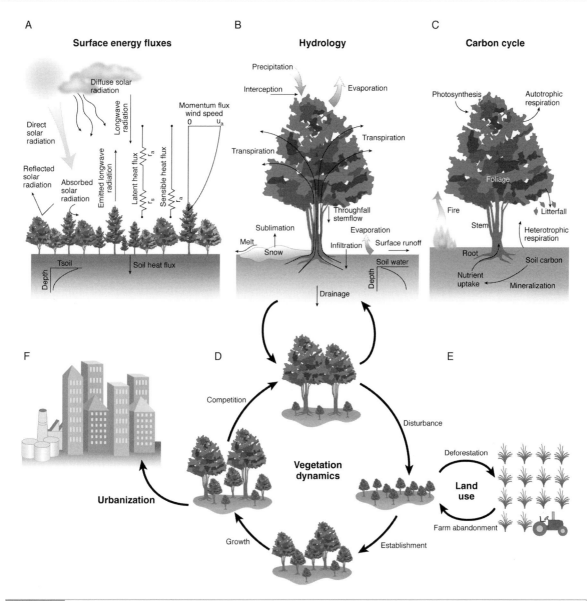

Figure 1.7 Scope of a terrestrial biosphere model for ecosystem–atmosphere coupling. Shown are the near-instantaneous fluxes of (a) energy, (b) water, and (c) carbon exchanged with the atmosphere at every model time step. Over longer timescales of decades to centuries, vegetation changes in response to (d) disturbance, establishment, growth, and competition and (e) land use. Some models also consider urban land cover. Adapted from Bonan (2008)

originally developed for deciduous oak–hickory forest and later generalized to other canopies. With greater computational power, there is renewed interest in including multilayer canopies in land surface models to better simulate the canopy environment (Bonan et al. 2014, 2018; Chen et al. 2016b; Ryder et al. 2016).

A variant of the multilayer model is the canopy–chemistry model. These models combine plant canopy models with atmospheric chemistry to simulate vertical profiles of gas concentrations in addition to temperature, water vapor, and wind. The gas concentrations result from emission and deposition of various chemical compounds at each layer in the canopy, chemical production and loss, and turbulent transport. Example models include the Canopy Atmospheric CHemistry Emission model (CACHE; Forkel et al. 2006; Bryan et al. 2012), the Chemistry of Atmosphere–Forest Exchange model (CAFE; Wolfe and Thornton 2011), the Atmospheric Chemistry and Canopy Exchange Simulation System (ACCESS; Saylor 2013), and the FORest Canopy Atmosphere Transfer model (FORCAsT; Ashworth et al. 2015).

1.4 | The Hydrological Ecosystem

Hydrologic modeling has primarily been concerned with rainfall–runoff models to simulate streamflow from observed precipitation and can be broadly classified into two groups (O'Connell 1991; Beven 2000; Todini 2007; Hrachowitz and Clark 2017). One of the first computer models was the Stanford Watershed Model in the early 1960s (Crawford and Linsley 1966; Crawford and Burges 2004; Donigian and Imhoff 2010). This is an example of a lumped parameter water balance model. It represents the effective hydrologic response of an entire watershed without explicitly accounting for spatial heterogeneity (e.g., soils, topography, vegetation) and lateral flows that influence runoff generation. Such models describe the relevant hydrological processes (e.g., rainfall, snowmelt, infiltration, evapotranspiration, runoff, drainage) – but in a simplified manner, often by semiempirical equations – and model the mass balance of several interconnected reservoirs of water. The particular parameters in this class of models do not necessarily have physical counterparts that can be measured. Instead, the model is calibrated for an individual watershed. Such models are also called conceptual storage models.

In contrast, physically based, distributed models discretize a watershed into small grid cells at high spatial resolution connected by vertical and lateral water flows (Figure 1.8). These models directly resolve surface and subsurface flows for the unsaturated and saturated zones and estimate catchment runoff as the numerical integration of fine-scale processes. Partial differential equations describe state variables at each grid cell using concepts of mass and momentum continuity. The models are spatially explicit, and so the partial differential equations are solved numerically as finite difference equations on a three-dimensional grid. Many of the parameters in these models have physical meaning and can be measured (e.g., saturated hydraulic conductivity, porosity). Freeze and Harlan (1969) outlined the basic principles of physically based, distributed models, and the Système Hydrologique Européen (SHE) model is an example (Abbott et al. 1986).

A third class of models, termed simplified distributed models (Beven 2000), represents spatial variability in runoff generation using distribution functions rather than directly resolving fine-scale heterogeneity. Example models include TOPMODEL (Beven and Kirby 1979), which relates runoff to a topographic index to characterize the saturated fraction of the surface, and the variable infiltration capacity (VIC) model (Liang et al. 1994, 1996), which uses an assumed distribution of infiltration capacity to account for spatial heterogeneity. VIC has also been termed a macroscale hydrologic model because it simulates hydrology at large spatial scales. The model simulates the surface energy balance, snowmelt, and hydrometeorology similarly to land surface models, but with an emphasis on runoff generation and streamflow. The model separates runoff to rivers into a fast response from saturation-excess surface runoff and a slower response via subsurface flow. It has been implemented and evaluated for particular river basins as well as continental and global simulations (Nijssen et al. 2001; Maurer et al. 2002).

Land surface models have been criticized for lacking hydrology, at least as hydrologists would recognize it (Clark et al. 2015a,b; Weiler and Beven 2015). The models focus on the surface energy balance and vertical water movement in the soil–plant–atmosphere system. Hydrology is considered mostly in the context of the surface energy budget (e.g., soil moisture control of evapotranspiration). Hillslope hydrology, lateral flows, groundwater dynamics, and channel routing – processes that hydrologists consider to be defining features of a hydrologic model – are difficult to represent at the large spatial scales of a global model (e.g., 1° or about 100 km). Recent efforts in the hydrologic community have advocated to include these processes in large-scale global models (Clark et al.

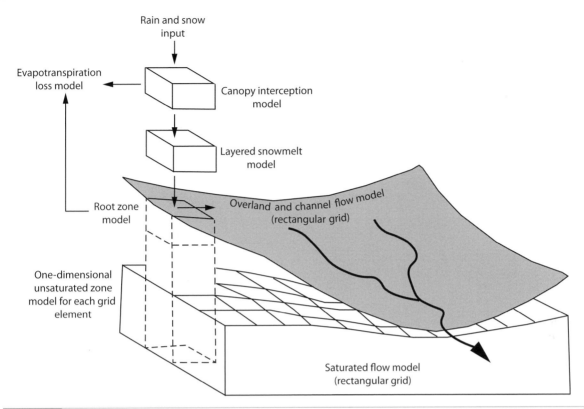

Figure 1.8 Hydrological processes in the Système Hydrologique Européen (SHE) model. The water balance consists of canopy interception, snowmelt, evapotranspiration, overland and channel flow, and unsaturated and saturated subsurface flow. The model discretizes a watershed into a three-dimensional grid in the horizontal and vertical dimensions. Flow in the unsaturated subsurface zone is one-dimensional (vertical) and links to two-dimensional (horizontal) overland flow and saturated flow models. Adapted from Abbott et al. (1986)

2015a,b). Some of the distinction between land surface and hydrologic models is disciplinary. The emphasis in many hydrologic models is on runoff generation, not hydrometeorological coupling with the atmosphere as in land surface models; and while evapotranspiration is needed for the water balance calculations, it is not necessarily the focus of hydrologic models. Hydrologic models also may not account for long-term changes in land use, vegetation cover, and stomatal conductance (e.g., in response to elevated CO_2), which are key drivers of changes in runoff and water availability. The ecological aspects of the hydrologic cycle and the interactions between water and ecosystems are considered a subset of hydrology known as ecohydrology, which emerged as a discipline in the late 1990s (Rajaram et al. 2015). The Regional Hydro-Ecologic Simulation System (RHESSys) is an example ecohydrology model that links hillslope hydrology, lateral water flows, and biogeochemistry (Band et al. 1993; Tague and Band 2004).

1.5 | One Biosphere

As seen in the preceding sections, the ecological, atmospheric, and hydrological communities have different ways of modeling terrestrial ecosystems depending on the scope of the model and the overarching science objectives. However, it is unmistakable that the biosphere is central to numerous aspects of global change research and that the biosphere integrates many Earth system processes (Figure 1.9). Climate models require surface fluxes of energy, moisture, and momentum. Models of atmospheric chemistry and air quality require surface fluxes of chemicals and particulate matter. The

(a)   (b)

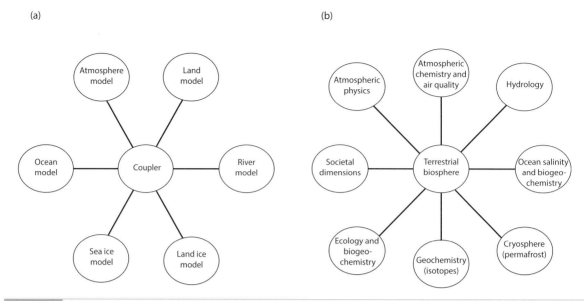

Figure 1.9 Two depictions of the biosphere in the Earth system. (a) A typical Earth system model includes component atmosphere, land, ocean, and sea ice models interacting through a flux coupler. Some models include river transport and glaciers. (b) The terrestrial biosphere is itself a coupler that integrates many disciplines and is central to understanding global change.

hydrologic cycle is affected by the type of vegetation, the amount of leaf area, and stomatal conductance. Transport of freshwater, carbon, and nutrients from land to ocean influences ocean salinity and chemistry. The occurrence of permafrost and its spatial extent depends on characteristics such as vegetation cover, leaf area, mosses, and peaty soils. Isotopes are important diagnostics of the hydrologic and carbon cycles because the abundance of water and carbon isotopes is affected by photosynthesis and evaporation. Biogeochemical cycles regulate greenhouse gases such as CO_2, CH_4, and N_2O and reactive nitrogen, BVOCs, dust, and smoke affect air quality and short-lived climate forcings. Understanding the biological impacts of climate change requires models of organisms, their physiology, how they are arranged into communities and ecosystems, and how ecosystems function. The societal dimensions of climate change manifest in terms of food, fiber, water resources, habitat, and biological diversity.

Is it possible to find a unified representation of terrestrial ecosystems to span these various disciplines? The advent of Earth system models, with their component models of the terrestrial biosphere, suggests that this can be done and that there can be a common basis to study terrestrial ecosystems in the Earth system. This book uses the terminology terrestrial biosphere model to describe such models. Terrestrial biosphere is meant to emphasize the living system; the central tenant is to model a living system of plants and microbes that connects the atmosphere, hydrosphere, and geosphere. Terrestrial biosphere model as used here is taken to mean a class of models that simulate the world's terrestrial ecosystems and their physical, chemical, and biological functioning at globally relevant spatial and temporal scales for coupling with Earth system models.

1.6 | Common Language

A cohesive modeling framework across disciplines requires a common language, but the various scientific communities have adopted their own vocabulary. Consider, for example, the simplest descriptor of a model – its general name. The terrestrial component of climate models has historically been known as a land surface model, but this is an uninformative descriptor of model capabilities. An alternative is the more specific terminology of soil–vegetation–atmosphere–transfer (SVAT) model, but this is cumbersome and has not been generally

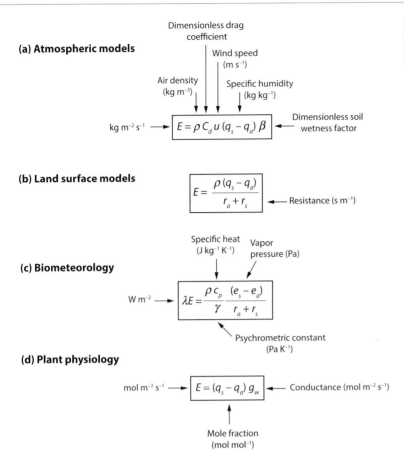

(a) Atmospheric models

Dimensionless drag coefficient

Wind speed (m s⁻¹)

Air density (kg m⁻³)

Specific humidity (kg kg⁻¹)

kg m⁻² s⁻¹ → $E = \rho\, C_d\, u\, (q_s - q_a)\, \beta$ ← Dimensionless soil wetness factor

(b) Land surface models

$$E = \frac{\rho\,(q_s - q_a)}{r_a + r_s}$$ ← Resistance (s m⁻¹)

(c) Biometeorology

Specific heat (J kg⁻¹ K⁻¹)

Vapor pressure (Pa)

W m⁻² → $$\lambda E = \frac{\rho\, c_p}{\gamma}\, \frac{(e_s - e_a)}{r_a + r_s}$$

Psychrometric constant (Pa K⁻¹)

(d) Plant physiology

mol m⁻² s⁻¹ → $E = (q_s - q_a)\, g_w$ ← Conductance (mol m⁻² s⁻¹)

Mole fraction (mol mol⁻¹)

Figure 1.10 Common parameterizations of evapotranspiration in various classes of models.

adopted. Neither descriptor accounts for the fact that nowadays land surface models simulate terrestrial biogeochemical fluxes, vegetation dynamics, and human management of the biosphere and hydrosphere.

Notation and scientific units present a challenge to interdisciplinary modeling of the biosphere. In a science that spans boundary layer meteorology, micrometeorology, atmospheric chemistry, plant physiology, ecosystem ecology, biogeochemistry, soil science, hydrology, and geochemistry, it is difficult to find a common language to express similar concepts. Each scientific community has its preferred notation. The forms of flux equations, the mathematical symbols used, and the units for the various terms in the equations are particularly vexing. Figure 1.10, for example, shows several equations used to calculate evapotranspiration in various models. Each equation describes evapotranspiration as a diffusive flux proportional to a driving force (the difference in water vapor between the

surface and atmosphere), but they differ in their implementation of the basic diffusion equation. Models use specific humidity, vapor pressure, or mole fraction depending on discipline. Atmospheric scientists relate the flux to a dimensionless drag coefficient and wind speed. Micrometeorologists might write an equation for the same flux but replace the drag coefficient with a resistance that combines turbulent transport and canopy physiology. Plant physiologists often use conductance rather than resistance in their equations of leaf fluxes.

Nor do the various communities agree on notation. The symbol L might mean the Obukhov length to a boundary layer meteorologist, but leaf area index to an ecologist. The symbol λ can mean wavelength, climate feedback, soil thermal conductivity, a soil pore-size distribution index, latent heat of vaporization, leaf water-use efficiency, a topographic index of hydrologic similarity, or Lagrange multiplier, depending on an atmospheric, soil science, plant physiology, hydrology, or mathematical perspective. The symbol ρ is often used to represent leaf reflectance but is more commonly used in atmospheric science to denote the density of air. Likewise, an atmospheric scientist will recognize q as specific humidity (kg H_2O kg^{-1} air), but it could equally mean mole fraction (mol H_2O mol^{-1} air) depending on how the flux equations are written. The symbol ε can mean emissivity in a surface flux model and isotope enrichment in the geochemistry community. Commonly accepted notation is used wherever possible in this book, within the constraint of minimizing duplication across chapters.

The scientific units are standard (Table A.1), though various metric prefixes are used for convenience (Table A.2), and some units are preferred to clarity. The most prominent example pertains to diffusive fluxes. Diffusive fluxes are commonly formulated in terms of a resistance with units s m^{-1} (the inverse of conductance). This can be cumbersome, and it hides the fact that the resistance pertains to the flux of a particular gas. For example, an evaporative resistance of 100 s m^{-1} has a conductance of 0.01 m s^{-1} (or 10 mm s^{-1}). A latent heat flux of 444 W m^{-2} is equivalent to a mass flux of about 0.18 g H_2O m^{-2} s^{-1}. A more straightforward approach is to define conductance in the same units as the flux (e.g., mol H_2O m^{-2} s^{-1}). A conductance of 10 mm s^{-1} is equivalent to about 0.4 mol m^{-2} s^{-1}, and an evaporative flux of 0.18 g H_2O m^{-2} s^{-1} is equal to 10 mmol H_2O m^{-2} s^{-1}. In most cases, this leads to only minor changes in the equations from their traditional disciplinary forms, and the equations will be familiar to readers. An exception is water potential. Soil scientists and hydrologists commonly use the dimension length (m), but plant physiologists use units of pressure (Pa). This distinction is retained to preserve forms of the equations recognizable to the respective disciplines.

1.7 Constructing a Terrestrial Biosphere Model

Physical interactions between the biosphere and atmosphere relate to exchanges of energy, water, and momentum and storage of heat and water in soil. Mathematical representation of these processes in atmospheric models requires resolving the diurnal cycle and coupling with the atmosphere at high temporal resolution (typically 30 minutes or less, but sometimes 10 minutes or less). Biogeochemical fluxes for atmospheric chemistry and composition also must be resolved at these short timescales. Yet the energy, water, and biogeochemical fluxes must also be integrated forward in time for periods of decades to millennia to simulate the changing ecological and hydrologic state of the land. Computational speed has restricted the detail with which terrestrial processes can be represented in global models, particularly in early generation models. While models that simulate the micrometeorology and physiology of a particular forest canopy, the temporal dynamics of an individual forest stand, or the hydrology and biogeochemistry of a watershed can be rich in process detail, global models have tended to be more simplistic. In practice, however, the distinction among models is becoming increasingly less apparent as the computational power of supercomputers increases. With performances allowing tens of petaflops (10^{15} floating point operations per second), the simplifications used in global models to enhance computational efficiency can be abandoned for the greater complexity of detailed process models.

In constructing a terrestrial biosphere model for climate and Earth system simulation, one must recognize the role the land surface and terrestrial ecosystems play in the fully coupled model and the various timescales of coupling (Figure 1.7). A terrestrial biosphere model provides the surface albedo (typically for direct beam and diffuse radiation and visible and near-infrared wavebands) and upward longwave radiation needed for atmospheric radiative transfer. It also provides sensible heat flux, latent heat flux, the corresponding water vapor flux, and zonal and meridional momentum fluxes for the atmospheric boundary layer. This requires determining the radiation absorbed at the surface and its partitioning into longwave emission, sensible and latent heat, soil heat storage, canopy storage, and snowmelt. The land provides runoff as freshwater input to the ocean model. That is obtained by partitioning precipitation into runoff, evapotranspiration, and water storage. The model also provides the net CO_2 exchange with the atmosphere, with component fluxes of gross primary production, autotrophic and heterotrophic respiration, fire, and land use. Additional biogeochemical fluxes include CH_4, nitrogen gas fluxes, BVOCs, dry deposition, dust, and wildfire chemical emissions. The biogeophysical, biogeochemical, and hydrologic fluxes must be solved at every model time step. These fluxes depend on the hydrologic and thermal state of the land, and so the models simulate canopy intercepted water, snow cover, soil moisture, and soil temperature. The ecological state is updated on timescales ranging from daily to seasonal leaf phenology to longer-term ecosystem structure and community composition over years to centuries. Key ecological states are vegetation cover, leaf area index, canopy height, and vegetation and soil carbon and nitrogen pools.

A terrestrial biosphere model, as with any model, is defined by the system boundary, state variables, model structure, parameters, forcing inputs to the model, initial conditions, and model outputs (Table 1.3). For a terrestrial biosphere model, the system of interest is a vertical column extending from deep in the soil (often taken as bedrock), upward through the plant canopy, and into the atmospheric boundary layer. Meteorological, ecological, biogeochemical, and hydrologic processes within this column are treated as one-dimensional (vertical only), and there are no lateral exchanges of water, carbon, or nutrients among columns (though some models are beginning to include lateral water flows). The models operate on a global domain represented as a three-dimensional grid so that all calculations are performed in every grid cell (Figure 1.11). A typical horizontal resolution for a global model is 1° in latitude and longitude (about 100 km), though some models have higher resolution. The vertical resolution in soil can vary from layers with a thickness of a few millimeters near the surface to several tens of centimeters deeper in the soil. Some models may also be vertically resolved aboveground in the canopy.

Model structure refers to the equations used to formulate specific processes as well as the interactions and connections among the various processes. This consists of prognostic equations that describe the time change in a state variable, often formulated as a conservation equation; flux parameterizations for the conservation equation; diagnostic equations that calculate variables from prognostic variables; and the numerical methods used to solve the state equations. Terrestrial biosphere models

Table 1.3 General model components

Model component	Definition
System boundary	Defines the system of interest; separates processes internal to the system from those external to the system
State variables	Time-varying quantities that appear on the left-hand side of a prognostic equation as a time derivative; also known as prognostic variables
Model structure	The equations used to formulate model variables and processes; the interactions and connections among variables and processes
Parameters	Constants used in process parameterizations
Forcing inputs	Quantities needed to evolve the model state
Initial conditions	State variables that need to be initialized for model start-up
Model outputs	Quantities simulated by the model; state variables as well as fluxes and other diagnostic variables

Source: Liu and Gupta (2007); Williams et al. (2009).

simulate many meteorological, hydrological, ecological, and biogeochemical processes at timescales from near-instantaneous fluxes to millennial changes in ecosystems (Table 1.4). One key constraint for the structure of a terrestrial biosphere model is that the fluxes must be solved on short time intervals and resolve the diurnal cycle. A second requirement is conservation of energy and mass, or else the atmosphere will spuriously gain or lose heat, water, CO_2, or other materials. Here, conservation is used in a physical sense and is a fundamental principle of global climate models. In contrast, ecologists often use conservation to mean preserved from change; a quantity (e.g., the ratio between two variables) is said to be conserved or is conservative if it is unchanged and is general over many environmental conditions. Additionally, there is no knowledge of future climate,

Table 1.4	Specific components of terrestrial biosphere models
State variables	Temperature, water, carbon, nitrogen
Energy	Solar radiation, longwave radiation, sensible heat, latent heat, biomass storage, soil heat, snow melt
Water	Interception, throughfall, stemflow, infiltration, surface runoff, soil water redistribution, subsurface runoff, snow melt, evaporation, transpiration, plant water uptake, stomatal conductance
Turbulent transport	Momentum flux, aerodynamic conductance, leaf boundary layer
Canopy scaling	Big leaf, sunlit and shaded leaves, nitrogen profile
Carbon	Photosynthesis, autotrophic respiration, allocation, litterfall, phenology, heterotrophic respiration, soil organic matter
Nitrogen	Mineralization, immobilization, plant uptake, denitrification, volatilization, leaching
Trace gases and aerosols	Dust, BVOCs, Nr, CH_4, aerosols, wildfire emissions
Demography	Establishment, competition, mortality
Landscape dynamics	Wildfire, harvest, land use and land-cover change, biogeography
Forcing inputs	Air temperature, humidity, wind speed, solar radiation, longwave radiation, air pressure, precipitation, CO_2, nitrogen deposition
Surface datasets	Soil texture, soil color, soil depth; leaf area index, canopy height, and vegetation type (for land surface models without dynamic vegetation)

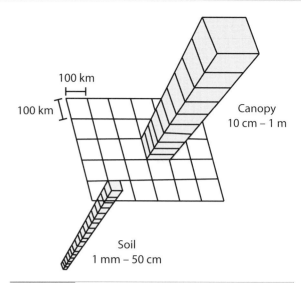

Figure 1.11 Terrestrial biosphere models utilize a three-dimensional grid that is structured in terms of longitude × latitude × level. A typical spatial resolution is 1° in longitude and latitude (equivalent to about 100 km in mid-latitudes). The vertical resolution in soil is variable, with thin layers near the surface and thicker layers deeper in the soil. If a multilayer plant canopy is included in the model, the vertical resolution varies depending on the height of the canopy.

only information of the immediate present state and the past. This is important when modeling leaf phenology or when using bioclimatic rules to determine where plants can grow.

Atmospheric models are formulated from fundamental physical principles that combine conservation of mass, heat, and momentum with diffusive flux transport, and the component land models follow the same approach. This is in contrast with ecological and hydrological models that may use semiempirical flux parameterizations for evapotranspiration or photosynthesis. A conservation equation (also known as a continuity equation) describes how the quantity of interest is conserved. For example, the change in the amount of mass can be described as the balance between fluxes into and out of a reservoir and a source or sink within the reservoir so that a conservation equation has the form:

change in storage = flux in − flux out + source.

$$(1.1)$$

A simple conservation equation for soil water is

$$\frac{dW}{dt} = P_R - E - D_R. \qquad (1.2)$$

In this equation, the change in the amount of water per unit area W is equal to the flux of water that infiltrates into the soil (taken here to be the same as precipitation P_R) minus losses from evaporation E

and drainage D_R. The evaporation and drainage fluxes themselves depend on soil water and must be specified. A simple equation is

$$E = \begin{cases} E_p(W/W_0) & W < W_0 \\ E_p & W \geq W_0 \end{cases}, \tag{1.3}$$

in which a potential evapotranspiration rate E_p is reduced if soil water is less than some amount W_0. The drainage loss can be represented as

$$D_R = a(W/W_{\max})^b, \tag{1.4}$$

in which $W_{\max}$ is the maximum water storage. Together, (1.2)–(1.4) provide equations to calculate the time evolution of soil water. Equation (1.2) is the prognostic equation. Prognostic equations use time derivatives to describe the change in a state variable and are integrated forward in time from some initial condition. In this example, W is the prognostic, or state, variable. Equations (1.3) for E and (1.4) for D_R are known as flux parameterizations. P_R and E_p are the model forcing (i.e., the input to the soil water model), and W_0, $W_{\max}$, a, and b are model parameters. State variables, prognostic equations, flux parameterizations, parameters, forcing inputs, and initial conditions are the main elements of all models. Some models additionally use diagnostic equations. While prognostic equations directly calculate the time change in a variable, diagnostic equations calculate variables from prognostic variables.

In atmospheric models, the term parameterization is used to describe a mathematical representation of a process that is too small-scale or complex to be directly resolved at the spatial and temporal scales of the model. Parameterizations approximate processes that cannot be explicitly represented from fundamental physical principles. One speaks of clouds, convection, radiative transfer, or boundary layer processes in terms of a parameterization. Clouds and precipitation are not modeled as individual droplets but rather as bulk processes; nor is radiative transfer modeled as individual photons; and turbulence transport in the boundary layer is not modeled with respect to individual eddies. Processes in a terrestrial biosphere model must be similarly parameterized. A model of leaf photosynthesis or stomatal conductance is a parameterization of complex biochemical and biophysical processes within a leaf that cannot be explicitly resolved. A radiative transfer parameterization simplifies the complexities of photon transport, typically by representing the canopy as a plane-parallel, horizontally homogeneous, turbid medium.

Prognostic variables must be numerically stepped forward from time n to $n + 1$ over the time interval Δt. Consider, for example, the water balance given by (1.2). A common method is to represent the left-hand side of the equation as a forward difference approximation using $(W^{n+1} - W^n)/\Delta t$ and to represent the fluxes on the right-hand side using W at time n. This is known in numerical methods as an explicit solution, and the value at time $n + 1$ depends on the state at time n. A more numerically robust method is the implicit solution in which the fluxes on the right-hand side are also specified in terms of W^{n+1}. Many modelers strive to write their equations in an implicit solution. Not all calculations, however, facilitate the implicit method, and many require iterative solutions because of numerous dependencies among the various equations to be solved.

Conservation equations combined with flux parameterizations are used to describe soil water, plant water storage, soil temperature, scalar quantities in the atmosphere, and biogeochemical cycles. Conservation equations facilitate assessment of conservation of mass and energy as a check of the model implementation and numerical solution. State equations such as (1.2) can also be checked for consistency of units. While conservation is a fundamental concept of the models, not all models utilize the same conservation equations. The surface energy budget equation may, for example, be missing heat storage in biomass or heat input from precipitation. The water balance equation may exclude lateral inflows and transfers between surface water and groundwater. The carbon balance may not account for land-use and land-cover change or loss of carbon in lateral water flows. With the addition of nitrogen and phosphorus cycles in a model, these, too, must balance but may neglect the complexities of gaseous and aqueous losses or geochemical inputs.

Model parameters are specified in two ways. Some parameters are considered to be physical constants (e.g., the diffusivity of water vapor in air, the thermal conductivity of water; Tables A.3 and A.4). Other parameters vary geographically and are input to the model as gridded (latitude by longitude) surface datasets. Soil texture, soil depth, and soil color are common inputs to a terrestrial biosphere model. If vegetation is not simulated, the type of vegetation, leaf area index, and canopy height are specified from gridded datasets.

An additional component of a model is the initial conditions. These are the state variables that need to be initialized for model start-up. Oftentimes these are not precisely known. If the initial state is incorrect, the state variables have to adjust to the forcing, and spurious trends will arise in the model output. The time required for the model to come into balance with the forcing is known as model spin-up. Many state variables in terrestrial biosphere models are, in fact, poorly known and are initialized from arbitrary values. An example is the initial carbon and nitrogen states. The exact values cannot be specified globally. Instead, the model is initialized with bare ground, and a constant annual forcing of climate and atmospheric CO_2 is cycled for many years until the model attains a state in equilibrium with the forcing. From these initial conditions, the model response to some perturbation (e.g., changing climate or elevated CO_2) can be simulated. A common example is to use atmospheric forcings for preindustrial conditions to spin up the model. Once the carbon and nitrogen pools have come into equilibrium, a transient simulation is performed for historical forcings through the nineteenth and twentieth centuries and for scenarios of the twenty-first century. Similar spin-up is required for soil temperature and soil moisture. Spin-up time can vary from a few years for soil temperature and soil moisture to several hundred years or longer for carbon and nitrogen.

1.8 | The Goals of Modeling

A primary use of models is to study system behavior in conditions beyond which measurements can be made; to allow predictions of system behavior, especially in response to some imposed perturbation; and to inform management and policy decisions. These usages of models are particularly important in the context of global change. Without an ability to conduct controlled experiments on climate, atmospheric scientists have long embraced models as a necessary tool to study climate and climate change. Models are required to make projections of Earth system response to anthropogenic forcings such as elevated concentrations of CO_2 and other greenhouse gases. It

is in this context that terrestrial biosphere and Earth system models are commonly used. The same is true for ecosystems, where experimental manipulations can only be made at small scales and for limited periods of time.

A model can also be seen as a formal organization of understanding; it originates from the knowledge of its developers about how the system operates. One purpose of modeling, then, is to identify the processes needed to adequately simulate the system. If a model replicates some observations, a scientist must ask why the model works correctly. If the model performs poorly, then the scientist must ask what is missing. Various parameter and parameterization sensitivity tests are performed with the model to understand its behavior. Such model experiments are very much like real-world experiments. Science advances by formulating a hypothesis, collecting data to test the hypothesis, analyzing the data and making inferences, and performing additional experiments as needed to refine the understanding. Models work in a similar way to increase understanding. Inaccuracies in a model lead to the formulation of hypotheses about processes that are missing from the model or are poorly known. In this sense, models are used to better understand observations of patterns and processes in the natural world. They can be used to test specific hypotheses of how a system functions and, in the context of global models, to test the generality of theory across a variety of climates and ecosystems. A key reason for developing terrestrial biosphere models, therefore, is to develop a process-level understanding across multiple ecosystems and at multiple timescales.

A third use of models is as a research tool to guide data collection. What are the critical parameters that need to be measured? How precisely must these parameters be measured to reduce model uncertainty? What new observations are needed to test the model? In this context, models inform data collection and experimental design to both test the model and advance process understanding.

1.9 | Complexity and Uncertainty

Models are simplifications of complex real-world systems, and terrestrial biosphere models are no exception. It is impossible, for example, to capture all the variability in the physiology of leaves or plants or the diversity of soils and microorganisms within a single forest stand, let alone the global biosphere. Mathematical equations can be formulated to describe the relevant processes, but these equations are generalizations for which there will always be exceptions. Terrestrial biosphere models attempt to represent the essential components of an ecosystem and the necessary processes to characterize ecosystem functioning. A key question that must be addressed during model development concerns the level of detail and process complexity required to adequately represent the biosphere. This problem is common to all models, and there is no single correct resolution. The answer is found in the scope of the scientific question that is driving the model development in the first place, the extent to which the functioning of the biosphere is understood, and by the computing resources available to run the model. One long-standing debate concerns the appropriate way to model plant canopies. Big-leaf models can be criticized as being physically incorrect for their simplistic representation of plant canopies, but multilayer models are equally criticized as inordinately complex (Raupach and Finnigan 1988). Which model is useful depends on the scope of the research: big-leaf models may be appropriate to calculate surface fluxes to the atmosphere; multilayer models also allow study of the canopy microclimate.

Ecosystem modelers routinely face fundamental questions concerning the balance between simplification and complexity – seen, for example, in the distinction between biogeochemical models and individual-based models. Do individual plants need to be recognized; or can biomass be lumped into aggregate pools of leaves, stems, and roots? Do individual species need to be explicitly represented, or can they be combined into a reduced number of plant functional types? What types of decomposing litter and soil carbon pools need to be represented? Do these pools need to be vertically resolved in the soil? Plant productivity can be represented by empirical relationships with temperature and rainfall; it can be modeled in relation to light-use efficiency as modified by environmental factors, or it can be modeled from leaf photosynthesis scaled to the canopy. Similarly, evapotranspiration can be simulated from simple empirical relationships with temperature and other factors, or it can be calculated from principles of diffusion in combination with surface energy balance constraints. Empirical and process models both try to represent and explain the same phenomena, though with different levels of understanding. The most sophisticated models rely on process understanding in which system behavior emerges from the process detail. Empirical models are convenient for their ease of use and the ability to calibrate the model to certain conditions, but may be restricted in their applicability. Process models may be sounder outside the range of conditions used to calibrate an empirical model, but they come with other sources of uncertainty.

An important point to remember is that greater complexity does not necessarily increase scientific understanding or lead to more precise model output. Demonstrating that a model can successfully reproduce some observations in nature is a necessary step in model development, but the more critical step is to understand why the model was successful. The more complex the model is in representing process detail, the more difficult it can be to understand why a certain outcome is obtained. This is particularly true in Earth system models – where terrestrial ecosystems, the atmosphere, and the ocean are coupled in an interdependent system. The simulations are computationally intensive, and experimentation with the model to isolate critical processes is often not possible. It is not practical to conduct numerous sensitivity experiments to understand model behavior. Instead, the models are limited to a few long climate simulations from which one must diagnose the reason for the simulated outcome. As a model becomes more process rich, it may be more mechanistically realistic, but uncertainty in the simulated outcome may increase because of a larger number of poorly known model parameters and as a result of complex interactions among various model compartments.

The ever-increasing complexity of terrestrial biosphere models is seen in the history of such models at the National Center for Atmosphere Research (Boulder, Colorado). As documented in the technical description of these models over 35 years, the number of scientific contributors to the models, the number of equations required to describe the models, and the length of the technical description have each increased dramatically so that successive model versions involve more processes and more computer code than previous versions (Figure 1.12). For example, four authors needed 79 equations to

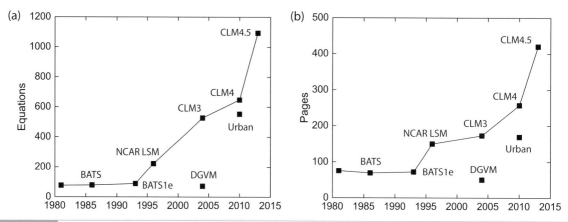

Figure 1.12 The increasing breadth and complexity of land surface models as documented by NCAR technical notes in terms of (a) the number of equations and (b) the number of pages (Dickinson et al. 1981, 1986, 1993; Bonan 1996; Levis et al. 2004; Oleson et al. 2004, 2010a,b, 2013).

describe the first version of BATS in 69 pages (Dickinson et al. 1986). Version 4.5 of the Community Land Model (CLM4.5) required 26 authors, 1093 equations, and 420 pages to describe the model (Oleson et al. 2013). Additional technical notes described the dynamic global vegetation model, introduced in 2004 (Levis et al. 2004), and the urban land cover parameterization, introduced in 2010 (Oleson et al. 2010a).

An additional consideration is the recognition that various alternative model structures and parameter sets can give equally good fits to observations. There are often many different ways in which models can achieve the same result, and various sets of parameters may be optimal within the same model structure. This concept has been termed equifinality, as discussed in the context of hydrologic models (Beven 1993, 2002, 2006; Beven and Freer 2001). It implies embracing multiple feasible models of natural systems rather than converging on one single correct answer. Yet too often, uncertainty is seen as failure. Beven (2006) addressed this point in the context of hydrologic models, but the same concept applies to models in general: "Science ... is supposed to be an attempt to work towards a single correct description of reality. It is not supposed to conclude that there must be multiple feasible descriptions of reality. The users of research also do not (yet) expect such a conclusion and might then interpret the resulting ambiguity of predictions as a failure (or at least an undermining) of the science."

With ever-increasing process complexity, new practices are needed to advance terrestrial biosphere models. A fundamental question for any modeling study is whether one should believe the results, but there is no simple means to assess this. There is an expectation that more data and better use of data through mathematical techniques of data assimilation and parameter optimization will improve the models. This is seen in the idea of benchmarking models in comparison with multiple observational datasets (Randerson et al. 2009; Luo et al. 2012). While such benchmarking is necessary to assess model performance, more fundamental, perhaps, is a need to reconsider the philosophy behind modeling complex environmental systems in light of uncertainty and equifinality and to embrace multiple model outcomes rather than converge on a single correct model (Beven 2002). In addition, model development must balance multiple goals of reliability, meaning the model must produce correct output; robustness, meaning the results are not overly sensitive to specific assumptions, approximations, or parameters; and realism, meaning the necessary processes are represented in sufficient detail (Prentice et al. 2015). Greater process richness and complexity may increase realism but reduce reliability and robustness. A correct model that produces acceptable results for theoretically sound reasons may not be a useful model (Raupach and Finnigan 1988). In this context, the development and evaluation of terrestrial biosphere models must embrace a synergy of ecological

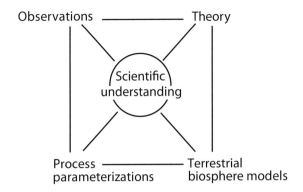

Figure 1.13 A balance among observations, theory, process parameterizations, and modeling studies are needed to advance our understanding of terrestrial ecosystems and their role in the Earth system. Adapted from Bonan (2014)

observations, sound theory to explain the observations, numerical parameterizations to mathematically implement the theory, and simulations to evaluate the parameterizations across scales, from leaf to canopy to global (Figure 1.13).

1.10 | Intent of the Book

This book bridges the gap between traditional ecosystem models developed with an ecological perspective and the global models developed by geophysicists for Earth system simulation. The book shows the commonality between the different modeling communities and provides a foundation for exchange of knowledge between ecological and geophysical scientists. Specific chapters describe the theory underlying particular process parameterizations and give the equations needed to numerically implement the parameterizations in a model. Supplemental programs allow readers to see how the equations are coded and solved. To the bane of some hydrologists, the book focuses on modeling land–atmosphere coupling with an emphasis on plants and ecosystems. The lateral water flows and surface–groundwater interactions that form the core of many hydrologic models are not considered.

Many processes have no agreed-upon way to model them, and even if scientists did agree on a perceptual understanding, they may not use the same numerical implementation or parameters. As

an example, consider the choices in modeling radiative transfer and the accompanying uncertainty arising from those choices. First, one must decide how to represent the canopy: plane-parallel, homogenous; big-leaf or vertically resolved; or three-dimensional. Many models distinguish sunlit and shaded leaves, especially for their differences in leaf gas exchange, but some models do not. Another choice is the leaf angle distribution and how this is used to calculate the light extinction coefficient. Does one represent a single leaf angle for the canopy or a distribution of leaf angle classes? Does leaf angle vary among types of plants? Is clumping of foliage considered? Leaf optical properties (reflectance, transmittance) need to be specified, as does the albedo of the underlying ground. Are leaf properties uniform throughout the canopy, or do they vary with depth in the canopy? The way in which brown leaves, wood, and other phytoelements are accounted for in addition to green leaves differs among models; and so, too, does the treatment of intercepted snow in the canopy.

Modeling entails a series of choices, and the final model is the outcome of those choices. In many instances, a particular choice for one model component will dictate how another component is implemented. Oftentimes, these choices or process implementations are a matter of convenience or expediency to fix a particular problem. Over time, such ad hoc coding accumulates in code that can be unwieldy and in which the implementation of new process parameterizations is limited by the existing code structure. The amassing of such code is termed technical debt and can become problematic in large models with hundreds of thousands of lines of code and a long history of development.

The book is written to help readers understand the choices faced in developing a terrestrial biosphere model. It takes the viewpoint that there is no one single correct way to model the biosphere–atmosphere system. Incertitude in a conceptual understanding of ecosystems, the parameterizations used to mathematically formulate processes, the parameters needed for the model, the forcing data to run the model, and the many different methods to numerically implement and solve the equations introduces uncertainty that propagates from the model input through the model structure to the model output. The book is not a statement of the correct way to model ecosystems or to argue about what processes may be missing from or are poorly represented in models. Such a book would quickly become dated. Instead, the book challenges readers to think about the underlying theory behind a particular parameterization, how that theory is numerically implemented, and whether it sufficiently replicates observations, and to assess how the numerical implementation affects the results – the synergies shown in Figure 1.13. Rather than asserting what modelers should do, the book illustrates a range of processes and highlights the choices that must be made to implement process parameterizations. It provides the details so that model developers can themselves decide what to do.

There is a big leap between writing the fundamental equations that need to be solved and their numerical implementation. That leap can involve many simplifications, approximations, or even ad hoc adjustments to get numerically efficient but stable solutions. Many equations used in models are in fact approximations of more complex, mathematically correct forms of the equations. These approximations are often justified in that the resulting imprecision is less than or comparable to the measurement error; the equations are more simple to use; and the approximations can, in some cases, produce an analytical solution to a complex equation. However, the necessary simplifications can change the solution of the equation. In contrast, the full complexity of the equation set can be retained but may require iterative numerical methods to solve. This is an important model distinction, as some models are written on the basis that analytical solutions are preferred while others are written with a preference for numerical methods. It is part of a fundamental modeling debate about model structural uncertainty and parameter uncertainty, in which a desire for mathematical exactness in process equations and their numerical solution must be balanced by inexactness in parameter choices. This book provides the necessary detail to understand the parameterizations. Where appropriate, approximations are highlighted so that

readers can understand the full derivation of the equations and assess the importance of simplifications.

Some topics are covered in great mathematical detail, while others have a more qualitative description. This relates to two points. First is a reflection of the scientific maturity of the topic. There is no such thing as a settled model, but some processes have converged on a common framework used across models. Photosynthesis of C_3 plants, for example, is modeled following Farquhar et al. (1980). However, various models differ greatly in their implementation of these equations; and so, though the models may agree on the same conceptual formulation of photosynthesis, they produce widely divergent results (Rogers et al. 2017). Similarly, many biogeochemical models use a CENTURY-like framework to represent soil organic matter but differ in the number of organic matter pools, turnover rates, and transfers among pools. A second purpose of model descriptions, then, is to provide an appreciation for the mathematical details needed to model ecosystems. Equations are necessary; what form do those equations have?

As a final word of caution, mathematical mistakes can – and indeed are – made. The derivation of equations can be wrong, as can be their numerical implementation, and incorrect equations can be provided in publications (and in models too). One can find examples of this throughout the modeling literature. It is prudent to re-derive equations as much as possible rather than treating them as gospel. This book provides sufficient detail to do so. Remember, too, that better mathematics and more data may not ultimately lead to an improved simulation of the biosphere, and disciplinary chauvinism often leads to criticisms of models for perceived lack of some detail thought to be critically important or some numerical method that is favored over another. Perhaps when modeling nature, less hubris and more humility are required.

Quantitative Description of Ecosystems

Chapter Overview

Terrestrial biosphere models characterize ecosystems by features that control biogeochemical cycles and energy, mass, and momentum fluxes with the atmosphere. These include leaf area index and its vertical profile in the canopy; the orientation of leaves, or leaf angle distribution; the vertical profile of leaf mass and leaf nitrogen in the canopy; the profile of roots in the soil; the size structure of plants; and the distribution of carbon within an ecosystem. This chapter defines these descriptors of ecosystems.

2.1 | Leaf Area Index

Leaves and other phytoelements exert strong control on biosphere–atmosphere fluxes. The transmission of radiation through a plant canopy depends on the amount of leaves and other plant materials (e.g., twigs, branches, and boles in forests) that obstruct solar radiation and on their spatial distribution and orientation. Leaves are also the sources of heat, moisture, and chemical exchanges between the biosphere and atmosphere and are sinks for other materials such as CO_2 and ozone. The foliage and plant materials protruding skyward exert drag on airflow, seen in the swaying of tall trees in strong winds, and cause wind speed to decrease near the surface. The amount of foliage in a plant canopy is denoted by leaf area index. The conventional definition of leaf area index is the projected, or one-sided, area of leaves per unit area of ground. Projected, or one-sided, leaf area is different from the total surface area of a leaf. Thin, flat leaves have a total surface area that is twice the projected leaf area (both sides of the leaf are included). Needles have a total surface area that is more than twice the projected leaf area. A more standard definition of leaf area index is one-half the total leaf surface area per unit ground area (Chen and Black 1992). A typical leaf area index for a productive forest is 4–6 $m^2 \, m^{-2}$ (Asner et al. 2003).

The distinction between total and projected leaf area is particularly important for conifer needles, which do not necessarily have a clearly defined upper and lower surface. Needles have a total surface area that is about 2.5 times the projected leaf area (Niinemets and Kull 1995; Waring and Running 2007, p. 30), but needles have complex shapes, and the exact ratio varies with needle shape (Figure 2.1). For needles shaped like a cylinder with thickness T (which is also the diameter) and length L_n, the surface area is $A_T = \pi \cdot T \cdot L_n$. The projected area viewed from above (also known as the planform area) is $A_P = T \cdot L_n$ so that $A_T = \pi A_P$. Needles shaped like a rhomboid have both a thickness T and a width W along the diagonals. Niinemets and Kull (1995) measured $T = 1.2$ mm and $W = 1.0$ mm in Norway spruce (*Picea abies*) needles, in which case $A_T = 3.4 A_P$. This ratio is smaller in needles with a thickness that is substantially greater than their width; e.g., $A_T = 2.6 A_P$ for a needle with $T = 1.1$ mm

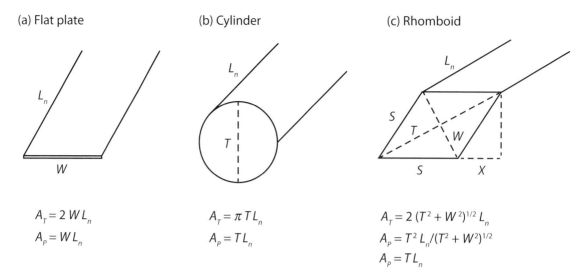

(a) Flat plate

$A_T = 2\,W\,L_n$
$A_P = W\,L_n$

(b) Cylinder

$A_T = \pi\,T\,L_n$
$A_P = T\,L_n$

(c) Rhomboid

$A_T = 2\,(T^2 + W^2)^{1/2}\,L_n$
$A_P = T^2\,L_n/(T^2 + W^2)^{1/2}$
$A_P = T\,L_n$

Figure 2.1 Geometric models of leaves for (a) a broadleaf shaped as a flat, thin plate with width W and length L_n, (b) a needle shaped as a cylinder with thickness (diameter) T and length L_n, and (c) a needle shaped as a rhomboid with thickness T, width W, and length L_n. Shown are calculations of total leaf area A_T and projected leaf area A_P. Two different calculations of projected leaf area are given for the rhomboidal needle. The first uses the projected needle thickness $S + X$, in which S is the side length and X is the additional length due to the projection from above. The second uses the needle thickness T. The rhomboid geometry is adapted from (Niinemets and Kull 1995)

and $W = 0.6$ mm. However, spruce needles do not conform exactly to a rhomboid, and an alternative calculation of the projected area is $A_p = T \cdot L_n$. In this case, the ratio of total to projected area is smaller than given in the previous examples ($A_T/A_P = 2.6$ and 2.3, respectively). These calculations demonstrate that the ratio of total to projected area varies with needle cross-sectional geometry, and observations show that this geometry is highly variable. For example, Steele et al. (1989) measured values of 2.5–3.0 for the ratio of total to projected area in Sitka spruce (*Picea sitchensis*) needles depending on tree age, and Niinemets and Kull (1995) found values of 2.5–3.5 in Norway spruce depending on the light environment.

The amount of foliage varies with height in the canopy, and cumulative leaf area index increases progressively with greater depth from the top of the canopy. Leaf area index measured below the canopy represents the total leaf area in the canopy. The leaf area density function, denoted $a(z)$ with units m^{-1}, is the foliage area (m^2) per unit volume of canopy space (m^3) at height z and describes the vertical distribution of leaf area. The total leaf area index L over a canopy of height h_c is

$$L = \int_0^{h_c} a(z)\,dz \qquad (2.1)$$

and the cumulative leaf area index at height z is

$$L(z) = \int_z^{h_c} a(z)\,dz. \qquad (2.2)$$

Cumulative leaf area index defines depth in the canopy in relation to the leaf area profile. It transforms the vertical coordinate system from physical height above the ground to one of cumulative leaf area. This transformed coordinate is commonly used to study vertical profiles of light availability and leaf nitrogen in plant canopies. Figure 2.2 shows the profile of leaf area in a deciduous forest of oak and hickory trees and in a Scots pine forest. The deciduous forest has a canopy height of about 23 m and a leaf area index of about 5 m^2 m^{-2}. Most leaf area is in the upper 5 m of the canopy. Over 75% of the total leaf area is located in the upper 25% of the canopy. The Scots pine forest is about 16 m tall. Most of the needles are located in the mid-canopy.

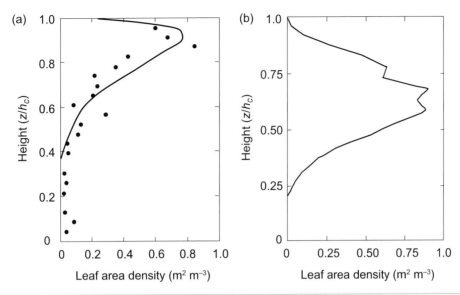

Figure 2.2 Profile of leaf area in (a) a deciduous oak–hickory (*Quercus–Carya*) forest (Baldocchi and Meyers 1988) and (b) a conifer Scots pine (*Pinus sylvestris*) forest (Halldin 1985). Height is given as a fraction of canopy height (z/h_c).

In practice, leaf area density is evaluated by measuring the amount of leaf area ΔL_i between two heights in the canopy separated by the distance Δz. This provides a discrete form of $a(z)$ in which the leaf area density at some level in the canopy is $a(z_i) = \Delta L_i/\Delta z$. In many canopies, the beta distribution probability density function describes profiles of leaf area density, and it provides a continuous representation of leaf area for use with multilayer canopy models(Massman 1982; Meyers and Paw U 1986; Wang et al. 1990; van den Hurk and McNaughton 1995; Markkanen et al. 2003; Wu et al. 2003; Boy et al. 2011; Bonan et al. 2018). In the beta distribution, the probability density function(Appendix A11) for a variable x, which has values between zero and one ($0 \leq x \leq 1$), is defined by two shape parameters p and q as

$$f(x) = \frac{x^{p-1}(1-x)^{q-1}}{B(p,q)}, \qquad (2.3)$$

in which the beta function $B(p,q)$ (Appendix A10) is a normalization constant to ensure that $f(x)$ integrates to one. As applied to leaf area density, $x = z/h_c$ is the relative height in the canopy, and

$$a(z) = \frac{L}{h_c}\frac{(z/h_c)^{p-1}(1-z/h_c)^{q-1}}{B(p,q)}. \qquad (2.4)$$

Equation(2.4) has a symmetric distribution about $z/h_c = 0.5$ for $p = q$. Larger values of p (so that $p > q$) skew the distribution towards the upper canopy, and the maximum leaf area density occurs at $z/h_c > 0.5$. The profiles in Figure 2.3 describe a variety of crop, grass, and forest canopies. In forests and shrublands, one additionally needs to account for the surface area of branches, trunks, and other woody surfaces. Figure 2.4 illustrates this for a deciduous forest. The leaf area index is 4.9 m^2 m^{-2}, and woody biomass contributes an additional 0.6 m^2 m^{-2} of surface area. Plant area index is the sum of leaf and wood area.

Question 2.1 The leaf area density profile in a plant canopy is $a(z) = 0.3(z/h_c)^{2.5}(1-z/h_c)/B(3.5,2.0)$. Calculate the leaf area index in a 20 m tall canopy.

Question 2.2 Five plant canopies have different leaf area density profiles described by the beta distribution with (p, q) equal to (1.0,1.0); (3.0,3.0); (8.0,8.0); (11.5,2.5); and (3.5,11.5). Which canopy has foliage in the upper canopy; an open overstory with a dense understory; a symmetric distribution; a symmetric

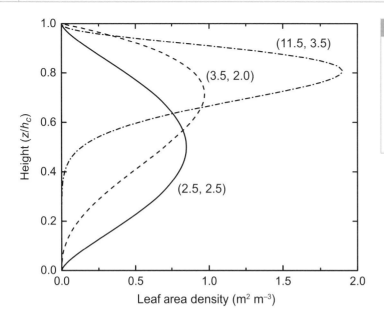

Figure 2.3 Generalized profiles of leaf area density in plant canopies. Shown are three different canopy profiles for (i) grass, soybean, maize, and wheat with $p = q = 2.5$; (ii) deciduous trees and spruce trees with $p = 3.5$ and $q = 2.0$; and (iii) pine trees with $p = 11.5$ and $q = 3.5$. These profiles are from Meyers et al. (1998) and Wu et al. (2003) and are shown here with $L/h_c = 0.5$ m^2 m^{-3}.

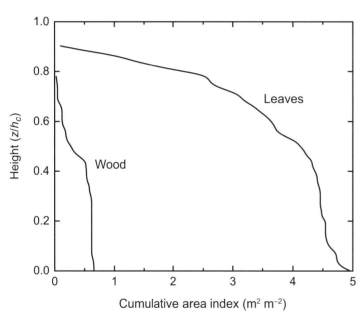

Figure 2.4 Cumulative leaf area index and wood area index in a deciduous oak–hickory forest. Adapted from Hutchinson et al. (1986)

distribution with a distinct layer of leaves at mid-canopy; a uniform distribution? Describe the effect of p and q on the leaf area density profile. Use Supplemental Program 2.1.

Question 2.3 Discuss the distinction between leaf area index and plant area index. Is it important to distinguish green leaves and wood in a forest canopy?

2.2 | Leaf Angle Distribution

The foliage in a plant canopy is oriented at many angles – varying among horizontal, semihorizontal, semivertical, and other angles – and with the leaf surface facing various azimuth directions. These different leaf orientations affect the reflection, transmission, and absorption of light through the plant canopy. The leaf inclination angle Θ_ℓ is the angle of the leaf from horizontal, which is also the angle between vertical and a line normal (perpendicular) to the upper leaf surface (Figure 2.5). The leaf surface also has an azimuth angle A_ℓ that describes its orientation in the compass directions north, east, south, west, etc. If the values of Θ_ℓ and A_ℓ are independent, a probability density function can be defined that describes the frequency distribution of these angles (Lemeur 1973). The leaf angle probability density function characterizes the fraction of leaf area in the canopy per unit leaf inclination angle. It is denoted by $f(\Theta_\ell)$ such that $f(\Theta_\ell)\,d\Theta_\ell$ is the fraction of the canopy foliage that has an inclination angle within the interval $\Theta_\ell \pm 0.5\,d\Theta_\ell$. Mathematically, this means that

$$\int_0^{\pi/2} f(\Theta_\ell)d\Theta_\ell = 1 \tag{2.5}$$

and the mean inclination angle $\bar{\Theta}_\ell$ is

$$\bar{\Theta}_\ell = \int_0^{\pi/2} \Theta_\ell f(\Theta_\ell)d\Theta_\ell. \tag{2.6}$$

Similarly, the distribution of azimuth angles is described by the probability density function:

$$\int_0^{2\pi} g(A_\ell)dA_\ell = 1. \tag{2.7}$$

If leaf azimuth is distributed randomly such that there is no preferred azimuth orientation, the azimuth probability density function is uniform and $g(A_\ell) = 0.5/\pi$. This is a common assumption in many analyses.

Leaf inclination angles can be measured in the field to derive the probability density function. Figure 2.6 shows common leaf angle distributions as described in Table 2.1 using the terminology of de Wit (1965). Planophile leaves are mostly horizontal; erectophile leaves are mostly vertical; plagiophile leaves are midway between vertical and horizontal; uniform leaves have the same abundance at any angle; and a spherical leaf angle distribution has an equal probability of any orientation (i.e., foliage orientation is random). In the spherical leaf distribution, leaves can be thought of as being distributed along the surface of a sphere (i.e., the angular distribution of leaves in the canopy is similar to the distribution of area on the surface of a sphere). Visualized this way, it is evident that erect leaves are more common than horizontal leaves, the latter

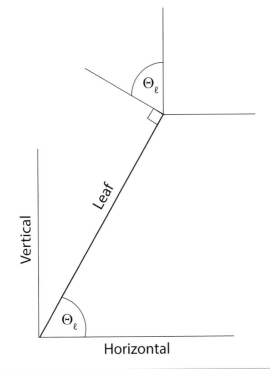

Figure 2.5 Illustration of a leaf (thick line) oriented at an angle Θ_ℓ to horizontal.

Table 2.1	Probability density functions for five leaf angle distributions and corresponding parameters for the equivalent beta distribution

Leaf type	$f(\Theta_\ell)$	Beta distribution	
		$\bar{\Theta}_\ell$	s
Planophile	$\frac{2}{\pi}(1 + \cos 2\Theta_\ell)$	26.76°	18.51
Erectophile	$\frac{2}{\pi}(1 - \cos 2\Theta_\ell)$	63.24°	18.50
Plagiophile	$\frac{2}{\pi}(1 - \cos 4\Theta_\ell)$	45.00°	16.27
Uniform	$\frac{2}{\pi}$	45.00°	25.98
Spherical	$\sin \Theta_\ell$	57.30°	21.55

Note: $\bar{\Theta}_\ell$ and s are given for units of degrees. Multiply by $\pi/2$ to convert to radians.
Source: From Goel and Strebel (1984)

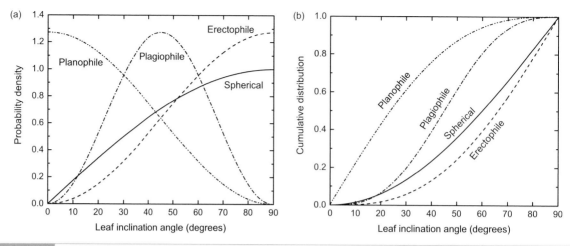

Figure 2.6 Planophile, erectophile, plagiophile, and spherical leaf angle distributions showing (a) the probability density function $f(\Theta_\ell)$ and (b) the cumulative distribution $F(\Theta_\ell)$. Equations for $f(\Theta_\ell)$ are from Table 2.1.

which are found only along the top and bottom of the sphere.

A two-parameter beta distribution fits a variety of leaf angle distributions (Goel and Strebel 1984). In this formulation, given here in units of radians, the leaf angle probability density function is

$$f(\Theta_\ell) = \frac{2}{\pi} \left(\frac{\Theta_\ell}{\pi/2} \right)^{p-1} \left(1 - \frac{\Theta_\ell}{\pi/2} \right)^{q-1} \frac{1}{B(p,q)}. \qquad (2.8)$$

The parameters p and q are defined by the mean leaf angle $\bar{\Theta}_\ell$ and the standard deviation s, with

$$p = \left(1 - \frac{s^2 + \bar{\Theta}_\ell^2}{\bar{\Theta}_\ell \pi/2} \right) \Big/ \left(\frac{s^2 + \bar{\Theta}_\ell^2}{\bar{\Theta}_\ell^2} - 1 \right) \qquad (2.9)$$

$$q = \left(\frac{\pi/2}{\bar{\Theta}_\ell} - 1 \right) p. \qquad (2.10)$$

Equation (2.8) matches the analytical functions with the appropriate values for $\bar{\Theta}_\ell$ and s (Table 2.1).

The ellipsoidal leaf angle probability density function is a generalized function that matches many leaf angle distributions (Campbell 1986, 1990). This function describes leaf orientations as arranged on the surface of an ellipsoid in which:

$$f(\Theta_\ell) = \frac{2x^3 \sin \Theta_\ell}{l(\cos^2\Theta_\ell + x^2 \sin^2\Theta_\ell)^2}, \qquad (2.11)$$

with x defined as the ratio of the horizontal semi-axis length to the vertical semiaxis length of the ellipsoid. The parameter l relates to x and has different expressions for $x < 1$, $x = 1$, and $x > 1$. For an ellipsoid that is taller than it is wide ($x < 1$):

$$l = x + \frac{\sin^{-1}e_1}{e_1}, \qquad (2.12)$$

with $e_1 = \sqrt{1 - x^2}$. A value $x = 0$ provides a vertical leaf distribution with all leaves oriented at 90°. In the special case of $x = 1$, the ellipsoid is a sphere in which $l = 2$ and (2.11) is the spherical leaf angle distribution. For an ellipsoid that is wider than it is tall ($x > 1$):

$$l = x + \frac{\ln\left[(1 + e_2)/(1 - e_2)\right]}{2e_2 x}, \qquad (2.13)$$

with $e_2 = \sqrt{1 - x^{-2}}$. A large value for x (as $x \to \infty$) gives a horizontal leaf angle distribution with all leaves oriented at 0°. Campbell (1990) derived a simpler equation that gives a continuous expression whereby the parameter l is approximated by

$$l = x + 1.774(x + 1.182)^{-0.733}. \qquad (2.14)$$

Many canopies can be represented with $0.1 < x < 10$ (Campbell 1986), and Figure 2.7 illustrates various distributions. The ellipsoidal function is constrained to a probability density of zero at an inclination angle of 0°. A rotated ellipsoidal distribution

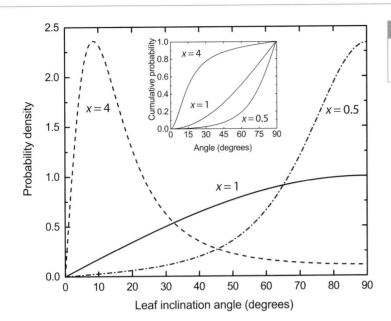

Figure 2.7 Ellipsoidal leaf angle distribution for spherical ($x = 1$), semihorizontal ($x = 4$), and semivertical ($x = 0.5$) canopies. The inset panel shows the cumulative distribution.

function allows for non-zero probability density with an inclination of zero (Thomas and Winner 2000).

To calculate the attenuation of radiation through a plant canopy, radiative transfer concepts must be integrated over the leaf angle distribution to obtain the light extinction coefficient K_b. Analytical solutions can be found for particular leaf angle distributions such as the spherical and ellipsoidal distributions, as well as for horizontal and vertical leaves (Chapter 14). Numerical methods are needed for other probability density functions, such as the beta distribution. In numerical applications, the leaf angle distribution can be represented as nine discrete angle classes of $10°$ intervals (Goudriaan 1977) or even three classes of $30°$ (Goudriaan 1988). The relative frequency of leaves in the ith class between angles $\Theta_{\ell,1}$ and $\Theta_{\ell,2}$ is

$$F_i = \int_{\Theta_{\ell,1}}^{\Theta_{\ell,2}} f(\Theta_\ell) d\Theta_\ell. \tag{2.15}$$

For example, F_i equals 0.015, 0.045, 0.074, 0.100, 0.123, 0.143, 0.158, 0.168, and 0.174 for the spherical leaf angle distribution, calculated as $F_i = \cos \Theta_{\ell,1} - \cos \Theta_{\ell,2}$ for the nine angle classes $0°–10°, \ldots, 80°–90°$. For a small interval, $\Delta\Theta_\ell$, F_i can be approximated by $f(\Theta_\ell)\Delta\Theta_\ell$.

The index χ_ℓ characterizes the departure of the leaf angle distribution from spherical (Ross 1975, 1981). This index is

$$\chi_\ell = \pm \frac{1}{2} \int_0^{\pi/2} |\sin \Theta_\ell - f(\Theta_\ell)| d\Theta_\ell \tag{2.16}$$

and equals +1 for horizontal leaves, 0 for spherical leaves, and −1 for vertical leaves. In practice, (2.16) can be approximated by

$$\chi_\ell = \pm \frac{1}{2}(|0.134 - F_1| + |0.366 - F_2| + |0.5 - F_3|), \tag{2.17}$$

with F_1, F_2, and F_3 representing the fractional abundance of leaves in inclination angle classes $0°–30°$, $30°–60°$, and $60°–90°$; and 0.134, 0.366, and 0.5 are the corresponding weights for the spherical distribution (Goudriaan 1977; Ross 1981). The sign of χ_ℓ is taken from the sign of the last term.

Question 2.4 Pisek et al. (2013) measured leaf angles for several broadleaf deciduous tree species and obtained parameters for the beta distribution. Use Supplemental Program 2.2 to calculate and graph the leaf angle distributions for the following seven species using the beta distribution and classify each by its appropriate distribution (planophile, erectophile, plagiophile, spherical).

Species	$\bar{\Theta}_\ell$	s
Norway maple	40.90°	13.65
Red maple	29.36°	14.93
Alder	51.03°	21.03
Gingko	51.85°	16.91
Tibetan cherry	65.97°	15.05
English oak	35.82°	19.43
Lilac	56.63°	18.66

Question 2.5 Three common leaf angle distributions are planophile, erectophile, and spherical. Graph these distributions and then calculate and graph the ellipsoidal leaf angle distribution. Find values of the parameter x for the ellipsoidal distribution that most closely match these leaf angle distributions. How well does the ellipsoidal distribution match these distributions?

Question 2.6 Calculate the Ross index χ_ℓ and the mean leaf angle $\bar{\Theta}_\ell$ for the five leaf angle distributions in Table 2.1 and for the ellipsoidal distribution with $x = 0.1, 0.5, 1, 4$, and 10. Order the leaf angle distributions from smallest to largest mean leaf angle. How does χ_ℓ vary in relation to $\bar{\Theta}_\ell$?

2.3 | Leaf Mass per Area

The biomass needed to support a unit of leaf area, termed leaf mass per area and commonly abbreviated as LMA (given here as m_a; kg m^{-2}), is an important leaf trait. Its inverse is the leaf area per unit mass and is known as specific leaf area. Species with low m_a produce a unit leaf area from a small amount of carbon. Leaves are thin and have a large area per unit mass, such as in broadleaved species. Species with high m_a require large amounts of carbon to produce a unit leaf area. Leaves are thick and have low area per unit mass,

Table 2.2	Leaf mass per area (LMA) for 257 species of forbs, shrubs, broadleaf trees, and needleleaf trees

Functional group	LMA (g m^{-2})
Forb	51
Broadleaf shrub	
Deciduous	71
Evergreen	141
Broadleaf tree	
Deciduous	73
Evergreen	112
Needleleaf tree	
Deciduous	100
Evergreen	263

Source: From Reich et al. (1998).

such as in conifer needles. Leaf mass per area describes carbon investment in photosynthesizing leaf area. It also relates to photosynthesis directly because maximum leaf photosynthetic capacity decreases and leaf longevity increases with higher m_a (Wright et al. 2004).

Leaf mass per area varies among types of plants (Table 2.2). Broadleaf species have smaller m_a than do needleleaf species. Deciduous species have smaller m_a compared with evergreen species. However, m_a is strongly influenced by growth conditions such as irradiance, CO$_2$ concentration, temperature, soil moisture, and nutrient availability and varies with leaf age (Poorter et al. 2009). In trees, m_a decreases with greater depth from the top of the canopy in relation to the decrease in light (Niinemets and Tenhunen 1997; Meir et al. 2002; Lloyd et al. 2010; Niinemets et al. 2015; Keenan and Niinemets 2016).

Leaf mass per area is calculated by dividing leaf dry mass by the projected leaf area and can be equivalently obtained as the product of leaf density (dry mass per volume) and leaf thickness (Witkowski and Lamont 1991) so that

$$m_a = \frac{\text{mass}}{\text{area}} = \frac{\text{mass}}{\text{area} \times \text{thickness}} \times \text{thickness}$$

$$= \frac{\text{mass}}{\text{volume}} \times \text{thickness} = \text{density} \times \text{thickness}.$$

(2.18)

This expression is for a thin, flat leaf with width W, length L_n, and thickness T, where the volume is $V = W \cdot L_n \cdot T$, the projected area is $A_P = W \cdot L_n$, and the thickness is $T = V/A_P$. For a needle shaped like a cylinder, where the needle diameter is the thickness, the volume is $L_n(\pi/4)T^2$ and the projected area is $T \cdot L_n$. In this case, $V/A_P = (\pi/4)T$ and

$$m_a = (\pi/4) \times \text{density} \times \text{thickness}. \qquad (2.19)$$

A more general expression for m_a replaces thickness with the ratio of volume to area (Poorter et al. 2009) whereby

$$m_a = \text{density} \times \frac{\text{volume}}{\text{area}}. \qquad (2.20)$$

The ratio of volume to area is the thickness T for a broadleaf and is equal to $(\pi/4)T$ for a cylindrical needle with diameter T. In this sense, m_a is the product of leaf density and thickness in flat broadleaves but is the product of density and volume-to-area ratio in needles.

Question 2.7 Witkowski and Lamont (1991) measured the foliage of two sclerophyllous shrubs – the broadleaf *Hakea lasianthoides* and the needleleaf *Hakea psilorrhyncha*. A sample broadleaf with a projected area of 484 mm^2 and thickness 0.332 mm has a dry mass of 50 mg. A cylindrical-shaped needle with diameter 0.66 mm and length 18.5 cm has a dry mass of 79 mg. Calculate leaf density and leaf mass per area for (a) the broadleaf and (b) the needleleaf. (c) Redo the calculations for the needleleaf to obtain the leaf mass per area on the basis of total leaf area.

2.4 | Canopy Nitrogen Profile

The amount of nitrogen in a plant canopy determines carbon uptake during photosynthesis. Nitrogen is an essential component of Rubisco and chlorophyll used in photosynthesis. Consequently, the photosynthetic capacity of a leaf scales linearly with the amount of nitrogen in the leaf (per unit leaf area), and photosynthetic rates increase with higher nitrogen concentrations (Field and Mooney 1986; Evans 1989; Wright et al. 2004). The vertical distribution of nitrogen in the canopy is important for modeling plant productivity. The investment of resources in photosynthetic capacity is expensive because it requires large amounts of nitrogen. Theory shows that canopy photosynthesis is maximized by allocating nitrogen such that leaves that receive high levels of light also have high photosynthetic capacity (i.e., high nitrogen concentration) to utilize that light (Chapter 15). Consequently, plant canopies have a nonuniform vertical distribution of nitrogen. A vertical profile in leaf nitrogen concentration per unit leaf area n_a develops in parallel with the profile of light in dense canopies so that leaf photosynthetic capacity decreases from top to bottom of the canopy. The original theory held that leaf nitrogen scales with the time mean profile of photosynthetically active radiation, but it is now recognized that the nitrogen gradient is not as steep as the light gradient.

Vertical gradients in n_a have been observed in many forest communities including: temperate broadleaf deciduous forest (Ellsworth and Reich 1993; Niinemets and Tenhunen 1997; Wilson et al. 2000; Meir et al. 2002; Lloyd et al. 2010); temperate broadleaf evergreen forest (Hollinger 1989, 1996); boreal conifer forests (Dang et al. 1997b); and tropical rainforest (Carswell et al. 2000; Domingues et al. 2005; Lloyd et al. 2010). Leaf nitrogen n_a (kg N m^{-2}) is the product of a leaf's mass-based foliage nitrogen concentration n_m (kg N kg^{-1} C) and its leaf mass per area m_a (here with the units kg C m^{-2}; carbon is typically 50% of dry biomass so that leaf C is one-half of leaf dry mass). In many forests, n_m is constant throughout the canopy, but n_a decreases because m_a decreases with depth in the canopy. The gradient in leaf nitrogen is generally found to be shallower than the light gradient (Hollinger 1996; Carswell et al. 2000; Meir et al. 2002; Niinemets 2007; Lloyd et al. 2010; Niinemets et al. 2015).

Vertical gradients in photosynthetic capacity and n_a in relation to the light profile have been found in many herbaceous plant communities including tall grass prairie (Schimel et al. 1991), tropical savanna (Anten et al. 1998b), wet meadow (Hirose and Werger 1994), and other plant assemblages (Hirose and Werger 1987; Hirose et al. 1988, 1989; Werger and Hirose 1991; Schieving et al. 1992; Anten et al. 1998a). A decline in m_a with low light has been observed in herbaceous plants, but in many studies the decline in

n_a within a canopy results from gradients of n_m, m_a, or both. Similar to forests, leaf nitrogen has a shallower gradient than the light profile (Hirose and Werger 1987; Werger and Hirose 1991; Schieving et al. 1992; Anten et al. 1995, 1998a,b). Vertical gradients of leaf nitrogen are also found in agricultural crops (Lemaire et al. 1991; Evans 1993a,b; Shiraiwa and Sinclair 1993; Anten et al. 1995; Connor et al. 1995; Grindlay 1997; Drouet and Bonhomme 1999, 2004; Dreccer et al. 2000a,b; Gastal and Lemaire 2002).

The decline in leaf nitrogen can be described by an exponential relationship with cumulative leaf area index x from the top of the canopy whereby the amount of leaf nitrogen per unit leaf area at the depth x is

$$n_a(x) = n_{a0}e^{-K_n x}. \tag{2.21}$$

Here, n_{a0} is the leaf nitrogen at the top of the canopy, and K_n is a parameter that describes the exponential decline in nitrogen with greater depth in the canopy. Some canopies have a shallow vertical profile of nitrogen; others have a steeper profile (Figure 2.8). The total nitrogen in the canopy, n_c (per unit ground area), is obtained by integrating the leaf nitrogen profile over the canopy whereby

$$n_c = \int_0^L n_a(x)\,dx = \frac{n_{a0}\left(1 - e^{-K_n L}\right)}{K_n}. \tag{2.22}$$

For a canopy with a leaf area index of 5 m² m⁻², a shallow profile with $K_n = 0.10$ contains almost twice as much nitrogen ($n_c = 3.93n_{a0}$) as does a steep profile with $K_n = 0.45$ ($n_c = 1.99n_{a0}$) for the same n_{a0}.

Question 2.8 Calculate the mass of nitrogen n_c in the following canopies:

n_{a0} (g N m⁻²)	K_n	L (m² m⁻²)
1.5	0.1	5
1.5	0.3	5
2.2	0.4	4
2.2	0.215	4
2.2	0.281	5

Question 2.9 Anten et al. (1995) calculated leaf nitrogen using the equation $n_a(x) = (n_{a0} - n_b)\exp(-K_n x) + n_b$ in which n_b is a base nitrogen content not associated with photosynthesis. Derive the equation for canopy nitrogen n_c.

2.5 | Root Profile

Fine roots (≤ 2 mm diameter) govern plant water uptake, and their vertical profile in soil is needed

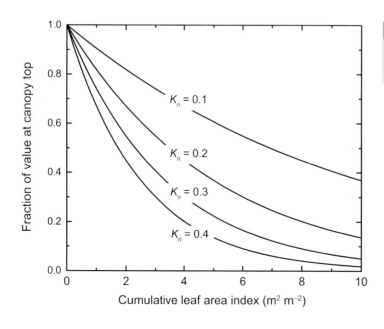

Figure 2.8 Canopy profiles of area-based leaf nitrogen in relation to cumulative leaf area index. Shown are exponential profiles for $n_a(x)/n_{a0}$ using four representative values of K_n.

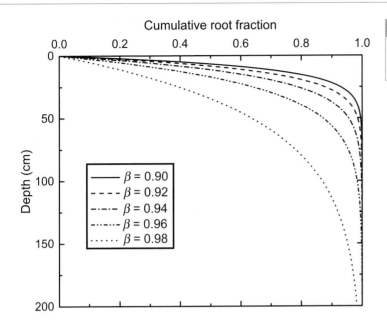

Cumulative root fraction

Depth (cm)

- —— $\beta = 0.90$
- - - - $\beta = 0.92$
- -·-· $\beta = 0.94$
- -··-·· $\beta = 0.96$
- ······ $\beta = 0.98$

Figure 2.9 Relative root abundance profiles commonly found in terrestrial ecosystems. Graphs show cumulative root distribution in relation to soil depth for five values of β.

to model plant hydraulics for transpiration. The root system is described by biomass M_r (kg m^{-2}) and its distribution with depth in soil. A general equation from Jackson et al. (1996) to describe the cumulative root fraction (0–1) at depth z (m) is

$$F(z) = 1 - \beta^{100z}. \tag{2.23}$$

Low values of β give a greater proportion of roots near the surface (Figure 2.9). Typical values are $\beta = 0.91$–0.98 (Table 2.3); values of 0.91 and 0.98 mean that 94% and 45% of root biomass is in the top 30 cm of soil, respectively. Tundra, boreal forest, temperate grassland, and shrubland have shallow root profiles, with about 80% or more of fine root biomass in the top 30 cm of soil (Table 2.3).

Root length density L_r (root length per unit volume of soil, m m^{-3}) is used in plant hydraulics models to calculate a soil-to-root hydraulic conductance (Chapter 13). It is obtained from the profile of root biomass as

$$L_{r,i} = r_l \left(\frac{M_r \Delta F_i}{\Delta z_i} \right). \tag{2.24}$$

The term in parentheses is the root biomass density (root biomass per unit soil volume, kg m^{-3}) in soil layer i that is Δz_i (m) thick and that contains ΔF_i of the total root biomass. $L_{r,i}\Delta z_i$ is the root length per unit area of soil (root length index, m m^{-2}) in the layer. Specific root length r_l is the length of root per

unit mass of root (m kg^{-1}) and is determined from the relationship:

$$r_l = \frac{1}{\rho_r \pi r_r^2}, \tag{2.25}$$

where ρ_r is the root tissue density (mass per unit volume of root tissue, kg m^{-3}) and πr_r^2 is the root cross-sectional area calculated from mean fine root radius r_r. Jackson et al. (1997) reviewed the properties of fine roots (Table 2.3). The mean fine root radius of trees is $r_r = 0.29$ mm and the specific root length is $r_l = 12.2$ m g^{-1} so that $\rho_r = 0.31$ g cm^{-3}. For shrubs, $r_r = 0.22$ mm, $r_l = 30$ m g^{-1}, and $\rho_r = 0.22$ g cm^{-3}. For grasses, $r_r = 0.11$ mm, $r_l = 118$ m g^{-1}, and $\rho_r = 0.22$ g cm^{-3}.

Although most fine root biomass is located in the upper 50–100 cm of soil, roots can extend much deeper (Canadell et al. 1996). Tundra plants are the shallowest rooted, extending to a depth of 50 cm on average. Desert and tropical savanna plants have roots 10–15 m deep. Deep-rooted plants are common in woody and herbaceous species across most terrestrial biomes (Table 2.3). On average, roots of trees extend 7.0 m deep, shrubs 5.1 m, and herbaceous perennials 2.6 m. These deep roots are a small portion of plant biomass, but are hydrologically important. They provide access to water reserves deep in the ground and help protect against intermittent droughts.

Table 2.3 Characteristics of live fine roots and maximum rooting depth specified by biome

Biome	Fine root biomass (g m^{-2})	Fine root length index (km m^{-2})	Fine root area index (m^2 m^{-2})	β	Percentage of fine root biomass in upper 30 cm	Maximum rooting depth (m)
Boreal forest	230	2.6	4.6	0.943	83	2.0
Temperate coniferous forest	500	6.1	11.0	0.980	45	3.9
Temperate deciduous forest	440	5.4	9.8	0.967	63	2.9
Tropical deciduous forest	280	3.5	6.3	0.982	42	3.7
Tropical evergreen forest	330	4.1	7.4	0.972	57	7.3
Shrubland	280	8.4	11.6	0.950	79	5.2
Temperate grassland	950	112	79.1	0.943	83	2.6
Tropical grassland & savanna	510	60.4	42.5	0.972	57	15.0
Desert	130	4.0	5.5	0.970	60	9.5
Tundra	340	7.4	5.2	0.909	94	0.5

Source: From Jackson et al. (1997) with maximum rooting depth from Canadell et al. (1996).

Question 2.10 A soil has the following profile of volumetric soil water θ_i at depth z_i. Calculate the average soil moisture content (weighted by the abundance of roots) of the soil column for a shallow root distribution ($\beta = 0.90$) and a deep root distribution ($\beta = 0.97$).

z_i (cm)	Δz_i (cm)	θ_i	z_i (m)	Δz_i (m)	θ_i
2.5	5	0.03	80	20	0.30
7.5	5	0.07	105	30	0.34
15	10	0.15	140	40	0.36
25	10	0.22	180	40	0.40
40	20	0.24	225	50	0.42
60	20	0.26			

2.6 | Ecosystem Structure

Vegetation can be described by the species present in a stand (i.e., community composition), the size of individual plants, and the amount of biomass in the stand. Biogeochemical models ignore the demographic characteristics of ecosystems and instead simulate the aggregate carbon pools in a stand without representing individual plants or species. Plant carbon is typically given by foliage, stem, and root pools. In contrast, individual plant models simulate vegetation in terms of the species present and the size of the individuals. Most vegetation stands show a pronounced hierarchy of plant sizes in which there are a few large, dominant individuals and many smaller, subordinate plants. Such an uneven size distribution is commonly seen in forests and is apparent in stem diameter. Other forest stands may have a more symmetric and narrow distribution of stem diameters, indicative of an even-age forest. The frequency distribution of stem diameter is an important descriptor of forest stands because stem diameter relates to biomass and height.

The three-parameter Weibull probability density function describes the frequency distribution of tree diameters (Bailey and Dell 1973). In this distribution, the probability density of a random variable x is

$$f(x) = \frac{c}{b}\left(\frac{x-a}{b}\right)^{c-1}\exp\left[-\left(\frac{x-a}{b}\right)^c\right] \qquad (2.26)$$

for $x > a$ and otherwise $f(x) = 0$. The parameter a is referred to as the location parameter, b is the scale parameter, and c is the shape parameter. The cumulative distribution is

$$F(x) = 1 - \exp\left[-\left(\frac{x-a}{b}\right)^c\right] \qquad (2.27)$$

and gives the probability of a diameter less than x. Figure 2.10 shows common diameter distributions found in forests.

Individual trees can be characterized in terms of diameter, height, and biomass in living (leaves, fine roots, sapwood) and structural (stem wood) components. These characteristics vary in relation to stem diameter D, as described by allometric relationships.

For example, biomass increases with larger stem diameter. A common relationship is

$$M_i = a_i D^{b_i}. \qquad (2.28)$$

The parameter a_i is known as the allometric constant; and b_i is the scaling exponent. These differ with the biomass compartment. Tree height follows a similar functional relationship with diameter. Ter-Mikaelian and Korzukhin (1997) summarized allometric relationships for many tree species. The total biomass of a forest stand is obtained by measuring stem diameter and applying species-specific allometric relationships to each individual found in the stand. Figure 2.11 illustrates these relationships for four common tree species in northern hardwood forests of North America using the parameters in Table 2.4. These allometric relationships are used in individual tree or cohort-based models of forest dynamics (Botkin et al. 1972; Shugart and West 1977; Shugart 1984; Pacala et al. 1996; Moorcroft et al. 2001; Fisher et al. 2015).

Terrestrial ecosystems contain large amounts of carbon belowground in soil. Soil carbon has a distinct vertical distribution in the first meter (Figure 2.12). Forests contain 50% of this carbon in the top 20 cm; grasslands have slightly less (42%). Additional carbon is found at depths below 1 m. In forests, an amount of soil organic carbon equal to 56% of that found in

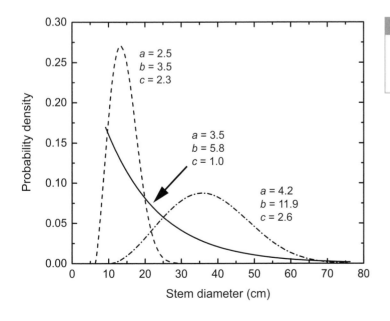

Figure 2.10 Weibull distribution of tree diameters in three forest stands illustrating common distributions. Values for the fitted parameters a, b, and c are from Bailey and Dell (1973) and given for diameter in inches.

a = 2.5
b = 3.5
c = 2.3

a = 3.5
b = 5.8
c = 1.0

a = 4.2
b = 11.9
c = 2.6

the first meter is stored in the second and third meters of soil. In grasslands, this deep soil carbon is 43% of that of the first meter. Most biogeochemical models follow a paradigm in which soil carbon is represented as aggregate pools of varying quality without specification of depth, but some models resolve the soil carbon profile (Chapter 18).

Question 2.11 The following table from Whittaker et al. (1974) gives the size distribution of sugar maple, yellow birch, and beech trees at the Hubbard Brook Experimental forest. Use the allometric relationships in Table 2.4 to calculate the total aboveground biomass in g m^{-2}.

Diameter (cm)	Density (ha^{-1})		
	Sugar maple	Yellow birch	Beech
2.5	180	67	373
7.5	122	55	154
12.5	72	38	65
17.5	45	31	27
22.5	35	21	19
27.5	23	16	12
32.5	9	9	8
37.5	3	6	0
42.5	2	0	6
47.5	1	1	2
52.5	0	0	2

Table 2.4 Coefficients for allometric equations of the form $y = aD^b$ for stem diameter D in cm

Species	Aboveground biomass (kg)		Stem biomass (kg)		Height (m)	
	a	b	a	b	a	b
Sugar maple (Acer saccharum)	0.1641	2.4209	0.1224	2.3718	3.4435	0.5317
Yellow birch (Betula alleghaniensis)	0.1684	2.4150	0.1385	2.2683	3.2248	0.5273
Beech (Fagus grandifolia)	0.1957	2.3916	0.1067	2.3981	2.8255	0.5753
Red spruce (Picea rubens)	0.2066	2.1830	0.0979	2.2046	1.3059	0.7408

Source: From Whittaker et al. (1974).

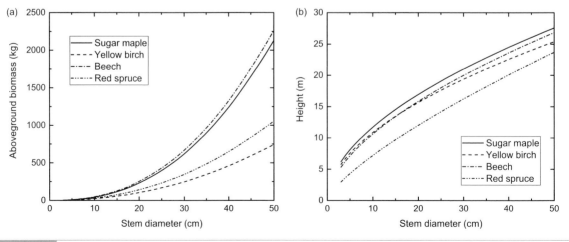

Figure 2.11 Relationships between stem diameter and (a) aboveground biomass and (b) height for sugar maple, yellow birch, beech, and red spruce trees at the Hubbard Brook Experimental Forest, New Hampshire using the allometric relationships in Table 2.4.

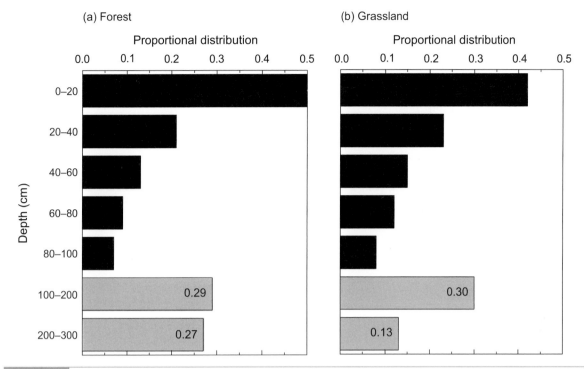

(a) Forest   (b) Grassland

Figure 2.12 Profiles of soil organic carbon distribution for (a) forest and (b) grassland. Black bars show the proportional distribution of soil organic carbon in the first meter of soil and sum to one. Gray bars show the additional carbon at depths of 100–200 cm and 200–300 cm, relative to the first meter. Data from Jobbágy and Jackson (2000) and reproduced from Bonan (2016).

2.7 | Supplemental Programs

2.1 Leaf Area Density: This program uses the beta distribution probability density function to calculate the leaf area density profile (2.4). It calculates and graphs the profiles in Figure 2.3 for different values of p and q with L/h_c specified as input.

2.2 Leaf Angle Distribution: This program uses the beta distribution to calculate leaf angle distribution (2.8). It uses the mean and standard deviations for the five leaf angle distributions in Table 2.1 and compares the numerical solution obtained with nine leaf angle classes (10° increments) to the analytical solutions.

3

Fundamentals of Energy and Mass Transfer

Chapter Overview

Energy and materials cycle throughout the Earth system. Heat in the atmosphere is exchanged by radiation, conduction, and convection. These fluxes determine the balance of energy gained, lost, or stored in a system and relate to temperature. A system that gains energy increases in temperature; a system that loses energy decreases in temperature. Convection similarly transfers materials in the movement of air. This transfer of materials is measured by the amount (mass or moles) and is commonly seen in gas diffusion (Fick's law). This chapter introduces the fundamental scientific concepts of energy and mass transfer needed to understand biosphere–atmosphere coupling. Table 3.1 summarizes the key flux equations derived in this chapter.

3.1 | Heat Storage

Heat and materials flow through the biosphere–atmosphere system. These exchanges are measured by the rate of flow. The transfer of materials is typically given by mass per unit time ($kg\ s^{-1}$) or equivalently by moles per unit time ($mol\ s^{-1}$). Flux density measures the flow of material per unit surface area per unit time ($kg\ m^{-2}\ s^{-1}$ or $mol\ m^{-2}\ s^{-1}$). The equivalent term for energy is $J\ m^{-2}\ s^{-1} = W\ m^{-2}$.

While flux density is the more precise term, it is commonly referred to simply as flux.

The first law of thermodynamics describes the conservation of energy. It states that in a closed system, energy can change from one form to another, but it cannot be created or destroyed; the total amount of energy in the system is conserved. Consider, for example, a system in which energy input is balanced by energy output and change in stored energy so that

$$\text{energy input} = \text{energy output} + \Delta Q, \qquad (3.1)$$

with ΔQ being the heat gain or loss in the volume. A volume with a unit of area and Δz (m) thickness stores the energy:

$$\Delta Q = c_v \left(\frac{\Delta T}{\Delta t} \right) \Delta z, \qquad (3.2)$$

where ΔT is the change in temperature (K) over the time period Δt (s) and c_v is heat capacity ($J\ m^{-3}\ K^{-1}$). Heat capacity is the amount of energy needed to raise the temperature of a unit volume of material by one degree. If the system loses the same amount of energy that it gains, there is no change in storage ($\Delta Q = 0$) and temperature remains constant ($\Delta T = 0$). If the system gains more energy than it loses, the excess energy is stored in the system as thermal energy, raising the temperature. Conversely, temperature decreases if the system loses more energy than it gains. Radiation, conduction, and convection are three means of energy transfer in the Earth system.

3.2 | Radiation

All materials with temperature greater than absolute zero (0 K, or −273.15°C) emit electromagnetic radiation. Electromagnetic radiation is called radiant energy and has the units joules (J). Radiant energy emitted or received per unit time is called the radiant flux and has the units of joules per second, or watts (1 W = 1 J s^{-1}). Radiant flux density is the radiant flux per unit surface area (W m^{-2}). Irradiance is the radiant flux density incident on a surface; emittance is that emitted by a surface.

The radiant flux density emitted by a body varies with its temperature and in relation to wavelength as defined by Planck's law (Figure 3.1). Objects with a higher temperature emit radiation at a greater rate than objects with a lower temperature. Additionally, the spectral distribution of radiant energy depends on the temperature of the object. The Sun, with a temperature of 6000 K, emits radiation in short wavelengths between 0.2 to 4 μm. Radiation in these wavelengths is known as solar, or shortwave, radiation. Solar radiation is divided into ultraviolet radiation with a wavelength of less than about 0.4 μm (containing 10% of the Sun's energy), visible radiation between about 0.4 and 0.7 μm (40% of the Sun's energy), and near-infrared radiation at wavelengths of greater than about 0.7 μm (50% of the Sun's energy). The Sun has peak emission at about 0.5 μm. Terrestrial objects, with an effective temperature of about 288 K, emit less radiation than the Sun and in longer wavelengths of 3–100 μm. Radiation at these wavelengths is called infrared, or longwave, radiation. Terrestrial objects have peak emission at about 10 μm.

The Stefan–Boltzmann law relates the radiant flux density, integrated over all wavelengths, emitted by an object to its temperature T (K) as

Table 3.1	Key equations derived in this chapter	
Process	Equation	Flux
Heat storage	(3.2)	$\Delta Q = c_v (\Delta T / \Delta t) \Delta z$
Longwave emittance for a gray body	(3.3)	$L^{\uparrow} = \varepsilon \sigma T^4$
Net radiation	(3.5)	$R_n = (1 - \rho) S^{\downarrow} + \varepsilon L^{\downarrow} - \varepsilon \sigma T^4$
Heat conduction	(3.7)	$F = -\kappa (\Delta T / \Delta z)$
Gas flux	(3.38)	$F_j = \Delta c_j g_j$
Evaporation	(3.39)	$E = (\Delta e / P) g_w$
Sensible heat flux	(3.43)	$H = c_p \Delta T g_h$

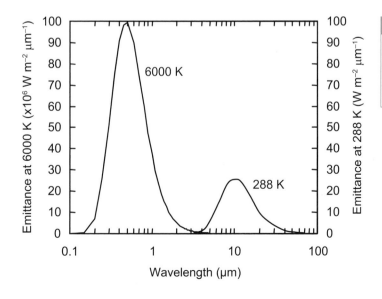

$$L^{\uparrow} = \varepsilon\sigma T^4, \qquad (3.3)$$

in which $L^{\uparrow}$ is emittance (W m^{-2}), $\sigma = 5.67 \times 10^{-8}$ W m^{-2} K^{-4} is the Stefan–Boltzmann constant, and ε is broadband emissivity. The Stefan–Boltzmann law was originally formulated for a blackbody. A blackbody is an object that is a perfect absorber of radiation at all wavelengths and that emits the maximum possible energy at all wavelengths for a given temperature. Most objects are not blackbodies and emit less radiation. Instead, blackbody emittance is reduced by the object's emissivity ε. Emissivity is defined as the ratio of the actual emittance to the blackbody emittance. Most objects have a broadband emissivity of 0.95–0.98 when integrated over all wavelengths.

An object absorbs radiant energy from the Sun. The radiation incident on an object is absorbed, reflected, or transmitted through the material so that

$$a_{\Lambda} + \rho_{\Lambda} + \tau_{\Lambda} = 1. \qquad (3.4)$$

In this equation, a_{Λ} is the fraction of incident radiation that is absorbed (absorptivity, or absorptance); ρ_{Λ} is the fraction of incident radiation that is reflected (reflectivity, or reflectance); and τ_{Λ} is the fraction of incident radiation that is transmitted (transmissivity, or transmittance). These optical properties vary with wavelength, denoted by Λ. Green leaves, for example, typically absorb 85% of the solar radiation in the visible waveband but absorb less than 50% of the radiation in the near-infrared waveband. Objects that are sufficiently thick are opaque ($\tau_{\Lambda} = 0$), and absorptivity is $a_{\Lambda} = 1 - \rho_{\Lambda}$. The solar radiation absorbed by an object is commonly calculated from this relationship. Terrestrial bodies also absorb longwave radiation, and (3.4) similarly describes absorptance, reflectance, and transmittance of longwave radiation. Additionally, Kirchhoff's law states that the absorptivity at a specified wavelength equals its emissivity at that wavelength – i.e., $a_{\Lambda} = \varepsilon_{\Lambda}$. A blackbody is a perfect absorber and emitter of radiation so that $a_{\Lambda} = \varepsilon_{\Lambda} = 1$ and $\rho_{\Lambda} = \tau_{\Lambda} = 0$. An opaque gray body ($\tau_{\Lambda} = 0$) has a reflectance $\rho_{\Lambda} = 1 - a_{\Lambda} = 1 - \varepsilon_{\Lambda}$.

Figure 3.2 illustrates the radiative balance of an opaque gray body receiving solar and longwave radiation from the atmosphere. Of the solar radiation incident on the body, an amount equal to $\rho S^{\downarrow}$ is

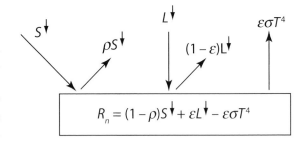

Figure 3.2 Radiative balance of an opaque gray body receiving downwelling solar $S^{\downarrow}$ and longwave $L^{\downarrow}$ radiation.

reflected; the net solar radiation absorbed is $(1 - \rho)S^{\downarrow}$. The longwave radiative transfer between the body and its surrounding environment is the balance between the energy radiated by the body and that absorbed from the environment. The longwave radiation incident on the body is $L^{\downarrow}$. An amount equal to $(1 - \varepsilon)L^{\downarrow}$ is reflected, and the body absorbs $\varepsilon L^{\downarrow}$. The body emits longwave radiation in relation to its temperature. The net radiation R_n absorbed is

$$R_n = (1 - \rho)S^{\downarrow} + \varepsilon L^{\downarrow} - \varepsilon\sigma T^4. \qquad (3.5)$$

Here, ρ is the reflectance integrated over all wavelengths (commonly called surface albedo), and ε is the broadband emissivity.

Electromagnetic radiation transfers energy in discrete units called quanta or photons. For photosynthesis, the number of photons, not energy, is important. The energy (J) of a photon with wavelength Λ (m) is hc/Λ, in which $h = 6.626 \times 10^{-34}$ J s is Planck's constant and $c = 3 \times 10^8$ m s^{-1} is the speed of light. The longer the wavelength is, the less the energy of the photon. The energy of a photon of blue light with wavelength 0.450 μm is 4.42×10^{-19} J. A photon of red light (0.680 μm) has 2.92×10^{-19} J. A photon of light with blue wavelength has more energy than a photon of light with red wavelength, but both have the same effect on photosynthesis. Only radiation with wavelengths between 0.4 and 0.7 μm, known as photosynthetically active radiation, is used during photosynthesis. The energy of a mole of photons is obtained by multiplying the energy per photon by Avogadro's number (6.022×10^{23} mol^{-1}). A mole of photons with wavelength 0.55 μm has the energy:

$$\frac{\left(6.626\times10^{-34}\,\mathrm{Js}\right)\left(3\times10^{8}\,\mathrm{ms}^{-1}\right)}{0.55\times10^{-6}\,\mathrm{m}}\left(6.022\times10^{23}\,\mathrm{mol}^{-1}\right)$$

$$= 0.218\times10^{6}\,\mathrm{Jmol}^{-1}.$$

Therefore, photosynthetically active radiation (with an average wavelength of 0.55 μm) is converted from W m^{-2} (J s^{-1} m^{-2}) to photosynthetic photon flux density with units μmol photon m^{-2} s^{-1} using the factor 4.6 μmol J^{-1}.

Question 3.1 On a winter night, a person sits next to a window with an effective temperature of 6°C. When the curtain is closed, the effective temperature is 18°C. How does the radiative balance of the person change when the curtain is closed? Assume an emissivity of one. Under which condition does the person feel warmer?

Question 3.2 Calculate the change in temperature over one hour for a volume of soil 50 cm thick that receives incoming solar and longwave radiation equal to 800 W m^{-2} and 400 W m^{-2}, respectively. The soil temperature is 30°C, emissivity is 0.96, reflectance is 0.20, and heat capacity is 2 MJ m^{-3} K^{-1}. Compare the temperature change to that for water, with heat capacity 4.18 MJ m^{-3} K^{-1}. Which body has the greatest change in temperature?

Question 3.3 The downwelling solar radiation is $S^{\downarrow} = 400$ W m^{-2}. What is the photosynthetic photon flux density?

3.3 | Conduction

Conduction is the transfer of heat within a material or between materials arising from molecular vibration without any motion of the material itself. Fourier's law gives the rate of heat transfer by conduction F (W m^{-2}) as:

$$F = -\kappa\frac{\partial T}{\partial z},\tag{3.6}$$

where κ is the thermal conductivity of the material (W m^{-1} K^{-1}) and $\partial T/\partial z$ is the temperature gradient (K m^{-1}) in the z-direction. The negative sign denotes

that the flux is positive for a negative temperature gradient. In numerical form:

$$F = -\kappa\frac{\Delta T}{\Delta z},\tag{3.7}$$

with ΔT the temperature difference between two points separated by the distance Δz. This is the form of the heat conduction equation commonly used in models. An equivalent equation uses a conductance defined by $\kappa/\Delta z$, with units W m^{-2} K^{-1}. Conductivity is a property of the material, whereas conductance additionally depends on the length Δz. The type of material affects the rate of heat transfer by conduction. Air is a very poor conductor of heat (0.02 W m^{-1} K^{-1}). Water has a higher thermal conductivity (0.57 W m^{-1} K^{-1}). Thermal conductivity for soils typically ranges from 0.3 to 2 W m^{-1} K^{-1}, depending on mineral composition, the amount of organic material, and soil water.

Question 3.4 Calculate heat loss by conduction between a body and overlying air if the air is 5°C colder than the body over a distance of 10 mm. How much heat is lost in water that is 5°C colder than body temperature over a distance of 10 mm?

3.4 | Molecular Diffusion

Diffusion is the process of mixing fluid properties by molecular motions. In still air, gases such as H_2O and CO_2 flow along a gradient from high to low concentration. Fick's law describes the rate of mass transfer through diffusion. The vertical transport and mixing of heat and mass through the movement of air is a form of diffusion termed convection and is similarly described by Fick's law but with a diffusion coefficient, or conductance, that reflects turbulent mixing rather than molecular diffusion. The meteorological community generally describes diffusion in terms of mass fluxes with units kg m^{-2} s^{-1} and conductance with units m s^{-1}, or more commonly resistance (the inverse of conductance) with units s m^{-1}. With molar units, flux and conductance have the same units (mol m^{-2} s^{-1}), as advocated by Campbell and Norman (1998). The appendix to this chapter gives the relationship between these different units.

Fick's law relates the diffusive flux to the concentration gradient times a proportionality constant known as the diffusion coefficient. The flux per unit area F_j (mol m^{-2} s^{-1}) of a gas with a concentration in a mixture c_j (mol mol^{-1}) is

$$F_j = -\rho_m D_j \frac{\partial c_j}{\partial z}, \tag{3.8}$$

in which D_j is the molecular diffusion coefficient (m^2 s^{-1}), ρ_m is the number of moles per unit volume (referred to as molar density, mol m^{-3}), and $\partial c_j / \partial z$ (m^{-1}) is the concentration gradient in the z-direction. The negative sign denotes that the flux is positive in the direction of decreasing concentration. Integrating between heights z_1 and z_2:

$$-F_j \int_{z_1}^{z_2} \frac{\partial z}{\rho_m D_j} = \int_{z_1}^{z_2} \partial c_j \tag{3.9}$$

and

$$F_j = -\rho_m D_j \left(\frac{c_{j,2} - c_{j,1}}{z_2 - z_1} \right), \tag{3.10}$$

with $c_{j,1}$ and $c_{j,2}$ the concentration at z_1 and z_2, respectively. More generally, the diffusive flux is expressed as a concentration difference multiplied by a conductance g_j (mol m^{-2} s^{-1}), or divided by a resistance r_j (s m^2 mol^{-1}), so that

$$F_j = \Delta c_j g_j = \frac{\Delta c_j}{r_j}. \tag{3.11}$$

In this form, the concentration difference Δc_j is specified from high to low such that the flux is positive in the direction of decreasing concentration. Comparison with (3.10) shows that $g_j = \rho_m D_j / \Delta z$. Molar density converts between conductance in m s^{-1} ($D_j / \Delta z$) and mol m^{-2} s^{-1}. The precise definition of conductance is that it is integrated over the length Δz. This is seen in the integration of F_j with respect to z given by (3.9) in which

$$g_j = \frac{1}{r_j} = \left(\int_{z_1}^{z_2} \frac{\partial z}{\rho_m D_j} \right)^{-1} = \rho_m \frac{D_j}{\Delta z}. \tag{3.12}$$

The comparable equation for heat transfer H (W m^{-2}) replaces c_j with $c_p T$, and

$$H = -\rho_m c_p D_h \frac{\partial T}{\partial z}, \tag{3.13}$$

with c_p the molar specific heat of moist air at constant pressure (J mol^{-1} K^{-1}) and D_h the molecular diffusion coefficient for heat (m^2 s^{-1}), commonly termed thermal diffusivity. The integrated flux equation is

$$H = c_p \Delta T g_h, \tag{3.14}$$

where $g_h = \rho_m D_h / \Delta z$ is the conductance for heat (mol m^{-2} s^{-1}).

Question 3.5 The diffusivity for heat is 20.8×10^{-6} m^2 s^{-1} at 15°C and at sea level. The corresponding molar density is 42.3 mol m^{-3}. Calculate the conductance across 1 mm. What is the corresponding conductivity? Compare this with values for H$_2$O (24.0×10^{-6} m^2 s^{-1}) and CO$_2$ (15.2×10^{-6} m^2 s^{-1}). What are the resistances in the units s m^{-1}?

3.5 | Gas Relationships

For the diffusive flux given by (3.11), the gas concentration c_j is defined by the number of moles as a fraction of the total mixture. The ideal gas law describes this and other measures of concentration. The ideal gas law relates the volume V (m^3) occupied by n moles of a gas at pressure P (Pa = N m^{-2} = J m^{-3}) and temperature T (K) as

$$PV = n\Re T, \tag{3.15}$$

with the universal gas constant $\Re = 8.314$ J K^{-1} mol^{-1}. One mole of gas (e.g., air) occupies a volume $V = 0.0236$ m^3 for a standard atmosphere at sea level. (The standard atmosphere defines sea level as 15°C and 1013.25 hPa, which is denoted STP for standard temperature and pressure). The inverse (the number of moles per unit volume) is the molar density:

$$\rho_m = \frac{n}{V} = \frac{P}{\Re T}. \tag{3.16}$$

Molar density at a given pressure and temperature is constant for all gases and equals 42.3 mol m^{-3} at STP. The density of a gas at pressure P and temperature T is equal to its mass m_j (kg) divided by the volume occupied by the gas. An equivalent form of the ideal gas law is

Table 3.2 Chemical composition of the atmosphere

Gas	Chemical symbol	Molecular mass (g mol^{-1})	Density (kg m^{-3})	Mole fraction (mol mol^{-1})	Mass concentration (kg m^{-3})
Nitrogen	N_2	28.01	1.185	0.7808	0.925
Oxygen	O_2	32.00	1.353	0.2095	0.284
Argon	Ar	39.95	1.690	0.00934	0.016
Carbon dioxide	CO_2	44.01	1.861	390×10^{-6}	7.3×10^{-4}
Neon	Ne	20.18	0.854	18.18×10^{-6}	1.6×10^{-5}
Helium	He	4.00	0.169	5.24×10^{-6}	8.9×10^{-7}
Methane	CH_4	16.04	0.678	1.80×10^{-6}	1.2×10^{-6}
Krypton	Kr	83.80	3.544	1.14×10^{-6}	4.0×10^{-6}
Hydrogen	H_2	2.02	0.085	0.5×10^{-6}	4.3×10^{-8}
Nitrous oxide	N_2O	44.01	1.861	0.32×10^{-6}	6.0×10^{-7}
Xenon	Xe	131.30	5.553	0.087×10^{-6}	4.8×10^{-7}
Dry air	–	28.97	1.225	1.00	1.225

Note: Density is calculated for a standard atmosphere at sea level (1013.25 hPa and 15°C) using (3.17). The mole fractions of CO_2, CH_4, and N_2O are for 2011. Other constituents are from Hartmann (1994, p. 8). Mass concentration is calculated using (3.21).

$$\text{density} = \frac{m_j}{V} = \frac{nM_j}{V} = \frac{P}{\Re T}M_j = \rho_m M_j, \quad (3.17)$$

in which M_j is molecular mass (kg mol^{-1}). Table 3.2 gives the molecular mass and density for various atmospheric gases at STP.

Air is composed of N_2, O_2, Ar, H_2O, CO_2, and other gases in trace amounts. Dalton's law of partial pressures states that the total pressure of a mixture of gases equals the sum of the partial pressures of each gas in the mixture. Partial pressure is the pressure that a gas would exert if it alone occupied the same volume as the mixture at the same temperature. The total pressure of air is

$$P = P_{N2} + P_{O2} + P_{Ar} + P_{H2O} + P_{CO2} + \dots. \quad (3.18)$$

Each individual gas follows the ideal gas law and has the mass concentration:

$$\rho_j = \frac{m_j}{V} = \frac{n_j M_j}{V} = \frac{P_j}{\Re T}M_j, \quad (3.19)$$

with m_j the mass, n_j the number of moles, M_j the molecular mass, and P_j the partial pressure of the gas.

Volume changes with temperature and pressure so that changes in ρ_j can occur independently of changes in mass. An alternative measure of concentration, independent of volume, is the mole fraction c_j.

This is equal to the number of moles n_j as a fraction of the total number of moles n in the mixture. Mole fraction is also referred to as the volume fraction and relates to the partial pressure of the gas. If n moles of air at pressure P and temperature T occupy a volume given by $PV = n\Re T$, the partial pressure for an individual gas in air at the same volume and temperature is $P_j V = n_j \Re T$ and the mole fraction is

$$c_j = \frac{n_j}{n} = \frac{P_j}{P}. \quad (3.20)$$

An equivalent expression for ρ_j, obtained by substituting (3.20) into (3.19), is

$$\rho_j = \frac{P}{\Re T}M_j c_j = \rho_m M_j c_j. \quad (3.21)$$

The mass concentration of a gas at a given temperature and pressure is equal to its density $\rho_m M_j$ multiplied by the mole fraction. For example, CO_2 (molecular mass, 44.01 g mol^{-1}) with mole fraction $c_j = 390$ µmol mol^{-1} has partial pressure $P_j = 39.5$ Pa and mass concentration $\rho_j = 0.73$ g m^{-3} at STP.

For many applications, it is convenient to represent air separately as water vapor and dry air. The latter is a general term for all gases other than H_2O, and the sum of these two components is the moist air. A mixture of gases is described by

$$\sum_j \frac{\rho_j}{M_j} = \frac{P}{\Re T} = \rho_m \qquad (3.22)$$

or, equivalently,

$$\sum_j \rho_j = \frac{P}{\Re T} \bar{M} = \rho_m \bar{M}, \qquad (3.23)$$

with $\bar{M}$ the mean molecular mass of the mixture. The mixture of gases behaves like a single gas with a mean molecular mass for the mixture. Substituting (3.21) into (3.23), the mean molecular mass of a mixture of gases is

$$\bar{M} = \sum_j M_j c_j. \qquad (3.24)$$

That is, the mean molecular mass is the average of the molecular masses of the individual gases weighted by the volume fraction of each gas. The molecular mass of dry air M_a is 28.97 g mol^{-1}, and the density of dry air is

$$\rho_a = \frac{P}{\Re T} M_a = \rho_m M_a. \qquad (3.25)$$

At STP, ρ_a = 1.225 kg m^{-3}. Table 3.2 gives the mass concentration of various constituents of dry air at STP. The density of dry air is the sum of the mass concentration of each individual gas. Three gases (N_2, O_2, Ar) comprise over 99.9% of the mass of dry air.

Question 3.6 What is the concentration in molecules per cm^3 equivalent to 1 ppm at STP?

Question 3.7 The mole fraction of CO_2 is expected to exceed 600 μmol mol^{-1} at some point in the future. Calculate the partial pressure P_j and mass concentration ρ_j at STP.

Question 3.8 In a sample of air, the mass concentration of sulfur dioxide (SO_2) is 406.4 μg m^{-3}. What is the mole fraction at STP?

Question 3.9 In developing a terrestrial biosphere model, you write an equation for photosynthesis in terms of the partial pressure of CO_2. The atmospheric model you couple to provides atmospheric CO_2 as a mass concentration. Write an equation to convert this to partial pressure.

3.6 | Atmospheric Humidity

Air pressure is the sum of dry air and water vapor so that

$$P = P_d + e, \qquad (3.26)$$

with P_d and e the partial pressure of dry air and the partial pressure of water vapor, respectively. The latter term is also known as vapor pressure. The dry air component follows the ideal gas law with partial pressure $P_d = P - e$, and its density is

$$\rho_d = \frac{P - e}{\Re T} M_a. \qquad (3.27)$$

The density of water vapor, also termed absolute humidity, is

$$\rho_v = \frac{e}{\Re T} M_w, \qquad (3.28)$$

with the molecular mass of water M_w equal to 18.02 g mol^{-1}.

The density of moist air is the sum of that for dry air and water vapor whereby

$$\rho = \rho_d + \rho_v = \frac{P}{\Re T} M_a \left[1 - \left(1 - \frac{M_w}{M_a} \right) \frac{e}{P} \right]$$
$$= \frac{P}{\Re T} M_a \left(1 - 0.378 \frac{e}{P} \right), \qquad (3.29)$$

with $M_w/M_a = 0.622$. Table 3.3 gives the density of dry air and moist air when saturated. The density of moist air is less than the density of dry air at the same temperature and pressure. Because moist air is less dense than dry air, water vapor is a source of buoyancy in the atmosphere. An equivalent expression for the density of moist air reverts to the standard gas equation using a virtual temperature T_v and

$$\rho = \frac{P}{\Re T_v} M_a. \qquad (3.30)$$

The virtual temperature of moist air is the temperature at which dry air would have the same density as the moist air. It is evident from (3.29) that

$$T_v = \frac{T}{1 - 0.378 e/P}. \qquad (3.31)$$

Table 3.3	Saturation vapor pressure over water and density of air in relation to temperature			
T (°C)	e_{sat} (Pa)	de_{sat}/dT (Pa K^{-1})	Density (kg m^{-3})	
			Dry	Saturated
−5	421	32	1.317	1.315
0	611	44	1.293	1.290
5	872	61	1.269	1.265
10	1227	82	1.247	1.241
15	1704	110	1.225	1.217
20	2337	146	1.204	1.194
25	3167	189	1.184	1.170
30	4243	243	1.165	1.146
35	5624	311	1.146	1.122
40	7378	393	1.127	1.096

Note: Density is calculated at standard pressure (1013.25 hPa).

An alternative equation uses specific humidity q (kg kg^{-1}), and

$$T_v = T\left[1 + \left(\frac{M_a}{M_w} - 1\right)q\right] = T(1 + 0.608q). \quad (3.32)$$

The density of moist air can be written as the product of molar density and molecular mass so that $\rho = \rho_m \bar{M}$, and it is evident that the molecular mass of moist air is

$$\bar{M} = \frac{\rho}{\rho_m} = M_a\left(1 - 0.378\frac{e}{P}\right) = \frac{M_a}{1 + 0.608q}. \quad (3.33)$$

This equation adjusts the molecular mass of dry air for water vapor, and the molecular mass of moist air is less than that of dry air.

The moisture content of air can be expressed in several ways. Mass mixing ratio (kg kg^{-1}) is the ratio of the mass of water vapor m_v in a parcel of air to the mass of dry air m_d in the parcel:

$$\chi_v = \frac{m_v}{m_d} = \frac{\rho_v}{\rho_d} = \frac{0.622e}{P - e}. \quad (3.34)$$

Specific humidity (kg kg^{-1}) is the ratio of the mass of water vapor in a parcel of air to the total mass of the air:

$$q = \frac{m_v}{m_d + m_v} = \frac{\rho_v}{\rho_d + \rho_v} = \frac{0.622e}{P - 0.378e}. \quad (3.35)$$

The mole fraction (mol mol^{-1}) is e/P. This is also known as the mole mixing ratio in atmospheric chemistry. Specific humidity and mole fraction are related by

$$q = \frac{M_w}{\bar{M}}\frac{e}{P}. \quad (3.36)$$

Relative humidity is a measure of how saturated the air is with water. It is the ratio of vapor pressure to saturated vapor pressure, expressed as a percentage, given as $100e/e_{sat}(T)$ and in which the term $e_{sat}(T)$ signifies that saturation vapor pressure varies with temperature. Vapor pressure deficit is the difference between the saturation vapor pressure (i.e., the maximum amount of water vapor that can be held in the air) and the actual vapor pressure – i.e., $e_{sat}(T) - e$.

Saturation vapor pressure is the maximum amount of water vapor that a parcel of air can hold. Saturation vapor pressure increases exponentially with warmer temperature, and warm air can hold considerably more water vapor when saturated than can cold air (Figure 3.3). A general equation to calculate saturation vapor pressure in relation to temperature is the polynomial:

$$e_{sat}(T) = 100\Big[a_0 + T\Big(a_1 + T\Big(a_2 + T\Big(a_3 + T\Big(a_4$$
$$+ T\big(a_5 + T(a_6 + T(a_7 + Ta_8))\big)\Big)\Big)\Big)\Big)\Big].$$
$$(3.37)$$

Temperature is given here in °C, and the coefficients a_0–a_8 vary over water or ice (Table 3.4). A similar polynomial gives de_{sat}/dT.

Question 3.10 In processing a sample of moist air (3% wet air), you obtain a mole fraction of CO_2 equal to 378.3 µmol mol^{-1}. You compare this to a previously reported mole fraction in dry air of 390 µmol mol^{-1} and conclude that the CO_2 concentration has deceased. However, you did not correct your measurement for the water vapor. Does your measurement differ from the prior value? Derive the necessary equation to explain this. Remember that the mole fraction of the gas is n_j/n and the mole fraction of water vapor is n_v/n.

Table 3.4	Coefficients to calculate saturation vapor pressure and its temperature derivative over water and ice using (3.37)				
	Water	Ice		Water	Ice
a_0	6.11213476	6.11123516	b_0	0.444017302	0.503277922
a_1	0.444007856	0.503109514	b_1	$0.286064092 \times 10^{-1}$	$0.377289173 \times 10^{-1}$
a_2	$0.143064234 \times 10^{-1}$	$0.188369801 \times 10^{-1}$	b_2	$0.794683137 \times 10^{-3}$	$0.126801703 \times 10^{-2}$
a_3	$0.264461437 \times 10^{-3}$	$0.420547422 \times 10^{-3}$	b_3	$0.121211669 \times 10^{-4}$	$0.249468427 \times 10^{-4}$
a_4	$0.305903558 \times 10^{-5}$	$0.614396778 \times 10^{-5}$	b_4	$0.103354611 \times 10^{-6}$	$0.313703411 \times 10^{-6}$
a_5	$0.196237241 \times 10^{-7}$	$0.602780717 \times 10^{-7}$	b_5	$0.404125005 \times 10^{-9}$	$0.257180651 \times 10^{-8}$
a_6	$0.892344772 \times 10^{-10}$	$0.387940929 \times 10^{-9}$	b_6	$-0.788037859 \times 10^{-12}$	$0.133268878 \times 10^{-10}$
a_7	$-0.373208410 \times 10^{-12}$	$0.149436277 \times 10^{-11}$	b_7	$-0.114596802 \times 10^{-13}$	$0.394116744 \times 10^{-13}$
a_8	$0.209339997 \times 10^{-15}$	$0.262655803 \times 10^{-14}$	b_8	$0.381294516 \times 10^{-16}$	$0.498070196 \times 10^{-16}$

Note: Coefficients for water are valid for the temperature range 0°C to 100°C, and coefficients for ice are valid for −75°C to 0°C. a_0–a_8 are valid for e_{sat}. b_0–b_8 are valid for de_{sat}/dT.
Source: From Flatau et al. (1992) as used in the Community Land Model.

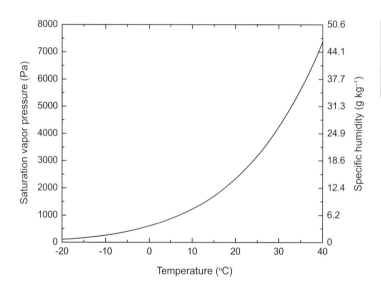

Figure 3.3 Saturation vapor pressure as a function of temperature. The left-hand axis shows vapor pressure in pascals. The right-hand axis shows specific humidity. Reproduced from Bonan (2016)

Question 3.11 Equation (3.24) gives the molecular mass of a mixture of gases. Use this formula to derive an expression for the molecular mass of moist air.

3.7 | Diffusive Flux Equations

Fick's law is formulated for molecular diffusion but also provides an expression for convective mass and heat fluxes in the atmosphere. Here, the conductance g_j represents turbulent mixing rather than molecular diffusion. The equation

$$F_j = \Delta c_j g_j \qquad (3.38)$$

describes fluxes of H_2O, CO_2, or other gases. For example, a typical CO_2 gradient between the stomatal cavity and air is $\Delta c_j = 100 \mu mol \, mol^{-1}$ and a representative conductance is $g_j = 0.2 \, mol \, CO_2 \, m^{-2} \, s^{-1}$. The resulting photosynthetic flux is $F_j = 20 \, \mu mol \, CO_2 \, m^{-2} \, s^{-1}$. A specific form of (3.38) describes evaporation by

Table 3.5	Common forms of evaporation diffusive flux equations

Units of measure	Flux
Absolute humidity: ρ_v (kg m^{-3})	$E = \Delta\rho_v g_w$
Mixing ratio: $\chi_v = \frac{\rho_v}{\rho_d}$ (kg kg^{-1})	$E = \rho_d \Delta\chi_v g_w$
Specific humidity: $q = \frac{\rho_v}{\rho}$ (kg kg^{-1})	$E = \rho\Delta q g_w$
Vapor pressure: e (Pa)	$E = \frac{M_w}{\Re T}\Delta e\, g_w$
Mole fraction: e/P (mol mol^{-1})	$E = M_w\rho_m \frac{\Delta e}{P} g_w$
Mole fraction: e/P (mol mol^{-1})	$E = \frac{M_w}{M_a}\rho_a \frac{\Delta e}{P} g_w$
Mole fraction: e/P (mol mol^{-1})	$E \approx \frac{M_w}{M_a}\rho \frac{\Delta e}{P} g_w$

Note: In this table, E has mass units (kg m^{-2} s^{-1}), and conductance g_w is in m s^{-1}.

$$E = \frac{\Delta e}{P} g_w. \tag{3.39}$$

A typical water vapor gradient in the surface boundary layer is 0.01 mol mol^{-1}, and a representative conductance is $g_w = 1$ mol m^{-2} s^{-1}. This gives an evaporative flux $E = 10$ mmol H$_2$O m^{-2} s^{-1}. Various common equations express evaporation in terms of absolute humidity ρ_v, mass mixing ratio χ_v, specific humidity q, or vapor pressure e (Table 3.5). The second-to-last equation in Table 3.5 is a form that uses the density of dry air given by (3.25). Multiplying by the ratio M_w/M_a corrects for the molecular mass of the water relative to dry air. Because $P >> e$, specific humidity in (3.35) is commonly approximated as $q \approx 0.622e/P$. This is the last equation in Table 3.5.

Evaporation is a flux of water measured by mass or moles. It is also an exchange of energy between the evaporating surface and air. Considerable amounts of energy are required to change water among its solid, liquid, and vapor states (Table 3.6). The energy absorbed in the evaporation of water is called the latent heat of vaporization. At 15°C, 2466 J are required to change 1 g of water from liquid to vapor (equivalent to 44.44 kJ mol^{-1}). Evaporation, therefore, is expressed as a mass flux (kg m^{-2} s^{-1}, or mol m^{-2} s^{-1}) or energy flux (J m^{-2} s^{-1}). The latter flux is termed latent heat flux and is

$$\lambda E = \lambda \frac{\Delta e}{P} g_w, \tag{3.40}$$

where λ is the molar latent heat of vaporization (J mol^{-1}). A common form of this equation uses the psychrometric constant γ (Pa K^{-1}) defined by

Table 3.6	Latent heat in relation to temperature

Temperature (°C)	Latent heat (J g^{-1})		
	Vaporization	Fusion	Sublimation
−20	2550	289	2839
−10	2525	312	2837
0	2501	334	2835
5	2490		
10	2478		
15	2466		
20	2454		
25	2442		
30	2430		
35	2419		
40	2407		

Note: Multiply by the molecular mass of water (M_w = 18.02 g mol^{-1}) to convert to J mol^{-1}.

$$\gamma = c_p P/\lambda \tag{3.41}$$

so that

$$\lambda E = \frac{c_p}{\gamma} \Delta e\, g_w. \tag{3.42}$$

An evaporative flux of 10 mmol H$_2$O m^{-2} s^{-1} is equivalent to a latent heat flux of 444 W m^{-2}.

The comparable equation for heat transfer by convection, commonly termed sensible heat flux, is

$$H = c_p \Delta T g_h. \tag{3.43}$$

A representative specific heat of moist air at constant pressure is $c_p = 29.2$ J mol^{-1} K^{-1} (derived in the chapter appendix). A typical conductance in the surface boundary layer is $g_h = 2$ mol m^{-2} s^{-1} so that a temperature difference $\Delta T = 5$ K produces a sensible heat flux $H = 292$ W m^{-2}.

Question 3.12 Calculate how much energy at STP is required to evaporate (a) 1 kg H$_2$O, (b) 1 mol H$_2$O, and (c) 1 mm H$_2$O (hint: the density of water is 1000 kg m^{-3}).

Question 3.13 Latent heat flux is 300 W m^{-2}. Give the equivalent water flux in (a) kg m^{-2} s^{-1}, (b) mol m^{-2} s^{-1}, and (c) mm H$_2$O s^{-1}.

Question 3.14 Derive any two equivalent expressions for evaporation using absolute humidity, mass mixing ratio, specific humidity, or vapor pressure. Show that these are the same algebraically.

Question 3.15 Momentum flux is commonly given as $\tau = \rho u g_{am}$ where u is wind speed (m s^{-1}) and τ has units kg m^{-1} s^{-2}. What units does the conductance g_{am} have? What is the equivalent equation in molar notation?

Table 3.7	Representative values of conductances in the biosphere–atmosphere system	
Conductance		Value (mol m^{-2} s^{-1})
Stomata		
Open		0.1–0.4
Closed		0.01
Leaf boundary layer		
Broad leaf, calm wind		0.4
Narrow leaf, high wind		4
Aerodynamic		
Short vegetation, calm wind		0.2
Tall vegetation, high wind		4

3.8 | Conductance and Resistance Networks

Fick's law represents the diffusive flux analogous to an electrical network. Ohm's law states that the current through a conductor between two points is directly proportional to the potential difference across the two points, or

$$\text{current} = \frac{\text{voltage}}{\text{resistance}}. \tag{3.44}$$

The diffusive flux analog is

$$\text{flux} = \frac{\text{potential difference}}{\text{resistance}}. \tag{3.45}$$

The electrical network analogy is important because the fluxes of heat, water vapor, and CO_2 between vegetation and the atmosphere can be represented by a network of resistances connected in series or in parallel. For example, gas exchange from one surface of a leaf is commonly described by epidermal, stomatal, and boundary layer processes connected in series. The total leaf gas exchange is from an upper and lower leaf surface connected in parallel. Canopy gas exchange includes an additional aerodynamic flux between the canopy and air above the canopy. Table 3.7 lists key conductances in the biosphere–atmosphere system and their representative values.

Resistances are additive in series while conductances are additive in parallel (Figure 3.4). For two resistances r_1 and r_2 in series (connected end to end, such as from the stomatal cavity to the leaf surface and the leaf surface to the air surrounding the leaf), the total resistance is

(a) Series

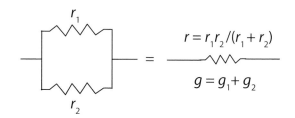

(b) Parallel

Figure 3.4 Depiction of a network of two resistances r_1 and r_2 connected (a) in series and (b) in parallel and the total resistance r. Also shown are conductances, denoted g_1 and g_2, and total conductance g.

$$r = r_1 + r_2. \tag{3.46}$$

Conductance is the reciprocal of resistance ($g = 1/r$), and the total conductance when connected in series is

$$\frac{1}{g} = \frac{1}{g_1} + \frac{1}{g_2} \Leftrightarrow g = \frac{g_1 g_2}{g_1 + g_2}. \tag{3.47}$$

For two resistances in parallel (side by side, such as for the sensible heat flux from both sides of the leaf), the total resistance is

$$\frac{1}{r} = \frac{1}{r_1} + \frac{1}{r_2}. \tag{3.48}$$

The conductance for a parallel network is

$$g = g_1 + g_2. \tag{3.49}$$

3.9 | Chapter Appendix – Diffusive Flux Notation

Mass fluxes with units kg m^{-2} s^{-1} and conductance with units m s^{-1}, or more commonly resistance with units s m^{-1}, are regularly used in the boundary layer meteorology literature and in atmospheric models. In the text that follows, variables with these units are shown in bold. A common form of Fick's law is given as the mass flux density $\mathbf{F}_j$ (kg m^{-2} s^{-1}) in relation to the concentration of the gas in mass per unit volume ρ_j (kg m^{-3}). In this form, conductance $\mathbf{g}_j$ has the units m s^{-1}. This notation is given in the following left-hand equation, and the corresponding molar notation is given in the right-hand equation:

$$\mathbf{F}_j = -D_j \frac{\partial \rho_j}{\partial z}, \qquad F_j = \frac{\mathbf{F}_j}{M_j} = -\rho_m D_j \frac{\partial c_j}{\partial z}. \tag{3.50}$$

The integrated flux equations are

$$\mathbf{F}_j = \Delta\rho_j \mathbf{g}_j, \qquad F_j = \frac{\mathbf{F}_j}{M_j} = \rho_m \Delta c_j \mathbf{g}_j = \Delta c_j g_j. \tag{3.51}$$

The left-hand and right-hand equations are related by $\rho_j = \rho_m M_j c_j$, from (3.21). Equation (3.51) shows that the conductances $\mathbf{g}_j$ and g_j are related by the volume occupied by one mole of gas as specified by the molar density $\rho_m = P/\Re T$. The equivalent conductances are

$$g_j = \rho_m \mathbf{g}_j. \tag{3.52}$$

A conductance $\mathbf{g}_j$ with units m s^{-1} is converted to g_j with units mol m^{-2} s^{-1} by multiplying by the molar density with units mol m^{-3}.

The comparable equations for sensible heat flux are

$$H = -\rho \mathbf{c_p} D_h \frac{\partial T}{\partial z}, \qquad H = -\rho_m c_p D_h \frac{\partial T}{\partial z}, \tag{3.53}$$

with ρ the density of moist air (kg m^{-3}) and $\mathbf{c_p}$ the mass specific heat of moist air at constant pressure (J kg^{-1} K^{-1}). Heat transfer by conduction, given from (3.6), is similar to Fick's law, with $D_h = \kappa/\rho \mathbf{c_p}$. Indeed, (3.53) is used to describe heat conduction in still air. The integrated flux equations are

$$H = \rho \mathbf{c_p} \Delta T \mathbf{g_h}, \qquad H = c_p \Delta T g_h, \tag{3.54}$$

with $\mathbf{g_h}$ (m s^{-1}) and g_h (mol m^{-2} s^{-1}) the conductance for heat.

The left-hand equations in (3.53) and (3.54) give the flux in relation to the density of moist air ρ and the mass specific heat of moist air at constant pressure $\mathbf{c_p}$. The term $\rho \mathbf{c_p}$ is the sum of the specific heat of dry air at constant pressure ($\mathbf{c_{pd}}$ = 1005 J kg^{-1} K^{-1}) and the specific heat of water vapor at constant pressure ($\mathbf{c_{pw}}$ = 1846 J kg^{-1} K^{-1}) multiplied by their respective densities (Webb et al. 1980; Brutsaert 1982, p. 43; Brutsaert 2005, p. 29) so that

$$\rho \mathbf{c_p} = \rho_d \mathbf{c_{pd}} + \rho_v \mathbf{c_{pw}}. \tag{3.55}$$

Dividing by ρ gives the equivalent expression:

$$\begin{aligned} \mathbf{c_p} &= (1-q)\mathbf{c_{pd}} + q\mathbf{c_{pw}} = \mathbf{c_{pd}}\left[1 + \left(\frac{\mathbf{c_{pw}}}{\mathbf{c_{pd}}} - 1\right)q\right] \\ &= \mathbf{c_{pd}}(1 + 0.84q). \end{aligned} \tag{3.56}$$

(Jacobson 2005, pp. 48–49, provides an alternative derivation of the same relationship.) Typical values are $\mathbf{c_p}$ = 1010 J kg^{-1} K^{-1} with q = 6 g kg^{-1} and 1020 J kg^{-1} K^{-1} with q = 18 g kg^{-1}. Using (3.56) for $\mathbf{c_p}$, (3.30) for ρ, and (3.32) for T_v gives

$$\begin{aligned} \rho \mathbf{c_p} &= \frac{P}{\Re T} M_a \mathbf{c_{pd}} \left(\frac{1 + 0.84q}{1 + 0.61q}\right) \\ &\approx \frac{P}{\Re T} M_a \mathbf{c_{pd}}(1 + 0.23q). \end{aligned} \tag{3.57}$$

The term $(P/\Re T)M_a$ is the dry air density, or ρ_a in (3.25); and because q is small (less than 10–20 g kg^{-1}), the term in parentheses is commonly neglected (over the range 0–20 g kg^{-1}, this term is 1.0–1.0045). Consequently, $\rho \mathbf{c_p}$ is approximated by $\rho_a \mathbf{c_{pd}}$. This gives the often-used equation for sensible heat flux in which

$$H = \rho_a \mathbf{c_{pd}} \Delta T \mathbf{g_h}. \tag{3.58}$$

c_p is the molar specific heat of moist air at constant pressure (J mol^{-1} K^{-1}). It is equal to

$$c_p = \mathbf{c_p}\left(\frac{M_a}{1 + 0.61q}\right),$$ (3.59)

with the term in brackets the molecular mass of moist air. This is equivalent to

$$c_p = \mathbf{c_{pd}}M_a\left(\frac{1 + 0.84q}{1 + 0.61q}\right) \approx \mathbf{c_{pd}}M_a(1 + 0.23q).$$ (3.60)

$c_p = 29.2$ J mol^{-1} K^{-1} is representative for a typical range of q.

Question 3.16 The psychrometric constant γ is commonly given in biometeorology studies as $c_pP/(0.622\lambda)$, which differs from (3.41) by the factor $1/0.622$. Equation (3.41) uses molar notation for c_p (J mol^{-1} K^{-1}) and λ (J mol^{-1}). The other equation uses the units J kg^{-1} K^{-1} and J kg^{-1}, respectively. Show that the two equations are equivalent.

4

Mathematical Formulation of Biological Flux Rates

Chapter Overview

Fick's law describes many rate processes in environmental physics, including diffusive fluxes, conduction, and water flow. Many biological fluxes are biochemical rather than biophysical and require formulations other than Fick's law. For example, the rate of photosynthesis increases with higher irradiance and higher CO_2 concentration. The rate of carbon loss during respiration increases with higher temperature. Rates of plant productivity and soil organic matter decomposition vary with temperature, soil moisture, and other factors. Common formulations for these processes are the Michaelis–Menten equation for a biochemical reaction; the Arrhenius equation to describe the temperature dependence of a biochemical reaction; minimum, multiplicative, and co-limiting rate multipliers; and first-order, linear differential equations to describe mass transfers within an ecosystem. In addition, mathematical principles of optimization provide a formal method to frame many ecological processes.

4.1 | Michaelis–Menten Equation

The Michaelis–Menten equation is a formulation of enzyme kinetics for chemical reactions. It relates the rate of an enzymatic reaction v to the concentration of a substrate $[S]$ as

$$v = \frac{v_{max}[S]}{[S] + K_m}. \tag{4.1}$$

In this equation, v_{max} is the maximum rate with saturating substrate concentration whereby $v = v_{max}$ when $[S] = \infty$, and K_m is the Michaelis–Menten constant, which defines the substrate concentration at which the reaction rate is one-half v_{max} (Figure 4.1).

The Michaelis–Menten equation is widely used to represent plant physiological processes. For example, the equation

$$V_c = \frac{V_{c\,max}\,c_i}{c_i + K_c(1 + o_i/K_o)} \tag{4.2}$$

describes the rate of carboxylation during photosynthesis when the enzyme Rubisco is limiting (Chapter 11). Here, the intercellular CO_2 concentration c_i ($\mu mol\ mol^{-1}$) is the substrate concentration, the Michaelis–Menten constant is $K_m = K_c(1 + o_i/K_o)$ with the units $\mu mol\ mol^{-1}$, and $V_{c\,max}$ is the maximum rate of carboxylation ($\mu mol\ m^{-2}\ s^{-1}$). A variant of the equation describes the photosynthetic response to light as

$$A = \frac{A_{max}E I^{\downarrow}}{E I^{\downarrow} + A_{max}}, \tag{4.3}$$

where $I^{\downarrow}$ is photosynthetically active radiation ($\mu mol\ photon\ m^{-2}\ s^{-1}$), E is light-use efficiency (mol CO_2 mol^{-1} photon) and represents the initial slope of the light-response curve, and A_{max} is the maximum photosynthetic rate at light saturation ($\mu mol\ CO_2$ $m^{-2}\ s^{-1}$). In this equation, the substrate

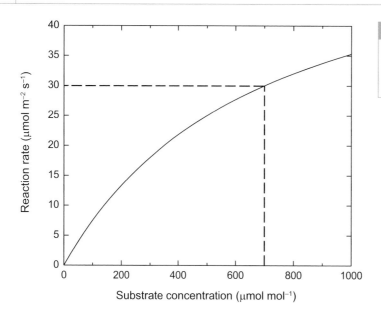

Figure 4.1 Michaelis–Menten equation for $v_{max} = 60$ µmol m^{-2} s^{-1} and $K_m = 700$ µmol mol^{-1}. Shown is the reaction rate v in relation to substrate concentration $[S]$. The dashed lines show v at $[S] = K_m$.

concentration is the irradiance $I^{\downarrow}$, the maximum rate is A_{max}, and $K_m = A_{max}/E$.

Question 4.1 The rate of a chemical reaction is commonly modeled with the Michaelis–Menten equation. Laboratory measurements show that v = 7.5 µmol m^{-2} s^{-1} at a substrate concentration $[S]$ = 100 µmol and v = 18.0 µmol m^{-2} s^{-1} at a substrate concentration $[S]$ = 300 µmol. Calculate the maximum rate v_{max} and the Michaelis–Menten constant K_m.

Question 4.2 The parameters v_{max} and K_m in the Michaelis–Menten equation can be estimated from rate measurements, as illustrated in the preceding question or more generally using nonlinear regression. An independent estimate of v_{max} is obtained from the initial slope of the response curve. Calculate $dv/d[S]$ using (4.1) and then relate v_{max} to the initial slope obtained with $[S] = 0$.

Question 4.3 Equation (4.3) gives the photosynthetic light-response curve as a Michaelis–Menten equation. Prove that E is the initial slope of the light response curve.

4.2 | Arrhenius Equation

The rate of biochemical reactions increases with higher temperatures. The Arrhenius equation describes the temperature dependence of a chemical reaction. For a rate v, its general form is

$$v = v_0 \exp\left(-\frac{\Delta H_a}{\Re T}\right), \qquad (4.4)$$

where v_0 is a constant with the same units as v, T is temperature (K), $\Re = 8.314$ J K^{-1} mol^{-1} is the universal gas constant, and ΔH_a is the activation energy (J mol^{-1}). The exponential term provides the temperature dependence. For example, a typical activation energy is $\Delta H_a = 50$ kJ mol^{-1}. At 20°C, $v = 1.23 \times 10^{-9} v_0$; at 30°C, $v = 2.42 \times 10^{-9} v_0$. This is a doubling for a 10°C increase in temperature. Activation energy is the threshold energy that reactants must acquire before the reaction can proceed and form products. A reaction with large activation energy requires much energy for the reaction to proceed; a reaction with small activation energy requires less energy for the reaction to proceed. In (4.4), the rate v at a given temperature increases as activation energy decreases. The constant v_0 can be eliminated from (4.4) if the rate v_1 at temperature T_1 and the rate v_2 at T_2 are known. Then,

$$v_1 = v_2 \ \exp\left[-\frac{\Delta H_a}{\mathfrak{R}}\left(\frac{1}{T_1}-\frac{1}{T_2}\right)\right]. \tag{4.5}$$

A typical reference temperature for biological reactions is 25°C so that v at temperature T is calculated from the rate v_{25} at 25°C as

$$v = v_{25} \ \exp\left[\frac{\Delta H_a}{298.15\mathfrak{R}}\left(1-\frac{298.15}{T}\right)\right]. \tag{4.6}$$

Another common expression used in ecological models to represent the temperature dependence of a physiological process is the Q_{10} function. Referenced to 25°C, this function is

$$v = v_{25} \ Q_{10}^{(T-298.15)/10}. \tag{4.7}$$

A value $Q_{10} = 2$ means that the rate doubles for every 10°C increase in temperature. At 25°C, (4.6) and (4.7) are related by

$$Q_{10} = \exp\left(\frac{10\Delta H_a}{298.15\mathfrak{R}T}\right). \tag{4.8}$$

In a common application, the Michaelis–Menten equation provides the rate of a biochemical reaction in relation to substrate concentration, and the Arrhenius equation adjusts the rate constants K_m and v_{max} for temperature. Photosynthesis models use this approach (Chapter 11).

Question 4.4 Calculate the value of ΔH_a for which $Q_{10} = 2$ at 25°C. Then graph and

compare the Arrhenius function (4.6) and the Q_{10} function (4.7).

Question 4.5 Using the results of the preceding question, calculate the Q_{10} of the Arrhenius function for various temperatures (0, 10, 20, 30°C). Is the Q_{10} constant? How does this compare with the Q_{10} function?

4.3 | Rate Modifiers

The rate of many ecological processes depends on environmental conditions that are not described by Michaelis–Menten kinetics. For example, the rate of carbon loss in foliage respiration increases with higher temperature. This and other ecological processes are commonly represented by an equation with the form

$$v = [S]v_{base}f(x). \tag{4.9}$$

This equation specifies the flux in relation to the pool size $[S]$ times a base rate per unit mass v_{base}, as adjusted for some environmental factor specified by $f(x)$. For example, the equation

$$v = [S]v_{25} \exp\left[\frac{\Delta H_a}{298.15\mathfrak{R}}\left(1-\frac{298.15}{T}\right)\right] \tag{4.10}$$

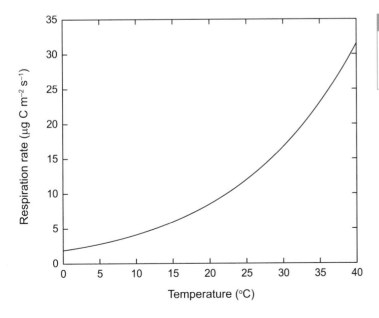

Figure 4.2 Respiration rate in relation to temperature illustrating the Arrhenius function with a base rate $v_{25} = 0.3$ μg C g^{-1} C s^{-1} at 25°C and pool size $[S] = 40$ g C m^{-2}. In this example, $\Delta H_a = 50$ kJ mol^{-1}.

describes foliage respiration using the Arrhenius equation. v is the leaf respiration rate (kg C m^{-2} s^{-1}), v_{25} is the base rate per unit mass (kg C kg^{-1} C s^{-1}) at 25°C, and $[S]$ is the foliage mass (kg C m^{-2}). Figure 4.2 illustrates this equation. Biogeochemical models routinely use this approach (Chapter 17).

Other processes are better conceived in terms of a maximum, or potential, rate for optimal environmental conditions that decreases to the extent that conditions are suboptimal. In this formulation, the flux rate is

$$v = [S]v_{\max}f(x),\qquad(4.11)$$

in which $f(x)$ is a rate-limiting factor that has a value between zero and one. The particular form of the rate-limiting factor can vary (Figure 4.3).

A common response function is a linear decline as a resource decreases below some critical abundance. In this method, the rate-limiting function is proportional to the amount of resource x when the resource level is below some threshold x'. Mathematically, this is given by the equation

$$f(x) = \begin{cases} x/x' & x < x' \\ 1 & x \geq x' \end{cases}.\qquad(4.12)$$

The response is linear for $x < x'$ and is constant for $x \geq x'$. Equation (4.12) is discontinuous at x'. Another function, which is continuous, is one where the rate increases linearly at low resource abundance and saturates at high resource

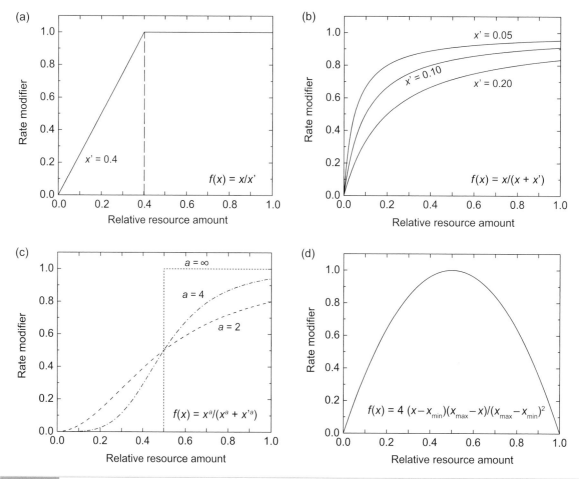

Figure 4.3 Rate-limiting factors $f(x)$ in relation to an environmental resource x. In these graphs, x is the relative resource abundance and x' defines a critical resource amount. (a) Linear response with $x' = 0.4$. (b) Michaelis–Menten response with $x' = 0.05$, 0.10, and 0.20. (c) Sigmoid response with $x' = 0.5$ and $a = 2$, 4, and ∞. (d) Parabolic response with $x_{\min} = 0$ and $x_{\max} = 1$.

abundance in a Michaelis–Menten response. In this formulation, the rate modifier is

$$f(x) = \frac{x}{x + x'},\qquad(4.13)$$

with x' now the resource amount where $f(x) = 0.5$, known as the half-saturation constant. x' determines how rapidly the response approaches the saturation rate. Low values for x' produce a steep initial slope, whereas high values produce a more gradual initial slope. A third response function gives a sigmoid relationship in which the rate increases rapidly with low resource abundance but saturates at high resource amount. A general function is

$$f(x) = \frac{x^a}{x^a + (x')^a},\qquad(4.14)$$

where a is a shape factor. This function produces a family of curves in which a larger value for the parameter a produces a steeper response. For very large values of a, $f(x)$ has a step response at $x = x'$. Some ecological processes are better described by a parabolic response curve in which more of a resource increases the rate up to some optimum, beyond which the rate decreases with greater resource amount. Such a response is given by

$$f(x) = \frac{4(x - x_{min})(x_{max} - x)}{(x_{max} - x_{min})^2}\qquad(4.15)$$

for $x_{min} \leq x \leq x_{max}$. This particular function has values of zero at the minimum x_{min} and maximum x_{max} resource amount and one-half at $(x_{max} + x_{min})/2$.

Ecological processes commonly respond to not one but several environmental factors. For example, the rate of carbon loss during decomposition increases with higher soil temperature and greater soil moisture but decreases with excessive soil moisture. Foliage respiration increases with higher temperature and greater amounts of leaf nitrogen. Photosynthesis increases with higher irradiance, higher CO_2 concentration, and greater amounts of foliage nitrogen and additionally varies with temperature and soil moisture. Stomata respond to light, CO_2, vapor pressure deficit, and soil moisture. One way to represent multiple resource limitations is to assume that the rate is constrained by the most limiting of the resources. This is known as Liebig's law of the minimum. If v_1, v_2, and v_3 are the rates

calculated separately for three different resources (x_1, x_2, and x_3, respectively), the realized rate is

$$v = \min(v_1, v_2, v_3).\qquad(4.16)$$

Some photosynthesis models use this to represent CO_2 assimilation as the lesser of a Rubisco-limited and light-limited rate (Chapter 11). An alternative is to allow multiple factors to concurrently limit the rate. This is done in a multiplicative manner whereby

$$v = [S]v_{max}f(x_1)f(x_2)f(x_3).\qquad(4.17)$$

Models of stomatal conductance (Chapter 12), soil organic matter decomposition (Chapter 18), BVOCs (Chapter 20), and nitrogen gas emissions (Chapter 20) use this method. Another common application is to represent carbon uptake during gross primary production (GPP) in photosynthetic light-use efficiency models. In these models, carbon assimilation is proportional to photosynthetically active radiation $I^\downarrow$ times a light-use efficiency E, which relates carbon gain to light absorbed (Prince and Goward 1995; Running et al. 2000, 2004; Zhao et al. 2005; Yuan et al. 2007). This potential productivity is reduced by environmental constraints using multiplicative factors so that

$$GPP = EI^\downarrow f_1(T)f_2(\theta)f_3(D),\qquad(4.18)$$

where $f_1(T)$, $f_2(\theta)$, and $f_3(D)$ are empirical functions scaled from zero to one that adjust photosynthesis for temperature, soil water, and vapor pressure deficit, respectively.

Where two factors limit a rate, another approach is to allow the rates v_1 and v_2 to co-limit the realized rate v. Co-limitation is obtained using the function

$$\Theta v^2 - (v_1 + v_2)v + v_1 v_2 = 0,\qquad(4.19)$$

and the co-limited rate is the smaller root given by

$$v = \frac{(v_1 + v_2) - [(v_1 + v_2)^2 - 4\Theta v_1 v_2]^{1/2}}{2\Theta}.\qquad(4.20)$$

The parameter Θ is a curvature factor that governs the degree of co-limitation. When $\Theta = 1$, this is equivalent to $v = \min(v_1, v_2)$. With $\Theta = 0$, (4.19) simplifies to $v = v_1 v_2/(v_1 + v_2)$. Intermediate values of Θ give response curves between these two extremes. Small values of Θ (e.g., 0.8) produce a gradual transition between rates with the result that v is less than $\min(v_1, v_2)$. Larger values (e.g., 0.98) give a more abrupt transition, and v

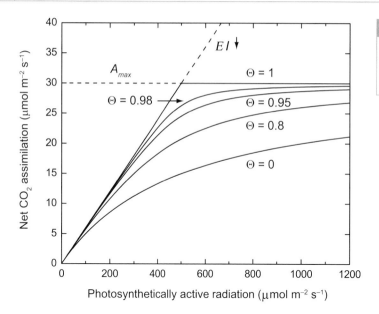

Figure 4.4 Co-limitation illustrated for photosynthetic response to light. The two dashed lines show the rates A_{max} and $EI^{\downarrow}$. The solid lines show the co-limited rate with $\Theta = 1$, 0.98, 0.95, 0.8, and 0.

approaches the rate determined by $\min(v_1, v_2)$. Some photosynthesis models use co-limitation to determine the photosynthetic rate in response to multiple limiting factors (Chapter 11).

A common application of (4.19) is to model photosynthesis in response to light. The photosynthetic light-response curve is governed by two rates: a maximum rate at light saturation A_{max}; and a linear increase with greater irradiance specified by $EI^{\downarrow}$. The co-limited photosynthetic rate A is

$$\Theta A^2 - \left(EI^{\downarrow} + A_{max}\right)A + EI^{\downarrow}A_{max} = 0. \qquad (4.21)$$

This equation has three parameters (E, A_{max}, Θ) and is sometimes referred to as a non-rectangular hyperbola (Figure 4.4). Equation (4.21) reduces to $A = \min\left(EI^{\downarrow}, A_{max}\right)$ with $\Theta = 1$. This is the same as the law of minimum (4.16) and is also referred to as a Blackman limiting response (Blackman 1905). With $\Theta = 0$, (4.21) simplifies to the Michaelis–Menten equation given by (4.3) and is also called a rectangular hyperbola. Some photosynthesis models use a variant of (4.21) to calculate electron transport in relation to irradiance (Chapter 11).

Question 4.6 The leaf photosynthesis model given by (4.3) can also be applied to a canopy, where here A_{max} and E are canopy-scale parameters and $I^{\downarrow}$ is canopy irradiance. Contrast this model with the production efficiency model given by (4.18). What is a key difference between these two models in their light response? Do they differ in the timescale at which they are valid?

Question 4.7 Foliage nitrogen can limit carbon uptake during gross primary production (GPP). In one model, GPP is calculated as $GPP = GPP_{max}f(N)$. Here, maximum GPP is obtained from a mechanistic photosynthesis model in response to light, temperature, and other factors, and $f(N)$ has a value from zero to one and decreases the maximum GPP for nitrogen availability. In this approach, nitrogen availability downregulates GPP if there is not sufficient nitrogen to meet the plant demand. Another model uses the relationship $GPP = (a + bN)f_1(T)f_2(\theta)f_3(D)$, where $a + bN$ represents a linear increase in maximum GPP in response to nitrogen and $f_1(T), f_2(\theta)$, and $f_3(D)$ decrease the maximum rate for temperature, soil water, and vapor pressure deficit, respectively. Describe a key conceptual difference in the response of GPP to the addition of nitrogen in these two models.

Question 4.8   Explain the difference between the non-rectangular hyperbola and rectangular hyperbola as used to model photosynthetic response to light.

Question 4.9   Equation (4.19) is used to calculate photosynthesis as the co-limited rate of two rates. Some models include a third rate. Explain how to co-limit three rates.

4.4 | First-Order, Linear Differential Equations

Many biological processes can be described by a first-order, linear differential equation. For example, the accumulation of material in an ecological system equals the rate of input minus the rate of loss. Over some time interval Δt, the change in mass of material m_1 is

$$\frac{\Delta m_1}{\Delta t} = \text{input rate} - \text{loss rate}. \quad (4.22)$$

The instantaneous rate of change is given by the first-order, linear differential equation

$$\frac{dm_1}{dt} = U - k_1 m_1. \quad (4.23)$$

In this equation, U is the input rate, and the loss rate is the product of the mass m_1 and the instantaneous fractional loss rate, or turnover rate, k_1. If U is an annual flux with units kg C m^{-2} y^{-1}, then m_1 has units kg C m^{-2}, k_1 has units y^{-1}, and time t is in years (y). Integration of (4.23) gives the mass at time t:

$$m_1(t) = \frac{U}{k_1}\left(1 - e^{-k_1 t}\right) + m_0 e^{-k_1 t}, \quad (4.24)$$

where m_0 is the initial mass at $t = 0$. At steady state, $dm_1/dt = 0$ and $m_1 = U/k_1$. This is also equal to $m_1(t)$ at $t = \infty$, as seen in (4.24). Figure 4.5a illustrates the behavior of (4.24). In this example, $U = 1000$ g C m^{-2} y^{-1} is the annual carbon input from net primary production, $k_1 = 0.1$ y^{-1} represents annual carbon loss, and the system accumulates 10 kg C m^{-2} at steady state. Equation (4.23) is simple and can be easily integrated to obtain an analytical solution. Other equations are more complex and must be solved using numerical methods (Appendix A3).

Many ecosystem processes can be described by a system of first-order, linear differential equations. For example, a simple biogeochemical model represents an ecosystem as two separate pools that describe vegetation and soil carbon (Figure 4.5b). The equation

$$\frac{dm_1}{dt} = U - k_1 m_1 \quad (4.25)$$

describes carbon accumulation in vegetation m_1 as the difference between net primary production U and litterfall $k_1 m_1$. The equation

$$\frac{dm_2}{dt} = k_1 m_1 - k_2 m_2 \quad (4.26)$$

represents soil carbon m_2 as the difference between litterfall input $k_1 m_1$ and decomposition loss $k_2 m_2$. Biogeochemical models use similar equations to describe plant (Chapter 17) and soil (Chapter 18) carbon. In these models, the ecosystem is represented by multiple pools of carbon, and a system of equations describes carbon turnover and flows among pools. The turnover rates vary with environmental factors such as temperature and soil moisture, represented either as a minimum limiting factor as in (4.16) or in a multiplicative manner as in (4.17).

Question 4.10   Using the one-pool biogeochemical model given by (4.23), calculate and graph the mass for $t = 0$ to $t = 400$ y for three different systems with initial mass zero and $U_1 = 1000$ g C m^{-2} y^{-1} and $k_1 = 0.1$ y^{-1}; $U_1 = 500$ g C m^{-2} y^{-1} and $k_1 = 0.05$ y^{-1}; and $U_1 = 500$ g C m^{-2} y^{-1} and $k_1 = 0.04$ y^{-1}. Calculate the mass at equilibrium. Which system accumulates the most carbon? How do the parameters U_1 and k_1 relate to the differences in carbon accumulation?

Question 4.11   For the two-pool biogeochemical model given by (4.25) and (4.26), derive equations for the pools m_1 and m_2 at steady state. Does a two-pool model accumulate more carbon at steady state than a one-pool model?

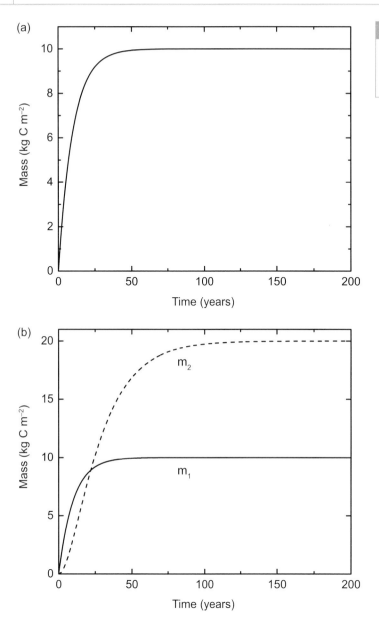

Figure 4.5 Representation of carbon pools using first-order, linear differential equations. (a) A one-pool model with $U = 1000$ g C m^{-2} y^{-1} and $k_1 = 0.1$ y^{-1}. (b) A two-pool model with $U = 1000$ g C m^{-2} y^{-1}, $k_1 = 0.1$ y^{-1}, and $k_2 = 0.05$ y^{-1}.

4.5 | Optimality Theory

Some biological processes can be construed in the context of optimization. If the process can be formulated in terms of a benefit and a cost, the optimal behavior is that which maximizes the difference between the benefit and cost. A model that utilizes this premise is called an optimality model. The construction of such a model requires a variable to be optimized, a benefit and cost associated with that variable, and a mathematical

formulation of the costs and benefits in relation to the variable of interest. A common premise in ecological studies is to maximize net carbon uptake over some time period. Consider, for example, plant productivity. Carbon uptake during photosynthesis increases with higher leaf area index, but carbon loss during respiration also increases as biomass increases. Monsi and Saeki (1953; reprinted as Monsi and Saeki 2005) used optimality theory to find the leaf area index that maximizes productivity. Equation (4.3) describes the photosynthetic rate of an individual leaf in relation to

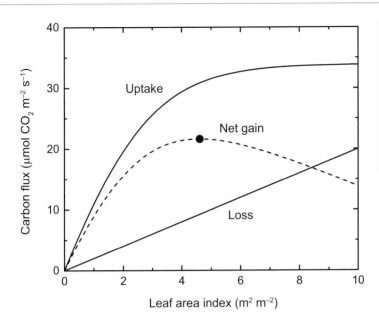

Figure 4.6 Productivity optimization in a system in which carbon uptake and carbon loss both depend on leaf area index. Carbon uptake is given by (4.27) with $A_{\max} = 15$ μmol CO_2 m^{-2} s^{-1}, $E = 0.06$ mol CO_2 mol^{-1} photon, $K_b = 0.7$, and $I_0^{\downarrow} = 1380$ μmol photon m^{-2} s^{-1}. Carbon loss is given by (4.28) with $R_d = 2$ μmol CO_2 m^{-2} s^{-1}. The optimal leaf area index that maximizes net carbon gain is $x = 4.6$ m^2 m^{-2} (denoted by the filled circle).

photosynthetically active radiation $I^{\downarrow}$. Integrating this leaf rate over the light profile in the canopy, the gross carbon uptake by a plant canopy with leaf area index x is

$$f(x) = \frac{A_{\max}}{K_b} \ln \left(\frac{A_{\max} + EK_b I_0^{\downarrow}}{A_{\max} + EK_b I_0^{\downarrow} e^{-K_b x}} \right). \tag{4.27}$$

The variables in this equation are the same as before, but additionally with $I_0^{\downarrow}$ the radiation incident on the top of the canopy and K_b the light extinction coefficient. Chapter 15 provides the derivation of this equation, in which carbon uptake increases up to some saturating value with greater leaf area index x as the canopy absorbs more radiation. Figure 4.6 illustrates this behavior for common values of $A_{\max}$, E, K_b, and $I_0^{\downarrow}$. The cost of leaves is carbon loss during respiration, which increases linearly with more leaves, as given by

$$g(x) = R_d x, \tag{4.28}$$

with R_d the respiration rate per unit leaf area (μmol CO_2 m^{-2} s^{-1}). The net carbon gain is $F(x) = f(x) - g(x)$. In this system, more leaf area results in higher carbon uptake, but the respiration cost is a constraint, or limitation, placed on the amount of leaf area (Figure 4.6). The maximum net carbon gain is found when $dF/dx = 0$, so that $df/dx = dg/dx$. The value of x that satisfies this relationship is the optimal leaf area index that

maximizes net carbon gain. In this example, the optimal amount of x is obtained when

$$x = \frac{1}{K_b} \ln \left[\frac{(A_{\max} - R_d) EK_b I_0^{\downarrow}}{A_{\max} R_d} \right]. \tag{4.29}$$

In the example shown in Figure 4.6, the optimal leaf area index is 4.6 m^2 m^{-2}, and the maximum net carbon gain is 21.6 μmol CO_2 m^{-2} s^{-1}.

The method of Lagrange multipliers provides a general mathematical means to maximize or minimize any function subject to the constraint that only points that satisfy a certain restriction are considered. Consider, for example, a function $f(x, y)$ that depends on two variables x and y with the constraint that $g(x, y) = 0$. Here, f and g are any two mathematical functions that both depend on x and y. The maxima and minima are found by introducing a new variable λ and analyzing the function F defined by

$$F(x, y, \lambda) = f(x, y) - \lambda g(x, y). \tag{4.30}$$

The values of x and y that satisfy this equation are those that maximize and/or minimize $f(x, y)$ subject to $g(x, y) = 0$. Here, λ is known as the Lagrange multiplier. It is a dummy variable needed to find x and y, but its exact value does not matter. The critical values of x and y at maximization or minimization are found by setting the partial derivatives of F with respect to x, y, and λ to zero

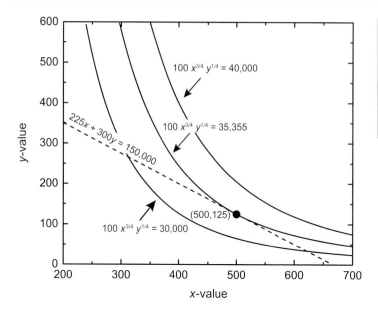

Figure 4.7 Graphical illustration of optimization in two dimensions. The dashed line shows the constraint $225x + 300y = 150,000$. The solid lines show three curves for $100x^{3/4}y^{1/4}$ equal to 30,000, 35,355, and 40,000. The optimal values $x = 500$ and $y = 125$ occur when the curve shares a tangent with the constraint function. This occurs for $100x^{3/4}y^{1/4} = 35,355$.

and solving a system of three simultaneous equations in which

$$\partial F/\partial x = 0 \quad \Rightarrow \quad \partial f/\partial x - \lambda \partial g/\partial x = 0 \tag{4.31}$$

$$\partial F/\partial y = 0 \quad \Rightarrow \quad \partial f/\partial y - \lambda \partial g/\partial y = 0 \tag{4.32}$$

$$\partial F/\partial \lambda = 0 \quad \Rightarrow \quad -g(x,y) = 0. \tag{4.33}$$

The solution is obtained by first solving for λ in terms of x and y, thereby removing λ from the equations, and then solving for x and y. This approach can be generalized to any number of variables.

A classic application of Lagrange multipliers is to maximize production in relation to cost in economics. The Cobb–Douglas production function relates the production of a good to labor and capital costs. In the example shown in Figure 4.7,

$$f(x,y) = 100x^{3/4}y^{1/4}. \tag{4.34}$$

Here, $f(x,y)$ is the total production in a year, x is the labor input (e.g., the total number of person-hours worked in a year), and y is the capital input (e.g., the real value of machinery, buildings, etc., used in production). The constraint is that total costs for labor and capital cannot exceed an amount equal to 150,000. If each unit of labor costs 225 and each unit of capital costs 300, the constraint function is

$$g(x,y) = 225x + 300y - 150,000 = 0. \tag{4.35}$$

This is the dashed straight line in Figure 4.7. The system of equations to solve is

$$75x^{-1/4}y^{1/4} - 225\lambda = 0 \tag{4.36}$$

$$25x^{3/4}y^{-3/4} - 300\lambda = 0 \tag{4.37}$$

$$225x + 300y = 150,000. \tag{4.38}$$

The solution is $x = 500$ units of labor and $y = 125$ units of capital, and the maximum production is 35,355 units. In this example, the equation $\lambda = (y/x)^{1/4}/3$ is obtained, but the Lagrange multiplier is not needed, other than to find x and y. However, λ does have economic meaning, and here it represents the marginal productivity of money. It is the increase in the maximum production when total cost is increased by one unit. In this example, $\lambda = 0.236$ so that an additional expenditure in cost of 10,000 increases production by 2360 units.

Optimality theory can be used to construct ecological models. One application is to derive a model of stomatal conductance from theoretical principles rather than an empirical model obtained from observations. Stomata open for carbon uptake during photosynthesis A_n (a benefit), but water is lost in transpiration E when stomata are open (a cost). In an optimality framework, stomata function so as to minimize the water cost of carbon gain or, equivalently, to maximize the carbon gain of water loss (Chapter 12). The optimal stomatal conductance is that in which the marginal water cost of carbon gain $\partial E / \partial A_n$ is constant at a value equal to λ. Here, λ is the Lagrange multiplier and relates transpiration water loss to carbon (mol H_2O mol^{-1} CO_2). The interpretation is that stomatal conductance varies such that $\partial E / \partial A_n = \lambda$. Analytical expressions for stomatal conductance can be obtained if λ is specified. In this optimization problem, the value for λ is a critical term in the solution, in contrast with the general mathematical use of Lagrange multipliers.

Another application of optimality theory has been to explain the vertical distribution of nitrogen in plant canopies (Figure 2.8). Daily canopy photosynthesis is maximized when nitrogen is allocated such that leaves that receive high levels of light also have high nitrogen concentration so as to have high photosynthetic capacity (Chapter 15). This is achieved when the marginal increase in photosynthesis for a unit increase in leaf nitrogen n_a is constant throughout the canopy (i.e., $\partial A_n / \partial n_a$ is constant at every level in the canopy).

Optimal stomatal conductance and canopy nitrogen profiles use formal concepts of mathematical optimization in which the Lagrange multipliers represent the marginal water cost of carbon gain and the marginal carbon gain of nitrogen, respectively. Other applications of optimality theory in ecological contexts invoke the principle of optimization without the formal mathematics. For example, because photosynthesis, transpiration, and leaf temperature vary with leaf size, optimality theory can be used to find the leaf size that maximizes water-use efficiency in a given environment (Parkhurst and Loucks 1972; Givnish and Vermeij 1976). Optimality theory provides insights to plant canopy organization. Under the premise of maximizing canopy photosynthesis, an optimal distribution of foliage in a plant canopy can be found (Horn 1971). Optimization of carbon gain also provides a framework to represent carbon allocation in biogeochemical models (Thomas and Williams 2014).

Question 4.12 Find the dimensions x, y, and z of a box with the largest volume if the total surface area is 150 cm^2.

5

Soil Temperature

Chapter Overview

The diurnal cycle of soil temperature and seasonal variation over the course of a year are important determinants of land surface climate. This chapter reviews the physics of soil heat transfer. Heat flows from high to low temperature through conduction. Thermal conductivity and heat capacity are key soil properties that determine heat transfer. These vary with the mineral composition of soil and also with soil moisture. In seasonally frozen soils, it is necessary to account for the different thermal properties of water and ice. Additionally, the change in phase of water consumes and releases heat during melting and freezing, respectively.

5.1 | Introduction

The starting point to understanding energy exchanges between the land and atmosphere is not the fluxes themselves, but rather soil temperature and the storage of heat in soil. During the day, when solar radiation heats the ground, the surface is typically warmer than the underlying soil, and heat flows into the soil. This transfer of heat into the soil cools the land surface. At night, the surface is typically cooler than the soil, and heat flows out of the soil. This loss of energy from the soil warms the land surface. The same behavior occurs annually, when soils store heat during warm months and lose heat during cold months. This heat storage and release moderates the surface temperature, reducing its diurnal and seasonal temperature range. Soil temperature also regulates the rate of many biological processes. In particular, the rate of heterotrophic respiration increases as an exponential function of soil temperature.

Figure 5.1a illustrates a typical diurnal cycle for soil temperature during the warm season. In the upper 20 cm or so, temperature increases with depth at night. The opposite occurs during the day, when heat flows into the soil, the upper soil warms, and temperature decreases with depth. The near-surface soil warms the most; the deep soil hardly warms at all. A similar pattern occurs over the course of a full year (Figure 5.1b). Temperature decreases with depth during the summer, but this pattern reverses during winter. The near-surface soil has a larger annual range in temperature than the deeper soil.

Early-generation models of Earth's land surface developed for use with climate models neglected soil heat storage (Manabe et al. 1965). Later models represented the soil as two layers (shallow and deep soil) to simulate diurnal and seasonal variation in temperature (Deardorff 1978; Dickinson et al. 1986, 1993). This approach is known as the force–restore method because the forcing of soil temperature by the soil heat flux is modified by a restoring term associated with deep soil temperature (Dickinson 1988). Two layers differentiate near-surface soil, which responds to the diurnal cycle, from deeper soil, which responds to the annual cycle.

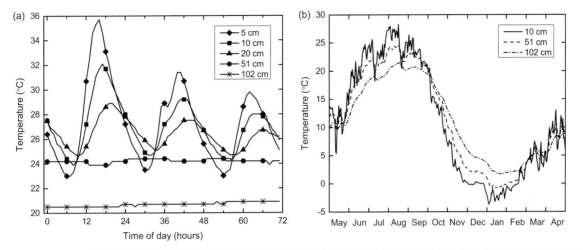

Figure 5.1 Soil temperature at various depths in an open field in Colorado, USA. (a) Diurnal cycle over a three-day period (1–3 August 2015). (b) Annual cycle (1 May 2015–30 April 2016). Data are for SCAN Site CPER (40°49′ N, 104°43′ W) from the Soil Climate Analysis Network (Natural Resources Conservation Service, US Department of Agriculture).

The current generation of models explicitly represents soil physics and multilayer heat transfer. This is an outcome of greater computational power, which allows for more process detail, but also because of the expanding scientific scope of the models, such as to study permafrost thawing and permafrost–carbon feedbacks with a warmer climate.

5.2 Transient, One-Dimensional Heat Conduction

Fourier's law describes heat transfer by conduction. Heat flows from high temperature to lower temperature at a rate that is equal to the product of the temperature gradient and the thermal conductivity of the material. For vertical heat transfer in a column of soil, the heat flux by conduction F (W m^{-2}) is

$$F = -\kappa \frac{\partial T}{\partial z}, \tag{5.1}$$

where κ is thermal conductivity (W m^{-1} K^{-1}), T is temperature (K), and z is depth (m). The term $\partial T/\partial z$ is the vertical temperature gradient (K m^{-1}) and is positive when temperature decreases with depth in the soil. The negative sign in (5.1) denotes that heat transfer is negative in the downward direction and positive in the upward direction.

A volume of soil warms if it gains more heat than it loses and cools if it has a net loss of heat. One-dimensional energy conservation requires that

$$c_v \frac{\partial T}{\partial t} = -\frac{\partial F}{\partial z} = \frac{\partial}{\partial z}\left(\kappa \frac{\partial T}{\partial z}\right). \tag{5.2}$$

In this equation, c_v is volumetric heat capacity (J m^{-3} K^{-1}), which is the product of soil density (kg m^{-3}) and mass specific heat (J kg^{-1} K^{-1}) ; t is time (s); $\partial T/\partial t$ is the change in temperature with time (K s^{-1}); and $\partial F/\partial z$ is the change in heat flux with depth (J m^{-3} s^{-1}). The negative sign ensures that temperature increases when there is a net gain of heat. If heat capacity and thermal conductivity do not vary with depth, (5.2) simplifies to

$$\frac{\partial T}{\partial t} = \frac{\kappa}{c_v} \frac{\partial^2 T}{\partial z^2}. \tag{5.3}$$

Thermal conductivity measures the heat flow in unit time by conduction through a unit thickness of a unit area of material across a unit temperature gradient. Heat capacity indicates the amount of heat required to change the temperature of a unit volume of material by 1 K. Thermal conductivity determines the rate of heat transfer for a unit temperature gradient, and heat capacity determines the temperature change resulting from this heat transfer. Soils with a large thermal conductivity gain and lose energy faster than soils with a small thermal conductivity. Soils with a small heat capacity warm and cool faster, for a given heat flux, than soils with a large heat capacity.

The derivation of (5.2) is understood by considering the energy balance of a volume of soil with horizontal area $\Delta x \Delta y$ and thickness Δz and in which energy is transferred only in the vertical dimension (Figure 5.2). The flux of energy into the soil across the cross-sectional area $\Delta x \Delta y$ is $F_{in} \Delta x \Delta y$ (J s^{-1}). Similarly, the energy flow out of the soil is similarly $F_{out} \Delta x \Delta y$. Energy conservation requires that the difference between the rate at which energy enters the soil volume at the top and exits the soil volume at the bottom must equal the rate of change in heat storage. The change in heat storage in the volume $\Delta x \Delta y \Delta z$ over the time interval Δt is therefore equal to

$$c_v \frac{\Delta T}{\Delta t} \Delta x \Delta y \Delta z = -(F_{in} - F_{out}) \Delta x \Delta y, \qquad (5.4)$$

so that

$$c_v \frac{\Delta T}{\Delta t} = -\frac{\Delta F}{\Delta z}. \qquad (5.5)$$

This is the finite difference approximation of (5.2). The left-hand side of the equation denotes the change in heat storage in a unit volume of soil, and the right-hand side is the vertical change in heat flux.

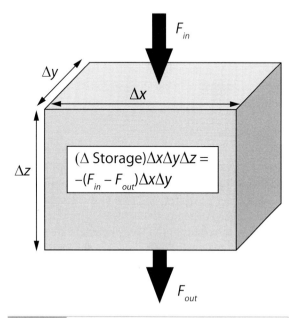

Figure 5.2 Energy balance for a soil volume with heat fluxes F_{in} entering the volume and F_{out} exiting the volume.

Question 5.1 The heat flux at depth $z_i = -5$ cm is $F_i = -150$ W m^{-2}, and the heat flux below this layer at $z_{i+1} = -10$ cm is $F_{i+1} = -120$ W m^{-2}. Calculate the change in temperature over one hour if heat capacity is $c_v = 2.5$ MJ m^{-3} K^{-1}. Compare this result with the temperature change if the soil has the heat capacity of water ($c_v = 4.2$ MJ m^{-3} K^{-1}).

5.3 | Numerical Implementation

For most applications, (5.2) is solved using finite difference methods. This numerical technique replaces the differential equation with difference equations in which the derivatives are approximated by finite differences (Appendix A4). These are obtained by representing the soil as a network of discrete nodal points that vary in space and time. Consider a soil as described in Figure 5.3. The soil consists of N layers, each with thickness Δz. Temperature, thermal conductivity, and heat capacity are defined at the midpoint of each soil layer. These properties are uniform over the layer. Depth decreases in the downward direction so that the surface is $z_0 = 0$ and $z_{i+1} < z_i$ (i.e., depths are negative distances from the surface). Temperature is

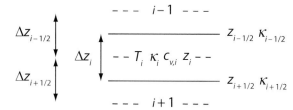

Figure 5.3 Cell-centered representation of soil as discrete layers. Shown is a single soil layer with thickness Δz_i. The depth $z_{i-1/2}$ is the interface between layer i and the adjacent layer $i-1$ above, and $z_{i+1/2}$ is the interface between i and $i+1$ below. Depths are negative distances from the surface so that $z_{i+1/2} = z_{i-1/2} - \Delta z_i$. The depth z_i is defined at the center of the layer so that $z_i = (z_{i-1/2} + z_{i+1/2})/2$. $\Delta z_{i\pm1/2}$ is the grid spacing between i and $i \pm 1$. Temperature T_i, thermal conductivity κ_i, and heat capacity $c_{v,i}$ are defined at the center of the layer at depth z_i and are uniform over the layer. An effective thermal conductivity $\kappa_{i\pm1/2}$ is defined at the interfaces between soil layers at depths $z_{i\pm1/2}$.

stepped forward in time from T^n at time n to T^{n+1} at time $n + 1$ over the time interval Δt.

Using a forward difference approximation at time n to represent the time derivative and a central difference approximation to represent the spatial derivative, (5.3) can be written as

$$\frac{T_i^{n+1} - T_i^n}{\Delta t} = \frac{\kappa}{c_v}\left(\frac{T_{i-1}^n - 2T_i^n + T_{i+1}^n}{\Delta z^2}\right). \qquad (5.6)$$

In this solution, κ, c_v, and Δz are invariant with soil depth. Rearranging terms gives an equation for temperature at time $n + 1$ in which

$$T_i^{n+1} = MT_{i-1}^n + (1 - 2M)T_i^n + MT_{i+1}^n, \qquad (5.7)$$

with

$$M = \frac{\kappa}{c_v}\frac{\Delta t}{\Delta z^2}. \qquad (5.8)$$

This is known as an explicit solution. The temperature T_i^{n+1} is calculated from T_{i-1}^n, T_i^n, and T_{i+1}^n known from the previous time step (Figure 5.4a). The explicit solution has only one unknown temperature at time $n + 1$ for each soil layer. For a soil with N layers, this gives N linear equations, each of which is solved independently for temperature. The

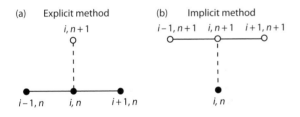

(a) Explicit method   (b) Implicit method

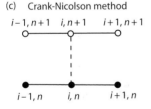

(c) Crank-Nicolson method

Figure 5.4 Geometric arrangement of nodal groups for (a) the explicit method, (b) the implicit method, and (c) the Crank–Nicolson method. The symbols i and n denote space and time, respectively. Open circles show the new point at which the solution is obtained. Filled circles show known points whose values are used to calculate the new point. Dashed lines connect points that are used to calculate time derivatives. Solid lines connect points that are used to calculate spatial derivatives.

equation is first-order accurate in time, and the solution is numerically stable and converges to the exact solution for $M \leq 1/2$. This constraint means that the time step Δt must decrease as Δz decreases.

A better solution is obtained from the implicit method, which avoids the numerical limitations of the explicit method. It evaluates the spatial derivatives using the unknown temperatures at time $n + 1$ and uses a backward difference approximation at time $n + 1$ for the time derivative so that

$$\frac{T_i^{n+1} - T_i^n}{\Delta t} = \frac{\kappa}{c_v}\left(\frac{T_{i-1}^{n+1} - 2T_i^{n+1} + T_{i+1}^{n+1}}{\Delta z^2}\right) \qquad (5.9)$$

and

$$-MT_{i-1}^{n+1} + (1 + 2M)T_i^{n+1} - MT_{i+1}^{n+1} = T_i^n. \qquad (5.10)$$

The implicit method is also first-order accurate in time but is numerically stable and convergent for $M > 0$. Equation (5.10) has three unknown temperatures (Figure 5.4b), and it requires solving a system of N simultaneous linear equations. This is evident when (5.10) is written in matrix form (Appendix A6) as:

$$
\begin{bmatrix}
1+3M & -M & 0 & 0 & 0 & 0 \\
-M & 1+2M & -M & 0 & 0 & 0 \\
0 & -M & 1+2M & -M & 0 & 0 \\
0 & 0 & \ddots & \ddots & \ddots & 0 \\
0 & 0 & 0 & -M & 1+2M & -M \\
0 & 0 & 0 & 0 & -M & 1+M
\end{bmatrix}
\times
\begin{bmatrix}
T_1^{n+1} \\
T_2^{n+1} \\
T_3^{n+1} \\
\vdots \\
T_{N-1}^{n+1} \\
T_N^{n+1}
\end{bmatrix}
=
\begin{bmatrix}
T_1^n + 2MT_0^{n+1} \\
T_2^n \\
T_3^n \\
\vdots \\
T_{N-1}^n \\
T_N^n
\end{bmatrix}
\qquad (5.11)
$$

in which T_0 is a specified boundary condition and $F_N = 0$. Equation (5.11) is a tridiagonal system of equations and can be solved efficiently with numerical methods (Appendix A8).

The Crank–Nicolson method (Crank and Nicolson 1947) increases the accuracy of the numerical solution by combining the explicit method with fluxes evaluated at time n and the implicit method with fluxes evaluated at $n + 1$ (Figure 5.4c). The actual flux is taken as the average of these two times with

$$\frac{T_i^{n+1} - T_i^n}{\Delta t} = \frac{\kappa}{c_v}\left(\frac{T_{i-1}^n - 2T_i^n + T_{i+1}^n}{2\Delta z^2} + \frac{T_{i-1}^{n+1} - 2T_i^{n+1} + T_{i+1}^{n+1}}{2\Delta z^2}\right).$$

(5.12)

This is a tridiagonal system of linear equations with the form

$$-\frac{M}{2}T_{i-1}^{n+1} + (1+M)T_i^{n+1} - \frac{M}{2}T_{i+1}^{n+1} = \frac{M}{2}T_{i-1}^n + (1-M)T_i^n$$
$$+ \frac{M}{2}T_{i+1}^n.$$

(5.13)

The Crank–Nicolson method is second-order accurate in both space and time. It combines the numerical stability of the implicit method with the accuracy of a second-order solution.

The preceding equations require constant thermal conductivity, heat capacity, and thickness in all soil layers. In models, however, it is desirable to have variable layer thickness (e.g., higher vertical resolution near the surface to resolve the diurnal cycle); and thermal properties may vary with depth because of changes in mineral composition, soil structure, and soil moisture. Derivation of heat transfer from an energy balance approach provides a general finite difference equation that accounts for variable thermal properties and soil layer thickness. Here, such a derivation is given for the implicit method. With reference to Figure 5.5b, the heat flux F_i from soil layer i to $i+1$ is

$$F_i = -(T_i - T_{i+1})\bigg/\left(\frac{\Delta z_i}{2\kappa_i} + \frac{\Delta z_{i+1}}{2\kappa_{i+1}}\right).$$

(5.14)

This equation is derived so that the energy flux from z_i to $z_{i+1/2}$ equals the flux from $z_{i+1/2}$ to z_{i+1} (so as to eliminate $T_{i+1/2}$). It is mathematically equivalent to

$$F_i = -\frac{\kappa_{i+1/2}}{\Delta z_{i+1/2}}(T_i - T_{i+1}),$$

(5.15)

in which $\Delta z_{i+1/2} = z_i - z_{i+1}$ is the distance between nodes i and $i+1$ and thermal conductivity is

$$\kappa_{i+1/2} = \frac{\kappa_i \kappa_{i+1}(z_i - z_{i+1})}{\kappa_i(z_{i+1/2} - z_{i+1}) + \kappa_{i+1}(z_i - z_{i+1/2})}.$$

(5.16)

This is the harmonic mean of the adjacent nodal conductivities κ_i and κ_{i+1}, and for constant Δz simplifies to $\kappa_{i+1/2} = \left[(\kappa_i^{-1} + \kappa_{i+1}^{-1})/2\right]^{-1}$ (i.e., the harmonic mean is the reciprocal of the arithmetic mean of the reciprocals). The energy balance for soil layer i is

$$\frac{c_{v,i}\Delta z_i}{\Delta t}(T_i^{n+1} - T_i^n) = -F_{i-1}^{n+1} + F_i^{n+1}$$

(5.17)

and

$$\frac{c_{v,i}\Delta z_i}{\Delta t}(T_i^{n+1} - T_i^n) = \frac{\kappa_{i-1/2}}{\Delta z_{i-1/2}}(T_{i-1}^{n+1} - T_i^{n+1})$$
$$- \frac{\kappa_{i+1/2}}{\Delta z_{i+1/2}}(T_i^{n+1} - T_{i+1}^{n+1}).$$

(5.18)

This is a tridiagonal system of equations that, after rearranging terms, can be written as

$$-\frac{\kappa_{i-1/2}}{\Delta z_{i-1/2}}T_{i-1}^{n+1} + \left(\frac{c_{v,i}\Delta z_i}{\Delta t} + \frac{\kappa_{i-1/2}}{\Delta z_{i-1/2}} + \frac{\kappa_{i+1/2}}{\Delta z_{i+1/2}}\right)T_i^{n+1}$$
$$- \frac{\kappa_{i+1/2}}{\Delta z_{i+1/2}}T_{i+1}^{n+1} = \frac{c_{v,i}\Delta z_i}{\Delta t}T_i^n$$

(5.19)

or, more generally,

$$a_i T_{i-1}^{n+1} + b_i T_i^{n+1} + c_i T_{i+1}^{n+1} = d_i.$$

(5.20)

Table 5.1 gives the matrix elements a, b, c, and d. The solution is equivalent to (5.11) for constant values of c_v, κ, and Δz. Sometimes it is more convenient to solve for the change in temperature rather than for temperature directly (e.g., as in Chapter 7). In this form, (5.19) is

$$-\frac{\kappa_{i-1/2}}{\Delta z_{i-1/2}}\Delta T_{i-1} + \left(\frac{c_{v,i}\Delta z_i}{\Delta t} + \frac{\kappa_{i-1/2}}{\Delta z_{i-1/2}} + \frac{\kappa_{i+1/2}}{\Delta z_{i+1/2}}\right)\Delta T_i$$
$$- \frac{\kappa_{i+1/2}}{\Delta z_{i+1/2}}\Delta T_{i+1} = \frac{\kappa_{i-1/2}}{\Delta z_{i-1/2}}(T_{i-1}^n - T_i^n) - \frac{\kappa_{i+1/2}}{\Delta z_{i+1/2}}(T_i^n - T_{i+1}^n),$$

(5.21)

with $\Delta T_i = T_i^{n+1} - T_i^n$. The left-hand side of the equation is comparable to (5.19), but now given in terms of ΔT. The right-hand side uses the fluxes $-F_{i-1}^n$ and F_i^n evaluated with temperatures at time n.

Special care must be taken for the first and last soil layers. The boundary condition at the bottom of the soil column ($i = N$) is typically taken as zero heat flux so that $F_N^{n+1} = 0$ (Figure 5.5c). This requirement necessitates use of soil that is several meters deep when simulating the annual cycle. Two different boundary conditions are possible for the first soil layer ($i = 1$). The surface temperature T_0^{n+1} can be specified as the boundary condition (this is known as a Dirichlet boundary condition). In this case, the

Table 5.1 Terms in (5.20) for the implicit formulation of soil temperature

Layer	a_i	b_i	c_i	d_i
$i = 1$	0	$\dfrac{c_{v,i}\Delta z_i}{\Delta t} + \dfrac{\kappa_{1/2}}{\Delta z_{1/2}} - c_i$	$-\dfrac{\kappa_{i+1/2}}{\Delta z_{i+1/2}}$	$\dfrac{c_{v,i}\Delta z_i}{\Delta t}T_i^n + \dfrac{\kappa_{1/2}}{\Delta z_{1/2}}T_0^{n+1}$
$1 < i < N$	$-\dfrac{\kappa_{i-1/2}}{\Delta z_{i-1/2}}$	$\dfrac{c_{v,i}\Delta z_i}{\Delta t} - a_i - c_i$	$-\dfrac{\kappa_{i+1/2}}{\Delta z_{i+1/2}}$	$\dfrac{c_{v,i}\Delta z_i}{\Delta t}T_i^n$
$i = N$	$-\dfrac{\kappa_{i-1/2}}{\Delta z_{i-1/2}}$	$\dfrac{c_{v,i}\Delta z_i}{\Delta t} - a_i$	0	$\dfrac{c_{v,i}\Delta z_i}{\Delta t}T_i^n$

Note: Boundary conditions are T_0^{n+1} for the first layer ($i = 1$) and zero heat flux for the bottom layer ($i = N$).

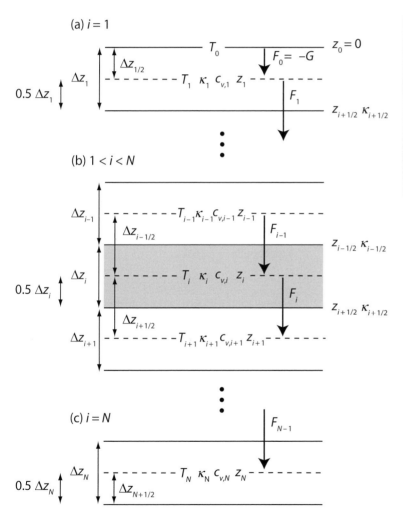

(a) $i = 1$

(b) $1 < i < N$

(c) $i = N$

Figure 5.5 Multilayer soil heat transfer in a cell-centered grid oriented such that $i = 1$ is the top soil layer at the surface and $i = N$ is the bottom soil layer. Shown are (a) the first soil layer ($i = 1$), (b) layers $1 < i < N$ depicted generally as three soil layers denoted $i - 1$, i, and $i + 1$, and (c) the bottom soil layer ($i = N$). The surface temperature T_0 or the flux F_0 provides the upper soil boundary condition, with zero heat flux at the bottom of the soil as the lower boundary condition.

heat flux into the soil G (W m^{-2}; taken as positive into the soil), with reference to Figure 5.5a, is

$$G = -F_0^{n+1} = \frac{\kappa_1}{\Delta z_{1/2}}\left(T_0^{n+1} - T_1^{n+1}\right), \tag{5.22}$$

and this equation is substituted for $-F_{i-1}^{n+1}$ in (5.17). Alternatively, the flux of energy into the soil can be specified directly (this is known as a Neumann boundary condition) so that $-F_{i-1}^{n+1} = G$. The total change in heat in the soil column equals the energy flux into the soil, and the numerical solution conserves energy according to

$$\sum_{i=1}^{N} \frac{c_{v,i} \Delta z_i}{\Delta t} \left(T_i^{n+1} - T_i^n \right) = G. \tag{5.23}$$

Question 5.2 Derive the soil heat flux given by (5.14), and then show that this is algebraically equivalent to (5.15) with thermal conductivity from (5.16).

Question 5.3 Derive (5.21) so that (5.18) is given in terms of $\Delta T_i = T_i^{n+1} - T_i^n$ rather than T_i^{n+1}.

Question 5.4. Table 5.1 gives the tridiagonal coefficients a_i, b_i, c_i, and d_i for the implicit method. Derive the same coefficients for the Crank–Nicolson method.

5.4 | Soil Thermal Properties

Thermal conductivity and heat capacity vary depending on mineral composition, organic matter content, porosity, and the water content of soils. Soils consist of solid particles and the voids around these particles. The total volume of voids is the pore space, or porosity. These pores can fill with water when the soil is wet or are mostly air when the soil is dry. Most soils are a mixture of air and water. The water fraction can consist of liquid water and ice. The overall thermal conductivity of a soil is an average of the conductivity of its solid, air, liquid water, and ice fractions (Table 5.2). Quartz has a thermal conductivity three times greater than clay minerals, and soils such as sands with high quartz content have a greater thermal conductivity than do clay soils. Organic material has an extremely low thermal conductivity, and peat soils with high organic matter content have a thermal conductivity that is one-quarter to one-third that of mineral soils. Air and water have a lower thermal conductivity than do mineral particles, but the thermal conductivity of water is more than 20 times that of air. Consequently, the thermal conductivity of soil increases as soil moisture increases. The thermal conductivity of frozen soil is larger than

Table 5.2 Thermal conductivity and heat capacity for soil components

	Thermal conductivity (W m^{-1} K^{-1})	Heat capacity (MJ m^{-3} K^{-1})
Quartz	8.80	2.12
Clay minerals	2.92	2.44
Organic matter	0.25	2.50
Liquid water	0.57	4.19
Ice	2.18	1.88
Air	0.025	0.0012

Source: From de Vries (1963).

that of unfrozen soil because ice conducts heat at a greater rate than does liquid water. Similarly, heat capacity increases with greater soil moisture because the heat capacity of water is three orders of magnitude greater than that of air.

Farouki (1981) described the Johansen method to calculate thermal conductivity. In this approach, thermal conductivity varies between dry and saturated values depending on soil moisture and accounting for differences in frozen and unfrozen states. Thermal conductivity varies linearly between dry and saturated values – denoted as κ_{dry} and κ_{sat}, respectively – so that

$$\kappa = \kappa_{dry} + \left(\kappa_{sat} - \kappa_{dry} \right) K_e. \tag{5.24}$$

The weighting factor is the dimensionless Kersten number K_e, which has a value of 0–1 depending on relative soil moisture S_e. For unfrozen soil:

$$K_e = \begin{cases} 1.0 + 0.7 \log_{10} S_e & \text{coarse-texture soil } (S_e > 0.05 \\ 1.0 + \log_{10} S_e & \text{fine-texture soil } (S_e > 0.1) \end{cases}, \tag{5.25}$$

and fine-texture soil has a sand content of less than 50%. For frozen soil:

$$K_e = S_e. \tag{5.26}$$

In these equations, $S_e = \theta/\theta_{sat}$ is relative wetness, with θ volumetric water content (m^3 H$_2$O m^{-3} soil) and θ_{sat} volumetric water content at saturation (also equal to porosity).

The dry and saturated thermal conductivities depend on soil properties. The thermal conductivity of dry soil varies with bulk density ρ_b (kg m^{-3}) according to

$$\kappa_{dry} = \frac{0.135\rho_b + 64.7}{2700 - 0.947\rho_b}, \qquad (5.27)$$

where $\rho_b = 2700(1 - \theta_{sat})$ and the density of soil solids is 2700 kg m^{-3}. The thermal conductivity of saturated soil is calculated from the thermal conductivity of the components (solids κ_{sol}, water κ_{wat}, ice κ_{ice}) and their respective volume fractions. For unfrozen soil:

$$\kappa_{sat} = \kappa_{sol}^{1-\theta_{sat}} \kappa_{wat}^{\theta_{sat}}. \qquad (5.28)$$

And for frozen soil:

$$\kappa_{sat} = \kappa_{sol}^{1-\theta_{sat}} \kappa_{ice}^{\theta_{sat}}. \qquad (5.29)$$

A general expression allowing a mixture of liquid water and ice is

$$\kappa_{sat} = \kappa_{sol}^{1-\theta_{sat}} \kappa_{wat}^{f_u \theta_{sat}} \kappa_{ice}^{(1-f_u)\theta_{sat}}. \qquad (5.30)$$

This equation recognizes that even at temperatures below freezing, the total water consists of unfrozen θ_{liq} and frozen θ_{ice} water $(\theta = \theta_{liq} + \theta_{ice})$, and $f_u = \theta_{liq}/\theta$ is the fraction of the total water that is unfrozen. Representative values are $\kappa_{wat} = 0.57$ and $\kappa_{ice} = 2.29$ W m^{-1} K^{-1}. The thermal conductivity of soil solids varies with the quartz content of soil. The Johansen method described by Farouki (1981) approximates κ_{sol} as

$$\kappa_{sol} = \kappa_q^q \kappa_o^{1-q}, \qquad (5.31)$$

where here q is the quartz content as a fraction of the total soil solids; $\kappa_q = 7.7$ W m^{-1} K^{-1} is the thermal conductivity of quartz; and $\kappa_o = 2.0$ W m^{-1} K^{-1} is the thermal conductivity of other minerals for soils with $q > 0.2$ and $\kappa_o = 3.0$ W m^{-1} K^{-1} for $q \leq 0.2$. The accuracy of this equation depends on the specified thermal conductivity of quartz and the quartz fraction. The quartz content can be approximated by the sand content (Peters-Lidard et al. 1998), though this is not necessarily correct (Lu et al. 2007).

The heat capacity of air is negligible (Table 5.2) so that the heat capacity of soil is given by the weighted fraction of the heat capacity of solids, water, and ice (de Vries 1963), and

$$c_v = (1 - \theta_{sat})c_{v,sol} + \theta_{liq}c_{v,wat} + \theta_{ice}c_{v,ice}. \qquad (5.32)$$

The heat capacity of water is $c_{v,wat} = \rho_{wat}c_{wat} = 4.19$ MJ m^{-3} K^{-1} and for ice is $c_{v,ice} = \rho_{ice}c_{ice} = 1.94$ MJ m^{-3} K^{-1} (Table A.4), and a representative value for soil solids is $c_{v,sol} = 1.926$ MJ m^{-3} K^{-1} (de Vries 1963) .

Equation (5.32) can be applied assuming all water is either liquid or ice to calculate the unfrozen and frozen heat capacity, respectively.

Figure 5.6 shows thermal conductivity for sand, sandy loam, loam, clay loam, and clay soils as a function of soil moisture using representative values for θ_{sat} and sand content from Clapp and Hornberger (1978). For all texture classes, thermal conductivity increases with greater soil moisture. Variation in thermal conductivity with soil texture arises from differences in θ_{sat}, which affects κ_{dry} and κ_{sat}, and from differences in sand content, which determine the quartz fraction (and through this κ_{sat}) and K_e. For similar soil moisture, unfrozen thermal conductivity decreases from sandy soils to loamy soils to clay soils. Quartz content is the primary determinant of this variation in thermal conductivity; differences in K_e between fine and coarse soil are less important except at low soil moisture. Frozen soils generally have a higher thermal conductivity than unfrozen soil at similar soil moisture because of the larger thermal conductivity of ice. Heat capacity also increases with greater soil moisture because of the high specific heat of water (Figure 5.7). Differences in heat capacity among soil texture classes are relatively minor and arise from variation in θ_{sat}. The heat capacity of frozen soil is less than that of unfrozen soil because of the smaller heat capacity of ice compared with water.

Snow has a much lower thermal conductivity compared with soil. Thermal conductivity increases with the bulk density of snow. A general relationship used in many models is

$$\kappa_{snow} = \kappa_{air} + \left(7.75 \times 10^{-5}\rho_{snow} + 1.105 \times 10^{-6}\rho_{snow}^2\right)$$
$$\times (\kappa_{ice} - \kappa_{air}), \qquad (5.33)$$

where ρ_{snow} is bulk density (kg m^{-3}) and κ_{air} is the thermal conductivity of air (Jordan 1991). The thermal conductivity of snow varies from about 0.05 W m^{-1} K^{-1} for fresh snow $(\rho_{snow} < 100$ kg m$^{-3})$ to 0.7 W m^{-1} K^{-1} for melting snow $(\rho_{snow} = 500$ kg m$^{-3})$.

Question 5.5 Quartz has a thermal conductivity that is three times that of clay minerals (Table 5.2). A soil with high sand content might, therefore, be expected to have a higher thermal conductivity compared with a soil with

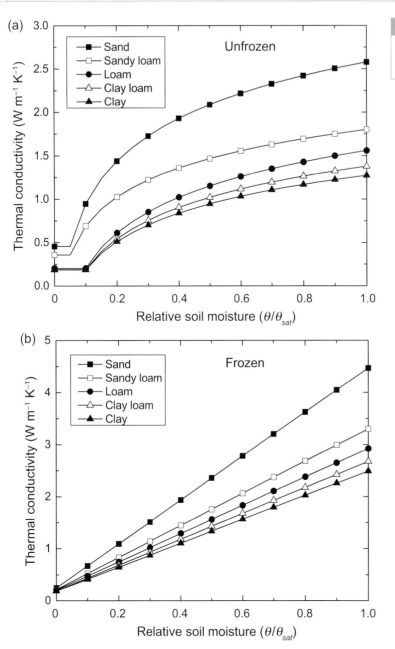

(a) Unfrozen — Sand, Sandy loam, Loam, Clay loam, Clay

(b) Frozen — Sand, Sandy loam, Loam, Clay loam, Clay

Figure 5.6 Thermal conductivity for (a) unfrozen and (b) frozen sand, sandy loam, loam, clay loam, and clay soils in relation to soil moisture.

less sand. (a) Use the Johansen method to calculate the thermal conductivity of a clay loam (32% sand, θ_{sat}= 0.476) with a volumetric water content $\theta = 0.75\theta_{sat}$. Assume all the water is unfrozen. Repeat the calculations for a silty clay loam (10% sand, θ_{sat}= 0.477). How does thermal conductivity differ between the two soils? Does thermal conductivity increase or decrease with greater percentage of sand? Explain why. (b) Graph the relationship between κ and κ_{sol} for water contents equal to θ_{sat}, $0.75\theta_{sat}$, $0.50\theta_{sat}$, and $0.25\theta_{sat}$. How does this explain the previous results in (a)?

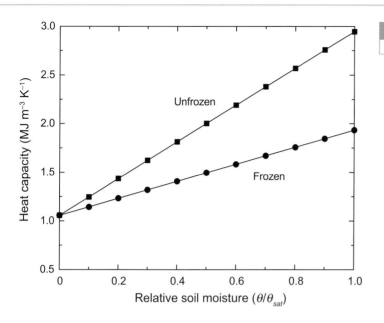

Figure 5.7 Unfrozen and frozen heat capacity for a loam soil in relation to soil water content.

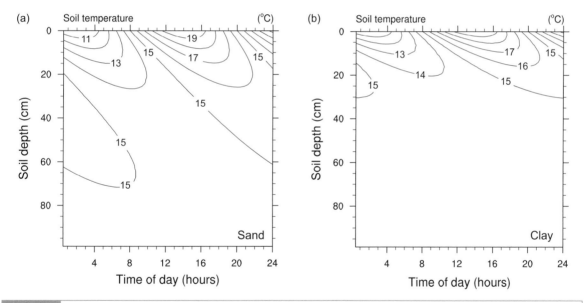

Figure 5.8 Simulated diurnal cycle of soil temperature for (a) sand and (b) clay soils with a periodic forcing in which surface temperature has a mean of 15°C and a diurnal range of 10°C. Minimum surface temperature occurs at 0200, and the maximum occurs at 1400. Thermal properties are invariant with depth. Sand: $\kappa = 2.42$ W m^{-1} K^{-1}, $c_v = 2.49$ MJ m^{-3} K^{-1}. Clay: $\kappa = 1.17$ W m^{-1} K^{-1}, $c_v = 2.61$ MJ m^{-3} K^{-1}. These are for unfrozen soil with $\theta = 0.8\theta_{sat}$. Here, $\Delta z = 2.5$ cm and $\Delta t = 1800$ s. Results are shown after soil temperature has spun up from initial conditions.

5.5 | Diurnal and Annual Cycles

Figure 5.8 illustrates the diurnal cycle of soil temperature for sand and clay soils. Surface temperature is represented by a sine wave with a mean of 15°C and a diurnal range of 10°C. Both soils warm during the day and cool at night, but the amplitude of the periodic forcing decreases with greater soil depth. Consider, for example, the sand.

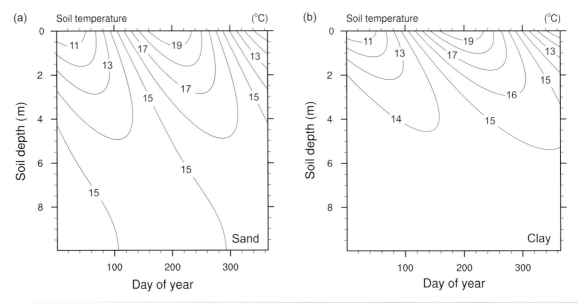

Figure 5.9 As in Figure 5.8, but for an annual cycle with a periodic forcing in which surface temperature has an annual mean of 15°C and an annual range of 10°C. Minimum surface temperature occurs on day 15, and the maximum is day 197.5. Here, $\Delta z = 5$ cm to accommodate more soil layers.

Temperature at a depth of 10 cm varies by about 5°C over the day; i.e., the surface amplitude has been reduced by about one-half. Temperature at a depth of 40 cm varies by less than 1°C over the day. Diurnal variation is minimal at deeper depths, and temperature is approximately equal to the diurnal mean. The amplitude of the surface forcing penetrates to a greater depth in sand than in clay because of the higher thermal conductivity of sand. It is also evident that the time of maximum temperature occurs later in the day as soil depth increases. In sand, for example, maximum temperature occurs at 1400 h at the surface. The maximum temperature occurs at about 1630 h at a depth of 10 cm and at about 1900 h at a depth of 20 cm.

Similar behavior is seen over the course of a year for sand and clay exposed to a mean surface temperature of 15°C and an annual range of 10°C (Figure 5.9). Soil temperature increases with depth during the cold season and decreases with depth during the warm season. Maximum soil temperature occurs later in the year as soil depth increases. The annual amplitude penetrates to a greater depth in sand than in clay. Temperature at a depth of 6 m for clay and 10 m or more for sand is nearly constant at the annual mean temperature. Over the course of a year, the surface amplitude penetrates to greater depth than that of the diurnal cycle. When simulating the annual cycle, deep soils are needed to satisfy the requirement of zero heat flux at the bottom of the soil column. There is a compromise between soil layer thickness and total soil depth. Models often utilize variable soil layer thickness with more soil layers near the surface (thin Δz) to resolve the diurnal cycle and progressively thicker layers (large Δz) with increasing depth to resolve the annual cycle.

Question 5.6 Many soils have a layer of organic material on the surface. The thermal conductivity of organic material is considerably less than that of mineral soil (a representative value is $\kappa = 0.5$ W m^{-1} K^{-1}). Describe the effect of an organic layer on the diurnal cycle of soil temperature.

Question 5.7 Describe the effects of snow cover on soil temperature.

5.6 | Phase Change

The freezing of soil water or melting of soil ice releases or absorbs energy, respectively. Formation of ice releases latent heat, and temperature remains constant at the freezing point while soil water freezes. Similarly, melting of ice consumes energy, during which temperature does not increase. Latent heat of fusion is the amount of energy required to convert a unit mass of frozen water to liquid. This transition requires 334 J g^{-1} (Table 3.6). Freezing liquid water to ice releases a similar amount of energy. The total energy involved in phase change depends on soil moisture. For a volumetric water content θ, the energy (J m^{-3}) required to freeze soil is $L_f\rho_{wat}\theta$, where $L_f = 0.334$ MJ kg^{-1} is the latent heat of fusion of water and $\rho_{wat} = 1000$ kg m^{-3} is the density of water. In coarse-grain soil such as sands, all the water present in the soil changes phase at a temperature close to 0°C ($T_f = 273.15$ K). These soils may be treated with good approximation as either completely frozen or unfrozen. In fine-grain soils such as silts and clays, some soil water remains unfrozen even at temperatures below freezing. The amount of unfrozen water decreases as temperature decreases, and the latent heat release occurs over some temperature range $T_f - \Delta T_f$.

A simple way to account for freezing and thawing is to add the latent heat associated with phase change to the heat conduction equation to yield an apparent heat capacity (Lunardini 1981). Including a latent heat source term as the unfrozen water θ_{liq} freezes, the heat conduction equation becomes

$$c_v \frac{\partial T}{\partial t} = \frac{\partial}{\partial z}\left(\kappa \frac{\partial T}{\partial z}\right) - L_f\rho_{wat}\frac{\partial \theta_{liq}}{\partial t}. \tag{5.34}$$

The second term on the right-hand side of the equation is a source of energy during freezing ($\partial\theta_{liq}/\partial t < 0$) and a sink of energy during melting ($\partial\theta_{liq}/\partial t > 0$). The change in liquid water can be expressed as $\partial\theta_{liq}/\partial t = (\partial\theta_{liq}/\partial T)(\partial T/\partial t)$, and (5.34) can be rewritten as

$$\left(c_v + L_f\rho_{wat}\frac{\partial\theta_{liq}}{\partial T}\right)\frac{\partial T}{\partial t} = \frac{\partial}{\partial z}\left(\kappa\frac{\partial T}{\partial z}\right). \tag{5.35}$$

Written this way, the term in parentheses on the left-hand side of the equation represents an effective heat capacity. Solution of this equation requires an expression for $\partial\theta_{liq}/\partial T$ to relate the amount of liquid water to temperature. In practice, however, the entire latent heat of fusion can be associated with a small finite temperature range between the freezing point and $T_f - \Delta T_f$ so that the effective heat capacity for a soil with water content θ is

$$c_v = \begin{cases} c_{vu} & T > T_f \\ c_{vf} + \dfrac{L_f\rho_{wat}\theta}{\Delta T_f} & T_f - \Delta T_f \leq T \leq T_f, \\ c_{vf} & T < T_f - \Delta T_f \end{cases} \tag{5.36}$$

where c_{vu} and c_{vf} are the unfrozen and frozen volumetric heat capacity, respectively. The apparent heat capacity is the heat capacity of soil constituents plus a term that accounts for the latent heat of fusion.

Lunardini (1981) described the apparent heat capacity formulation of Bonacina et al. (1973), in which the latent heat of fusion is added to the heat capacity over the temperature range $T_f \pm \Delta T_f$. With reference to Figure 5.10, the area given by $c_v 2\Delta T_f$ is equal to the area given by $c_{vf}\Delta T_f + c_{vu}\Delta T_f$ plus the latent heat of fusion so that

$$c_v 2\Delta T_f = \int_{T_f-\Delta T_f}^{T_f} c_{vf}dT + L_f\rho_{wat}\theta + \int_{T_f}^{T_f+\Delta T_f} c_{vu}dT, \tag{5.37}$$

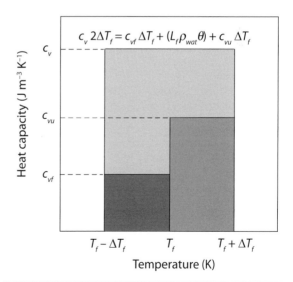

Figure 5.10 Derivation of the apparent heat capacity c_v over the temperature range $T_f \pm \Delta T_f$ using frozen c_{vf} and unfrozen c_{vu} heat capacity and latent heat of fusion $L_f\rho_{wat}\theta$.

and the apparent heat capacity is

$$c_v = \begin{cases} c_{vu} & T > T_f + \Delta T_f \\ \dfrac{c_{vf} + c_{vu}}{2} + \dfrac{L_f \rho_{wat} \theta}{2\Delta T_f} & T_f - \Delta T_f \leq T \leq T_f + \Delta T_f \\ c_{vf} & T < T_f - \Delta T_f \end{cases}.$$

$$(5.38)$$

Similarly, thermal conductivity is expressed as a linear function of frozen κ_f and unfrozen κ_u conductivities over the temperature range $T_f \pm \Delta T_f$ so that

$$\kappa = \begin{cases} \kappa_u & T > T_f + \Delta T_f \\ \kappa_f + \dfrac{\kappa_u - \kappa_f}{2\Delta T_f}\left(T - T_f + \Delta T_f\right) & T_f - \Delta T_f \leq T \leq T_f + \Delta T_f \\ \kappa_f & T < T_f - \Delta T_f \end{cases}.$$

$$(5.39)$$

The accuracy of the apparent heat capacity method depends on the layer thickness Δz, the time step Δt, and the freezing temperature range ΔT_f. The numerical scheme simulates phase change depth and soil temperatures that evolve over time in a stepwise fashion. The width of these steps varies with Δz. The latent heat effect may be missed completely if the grid spacing is too small so that the T_f isotherm moves across the soil layer in less than one time step. Greater smoothing can be achieved with larger ΔT_f. Bonan (1991) showed that $\Delta T_f = 0.5$ K works well with $\Delta z = 0.1$ m and $\Delta t = 3600$ s, and Bonan (1996)

used the method in a global land surface model. The numerical solution using (5.38) for heat capacity and (5.39) for thermal conductivity compares favorably with an analytical solution. In this example, the initial soil temperature is 2°C, the boundary condition is $T_0 = -10°C$, $L_f \rho_{wat} \theta = 110.45$ MJ m^{-3}, $c_{vu} = 2.862$ MJ m^{-3} K^{-1}, $c_{vf} = 1.966$ MJ m^{-3} K^{-1}, $\kappa_u = 1.860$ W m^{-1} K^{-1}, and $\kappa_f = 2.324$ W m^{-1} K^{-1}. The frost penetration depth closely matches the analytical solution (Figure 5.11). The soil cools too much when phase change is excluded (Figure 5.12). With phase change, soil temperature follows the analytical solution. Soil temperature advances over time in a stepwise manner, but the error is small compared with the error arising from not accounting for phase change.

Another numerical method to account for phase change is through the concept of excess energy (Lunardini 1981). In this approach, the usual finite difference equations are written for soil temperature. During melting, temperature is set to the freezing point, and the excess energy that allows temperature to exceed the freezing point is used to melt ice. Conversely, the energy released from freezing increases temperature to the freezing point. The temperature of a soil layer undergoing phase change is held constant at the freezing point until the heat gain or loss equals the latent heat of that soil layer. Niu and Yang (2006) described such an approach, which is summarized

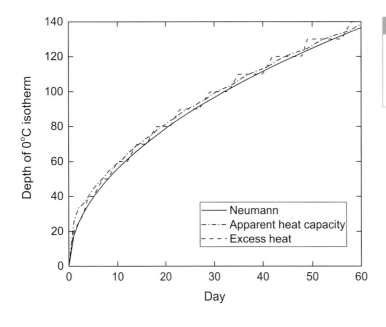

Figure 5.11 Frost penetration depth simulated with the apparent heat capacity method and the excess energy method compared with the Neumann analytical solution. The analytical solution is from Jumikis (1966) and Lunardini (1981), as described in Supplemental Program 5.3.

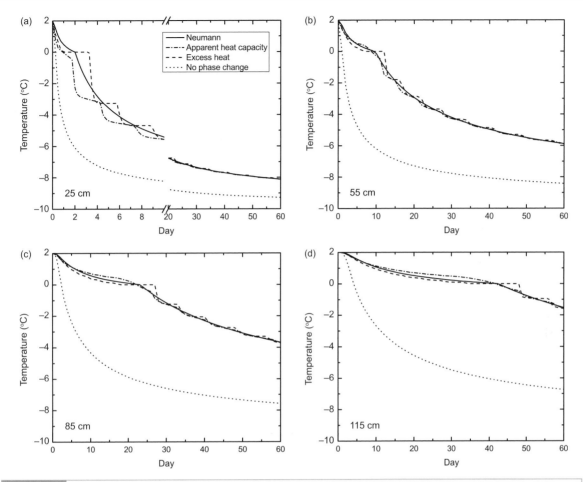

Figure 5.12 As in Figure 5.11, but for soil temperatures at (a) 25 cm, (b) 55 cm, (c) 85 cm, and (d) 115 cm. Also shown is the temperature solution without accounting for phase change. Panel (a) shows temperatures for days 0–10 and days 20–60.

as follows, and a variant of this is used in the Community Land Model.

For convenience, liquid and frozen water are expressed in the units kg H_2O m^{-2} with $W_{liq} = \rho_{wat}\theta_{liq}\Delta z$ and $W_{ice} = \rho_{ice}\theta_{ice}\Delta z$. The energy balance is

$$c_v \frac{\partial T}{\partial t} = \frac{\partial}{\partial z}\left(\kappa \frac{\partial T}{\partial z}\right) + L_f \frac{\partial W_{ice}}{\partial t}\frac{1}{\Delta z}. \qquad (5.40)$$

The second term on the right-hand side of the equation is the energy released from freezing ($\partial W_{ice}/\partial t > 0$) or consumed in thawing ($\partial W_{ice}/\partial t < 0$). Soil temperatures are first updated from time n to $n+1$ without accounting for phase change. Melting occurs in soil layers that have ice

and in which temperature at $n+1$ is above freezing. Freezing occurs in layers with liquid water and in which temperature at $n+1$ is below freezing. The temperature of a soil layer undergoing phase change is reset to T_f. The latent heat for freezing (+) or melting (−) is determined from the energy that must be added (freezing) or removed (thawing) to change temperature to the freezing point. For a layer undergoing phase change, this energy flux (W m^{-2}) is

$$H'_f = \frac{c_v \Delta z}{\Delta t}\left(T_f - T^n\right) \qquad (5.41)$$

and is subject to two constraints. In the freezing phase, it cannot exceed the latent heat released from freezing all the liquid water. The latent heat consumed by melting all the ice provides a similar constraint during the melting phase. The actual energy for phase change is

$$H_f = \begin{cases} \min\left(H'_f, \dfrac{L_f W_{liq}}{\Delta t}\right) & \text{freezing} \\[2ex] \max\left(H'_f, -\dfrac{L_f W_{ice}}{\Delta t}\right) & \text{melting} \end{cases}. \quad (5.42)$$

This energy is used to freeze or melt ice so that the change in ice is

$$\frac{\Delta W_{ice}}{\Delta t} = \frac{H_f}{L_f}, \quad (5.43)$$

and the corresponding change in unfrozen water is $\Delta W_{liq} = -\Delta W_{ice}$. In practice, (5.42) restricts melting to not exceed the amount of ice present in the soil layer ($W_{ice}^{n+1} \geq 0$); nor can the amount of ice exceed the total water (ice + liquid) from the previous time step upon freezing ($W_{ice}^{n+1} \leq W_{ice}^n + W_{liq}^n$). An alternative to (5.42) and (5.43) is to calculate the change in ice from

$$\frac{\Delta W_{ice}}{\Delta t} = \frac{H'_f}{L_f} \quad (5.44)$$

and to limit W_{ice}^{n+1} for the preceding two numerical constraints. The latent heat from phase change is then

$$H_f = \frac{L_f}{\Delta t}\left(W_{ice}^{n+1} - W_{ice}^n\right). \quad (5.45)$$

The two methods are mathematically equivalent. The residual energy is used to cool (freezing) or warm (melting) the soil layer so that soil temperature is

$$T^{n+1} = T_f - \frac{\Delta t}{c_v \Delta z}\left(H'_f - H_f\right). \quad (5.46)$$

This numerical scheme conserves energy as

$$\sum_{i=1}^{N} \frac{c_{v,i}\Delta z_i}{\Delta t}\left(T_i^{n+1} - T_i^n\right) = G + \sum_{i=1}^{N} H_{f,i}. \quad (5.47)$$

The first term on the right-hand side of the equation is the heat flux into the soil, and the second term is the total latent heat from phase change.

The excess energy method provides a reasonable approximation to phase change and allows for co-occurrence of mixed phases of water in the soil. Depth of frost penetration (Figure 5.11) and soil temperature (Figure 5.12) match an analytical solution but advance in a stepwise fashion as soil layers undergo phase change. A limitation of this approach as described here is that it does not allow for supercooled soil water. Supercooled water is the liquid water that co-occurs with ice over a range of temperatures below freezing. The excess energy method can be extended to account for supercooled water using a relationship between unfrozen water and temperature (Cherkauer and Lettenmaier 1999; Cox et al. 1999; Niu and Yang 2006; Barman and Jain 2016).

Question 5.8 Calculate the latent heat of fusion for a soil with volumetric water content $\theta = 0.3$ and another with $\theta = 0.4$. Which soil warms faster during thawing? Which soil freezes more rapidly?

5.7 | Supplemental Programs

5.1 Soil Thermal Properties: This program calculates thermal conductivity using the Johansen method, as in Figure 5.6.

5.2 Soil Temperature: This program calculates soil temperature over a diurnal cycle, as in Figure 5.8, using the implicit solution with (5.20) and Table 5.1. The diurnal cycle of temperature at the soil surface is represented as a sine wave in which surface temperature varies periodically between some maximum and minimum value. The temperature of the soil surface at time t (here in hours of the day) is

$$T_0(t) = \bar{T}_0 + A_0 \sin\left[2\pi(t-8)/24\right], \quad (5.48)$$

in which 24 hours is the period of oscillation for a diurnal cycle, $\bar{T}_0$ is the average temperature over this period, and A_0 is the amplitude (i.e., one-half the difference between maximum and minimum temperatures) over the same time period. $(t-8)/24$ is time in fraction of a day, and the 8 hour offset gives a minimum temperature at $t = 2$ h and maximum temperature at 14 h. The program calculates soil thermal properties based on soil texture and water content using the Johansen method and phase change using either the apparent heat capacity or excess energy methods.

5.3 Phase Change: This program compares the apparent heat capacity and excess energy methods to calculate phase change with the Neumann analytical solution (Lunardini 1981). For a soil that has an initial temperature $T_b > T_f$ and in which the temperature of the surface is reduced to $T_0 < T_f$ and held constant, the temperature at depth z (m) and time t (s) is

$$T(z,t) = T_0 + (T_f - T_0) \frac{\text{erf}\left[\dfrac{z}{2\sqrt{\alpha_f t}}\right]}{\text{erf}[v]} \tag{5.49}$$

for the frozen zone, and

$$T(z,t) = T_b - (T_b - T_f) \frac{1 - \text{erf}\left[\dfrac{z}{2\sqrt{\alpha_u t}}\right]}{1 - \text{erf}\left[v\sqrt{\dfrac{\alpha_f}{\alpha_u}}\right]} \tag{5.50}$$

for the unfrozen zone. In this solution, z is positive downward; α_u and α_f are the unfrozen and frozen thermal diffusivities (κ/c_v), respectively; and erf denotes the error function (Appendix A10). The frost penetration depth (m; positive downward) is $2v\sqrt{\alpha_f t}$. v varies with diffusivity, latent heat of fusion, T_b, and T_0. In this example, $T_b = 2°C$, $T_0 = -10°C$, $L_f \rho_{wat} \theta = 110.45$ MJ m^{-3}, $c_{vu} = 2.862$ MJ m^{-3} K^{-1}, $c_{vf} = 1.966$ MJ m^{-3} K^{-1}, $\kappa_u = 1.860$ W m^{-1} K^{-1}, and $\kappa_f = 2.324$ W m^{-1} K^{-1}, for which $v = 0.0006/\left(2\sqrt{\alpha_f}\right)$ (Jumikis 1966). This program was used for Figures 5.11 and 5.12.

5.8 | Modeling Projects

1. The code in Supplemental Program 5.2 calculates the diurnal cycle of soil temperature. Compare results after one simulation day, 10 simulations days, and 100 days. How does soil temperature differ between days? When does soil temperature reach equilibrium?

2. Figure 5.8a shows the diurnal cycle of soil temperature for sand. Contrast this when the soil is covered by 5 cm of organic material with $\kappa = 0.5$ W m^{-1} K^{-1}. What is the effect of an organic layer on the diurnal cycle?

3. Modify the code in Supplemental Program 5.2 to calculate the annual cycle of soil temperature with a minimum surface temperature of $-10°C$ on day 15 and a maximum of $30°C$ on day 197.5. In this case, $T_0(t) = \bar{T}_0 + A_0 \sin\left[2\pi(t - 106.25)/365\right]$ for t in days of the year (including fractions of the current day). Set the depth of snow to 10 cm during November through February; otherwise, there is no snow. Use $\kappa_{snow} = 0.3$ W m^{-1} K^{-1}. Compare soil temperatures with and without the snow. Describe the effect of snow on soil temperature.

4. Investigate the effects of phase change on the annual cycle of soil temperature. Compare simulations with and without phase change.

Turbulent Fluxes and Scalar Profiles in the Surface Layer

Chapter Overview

Monin–Obukhov similarity theory provides mathematical expressions for the momentum, sensible heat, and evaporation fluxes between the land surface and the atmosphere; the corresponding vertical profiles of wind speed, temperature, and water vapor; and the aerodynamic conductances that regulate surface fluxes. Above rough plant canopies, however, these fluxes and profiles deviate from Monin–Obukhov similarity theory due to the presence of the roughness sublayer. This chapter develops the meteorological theory and mathematical expressions to model turbulent fluxes and scalar profiles in the surface layer of the atmosphere.

6.1 Introduction

Fluxes of sensible heat and evaporation between the land and atmosphere occur because of turbulent mixing of air and the resultant transport of heat and water vapor, generally away from the surface during the day. Turbulence occurs when wind flows over Earth's surface and the ground, vegetation, and other objects retard the fluid motion of air. The reduction in wind speed near the surface creates turbulence that mixes air from above downward and from below upward and transports the temperature, water vapor, and momentum (mass ×

velocity) of the parcels of air being mixed. Vertical turbulent motion can also occur due to surface heating and buoyancy. In the daytime, the land surface is typically warmer than the overlying air, and strong solar heating provides buoyant energy. Warm air is less dense than cold air, and rising air enhances mixing and the transport of heat and water vapor away from the surface. In this unstable atmosphere, sensible heat flux is positive, temperature decreases with greater height above the surface, and vertical transport increases with strong surface heating. At night, the land surface cools. The lowest levels of the atmosphere become stable – with cold, dense air trapped near the surface and warmer air above. Sensible heat flux is negative (i.e., toward the surface). These conditions suppress vertical motion and reduce transport.

The turbulent fluxes of momentum, sensible heat, and water vapor are nearly constant with height in the layer of the atmosphere near the surface extending to a height of 50–100 m or so. This region of the atmosphere is known as the surface, or constant flux, layer. Within it, the fluxes of momentum, sensible heat, and water vapor are related to mean vertical gradients of wind speed, potential temperature, and water vapor, which follow a logarithmic profile with height, as shown in Figure 6.1 for wind speed. The use of flux–gradient relationships in boundary layer meteorology is key to understanding turbulent fluxes and scalar profiles, and Monin–Obukhov similarity theory provides one set of such relationships. This theory provides mathematical expressions for the surface fluxes, the

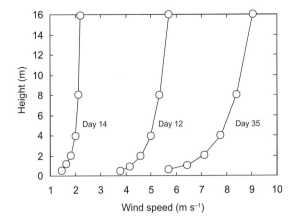

Figure 6.1 Wind profile above sparse grassland in southeastern Australia under near-neutral conditions measured at 1630 hours on days 12, 14, and 35 of the Wangara experiment. Data from Clarke et al. (1971) and reproduced from Bonan (2016)

associated scalar profiles, and additionally the aerodynamic conductances that regulate the fluxes (Figure 6.2).

6.2 | Turbulent Fluxes

A parcel of air has scalar properties such as temperature, water vapor, or CO_2 concentration that are carried with the parcel as it is mixed in the atmosphere. (A scalar quantity has only magnitude. A vector quantity has both magnitude and direction. Temperature and gas concentration are scalars. Velocity is a vector quantity.) Scalars of interest in the boundary layer include u, the velocity component in the x-direction, or zonal wind (m s^{-1}); v, the velocity component in the y-direction, or meridional wind (m s^{-1}); w, the vertical velocity (m s^{-1}); θ, potential temperature (K); and c, gas

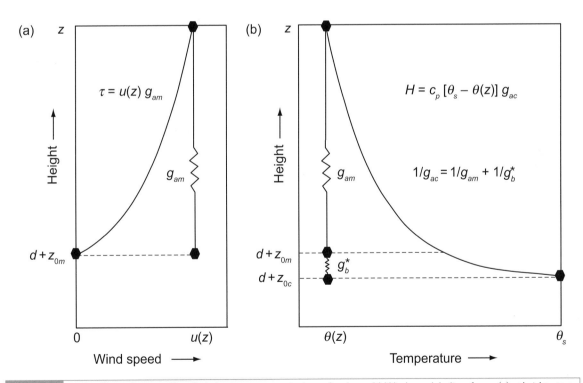

Figure 6.2 Idealized logarithmic profiles of wind and temperature in the surface layer. (a) Wind speed declines from $u(z)$ at height z to a surface value of $u = 0$ at $d + z_{0m}$. The conductance g_{am} governs the wind profile and the momentum flux τ. (b) For an unstably stratified surface layer, temperature increases from $\theta(z)$ at height z to the surface with θ_s at $d + z_{0c}$. The conductance g_{ac} governs the temperature profile and the sensible heat flux H. This conductance consists of two conductances in series: g_{am} between z and $d + z_{0m}$ and g_b^* arising from the difference between z_{0m} and z_{0c}.

concentration given as mole fraction (mol mol^{-1}). For water vapor, the mole fraction is $q = e/P$. (In this chapter, q is mole fraction with units mol mol^{-1} – rather than specific humidity with units kg kg^{-1}, as commonly used in boundary layer meteorology. These are related by molecular mass using (3.33). The distinction relates to ρ_m rather than ρ in the flux equations.) The turbulent flux of a scalar c is denoted by $\overline{w'c'}$, which is equal to the covariance of c and w. Of importance in boundary layer meteorology are the covariances $\overline{u'w'}$, $\overline{v'w'}$, $\overline{w'\theta'}$, and $\overline{w'q'}$, which are related to the turbulent fluxes of momentum, heat, and water vapor.

Boundary layer meteorology theory describes sensible heat flux H (W m^{-2}) and water vapor flux E (mol m^{-2} s^{-1}) by

$$H = \rho_m c_p \overline{w'\theta'} \tag{6.1}$$

$$E = \rho_m \overline{w'q'}, \tag{6.2}$$

where $\rho_m = P/\Re T$ is molar density (42.3 mol m^{-3} is a representative value; Eq. 3.16) and c_p is the specific heat of moist air at constant pressure (29.2 J mol^{-1} K^{-1} is a representative value; Eq. 3.59). The momentum flux τ has zonal and meridional components given by

$$\begin{aligned} \tau_x &= -\rho_m \overline{u'w'} \\ \tau_y &= -\rho_m \overline{v'w'} \end{aligned} \tag{6.3}$$

and

$$\tau = -\rho_m \left[\left(\overline{u'w'} \right)^2 + \left(\overline{v'w'} \right)^2 \right]^{1/2}. \tag{6.4}$$

For convenience, the x-axis is oriented along the direction of surface wind so that $v = 0$, and

$$\tau = -\rho_m \overline{u'w'}. \tag{6.5}$$

As given here, τ has units mol m^{-1} s^{-2}. Momentum flux more commonly is given using the density of air (ρ; kg m^{-3}) with units kg m^{-1} s^{-2}. These units are related by molecular mass (kg mol^{-1}) through (3.33).

Boundary layer meteorology defines the scales u_*, θ_*, and q_* in relation to turbulent fluxes using the expressions

$$u_* u_* = -\overline{u'w'} = \frac{\tau}{\rho_m} \tag{6.6}$$

$$\theta_* u_* = -\overline{w'\theta'} = -\frac{H}{\rho_m c_p} \tag{6.7}$$

$$q_* u_* = -\overline{w'q'} = -\frac{E}{\rho_m}, \tag{6.8}$$

where u_* (m s^{-1}) is the characteristic scale for velocity, commonly known as friction velocity, and θ_* (K) and q_* (mol mol^{-1}) are similar characteristic scales for temperature and water vapor. Table 6.1 summarizes various definitions of fluxes used in this chapter.

Sensible heat flux is specified in relation to potential temperature rather than air temperature. Potential temperature is used in boundary layer meteorology because it is independent of pressure or volume and is conserved in vertical motion. Potential temperature (commonly designated θ, not to be confused with the same symbol for volumetric soil moisture) is the temperature a parcel of air at some height above the surface would have if it was brought adiabatically from its actual pressure P to a reference pressure P_0 (equal to 1000 hPa). This is given by

$$\theta = T(P_0/P)^{0.286}. \tag{6.9}$$

Potential temperature at some height z (m) above the surface can be approximated from air temperature using the dry adiabatic lapse rate ($\Gamma_d = 0.0098$ K m^{-1}) in which

Table 6.1	Various flux definitions used in this chapter		
Quantity		Symbol	Units
Momentum			
Surface flux		τ	mol m^{-1} s^{-2}
Surface kinematic flux		τ/ρ_m	m^2 s^{-2}
Turbulent flux		$\overline{u'w'}$	m^2 s^{-2}
Sensible heat			
Surface energy flux		H	W m^{-2}
Surface kinematic flux		$H/\rho_m c_p$	K m s^{-1}
Turbulent flux		$\overline{w'\theta'}$	K m s^{-1}
Water vapor			
Surface molar flux		E	mol m^{-2} s^{-1}
Surface kinematic flux		E/ρ_m	mol mol^{-1} m s^{-1}
Turbulent flux		$\overline{w'q'}$	mol mol^{-1} m s^{-1}

$$\theta = T + \Gamma_d z. \tag{6.10}$$

The distinction is small within the first tens of meters above the surface. If air temperature is measured close to the surface (e.g., 20–30 m), the two quantities are similar and are often interchanged. By definition, $\theta_s = T_s$ at the surface.

6.3 | Gradient–Diffusion Theory

The equations for τ, H, and E developed in the preceding section require the covariances $\overline{u'w'}$, $\overline{w'\theta'}$, and $\overline{w'q'}$, respectively. Eddy covariance is a methodology to measure these terms (Baldocchi et al. 1988; Aubinet et al. 2000, 2012; Baldocchi 2003). In models, however, they must be parameterized. A common approach, analogous to molecular diffusion, is to assume that vertical turbulent transport in the surface layer occurs by diffusion along the mean concentration gradient (gradient–diffusion theory, or K-theory). Fick's law relates the diffusive flux to the molecular diffusion coefficient D_j (m^2 s^{-1}), as given in (3.8) for a gas and (3.13) for temperature. K-theory similarly describes turbulent fluxes in the surface layer by

$$\tau = \rho_m K_m(z) \frac{\partial \bar{u}}{\partial z} \tag{6.11}$$

$$H = -\rho_m c_p K_c(z) \frac{\partial \bar{\theta}}{\partial z} \tag{6.12}$$

$$E = -\rho_m K_c(z) \frac{\partial \bar{q}}{\partial z}, \tag{6.13}$$

with K_m and K_c (m^2 s^{-1}) the eddy diffusivity for momentum and scalars, respectively. (Many textbooks denote separate diffusivities for heat and water vapor but then assume they are equal.) These are not the molecular diffusivity but rather represent turbulent motion, and the complexities of turbulent transport are collectively represented by the diffusivities. $\bar{u}$, $\bar{\theta}$, and $\bar{q}$ are the mean wind speed, temperature, and water vapor concentration, respectively, and (6.11)–(6.13) are known as a first-order turbulence closure because the turbulent fluxes (i.e., the covariance, or second-order, terms) are parameterized using the mean (first-order)

vertical gradient terms (e.g., $\overline{u'w'}$ and $\partial \bar{u}/\partial z$ for τ). For convenience, the specific notation of the mean using an overbar is dropped for subsequent equations so that u, θ, and q are taken to denote the mean.

Similar to molecular diffusion, the vertically integrated flux equations between heights z_1 and z_2 relate the turbulent fluxes to an aerodynamic conductance rather than a diffusivity. In conductance notation, the turbulent fluxes are

$$\tau = u(z)g_{am}(z) \tag{6.14}$$

$$H = -c_p[\theta(z) - \theta_s]g_{ac}(z) \tag{6.15}$$

$$E = -[q(z) - q_s]g_{ac}(z). \tag{6.16}$$

Whereas the conductance in (3.11) represents that for molecular diffusion in still air, here g_{am} and g_{ac} are the equivalent conductances for turbulent motions. Equations (6.14)–(6.16) are used in models to calculate surface fluxes. Expressions for the conductances are obtained from profiles of u, θ, and q in the surface layer.

Question 6.1 Boundary layer meteorology commonly uses specific humidity ($q_{(mass)}$; kg kg^{-1}) to calculate E as a mass flux (kg m^{-2} s^{-1}) rather than mole fraction ($q_{(mol)}$; mol mol^{-1}) with E in molar units (mol m^{-2} s^{-1}). In this form, evaporation is related to $\rho q_{(mass)}$. Show that $\rho_m M_w q_{(mol)}$ is the equivalent expression to give E as a mass flux.

Question 6.2 Boundary layer meteorology commonly formulates turbulent fluxes in relation to a dimensionless drag coefficient rather than a conductance. In this form, the momentum flux is $\tau/\rho_m = C_m(z)u^2(z)$, and the sensible heat flux is $H/\rho_m c_p = C_h(z)u(z)[\theta_s - \theta(z)]$. Derive equations for the drag coefficients $C_m(z)$ and $C_h(z)$ in relation to u_* and θ_*. How do the drag coefficients relate to the aerodynamic conductances g_{am} and g_{ac}?

6.4 | Logarithmic Profiles

In the surface layer, wind speed, temperature, water vapor, and other scalars typically have logarithmic profiles in relation to height. These profiles are derived from similarity theory that relates the momentum, sensible heat, and water vapor fluxes with the mean vertical wind speed, temperature, and water vapor gradients, respectively. The profile of wind speed in a neutral surface layer is described by the relationship

$$\frac{(z-d)}{u_*}\frac{\partial u}{\partial z} = \frac{1}{k},\tag{6.17}$$

in which z is height (m) above the ground, d is the zero-plane displacement height (m), and k is the von Karman constant. The terms on the left-hand side of the equation form a single dimensionless quantity that is constant and equal to $1/k$. Measurements find that the von Karman constant is approximately equal to 0.4. Over some surfaces, the protrusion of roughness elements above the surface displaces turbulent flow upward. Trees, for example, extend upward into the atmosphere from the ground. The displacement height is the vertical displacement caused by surface elements. It is zero for bare ground and greater than zero for vegetation. Equation (6.17) shows that when scaled by u_*, the mean vertical wind gradient $\partial u/\partial z$ depends only on height $z-d$ above the surface. From the definition of u_* in (6.6), it is evident that (6.17) relates $\partial u/\partial z$ to τ so that the vertical wind gradient depends on the surface forcing and height.

Integrating $\partial u/\partial z$ between two heights ($z_2 > z_1$) gives

$$u_2 - u_1 = \frac{u_*}{k}\ln\left(\frac{z_2-d}{z_1-d}\right).\tag{6.18}$$

The surface is defined at $z_1 = d + z_{0m}$ where $u_1 = 0$. Substituting these for z_1 and u_1 in (6.18) gives an expression for wind speed at height z whereby

$$u(z) = \frac{u_*}{k}\ln\left(\frac{z-d}{z_{0m}}\right).\tag{6.19}$$

The term z_{0m} (m) is the roughness length for momentum. It is the theoretical height at which wind speed is zero. In this context, z_{0m} is the height above the displacement height where the wind profile

extrapolates to zero, and the height $d + z_{0m}$ is known as the apparent sink of momentum. This roughness length is generally less than 0.01 m for soil and increases to more than 1 m for tall vegetation.

Similar equations pertain to temperature and water vapor with

$$\frac{(z-d)}{\theta_*}\frac{\partial \theta}{\partial z} = \frac{1}{k}\tag{6.20}$$

$$\frac{(z-d)}{q_*}\frac{\partial q}{\partial z} = \frac{1}{k}.\tag{6.21}$$

Integrating $\partial\theta/\partial z$ and $\partial q/\partial z$ gives the temperature and water vapor at height z:

$$\theta(z) - \theta_s = \frac{\theta_*}{k}\ln\left(\frac{z-d}{z_{0c}}\right)\tag{6.22}$$

$$q(z) - q_s = \frac{q_*}{k}\ln\left(\frac{z-d}{z_{0c}}\right).\tag{6.23}$$

The surface values are θ_s and q_s at $z_1 = d + z_{0c}$ with z_{0c} the scalar roughness length. Similar to momentum, $d + z_{0c}$ is the effective height at which heat and water vapor are exchanged with the atmosphere and is known as the apparent source of heat and water vapor. These equations relate H and E to $\partial\theta/\partial z$ and $\partial q/\partial z$, respectively, through θ_* and q_*.

From these relationships, the eddy diffusivity for momentum in (6.11) is

$$K_m(z) = ku_*(z-d)\tag{6.24}$$

and has the dimensions of velocity × length. Similar expressions can be obtained for heat and water vapor whereby $K_m = K_c$ for a neutral surface layer in the absence of surface heating. The aerodynamic conductance for momentum in (6.14) is

$$g_{am}(z) = \rho_m k^2 u(z)\left[\ln\left(\frac{z-d}{z_{0m}}\right)\right]^{-2};\tag{6.25}$$

and, similarly, the conductance for scalars in (6.15) and (6.16) is

$$g_{ac}(z) = \rho_m k^2 u(z)\left[\ln\left(\frac{z-d}{z_{0m}}\right)\right]^{-1}\left[\ln\left(\frac{z-d}{z_{0c}}\right)\right]^{-1}.\tag{6.26}$$

Mixing length theory provides the same form for the wind profile but defines a mixing length

l_m, which is the length scale of turbulent motion. This is the distance that a parcel can travel vertically while retaining its momentum unchanged. It is the distance the parcel moves in the mixing process that generates the momentum flux. In mixing length theory, the equivalent expression for the momentum flux is

$$\frac{\tau}{\rho_m} = l_m^2 \left(\frac{\partial u}{\partial z}\right)^2, \tag{6.27}$$

with $l_m = k(z - d)$ whereby $K_m = l_m^2 \partial u/\partial z$. Similar derivations for temperature and water vapor give

$$\frac{H}{\rho_m c_p} = -l_c l_m \frac{\partial u}{\partial z}\frac{\partial \theta}{\partial z} \tag{6.28}$$

$$\frac{E}{\rho_m} = -l_c l_m \frac{\partial u}{\partial z}\frac{\partial q}{\partial z}, \tag{6.29}$$

with $l_m = l_c$ for a neutral surface layer and $K_c = l_c l_m \partial u/\partial z$. In this sense, τ, H, and E can be viewed in terms of an eddy diffusivity that is obtained from mixing length theory.

Question 6.3 Derive the expressions for K_m and K_c as given by (6.24) and g_{am} and g_{ac} as given by (6.25) and (6.26).

Question 6.4 Textbooks and models traditionally use the units kg m^{-1} s^{-2} for τ and specific humidity (kg kg^{-1}) for q so that E has the units kg m^{-2} s^{-1}, and they express fluxes in resistance notation with the units s m^{-1}. Write the flux equations using this notation and also the corresponding flux–profile equations and resistances. What are the differences compared with the molar notation used here?

6.5 | Monin–Obukhov Similarity Theory

The equations for u, θ, and q in the preceding section pertain to a surface layer in the absence of buoyancy. Monin–Obukhov similarity theory extends the profile equations to account for buoyancy. In particular, strong surface heating in unstable conditions enhances upward turbulent motion and alters the flux–gradient relationships. Conversely, stable conditions suppress turbulent motion. The effect of buoyancy is accounted for by the dimensionless parameter ζ, which is defined as

$$\zeta = \frac{z - d}{L_{MO}}, \tag{6.30}$$

where L_{MO} (m) is the Obukhov length scale. This length scale is

$$L_{MO} = \frac{-u_*^3 \theta_v}{kg(H_v/\rho_m c_p)} = \frac{u_*^2 \theta_v}{kg\theta_{v*}}, \tag{6.31}$$

with $L_{MO} < 0$ for unstable conditions and $L_{MO} > 0$ for stable conditions. $|L_{MO}|$ approaches ∞ for neutral conditions, and smaller absolute values indicate larger deviations from neutral conditions. For example, with $u_* = 0.25$ m s^{-1} and $\theta_v = 288$ K, $L_{MO} = -4.7$ m for a heat flux $H_v = 300$ W m^{-2}. A smaller heat flux H_v gives a correspondingly more negative L_{MO}. Negative values of ζ indicate an unstable surface layer, positive values indicate a stable surface layer, and $\zeta = 0$ for neutral conditions.

The length scale L_{MO} incorporates three variables. Turbulence is directly related to surface heating, and the term $-H_v/\rho_m c_p = \theta_{v*}u_*$ is the surface buoyancy heat flux. When H_v is positive (so that the surface is warmer than the overlying air), the atmosphere is unstable and mixing is enhanced. A negative heat flux (surface is cooler than air) indicates a stable atmosphere with diminished mixing. The term g/θ_v relates to the static stability of the atmosphere (with g = 9.81 m s^{-2} the gravitational acceleration), which also accounts for buoyancy. The term u_*^2 is the momentum flux τ/ρ_m. Moist air is less dense than dry air. The effect of water vapor on buoyancy is accounted for through virtual potential temperature θ_v and virtual heat flux H_v. Virtual potential temperature is similar to virtual temperature (3.32) and increases with greater water vapor. Virtual heat flux increases with greater evaporation. Garratt (1992, pp. 10, 36, 38) defined this flux as

$$H_v = H + 0.61 c_p \theta E (M_w/\bar{M}) \tag{6.32}$$

so that

$$\theta_{v*} = \theta_* + 0.61\theta q_* (M_w/\bar{M}). \tag{6.33}$$

The term $M_w/\bar{M}$ is the ratio of the molecular mass of water to the molecular mass of moist air (3.33) and appears because q as used here is expressed as mole fraction (mol mol^{-1}) instead of specific humidity (kg kg^{-1}).

Monin–Obukhov similarity theory states that, when scaled appropriately, the dimensionless mean vertical gradients of wind speed, temperature, and water vapor at some height z above the ground are unique functions of ζ. For wind,

$$\frac{k(z-d)}{u_*}\frac{\partial u}{\partial z} = \phi_m\left(\frac{z-d}{L_{MO}}\right) \tag{6.34}$$

and, similarly, for temperature and water vapor,

$$\frac{k(z-d)}{\theta_*}\frac{\partial \theta}{\partial z} = \phi_c\left(\frac{z-d}{L_{MO}}\right) \tag{6.35}$$

$$\frac{k(z-d)}{q_*}\frac{\partial q}{\partial z} = \phi_c\left(\frac{z-d}{L_{MO}}\right). \tag{6.36}$$

These equations are comparable to (6.17), (6.20), and (6.21) but include the similarity functions ϕ_m and ϕ_c. These functions relate τ and $\partial u/\partial z$, H and $\partial \theta/\partial z$, and E and $\partial q/\partial z$. The similarity functions are obtained from measurements, commonly taken over flat, homogeneous surfaces with short vegetation. Representative functions are

$$\phi_m(\zeta) = \begin{cases} (1 - 16\zeta)^{-1/4} & \zeta < 0 \\ 1 + 5\zeta & \zeta \geq 0 \end{cases} \tag{6.37}$$

for momentum, and

$$\phi_c(\zeta) = \begin{cases} (1 - 16\zeta)^{-1/2} & \zeta < 0 \\ 1 + 5\zeta & \zeta \geq 0 \end{cases} \tag{6.38}$$

for scalars (Dyer and Hicks 1970; Dyer 1974; Brutsaert 1982, pp. 68–71; Garratt 1992, pp. 52–54). They have values less than one when the atmosphere is unstable, equal to one for neutral conditions, and greater than one when the atmosphere is stable (Figure 6.3a).

The wind profile is obtained by integrating $\partial u/\partial z$ between two heights ($z_2 > z_1$) to give

$$u_2 - u_1 = \frac{u_*}{k}\left[\ln\left(\frac{z_2-d}{z_1-d}\right) - \psi_m\left(\frac{z_2-d}{L_{MO}}\right) + \psi_m\left(\frac{z_1-d}{L_{MO}}\right)\right], \tag{6.39}$$

and the wind speed at height z is

$$u(z) = \frac{u_*}{k}\left[\ln\left(\frac{z-d}{z_{0m}}\right) - \psi_m\left(\frac{z-d}{L_{MO}}\right) + \psi_m\left(\frac{z_{0m}}{L_{MO}}\right)\right]. \tag{6.40}$$

This is the familiar logarithmic profile in which wind speed decreases near the surface, and the term $-[\psi_m(\zeta) - \psi_m(z_{0m}/L_{MO})]$ gives the deviation from the log profile due to atmospheric stability. Similarly, $\partial \theta/\partial z$ and $\partial q/\partial z$ are integrated to give

$$\theta_2 - \theta_1 = \frac{\theta_*}{k}\left[\ln\left(\frac{z_2-d}{z_1-d}\right) - \psi_c\left(\frac{z_2-d}{L_{MO}}\right) + \psi_c\left(\frac{z_1-d}{L_{MO}}\right)\right] \tag{6.41}$$

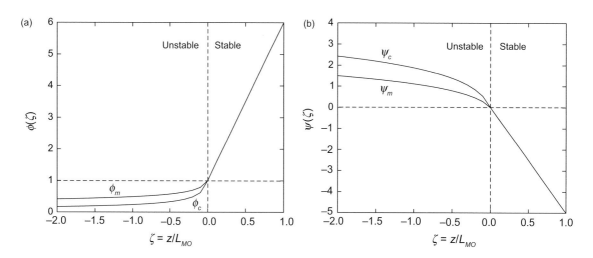

Figure 6.3 Similarity functions (a) $\phi_m(\zeta)$ and $\phi_c(\zeta)$ and (b) $\psi_m(\zeta)$ and $\psi_c(\zeta)$ in relation to $\zeta = z/L_{MO}$.

$$q_2 - q_1 = \frac{q_*}{k}\left[\ln\left(\frac{z_2 - d}{z_1 - d}\right) - \psi_c\left(\frac{z_2 - d}{L_{MO}}\right) + \psi_c\left(\frac{z_1 - d}{L_{MO}}\right)\right]$$

(6.42)

or, more commonly,

$$\theta(z) - \theta_s = \frac{\theta_*}{k}\left[\ln\left(\frac{z - d}{z_{0c}}\right) - \psi_c\left(\frac{z - d}{L_{MO}}\right) + \psi_c\left(\frac{z_{0c}}{L_{MO}}\right)\right]$$

(6.43)

$$q(z) - q_s = \frac{q_*}{k}\left[\ln\left(\frac{z - d}{z_{0c}}\right) - \psi_c\left(\frac{z - d}{L_{MO}}\right) + \psi_c\left(\frac{z_{0c}}{L_{MO}}\right)\right].$$

(6.44)

When sensible heat flux and evaporation are positive, so that heat and water vapor are exchanged into the atmosphere, temperature and water vapor decrease with height above the surface. This typically occurs during the day. At night, when sensible heat flux is negative (i.e., toward the surface), temperature increases with height.

The functions ψ_m and ψ_c account for the influence of atmospheric stability on turbulent fluxes and are obtained with the integration of $\partial u/\partial z$, $\partial \theta/\partial z$, and $\partial q/\partial z$ in (6.34)–(6.36). These are formally defined as

$$\psi_m(\zeta) = \int_0^\zeta \frac{1 - \phi_m(\zeta')}{\zeta'}d\zeta'$$

(6.45)

and similarly for ψ_c. For momentum,

$$\psi_m(\zeta) = \begin{cases} 2\ln\left(\frac{1+x}{2}\right) + \ln\left(\frac{1+x^2}{2}\right) - 2\tan^{-1}x + \frac{\pi}{2} & \zeta < 0 \\ -5\zeta & \zeta \geq 0 \end{cases}$$

(6.46)

and for scalars,

$$\psi_c(\zeta) = \begin{cases} 2\ln\left(\frac{1+x^2}{2}\right) & \zeta < 0 \\ -5\zeta & \zeta \geq 0 \end{cases},$$

(6.47)

with $x = (1 - 16\zeta)^{1/4}$. These functions have large positive values when the atmosphere is very unstable, decreasing toward zero as the atmosphere become less unstable and with negative values in a stable atmosphere (Figure 6.3b). As a result, the vertical profiles of u and θ are steep in an unstable surface layer, when turbulence efficiently mixes air, and become shallower as the surface layer becomes more neutrally stratified (Figure 6.4a,b).

Equations (6.40), (6.43), and (6.44) allow calculation of momentum, sensible heat, and evaporation fluxes from wind speed, temperature, and water vapor measured at a reference height z and at the surface. However, use of Monin–Obukhov similarity theory has limitations (Foken 2006). The similarity functions are valid for moderate values of ζ from about −2 to 1; they are applicable to homogeneous surfaces and must be modified over tall vegetation to account for the roughness sublayer; and they have errors of about 10%–20% even under ideal conditions.

Calculation of fluxes using Monin–Obukhov similarity theory is numerically challenging because L_{MO} depends on u_* and θ_* (and additionally q_* if the effects of water vapor on buoyancy are included), yet these themselves depend on L_{MO}. This is more evident when the terms in (6.31) are expanded. Ignoring, for simplicity, the effects of water vapor, u_* and θ_* can be replaced with (6.40) and (6.43), respectively, so that (6.31) is equivalently expressed as

$$Ri_B = \zeta\left[\ln\left(\frac{z - d}{z_{0c}}\right) - \psi_c(\zeta) + \psi_c\left(\frac{\zeta z_{0c}}{z - d}\right)\right]\left[\ln\left(\frac{z - d}{z_{0m}}\right)\right.$$

$$\left. - \psi_m(\zeta) + \psi_m\left(\frac{\zeta z_{0m}}{z - d}\right)\right]^{-2},$$

(6.48)

with $\zeta = (z - d)/L_{MO}$ and in which Ri_B is the bulk Richardson number defined by

$$Ri_B = \frac{g(z - d)}{u^2(z)}\frac{[\theta(z) - \theta_s]}{\theta(z)}.$$

(6.49)

Equation (6.48) also replaces z_{0m}/L_{MO} with the equivalent expression $\zeta z_{0m}/(z - d)$ and similarly for z_{0c}/L_{MO}. More generally, (6.48) can be written as $Ri_B = \zeta F_c(\zeta)F_m^{-2}(\zeta)$, with F_c and F_m evident from (6.48). The effect of water vapor is included by replacing θ with θ_v to calculate Ri_B. However, the relationship between Ri_B and ζ is complex; and the solution for L_{MO} requires iterative numerical algorithms for root finding such as bisection, secant, or Brent's methods (Appendix A5). Newton–Raphson iteration is also possible with the derivative:

$$\frac{\partial Ri_B}{\partial \zeta} = \frac{Ri_B}{\zeta}\left\{1 + \left[\phi_c(\zeta) - \phi_c\left(\frac{\zeta z_{0c}}{z - d}\right)\right]F_c^{-1}(\zeta)\right.$$

$$\left. - 2\left[\phi_m(\zeta) - \phi_m\left(\frac{\zeta z_{0m}}{z - d}\right)\right]F_m^{-1}(\zeta)\right\}.$$

(6.50)

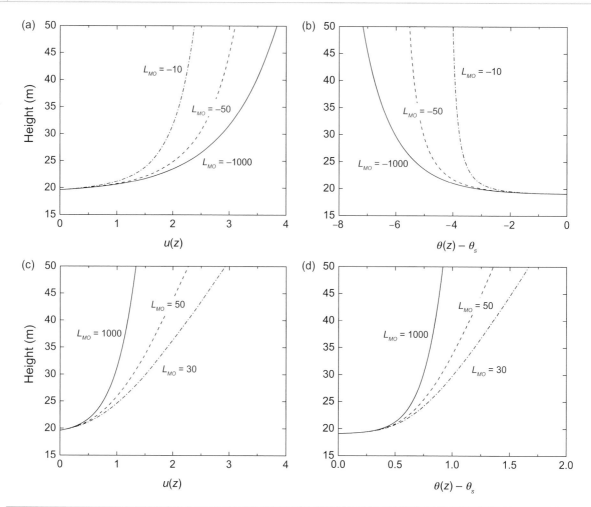

Figure 6.4 Vertical profiles of wind speed and temperature for different conditions of atmospheric stability. The top panels show unstable conditions for wind speed and temperature, and the bottom panels are for stable conditions. Calculations follow Physick and Garratt (1995) with $d = 19$ m, $z_{0m} = 0.6$ m, and $z_{0c} = 0.135 z_{0m}$. For unstable conditions, $u_* = 0.4$ m s^{-1} and $\theta_* = -0.5$ K ($H = 247$ W m^{-2}). Profiles are calculated for $L_{MO} = -1000, -50,$ and -10 m (i.e., with increasing instability). For stable conditions, $u_* = 0.13$ m s^{-1} and $\theta_* = 0.06$ K ($H = -10$ W m^{-2}). Profiles are calculated for $L_{MO} = 1000, 50,$ and 30 m (i.e., with increasing stability).

In many cases, it is better to use a generalized numerical solution that can be modified for different flux–profile relationships, such as to account for the roughness sublayer (Section 6.8). Figure 6.5 illustrates the most general iterative procedure, in which u_* and θ_{v*} are calculated for an initial estimate of L_{MO}, and L_{MO} is then updated for the current values of u_* and θ_{v*}. The iteration continues until the change in L_{MO} is less than some small amount. Numerical algorithms for root finding such as bisection, secant, or Brent's methods can be used.

Question 6.5 In Monin–Obukhov similarity theory, what are the equations that relate τ, H, and E to u, $\theta_s - \theta$, and $q_s - q$, respectively?

Question 6.6 Panofsky (1963) introduced an equation for wind speed in the surface layer that differs from (6.40). In this form:

$$u(z) = \frac{u_*}{k}\left[\ln\left(\frac{z-d}{z_{0m}}\right) - \psi_m(\zeta)\right],$$

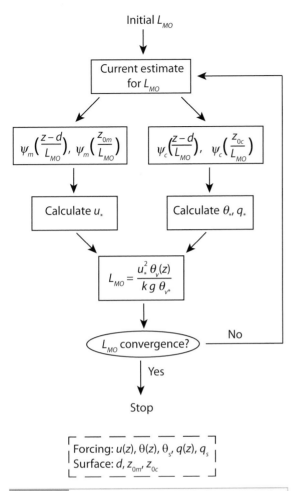

Figure 6.5 Illustration of the iterative calculations required for the Obukhov length L_{MO}, given a specified surface temperature θ_s and with u and θ at height z. Including the effects of water vapor on buoyancy requires q_s and q to calculate q_*. Surface parameters are d, z_{0m}, and z_{0c}.

and $\psi_m(\zeta)$ is

$$\psi_m(\zeta) = \int_{z_{0m}/L_{MO}}^{\zeta} \frac{1 - \phi_m(\zeta')}{\zeta'} d\zeta',$$

which differs from (6.45) in that the lower limit of the integration is z_{0m}/L_{MO} rather than zero. (a) Show that this calculation of wind speed is equivalent to (6.40) with $\psi_m(\zeta)$ given by (6.45). (b) Some publications give the wind profile as above but then calculate $\psi_m(\zeta)$ using (6.46), which is the solution to (6.45) as introduced by Paulson (1970). What error arises from this approximation?

Question 6.7 Use Supplemental Program 6.1 to calculate u_*, θ_*, and H for the following measurements made over a forest where the displacement height is $d = 19$ m:

z (m)	u (m s^{-1})	θ (°C)
21	1.0	29.0
29	2.1	28.1

Repeat the calculations but with the temperature at 21 m increased to 29.5°C. Why is u_* different from the previous calculation?

6.6 | Aerodynamic Conductances

Equations (6.11)–(6.13) use gradient–diffusion theory, or K-theory, to express τ, H, and E in relation to diffusion along the mean vertical concentration gradient with K_m and K_c the eddy diffusivity for momentum and scalars, respectively. Comparison with (6.34)–(6.36) show that these diffusivities are given by

$$K_m(z) = ku_*(z - d)\phi_m^{-1}\left(\frac{z - d}{L_{MO}}\right) \tag{6.51}$$

$$K_c(z) = ku_*(z - d)\phi_c^{-1}\left(\frac{z - d}{L_{MO}}\right). \tag{6.52}$$

The diffusivities represent turbulent motion. They increase with height above the surface, which characterizes the scale of the eddies. Dividing (6.51) and (6.52) by u_* gives the mixing lengths l_m and l_c, respectively. Larger eddies at greater height increase turbulent transport. To maintain a constant flux with respect to height, increases in transport with greater height (e.g., K_m) are balanced by corresponding decreases in the vertical gradient (e.g., $\partial u/\partial z$). Enhanced mixing under unstable conditions is represented by the similarity functions ϕ_m and ϕ_c.

These functions have values less than one when the atmosphere is unstable and greater than one when the atmosphere is stable so that the diffusivities increase with an increasingly unstable atmosphere. To maintain a constant flux, the vertical gradient ($\partial u/\partial z$, $\partial\theta/\partial z$) must decrease with increasing instability, and the vertical profiles of u and θ become steeper (Figure 6.4a,b).

Equations (6.14)–(6.16) are the vertically integrated form of (6.11)–(6.13) and express turbulent fluxes as the bulk transfer between the atmosphere at some height z with u, θ, and q and the surface with θ_s and q_s. In this form, g_{am} and g_{ac} are aerodynamic conductances (mol m^{-2} s^{-1}) for momentum and scalars, respectively, between the height z and the surface. These conductances are evident from the flux–profile equations – using (6.40), (6.43), and (6.44) – but more formally are obtained from the diffusivity integrated between two heights. For momentum,

$$\frac{1}{g_{am}} = \int_{z_1}^{z_2} \frac{dz}{\rho_m K_m} \qquad (6.53)$$

in a manner analogous to (3.12) for molecular diffusion. Similar integration gives g_{ac}. These conductances are

$$g_{am}(z) = \rho_m k^2 u(z) \left[\ln\left(\frac{z-d}{z_{0m}}\right) - \psi_m(\zeta) + \psi_m\left(\frac{z_{0m}}{L_{MO}}\right) \right]^{-2} \qquad (6.54)$$

$$g_{ac}(z) = \rho_m k^2 u(z) \left[\ln\left(\frac{z-d}{z_{0m}}\right) - \psi_m(\zeta) + \psi_m\left(\frac{z_{0m}}{L_{MO}}\right) \right]^{-1}$$
$$\left[\ln\left(\frac{z-d}{z_{0c}}\right) - \psi_c(\zeta) + \psi_c\left(\frac{z_{0c}}{L_{MO}}\right) \right]^{-1}, \qquad (6.55)$$

with $\zeta = (z-d)/L_{MO}$. Typical values for a neutral surface layer are 0.4–4 mol m^{-2} s^{-1} (Figure 6.6).

The aerodynamic conductances are greater than the neutral value when the surface layer is unstable, resulting in a large flux for a given surface-to-air difference, and less than the neutral value when the surface layer is stable, resulting in a smaller flux for a given difference. This is seen in Figure 6.4. In unstable conditions, sensible heat flux is positive (in this example, $H = 247$ W m^{-2}), and the surface is warmer than the air. A smaller temperature difference between the surface and air arises from the same sensible heat flux as the atmosphere becomes more unstable (as $L_{MO} \to 0$). This is because g_{ac} increases so that $\theta_s - \theta$ must decrease for the same H. For stable conditions, sensible heat flux is

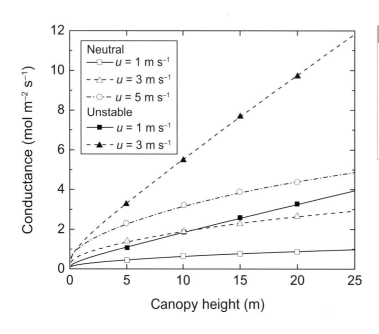

Figure 6.6 Aerodynamic conductance for scalars g_{ac} in relation to canopy height. For this example, $d = 0.67h_c$, $z_{0m} = 0.13h_c$, and $z_{0c} = 0.1z_{0m}$. Conductance is calculated at a height 10 m above the canopy ($z = h_c + 10$). Shown are conductances for three wind speeds ($u = 1$, 3, and 5 m s^{-1}) and for neutral ($L_{MO} = -\infty$) and unstable ($L_{MO} = -10$ m) conditions.

negative and typically small (in this example, $H = -10 \text{ W m}^{-2}$). Temperature increases with height, and a larger temperature difference is needed in increasingly stable conditions to achieve the same H because g_{ac} decreases with greater stability. Similar results are seen with wind speed.

Models calculate τ, H, and E from atmospheric and surface variables provided the aerodynamic conductances are specified. These conductances require ζ, which must be obtained using numerical methods. To simplify the calculation, some models parameterize the conductances using Ri_B rather than ζ. The iterative solution to ζ given by (6.48) can be replaced with analytical functions of Ri_B (Louis 1979). Then, the conductances can be written in the form:

$$g_{am}(z) = g_{am}(z)|_{\text{neutral}} \cdot F_m(\text{Ri}_B) \tag{6.56}$$

$$g_{ac}(z) = g_{ac}(z)|_{\text{neutral}} \cdot F_c(\text{Ri}_B). \tag{6.57}$$

The neutral value of the conductance is obtained as before, but without stability corrections. The functions $F(\text{Ri}_B)$ modify the neutral value to account for stability. In this form, the effects of atmospheric stability can be obtained directly from model variables (temperature and wind speed) using (6.49) for Ri_B. The functions $F(\text{Ri}_B)$ are formally related to the stability functions ψ_m and ψ_c but avoid the need for iteration. Several forms for these functions have been proposed, especially for very stable conditions that develop at night with clear sky and low wind speed (Louis et al. 1982; Mahrt 1987; Holtslag and Beljaars 1989; Beljaars and Holtslag 1991). For these conditions, evaluation of surface fluxes using Monin–Obukhov similarity theory is challenged (Holtslag et al. 2013).

Question 6.8 Give formulas for the aerodynamic conductances g_{am} and g_{ac} between heights z_1 and z_2 in the surface layer. Calculate these conductances for Question 6.7. What are the equivalent resistances in units of s m^{-1}?

6.7 | Roughness Length

In applying surface flux theory, roughness length and displacement height must be specified.

Roughness length can be eliminated by evaluating the profile equations at the canopy height h_c. Equation (6.39) for wind, for example, can be equivalently expressed as

$$u(z) - u(h_c) = \frac{u_*}{k}\left[\ln\left(\frac{z-d}{h_c-d}\right) - \psi_m\left(\frac{z-d}{L_{MO}}\right)\right.$$
$$\left. + \psi_m\left(\frac{h_c-d}{L_{MO}}\right)\right]. \tag{6.58}$$

The ratio of friction velocity to wind speed at the canopy top can be defined as $\beta = u_*/u(h_c)$ so that (6.58) can be rewritten in the form:

$$u(z) = \frac{u_*}{k}\left[\ln\left(\frac{z-d}{h_c-d}\right) - \psi_m\left(\frac{z-d}{L_{MO}}\right) + \psi_m\left(\frac{h_c-d}{L_{MO}}\right) + \frac{k}{\beta}\right]. \tag{6.59}$$

Comparing (6.59) and (6.40) shows that

$$z_{0m} = (h_c-d)\exp\left[-\frac{k}{\beta}\right]\exp\left[-\psi_m\left(\frac{h_c-d}{L_{MO}}\right) + \psi_m\left(\frac{z_{0m}}{L_{MO}}\right)\right]. \tag{6.60}$$

This equation for z_{0m} is not helpful by itself because z_{0m} appears also in the right-hand side of the equation. Equation (6.59), however, is very useful. It eliminates z_{0m} and though it introduces a new parameter β, that parameter can be calculated from physical principles, as described in Section 6.8.

Displacement height and roughness length are, in fact, complex functions of plant canopy structure (Shaw and Pereira 1982). Both increase with height of the canopy; and both depend on the amount of leaves, branches, and stems and their distribution in the canopy. Choudhury and Monteith (1988) simplified the relationships of Shaw and Pereira (1982) to obtain

$$\frac{d}{h_c} = 1.1\ln\left[1 + (c_d L)^{1/4}\right] \tag{6.61}$$

and:

$$\frac{z_{0m}}{h_c} = \begin{cases} \frac{z_0'}{h_c} + 0.3\sqrt{c_d L} & c_d L \leq 0.2 \\ 0.3(1 - d/h_c) & 0.2 < c_d L \leq 1.5 \end{cases}, \tag{6.62}$$

where L is the plant area index of the canopy (foliage, stem, wood) and c_d is the leaf aerodynamic drag coefficient (a common value is $c_d = 0.2$). For sparse canopies with low plant area index, roughness length has a different relationship and additionally depends on the roughness length of the soil surface z_0'. Raupach (1994, 1995) also provided equations for d/h_c and z_{0m}/h_c in relation to the amount of plant material. In this parameterization:

$$\frac{d}{h_c} = 1 - \frac{1 - \exp\left(-\sqrt{7.5L}\right)}{\sqrt{7.5L}} \tag{6.63}$$

and

$$\frac{z_{0m}}{h_c} = \left(1 - \frac{d}{h_c}\right) \exp\left[-\frac{k}{\beta} + 0.193\right]. \tag{6.64}$$

The latter equation for z_{0m}/h_c is comparable to (6.60) but is simplified in that the factor 0.193 accounts for the departure of the wind profile from the logarithmic relationship. The term $\beta = u_*/u(h_c)$ is defined by

$$\beta = (0.003 + 0.3L/2)^{1/2}, \tag{6.65}$$

with a maximum value whereby $\beta \leq 0.3$. The five constants in (6.63)–(6.65) are free parameters but are given here with recommended values (Raupach 1994, 1995).

The roughness length of scalars such as heat and water vapor for plant canopies is typically less than that for momentum. That is, the apparent sources of scalars occur at a lower height in the canopy than the apparent sink of momentum, and the aerodynamic conductances for sensible heat and water vapor are less than that for momentum (i.e., the resistances are larger). This is because momentum transport occurs from the drag of plant elements on wind flow, whereas scalar transport depends on diffusion from the plant elements. For a scalar, the smaller roughness length can be considered as an excess resistance (Figure 6.2), as introduced by Chamberlain (1966). This resistance is easily seen for neutral stability, when

$$\frac{1}{g_b^*} = \frac{1}{\rho_m k u_*} \ln\left(\frac{z_{0m}}{z_{0c}}\right). \tag{6.66}$$

The term $\ln(z_{0m}/z_{0c})$ is denoted as kB^{-1}, where B is a measure of the difference in rate of transport of a scalar and momentum and is known as the Stanton number. A typical value for kB^{-1} over vegetation is approximately two (Thom 1972; Garratt and Hicks 1973; Brutsaert 1982, pp. 109–110) so that z_{0c} is about 10% that for momentum. However, larger values ($kB^{-1} > 10$) are common, especially for sparse canopies (Qualls and Brutsaert 1996; Verhoef et al. 1997; Massman 1999). This excess resistance is a manifestation of leaf-level boundary layer processes integrated over the whole canopy and so depends on canopy characteristics such as the size of leaves and leaf area index (McNaughton and van den Hurk 1995; Massman 1999).

Despite these complexities, a simplification is to specify roughness length in relation to the canopy height by $z_{0m} = 0.13h_c$ and, similarly, displacement height by $d = 0.67h_c$ (Cowan 1968; Brutsaert 1982, pp. 113–116; Campbell and Norman 1998, p. 71; Shuttleworth 2012, p. 343; Monteith and Unsworth 2013, p. 304). A further simplification is to use $z_{0c} = 0.1z_{0m}$. Figure 6.6 shows g_{ac} in relation to canopy height using these simple relationships. Aerodynamic conductance increases with increasing vegetation height due to greater roughness length. Short grass is a smoother surface than tall grass. Forests are aerodynamically rougher than crops or grasses.

Question 6.9 Use (6.61) and (6.62) from Choudhury and Monteith (1988) to graph d/h_c and z_{0m}/h_c for a range of canopies with a plant area index from 0.1 to 10 m^2 m^{-2}. Compare these with the Raupach (1994, 1995) parameterization. How do the results compare with the simplifications $d = 0.67h_c$ and $z_{0m} = 0.13h_c$? How sensitive are the results to leaf drag coefficient?

Question 6.10 Prove that g_{ac} obtained from (6.55) is equal to the aerodynamic conductance for momentum together with the excess conductance. Use neutral conditions for simplicity.

Question 6.11 A model uses the equations $g_{am} = \rho_m u_*^2/u$ and $g_{ac} = \rho_m u_* \theta_*/(\theta - \theta_s)$ to calculate τ and H. (a) Are these the correct equations for g_{am} and g_{ac}? (b) The code is revised to use the equation $\rho_m/g_{ac} = u/u_*^2 + B^{-1}/u_*$. Explain how this equation relates to the previous expression for g_{ac}.

6.8 | Roughness Sublayer

The presence of tall vegetation significantly impacts turbulence and alters the vertical gradients of wind speed, temperature, and water vapor compared with the absence of vegetation. The empirical similarity functions ϕ_m and ϕ_c in (6.37) and (6.38) were obtained over smooth surfaces (e.g., short grassland) with small aerodynamic roughness. Deviations from the expected logarithmic profiles have been observed over tall crops, forests, and other rough surfaces (Figure 6.7). In the air immediately above these plant canopies and extending to two to three times the canopy height, known as the roughness sublayer, values of ϕ_m and ϕ_c are reduced compared to that predicted from Monin–Obukhov similarity theory so that the diffusivities K_m and K_c are larger (Garratt 1978; Raupach 1979).

Garratt (1980, 1983) and Physick and Garratt (1995) developed a roughness sublayer parameterization. The influence of the roughness sublayer can be accounted for by modifying the flux–gradient relationship of (6.34) so that

$$\frac{k(z-d)}{u_*}\frac{\partial u}{\partial z} = \Phi_m(z), \qquad (6.67)$$

with Φ_m an effective similarity function that additionally accounts for the roughness sublayer. A similar equation applies to temperature in which

$$\frac{k(z-d)}{\theta_*}\frac{\partial \theta}{\partial z} = \Phi_c(z). \qquad (6.68)$$

Φ_m and Φ_c can take many forms. In their approach,

$$\Phi_m(z) = \phi_m\left(\frac{z-d}{L_{MO}}\right)\hat{\phi}_m\left(\frac{z-d}{z_* - d}\right). \qquad (6.69)$$

ϕ_m is the Monin–Obukhov similarity function as before and uses the dimensionless parameter $\zeta = (z-d)/L_{MO}$, in which the length scale is the Obukhov length L_{MO}. $\hat{\phi}_m$ accounts for the departure from Monin–Obukhov similarity theory because of the roughness sublayer. This adjusts the vertical gradient of wind speed for an additional length

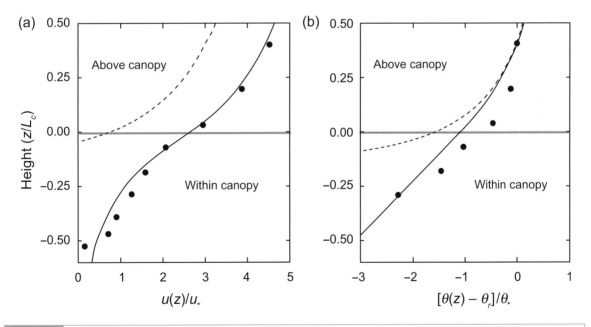

Figure 6.7 Vertical profiles of (a) mean wind speed (with $L_c/L_{MO} = -0.76$ and $\beta = 0.38$) and (b) mean potential temperature (with $L_c/L_{MO} = -0.71$, $\beta = 0.39$, and $Pr = 0.28$) above and within a forest canopy. Observations are for the Tumbarumba forest in unstable conditions. The dashed line is the surface layer profile extrapolated to the canopy. The solid line is the profile predicted from roughness sublayer theory. Height z is normalized by the canopy length scale L_c. Wind speed is normalized by u_*. Potential temperature is given as the deviation from the reference height temperature θ_r and normalized by θ_*. The horizontal gray line denotes the canopy height. Adapted from Harman and Finnigan (2007, 2008).

scale, which is the height of the roughness sublayer (denoted by z_*). An expression for $\hat{\phi}_m$ is

$$\hat{\phi}_m\left(\frac{z-d}{z_*-d}\right) = \exp\left[-0.7\left(1-\frac{z-d}{z_*-d}\right)\right] \quad (6.70)$$

for $z \leq z_*$, in which $\hat{\phi}_m$ increases from 0.5 at $z = d$ to 1 at $z \geq z_*$. In this formulation, $\Phi_m \leq \phi_m$ and the influence of the roughness sublayer decreases exponentially with greater height. Other variants have been proposed for $\hat{\phi}_m$, and with separate functions for momentum and temperature, but still using the height scale z_* (Cellier and Brunet 1992; Mölder et al. 1999; De Ridder 2010). A challenge with these parameterizations is to define the roughness sublayer height z_*.

The wind speed profile is obtained by integrating $\partial u/\partial z$ from $d + z_{0m}$ to z so that

$$\frac{k}{u_*}[u(z) - u(d + z_{0m})] =$$

$$\ln\left(\frac{z-d}{z_{0m}}\right) - \psi_m\left(\frac{z-d}{L_{MO}}\right) + \psi_m\left(\frac{z_{0m}}{L_{MO}}\right) - \int_{d+z_{0m}}^{z} \phi_m\left(\frac{z'-d}{L_{MO}}\right)$$

$$\left[1 - \hat{\phi}_m\left(\frac{z'-d}{z_*-d}\right)\right]\frac{dz'}{z'-d}. \quad (6.71)$$

This equation differs from (6.40) obtained using Monin–Obukhov similarity theory in two respects. The integral on the right-hand side of the equation is an extra term that accounts for the roughness sublayer. Additionally, wind speed is taken as zero at $d + z_{0m}$ in (6.40). With roughness sublayer theory, however, wind speed is non-zero at $d + z_{0m}$. This is shown by equating (6.40) and (6.71) at z_*, at which height the roughness sublayer produces no deviation from Monin–Obukhov similarity theory because $\hat{\phi}_m = 1$. Substituting (6.40) for $u(z)$ in (6.71) and evaluating at $z = z_*$ shows that the surface wind speed is

$$u(d+z_{0m}) = \frac{u_*}{k}\int_{d+z_{0m}}^{z_*}\phi_m\left(\frac{z'-d}{L_{MO}}\right)\left[1-\hat{\phi}_m\left(\frac{z'-d}{z_*-d}\right)\right]\frac{dz'}{z'-d}. \quad (6.72)$$

Equation (6.71) then becomes

$$u(z) = \frac{u_*}{k}\left[\ln\left(\frac{z-d}{z_{0m}}\right) - \psi_m\left(\frac{z-d}{L_{MO}}\right) + \psi_m\left(\frac{z_{0m}}{L_{MO}}\right) + \hat{\psi}_m(z)\right]. \quad (6.73)$$

The term $\hat{\psi}_m(z)$ accounts for the roughness sublayer and is

$$\hat{\psi}_m(z) = \int_z^{z_*}\phi_m\left(\frac{z'-d}{L_{MO}}\right)\left[1-\hat{\phi}_m\left(\frac{z'-d}{z_*-d}\right)\right]\frac{dz'}{z'-d}. \quad (6.74)$$

$\hat{\psi}_m$ is positive and gives the deviation from the expected logarithmic profile at heights below z_*; its value decreases with greater height and equals zero for $z \geq z_*$. Equation (6.74) must be integrated by numerical methods, although other equations for $\hat{\phi}_m$ allow an analytic expression for $\hat{\psi}_m$ (De Ridder 2010). The corresponding equation for temperature is

$$\theta(z) - \theta_s = \frac{\theta_*}{k}\left[\ln\left(\frac{z-d}{z_{0c}}\right) - \psi_c\left(\frac{z-d}{L_{MO}}\right) + \psi_c\left(\frac{z_{0c}}{L_{MO}}\right) + \hat{\psi}_c(z)\right]. \quad (6.75)$$

Φ_c is calculated as in (6.69) for Φ_m, and $\hat{\psi}_c$ is calculated as in (6.74) for $\hat{\psi}_m$, but replacing ϕ_m with ϕ_c and with $\hat{\phi}_c = \hat{\phi}_m$.

Figure 6.8 illustrates vertical profiles calculated with and without the roughness sublayer. Wind speed and temperature at any height are larger with the roughness sublayer than in Monin–Obukhov similarity theory. Additionally, while $u = 0$ at $d + z_{0m}$ with Monin–Obukhov similarity theory, the wind speed at this height is non-zero with the roughness sublayer. Similarly, $\theta - \theta_s$ calculated with the roughness sublayer is non-zero at $d + z_{0c}$, in contrast to Monin–Obukhov similarity theory in which $\theta = \theta_s$. Another consequence of the roughness sublayer is larger aerodynamic conductances because of the additional terms $\hat{\psi}_m$ and $\hat{\psi}_c$. Sensible heat flux calculated from measured gradients $\partial\theta/\partial z$ and $\partial u/\partial z$ using roughness sublayer flux–gradient relationships is larger than that calculated using Monin–Obukhov similarity theory.

Question 6.12 Repeat Question 6.7 and Question 6.8 using Supplemental Program 6.1, but now accounting for the roughness sublayer using the Physick and Garratt (1995) parameterization. Use $z_* = 49$ m. How do τ, H, and the conductances g_{am} and g_{ac} compare with Monin–Obukhov similarity theory?

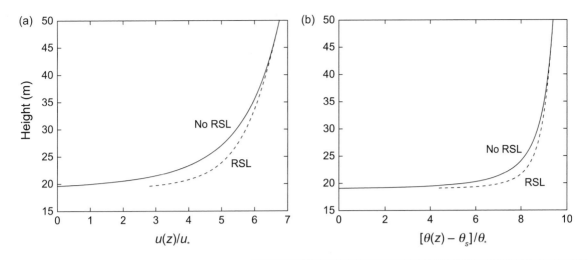

Figure 6.8 Vertical profiles of (a) u/u_* and (b) $(\theta - \theta_s)/\theta_*$ with (dashed lines) and without (solid lines) the influence of the roughness sublayer. Calculations are as in Figure 6.4, but with $L_{MO} = -20$ m and $z_* = 49$ m. See also Physick and Garratt (1995).

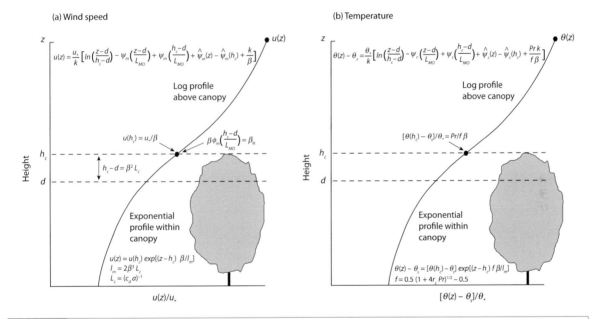

Figure 6.9 Illustration of the Harman and Finnigan (2007, 2008) roughness sublayer parameterization with coupling to the plant canopy for (a) wind speed and (b) temperature. The parameterization eliminates a priori specification of d, z_{0m}, and z_{0c} as input parameters but requires parameters for the canopy height h_c, leaf area density a, leaf drag coefficient c_d, leaf Nusselt number r_c, Prandtl number Pr, β_N, and uses $c_2 = 0.5$.

Harman and Finnigan (2007) developed a more detailed roughness sublayer parameterization that modifies Monin–Obukhov similarity theory for canopy-induced turbulence. Their approach introduces a new length scale and couples the above-canopy momentum and scalar fluxes with

equations for the momentum and scalar balances within a dense, horizontally homogenous canopy to obtain the necessary parameters. The resulting equation set produces logarithmic profiles above the canopy and exponential profiles within the canopy (Figure 6.9). This coupling introduces complexity arising from drag elements (foliage, branches, stems) in the canopy and introduces a dimensionless parameter $(z - d)/(l_m/\beta)$, which is the height $z - d$ normalized by the length scale l_m/β. This length scale relates to canopy structure and is equal to $u/(\partial u/\partial z)$ at the top of the canopy. It is specified by a mixing length in the canopy l_m and $\beta = u_*/u(h_c)$, which is the ratio of friction velocity to wind speed at the canopy height introduced previously in Section 6.7. The parameterization is valid for a uniform, dense, homogeneous canopy in which momentum is absorbed mostly by the canopy and in which the ground exerts minimal drag.

The wind profile above the canopy at heights $z \geq h_c$ is

$$\frac{\partial u}{\partial z} = \frac{u_*}{k(z - d)} \Phi_m(z) \tag{6.76}$$

so that

$$u(z) = \frac{u_*}{k} \left[\ln \left(\frac{z - d}{z_{0m}} \right) - \psi_m \left(\frac{z - d}{L_{MO}} \right) + \psi_m \left(\frac{z_{0m}}{L_{MO}} \right) + \hat{\psi}_m(z) \right], \tag{6.77}$$

as before. The profile equation can be rewritten in the form

$$u(z) = \frac{u_*}{k} \left[\ln \left(\frac{z - d}{h_c - d} \right) - \psi_m \left(\frac{z - d}{L_{MO}} \right) + \psi_m \left(\frac{h_c - d}{L_{MO}} \right) \right.$$
$$\left. + \hat{\psi}_m(z) - \hat{\psi}_m(h_c) + \frac{k}{\beta} \right], \tag{6.78}$$

in which the roughness length is

$$z_{0m} = (h_c - d) \exp \left[-\frac{k}{\beta} \right] \exp \left[-\psi_m \left(\frac{h_c - d}{L_{MO}} \right) \right.$$
$$\left. + \psi_m \left(\frac{z_{0m}}{L_{MO}} \right) \right] \exp \left[\hat{\psi}_m(h_c) \right]. \tag{6.79}$$

These latter two equations are similar to (6.59) and (6.60) introduced previously in Section 6.7 but with additional terms for the roughness sublayer. Within the canopy $(z \leq h_c)$, Harman and Finnigan (2007) specified an exponential wind profile in which wind speed varies with height (taken as the deviation from the canopy top) divided by the length scale l_m/β so that

$$u(z) = u(h_c) \exp \left(\frac{z - h_c}{l_m/\beta} \right). \tag{6.80}$$

The exponential profile is obtained from conservation of momentum as described in Chapter 16. This provides an expression for the mixing length l_m (m) whereby

$$l_m = 2\beta^3 L_c. \tag{6.81}$$

The canopy length scale L_c (m) is defined in relation to leaf area density and the leaf aerodynamic drag coefficient c_d. This length scale is assumed to be constant in the canopy so that $L_c = (c_d a)^{-1}$ with leaf area density a (m^2 m^{-3}) equal to plant area index divided by canopy height. Displacement height is specified by

$$h_c - d = \frac{l_m}{2\beta} = \beta^2 L_c. \tag{6.82}$$

The wind profiles are coupled at the canopy height through $u(h_c)$. For continuity, $\partial u/\partial z$ above the canopy from (6.76) must equal $\partial u/\partial z$ within the canopy from (6.80) at $z = h_c$. Mathematically, this means that

$$\frac{u_*}{k(h_c - d)} \Phi_m(h_c) = \frac{u(h_c)}{l_m/\beta}, \tag{6.83}$$

which simplifies to

$$\Phi_m(h_c) = \frac{k}{2\beta}. \tag{6.84}$$

In Harman and Finnigan (2007), the modified similarity function for momentum is

$$\Phi_m(z) = \phi_m \left(\frac{z - d}{L_{MO}} \right) \hat{\phi}_m \left(\frac{z - d}{l_m/\beta} \right). \tag{6.85}$$

This equation is similar to (6.69) introduced previously, but that equation uses the height of the roughness sublayer $z_* - d$ as the length scale, whereas the length scale for $\hat{\phi}_m$ in (6.85) is l_m/β. The roughness sublayer function is

$$\hat{\phi}_m \left(\frac{z - d}{l_m/\beta} \right) = 1 - c_1 \exp \left[-c_2 \left(\frac{z - d}{l_m/\beta} \right) \right]. \tag{6.86}$$

This function has a form such that the influence of the roughness sublayer decays exponentially with height above the canopy. The parameter c_1 is found by evaluating $\hat{\phi}_m$ at the canopy height h_c, and with $h_c - d = l_m/2\beta$ from (6.82):

$$c_1 = \left[1 - \hat{\phi}_m \left(\frac{h_c - d}{l_m/\beta} \right) \right] \exp \left(\frac{c_2}{2} \right). \tag{6.87}$$

This equation needs an expression for $\hat{\phi}_m$ evaluated at the canopy height h_c. That is obtained from (6.85), whereby $\hat{\phi}_m = \Phi_m/\phi_m$, and with $\Phi_m(h_c) = k/2\beta$ from (6.84):

$$\hat{\phi}_m \left(\frac{h_c - d}{l_m/\beta} \right) = \frac{k}{2\beta} \phi_m^{-1} \left(\frac{h_c - d}{L_{MO}} \right). \tag{6.88}$$

A simplification of the theory is to take $c_2 = 0.5$. $\hat{\phi}_m < 1$ denotes the influence of the roughness sublayer, and $\hat{\phi}_m$ increases to one with greater height above the canopy.

The roughness sublayer function $\hat{\psi}_m$ is

$$\hat{\psi}_m(z) = \int_z^\infty \phi_m \left(\frac{z' - d}{L_{MO}} \right) \left[1 - \hat{\phi}_m \left(\frac{z' - d}{l_m/\beta} \right) \right] \frac{dz'}{z' - d}. \tag{6.89}$$

This integration is to ∞ because here $\hat{\phi}_m$ is continuous with z, whereas the integration in (6.74) is to z_* because that expression for $\hat{\phi}_m$ is discontinuous at z_*. Equation (6.89) must be integrated by numerical methods. $\hat{\psi}_m$ is positive and gives the deviation of wind speed due to the roughness sublayer. In this approach, there is no specified roughness sublayer height (i.e., z_* is not a required parameter), and the roughness sublayer influence decreases with greater height ($\hat{\psi}_m \to 0$ as $z \to \infty$).

$\beta = u_*/u(h_c)$ is a critical unknown in the roughness sublayer parameterization. The squared valued represents the drag exerted by the canopy and is the drag coefficient for momentum evaluated at the top of the canopy – i.e., $C_m(h_c) = u_*^2/u^2(h_c)$. Harman and Finnigan (2007) derived an expression for β from a relationship in which

$$\beta \phi_m \left(\frac{h_c - d}{L_{MO}} \right) = \beta_N, \tag{6.90}$$

with β_N the value of $u_*/u(h_c)$ for neutral conditions (a representative value is $\beta_N = 0.35$). This is a nonlinear equation that can be solved to obtain β. Equation (6.90) can be written as

$$\beta \phi_m \left(\beta^2 L_c/L_{MO} \right) = \beta_N \tag{6.91}$$

because $h_c - d = \beta^2 L_c$. The exact form of the equation for β depends on whether ϕ_m is evaluated for unstable or stable conditions. With ϕ_m given by (6.37), the expanded form of (6.91) for unstable conditions ($L_{MO} < 0$) is a quadratic equation for β^2 in which

$$(\beta^2)^2 + 16 \frac{L_c}{L_{MO}} \beta_N^4 (\beta^2) - \beta_N^4 = 0. \tag{6.92}$$

The correct solution is the larger root. For stable conditions ($L_{MO} > 0$), a cubic equation is obtained for β whereby

$$5 \frac{L_c}{L_{MO}} \beta^3 + \beta - \beta_N = 0. \tag{6.93}$$

This equation has one real root.

Harman and Finnigan (2008) extended the theory to include temperature and other scalars (Figure 6.9b). The temperature profile above the canopy in the roughness sublayer is as obtained previously with

$$\frac{\partial \theta}{\partial z} = \frac{\theta_*}{k(z - d)} \Phi_c(z) \tag{6.94}$$

and

$$\theta(z) - \theta_s = \frac{\theta_*}{k} \left[\ln \left(\frac{z - d}{z_{0c}} \right) - \psi_c \left(\frac{z - d}{L_{MO}} \right) + \psi_c \left(\frac{z_{0c}}{L_{MO}} \right) + \hat{\psi}_c(z) \right]. \tag{6.95}$$

Temperature within the canopy has an exponential profile similar to wind. As described in Chapter 16, the equations of conservation of heat and momentum can be solved to give

$$\theta(z) - \theta_s = [\theta(h_c) - \theta_s] \exp \left[\frac{f(z - h_c)}{l_m/\beta} \right]. \tag{6.96}$$

The parameter f relates the length scale of heat to that of momentum and is given by

$$f = \frac{1}{2} (1 + 4 r_c \text{Pr})^{1/2} - \frac{1}{2}, \tag{6.97}$$

with r_c the leaf Nusselt number for heat (or Stanton number for other scalars; a representative value is $r_c = 0.1$–0.2) and Pr the Prandtl number (or Schmidt number Sc for other scalars). Similar to wind, continuity above and within the canopy means that at the canopy height h_c,

$$\frac{\partial \theta}{\partial z} = \frac{\theta_*}{k(h_c - d)} \Phi_c(h_c) = \frac{f[\theta(h_c) - \theta_s]}{l_m/\beta} \tag{6.98}$$

so that

$$\frac{\theta(h_c) - \theta_s}{\theta_*} = \frac{\text{Pr}}{f\beta}, \tag{6.99}$$

with $\text{Pr} = \Phi_c(h_c)/\Phi_m(h_c)$.

Equation (6.95) for temperature requires the scalar roughness length z_{0c}. Similar to wind, an alternative equation provides temperature between z and h_c as

$$\theta(z) - \theta(h_c) = \frac{\theta_*}{k}\left[\ln\left(\frac{z-d}{h_c-d}\right) - \psi_c\left(\frac{z-d}{L_{MO}}\right)\right.$$

$$\left. + \psi_c\left(\frac{h_c-d}{L_{MO}}\right) + \hat{\psi}_c(z) - \hat{\psi}_c(h_c)\right]. \quad (6.100)$$

This equation eliminates z_{0c} but requires the temperature $\theta(h_c)$ at the canopy height. An equivalent form of (6.100) is obtained by substituting (6.99) for $\theta(h_c)$ so that

$$\theta(z) - \theta_s = \frac{\theta_*}{k}\left[\ln\left(\frac{z-d}{h_c-d}\right) - \psi_c\left(\frac{z-d}{L_{MO}}\right) + \psi_c\left(\frac{h_c-d}{L_{MO}}\right)\right.$$

$$\left. + \hat{\psi}_c(z) - \hat{\psi}_c(h_c) + \frac{Prk}{f\beta}\right]. \quad (6.101)$$

This is the same as (6.95) with roughness length equal to

$$z_{0c} = (h_c - d)\exp\left[-\frac{Prk}{f\beta}\right]\exp\left[-\psi_c\left(\frac{h_c-d}{L_{MO}}\right)\right.$$

$$\left. + \psi_c\left(\frac{z_{0c}}{L_{MO}}\right)\right]\exp\left[\hat{\psi}_c(h_c)\right]. \quad (6.102)$$

For temperature, the modified similarity function is

$$\Phi_c(z) = \phi_c\left(\frac{z-d}{L_{MO}}\right)\hat{\phi}_c\left(\frac{z-d}{l_m/\beta}\right). \quad (6.103)$$

The roughness sublayer function $\hat{\phi}_c$ is the same as $\hat{\phi}_m$ for momentum given by (6.86), but with

$$c_1 = \left[1 - \hat{\phi}_c\left(\frac{h_c-d}{l_m/\beta}\right)\right]\exp\left(\frac{c_2}{2}\right) \quad (6.104)$$

and

$$\hat{\phi}_c\left(\frac{h_c-d}{l_m/\beta}\right) = \frac{Prk}{2\beta}\phi_c^{-1}\left(\frac{h_c-d}{L_{MO}}\right). \quad (6.105)$$

Equation (6.105) is similar to (6.88) for momentum, but with $\Phi_c(h_c) = Prk/2\beta$. The function $\hat{\psi}_c$ is evaluated similarly to $\hat{\psi}_m$ for momentum using (6.89), but with ϕ_c and $\hat{\phi}_c$.

The Prandtl number is the ratio of the diffusivities for momentum and heat (or equivalently the mixing lengths) so that

$$Pr = \frac{K_m(h_c)}{K_c(h_c)} = \frac{\Phi_c(h_c)}{\Phi_m(h_c)}. \quad (6.106)$$

It varies across the roughness sublayer from approximately 0.5 at the canopy height to one above the roughness sublayer. Harman and Finnigan (2008) defined it at the top of the canopy and parameterized it using the equation

$$Pr = 0.5 + 0.3\tanh(2L_c/L_{MO}), \quad (6.107)$$

which has a value of 0.5 with neutral conditions. The same equations apply to other scalars (e.g., water vapor) but using the Schmidt number Sc, which is similarly parameterized so that Sc = Pr.

Equation (6.78) describes wind speed above the canopy, similar to (6.40), but accounting for the roughness sublayer. Furthermore, the aerodynamic properties of the individual canopy elements determine the bulk aerodynamic characteristics of the canopy. The roughness sublayer parameterization replaces d and z_{0m} with the canopy structural parameters a, h_c, c_d and introduces the additional parameters β_N and c_2. Coupling of the wind profile above the canopy with the profile in the canopy makes wind speed depend on canopy density as represented by the leaf drag coefficient c_d, leaf area density a, and canopy height h_c. Leaf area density can be estimated as leaf area index divided by canopy height (L/h_c) or, more precisely, plant area index to include the area of all canopy elements (live leaves, dead leaves, stems, branches). The temperature profile given by (6.101) additionally requires Pr and the leaf parameter r_c. The roughness lengths z_{0m} and z_{0c} are no longer required. A further outcome of coupling with the canopy is that d is not constant but, rather, varies depending on atmospheric stability through β.

Implementation of these equations requires an iterative calculation for L_{MO} (Figure 6.10). β is obtained for an initial value of L_{MO}. This provides d, and Pr and f are also determined for the current L_{MO}. The functions ϕ_m and ϕ_c are evaluated at h_c to obtain c_1, needed to evaluate $\hat{\phi}_m$ and $\hat{\phi}_c$. The roughness sublayer functions $\hat{\psi}_m$ and $\hat{\psi}_c$ are evaluated at z and h_c, and u_* and θ_* are obtained from (6.78) and (6.101), respectively. q_* is obtained similarly to θ_*, but replacing Pr with Sc. A new estimate of L_{MO} is evaluated, and the iteration is repeated until convergence in L_{MO}. Supplemental Program 7.1 uses the roughness sublayer parameterization to calculate surface fluxes.

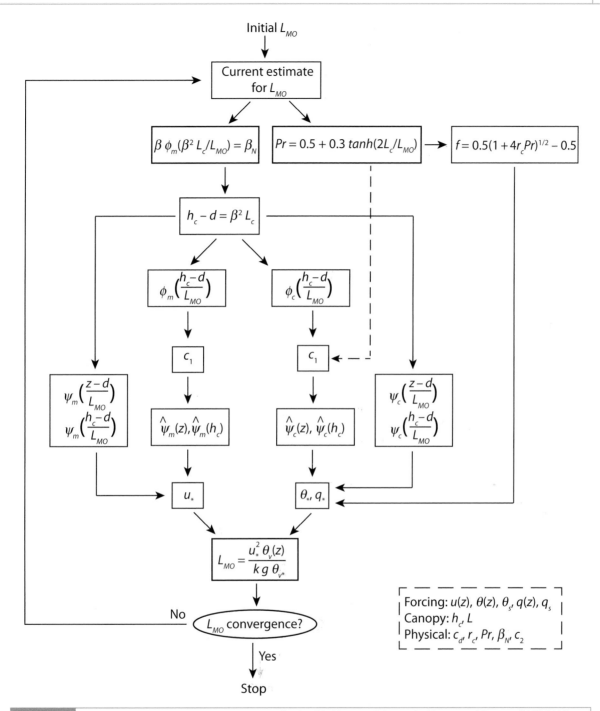

Figure 6.10 Illustration of the iterative calculations required for the Obukhov length L_{MO} in the Harman and Finnigan (2007, 2008) roughness sublayer parameterization.

Harman (2012) considered the consequences of the roughness sublayer parameterization. Because aerodynamic conductance increases, the same turbulent flux is obtained with smaller differences between the surface and atmosphere; or, conversely, a given surface-to-air difference produces a greater turbulent flux. This has greater effect on sensible heat flux than on latent heat flux because the latter is additionally regulated by canopy conductance. Consequently, available energy is preferentially partitioned to sensible heat, and the Bowen ratio $H/\lambda E$ increases. The surface also exerts greater drag for the same near-surface wind speed, as seen in larger u_*. Although the numerical implementation is complex, an advantage is that the theory utilizes observable parameters. Critical parameters (c_d, r_c, β_N, Pr) have physical meaning, have a well-defined range of observed values, and are not free parameters to fit the model to observations. The canopy length scale L_c is assumed to be constant, not either leaf area density or the leaf drag coefficient separately. The formulation asymptotes to Monin–Obukhov similarity theory with increasing height above the ground or for short canopies. Chapter 16 further uses the theory to calculate scalar concentrations in canopies.

The roughness sublayer theory as developed by Harman and Finnigan (2007, 2008) has implications for kB^{-1}, introduced previously in Section 6.7. Equations (6.79) for z_{0m} and (6.102) for z_{0c} relate roughness length to the aerodynamic characteristics of leaves and other structural elements in the canopy. With neutral stability and ignoring for the moment the $\hat{\psi}_m$ and $\hat{\psi}_c$ terms,

$$kB^{-1} = \ln \frac{z_{0m}}{z_{0c}} = \frac{k}{\beta}\left(\frac{\mathrm{Pr}}{f} - 1\right). \tag{6.108}$$

Representative values for neutral conditions are Pr $= 0.5$ and $\beta = 0.35$, in which case $kB^{-1} = 10.8$ with $r_c = 0.1$ and 5.1 with $r_c = 0.2$. r_c is the leaf Nusselt (or Stanton) number and is a measure of the difference in scalar and momentum transport across leaves. The smaller roughness length for scalars compared with momentum arises from low values of r_c, as discussed also by Kondo and Kawanaka (1986). While these estimates are illustrative, in fact, kB^{-1} varies with atmospheric stability in a complex relationship.

6.9 | Supplemental Programs

6.1 Flux Calculations and Obukhov Length: This code illustrates the calculation of friction velocity and sensible heat flux using Monin–Obukhov similarity theory given wind speed and temperature at two heights. Bisection is used to iteratively solve for L_{MO} given two initial estimates that bracket the solution. The program can optionally use the Physick and Garratt (1995) roughness sublayer parameterization.

6.2 Roughness Length: This code calculates z_{0m} and z_{0c} using the Harman and Finnigan (2007, 2008) roughness sublayer parameterization. Canopy parameters (h_c, a, c_d, r_c) and L_{MO} are specified inputs.

6.10 | Modeling Projects

1. Eddy covariance measurements show that $\overline{w'\theta'} = 0.24$ K m s^{-1}. Sensors placed at $z= 30$ m on the tower show that $\theta= 30°C$ and $u= 5$ m s^{-1}. The surface is forested with height $h_c = 20$ m. (a) Modify Supplemental Program 6.1 to calculate θ_s and g_{ac} using Monin–Obukhov similarity theory. Assume that $d = 0.67h_c$, $z_{0m} = 0.13h_c$, and $z_{0c} = 0.1z_{0m}$. (b) Compare these results to a surface with $h_c = 0.5$ m and with the reference height 10 m above the canopy ($z= 10.5$ m). (c) Then compare both temperatures to a lower wind speed (e.g., $u(z)= 2$ m s^{-1}). How do surface roughness and wind speed affect the temperature calculation?

2. Supplemental Program 6.1 uses bisection to solve for τ, H, and E. Compare this solution with that obtained from Newton–Raphson iteration using (6.48)–(6.50). Ignore the effects of water vapor in calculating L_{MO} and Ri_B. Do both programs give similar answers? Discuss the limitations in using Newton–Raphson.

3. Supplemental Program 6.2 calculates z_{0m} and z_{0c} using the Harman and Finnigan (2007, 2008) roughness sublayer parameterization. Examine the sensitivity of kB^{-1} to canopy parameters (h_c, a, c_d, r_c) and different values of L_{MO}.

Surface Energy Fluxes

Chapter Overview

Surface temperature is the temperature that balances net radiation, sensible heat flux, latent heat flux, soil heat flux, and storage of heat in biomass. This chapter develops the theory and mathematical expressions to model the surface energy balance and surface temperature. The surface energy balance is a nonlinear equation that must be solved for surface temperature using numerical methods. A more complex solution couples the surface energy balance with a model of soil temperature and a bucket model of soil hydrology. The Penman–Monteith equation is a linearization of the surface energy balance.

7.1 | Introduction

Net radiation is the solar and longwave radiation absorbed by the land surface after accounting for reflection of solar radiation and emission of longwave radiation. The overall energy balance at the land surface requires that the energy gained from net radiation be balanced by the fluxes of sensible and latent heat to the atmosphere and the storage of heat. These fluxes vary over the course of a day and throughout the year, principally in relation to soil moisture, the amount of leaf area, the physiology of leaves, and the diurnal and annual cycles of solar radiation. Figure 7.1 illustrates the diurnal cycle for an aspen forest in Canada. In the early morning, the land surface typically has a negative radiative balance because no solar radiation is absorbed but longwave radiation is lost. Sensible and latent heat fluxes are small. During daylight hours, solar radiation increases and there is a net gain of radiation at the surface. The land warms and some of this energy returns to the atmosphere as sensible and latent heat. The remainder warms the soil. Fluxes typically are strongest in the early to middle afternoon and decrease late in the afternoon when solar radiation diminishes. Figure 7.2 illustrates the annual cycle of fluxes for a temperate deciduous forest and a tropical rainforest. The tropical forest receives relatively constant solar radiation throughout the year. High amounts of rainfall keep the soil wet, and much of the available energy is used to evaporate water. The temperate deciduous forest has an annual cycle to latent heat flux that tracks solar radiation. Sensible heat flux exceeds latent heat flux in spring before leaves emerge. Thereafter, latent heat flux exceeds sensible heat flux during summer months as the foliage sustains high rates of transpiration.

7.2 | Energy Balance

The land surface energy balance is commonly written as

$$R_n = \left(S^{\downarrow} - S^{\uparrow}\right) + \left(L^{\downarrow} - L^{\uparrow}\right) = H + \lambda E + G + \text{storage},$$

(7.1)

in which all fluxes have the units W m^{-2}. Net radiation R_n at the surface is the energy gained from incoming solar radiation $S^{\downarrow}$ and longwave radiation

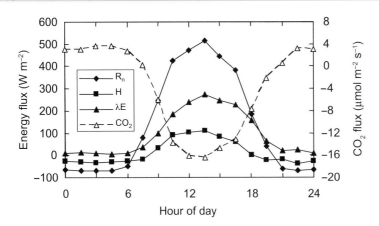

Figure 7.1 Diurnal cycle of net radiation R_n, sensible heat flux H, latent heat flux λE, and CO_2 for an aspen forest (*Populus tremuloides*) in Prince Albert National Park, Saskatchewan. The tower fluxes are described by Blanken et al. (1997) and are shown here for an average summer day as described by Bonan et al. (1997). Reproduced from Bonan (2016)

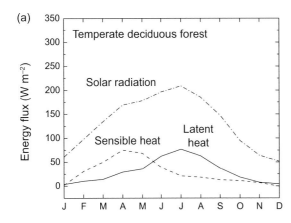

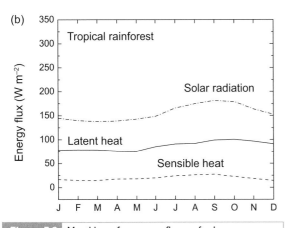

Figure 7.2 Monthly surface energy fluxes of solar radiation, latent heat flux, and sensible heat flux for (a) a temperate deciduous forest in central Massachusetts (Urbanski et al. 2007) and (b) a tropical rainforest in Brazil (Miller et al. 2011).

$L^{\downarrow}$ minus energy losses from reflected solar radiation $S^{\uparrow}$ and outgoing longwave radiation $L^{\uparrow}$ (positive denotes energy gain). This is balanced by sensible heat H and latent heat λE fluxes (positive to the atmosphere) and the heat flux that is stored in soil by conduction G (positive into the soil). Heat storage in the canopy is an additional energy flux. Latent heat flux is the product of the evaporative water flux E (mol m^{-2} s^{-1}) times latent heat of vaporization λ (J mol^{-1}). The various fluxes in the surface energy balance can be written in terms of a surface temperature θ_s, and (7.1) can be solved for the temperature that balances the energy budget. Such a model is known as a bulk surface model. It does not explicitly represent plant canopies or distinguish vegetation and soil but, rather, uses a bulk surface from which heat and moisture are exchanged with the atmosphere and in which the complexities of the land–atmosphere interface are greatly simplified (Figure 7.3a). First generation models of Earth's land surface for coupling with atmospheric models defined the surface in this manner (Manabe et al. 1965; Williamson et al. 1987).

Some of the incident solar radiation is reflected, and the remainder is absorbed. The amount reflected is

$$S^{\uparrow} = \sum_{\Lambda} \rho_{\Lambda} S_{\Lambda}^{\downarrow}, \qquad (7.2)$$

in which ρ_{Λ} is surface reflectance (commonly called surface albedo), defined as the fraction of $S_{\Lambda}^{\downarrow}$ that is reflected by the surface. The remainder, equal to $(1 - \rho_{\Lambda})S_{\Lambda}^{\downarrow}$, is the solar radiation absorbed by the surface. The subscript Λ denotes that reflectance

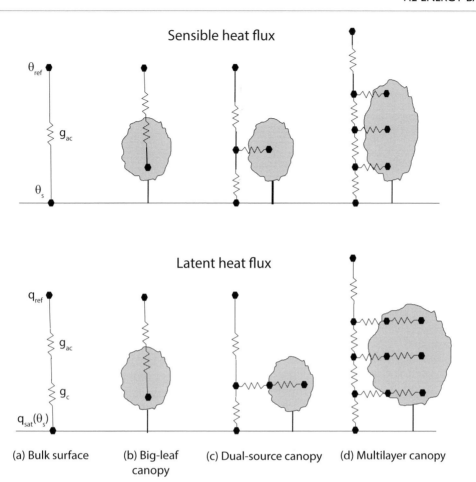

Sensible heat flux

Latent heat flux

(a) Bulk surface   (b) Big-leaf canopy   (c) Dual-source canopy   (d) Multilayer canopy

Figure 7.3 Conductance networks for sensible heat flux (top) and latent heat flux (bottom) for various depictions of the land surface. This chapter describes the bulk surface and big-leaf canopies. Chapter 15 provides further details of the big-leaf canopy, as well as the dual-source and multilayer canopy. Adapted from Bonan (2016)

and solar radiation vary with wavelength, and it is common to broadly define visible (less than 0.7 μm) and near-infrared (greater than 0.7 μm) wavebands because each accounts for approximately 50% of total solar radiation and because leaf reflectance varies greatly between these wavebands. The surface emits longwave radiation and additionally reflects some incoming longwave radiation. The outgoing longwave radiation is

$$L^{\uparrow} = \varepsilon\sigma\theta_s^4 + (1-\varepsilon)L^{\downarrow}, \qquad (7.3)$$

with θ_s surface temperature (K) and ε emissivity. The first term on the right-hand side of this equation is the emitted radiation as described by (3.3). The second term represents the incident longwave radiation that is reflected upward. The emissivity of a

surface is also its absorptivity so that $1-\varepsilon$ is the fraction of incoming longwave radiation that is reflected. Similar to (3.5), the net radiation at the surface is

$$R_n = \sum_\Lambda (1-\rho_\Lambda)S_\Lambda^{\downarrow} + \varepsilon L^{\downarrow} - \varepsilon\sigma\theta_s^4. \qquad (7.4)$$

Sensible heat flux is represented analogous to diffusive fluxes so that

$$H = c_p(\theta_s - \theta_{ref})g_{ac}. \qquad (7.5)$$

This equation is similar to (6.15) as described in Chapter 6, with θ_{ref} the potential temperature at some reference height above the surface and g_{ac} the aerodynamic conductance for scalars between the surface and the reference height. Evapotranspiration is also represented as a diffusive flux so that

$$E = \frac{q_{sat}(\theta_s) - q_{ref}}{g_c^{-1} + g_{ac}^{-1}}. \tag{7.6}$$

Here, q denotes mole fraction (e/P; mol mol^{-1}), and $q_{sat}(\theta_s) - q_{ref}$ is the water vapor deficit between the evaporating surface, which is saturated with moisture at the temperature θ_s, and the surrounding air.

The total conductance for evapotranspiration depends on the aerodynamic conductance and a surface conductance arranged in series, given by

$$g_w^{-1} = g_c^{-1} + g_{ac}^{-1}. \tag{7.7}$$

The conductance g_c represents the effects of plant canopies and surface wetness on evapotranspiration and decreases as the surface becomes drier so that a wet site has a higher evaporative flux than a dry site, all other factors being equal. The early land surface models used with climate models ignored the complexity of soil moisture and vegetation controls on turbulent fluxes (Manabe et al. 1965; Williamson et al. 1987). In these models, evapotranspiration was described by the bulk aerodynamic formulation as in (7.6), but without a surface conductance and with the flux instead adjusted for soil wetness through a factor β_w. This is a dimensionless wetness factor scaled from zero when soil is dry to one when soil is wet that adjusts potential flux for the extent that soil water limits evapotranspiration. In this approach, the evaporative conductance is $g_w = \beta_w g_{ac}$.

Some of the net radiation at the surface is used to warm soil. The rate at which heat is exchanged by conduction between the surface and underlying soil depends on the temperature gradient and thermal conductivity. As described in Chapter 5, the heat transfer between the surface with temperature θ_s and soil with temperature T_1 at depth Δz is

$$G = \frac{\kappa}{\Delta z}(\theta_s - T_1). \tag{7.8}$$

The flux G is the storage of energy in the soil column. Here, it is specified in relation to thermal conductivity κ and a known temperature at some depth in the soil. Energy conservation requires that the flux into the soil equal the change in heat content of the soil so that

$$G = \int_0^z c_v \frac{\partial T}{\partial t} dz. \tag{7.9}$$

The right-hand side of this equation is integrated to a depth z in soil with heat capacity c_v and temperature T.

Energy storage in the canopy includes heat stored in plant biomass (leaves, twigs, branches, trunks) and in the canopy air arising from changes in temperature and also latent heat storage in the canopy air. Canopy storage is small when averaged over long time periods and is commonly neglected from the surface energy balance, but can be a large term in the hourly energy balance (up to 50 W m^{-2} or more) in forests as energy is stored during the day and released at night (Aston 1985; McCaughey 1985; Moore and Fisch 1986; McCaughey and Saxton 1988; Blanken et al. 1997; Wilson and Baldocchi 2000; Oliphant et al. 2004; Michiles and Gielow 2008; Lindroth et al. 2010). Energy storage also contributes to the surface energy balance of croplands (Meyers and Hollinger 2004). Another term commonly neglected in the surface energy balance is energy storage during photosynthesis, which can be non-negligible (Blanken et al. 1997; Meyers and Hollinger 2004).

Thom (1975) outlined the principles with which to estimate storage fluxes. Sensible heat storage in a column of air with temperature θ extending from the ground to a height z is

$$S_h = \int_0^z \rho_m c_p \frac{\partial \theta}{\partial t} dz, \tag{7.10}$$

and latent heat storage is

$$S_q = \int_0^z \lambda \rho_m \frac{\partial q}{\partial t} dz. \tag{7.11}$$

Representative values are ρ_m = 42.3 mol m^{-3}, c_p = 29.2 J mol^{-1} K^{-1}, and λ = 44.44 kJ mol^{-1} so that sensible heat storage over one hour in a column of air is $S_h = 0.34 \cdot z \cdot \Delta\theta$ and latent heat storage simplifies to $S_q = 522 \cdot z \cdot \Delta q$. A temperature change of 5 K over one hour produces a storage flux of 34 W m^{-2} in a 20 m column of air. A change in water vapor of 0.005 mol mol^{-1} (5 hPa) over one hour results in a storage of 52 W m^{-2}. The energy storage in biomass is

$$S_v = \int_0^{h_c} \rho_{veg} c_{veg} \frac{\partial T_{veg}}{\partial t} \, dz. \qquad (7.12)$$

Here, ρ_{veg} is the biomass density of vegetation (kg m^{-3}), c_{veg} is the specific heat of plant material (J kg^{-1} K^{-1}), T_{veg} is the temperature of the plant material (K), and storage is integrated over the canopy height h_c. For vegetation, the integral of $\rho_{veg} dz$ over the height 0 to h_c is the mass of plant material per unit area m_{veg} (kg m^{-2}), and a further simplification is to take c_{veg} as 70% that of water (~2930 J kg^{-1} K^{-1}) so that energy storage over one hour is $S_v = 0.8 m_{veg} \Delta T_{veg}$. A forest with a biomass of 20 kg m^{-2} (fresh weight) storages 80 W m^{-2} for a 5 K change in temperature.

Determining the heat capacity of vegetation is necessary to calculate the biomass storage term. Studies have calculated energy storage in forests using a single heat capacity as in (7.12) with $c_{veg} = 2650$ J kg^{-1} K^{-1} (Moore and Fisch 1986), 2800 J kg^{-1} K^{-1} (Lindroth et al. 2010), 70% that of water (McCaughey 1985; McCaughey and Saxton 1988; Oliphant et al. 2004), or 3340 J kg^{-1} K^{-1} (Wilson and Baldocchi 2000). Other studies account for the different specific heat of foliage, twigs and branches, and tree stems. Aston (1985) used values of 2593, 2760, and 2802 J kg^{-1} K^{-1} for twigs and branches, leaves, and trunks respectively, in a eucalypt forest. Michiles and Gielow (2008) used values of 2457, 2398, and 2769 J kg^{-1} K^{-1} for twigs, branches, and leaves, respectively, in a tropical rainforest. A further difficulty is that energy storage in tree trunks is a complex calculation that can require modeling temperature and heat transfer in the bole (Moore and Fisch 1986; Michiles and Gielow 2008; Lindroth et al. 2010) and, additionally, the radiative balance and surface heat exchange of trunks (Haverd et al. 2007).

Eddy covariance measurements of sensible and latent heat fluxes are commonly used to evaluate terrestrial biosphere models. A challenge is that these measurements may fail to balance the net available energy $R_n - G$. The reasons for lack of closure include methodological concerns, failure to account for storage terms, and landscape heterogeneity (Foken 2008; Hendricks Franssen et al. 2010; Leuning et al. 2012; Stoy et al. 2013). One means to correct the observations to force closure is to assume that $R_n - G$ is representative of the measurement area and to adjust H and λE using the measured Bowen ratio $B = H/\lambda E$ (Twine et al. 2000). The Bowen ratio is assumed to be unaffected by energy closure so that the corrected latent heat flux is $\lambda E = (R_n - G)/(B + 1)$ and the corrected sensible heat flux is $H = B \cdot \lambda E$. A further challenge in comparing modeled with observed fluxes is that the model may not fully account for all the terms in the surface energy balance. This is particularly true for the canopy air and biomass storage fluxes, which typically are not included in models.

Question 7.1 An atmospheric modeler calculates latent heat flux using a wetness factor β_w to decrease the evaporative flux as soil becomes drier so that the evaporative conductance is $g_w = \beta_w g_{ac}$. A biometeorologist conceives of surface–atmosphere coupling in a different manner, using aerodynamic and surface conductances acting in series so that $g_w^{-1} = g_c^{-1} + g_{ac}^{-1}$. Relate the wetness factor β_w to the surface conductance g_c.

Question 7.2 If net radiation is 600 W m^{-2}, sensible heat flux is 250 W m^{-2}, and latent heat flux is 300 W m^{-2}, calculate how much a column of soil with heat capacity 2.5 MJ m^{-3} K^{-1} and depth 50 cm will warm in one hour.

7.3 | Surface Temperature

The net radiation that impinges on the land surface is balanced by energy lost or gained through sensible heat, latent heat, conduction, and canopy storage. This balance is maintained by the surface temperature. This is the temperature that balances the energy budget and is obtained, in the absence of canopy storage, from the equation:

$$\sum_\Lambda (1 - \rho_\Lambda) S_\Lambda^\downarrow + \varepsilon L^\downarrow = Q_a = \varepsilon \sigma \theta_s^4 + c_p \left(\theta_s - \theta_{ref} \right) g_{ac} + \frac{\lambda \left[q_{sat}(\theta_s) - q_{ref} \right]}{g_c^{-1} + g_{ac}^{-1}} + \frac{\kappa}{\Delta z} (\theta_s - T_1). \qquad (7.13)$$

The left-hand side of this equation constitutes the radiative forcing Q_a, which is the sum of absorbed solar radiation and longwave radiation, respectively. The right-hand side of the equation consists of emitted longwave radiation, sensible heat flux, latent heat flux, and heat storage in soil. This energy balance equation represents the surface as an infinitely thin layer without any heat capacity, and θ_s is referred to as a skin temperature. Equation (7.13) can be applied to plant canopies using representative parameters. In this case, ρ_Λ represents the albedo of vegetation. The aerodynamic conductance g_{ac} varies with roughness length and displacement, and both depend on canopy height and leaf area index as described in Chapter 6. The canopy conductance g_c is leaf stomatal conductance integrated over the canopy.

Equation (7.13) is nonlinear and requires numerical methods to solve for θ_s because of the dependence of emitted longwave radiation on θ_s^4 and of latent heat flux on $q_{sat}(\theta_s)$. One approach is Newton–Raphson iteration, as described in Appendix A5 on root finding. The root of (7.13) is found by rewriting it as

$$F_0(\theta_s) = Q_a - \varepsilon\sigma\theta_s^4 - c_p(\theta_s - \theta_{ref})g_{ac} - \frac{\lambda\left[q_{sat}(\theta_s) - q_{ref}\right]}{g_c^{-1} + g_{ac}^{-1}}$$
$$- \frac{\kappa}{\Delta z}(\theta_s - T_1) = 0. \tag{7.14}$$

Newton–Raphson iteration refines the estimate of θ_s until the change in value between successive iterations given by

$$\Delta\theta_s = -\frac{F_0(\theta_s)}{\partial F_0/\partial\theta_s} \tag{7.15}$$

is less than some convergence criterion. This criterion typically has to be small (e.g., less than 0.001 K) to ensure fluxes balance. Newton–Raphson iteration requires the derivative of (7.14) with respect to θ_s. If the aerodynamic conductance g_{ac} is specified, the derivative is

$$\frac{\partial F_0}{\partial\theta_s} = -4\varepsilon\sigma\theta_s^3 - c_p g_{ac} - \frac{\lambda}{g_c^{-1} + g_{ac}^{-1}}\frac{dq_{sat}(\theta_s)}{dT} - \frac{\kappa}{\Delta z}. \tag{7.16}$$

If fact, however, the aerodynamic conductance is itself a complex function of sensible and latent heat fluxes through the Obukhov length L_{MO} (Chapter 6) and so also depends on θ_s. Because of this dependence, the temperature derivative of (7.14) cannot be

directly obtained, and other numerical methods such as the secant method or Brent's method (Appendix A5) are required to solve for θ_s (Figure 7.4a). The solution is numerically complex and attempts to find a surface temperature that simultaneously satisfies the energy balance and the atmospheric stability requirement.

Equation (7.13) can be used in a model to calculate surface temperature and energy fluxes over the course of a day. Surface temperature at time $n + 1$ is determined based on the atmospheric forcing (e.g., solar radiation, longwave radiation, air temperature, humidity, wind speed) for that time step. Because there is no heat capacity in (7.13), the solution can, if one is not careful, give large temperature changes between successive time steps related to inferred stable ($\theta_s < \theta_{ref}$) and unstable ($\theta_s > \theta_{ref}$) conditions. In addition, calculation of surface temperature requires coupling the surface energy balance with equations for soil temperature. The difficulty is that soil temperature itself depends on the soil heat flux. One solution is to calculate surface temperature at time $n + 1$ using soil temperature from time n and to then update soil temperature for the calculated soil heat flux. This method must be utilized in a model that solves the surface energy balance according to (7.13), whereby θ_s^{n+1} is calculated using the soil temperature T_1^n at the previous time step.

An alternative approach is to directly couple the equations for the surface energy balance and soil temperature so that all temperatures are calculated at time $n + 1$ in an implicit solution. This requires conceptualizing the surface not as a skin temperature with no heat capacity but, rather, as a thin soil layer only a few millimeters thick so as to represent surface conditions. The surface layer has thickness Δz_1, heat capacity $c_{v,1}$, and temperature T_1 (to distinguish this from skin temperature θ_s). The surface energy balance is written so that

$$F_0(T_1) = Q_a - \varepsilon\sigma T_1^4 - c_p(T_1 - \theta_{ref})g_{ac} - \frac{\lambda\left[q_{sat}(T_1) - q_{ref}\right]}{g_c^{-1} + g_{ac}^{-1}} \tag{7.17}$$

is the heat flux into the soil. Whereas (7.13) has a direct term for soil heat flux, here the flux is calculated as a residual in the energy balance. Equation (7.17) cannot be directly solved for T_1, but it can be used as the boundary condition for the system of equations to calculate soil temperatures.

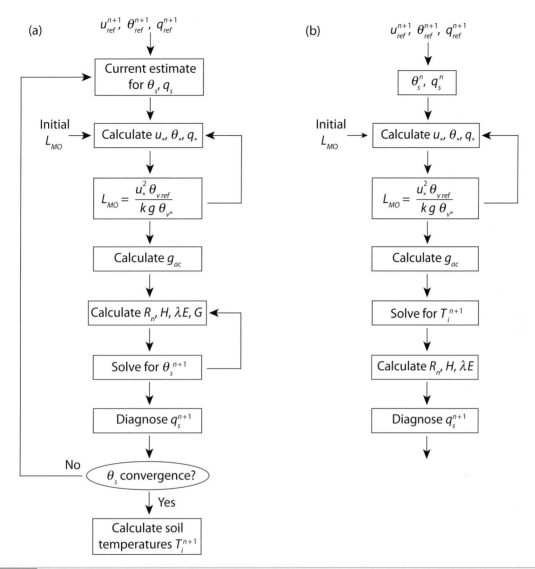

Figure 7.4 Illustration of the surface temperature calculation given u_{ref}, θ_{ref}, and q_{ref} at time $n+1$ at some reference height z_{ref}. (a) Iterative calculation of θ_s. For initial estimates of surface temperature θ_s and water vapor q_s, the Obukhov length L_{MO} is determined numerically. This is then used to obtain the aerodynamic conductance, from which a new surface temperature and the corresponding surface energy fluxes are calculated. The new values of θ_s and q_s are used to again calculate L_{MO}. The iteration continues until the change in θ_s is small. This method simultaneously finds values of L_{MO} and θ_s that are consistent at time $n+1$. (b) Implicit calculation of θ_s. In this method, θ_s and q_s for time n are used to calculate L_{MO} given the atmospheric forcing at $n+1$. This provides the aerodynamic conductance used in an implicit calculation of surface and soil temperatures at $n+1$. This approach uses surface values at n and atmospheric values at $n+1$ to determine L_{MO}.

As described in Chapter 5, energy conservation requires that for the first soil layer ($i = 1$),

$$\frac{c_{v,i}\Delta z_i}{\Delta t}\left(T_i^{n+1} - T_i^n\right) = F_0\left(T_i^{n+1}\right) - \frac{\kappa_{i+1/2}}{\Delta z_{i+1/2}}\left(T_i^{n+1} - T_{i+1}^{n+1}\right).$$

(7.18)

F_0 is a nonlinear expression with respect to T_1^{n+1} but can be linearized using a Taylor series approximation (Appendix A1) so that

$$F_0\left(T_1^{n+1}\right) = F_0\left(T_1^n\right) + \frac{\partial F_0\left(T_1^n\right)}{\partial T_1}\left(T_1^{n+1} - T_1^n\right).$$

(7.19)

Table 7.1 Terms in (7.21) for soil temperature

Layer	a_i	b_i	c_i	d_i
$i = 1$	0	$\dfrac{c_{v,i}\Delta z_i}{\Delta t} - c_i - \dfrac{\partial F_0(T_i^n)}{\partial T_i}$	$-\dfrac{\kappa_{i+1/2}}{\Delta z_{i+1/2}}$	$F_0(T_i^n) - \dfrac{\kappa_{i+1/2}}{\Delta z_{i+1/2}}\left(T_i^n - T_{i+1}^n\right)$
$1 < i < N$	$-\dfrac{\kappa_{i-1/2}}{\Delta z_{i-1/2}}$	$\dfrac{c_{v,i}\Delta z_i}{\Delta t} - a_i - c_i$	$-\dfrac{\kappa_{i+1/2}}{\Delta z_{i+1/2}}$	$\dfrac{\kappa_{i-1/2}}{\Delta z_{i-1/2}}\left(T_{i-1}^n - T_i^n\right) - \dfrac{\kappa_{i+1/2}}{\Delta z_{i+1/2}}\left(T_i^n - T_{i+1}^n\right)$
$i = N$	$-\dfrac{\kappa_{i-1/2}}{\Delta z_{i-1/2}}$	$\dfrac{c_{v,i}\Delta z_i}{\Delta t} - a_i$	0	$\dfrac{\kappa_{i-1/2}}{\Delta z_{i-1/2}}\left(T_{i-1}^n - T_i^n\right)$

This is an expansion of T_1 at $n + 1$ about its previous value at time n. Substituting this expression into (7.18) gives a linear equation in which the temperature of the surface layer is

$$\frac{c_{v,i}\Delta z_i}{\Delta t}\left(T_i^{n+1} - T_i^n\right) = F_0(T_i^n) + \frac{\partial F_0(T_i^n)}{\partial T_i}\left(T_i^{n+1} - T_i^n\right)$$
$$- \frac{\kappa_{i+1/2}}{\Delta z_{i+1/2}}\left(T_i^{n+1} - T_{i+1}^{n+1}\right). \qquad (7.20)$$

This is combined with the remaining soil layers to give a tridiagonal system of linear equations that can be solved for soil temperature as in Chapter 5. Table 7.1 gives terms for the tridiagonal equations using the notation $\Delta T = T^{n+1} - T^n$ in which

$$a_i\Delta T_{i-1} + b_i\Delta T_i + c_i\Delta T_{i+1} = d_i. \qquad (7.21)$$

Because it is implicit, the solution is more numerically stable than the previous iterative calculation. However, the Obukhov length is calculated using surface values at n and atmospheric values at $n + 1$ (Figure 7.4b). In this method, the surface fluxes of longwave radiation, sensible heat, and latent heat are not calculated directly, but rather are solved simultaneously with soil temperatures. It is important to remember that the fluxes encompassed in F_0^{n+1} have been linearized in (7.19) to calculate T_1^{n+1}, and therefore the fluxes at $n + 1$ must be calculated using the linearized form of the flux equations.

Figure 7.5 illustrates the diurnal cycle of surface fluxes on a warm summer day for tall forest and short grassland vegetation. Diurnal variation in air temperature and net radiation drives surface fluxes in these simulations. The forest is aerodynamically rougher than the grassland, seen in the larger friction velocity and aerodynamic conductance g_{ac}. Sensible heat exchange between the forest and the atmosphere is larger than for the grassland (to the atmosphere during the day; to the surface at night). Latent heat flux is also larger for the forest than for the grassland. Sensible heat flux is directly proportional to g_{ac}, and the larger conductance for forests results in more sensible heat exchange. Latent heat flux is additionally regulated by canopy conductance g_c acting in series with g_{ac}. The low canopy conductance (midday maximum 0.5 mol m^{-2} s^{-1}) produces a smaller change in total conductance $g_c^{-1} + g_{ac}^{-1}$. Because the forest is aerodynamically well-coupled with the atmosphere, its temperature is somewhat similar to air temperate. In contrast, the grassland is several degrees warmer than the air during the day and cooler than the air at night. A particular challenge, recognized early in the development of surface flux models, is that the surface can become energetically disconnected from the atmosphere in a stable boundary layer because of low aerodynamic conductance (Louis 1979; Holtslag et al. 2013). During this condition, the surface cools by radiation and unrealistically low nighttime temperatures may result.

A further difficulty in solving surface fluxes for coupling with atmospheric models is that θ_{ref} and q_{ref} at some height in the atmosphere above the surface must be specified, but these depend on surface fluxes. Various numerical methods are available to couple surface fluxes with the atmosphere (Polcher et al. 1998). Some models use a fully implicit solution in which atmospheric profiles of temperature and water vapor are solved simultaneously with surface and soil temperatures at time $n + 1$ in a set of governing equations from the top of the atmospheric boundary layer to the bottom of the soil column. In this method, the atmospheric profiles of temperature and water vapor are

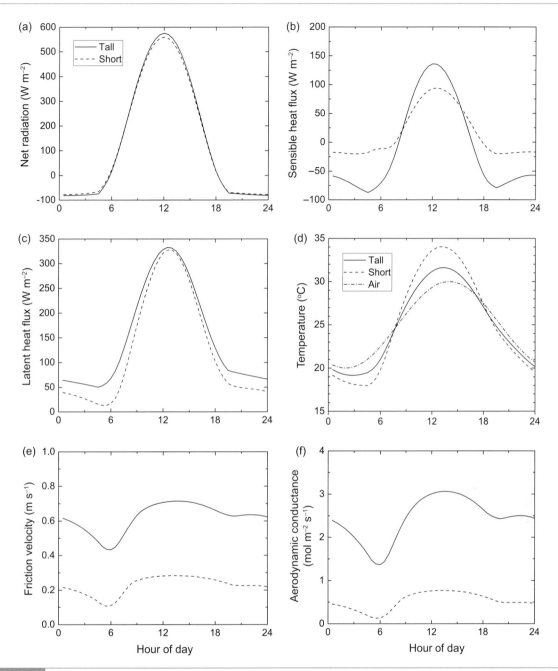

Figure 7.5 Simulated diurnal cycle of (a) net radiation, (b) sensible heat flux, (c) latent heat flux, (d) temperature, (e) friction velocity, and (f) aerodynamic conductance for a warm summer day. Simulations used Monin–Obukhov similarity theory with prescribed albedo, roughness length, displacement height, canopy conductance, and soil thermal properties.

synchronous with conditions at the land surface. Other models use an explicit numerical solution in which surface fluxes for $n+1$ are calculated from atmospheric states at n.

Question 7.3 Write a program that uses Newton–Raphson iteration to solve (7.13) for θ_s given specified conductances. Compare sensible heat and latent heat fluxes calculated using a stringent convergence criterion ($\Delta\theta_s \leq 0.001$ K) with answers obtained using $\Delta\theta_s \leq 0.1$ K. What is the difference in the surface energy balance between these two approaches? Why does this difference occur? Use these results to explain why the convergence criterion $\Delta\theta_s$ for Newton–Raphson iteration has to be small to ensure that the surface energy budget balances.

Question 7.4 A scientist writes a model that uses (7.21) to calculate soil temperatures. The temperature of the first soil layer T_1 is equated with the surface temperature θ_s; and longwave radiation, sensible heat flux, and latent heat flux are calculated from their standard flux definitions using (7.3), (7.5), and (7.6). The scientist finds that the resulting value for $Q_a - L^\uparrow - H - \lambda E$ is not the same as the energy used to warm the soil. Why is this?

7.4 | A Bucket Model Hydrologic Cycle

The hydrologic cycle is a critical determinant of surface fluxes and temperature because of several processes. First, soil moisture is a direct control of evapotranspiration, either through the empirical wetness factor β_w or more precisely through the surface conductance g_c. Second, the energy required to change water from liquid to vapor during evaporation cools the evaporating surface. In addition, the presence of snow on the ground influences the surface energy balance because the high albedo of snow reduces net energy at the surface, because some of the energy is used to melt the snow rather than warm the soil, and because the low thermal conductivity of snow insulates the underlying soil. Simulation of surface energy fluxes, therefore, requires an associated model of snow, soil moisture, and hydrologic fluxes. Hydrology models can be very complex in their representation of processes.

Chapter 8 describes soil moisture, and Chapter 9 provides further detail on hydrologic scaling.

The early land surface models used with climate models abstracted the complexity of the hydrologic cycle in a simplified approach analogous to a bucket (Manabe 1969). The bucket has a maximum water-holding capacity W_0. The amount of water W in the bucket is calculated by allowing the bucket to fill up to the amount W_0. Thereafter, precipitation in excess of evaporation runs off. There is no runoff for $W < W_0$, and the change in soil water is the difference between precipitation and evapotranspiration. In this method, the wetness factor is

$$\beta_w = \begin{cases} W/W_0 & W < W_0 \\ 1 & W \geq W_0 \end{cases}. \tag{7.22}$$

The maximum water-holding capacity W_0 is taken to be 75% of field capacity. Field capacity is set to 15 cm of water everywhere. Mintz and Walker (1993) used this methodology to calculate evapotranspiration and soil moisture in a global model.

Milly and Shmakin (2002) revised the bucket model to better represent surface processes, as used in the GFDL LM2 land model (Anderson et al. 2004). The single bucket was expanded to three pools so that water storage on land is composed of snow W_S, root zone soil water W_R, and groundwater W_G (with units kg H_2O m^{-2}). The snowpack is replenished by snowfall P_S and depleted by snowmelt M_S and loss of snow from sublimation E_S so that the water balance is

$$\frac{dW_S}{dt} = P_S - M_S - E_S. \tag{7.23}$$

The root zone water balance is similarly

$$\frac{dW_R}{dt} = P_R + M_S - E_R - D_R, \tag{7.24}$$

where P_R is rainfall precipitation, E_R is evapotranspiration loss, and drainage to groundwater D_R occurs when the water-holding capacity of the root zone is exceeded. Evapotranspiration is constrained by root zone soil water in a manner similar to β_w. Root zone drainage replenishes the groundwater so that

$$\frac{dW_G}{dt} = D_R - D_G, \tag{7.25}$$

where D_G is groundwater discharge.

When snow is on the ground, the surface temperature cannot exceed the freezing point while snow is melting. This physical principle provides a simple means to calculate snowmelt from a mass balance perspective. Snow melts if the surface temperature exceeds the freezing point. In this case, the surface temperature is set to freezing, and the energy imbalance is used to melt snow. The maximum snowmelt rate is $W_S/\Delta t$ (kg H_2O m^{-2} s^{-1}), which is converted to an energy flux (W m^{-2}) by multiplying by the latent heat of fusion L_f (J kg^{-1}). Any excess energy above $L_f W_S/\Delta t$ is used to warm the soil. The chapter appendix describes this approach as implemented in the LM2 model (Milly and Shmakin 2002). Much more complicated models treat the snowpack using multilayer heat transfer with compaction and refreezing of liquid water (Chapter 9).

Question 7.5 What are the atmospheric forcing data that need to be provided to solve the surface energy balance with a bucket hydrology? What are the site variables that need to be provided? What are the initial conditions?

7.5 | Soil Evaporation

Equation (7.6) describes evapotranspiration from a bulk surface. Evaporation from soil is more complicated. As evaporation proceeds, a dry soil layer develops above the wet soil and becomes progressively deeper with time. The rate of evaporation depends on diffusion of water vapor through this dry layer to the ground surface. The effect of soil drying on evaporation depends on physical principles of water movement in soil (e.g., Smits et al. 2012). More commonly, however, soil evaporation is described as a diffusion process in which

$$E = \frac{h_{s1}q_{sat}(T_1) - q_{ref}}{g_c^{-1} + g_{ac}^{-1}}. \tag{7.26}$$

This equation is similar to (7.6) but adjusts the saturation water vapor of the top soil layer for the relative humidity of the soil pore space so that $h_{s1}q_{sat}(T_1)$ is the water vapor concentration in the soil pores where water evaporates. Philip (1957a) gave the expression

$$h_{s1} = \exp\left(\frac{gM_w\psi_1}{\mathfrak{R}T_1}\right), \tag{7.27}$$

where ψ_1 is the matric potential of the first soil layer (here with units m), T_1 is the temperature of the first soil layer (K), g is gravitational acceleration (m s^{-2}), M_w is the molecular mass of water (kg mol^{-1}), and $\mathfrak{R}$ is the universal gas constant (J K^{-1} mol^{-1}).

For soil evaporation, the conductance g_c represents diffusion of water vapor between the water-filled soil pores and the ground surface. This conductance decreases as the surface becomes drier and is commonly parameterized using empirical relationships with the water content of the near-surface soil. In SiB2, for example,

$$g_c = \rho_m \exp\left(4.255 S_e - 8.206\right), \tag{7.28}$$

where S_e is the water content of the top soil layer relative to saturation (Sellers et al. 1996a). In this formulation, a typical conductance for a saturated soil is 0.8 mol m^{-2} s^{-1}, decreasing to less than 0.05 mol m^{-2} s^{-1} for a dry soil. Lee and Pielke (1992) devised a commonly used wetness factor for soil evaporation with the form

$$\beta_w = \frac{1}{4}\left[1 - \cos\left(\frac{\theta}{\theta_{fc}}\pi\right)\right]^2, \tag{7.29}$$

where here θ/θ_{fc} is soil moisture relative to field capacity. Empirical relationships such as this vary with soil texture and surface conditions. Other parameterizations are derived from gaseous diffusion in soil (Tang and Riley 2013), and some adjust surface conductance for the presence of litter (Sakaguchi and Zeng 2009).

A simple physically based approach is to formulate an expression for surface conductance using the depth of the dry surface layer. In this method, evaporation is represented by diffusion of water vapor from wet soil through an overlying dry surface layer. As soil becomes drier, the dry layer deepens and evaporation decreases. Choudhury and Monteith (1988) related g_c to the depth of the dry soil layer l_d (m) with the equation:

$$g_c = \rho_m \frac{\theta_{sat}t_f D_w}{l_d}. \tag{7.30}$$

(Note that Choudhury and Monteith (1988) described tortuosity as a factor greater than one and used $t_f = 2$. In (7.30), tortuosity has the more common convention with a factor less than one.)

This equation can be understood in terms of gas diffusion in soils. Diffusion of water vapor q (mol mol^{-1}) in soil is described by

$$F_w = -\rho_m \theta_{air} D \frac{\partial q}{\partial z}. \tag{7.31}$$

This is Fick's law as given by (3.8) but additionally includes the term θ_{air} for the air-filled pore space (m^3 air m^{-3} soil). Diffusion of a gas in soil is less than that in air because solid particles and liquids impede transport and because the flow path through pores is more tortuous than in air. The effective diffusivity of water vapor in soil is less than the diffusivity D_w in air so that $D = t_f D_w$ (Penman 1940; Millington 1959); t_f is commonly described as a tortuosity factor. Air-filled pore space is the soil porosity θ_{sat} (equal to the volumetric water content at saturation; Chapter 8) minus the volumetric water content. Some water is still held in soil when it is air-dry, but air-filled pore space can be approximated by θ_{sat} so that the conductance over a dry layer with depth l_d is equal to (7.30). Expressions for tortuosity (i.e., $t_f = D/D_w$) can be found elsewhere. One simple equation is $t_f = 0.66\theta_{air}$ (Penman 1940; Millington 1959). A more complex equation is used by Tang and Riley (2013) and Swenson and Lawrence (2014). The depth of the dry soil layer is a critical term in the conductance calculation. Swenson and Lawrence (2014) described an implementation of (7.30). A representative conductance is on the order of 0.01 mol m^{-2} s^{-1} for a 10 mm dry layer.

Question 7.6 Use (7.30) to calculate the evaporative resistance for a soil with a porosity of 0.45 if the top 5 mm is dry. What is the corresponding relative soil wetness if conductance is calculated using (7.28)?

7.6 | Penman–Monteith Equation

The surface energy balance as described in this chapter is known as a bulk surface without representation of plant canopies. The energy balance can be extended to plant canopies by treating the surface as a "big leaf" and is further extended by distinguishing vegetation and soil fluxes (Figure 7.3).

Chapter 15 provides more details on plant canopy models. The Penman–Monteith equation for evapotranspiration is an example of the application of the surface energy budget to a plant canopy using the big-leaf concept. Penman (1948) derived the equation by combining the thermodynamic and aerodynamic aspects of evaporation. The original derivation by Penman was for evaporation over open water (see review by Monteith 1981a) but was extended to a plant canopy by substituting the appropriate conductances and is commonly referred to as the Penman–Monteith equation (Monteith 1965; Jarvis and McNaughton 1986).

Evaporation of water from a saturated surface is a thermodynamic process in which energy is required to change water from liquid to vapor. Latent heat flux (in the absence of biomass storage) is

$$\lambda E = (R_n - G) - H = (R_n - G) - c_p(T_s - T_a)g_{ac}, \tag{7.32}$$

where here T_s and T_a are used to denote surface and air temperature, respectively. Evaporation is also an aerodynamic process related to turbulent transport of water vapor away from the surface. This is represented by (7.6), in which latent heat flux increases as evaporative demand, given by the water vapor deficit between the surface with $q_{sat}(T_s)$ and air with q_a, increases. Saturation vapor pressure is linearized using a Taylor series expansion so that

$$q_{sat}(T_s) = q_{sat}(T_a) + s(T_s - T_a) \tag{7.33}$$

in which $q_{sat}(T_a)$ is saturation vapor pressure evaluated at the air temperature T_a and $s = dq_{sat}/dT$ is evaluated at T_a. Substituting (7.33) into (7.6) gives

$$\lambda E = \frac{\lambda[q_{sat}(T_a) - q_a + s(T_s - T_a)]}{g_c^{-1} + g_{ac}^{-1}}. \tag{7.34}$$

Latent heat flux is obtained by finding the surface temperature that satisfies (7.32) and (7.34). From (7.32):

$$T_s - T_a = \frac{(R_n - G) - \lambda E}{c_p g_{ac}}. \tag{7.35}$$

Substituting (7.35) into (7.34) gives

$$\lambda E = \frac{s(R_n - G) + c_p[q_{sat}(T_a) - q_a]g_{ac}}{s + (1 + g_{ac}/g_c)c_p/\lambda}. \tag{7.36}$$

For a dense plant canopy, the surface conductance is the canopy conductance (Kelliher et al. 1995). By measuring evapotranspiration from a canopy, if all other terms are known, the Penman–Monteith equation can be rearranged to give

$$\frac{1}{g_c} = \frac{a+1}{g_{ac}}\left[\frac{a(R_n - G)}{(a+1)\lambda E} - 1\right] + \frac{[q_*(T_a) - q_a]}{E}, \quad (7.37)$$

with $a = s\lambda/c_p$. This conductance captures the effects of leaf area, canopy coverage, photosynthetic capacity, and soil moisture on the partitioning of net radiation into latent heat flux. The Penman–Montieth equation is known as a big-leaf model because it represents the canopy as a single leaf with an effective conductance g_c.

Question 7.7 Use the Penman–Monteith equation to calculate canopy conductance for latent heat flux equal to 300 W m^{-2} with $R_n - G = 400$ W m^{-2}, $T_a = 25°C$, 75% relative humidity, and $g_{ac} = 2$ mol m^{-2} s^{-1}.

Question 7.8 This chapter uses mole fraction to calculate latent heat flux. A more common expression instead uses vapor pressure with $\lambda E = (c_p/\gamma)[e_{sat}(T_s) - e_a]g_w$ and $g_w = 1/(g_c^{-1} + g_{ac}^{-1})$. Derive the Penman–Monteith equation using this notation.

7.7 | Chapter Appendix: Snowmelt

In the LM2 model (Milly and Shmakin 2002), surface fluxes and soil temperature are calculated implicitly at time $n+1$ using the tridiagonal equations given by (7.21). A common numerical algorithm to solve a tridiagonal system of equations has a forward sweep from the surface layer ($i = 1$) to the bottom soil layer ($i = N$) and then a backward sweep from N to 1 (Appendix A8). In this solution, the first temperature calculated is T_N^{n+1}. To limit the surface temperature to freezing when snow is melting, it is necessary to solve for T_1^{n+1} first. If the numerical algorithm is revised in such a manner, the temperature of the surface soil layer (in the absence of snowmelt) is

$$\hat{T}_1^{n+1} = T_1^n + \frac{d_1 - c_1 f_2}{b_1 - c_1 e_2}, \quad (7.38)$$

with b_1 (W m^{-2} K^{-1}), c_1 (W m^{-2} K^{-1}), and d_1 (W m^{-2}) terms in the tridiagonal equation (7.21) and e_2 (dimensionless) and f_2 (K) obtained from the numerical algorithm. The excess energy (W m^{-2}) above the freezing temperature T_f is $\left(\hat{T}_1^{n+1} - T_f\right)(b_1 - c_1 e_2)$. This is the maximum energy to melt snow. The actual energy is the lesser of this or the amount of snow so that

$$G_{snow} = \min\left[\left(\hat{T}_1^{n+1} - T_f\right)(b_1 - c_1 e_2), L_f \frac{W_s}{\Delta t}\right], \quad (7.39)$$

and the actual temperature increase is

$$T_1^{n+1} = T_1^n + \frac{(d_1 - c_1 f_2) - G_{snow}}{b_1 - c_1 e_2}. \quad (7.40)$$

If all the excess energy is used to melt snow, $T_1^{n+1} = T_f$. For lesser amounts of snowmelt, $T_1^{n+1} > T_f$. The energy available to warm the soil is

$$G_{soil} = F_0\left(T_1^n\right) + \frac{\partial F_0\left(T_1^n\right)}{\partial T_1}\left(T_1^{n+1} - T_1^n\right) - G_{snow}. \quad (7.41)$$

The derivation of these equations can be understood by considering changes in the soil energy balance upon setting $\hat{T}_1^{n+1}$ to any value T_1'. In this case, the soil heat flux is

$$G = F_0\left(T_1^n\right) + \frac{\partial F_0\left(T_1^n\right)}{\partial T_1}\left(T_1' - T_1^n\right). \quad (7.42)$$

However, setting $\hat{T}_1^{n+1}$ to T_1' itself changes the heat content of the soil by an additional amount equal to $\left(T_1' - \hat{T}_1^{n+1}\right)(b_1 - c_1 e_2)$ (W m^{-2}) so that the total flux of energy to the soil is

$$G_{soil} = G + \left(T_1' - \hat{T}_1^{n+1}\right)(b_1 - c_1 e_2). \quad (7.43)$$

This equation can be equivalently written as $G_{soil} = G - G'$. In this form, the energy flux $G' = \left(\hat{T}_1^{n+1} - T_1'\right)(b_1 - c_1 e_2)$ must be subtracted from the soil. In the case of snowmelt, $T_1' = T_f$ and G' is the energy subtracted from G to melt snow.

Question 7.9 Rewrite the tridiagonal algorithm given in Appendix A8 so that the surface layer temperature T_1^{n+1} is obtained first rather than the bottom layer temperature T_N^{n+1}.

7.8 | Supplemental Programs

7.1 Diurnal Cycle of Surface Fluxes: This program calculates the diurnal cycle of surface fluxes and soil temperatures in an implicit solution as in Figure 7.5. It combines principles of soil heat transfer (Chapter 5), turbulent fluxes (Chapter 6), and the surface energy balance described in this chapter. Air temperature varies sinusoidally between 20°C and 30°C. Other atmospheric inputs are relative humidity, wind speed, surface pressure, precipitation, and forcing height. Solar radiation is calculated for clear sky based on solar geometry, and longwave radiation varies with atmospheric temperature. Surface emissivity, albedo, canopy height, and leaf area index are also required inputs. Canopy conductance depends on leaf area index and the sunlit and shaded fractions of the canopy. The code uses Monin–Obukhov similarity theory with an optional roughness sublayer parameterization from Harman and Finnigan (2007, 2008) as described in Chapter 6. There is an optional bucket model hydrology in which latent heat flux is multiplied by a soil wetness factor. The code uses 10 soil layers of variable thickness in which thermal properties vary with texture and soil moisture. The time step is 30 minutes.

7.9 | Modeling Projects

1. Supplemental Program 7.1 calculates the diurnal cycle of surface fluxes for a warm summer day as in Figure 7.5. Use this program to calculate H, λE, and θ_s. Compare the diurnal cycle for a one-day simulation with that after 5, 10, and 30 days. Why do the results differ?

2. Forests generally have a lower albedo compared with grassland, are aerodynamically rougher, and have a higher canopy conductance. Examine the changes in surface fluxes and surface temperature in relation to these parameters. Which parameter has the largest impact?

3. Calculate the diurnal cycle of surface temperature and energy fluxes for a soil with low heat capacity ($c_v = 1.5$ MJ m^{-3} K^{-1}). Contrast this with a soil that has high heat capacity ($c_v = 3.0$ MJ m^{-3} K^{-1}). How does heat capacity affect the diurnal cycle of surface temperature?

4. Supplemental Program 7.1 has a bucket hydrology model. Use this to examine the effects of soil wetness on surface fluxes. How do fluxes vary in relation to the precipitation rate? How long is the model spin-up?

5. Use the Choudhury and Monteith (1988) and Raupach (1994, 1995) parameterizations for d and z_{0m} (Chapter 6) to examine the sensitivity of surface fluxes to leaf area index. Note that leaf area index also affects canopy conductance.

6. Supplemental Program 7.1 has the Harman and Finnigan (2007, 2008) roughness sublayer parameterization as described in Chapter 6. Use this to study the effects of the roughness sublayer on surface fluxes. Describe why surface fluxes are different compared with the model's implementation of Monin–Obukhov similarity theory. Hint: compare roughness lengths.

Soil Moisture

Chapter Overview

Water flows from high to low potential as described by Darcy's law. The Richards equation combines Darcy's law with principles of water conservation to calculate water movement in soil. Particular variants of the Richards equation are the mixed-form, head-based, and moisture-based equations. Water movement is determined by hydraulic conductivity and matric potential, both of which vary with soil moisture and additionally depend on soil texture. This chapter reviews soil moisture and the Richards equation. Numerical solutions are given for the various forms of the equation.

8.1 | Introduction

The region of soil between the ground surface and the water table is known as the unsaturated, or vadose, zone, and the water held in this zone is called soil moisture (Figure 8.1). The water content of the vadose zone is dynamic, ranging from saturation in the upper soil layers near the surface during infiltration to nearly dry in prolonged absence of rainfall as plant roots extract water during transpiration. The vertical profile of soil water is a particularly important determinant of land–atmosphere coupling. A dry surface layer develops in the absence of rainfall, and this dry layer impedes soil evaporation. Conversely, plant roots can extend deep in the soil to sustain transpiration during dry periods. Below the vadose zone lies saturated groundwater,

and soil moisture also controls the fluxes of water between the vadose zone and groundwater.

The first models of the land surface used in climate simulations ignored the complexity of the hydrologic cycle and instead abstracted it using the bucket analogy in which soil is treated as a bucket that fills from precipitation, empties from evapotranspiration, or spills over as runoff as described in Chapter 7. In fact, however, storage of water and its movement in soil is much more complex. When soil is wet, water is loosely held in soil and quickly drains due to the force of gravity. When soil is dry, water movement becomes more difficult, and at some critical amount the water is strongly bound to soil particles and can no longer be removed. Figure 8.2 illustrates the dynamics of water movement during infiltration into initially dry soil. A distinct wetting front moves progressively downward over time. The upper soil becomes saturated with water while the deeper soil remains dry. In the sandy soil shown in Figure 8.2, the upper 50–60 cm become saturated after 42 minutes (0.7 h). This dynamics is explained from physical principles using Darcy's law and the Richards equation.

8.2 | Measures of Soil Moisture

A typical soil consists of solid particles of varying size and shape that are interconnected by pores. Water completely fills these pores when the soil is saturated, or the pores consist mainly of air when the soil is dry. Most conditions in the field are in-between, and soil is a mix of solid particles, water,

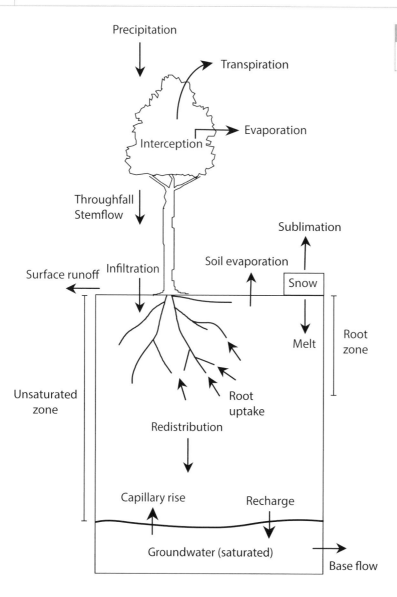

Figure 8.1 Water flows in a soil column extending from the ground surface to the water table.

and air (Figure 8.3). The bulk density of a soil ρ_b (kg m^{-3}) is the mass of soil solids per volume of soil so that

$$\rho_b = \frac{\text{mass of solids}}{\text{volume of soil}} = \frac{m_s}{V}. \qquad (8.1)$$

The bulk volume of soil consists of the total volume of solids and pore space ($V = V_s + V_p$). The particle density ρ_s (kg m^{-3}) is the mass of soil solids per volume of soil solids whereby

$$\rho_s = \frac{\text{mass of solids}}{\text{volume of solids}} = \frac{m_s}{V_s}. \qquad (8.2)$$

If a volume of soil 10 cm $\times$ 10 cm $\times$ 10 cm has a dry mass of 1.325 kg, its bulk density is 1325 kg m^{-3}. If the pore space comprises one-half of this volume, $V_s = 0.5V$, and the particle density is 2650 kg m^{-3}. A typical particle density is, in fact, 2650 kg m^{-3}. The fraction of the soil volume comprising pores, known as porosity, is $(V - V_s)/V$. When saturated, water fills all the pores so that porosity is also the volumetric water content at saturation θ_{sat}. Porosity is calculated from bulk density and particle density by

$$\theta_{sat} = 1 - \frac{\rho_b}{\rho_s}. \qquad (8.3)$$

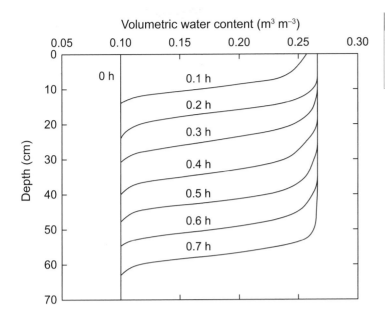

Volumetric water content (m³ m⁻³)

Depth (cm)

Figure 8.2 Soil water movement during infiltration into sand. Shown are the initial moisture profile ($\theta = 0.1$) and profiles in increments of 0.1 h. Adapted from Haverkamp et al. (1977)

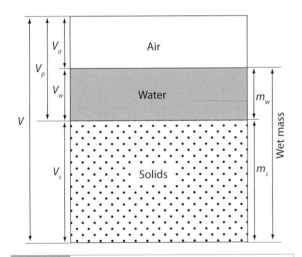

Figure 8.3 Depiction of soil with total volume V comprising soil solids V_s, water V_w, and air V_a. The total volume of pore space is $V_p = V_w + V_a$. The total soil mass consists of soil solids with mass m_s and water with mass m_w.

The amount of water can be measured by volume (m³ H₂O) and equivalently by mass per area (kg H₂O m⁻²) or depth (m H₂O). These are related by the density of water ($\rho_{wat} = 1000$ kg m⁻³) so that mass per area = depth × density. One kilogram of water spread over an area of one square meter (1 kg m⁻²) is equivalent to a depth of 1 mm and a volume of 0.001 m³. A common measure of soil moisture is volumetric water content (m³ H₂O m⁻³ soil). Volumetric water content is

$$\theta = \frac{\text{volume of water}}{\text{volume of soil}} = \frac{V_w}{V}. \qquad (8.4)$$

Volumetric water content is also the depth of water per unit depth of soil. A soil with thickness Δz m contains $\theta \Delta z$ m of water. The mass of water per area W (kg m⁻²) in a volume of soil with volumetric water content θ is

$$W = \theta \Delta z \rho_{wat}. \qquad (8.5)$$

Another measure of soil moisture is gravimetric (mass) water content (kg H₂O kg⁻¹ dry soil). Gravimetric water content is

$$\theta_m = \frac{\text{mass of water}}{\text{mass of dry soil}} = \frac{m_w}{m_s}. \qquad (8.6)$$

Volumetric water content is related to gravimetric water content by the density of water and the bulk density of soil as

$$\theta = \theta_m \frac{\rho_b}{\rho_{wat}}. \qquad (8.7)$$

Another measure is the effective saturation, which is defined as the soil moisture θ above some residual amount θ_{res} relative that at saturation:

$$S_e = \frac{\theta - \theta_{res}}{\theta_{sat} - \theta_{res}}. \qquad (8.8)$$

Table 8.1	Various measures of soil water for a volume of soil with dimensions 10 cm × 10 cm × 10 cm that has a mass of 1.7 kg when wet, 1.45 kg when dry, and particle density $\rho_s = 2650$ kg m^{-3}
Quantity	**Amount**
Mass of water	$m_w = 1.7$ kg $- 1.45$ kg $= 0.25$kg
Mass of soil solids	$m_s = 1.45$kg
Bulk volume of soil	$V = 0.1$ m $\times 0.1$ m $\times 0.1$ m $= 0.001$m^3
Bulk density	$\rho_b = m_s/V = 1450$ kg m^{-3}
Porosity	$\theta_{sat} = 1 - \rho_b/\rho_s = 0.453$
Volume of water	$V_w = m_w/\rho_{wat} = 0.00025$ m^3
Gravimetric water content	$\theta_m = m_w/m_s = 0.172$ kg kg^{-1}
Volumetric water content	$\theta = V_w/V = \theta_m\rho_b/\rho_{wat} = 0.25$ m^3 m^{-3}
Water content relative to saturation	$\theta/\theta_{sat} = 0.552$
Depth of water	$\theta\Delta z = 0.25 \times 0.1$ m $= 0.025$ m
Mass of water per area	$W = \dfrac{0.25 \text{ kg}}{0.1 \text{ m} \times 0.1 \text{ m}} = \theta\Delta z\rho_{wat} = 25$ kg m^{-2}

Table 8.1 provides example calculations for these various measures of soil moisture.

8.3 | Matric Potential and Hydraulic Conductivity

Water is tightly held by the surfaces of soil particles. This creates a negative pressure, or suction, called matric potential ψ that binds water to the soil. (The symbol h is commonly used in soil science and hydrology to denote this as pressure head.) The matric force is positive when represented as suction and negative when given as potential. Matric potential varies with soil moisture. Relatively weak suction is exerted on water when soil is saturated (matric potential is high), but suction increases sharply (matric potential decreases) as soil becomes drier and strong forces bind water in small pores. This relationship between ψ and θ is quite nonlinear and varies depending on soil texture (Figure 8.4). Water is loosely held in sandy soils (low suction) and tightly held in clay soils (high suction).

The dependence between ψ and θ is referred to as the soil moisture retention curve and is described mathematically by equations that relate θ to ψ or, equivalently, ψ to θ, denoted $\theta(\psi)$ or $\psi(\theta)$, respectively. Table 8.2 gives three common relationships. Brooks and Corey (1964, 1966) related ψ to the

effective saturation S_e. In this equation, ψ_b and c are empirical parameters used to fit the data; ψ_b is the air entry water potential and is the value of ψ at which $S_e = 1$; c is referred to as the pore-size distribution index. Campbell (1974) proposed a similar relationship but with $\theta_{res} = 0$, in which case ψ_b can be thought of as the matric potential at saturation. Van Genuchten (1980) developed another widely used soil moisture retention curve. In this relationship, the empirical parameter α is the inverse of the air entry potential ($\alpha = 1/|\psi_b|$), and n is the pore-size distribution index.

Parameter values vary depending on soil texture, and various so-called pedotransfer functions relate hydraulic parameters to discrete texture classes or as continuous functions of sand, clay, or other soil properties. The Brooks and Corey (1964, 1966) parameters, for example, can be related to sand, clay, and porosity (Rawls and Brakensiek 1985; Rawls et al. 1993). Clapp and Hornberger (1978) estimated parameters for various soil texture classes using the Campbell (1974) relationship, and Cosby et al. (1984) subsequently related θ_{sat}, ψ_{sat}, and b to the sand and clay content of soil. The van Genuchten (1980) parameters can be difficult to estimate (Carsel and Parrish 1988; Leij et al. 1996; Schaap et al. 1998, 2001). Table 8.3 gives representative values for soil texture classes.

Water held in soil is subjected to two forces, or potentials. The force of gravity pulls water

Table 8.2 | Soil moisture retention and hydraulic conductivity functions

$\theta(\psi)$	$K(\theta)$
(a) Brooks and Corey (1964, 1966)	

$$S_e = \frac{\theta - \theta_{res}}{\theta_{sat} - \theta_{res}} = \left(\frac{\psi}{\psi_b}\right)^{-c} \qquad K = K_{sat}S_e^{2/c+3}$$

$$\frac{d\theta}{d\psi} = \frac{-c(\theta_{sat} - \theta_{res})}{\psi_b}\left(\frac{\psi}{\psi_b}\right)^{-c-1} \qquad \frac{dK}{d\theta} = \frac{K_{sat}}{\theta_{sat} - \theta_{res}}(2/c + 3)S_e^{2/c+2}$$

(b) Campbell (1974)

$$\frac{\theta}{\theta_{sat}} = \left(\frac{\psi}{\psi_{sat}}\right)^{-1/b} \qquad K = K_{sat}\left(\frac{\theta}{\theta_{sat}}\right)^{2b+3}$$

$$\frac{d\theta}{d\psi} = \frac{-\theta_{sat}}{b\,\psi_{sat}}\left(\frac{\psi}{\psi_{sat}}\right)^{-1/b-1} \qquad \frac{dK}{d\theta} = \frac{K_{sat}(2b + 3)}{\theta_{sat}}\left(\frac{\theta}{\theta_{sat}}\right)^{2b+2}$$

(c) van Genuchten (1980)

$$S_e = \frac{\theta - \theta_{res}}{\theta_{sat} - \theta_{res}} = \left[1 + (\alpha|\psi|)^n\right]^{-m}, \qquad K = K_{sat}S_e^{1/2}\left[1 - \left(1 - S_e^{1/m}\right)^m\right]^2$$

$$m = 1 - 1/n$$

$$\frac{d\theta}{d\psi} = \frac{\alpha mn(\theta_{sat} - \theta_{res})(\alpha|\psi|)^{n-1}}{\left[1 + (\alpha|\psi|)^n\right]^{m+1}} \qquad \frac{dK}{d\theta} = \frac{K_{sat}}{(\theta_{sat} - \theta_{res})}\left[\frac{f^2}{2S_e^{1/2}} + \frac{2S_e^{1/m-1/2}f}{\left(1 - S_e^{1/m}\right)^{1-m}}\right],$$

$$f = 1 - \left(1 - S_e^{1/m}\right)^m$$

Note: (a) $S_e = 1$ and $K = K_{sat}$ for $\psi > \psi_b$. (b) $\theta/\theta_{sat} = 1$ and $K = K_{sat}$ for $\psi > \psi_{sat}$. (c) $S_e = 1$ and $K = K_{sat}$ for $\psi > 0$.

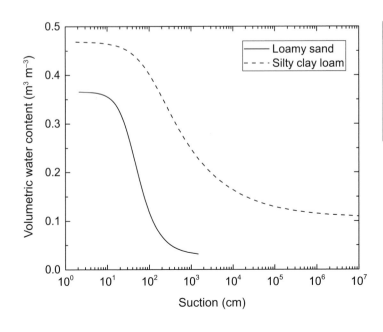

Figure 8.4 Water content in relation to matric potential given here as suction $(-\psi)$. Shown is the van Genuchten (1980) $\theta(\psi)$ relationship for Berino loamy fine sand $(\theta_{res} = 0.0286, \theta_{sat} = 0.3658, \alpha = 0.028 \text{ cm}^{-1}, n = 2.239)$ and Glendale silty clay loam $(\theta_{res} = 0.106, \theta_{sat} = 0.4686, \alpha = 0.0104 \text{ cm}^{-1}, n = 1.3954)$ from Hills et al. (1989).

Table 8.3 Parameter values for the Campbell (1974) and van Genuchten (1980) $\theta(\psi)$ and $K(\theta)$ functions arranged by soil texture

Soil type	Campbell				van Genuchten				
	θ_{sat}	ψ_{sat} (cm)	b	K_{sat} (cm h^{-1})	θ_{sat}	θ_{res}	α (cm^{-1})	n	K_{sat} (cm h^{-1})
Sand	0.395	−12.1	4.05	63.36	0.43	0.045	0.145	2.68	29.70
Loamy sand	0.410	−9.0	4.38	56.28	0.41	0.057	0.124	2.28	14.59
Sandy loam	0.435	−21.8	4.90	12.48	0.41	0.065	0.075	1.89	4.42
Silt loam	0.485	−78.6	5.30	2.59	0.45	0.067	0.020	1.41	0.45
Loam	0.451	−47.8	5.39	2.50	0.43	0.078	0.036	1.56	1.04
Sandy clay loam	0.420	−29.9	7.12	2.27	0.39	0.100	0.059	1.48	1.31
Silty clay loam	0.477	−35.6	7.75	0.61	0.43	0.089	0.010	1.23	0.07
Clay loam	0.476	−63.0	8.52	0.88	0.41	0.095	0.019	1.31	0.26
Sandy clay	0.426	−15.3	10.4	0.78	0.38	0.100	0.027	1.23	0.12
Silty clay	0.492	−49.0	10.4	0.37	0.36	0.070	0.005	1.09	0.02
Clay	0.482	−40.5	11.4	0.46	0.38	0.068	0.008	1.09	0.20

Note: Soils are arranged from least to most clay.
Source: Campbell (1974) parameters from Clapp and Hornberger (1978); van Genuchten (1980) parameters from Carsel and Parrish (1988) and Leij et al. (1996).

downward, denoted as gravitational potential. Gravitational potential is the height z above some arbitrary reference height. The second force is the matric potential, which holds water to the soil particles. The total potential, also known as hydraulic head, is $\psi + z$, and this is the work per unit weight required to move an amount of water at some elevation and matric potential to another position in the soil with a different potential. Soil water potential as used in this chapter has the units of energy per unit weight and has the dimension of length (m):

$$\frac{energy}{weight} = \frac{J}{N} = \frac{N \cdot m}{N} = m$$

Other units for potential are energy per unit mass (J kg^{-1}) or energy per unit volume (J m^{-3}), and this latter expression is also the units of pressure (Pa). Soil water potential is converted from m to J kg^{-1} by multiplying by gravitational acceleration g and to J m^{-3} by multiplying by $\rho_{wat}g$:

$$J\,kg^{-1} = g\psi$$

$$J\,m^{-3} = Pa = \rho_{wat}g\psi.$$

A wet soil with a matric potential of −0.01 MPa has a suction of approximately 1000 mm; a dry soil with −1.5 MPa has a suction of approximately 150 m.

Hydraulic conductivity governs the rate of water flow for a unit gradient in potential. Hydraulic conductivity decreases sharply as soil becomes drier because suction increases and because the pore space filled with water becomes smaller and discontinuous. This relationship is nonlinear and varies with soil texture (Figure 8.5). Sandy soil has a higher conductivity than clay soil. The relationship between K and θ is not independent of the relationship between θ and ψ, and the derivation of $K(\theta)$ requires an expression for $\theta(\psi)$ (Figure 8.4). This expression is given in terms of the effective saturation S_e, or θ/θ_{sat} in the Campbell (1974) equation, so that hydraulic conductivity can be equivalently expressed as $K(\psi)$. The expressions for hydraulic conductivity also require K_{sat}, the hydraulic conductivity at saturation. The term $S_e^{1/2}$ in the van Genuchten (1980) equation for hydraulic conductivity is a common form, but the exponent can vary (Schaap et al. 2001). Better fit to data can be achieved by replacing K_{sat} with a curve fitting parameter K_0, but this has a value that is usually less than K_{sat} so that $K(\theta) \neq K_{sat}$ at θ_{sat} (Schaap et al. 2001).

Question 8.1 Graph and compare the van Genuchten (1980) and Campbell (1974) relationships for $\theta(\psi)$ and $K(\theta)$

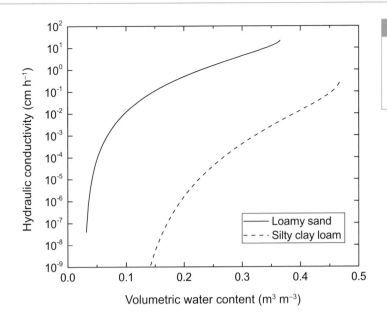

Figure 8.5 Hydraulic conductivity in relation to water content. Shown is the van Genuchten (1980) $K(\theta)$ relationship for Berino loamy fine sand and Glendale silty clay loam. Parameter values are as in Figure 8.4 and K_{sat}= 22.54 and 0.55 cm h^{-1}, respectively (Hills et al. 1989).

for sandy loam, loam, and clay loam using parameter values in Table 8.3. Describe differences in the shape of these relationships.

8.4 | Richards Equation

Darcy's law describes water flow. In the vertical dimension, the rate of water movement is

$$Q = -K(\theta)\frac{\partial(\psi + z)}{\partial z} = -K(\theta)\left[\frac{\partial\psi}{\partial z} + 1\right]$$

$$= -K(\theta)\frac{\partial\psi}{\partial z} - K(\theta). \tag{8.9}$$

This is a form of Fick's law and relates the rate of flow to the product of the hydraulic conductivity and the vertical gradient in water potential. The flux Q is the volume of water (m^3) flowing through a unit cross-sectional area (m^2) per unit time (s) and has the dimensions length per time (m s^{-1}). Hydraulic conductivity has the same units, and $K(\theta)$ denotes that hydraulic conductivity depends on soil moisture. The total potential $\psi + z$ has dimensions of length (m). It governs water movement so that water flows from high to low potential. The vertical depth z is taken as positive in the upward direction so that $z = 0$ is the ground surface and $z < 0$ is the elevation relative to the surface

with greater depth into the soil. The matric potential has values $\psi < 0$ for unsaturated soil and $\psi \geq 0$ for saturated soil. The negative sign in (8.9) ensures that downward water flow is negative and upward flow is positive.

The flow of water given by Darcy's law depends strongly on soil moisture. The dominant force causing water to move in a wet soil is the gravitation potential. Water near the surface has a higher gravitational potential than water deeper in the soil, and because water flows from high potential to low potential, it flows downward from the force of gravity. This is given by z in (8.9), which is the height relative to the ground surface. In drier soils, matric potential decreases, and the adsorptive force binding water to soil particles generally exceeds the gravitational force pulling water downward. This reduces the rate of water flow. The lower hydraulic conductivity in dry soils also restricts water movement.

An equation for the time rate of change in soil moisture is obtained from principles of conservation similar to that for soil temperature. Consider a volume of soil with horizontal area $\Delta x \Delta y$ and thickness Δz and in which water flows only in the vertical dimension (Figure 8.6). The mass flux of water (kg s^{-1}) entering the soil across the cross-sectional area $\Delta x \Delta y$ is $\rho_{wat}Q_{in}\Delta x \Delta y$, and the flux out of the soil is similarly $\rho_{wat}Q_{out}\Delta x \Delta y$. Conservation requires that the difference between the flux of

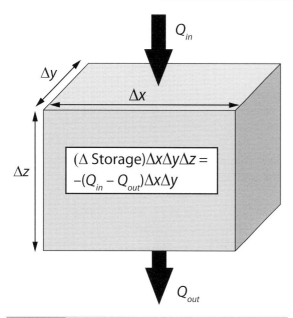

$$(\Delta \text{ Storage})\Delta x\Delta y\Delta z = -(Q_{in} - Q_{out})\Delta x\Delta y$$

Figure 8.6 Water balance for a soil volume with the fluxes Q_{in} entering the volume and Q_{out} exiting the volume.

water into and out of the soil equals the rate of change in water storage. The mass of water in the soil volume is $\rho_{wat}\theta\Delta x\Delta y\Delta z$ so that the change in soil water over the time interval Δt is

$$\rho_{wat}\frac{\Delta\theta}{\Delta t}\Delta x\Delta y\Delta z = -\rho_{wat}(Q_{in} - Q_{out})\Delta x\Delta y \qquad (8.10)$$

and

$$\frac{\Delta\theta}{\Delta t} = -\frac{\Delta Q}{\Delta z}. \qquad (8.11)$$

Equation (8.11) is the continuity equation for water, with the left-hand side the change in water storage and the right-hand side the flux divergence.

In the notation of calculus, the continuity equation is

$$\frac{\partial\theta}{\partial t} = -\frac{\partial Q}{\partial z}, \qquad (8.12)$$

and substituting Darcy's law for Q gives

$$\frac{\partial\theta}{\partial t} = \frac{\partial}{\partial z}\left[K(\theta)\frac{\partial\psi}{\partial z} + K(\theta)\right] = \frac{\partial}{\partial z}\left[K(\theta)\frac{\partial\psi}{\partial z}\right] + \frac{\partial K}{\partial z}. \qquad (8.13)$$

This is the Richards equation and describes the movement of water in an unsaturated porous medium (Richards 1931). Equation (8.13) is called

the mixed-form equation because it includes the time rate of change in θ on the left-hand side and the vertical gradient in ψ on the right-hand side. Other forms of the Richards equation use the dependence between ψ and θ to express the equation in terms of only one unknown variable. The head-based, or ψ-based, form transforms the storage term on the left-hand side of the equation from θ to ψ so that ψ is the dependent variable. This uses the chain rule to expand $\partial\theta/\partial t$ as

$$\frac{\partial\theta}{\partial t} = \frac{d\theta}{d\psi}\frac{\partial\psi}{\partial t} = C(\psi)\frac{\partial\psi}{\partial t}, \qquad (8.14)$$

in which $C(\psi) = d\theta/d\psi$ is known as the specific moisture capacity (m^{-1}) and is the slope of the soil moisture retention curve. Then (8.13) is rewritten as

$$C(\psi)\frac{\partial\psi}{\partial t} = \frac{\partial}{\partial z}\left[K(\theta)\frac{\partial\psi}{\partial z}\right] + \frac{\partial K}{\partial z}. \qquad (8.15)$$

The ψ-based form is applicable for unsaturated and saturated conditions and provides a continuous equation for water flow in the vadose zone and for groundwater. However, it is not mass conserving because the specific moisture capacity $d\theta/d\psi$ that appears in the storage term itself depends on ψ and so is not constant over a discrete time interval during which ψ changes value (Milly 1985; Celia et al. 1990). Whereas (8.14) is mathematically correct, its temporal discretization over some time interval Δt (as required in numerical methods) is not equivalent. The moisture-based, or θ-based, equation uses θ as the dependent variable with

$$\frac{\partial\theta}{\partial t} = \frac{\partial}{\partial z}\left[D(\theta)\frac{\partial\theta}{\partial z}\right] + \frac{\partial K}{\partial z}, \qquad (8.16)$$

in which $D(\theta) = K(\theta)/C(\psi)$ is referred to as the hydraulic diffusivity (m^2 s^{-1}). In this equation, the specific moisture capacity appears within the spatial derivative. The θ-based form is mass conserving but is restricted to the unsaturated zone because soil moisture does not vary within a saturated porous medium (soil moisture is bounded by $0 \leq \theta \leq \theta_{sat}$) whereas pressure head does vary. Furthermore, the equation is restricted to homogenous soils because θ is not continuous across soil layers with different $\theta(\psi)$ relationships. As a result, soils in which texture varies with depth have discontinuous vertical profiles of θ, whereas ψ is continuous even in inhomogeneous soils.

The Richards equation requires relationships for $K(\theta)$ and $\theta(\psi)$. Analytical solutions are difficult to obtain because these relationships are highly non-linear. Instead, numerical methods are used, which requires first writing the finite difference approximation of the partial differential equation and then linearizing the nonlinear terms involving $K(\theta)$ and $C(\psi)$. The accuracy of the solution very much depends on the details of the numerical methods, the form of the Richards equation, and the size of the time step and grid spacing. The literature on numerical methods to solve the Richards equation is enormous. The next two sections provide an introduction to this literature with the caveat that many more numerical techniques are available and the merits of particular methods are still being debated.

Question 8.2 Soil at a depth of 5 cm has a matric potential of −478 mm, and the matric potential 50 cm deeper is −843 mm. Calculate the vertical water flux with a hydraulic conductivity of 2 mm h^{-1}. What is the horizontal water flux if both locations are at the same depth in the soil but separated by 50 cm? Explain the difference between the two fluxes.

Question 8.3 In Darcy's law given by (8.9), water flux has the units m H$_2$O s^{-1}, hydraulic conductivity is m s^{-1}, and hydraulic head is m. Is
$$Q = -K(\theta)\frac{\partial}{\partial z}\left(\frac{\psi}{\rho_{wat}g} + z\right)$$ an equivalent equation? What are the units for ψ, $K(\theta)$, and Q in this equation? Derive the conversion factor for $K(\theta)$ from m s^{-1} to the same units as ψ.

Question 8.4 A model calculates soil moisture in the unsaturated zone using the mixed-form Richards equation and solves the equation $\frac{\partial\theta}{\partial t} = \frac{\partial}{\partial z}\left[K(\theta)\frac{\partial\psi}{\partial z}\right] - \frac{\partial K}{\partial z}$. Explain the difference between this equation and (8.13). Are the equations equivalent?

8.5 | Finite Difference Approximation

The finite difference approximation for the Richards equation represents the soil as a network of discrete nodal points that vary in space and time similar to that for soil temperature. In doing so, it is necessary to remember that hydraulic conductivity is not constant but, rather, varies with depth depending on soil moisture so that the mixed-form equation is expanded as

$$\frac{\partial\theta}{\partial t} = K(\theta)\frac{\partial^2\psi}{\partial z^2} + \frac{\partial K}{\partial z}\frac{\partial\psi}{\partial z} + \frac{\partial K}{\partial z}. \tag{8.17}$$

For reasons of numerical stability similar to soil temperature, (8.17) is solved using an implicit time discretization in which the spatial derivatives $\partial K/\partial z$, $\partial\psi/\partial z$, and $\partial^2\psi/\partial z^2$ are written numerically using a central difference approximation at time $n+1$, and the time derivative uses a backward difference approximation at $n+1$ (Appendix A4). For a vertical grid with discrete layers each equally spaced at a distance Δz and with z positive in the upward direction so that layer i is above layer $i+1$, the numerical form of (8.17) is

$$\frac{\theta_i^{n+1} - \theta_i^n}{\Delta t} = \frac{K_i^{n+1}}{\Delta z^2}\left(\psi_{i-1}^{n+1} - 2\psi_i^{n+1} + \psi_{i+1}^{n+1}\right)$$
$$+ \left(\frac{K_{i-1}^{n+1} - K_{i+1}^{n+1}}{2\Delta z}\right)\left(\frac{\psi_{i-1}^{n+1} - \psi_{i+1}^{n+1}}{2\Delta z}\right)$$
$$+ \frac{K_{i-1}^{n+1} - K_{i+1}^{n+1}}{2\Delta z}. \tag{8.18}$$

This is an implicit solution in which θ and ψ are expressed at $n+1$, and K is similarly evaluated with θ at $n+1$. Rearranging terms gives an equivalent form in which

$$\frac{\theta_i^{n+1} - \theta_i^n}{\Delta t} = \frac{K_{i-1/2}^{n+1}}{\Delta z^2}\left(\psi_{i-1}^{n+1} - \psi_i^{n+1}\right) - \frac{K_{i+1/2}^{n+1}}{\Delta z^2}\left(\psi_i^{n+1} - \psi_{i+1}^{n+1}\right)$$
$$+ \frac{K_{i-1/2}^{n+1} - K_{i+1/2}^{n+1}}{\Delta z}, \tag{8.19}$$

with

$$K_{i\pm1/2}^{n+1} = K_i^{n+1} \pm \frac{K_{i+1}^{n+1} - K_{i-1}^{n+1}}{4}. \tag{8.20}$$

The ψ-based form of the equation is obtained by replacing the left-hand side of (8.19) with $C_i^{n+1}\left(\psi_i^{n+1} - \psi_i^n\right)/\Delta t$.

A more general derivation of (8.19) is obtained by considering the mass balance of a soil layer. Figure 8.7 depicts the soil profile in a cell-centered grid of N discrete layers similar to that used for soil temperature. Soil layer i has a thickness Δz_i. Water content θ_i, matric potential ψ_i, and hydraulic

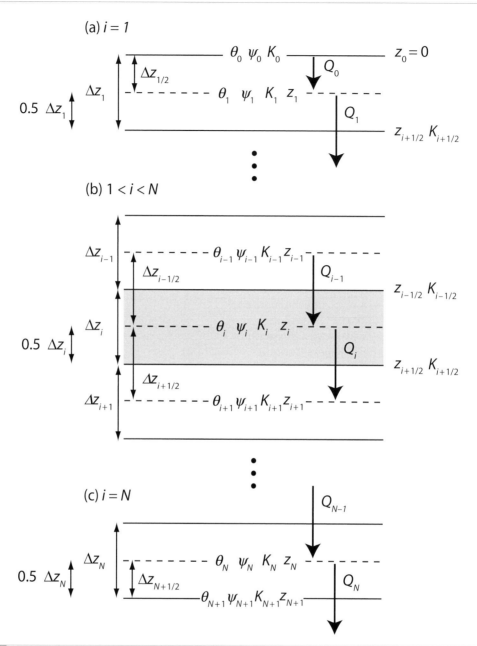

Figure 8.7 Multilayer soil water flow in a cell-centered grid oriented such that $i = 1$ is the top soil layer at the surface, and $i = N$ is the bottom soil layer. Each layer has a thickness Δz_i. The depth $z_{i-1/2}$ is the interface between adjacent layers $i - 1$ and i, and $z_{i+1/2}$ is the interface between i and $i + 1$. Depths are negative distances from the surface so that $\Delta z_i = z_{i-1/2} - z_{i+1/2}$ (i.e., $z_{i+1/2} = z_{i-1/2} - \Delta z_i$). The depth z_i is defined at the center of layer i so that $z_i = (z_{i-1/2} + z_{i+1/2})/2$. $\Delta z_{i\pm1/2}$ is the grid spacing between i and $i \pm 1$. Water content θ_i, matric potential ψ_i, and hydraulic conductivity K_i are defined at the center of layer i at depth z_i and are uniform over the layer. An effective hydraulic conductivity $K_{i\pm1/2}$ is defined at the interface between soil layers at depth $z_{i\pm1/2}$. Shown are (a) the first soil layer ($i = 1$); (b) layers $1 < i < N$ depicted generally as three soil layers denoted $i - 1$, i, and $i + 1$; and (c) the bottom soil layer ($i = N$). The surface matric potential ψ_0 or the flux Q_0 provide the upper soil boundary condition, and the lower boundary condition at the bottom of the soil is ψ_{N+1} or gravitational drainage Q_N.

conductivity K_i are defined at the center of the layer at depth z_i and are uniform over the layer. Depth decreases in the downward direction from the surface so that $z_0 = 0$ denotes the surface and $z_{i+1} < z_i$ (i.e., depths are negative distances from the surface). The flux of water Q_i between adjacent layers i and $i+1$ depends on the hydraulic conductivities of the two layers. The hydraulic conductivity $K_{i+1/2}$ replaces the vertically varying hydraulic conductivities in the two adjacent layers with an effective conductivity for an equivalent homogenous medium so that the flux Q_i is

$$Q_i = -\frac{K_{i+1/2}}{\Delta z_{i+1/2}} (\psi_i - \psi_{i+1}) - K_{i+1/2}, \qquad (8.21)$$

with $\Delta z_{i+1/2} = z_i - z_{i+1}$ the distance between nodes i and $i+1$. The flux Q_{i-1} at the top of soil layer i is similarly

$$Q_{i-1} = -\frac{K_{i-1/2}}{\Delta z_{i-1/2}} (\psi_{i-1} - \psi_i) - K_{i-1/2}, \qquad (8.22)$$

and $\Delta z_{i-1/2} = z_{i-1} - z_i$. With fluxes expressed for time $n+1$, the mass balance for soil layer i is

$$\frac{\theta_i^{n+1} - \theta_i^n}{\Delta t} = -\frac{Q_{i-1}^{n+1} - Q_i^{n+1}}{\Delta z_i} \qquad (8.23)$$

so that

$$\frac{\Delta z_i}{\Delta t} \left(\theta_i^{n+1} - \theta_i^n\right) = \frac{K_{i-1/2}^{n+1}}{\Delta z_{i-1/2}} \left(\psi_{i-1}^{n+1} - \psi_i^{n+1}\right) - \frac{K_{i+1/2}^{n+1}}{\Delta z_{i+1/2}}$$
$$\left(\psi_i^{n+1} - \psi_{i+1}^{n+1}\right) + K_{i-1/2}^{n+1} - K_{i+1/2}^{n+1}. \qquad (8.24)$$

For constant soil layer thickness, this is equivalent to (8.19).

Equation (8.24) describes the water balance of soil layers $1 < i < N$. Special equations are needed for the top ($i = 1$) and bottom ($i = N$) layers to account for boundary conditions. Boundary conditions at the surface are specified in terms of ψ_0 (Dirichlet boundary condition) or as a flux of water Q_0 into the soil (Neumann boundary condition). With the surface value ψ_0 specified, (8.24) is still valid, in which case the corresponding flux of water into the soil is

$$Q_0^{n+1} = -\frac{K_{1/2}^{n+1}}{\Delta z_{1/2}} \left(\psi_0^{n+1} - \psi_1^{n+1}\right) - K_{1/2}^{n+1}. \qquad (8.25)$$

Alternatively, Q_0 can be directly specified as the boundary condition (e.g., as an infiltration rate; negative into the soil), in which case the water balance for layer $i = 1$ is

$$\frac{\Delta z_i}{\Delta t} \left(\theta_i^{n+1} - \theta_i^n\right) = -Q_0^{n+1} - \frac{K_{i+1/2}^{n+1}}{\Delta z_{i+1/2}} \left(\psi_i^{n+1} - \psi_{i+1}^{n+1}\right) - K_{i+1/2}^{n+1}. \qquad (8.26)$$

The lower boundary condition at the bottom of the soil column can similarly be specified in terms of ψ or as a flux of water. When given as the matric potential ψ_{N+1}, the corresponding flux of water draining out of the soil column is

$$Q_N^{n+1} = -\frac{K_{N+1/2}^{n+1}}{\Delta z_{N+1/2}} \left(\psi_N^{n+1} - \psi_{N+1}^{n+1}\right) - K_{N+1/2}^{n+1}. \qquad (8.27)$$

A common flux boundary condition specifies free drainage at the bottom of the soil column in which $Q_N = -K_N$. This is referred to as a unit hydraulic gradient because $\partial\psi/\partial z = 0$ in the Darcian flux. The water balance of layer $i = N$ is then

$$\frac{\Delta z_i}{\Delta t} \left(\theta_i^{n+1} - \theta_i^n\right) = \frac{K_{i-1/2}^{n+1}}{\Delta z_{i-1/2}} \left(\psi_{i-1}^{n+1} - \psi_i^{n+1}\right) + K_{i-1/2}^{n+1} - K_i^{n+1}. \qquad (8.28)$$

Alternatively, the soil column can be coupled with a groundwater model to allow for the influence of water table dynamics on soil moisture. The total change in water in the soil column equals the net flux into the soil so that conservation is given by

$$\sum_{i=1}^{N} \left(\theta_i^{n+1} - \theta_i^n\right) \Delta z_i = \left(Q_N^{n+1} - Q_0^{n+1}\right) \Delta t. \qquad (8.29)$$

Equation (8.20) defines the effective conductivity $K_{i\pm1/2}$ from the central difference approximation for $(\partial K/\partial z)(\partial\psi/\partial z)$. Hydraulic conductivity can differ substantially between layers because of vertical gradients in soil moisture (e.g., during infiltration into a dry soil). The equation used to represent the effective conductivity affects the accuracy of the numerical solution, and other expressions can be used for $K_{i\pm1/2}$ (Haverkamp and Vauclin 1979; Warrick 1991, 2003). The effective conductivity can be defined as the arithmetic mean of the adjacent nodal conductivities in which

$$K_{i\pm1/2} = 0.5(K_i + K_{i\pm1}). \qquad (8.30)$$

Another expression is the geometric mean whereby

$$K_{i\pm1/2} = (K_iK_{i\pm1})^{1/2}. \tag{8.31}$$

The harmonic mean is the reciprocal of the arithmetic mean of the reciprocals with

$$K_{i\pm1/2} = \frac{1}{0.5\left(K_i^{-1} + K_{i\pm1}^{-1}\right)} = \frac{2K_iK_{i\pm1}}{K_i + K_{i\pm1}}. \tag{8.32}$$

This latter equation is similar to (5.16) for the effective thermal conductivity and is obtained for constant Δz from continuity of flow across the interface so that the flux of water from depth z_i to $z_{i+1/2}$ equals the flux from $z_{i+1/2}$ to z_{i+1}. While the harmonic mean ensures continuity of fluxes at the interface, it is weighted towards the lower value of the two hydraulic conductivities and is the smallest of the three means. The arithmetic mean has the largest value, and the geometric mean has an intermediate value. The arithmetic mean is commonly used – e.g., as in the numerical solutions of Haverkamp et al. (1977) and Celia et al. (1990).

The numerical form of the ψ-based Richards equation is similar to (5.18) for soil temperature. For a soil with N layers, this is a tridiagonal system of N equations with N unknown values of ψ at time $n + 1$. This is more obvious by rewriting the finite difference approximation of the mixed-form equation given by (8.24) in the ψ-based form and rearranging terms to get

$$-\frac{K_{i-1/2}^{n+1}}{\Delta z_{i-1/2}}\psi_{i-1}^{n+1} + \left(\frac{C_i^{n+1}\Delta z_i}{\Delta t} + \frac{K_{i-1/2}^{n+1}}{\Delta z_{i-1/2}} + \frac{K_{i+1/2}^{n+1}}{\Delta z_{i+1/2}}\right)\psi_i^{n+1}$$
$$-\frac{K_{i+1/2}^{n+1}}{\Delta z_{i+1/2}}\psi_{i+1}^{n+1} = \frac{C_i^{n+1}\Delta z_i}{\Delta t}\psi_i^n + K_{i-1/2}^{n+1} - K_{i+1/2}^{n+1}. \tag{8.33}$$

A general form for this equation is

$$a_i\psi_{i-1}^{n+1} + b_i\psi_i^{n+1} + c_i\psi_{i+1}^{n+1} = d_i, \tag{8.34}$$

or, in matrix notation (Appendix A6),

$$\begin{bmatrix} b_1 & c_1 & 0 & 0 & 0 & 0 \\ a_2 & b_2 & c_2 & 0 & 0 & 0 \\ 0 & a_3 & b_3 & c_3 & 0 & 0 \\ 0 & 0 & \ddots & \ddots & \ddots & 0 \\ 0 & 0 & 0 & a_{N-1} & b_{N-1} & c_{N-1} \\ 0 & 0 & 0 & 0 & a_N & b_N \end{bmatrix} \times \begin{bmatrix} \psi_1^{n+1} \\ \psi_2^{n+1} \\ \psi_3^{n+1} \\ \vdots \\ \psi_{N-1}^{n+1} \\ \psi_N^{n+1} \end{bmatrix} = \begin{bmatrix} d_1 \\ d_2 \\ d_3 \\ \vdots \\ d_{N-1} \\ d_N \end{bmatrix}, \tag{8.35}$$

in which the matrix elements a, b, c, and d are evident from (8.33), with special values for $i = 1$ and $i = N$ to account for boundary conditions (Table 8.4).

However, whereas soil temperature uses a linear equation, (8.33) is a nonlinear equation for ψ because of the complex dependence of K and C on ψ. This nonlinearity is evident when these terms are replaced with their various expressions given in Table 8.2. An additional complexity is that ψ at $n + 1$ also appears on the right-hand side of (8.33) through K and C. Solving a nonlinear equation is challenging, and solving the system of nonlinear equations required to represent N soil layers is especially challenging. The solution requires linearizing (8.33) with respect to ψ through various numerical methods. One simple linearization uses values of K and C obtained at the preceding time step (at time n rather than $n + 1$). This is an implicit solution for ψ but with explicit linearization of K and C (Haverkamp et al. 1977).

A better numerical technique is the predictor–corrector method. This is a two-step solution that

Table 8.4	Tridiagonal terms for the ψ-based Richards equation			
Layer	a_i	b_i	c_i	d_i
$i = 1$	0	$\frac{C_i^{n+1}\Delta z_i}{\Delta t} + \frac{K_{1/2}^{n+1}}{\Delta z_{1/2}} - c_i$	$-\frac{K_{i+1/2}^{n+1}}{\Delta z_{i+1/2}}$	$\frac{C_i^{n+1}\Delta z_i}{\Delta t}\psi_i^n + \frac{K_{1/2}^{n+1}}{\Delta z_{1/2}}\psi_0^{n+1} + K_{1/2}^{n+1} - K_{i+1/2}^{n+1}$
$1 < i < N$	$-\frac{K_{i-1/2}^{n+1}}{\Delta z_{i-1/2}}$	$\frac{C_i^{n+1}\Delta z_i}{\Delta t} - a_i - c_i$	$-\frac{K_{i+1/2}^{n+1}}{\Delta z_{i+1/2}}$	$\frac{C_i^{n+1}\Delta z_i}{\Delta t}\psi_i^n + K_{i-1/2}^{n+1} - K_{i+1/2}^{n+1}$
$i = N$	$-\frac{K_{i-1/2}^{n+1}}{\Delta z_{i-1/2}}$	$\frac{C_i^{n+1}\Delta z_i}{\Delta t} - a_i$	0	$\frac{C_i^{n+1}\Delta z_i}{\Delta t}\psi_i^n + K_{i-1/2}^{n+1} - K_N^{n+1}$

Note: Boundary conditions are ψ_0^{n+1} for the first layer ($i = 1$) and free drainage for the bottom layer ($i = N$).

solves the Richards equation twice. The predictor step uses explicit linearization to solve for ψ over one-half a full time step ($\Delta t/2$) at time $n + 1/2$ with K and C from time n. The resulting values for ψ are used to evaluate K and C at $n + 1/2$, and the corrector step uses these to obtain ψ over the full time step (Δt) at $n + 1$. The method can be applied to the ψ-based Richard equations (Haverkamp et al. 1977) or the θ-based equation (Hornberger and Wiberg 2005). These implementations use an implicit solution for the predictor step and the Crank–Nicolson method (Appendix A4) for the corrector step. The predictor equation for ψ at $n + 1/2$ is

$$-\frac{K_{i-1/2}^n}{\Delta z_{i-1/2}}\psi_{i-1}^{n+1/2} + \left(\frac{C_i^n \Delta z_i}{\Delta t/2} + \frac{K_{i-1/2}^n}{\Delta z_{i-1/2}} + \frac{K_{i+1/2}^n}{\Delta z_{i+1/2}}\right)\psi_i^{n+1/2}$$

$$-\frac{K_{i+1/2}^n}{\Delta z_{i+1/2}}\psi_{i+1}^{n+1/2} = \frac{C_i^n \Delta z_i}{\Delta t/2}\psi_i^n + K_{i-1/2}^n - K_{i+1/2}^n. \quad (8.36)$$

The corrector equation solves for ψ over a full time step using the Crank–Nicolson method with fluxes evaluated at time n and $n + 1$ whereby

$$-\frac{K_{i-1/2}^{n+1/2}}{2\Delta z_{i-1/2}}\psi_{i-1}^{n+1} + \left(\frac{C_i^{n+1/2}\Delta z_i}{\Delta t} + \frac{K_{i-1/2}^{n+1/2}}{2\Delta z_{i-1/2}} + \frac{K_{i+1/2}^{n+1/2}}{2\Delta z_{i+1/2}}\right)\psi_i^{n+1}$$

$$-\frac{K_{i+1/2}^{n+1/2}}{2\Delta z_{i+1/2}}\psi_{i+1}^{n+1} = \frac{C_i^{n+1/2}\Delta z_i}{\Delta t}\psi_i^n + \frac{K_{i-1/2}^{n+1/2}}{2\Delta z_{i-1/2}}\left(\psi_{i-1}^n - \psi_i^n\right)$$

$$-\frac{K_{i+1/2}^{n+1/2}}{2\Delta z_{i+1/2}}\left(\psi_i^n - \psi_{i+1}^n\right) + K_{i-1/2}^{n+1/2} - K_{i+1/2}^{n+1/2}. \quad (8.37)$$

Equations (8.36) and (8.37) are both a tridiagonal system of linear equations and are easily solved for ψ (Appendix A8).

Question 8.5 Soil temperature is commonly modeled with zero heat flux as the boundary condition at the bottom of the soil column. Explain how this is similar to the free drainage boundary condition for soil moisture.

Question 8.6 Compare the ψ-based Richards equation given by (8.33) with that for soil temperature given by (5.19). What are the similarities? What is a key difference? Why are iterative methods required to solve the Richards equation but not soil temperature?

Question 8.7 Table 8.4 gives the tridiagonal coefficients a_i, b_i, c_i, and d_i for the ψ-based Richards equation using the implicit method. Derive the same coefficients for the Crank–Nicolson method as used in the predictor–correct solution. What is the equation for the infiltration rate Q_0?

8.6 | Iterative Numerical Solutions

Other numerical methods use iterative calculations. These methods approach the correct solution by using successive approximations in which values for K and C from one iteration are used at the next iteration. Picard iteration, which is an example of fixed-point iteration (Appendix A5), is one such numerical algorithm. As applied to the ψ-based Richards equation, Picard iteration provides successive estimates for ψ using values of K and C evaluated with the previous value of ψ (Celia et al. 1990). The iteration repeats until ψ does not change value between iterations. With n denoting time and m denoting iteration, (8.33) is written as

$$-\frac{K_{i-1/2}^{n+1,m}}{\Delta z_{i-1/2}}\psi_{i-1}^{n+1,m+1} + \left(\frac{C_i^{n+1,m}\Delta z_i}{\Delta t} + \frac{K_{i-1/2}^{n+1,m}}{\Delta z_{i-1/2}} + \frac{K_{i+1/2}^{n+1,m}}{\Delta z_{i+1/2}}\right)\psi_i^{n+1,m+1}$$

$$-\frac{K_{i+1/2}^{n+1,m}}{\Delta z_{i+1/2}}\psi_{i+1}^{n+1,m+1} = \frac{C_i^{n+1,m}\Delta z_i}{\Delta t}\psi_i^n + K_{i-1/2}^{n+1,m} - K_{i+1/2}^{n+1,m}. \quad (8.38)$$

The values of K and C are obtained from iteration m so that (8.38) is a tridiagonal system of linear equations that is solved for ψ at time $n + 1$ and iteration $m + 1$. It is more convenient to rewrite this equation to solve for the change in ψ between iterations ($\delta^{m+1} = \psi^{n+1,m+1} - \psi^{n+1,m}$) rather than directly for ψ itself. With this modification, (8.38) becomes

$$-\frac{K_{i-1/2}^{n+1,m}}{\Delta z_{i-1/2}}\delta_{i-1}^{m+1} + \left(\frac{C_i^{n+1,m}\Delta z_i}{\Delta t} + \frac{K_{i-1/2}^{n+1,m}}{\Delta z_{i-1/2}} + \frac{K_{i+1/2}^{n+1,m}}{\Delta z_{i+1/2}}\right)\delta_i^{m+1}$$

$$-\frac{K_{i+1/2}^{n+1,m}}{\Delta z_{i+1/2}}\delta_{i+1}^{m+1} = f_i^{n+1,m}, \quad (8.39)$$

with

$$f_i^{n+1,m} = \frac{K_{i-1/2}^{n+1,m}}{\Delta z_{i-1/2}} \left(\psi_{i-1}^{n+1,m} - \psi_i^{n+1,m} \right)$$
$$- \frac{K_{i+1/2}^{n+1,m}}{\Delta z_{i+1/2}} \left(\psi_i^{n+1,m} - \psi_{i+1}^{n+1,m} \right)$$
$$+ K_{i-1/2}^{n+1,m} - K_{i+1/2}^{n+1,m}$$
$$- \frac{C_i^{n+1,m} \Delta z_i}{\Delta t} \left(\psi_i^{n+1,m} - \psi_i^n \right). \tag{8.40}$$

Terms are known at iteration m, and (8.39) is solved for δ at iteration $m+1$. The right-hand side of (8.39) is the ψ-based Richards equation evaluated at the mth iteration. As the solution converges, δ becomes small so that the left-hand side approaches zero, and (8.39) is the standard finite difference approximation with K and C expressed for $n+1$. Picard iteration is a simple procedure that evaluates K and C for the current estimate of ψ, uses these to solve for a new value of ψ, and repeats this calculation until convergence is achieved. However, convergence can require many iterations and is not always guaranteed.

A more complex numerical method uses Newton–Raphson iteration to linearize the system of N equations (Appendix A9). This method reformulates the solution in terms of finding the roots of the system of equations. Newton–Raphson iteration defines δ^{m+1} as given previously but uses a Taylor series approximation to linearize the Richards equation and solves for δ^{m+1} that satisfies the equation

$$\frac{\partial f_i}{\partial \psi_{i-1}} \delta_{i-1}^{m+1} + \frac{\partial f_i}{\partial \psi_i} \delta_i^{m+1} + \frac{\partial f_i}{\partial \psi_{i+1}} \delta_{i+1}^{m+1} = -f_i^{n+1,m}. \tag{8.41}$$

The right-hand side is the Richards equation evaluated at iteration m as in (8.40), and the left-hand side uses the partial derivatives $\partial f/\partial \psi$ evaluated at iteration m. The iteration proceeds until δ is less than some convergence criterion. Equation (8.41) is similar to Picard iteration; but whereas that method uses the standard terms in the Richards equation, Newton–Raphson iteration requires evaluating the partial derivatives with respect to ψ. In linear algebra, the partial derivatives are referred to as the Jacobian matrix. The two methods differ in the computational efficiency and robustness of the numerical solution (Paniconi et al. 1991; Paniconi and Putti 1994; Lehmann and Ackerer 1998). Picard iteration may fail to converge or may need many iterations to

converge. Newton–Raphson iteration requires evaluating a matrix of partial derivatives (the Jacobian) and is computationally more expensive per iteration but can converge in fewer iterations and provide a more robust solution (though it, too, can fail to converge).

The difficulty in using the ψ-based Richards equation is that it does not conserve mass because the specific moisture capacity $C(\psi)$ that appears in the water storage term is not constant over a time step (Milly 1985; Celia et al. 1990). Celia et al. (1990) devised a mass-conserving numerical solution for the mixed-form Richards equation that is a modified Picard iteration. The mixed-form equation is

$$\frac{\Delta z_i}{\Delta t} \left(\theta_i^{n+1,m+1} - \theta_i^n \right) = \frac{K_{i-1/2}^{n+1,m}}{\Delta z_{i-1/2}} \left(\psi_{i-1}^{n+1,m+1} - \psi_i^{n+1,m+1} \right)$$
$$- \frac{K_{i+1/2}^{n+1,m}}{\Delta z_{i+1/2}} \left(\psi_i^{n+1,m+1} - \psi_{i+1}^{n+1,m+1} \right)$$
$$+ K_{i-1/2}^{n+1,m} - K_{i+1/2}^{n+1,m}, \tag{8.42}$$

with n referring to time and m to iteration as before. Mass conservation is achieved by using a Taylor series approximation (Appendix A1) for $\theta_i^{n+1,m+1}$ in which

$$\theta_i^{n+1,m+1} = \theta_i^{n+1,m} + C_i^{n+1,m} \left(\psi_i^{n+1,m+1} - \psi_i^{n+1,m} \right), \tag{8.43}$$

with

$$C_i^{n+1,m} = \left. \frac{d\theta_i}{d\psi_i} \right|^{n+1,m}. \tag{8.44}$$

Substituting this expression into (8.42) and converting to residual form gives

$$- \frac{K_{i-1/2}^{n+1,m}}{\Delta z_{i-1/2}} \delta_{i-1}^{m+1} + \left(\frac{C_i^{n+1,m} \Delta z_i}{\Delta t} + \frac{K_{i-1/2}^{n+1,m}}{\Delta z_{i-1/2}} + \frac{K_{i+1/2}^{n+1,m}}{\Delta z_{i+1/2}} \right) \delta_i^{m+1}$$
$$- \frac{K_{i+1/2}^{n+1,m}}{\Delta z_{i+1/2}} \delta_{i+1}^{m+1} = f_i^{n+1,m}, \tag{8.45}$$

as with the ψ-based equation, but now with

$$f_i^{n+1,m} = \frac{K_{i-1/2}^{n+1,m}}{\Delta z_{i-1/2}} \left(\psi_{i-1}^{n+1,m} - \psi_i^{n+1,m} \right)$$
$$- \frac{K_{i+1/2}^{n+1,m}}{\Delta z_{i+1/2}} \left(\psi_i^{n+1,m} - \psi_{i+1}^{n+1,m} \right) + K_{i-1/2}^{n+1,m}$$
$$- K_{i+1/2}^{n+1,m} - \frac{\Delta z_i}{\Delta t} \left(\theta_i^{n+1,m} - \theta_i^n \right). \tag{8.46}$$

Table 8.5 | Tridiagonal terms for the modified Picard iteration of the mixed-form Richards equation

Layer	a_i	b_i	c_i	d_i
$i = 1$	0	$\dfrac{C_i^{n+1,m}\Delta z_i}{\Delta t} - c_i$	$-\dfrac{K_{i+1/2}^{n+1,m}}{\Delta z_{i+1/2}}$	$\dfrac{K_{1/2}^{n+1,m}}{\Delta z_{1/2}}\left(\psi_0^{n+1} - \psi_i^{n+1,m}\right)$ $-\dfrac{K_{i+1/2}^{n+1,m}}{\Delta z_{i+1/2}}\left(\psi_i^{n+1,m} - \psi_{i+1}^{n+1,m}\right)$ $+K_{1/2}^{n+1,m} - K_{i+1/2}^{n+1,m} - \dfrac{\Delta z_i}{\Delta t}\left(\theta_i^{n+1,m} - \theta_i^n\right)$
$1 < i < N$	$-\dfrac{K_{i-1/2}^{n+1,m}}{\Delta z_{i-1/2}}$	$\dfrac{C_i^{n+1,m}\Delta z_i}{\Delta t} - a_i - c_i$	$-\dfrac{K_{i+1/2}^{n+1,m}}{\Delta z_{i+1/2}}$	$\dfrac{K_{i-1/2}^{n+1,m}}{\Delta z_{i-1/2}}\left(\psi_{i-1}^{n+1,m} - \psi_i^{n+1,m}\right)$ $-\dfrac{K_{i+1/2}^{n+1,m}}{\Delta z_{i+1/2}}\left(\psi_i^{n+1,m} - \psi_{i+1}^{n+1,m}\right)$ $+K_{i-1/2}^{n+1,m} - K_{i+1/2}^{n+1,m} - \dfrac{\Delta z_i}{\Delta t}\left(\theta_i^{n+1,m} - \theta_i^n\right)$
$i = N$	$-\dfrac{K_{i-1/2}^{n+1,m}}{\Delta z_{i-1/2}}$	$\dfrac{C_i^{n+1,m}\Delta z_i}{\Delta t} - a_i$	0	$\dfrac{K_{i-1/2}^{n+1,m}}{\Delta z_{i-1/2}}\left(\psi_{i-1}^{n+1,m} - \psi_i^{n+1,m}\right) + K_{i-1/2}^{n+1,m} - K_N^{n+1,m}$ $-\dfrac{\Delta z_i}{\Delta t}\left(\theta_i^{n+1,m} - \theta_i^n\right)$

Note: Boundary conditions are ψ_0^{n+1} for the first layer ($i = 1$) and free drainage for the bottom layer ($i = N$).

Equation (8.45) is a tridiagonal system of linear equations that is solved for δ. Table 8.5 gives the various terms in the tridiagonal equations. The iteration is repeated until some convergence criterion is satisfied. This can be an absolute threshold in which $|\,\delta_i\,| \le \varepsilon_a$ for each layer or can also include both an absolute and a relative error term in which case $|\,\delta_i\,| \le \varepsilon_a + \varepsilon_r\,|\,\psi_i^{n+1,m}\,|$. As the iteration converges, δ_i approaches zero and (8.45) reduces to the mixed-form Richards equation. The key difference compared with the ψ-based Picard iteration is the use of a Taylor series approximation for θ, and this ensures mass conservation. At convergence, $\psi_i^{n+1,m+1} - \psi_i^{n+1,m}$ approaches zero, thereby eliminating inaccuracy in evaluating C_i.

Accurate solution of the Richards equation requires a small time step Δt and spatial increment Δz. This is particularly true during infiltration into dry soil, where there is a sharp wetting front. Some models utilize adaptive time stepping in which Δt is dynamically adjusted and varies between some minimum and maximum value based on specified criteria. One simple method is to adjust Δt at every time, based on the number of iterations required for convergence at the previous time (Paniconi et al. 1991; Paniconi and Putti 1994). The time step is likely to be too short if few iterations are needed to achieve convergence, but is likely to be too long if convergence requires many iterations. The time step is increased by a specified factor if convergence is achieved in fewer than some number of iterations, is decreased if some number of iterations is exceeded, or is otherwise left unchanged. If the solution fails to converge after a maximum number of iterations, Δt is decreased by some fraction and the iteration is restarted.

Global models must simulate tens of thousands of soil columns over hundreds of years, and small vertical or temporal step sizes pose a large computational burden. In these models, a linear form of the θ-based Richards equation is commonly used because it conserves mass for all step sizes. The linearization is attained using a Taylor series approximation and can be solved directly for θ at

$n + 1$ without iteration or also with Newton–Raphson iteration. The cost with larger step sizes and fewer iterations is less accuracy in the solution. With reference to Figure 8.7b, the water balance for soil layer i at time $n + 1$ and iteration $m + 1$ is

$$\frac{\Delta z_i}{\Delta t}\left(\theta_i^{n+1,m+1} - \theta_i^n\right) = -Q_{i-1}^{n+1,m+1} + Q_i^{n+1,m+1}. \quad (8.47)$$

The flux Q_{i-1} is linearized as

$$Q_{i-1}^{n+1,m+1} = Q_{i-1}^{n+1,m} + \frac{\partial Q_{i-1}}{\partial \theta_{i-1}}\delta_{i-1}^{m+1} + \frac{\partial Q_{i-1}}{\partial \theta_i}\delta_i^{m+1}, \quad (8.48)$$

and Q_i is

$$Q_i^{n+1,m+1} = Q_i^{n+1,m} + \frac{\partial Q_i}{\partial \theta_i}\delta_i^{m+1} + \frac{\partial Q_i}{\partial \theta_{i+1}}\delta_{i+1}^{m+1}, \quad (8.49)$$

with $\delta^{m+1} = \theta^{n+1,m+1} - \theta^{n+1,m}$. Substituting these expressions into (8.47), the water balance is

$$-\frac{\partial Q_{i-1}}{\partial \theta_{i-1}}\delta_{i-1}^{m+1} - \left(\frac{\Delta z_i}{\Delta t} + \frac{\partial Q_{i-1}}{\partial \theta_i} - \frac{\partial Q_i}{\partial \theta_i}\right)\delta_i^{m+1} + \frac{\partial Q_i}{\partial \theta_{i+1}}\delta_{i+1}^{m+1}$$

$$= Q_{i-1}^{n+1,m} - Q_i^{n+1,m} + \frac{\Delta z_i}{\Delta t}\left(\theta_i^{n+1,m} - \theta_i^n\right). \quad (8.50)$$

The solution becomes more accurate with multiple iterations as δ approaches zero. The complexity lies in the partial derivatives, which include expressions for $\partial \psi/\partial \theta$ and $\partial K/\partial \theta$.

Question 8.8 Contrast the predictor–corrector, modified Picard, and Newton–Raphson methods to solve the Richards equations. What are the main differences among these methods? What are the similarities?

Question 8.9 Use Newton–Raphson iteration to solve the system of equations:
$x_1^2 + x_2^2 = 4$ and $x_1 x_2 = 1$.

8.7 | Infiltration

The Richards equation can be used to model infiltration into soil. The boundary condition at the soil surface depends on the rate at which water is applied to the soil. When the supply rate is less than the saturated hydraulic conductivity, no water accumulates on the surface and a flux (Neumann) boundary condition is used. If sufficient water is provided so that the soil surface is saturated but water does not pond, a concentration (Dirichlet) boundary condition is used with $\theta_0 = \theta_{sat}$. If water ponds on the soil surface, the boundary condition is $\theta_0 = \theta_{sat}$ and with a small positive depth of water on the surface. Figure 8.8 shows soil moisture profiles during infiltration into sand and Yolo light clay using the predictor–corrector method, as in

(a) Sand

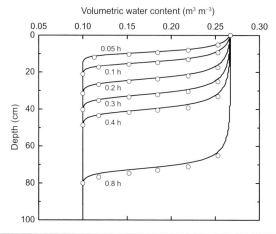

(b) Yolo light clay

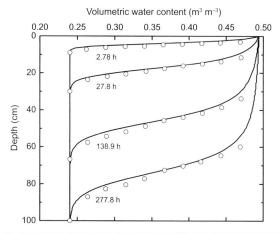

Figure 8.8 Soil moisture profiles for (a) sand and (b) Yolo light clay during infiltration with a specified θ_0 boundary condition. Simulations are as in Haverkamp et al. (1977) and use the predictor–correction method. Results are shown at various times up to 0.8 h (48 minutes) for sand and 277.8 h (11.6 days) for clay. The open circles show the analytical solution from Haverkamp et al. (1977).

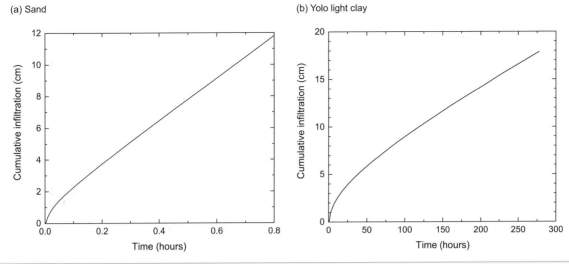

(a) Sand

(b) Yolo light clay

Figure 8.9 As in Figure 8.8, but for cumulative infiltration.

Haverkamp et al. (1977). Initial conditions for sand are $\theta = 0.1$ at $t = 0$, and the boundary condition is $\theta_0 = 0.267$ for $t > 0$. The simulated soil moisture closely matches the analytical solution. For clay, the initial and boundary conditions are $\theta = 0.24$ and $\theta_0 = 0.495$. Infiltration into the clay proceeds much slower than the sand. The sand absorbs almost 12 cm of water in 48 minutes while 18 cm of water infiltrates into the clay over 11.6 days (Figure 8.9).

8.8 Source and Sink Terms

The Richards equation, as discussed in this chapter, considers only the vertical Darcian fluxes of water. An additional source or sink term can be added (source) or subtracted (sink) to account for other water fluxes. Primary among these is evapotranspiration loss, which is accounted for by subtracting a root extraction, or sink, term in the Richards equation. Other plant-mediated water fluxes such as hydraulic redistribution can also be considered. In this case, the water balance is

$$\frac{\Delta z_i}{\Delta t}\left(\theta_i^{n+1} - \theta_i^n\right) = -Q_{i-1}^{n+1} + Q_i^{n+1} \pm S_{w,i}, \quad (8.51)$$

where $S_{w,i}$ (m s^{-1}) is the flux of water added or subtracted in soil layer i.

Evapotranspiration must be partitioned to root uptake in each soil layer. A common method uses the root profile weighted by a soil wetness factor β_w. For soil layer i, a simple wetness factor is

$$\beta_{w,i} = \begin{cases} \dfrac{\psi_i - \psi_{dry}}{\psi_{opt} - \psi_{dry}} & \psi_i > \psi_{dry} \\ 0 & \psi_i \le \psi_{dry} \end{cases}, \quad (8.52)$$

in which ψ_{dry} is the matric potential at which transpiration ceases and ψ_{opt} is the matric potential at which $\beta_w = 1$. Some models use volumetric moisture rather than matric potential; the difference relates to the nonlinearity of the $\theta(\psi)$ relationship. Total evapotranspiration E is partitioned to an individual soil layer in relation to the relative root fraction ΔF_i obtained from (2.23), and

$$S_{w,i} = E\left(\Delta F_i \beta_{w,i} \bigg/ \sum_{i=1}^{N} \Delta F_i \beta_{w,i}\right). \quad (8.53)$$

More complex models of plant hydraulics calculate root uptake from physiological principles (Chapter 13).

Hydraulic redistribution is a process by which roots move water upward and downward in the soil. At night, roots can transport water from wet, deep soil layers to dry, upper soil layers, thereby increasing the supply of water available to near-surface roots. Downward root-mediated transport can also occur from wet, upper layers to dry, lower layers after rainfall. Modeling studies show that hydraulic redistribution helps sustain photosynthesis and transpiration during the dry season (Lee et al. 2005; Zheng and Wang 2007; Baker et al. 2008; Wang

2011; Li et al. 2012b; Yan and Dickinson 2014). Many models include the water flux from hydraulic redistribution as a source or sink term in the Richards equation based on Ryel et al. (2002), though Amenu and Kumar (2008) provided an alternative parameterization. The flux is calculated from the difference in matric potential between two layers using Darcy's law. At night, the flux of water by hydraulic redistribution from layer j to layer i is

$$S_{w,j \to i} = -C_{RT}\left(\psi_j - \psi_i\right) \max\left(c_i, c_j\right) \frac{\Delta F_i \Delta F_j}{1 - \Delta F_j}, \quad (8.54)$$

where C_{RT} is a maximum radial soil–root conductance (m MPa^{-1} s^{-1}) and $\Delta \psi$ is the difference in water potential (MPa) between the uptake and release soil layers. The term c_i or c_j is a factor that reduces conductance for soil moisture in the source or sink layer and is given by

$$c_i = \frac{1}{1 + (\psi_i/\psi_{50})^b}, \quad (8.55)$$

in which ψ_{50} is the matric potential at which conductance is reduced by 50%. Representative parameters are $C_{RT} = 0.097$ cm MPa^{-1} h^{-1}, $\psi_{50} = -1$ MPa, and $b = 3.22$ (Ryel et al. 2002), as used also in some land surface models (Zheng and Wang 2007; Wang 2011; Li et al. 2012b; Yan and Dickinson 2014). The rightmost term in (8.54) accounts for root abundance. The denominator is $1 - \Delta F_j$ when $\theta_j > \theta_i$, but $1 - \Delta F_i$ when $\theta_i > \theta_j$. Models commonly do not allow hydraulic redistribution to the top soil layer. Otherwise, the soil surface is continually wetted, and excessive soil evaporation can occur.

Question 8.10 Write the mixed-form Richards equation (8.13) with a sink term. What are the units of S?

8.9 | Soil Heterogeneity

The Richards equation is commonly used in land surface models. It applies to homogenous soil columns in which soil hydraulic properties are horizontally uniform but can vary in the vertical dimension. Spatially homogenous soil columns are applicable for laboratory studies or at small scales, but field soils are, in fact, quite heterogeneous. Several texture classes can co-occur within a small footprint, hydraulic conductivity and specific moisture capacity can differ within a texture class, and the presence of macropores can alter water movement in soils. Soil heterogeneity can be addressed through stochastic methods applied to the Richards equation, such as treating hydraulic conductivity as a random variable (Gelhar 1986; Milly 1988). These methods parameterize spatial heterogeneity statistically rather than explicitly representing the variability. A goal of such parameterizations is to obtain not only the mean water flow and moisture profile, but also the variance. A secondary goal is to obtain effective hydraulic parameters at large scales for which the Richards equation can be used. One approach is to represent soil as independent columns that vary in hydraulic parameters such as K_{sat}, for which the Richards equation is solved. This characterizes soil heterogeneity by a series of independent, one-dimensional flow problems. The mean soil moisture and its variance are obtained from an ensemble of Monte Carlo simulations in which, for example, K_{sat} is drawn from a specified probability density function or by numerically integrating the solution over the distribution (Bresler and Dagan 1983; Clapp et al. 1983; Dagan and Bresler 1983). Another method is to solve the Richards equation in a probabilistic treatment to obtain a probability distribution for soil moisture with a mean and variance. This approach formulates the Richards equation as a stochastic partial differential equation with hydraulic conductivity taken as a random variable (Yeh et al. 1985a,b,c; Mantoglou and Gelhar 1987a,b,c; Chen et al. 1994). A recent example is Vrettas and Fung (2015), who applied the concept to a local watershed but also suggested its applicability to large-scale land surface models.

8.10 | Supplemental Programs

8.1 Predictor–Corrector Solution for the ψ-Based Richards Equation: This program implements the predictor–correction method given by (8.36) and (8.37). Boundary conditions are θ_0 and free drainage. The code uses either the Campbell (1974) or van Genuchten (1980) relationships for $\theta(\psi)$ and $K(\theta)$. Specific configurations match Haverkamp et al. (1977) for sand and Yolo light clay as in Figure 8.8 and Figure 8.9. $\theta(\psi)$ uses the van Genuchten relationships with $\theta_{res} = 0.075$, $\theta_{sat} = 0.287$, $\alpha = 0.027$ cm^{-1}, $n = 3.96$, and $m = 1$ (sand); and $\theta_{res} = 0.124$, $\theta_{sat} = 0.495$, $\alpha = 0.026$ cm^{-1}, $n = 1.43$, and $m = 0.3$ (clay). In Haverkamp et al. (1977), $K = K_{sat}A/\left(A + |\psi|^B\right)$ with $K_{sat} = 34$ cm h^{-1}, $A = 1.175 \times 10^6$, and $B = 4.74$ (sand); and $K_{sat} = 0.0443$ cm h^{-1}, $A = 124.6$, and $B = 1.77$ (clay).

8.2 Modified Picard Iteration for the Mixed-Form Richards Equation: This program is similar to the previous program but implements the modified Picard iteration (8.45) with Table 8.5. Critical parameters for the solution are the tolerance ε_a, which is the maximum allowable change in ψ between iterations for convergence. This parameter determines the accuracy of the water balance.

8.11 | Modeling Projects

1. Use the ψ-based predictor–corrector method (Supplemental Program 8.1) to calculate the amount of water that infiltrates into a sandy loam. Compare results using the van Genuchten (1980) and Campbell (1974) relationships for $\theta(\psi)$ and $K(\theta)$ with parameters from Table 8.3.

2. Repeat the previous problem, but using the modified Picard iteration (Supplemental Program 8.2). How do the results compare with the predictor–corrector method? What can be said about parameter uncertainty versus numerical methods?

Hydrologic Scaling and Spatial Heterogeneity

Chapter Overview

The hydrologic cycle on land includes infiltration, runoff, and evapotranspiration. This chapter reviews the principles of infiltration and runoff and discusses parameterizations to scale runoff over large regions with heterogeneous soils. These parameterizations rely on statistical distributions to characterize soil moisture variability. A similar statistical approach accounts for nonlinearity in soil moisture effects on evapotranspiration using either continuous or discrete statistical distributions. Snow cover is particularly patchy, and models divide land into snow-covered and snow-free areas when calculating surface fluxes. Many models additionally account for heterogeneity in vegetation and soils by dividing a model grid cell into several tiles.

9.1 | Introduction

The amount of water held in soil is the balance of several processes (Figure 8.1). Precipitation provides the input of water to replenish soils, but not all precipitation actually reaches the ground. Leaves, twigs, and branches of plants intercept some snow and rain, and this water quickly evaporates. The water that is not intercepted falls to the ground as throughfall or stemflow. Throughfall reaches the ground directly through openings in the plant canopy or by dripping down from leaves, twigs,

and branches. Stemflow reaches the ground by flowing down plant stems and tree trunks. Interception, thoughfall, and stemflow are described further in Chapter 15, on plant canopies. The water at the ground infiltrates into the soil, accumulates in small puddles on the surface, or runs off over the surface as overland flow. Overland flow is runoff generated when rainfall exceeds the infiltration capacity of soil, resulting first in ponding of water on the soil surface and then flow across the surface. Runoff and evapotranspiration are linked through soil moisture, and the ability to model evapotranspiration, therefore, also determines the generation of runoff (Koster and Milly 1997). Soil moisture varies considerably across the land surface, and accurate representation of the hydrologic cycle in models must account for soil heterogeneity. Spatial variability in soil texture and water-holding capacity is particularly important for infiltration and runoff. If the precipitation falls as snow, the water first accumulates in a snowpack before melting and becoming available to replenish the soil. Similar to soil moisture, snow is patchily distributed across the land surface.

9.2 | Infiltration

Infiltration is the vertical movement of water into soil, and the maximum amount of water that can infiltrate into soil is known as infiltration capacity. Infiltration capacity depends on soil moisture at the start of infiltration, the hydraulic properties of the

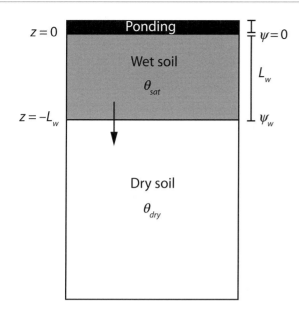

$z = 0$

Ponding

$\psi = 0$

Wet soil

θ_{sat}

L_w

$z = -L_w$

ψ_w

Dry soil

θ_{dry}

Figure 9.1 The Green–Ampt equation represents infiltration as a uniform wetting front moving downward in a dry soil. Ponding of water on the surface is considered to be negligible in deriving the equation from Darcy's law.

soil, and the length of the infiltration event. Infiltration can be modeled by solving the Richards equation (Figures 8.8 and 8.9), but analytical solutions are also available. One method to calculate infiltration capacity is the Green–Ampt equation, which is derived from Darcy's law in combination with a continuity equation for water (Green and Ampt 1911). The Green–Ampt equation uses the concept of a horizontal wetting front at some distance L_w into the soil (Figure 9.1). Soil above the wetting front is saturated with a water content θ_{sat}, whereas soil below the wetting front has a uniform water content θ_{dry}. Darcy's law provides the flux of water i_c infiltrating into soil from the surface with $z_1 = 0$ to the wetting front at the depth $z_2 = -L_w$ in which

$$i_c = K_{sat} \left[\frac{(\psi_2 + z_2) - (\psi_1 + z_1)}{z_2 - z_1} \right]$$

$$= K_{sat} \left(\frac{-\psi_w}{L_w} + 1 \right) = K_{sat} \left(\frac{|\psi_w|}{L_w} + 1 \right). \qquad (9.1)$$

The surface has matric potential $\psi_1 = 0$, and the wetting front has matric potential $\psi_2 = \psi_w$, and here Darcy's law is written so that i_c is positive downward into the soil. The total cumulative infiltrated water I_c at time t is the time integral of i_c, and

this is also equal to the depth of the wetting front multiplied by the change in water $\Delta\theta = \theta_{sat} - \theta_{dry}$ so that

$$I_c = \Delta\theta L_w. \qquad (9.2)$$

Another expression for infiltration rate is the time derivative of I_c, whereby

$$i_c = \frac{dI_c}{dt} = \Delta\theta \frac{dL_w}{dt}.$$

Equating (9.1) and (9.2) gives the time rate of change for L_w,

$$\Delta\theta \frac{dL_w}{dt} = K_{sat} \left(\frac{|\psi_w|}{L_w} + 1 \right), \qquad (9.3)$$

and substituting $I_c/\Delta\theta$ from (9.2) for L_w gives

$$\frac{dI_c}{dt} = K_{sat} \left(\frac{|\psi_w|\Delta\theta}{I_c} + 1 \right). \qquad (9.4)$$

An expression for $I_c(t)$ is obtained from (9.4) by rearranging terms and integrating

$$\int_0^{I_c(t)} \left[1 - \frac{|\psi_w|\Delta\theta}{|\psi_w|\Delta\theta + I_c} \right] dI_c = \int_0^t K_{sat} dt \qquad (9.5)$$

to obtain

$$I_c(t) = K_{sat}t + |\psi_w|\Delta\theta \ln\left[1 + \frac{I_c(t)}{|\psi_w|\Delta\theta} \right]. \qquad (9.6)$$

This equation defines cumulative infiltration to time t given K_{sat}, ψ_w, θ_{sat}, and θ_{dry}. Because I_c appears on both the left- and right-hand sides of the equation, (9.6) must be solved using numerical root finding methods (Appendix A5). Use of the Green–Ampt equation requires the matric potential at the wetting front. This is a complex function of hydraulic conductivity (Assouline 2013). A simplification from Rawls et al. (1993) uses the Campbell (1974) relationship in Table 8.2 with

$$\psi_w = \left(\frac{2b + 3}{2b + 6} \right) \psi_{sat}, \qquad (9.7)$$

and a further simplification is $\psi_w = 0.76\psi_{sat}$ (Dingman 1994, p. 243). Figure 9.2 shows infiltration capacity over one hour for three soils. Infiltration capacity is high during initial stages and decreases to a near constant rate over time. The sandy loam has the greatest infiltration capacity, the loam has an intermediate value, and the clay loam has the lowest infiltration capacity.

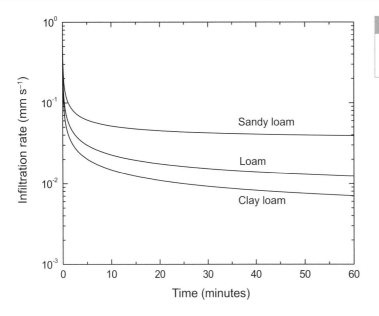

Figure 9.2 Green–Ampt infiltration capacity for sandy loam, loam, and clay loam with water content $\theta_{dry} = 0.3$ and hydraulic properties from Clapp and Hornberger (1978) as in Table 8.3.

Philip (1957b) derived another expression for infiltration from an analytical solution to the Richards equation in which

$$i_c(t) = 0.5St^{-1/2} + A, \qquad (9.8)$$

where S and A are soil properties that can be estimated by fitting the equation to infiltration measurements. The parameter A relates to saturated hydraulic conductivity, and the sorptivity S depends on soil texture and initial water. Simplified expressions can be used to relate these parameters to soil properties in which $A = 2K_{sat}/3$ and $S = \left[2\left(\theta_{sat} - \theta_{dry}\right)K_{sat}|\psi_w|\right]^{1/2}$ for K_{sat} with units cm h^{-1} and ψ_w with units cm (Rawls et al. 1993). Eagleson (1978) provided more complicated expressions for S and A, as described also by Bras (1990, pp. 360–361). The Green–Ampt and Philip equations represent infiltration into idealized soil columns. Vegetation cover, the presence of a litter layer, slope, soil compaction, and macropores all affect infiltration and ponding.

Question 9.1 Equation (9.6) for Green–Ampt infiltration gives the cumulative infiltration I_c at time t. Explain how this is used to calculate the infiltration rate i_c.

Question 9.2 Rewrite the Green–Ampt equation to obtain an expression for the time t required for an amount of water I_c to infiltrate. Then use this expression to calculate the time required for 25 mm of water to infiltrate in a soil with $K_{sat} = 0.007$ mm s^{-1}, $\psi_w = -478$ mm, $\theta_{sat} = 0.451$, and $\theta_{dry} = 0.3$. What is the instantaneous infiltration rate at this time? What is the depth of the wetting front?

Question 9.3 A typical model time step is 30 minutes. Use the Green–Ampt equation to calculate infiltration capacity for the soils in Figure 9.2 over this time interval.

Question 9.4 Use (9.8) to calculate the cumulative infiltration after 30 minutes for a soil with $S = 2.9$ cm h$^{-1/2}$ and $A = 0.4$ cm h^{-1}.

9.3 | Runoff

Surface runoff occurs when the amount of water reaching the ground exceeds the capacity of soil to gain water during infiltration. This is known as infiltration-excess runoff. The local runoff rate R at a particular location is the rainfall P_R in excess of infiltration capacity i_c so that

$$R = \begin{cases} P_R - i_c & P_R > i_c \\ 0 & P_R \leq i_c \end{cases}. \qquad (9.9)$$

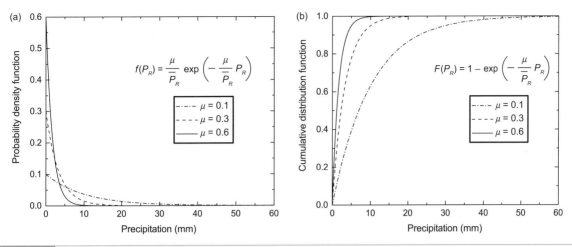

Figure 9.3 Exponential precipitation distribution showing (a) the probability density function $f(P_R)$ and (b) the cumulative distribution function $F(P_R)$ for μ = 0.1, 0.3, and 0.6 with $\bar{P}_R$ = 1 mm.

Infiltration-excess runoff occurs on soils with low infiltration capacity. However, many soils have high infiltration capacity, and precipitation on these soils is rarely sufficient to generate infiltration-excess runoff. Runoff also occurs where precipitation falls on saturated areas. This is known as saturation-excess runoff. It is common in low-lying areas with shallow water tables, such as wetlands or along stream channels. All precipitation runs off over these areas. An additional factor is that not all locations receive the same precipitation during a storm. This section describes several methods to integrate runoff generation over large scales. These methods use probability density functions (Appendix A11) to describe spatial variability and then integrate runoff generation over the statistical distribution.

9.3.1 Exponential Distribution of Precipitation and Soil Moisture

One approach to calculate runoff at large spatial scales is to account for spatial variability in precipitation (Shuttleworth 1988; Pitman et al. 1990; Dolman and Gregory 1992; Eltahir and Bras 1993). Rain is assumed to fall over a fraction of the land μ, and $1 - \mu$ receives no rain. Within the raining area, the local precipitation at a given point is exponentially distributed with the probability density function,

$$f(P_R) = \frac{\mu}{\bar{P}_R} \exp\left(-\frac{\mu}{\bar{P}_R} P_R\right), \qquad (9.10)$$

in which $\bar{P}_R$ is the precipitation averaged over the entire surface. The average precipitation over the raining area is $\bar{P}_R/\mu$. The cumulative distribution gives the probability that local precipitation is less than a particular value and shows that the occurrence of extreme high local rainfall increases as μ decreases. In Figure 9.3, $\bar{P}_R$ = 1 mm and 50% of the surface receives more than 1.2 mm with μ = 0.6; this increases to 2.3 mm with μ = 0.3 and 7 mm with μ = 0.1. In this example, 5% of the surface receives more than 30 mm of rainfall when μ = 0.1. If infiltration capacity is spatially invariant, the average large-scale runoff is obtained by integrating $P_R - i_c$ (which provides the runoff rate from the raining fraction μ) and recognizing that the fractional area $1 - \mu$ receives no rainfall so that

$$\bar{R} = \mu \int_{i_c}^{\infty} (P_R - i_c) f(P_R) dP_R = \bar{P}_R \exp\left(-\frac{\mu i_c}{\bar{P}_R}\right). \qquad (9.11)$$

In this equation, the amount of precipitation that becomes runoff ($\bar{R}/\bar{P}_R$) decreases as infiltration capacity increases relative to precipitation (i.e., as $i_c/\bar{P}_R$ increases), but locally concentrated rainfall (smaller μ) increases runoff (Figure 9.4).

This approach can be extended to account for soil moisture heterogeneity by assuming that soil moisture is also distributed spatially with a probability density function (Entekhabi and Eagleson 1989; Johnson et al. 1993). A simple assumption is

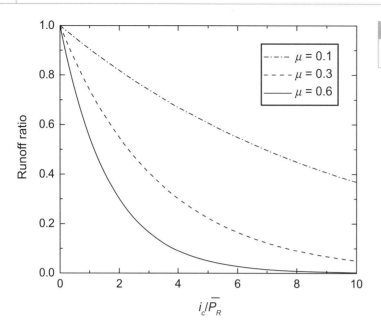

Figure 9.4 Effect of precipitation variability on runoff. Shown is the runoff ratio R/P_R in relation to i_c/P_R for $\mu = 0.1$, 0.3, and 0.6.

that the local relative soil wetness $S_e = \theta/\theta_{sat}$ is exponentially distributed with

$$f(S_e) = \frac{1}{\bar{S}_e} \exp\left(-\frac{S_e}{\bar{S}_e}\right), \tag{9.12}$$

and $\bar{S}_e$ is the spatial average wetness. Infiltration-excess runoff occurs over the fractional area that is not saturated and in which P_R exceeds i_c, and saturation-excess runoff occurs over the fractional area that is saturated whereby

$$R = \begin{cases} P_R - i_c & P_R > i_c, \ 0 \le S_e < 1 \\ P_R & 1 \le S_e < \infty \end{cases}. \tag{9.13}$$

The average runoff is obtained by integrating the equation:

$$\bar{R} = \mu \int_0^1 \int_{i_c}^\infty (P_R - i_c) f(P_R) dP_R f(S_e) dS_e$$
$$+ \mu \int_1^\infty \int_0^\infty P_R f(P_R) dP_R f(S_e) dS_e. \tag{9.14}$$

The first term on the right-hand side of the equation is the infiltration-excess runoff for the portion of the surface that is unsaturated, given by the integral of $f(S_e)$ below one, and where precipitation exceeds infiltration capacity. The second term is the saturation-excess runoff from precipitation that

falls onto the saturated portion of the surface given by the integral of $f(S_e)$ above one. Over the saturated fraction, all precipitation runs off.

Infiltration capacity depends on soil water, and the solution to (9.14) requires an expression for i_c in relation to S_e. Entekhabi and Eagleson (1989) derived such a relationship from Darcy's law in which

$$i_c = K_{sat}\left(\frac{\partial \psi}{\partial z} + 1\right) = K_{sat}\left(\frac{\partial \psi}{\partial S_e}\frac{\partial S_e}{\partial z} + 1\right). \tag{9.15}$$

The rightmost form of the equation uses $\partial S_e/\partial z$ rather than $\partial \psi/\partial z$ so as to be able to integrate i_c with respect to S_e. They approximated (9.15) using

$$i_c = K_{sat}\left(\frac{\partial |\psi|}{\partial S_e} \cdot \frac{S_e - 1}{\Delta z} + 1\right) = K_{sat} \nu S_e + K_{sat}(1 - \nu). \tag{9.16}$$

In this equation, the soil surface is assumed to be saturated (i.e., $S_e = 1$), the soil at depth Δz from the surface has water content S_e, and $\nu = (\partial |\psi|/\partial S_e)/\Delta z$ evaluated for $S_e = 1$. To be consistent with the notation of Entekhabi and Eagleson (1989), Δz is positive and $|\psi|$ denotes use of suction rather than potential. $\partial \psi/\partial S_e$ varies for different constitutive relationships (Table 8.2). Using the relation of Campbell (1974), for example, $\nu = -b\,|\psi_{sat}|/\Delta z$. Substituting (9.16) into (9.14) and integrating gives

$$\bar{R} = \bar{P}_R \left[\frac{\exp\left(-\mu I + \mu Iv\right) - \exp\left(-\mu I - 1/\bar{S}_e\right)}{1 + \mu Iv\bar{S}_e} \right]$$

$$+ \bar{P}_R \exp\left(-1/\bar{S}_e\right), \tag{9.17}$$

with $I = K_{sat}/\bar{P}_R$. The first term on the right-hand side of the equation is the infiltration-excess runoff from the area that is unsaturated, comparable to (9.11). This is evident when matric forces are ignored so that $v = 0$ and infiltration capacity depends only on gravitational potential ($i_c = K_{sat}$). In this case, the infiltration-excess term in (9.17) reduces to a rate $\bar{P}_R \exp(-\mu i_c/\bar{P}_R)$ times the unsaturated area $1 - \exp(-1/\bar{S}_e)$. Equation (9.17) has similar properties to (9.11) in that infiltration-excess runoff increases as precipitation is concentrated into a smaller area (i.e., as μ decreases) and decreases as hydraulic conductivity increases relative to precipitation (i.e., as I increases). However, (9.17) has the additional feature that the partial area that is saturated, given by $\exp(-1/\bar{S}_e)$, always generates runoff at the rate $\bar{P}_R$. Figure 9.5 illustrates the behavior of (9.17) for sandy loam and clay soils with $\mu = 0.3$ and precipitation that is 1/15 the saturated hydraulic conductivity of sandy loam. For both soils, the dominant mechanism generating runoff is saturation excess, and total runoff increases with wetter soil because of greater saturated area. For the sandy loam with its high K_{sat}, infiltration-excess runoff is negligible, and runoff is entirely generated by the partial area that is saturated. The clay has a smaller K_{sat} and some additional runoff occurs because the infiltration capacity is exceeded.

Question 9.5 Repeat the calculations for Figure 9.5, but for loam, and contrast its runoff with sandy loam and clay. Increase precipitation to 25 mm per hour. How does infiltration-excess runoff change with greater precipitation?

9.3.2 Variable Infiltration Capacity

Spatial variability in soil moisture and runoff generation can be represented by treating soil as a bucket with a specified water-holding capacity. The bucket fills with water until this capacity is reached, after which excess water runs off. Because the water-holding capacity is the maximum amount of water that can be stored, it can also be thought of as the infiltration capacity i_c. The water balance of a bucket with an initial amount of water W_0 that receives precipitation P_R over some time interval Δt is

$$R = \begin{cases} P_R \Delta t - (i_c - W_0) & P_R \Delta t > i_c - W_0 \\ 0 & P_R \Delta t \le i_c - W_0 \end{cases}, \tag{9.18}$$

and here W_0, i_c, and R have units of depth (mm H_2O). A large region can be represented by many individual buckets that vary in storage capacity. Runoff occurs only from buckets that are filled to capacity, and the smallest buckets overflow first. As precipitation adds more water, a greater number of buckets fill to capacity; more of the land is saturated and produces runoff. In this representation of runoff, rainfall infiltrates where there is capacity to absorb water and runs off where the infiltration capacity is exceeded.

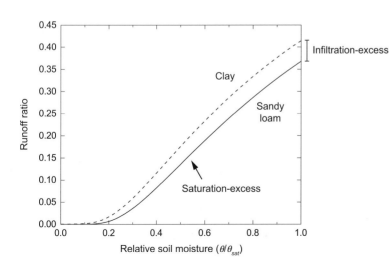

Figure 9.5 Runoff ratio $\bar{R}/\bar{P}_R$ in relation to soil moisture $\bar{S}_e$ for sandy loam and clay soils. Both soils have the same saturation-excess runoff. Infiltration-excess runoff is negligible for sandy loam so that the total runoff is from saturation-excess. The clay produces some infiltration-excess runoff so that the total runoff is larger than that from only saturation-excess (dashed line). Hydraulic properties are from Table 8.3 for Campbell (1974). For these graphs, $\mu = 0.3$, $\Delta z = 10$ cm, $I = 15$ (sandy loam), and $I = 0.55$ (clay).

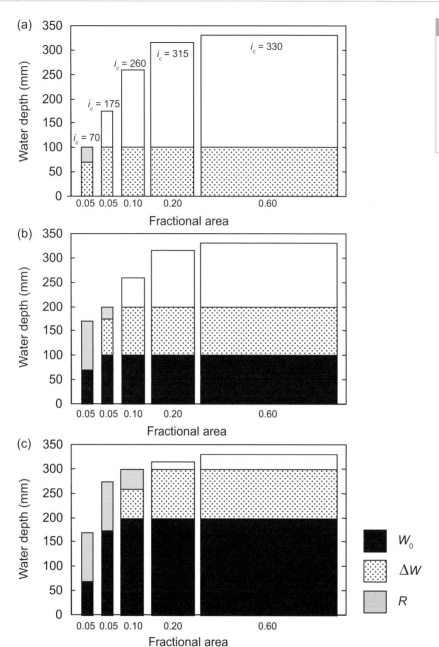

Figure 9.6 Example of the variable infiltration capacity runoff formulation. The region consists of five subareas that differ in infiltration capacity and area, as represented by the depth and width of the bucket. The three panels illustrate the water balance of each bucket after 100 mm of rainfall.

Figure 9.6 shows how (9.18) is used to calculate runoff. In this example, the land consists of five subareas that differ in infiltration capacity: 5% of the area has $i_c = 70$ mm; 5% has $i_c = 175$ mm; 10% has $i_c = 260$ mm; 20% has $i_c = 315$ mm; and 60% has $i_c = 330$ mm. The maximum water storage across the region is the area-weighted sum of i_c so that $W_{\max} = 299.25$ mm. Figure 9.6a illustrates this as five buckets that differ in depth (storage capacity) and width (area). If the initial water content of each bucket is $W_0 = 0$ and $P_R = 100$ mm, the four deepest buckets with infiltration capacity greater than 100 mm each store 100 mm of water and no precipitation runs off. The shallowest bucket with infiltration capacity $i_c = 70$ mm fills with water and then generates 30 mm runoff. The area-weighted change in soil water is $\Delta W = 98.5$ mm, and runoff is $R = 1.5$ mm. At the end of this rainfall event, only the shallowest bucket fills to capacity, and the fraction of the land that is saturated is $f_{sat} = 0.05$. Figure 9.6b illustrates the changes with another 100 mm of rainfall. The shallowest bucket is saturated, and all precipitation runs off. The second bucket stores an additional 75 mm of water to reach capacity and then generates 25 mm runoff. The three deepest buckets each absorb 100 mm of water. The area-weighted values are $\Delta W = 93.75$ mm, $R = 6.25$ mm, and $f_{sat} = 0.1$. If an additional 100 mm of rain falls, the two shallowest buckets (each saturated) generate 100 mm runoff while the third bucket fills to capacity and then generates 40 mm of runoff (Figure 9.6c). The two largest buckets each absorb 100 mm of rain and generate no runoff. The area-weighted values are $\Delta W = 86$ mm, $R = 14$ mm, and $f_{sat} = 0.2$.

The water balance shown in Figure 9.6 is an example of a probability distributed rainfall–runoff model (Moore 2007). In this approach, the storage capacity i_c at any point is distributed spatially with the probability density function $f(i_c)$. The portion of the land with storage capacity less than or equal to i_c is given by the cumulative distribution function

$F(i_c)$. Figure 9.7a shows a common distribution function. The maximum water storage is

$$W_{max} = \int_0^\infty i_c f(i_c)\,di_c = \int_0^\infty [1 - F(i_c)]\,di_c = \bar{i}_c. \tag{9.19}$$

As shown in Figure 9.7a, this is the area above the curve represented by $F(i_c)$. The area below the curve is $i_{c\,max} - W_{max}$, and this is the amount of water that cannot be stored. There is a critical storage capacity i_c^* below which stores are filled. The portion of land that contains stores with capacity less than or equal to i_c^* is

$$F(i_c^*) = \int_0^{i_c^*} f(i_c)\,di_c. \tag{9.20}$$

This is the area of the watershed that is saturated and generates runoff. The total amount of water stored relates to i_c^* by

$$W = \int_0^{i_c^*} [1 - F(i_c)]\,di_c. \tag{9.21}$$

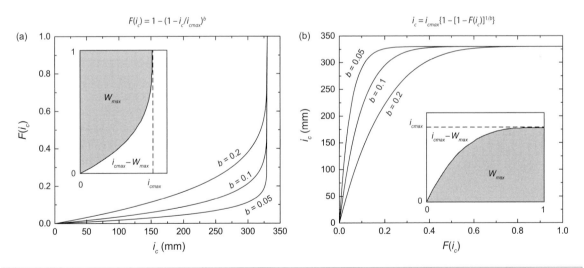

$$F(i_c) = 1 - (1 - i_c/i_{cmax})^b \qquad\qquad i_c = i_{cmax}\{1 - [1 - F(i_c)]^{1/b}\}$$

Figure 9.7 Statistical distribution of infiltration capacity. (a) Shown is the cumulative distribution $F(i_c)$ in relation to infiltration capacity i_c. The particular cumulative distribution function is for the Xinanjiang model. Curves are given for $i_{c\,max} = 330$ mm and with $b = 0.05$, 0.1, and 0.2. The inset panel shows that W_{max} is the area above $F(i_c)$ from 0 to $i_{c\,max}$. The area below the curve is $i_{c\,max} - W_{max}$. (b) Shown is the same distribution function, but rewritten so that infiltration capacity is expressed in relation to $F(i_c)$. Here, W_{max} is the area below i_c from 0 to 1.

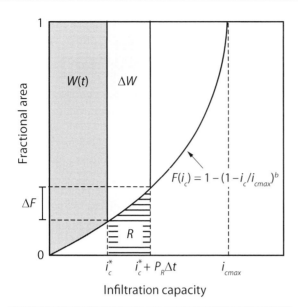

Figure 9.8 Runoff generation for the probability distributed rainfall–runoff method. Shown is the cumulative distribution function $F(i_c)$ for the Xinanjiang model. The precipitation $P_R\Delta t$ increases soil water storage by ΔW and increases the saturated fraction by ΔF. Runoff is the area below the cumulative distribution function such that $R = P_R\Delta t - \Delta W$. Adapted from Moore (2007).

Equation (9.21) can be solved to obtain i_c^* for a given W.

Figure 9.8 shows how these relationships are used to calculate runoff. The amount of water storage W at time t relates to i_c^* at time t and is the area above $F(i_c)$ between 0 and i_c^*. Associated with this water storage is the saturated fraction $F(i_c^*)$. Over the time interval Δt, an amount of precipitation equal to $P_R\Delta t$ falls over the land. Some portion of the water infiltrates into the soil, thereby increasing W by ΔW, and the remainder runs off from the saturated area so that

$$R = P_R\Delta t - \Delta W. \tag{9.22}$$

The change in water storage is $\Delta W = W(t + \Delta t) - W(t)$ and is obtained using (9.21) whereby

$$\Delta W = \int_{i_c^*}^{i_c^* + P_R\Delta t} [1 - F(i_c)]di_c. \tag{9.23}$$

As shown in Figure 9.8, ΔW is the area above $F(i_c)$ from i_c^* to $i_c^* + P_R\Delta t$ and R is the area below $F(i_c)$.

The Xinanjiang model uses this concept to represent spatial variability in soil water storage capacity and its effect on runoff (Zhao et al. 1980; Zhao 1992; Wood et al. 1992; Zhao and Liu 1995). The spatial variation of water-holding (or infiltration) capacity is described by

$$F(i_c) = 1 - \left(1 - \frac{i_c}{i_{c\,max}}\right)^b \tag{9.24}$$

and

$$f(i_c) = \frac{d}{di_c}F(i_c) = \frac{b}{i_{c\,max}}\left(1 - \frac{i_c}{i_{c\,max}}\right)^{b-1}, \tag{9.25}$$

where $i_{c\,max}$ is the maximum infiltration capacity and b is a shape parameter with typical values of 0.1–0.5. This is the distribution function shown in Figure 9.7a. The maximum storage is given from (9.19), and

$$W_{max} = \frac{i_{c\,max}}{b + 1}. \tag{9.26}$$

Equation (9.24) is also commonly written in the form

$$i_c = i_{c\,max}\left\{1 - [1 - F(i_c)]^{1/b}\right\}, \tag{9.27}$$

as shown in Figure 9.7b. In this form, it is easy to see that i_c becomes a constant value $i_{c\,max}$ as $b \to 0$. Spatial variation in infiltration capacity increases as b increases. Figure 9.7b illustrates the infiltration capacity curve for $i_{c\,max} = 330$ mm and $b = 0.1$, which was used to derive the infiltration capacities in the numerical example shown in Figure 9.6. About 20% of the land has infiltration capacity less than 295 mm; 10%, less than 215 mm; and 5%, less than 132 mm. It is evident from Figure 9.7b that the maximum storage is the area below the infiltration capacity curve. The area $i_{c\,max} - W_{max}$ above the curve is excess water that cannot be stored in soil and is lost as runoff. If infiltration capacity is spatially invariant (i.e., $b= 0$), the maximum amount of water is $i_{c\,max}$. As b becomes larger, the water storage capacity deceases below $i_{c\,max}$ and more water is lost to runoff.

Figure 9.8 illustrates the runoff calculations. The water storage at time t is obtained from (9.21) and is equal to

$$W(t) = W_{max}\left[1 - \left(1 - \frac{i_c^*}{i_{c\,max}}\right)^{b+1}\right]. \tag{9.28}$$

Rearranging terms in this equation gives the critical storage capacity:

$$i_c^* = i_{c\max} \left\{ 1 - \left[1 - \frac{W(t)}{W_{\max}} \right]^{1/(b+1)} \right\}. \tag{9.29}$$

The saturated fraction at time t is obtained by evaluating (9.24) with i_c^* so that

$$F(i_c^*) = 1 - \left[1 - \frac{W(t)}{W_{\max}} \right]^{b/(b+1)}. \tag{9.30}$$

The change in water storage over the time interval Δt after receiving the rainfall $P_R \Delta t$ is given by (9.23) as the area above $F(i_c)$ from i_c^* to $i_c^* + P_R \Delta t$. In practice, it is more convenient to calculate water storage at time $t + \Delta t$ using

$$W(t + \Delta t) = W_{\max} \left[1 - \left(1 - \frac{i_c^* + P_R \Delta t}{i_{c\max}} \right)^{b+1} \right] \tag{9.31}$$

so that runoff is

$$R = \begin{cases} P_R \Delta t - W_{\max} + W(t) + W_{\max} \left[1 - \dfrac{i_c^* + P_R \Delta t}{i_{c\max}} \right]^{b+1} & i_c^* + P_R \Delta t < i_{c\max} \\ P_R \Delta t - W_{\max} + W(t) & i_c^* + P_R \Delta t \ge i_{c\max} \end{cases} \tag{9.32}$$

Figure 9.9 is an example calculation for different initial water stores. The shape parameter b influences runoff generation. Small values of b give a small fractional area with low infiltration capacity, producing a small saturated area and low runoff. Larger values of b give a larger fractional area with low infiltration capacity, a larger saturated fraction, and greater runoff for a given amount of rainfall.

The variable infiltration capacity (VIC) model uses the Xinanjiang runoff parameterization (Liang et al. 1994, 1996). The model additionally calculates base flow from the deepest soil layer to separate the fast contribution of saturation-excess surface runoff to riverflow and a slower contribution of subsurface drainage. The model has been evaluated in continental and global simulations (Nijssen et al. 2001; Maurer et al. 2002). The Xinanjiang runoff has been used in some land surface models (Dümenil and Todini 1992; Stamm et al. 1994; Ducharne et al. 1998; Hagemann and Gates 2003; Wang et al. 2008; Balsamo et al. 2009; Li et al. 2011).

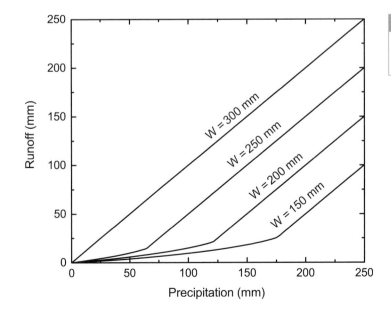

Figure 9.9 Runoff generation for the Xinanjiang model in relation to precipitation for $W_{\max} = 300$ mm, $b = 0.1$, and $W(t) = 150, 200, 250,$ and 300 mm.

Question 9.6 The numerical example shown in Figure 9.6 used $i_{c\,max} = 330$ mm and $b = 0.1$ to generate the infiltration capacity and area of each bucket. Use the analytical solutions to calculate W at the beginning of each time interval and also ΔW, R, and the saturated fraction after each time interval. How do the results compare with the numerical example? How much water infiltrates into the soil?

Question 9.7 Repeat the runoff calculations in Figure 9.9, but with $b = 0.2$. How does spatial variability in infiltration capacity affect runoff?

9.3.3 TOPMODEL Saturated Area

The hydrologic model TOPMODEL (Beven and Kirkby 1979; Beven et al. 1995; Hornberger et al. 1998; Beven 2000) is a physically based watershed model that uses the variable contributing area concept to calculate runoff in relation to topography. A key premise of the model is that the saturated zone is in equilibrium with recharge from an upslope contributing area. A second assumption is that the water table is parallel to the ground surface so that the local hydraulic gradient relates to the local slope angle β by $\tan\beta$ (Figure 9.10). Another important assumption is that saturated hydraulic conductivity decreases exponentially with greater depth into the soil so that

$$K_{sat}(z) = K_{sat,0}\exp\left(-f_k|z|\right), \tag{9.33}$$

where $K_{sat,0}$ is the saturated hydraulic conductivity at the surface and the parameter f_k describes the decrease in saturated hydraulic conductivity with

depth z. At any location in the watershed, the water table depth z_∇ (here taken as positive downward) is

$$z_\nabla = \bar{z}_\nabla + \frac{1}{f_k}(\bar{x} - \ln x), \tag{9.34}$$

with $\bar{z}_\nabla$ the average depth to water table, $x = \ln(a/\tan\beta)$ a local topographic index, and $\bar{x}$ the average topographic index. The topographic index relates to the upslope drainage area per unit contour width defined by $a = A/c$, where here A is the area of the hillslope segment and c is the hillslope width (Figure 9.10). High values of $\ln(a/\tan\beta)$ result from large contributing areas and shallow slopes, such as at the base of hillslopes and near streams. These topographic conditions are likely to have a shallow water table and greater area of saturation. Low values indicate relatively little upslope contributing area and steep slopes, typically along ridges and hilltops, with greater depth to water table and less saturation.

Equation (9.34) can be used to calculate the saturated area from which runoff occurs. Saturated locations are those with $z_\nabla \le 0$. By setting $z_\nabla = 0$, it can be shown that all locations in which

$$\ln x > \bar{z}_\nabla f_k + \bar{x} \tag{9.35}$$

are saturated. Given a probability density function $f(x)$ that characterizes the spatial distribution of the topographic index $\ln(a/\tan\beta)$, the saturated fraction of the watershed is the cumulative area in which the local topographic index exceeds (9.35) whereby

$$f_{sat} = \int f(x)dx \tag{9.36}$$

for $x \ge \bar{z}_\nabla f + \bar{x}$. One approach has been to represent $f(x)$ as a three-parameter gamma distribution (Sivapalan et al. 1987; Wolock 1993; Ducharne et al.

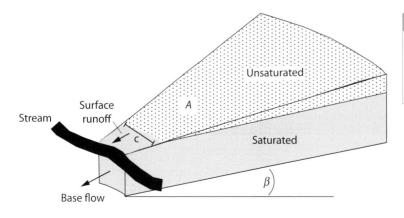

Figure 9.10 Topographically defined variable contributing area in TOPMODEL for a hillslope with an angle β, upslope area A, and width c. Shading denotes the saturated zone. Reproduced from Bonan (2016)

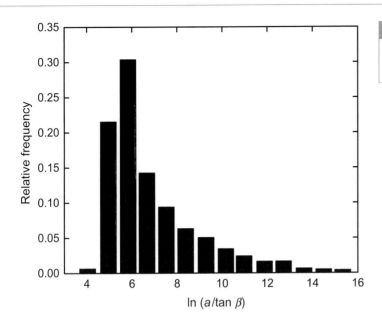

2000; Chen and Kumar 2001). In practice, the distribution of $\ln(a/\tan\beta)$ is discretized into some number of bins using high-resolution digital elevation data to estimate the local slope angle and area draining through that point (Figure 9.11). Each bin represents a portion of the watershed that is hydrologically similar. Equation (9.36) is then evaluated numerically as the sum of the area for grid cells where $z_i \leq 0$. All the precipitation that falls on the saturated fraction becomes runoff. There is also an associated base flow so that, similar to VIC, TOPMODEL separates total runoff to rivers into saturation-excess overland flow and subsurface base flow.

TOPMODEL is a distributed modeling framework in that hydrologic calculations are performed for increments of the topographic index over the entire watershed. By representing topography as discrete parcels of topographic index or as statistical distributions, runoff can be calculated and routed into streams during a storm. TOPMODEL and its variants have been applied to many watersheds and are used to represent hydrologic processes in some land surface models. Famiglietti and Wood (1994), for example, used the TOPMODEL approach to calculate energy and water fluxes associated with each topographic bin. This can be computationally expensive, depending on the number of bins used. Instead, Stieglitz et al. (1997) used analytical forms of the TOPMODEL equations to integrate over an entire watershed and coupled the resulting runoff to the

traditional soil column approach used in land surface models, and Chen and Kumar (2001) applied a similar approach over North America (Figure 9.12). Other models have utilized comparable methods, but applied to a rectangular grid instead of watersheds. Niu et al. (2005) simplified the TOPMODEL framework by representing the spatial variability of the topographic index as an exponential distribution. This approach is used in the Community Land Model, in which the saturated fraction is approximated by

$$f_{sat} = f_{max} \exp\left(-0.5\bar{z}_\nabla f_k\right), \tag{9.37}$$

with f_{max} the maximum fractional saturated area defined as the area within a grid cell with topographic index greater than or equal to the grid cell mean relative to the total grid cell area. For global simulations, $f_k = 0.5$ m^{-1}. The mean water table depth $\bar{z}_\nabla$ is diagnosed from the soil moisture profile.

Question 9.8 Calculate the local topographic index for a segment of a watershed with an upslope area $A = 500$ m^2, contour length at the base $c = 25$ m, and a local slope of $5°$. What is the topographic index for a smaller upslope area of 250 m^2? For a shallower slope of $1°$? For a smaller contour length $c = 5$ m? What do the topographic indices indicate about runoff?

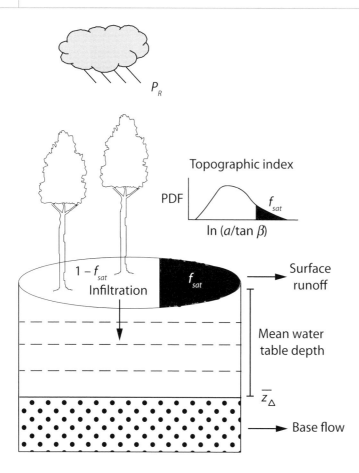

Figure 9.12 TOPMODEL saturated fraction as applied to a soil column model. A probability density function of the topographic index $\ln(a/\tan\beta)$ is used to calculate the saturated fraction f_{sat}. Surface runoff is $f_{sat}P_R$, and $(1 - f_{sat})P_R$ infiltrates into the soil. Adapted from Stieglitz et al. (1997) and Chen and Kumar (2001)

Question 9.9 The following table gives the distribution of the topographic index for the Sleepers River watershed shown in Figure 9.11. Calculate the saturated fraction if the mean water table depth is $\bar{z}_{\nabla} = 0.2$ m and $f_k = 3.3$, as in Niu et al. (2005). What is the saturated fraction for $\bar{z}_{\nabla} = 0.7$ m? How do these compare with the Niu et al. (2005) approximation given by (9.37)?

$\ln(a/\tan\beta)$	Relative area	$\ln(a/\tan\beta)$	Relative area	$\ln(a/\tan\beta)$	Relative area
4.0	0.006	8.4	0.064	12.7	0.017
5.0	0.219	9.3	0.051	13.6	0.007
5.8	0.309	10.2	0.035	14.5	0.005
6.7	0.145	11.0	0.025	15.3	0.004
7.5	0.096	11.9	0.017		

9.4 | Evapotranspiration

The spatial heterogeneity in soil moisture that affects runoff also affects evapotranspiration. A common approach used in models decreases evapotranspiration for dry soils according to

$$E = E_p f(S_e), \tag{9.38}$$

in which E_p is a potential rate in the absence of soil moisture limitation and $f(S_e)$ is a moisture stress function that has a value of zero when soil is dry and one when soil is wet. There can be considerable spatial variability in soil moisture, with wet locales generally having greater evapotranspiration than dry locales, all other factors being equal. Models that use a grid-average soil moisture to calculate evapotranspiration cannot account for this subgrid heterogeneity.

Entekhabi and Eagleson (1989) applied their subgrid soil moisture parameterization to also calculate evapotranspiration. They used the function,

$$f(S_e) = \begin{cases} 0 & 0 \leq S_e \leq S_w \\ \dfrac{S_e - S_w}{S_* - S_w} & S_w < S_e < S_* , \\ 1 & S_* \leq S_e \leq \infty \end{cases} \quad (9.39)$$

in which S_w and S_* are critical threshold values of relative soil wetness $S_e = \theta/\theta_{sat}$. If S_e varies spatially in a grid cell, the area-weighted average evapotranspiration is

$$\bar{E} = E_p \int_{S_w}^{S_*} \left(\frac{S_e - S_w}{S_* - S_w}\right) f(S_e) dS_e + E_p \int_{S_*}^{\infty} f(S_e) dS_e. \quad (9.40)$$

For exponentially distributed soil moisture, the solution is

$$\bar{E} = E_p \left(\frac{\bar{S}_e}{S_* - S_w}\right) \left[\exp\left(-\frac{S_w}{\bar{S}_e}\right) - \exp\left(-\frac{S_*}{\bar{S}_e}\right)\right]. \quad (9.41)$$

More complicated soil moisture stress functions require numerical solutions. Sellers et al. (2007) developed a statistical binning approach for subgrid soil moisture that is used in SiB3. Evapotranspiration is approximated by some number of discrete soil moisture bins that vary from dry to saturated so that

$$\bar{E} = E_p \sum_i f(S_{e,i}) a_i, \quad (9.42)$$

where $S_{e,i}$ is the relative wetness of the ith bin and a_i is the fractional area associated with the bin. Sellers et al. (2007) demonstrated the utility of soil moisture binning using the stress function

$$f(S_e) = \max\left[0, 0.25(S_e - 1), \tanh\left(0.0045\psi_{sat} S_e^{-8}\right) + 1\right], \quad (9.43)$$

with $\psi_{sat} = -0.5$ m the matric potential at saturation (typical of a loam). The subgrid binning parameterization has been implemented in SiB3 with 10 bins (Baker et al. 2017). A key to its implementation is to dynamically update the area of the bins.

Question 9.10 Graph $\bar{E}/E_p$ in relation to $\bar{S}_e$ for (9.41) with $S_w = 0.1$ and $S_* = 0.8$.

Question 9.11 Calculate area-average evapotranspiration $\bar{E}/E_p$ for the following soil wetness distribution using (9.43) for soil moisture stress.

Compare this estimate with that obtained using the average soil wetness $\bar{S}_e$. Repeat the calculations with the linear function $f(S_e) = S_e$. Contrast the utility of subgrid binning for linear and nonlinear stress functions.

$S_{e,i}$	a_i	$S_{e,i}$	a_i	$S_{e,i}$	a_i	$S_{e,i}$	a_i
0.05	0.003	0.35	0.115	0.65	0.192	0.95	0.008
0.15	0.010	0.45	0.207	0.75	0.119		
0.25	0.047	0.55	0.254	0.85	0.044		

9.5 | Snow Cover

The presence of snow on the ground affects the thermal state of land in several ways. Fresh snow has a much higher albedo compared with soil, leaves, and other plant material. In addition to reducing the absorption of solar radiation, some of the net radiation is used to melt snow rather than warm the soil. Snow also has a much lower thermal conductivity compared with soil (Chapter 5) so that the presence of a thick snowpack insulates the underlying soil. While snow is melting, the surface temperature cannot exceed freezing. This can create a strong temperature inversion on warm, sunny days between the cold snow surface and warmer air above.

The physical processes to include in a snow model are complex and include snowfall accumulation, melting and refreezing, compaction, liquid water flow, grain size growth, absorption of solar radiation, and turbulent fluxes with the atmosphere. The thermodynamics of snowmelt can be treated in a framework that combines the surface energy balance with multilayer heat transfer and phase change in a column consisting of snow and soil. The surface energy balance needs to account for sublimation of snow rather than evaporation of liquid water and may also include heat input from rain. Extreme stable conditions can develop over melting snow, and snowpack models may use stability corrections for turbulent fluxes that are unique for snow (in contrast with those in Chapter 6). The albedo of snow depends on solar zenith angle; the amount of aerosols such as black carbon, mineral dust, and organic carbon deposited onto and mixed into the snowpack; and snow aging as represented through changes in ice grain size. Grain size changes over time in relation to dry snow metamorphism, liquid water metamorphism, refreezing of liquid water, and the addition of fresh snow. Some models also account for vertical absorption of radiation in the snowpack. A model must account for liquid water movement in the snowpack and refreezing of meltwater. In the early phases of melt, the snowpack gains energy and temperature increases to 0°C, after which melting occurs. Initially, this water may be retained within the snowpack, but subsequently, as more energy is absorbed, the melt water is released from the snow. These processes can be represented in physically based snow models (Anderson 1976; Jordan 1991), which have been a starting point for snow cover parameterizations in many land surface models.

A critical requirement of snow models is to account for compaction. Snow has a density of 50–100 kg m^{-3} when fresh. Low density occurs during conditions of cold temperatures and low wind speed. Warmer temperatures and windy conditions increase density. Snow compacts over time, increasing to a density of 300–500 kg m^{-3} or so during springtime melt as a result of settling of new snow, the mass of overlying snow, densification during melting, and other processes. Compaction manifests in an increase in bulk density and a decrease in depth. This is seen in the relationship between snow water equivalent and snow depth in which

$$W_{snow} = h_{snow}\rho_{snow}, \qquad (9.44)$$

where W_{snow} is the mass of water per unit area (kg m^{-2}), h_{snow} is the depth of snow (m), and ρ_{snow} is density (kg m^{-3}). A complexity is that the number of layers and their thickness change over time as snow accumulates from snowfall, compacts, or melts.

The fraction of the ground covered by snow is also important. Snow rarely covers the ground uniformly and instead is distributed patchily across the surface. Models partition the land surface into snow-covered and snow-free fractions when calculating surface energy fluxes and the hydrologic cycle (Figure 9.13). The relationship between snow

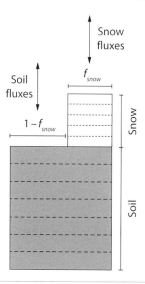

Figure 9.13 Depiction of subgrid snow cover. Snow covers the fraction f_{snow} of the ground surface, from which fluxes with the atmosphere are calculated for snow. The fraction $1 - f_{snow}$ is bare soil. Temperature and water are calculated in a multilayer snow and soil column.

depth and snow cover fraction is complex, differs between accumulation and melt because of the patchiness of snowmelt, and the parameterizations used in model vary considerably (Liston 2004; Clark et al. 2011b). Table 9.1 gives the snow cover fraction parameterization used in three versions of the Community Land Model, and these have very different relationships with snow depth (Figure 9.14). In CLM3, a snowpack 10 cm deep has $f_{snow} = 0.5$. Snow fraction depends on the density of snow in CLM4. A 10 cm snowpack has $f_{snow} = 1$ when density is low (50–100 kg m^{-3}), such as may be found in fresh snow, and a smaller snow fraction ($f_{snow} = 0.76$) when density is high (400 kg m^{-3}), such as may be found when snow is melting. The ground is mostly covered when snow depth is 20 cm regardless of density. In CLM4.5, different equations are used during the accumulation and melt phases. During melt, snow fraction varies depending on an index of topographic variability, defined by the standard deviation of elevation within a model grid cell. With low topographic variability ($n = 8$), the ground is completely covered by snow when depth is 50 cm. Greater topographic variability (lower values for n) give correspondingly lower snow fraction.

Table 9.1	Snow cover fraction parameterizations in three versions of the Community Land Model	
Model	Parameterization	Reference
CLM3	$f_{snow} = \dfrac{h_{snow}}{10z_0 + h_{snow}}$	Oleson et al. (2004)
CLM4	$f_{snow} = \tanh\left[\dfrac{h_{snow}}{2.5z_0}\left(\dfrac{\rho_{snow}}{\rho_{new}}\right)^{-m}\right]$	Niu and Yang (2007); Oleson et al. (2010b)
CLM4.5	$f_{snow} = 1 - \left[\dfrac{1}{\pi}\cos^{-1}\left(2\dfrac{W_{snow}}{W_{snow,\,max}} - 1\right)\right]^{n}$	Swenson and Lawrence (2012); Oleson et al. (2013)

Note: h_{snow}, snow depth (m); $z_0 = 0.01$ m, roughness length of ground; ρ_{snow}, density of snow (kg m^{-3}), $\rho_{new} = 100$ kg m^{-3}, density of fresh snow. In CLM4, $m = 1$. The equation for CLM4.5 is for melting only with W_{snow} snow water equivalent (kg m^{-2}), $W_{snow,\,max}$ maximum snow water equivalent, and $n = 200/\sigma_{elev}$ an index of topographic variability defined in relation to the standard deviation of elevation within a grid cell ($n = 1$ when $\sigma_{elev} = 200$ m).

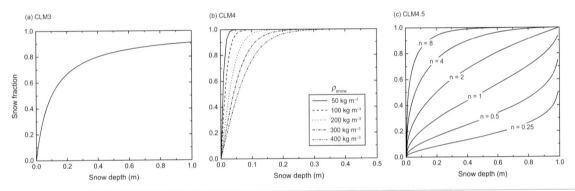

Figure 9.14 Fraction of the ground covered by snow as represented in three versions of the Community Land Model (Table 9.1). CLM4 is shown for $\rho_{snow} = 50$–400 kg m^{-3}. CLM4.5 is shown for $n = 0.25$–8 with $W_{snow,\,max} = 300$ kg m^{-2}, and snow depth is converted to snow water equivalent using $\rho_{snow} = 300$ kg m^{-3}.

Question 9.12 In CLM4.5, the parameterization of saturated fraction and snow fraction rely on topographic variability in a model grid cell as shown by (9.37) and Table 9.1. Discuss whether these parameterizations are appropriate when comparing the model to eddy covariance flux tower measurements with a typical spatial footprint of 1 km^2.

9.6 | Subgrid Tiling

Many models explicitly account for spatial heterogeneity in vegetation and soil within a grid cell using subgrid tiling (Koster and Suarez 1992; Seth et al. 1994; Bonan et al. 2002; Essery et al. 2003; Melton and Arora 2014). This approach divides a model grid cell into a mosaic of independent smaller patches of homogenous vegetation or soil (Figure 9.15). Each tile develops its own surface climate in response to the atmospheric forcing, and the atmosphere model, in turn, sees the weighted average fluxes of the individual patches. The subgrid tiling can be based on vegetation type, soil texture, soil depth, elevation, or some combination of these.

Tiling is particularly useful in mixed lifeform biomes such as savanna with a mixture of trees and grasses, a crop–forest mosaic, or a mixed forest with deciduous broadleaf and evergreen needleleaf trees (Bonan et al. 2002). The various plant types differ greatly in photosynthesis, phenology, carbon

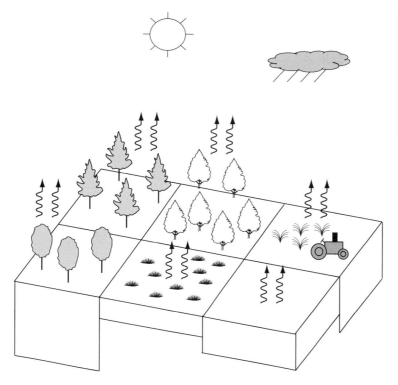

Figure 9.15 Depiction of subgrid tiling. The grid cell is represented by six tiles that differ in vegetation and soil depth. Each tile receives the same atmospheric forcing but calculates unique surface fluxes depending on the tile's vegetation and soil. The fluxes to the atmosphere are weighted by the fractional area of each tile.

allocation, and other physiological traits. A tiling approach is needed to separate the different life forms into distinct homogenous patches. While this accounts for varying physiology and life history characteristics, important questions remain about the scale at which the plant types interact. In savanna, for example, tiling may neglect forest–grass interactions that are important for shaping vegetation distributions and response to environmental change. Similarly, trees are comingled in a mixed forest and compete for light, water, and nutrients unlike in the tiled mosaic. Some models such as the Community Land Model allow aboveground tiling for plant canopies, but shared soil resources belowground. Such tiling affects surface fluxes compared with tiling in which each plant type has its own soil column (Schultz et al. 2016). An additional challenge is the representation of sparse or partial canopy cover. Models distinguish a bare ground fraction of the surface, but the effect on surface fluxes differs if this represents a small gap in the canopy on the order of 100–1000 m^2 or a larger opening on the order of 10–100 km^2 created by fire. Small canopy gaps can be adequately modeled using a homogeneous big-leaf canopy in which the leaf area index is reduced for the bare ground

fraction. Large canopy openings may be better modeled as separate patches of dense vegetation and bare soil. This is particularly true for widely spaced shrubland or desert scrub vegetation.

9.7 | Supplemental Programs

9.1 Infiltration Capacity: This program calculates infiltration rate using the Green–Ampt equation as in Figure 9.2. It uses bisection to solve (9.6) for I_c at time t.

9.8 | Modeling Projects

1. (a) Use Supplemental Program 9.1 to calculate Green–Ampt infiltration as in Figure 9.2. Compare this with results obtained for the parameters from Rawls et al. (1982, 1993), as used in the model of Stieglitz et al. (1997) and given in the following table. (b) Repeat (a), but using (9.8) with A and S calculated as in Rawls et al. (1993). (c) Using linear regression, fit (9.8) to the Green–Ampt infiltration data to obtain estimates for A and S. How do these compare with the Rawls

et al. (1993) estimates used in (b)? Discuss the importance of parameter uncertainty and model structural uncertainty.

Soil type	θ_{sat}	θ_{res}	ψ_{sat} (cm)	b	K_{sat} (cm h^{-1})
Sandy loam	0.453	0.041	−14.7	2.65	2.59
Loam	0.463	0.027	−11.2	3.97	1.32
Clay loam	0.464	0.075	−25.9	4.13	0.23

2. Use Supplemental Program 7.1 to investigate subgrid tiling. Calculate the diurnal cycle of surface energy fluxes for a tall forest (canopy height, 20 m) and a short grassland (canopy height, 0.5 m). Compare the average fluxes for these two sites with the fluxes calculated using the average canopy height (10.25 m). Try varying other site parameters such as the amount of soil water, maximum canopy conductance, and leaf area index.

Leaf Temperature and Energy Fluxes

Chapter Overview

Extension of the bulk surface energy balance described in Chapter 7 to include vegetation involves the formulation of leaf fluxes. Energy from the net radiation absorbed by a leaf is stored in biomass or is dissipated as sensible and latent heat, and the balance of these fluxes – as influenced by prevailing meteorological conditions, leaf biophysics, and leaf physiology – determines leaf temperature. This chapter develops the biophysical foundation and mathematical equations to describe leaf temperature and the leaf energy budget. A critical determinant of leaf fluxes and temperature is leaf boundary layer conductance, which depends on wind speed and leaf size.

10.1 | Introduction

A leaf absorbs solar and longwave radiation, and the dissipation of this radiation as sensible and latent heat determines the leaf microclimate. High radiative loading onto a leaf increases its temperature, and the leaf can be substantially warmer than the surrounding air unless turbulent processes are sufficient to remove the heat. Conditions such as high wind speed or small leaf size increase turbulent exchange with the surrounding air and help dissipate heat. The dynamics of leaf–air coupling is seen in leaf temperature. On a warm summer day, for example, the leaves of a deciduous tree can be 5°C–7°C warmer than air temperature when exposed

to full sunlight, but intermittent clouds cause rapid cooling to 2°C–3°C below air temperature (Figure 10.1). Wind has a similar effect on temperature, and a change from calm air to blowing air can produce a near-instantaneous large decrease in leaf temperature.

Mathematical models of leaf energy fluxes describe these changes in leaf temperature. The early theoretical development of these models was based on analysis of the energy balance of leaves and application of heat transfer theory for flat plates (Raschke 1960; Gates 1962, 1963, 1965, 1966; Gates et al. 1968; Gates and Papian 1971). Current theory formalizes the linkages among leaf energy, water, and CO_2 fluxes; and its mathematical implementation couples leaf temperature, the energy balance, photosynthesis, stomatal conductance, and the leaf boundary layer in a quasi-mechanistic representation of leaf biophysics and biochemistry. This theory represents a leaf in terms of its radiative environment, boundary layer processes, the physiology of stomata, and the biochemistry of photosynthesis (Figure 10.2). The radiative environment regulates leaf temperature by the absorption of solar and longwave radiation and the emission of longwave radiation, and this net radiation is dissipated through turbulent exchange. The boundary layer regulates transfer of heat, water vapor, and CO_2 between the ambient air and the leaf surface. Stomata provide additional regulation of water vapor and CO_2 diffusion between the leaf surface and intercellular spaces. Biochemistry is responsible for fixing CO_2 during photosynthesis in reactions catalyzed by Rubisco in C_3 plants and PEP carboxylase in C_4 plants, and the radiative environment further regulates leaf temperature and

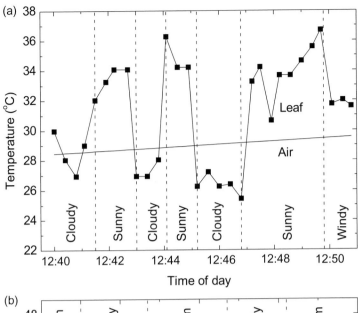

(a)

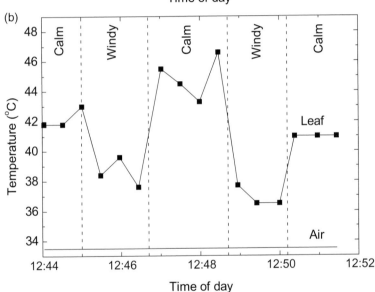

(b)

Figure 10.1 Measured leaf temperature of lanceleaf cottonwood (*Populus acuminata*) on a summer day in Boulder, Colorado, for (a) sunny and cloudy conditions over an 11-minute period and (b) calm and windy conditions over a 7-minute period. Also shown is air temperature. Adapted from Gates (1963)

fluxes through its effects on photosynthesis and stomata. This chapter considers boundary layer processes. The next two chapters examine photosynthesis and stomatal conductance.

10.2 | Leaf Energy Balance

The net energy stored by a leaf is the balance between net radiation R_n, convective (sensible) heat flux H, and latent heat flux λE so that

$$c_L \frac{\partial T_\ell}{\partial t} = R_n - H - \lambda E. \tag{10.1}$$

H and λE are positive away from the leaf surface, and fluxes are per unit leaf area with the units W m^{-2}. Leaf heat capacity c_L (J m^{-2} K^{-1}) is the product of leaf specific heat (J kg^{-1} K^{-1}) and leaf mass per area (kg m^{-2}). These fluxes depend on leaf temperature T_ℓ, and models of leaf energy fluxes solve for the temperature that balances the energy budget. It is convenient to separate net radiation into two terms. Radiative forcing Q_a is the sum of absorbed solar

(a) Radiative environment

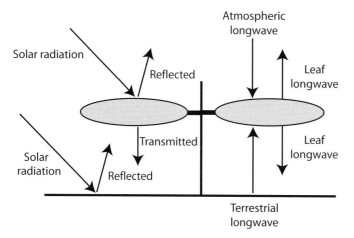

Figure 10.2 Biophysics and biochemistry of leaves. (a) The radiative environment consists of solar radiation (left) and longwave radiation (right). (b) Leaf fluxes include CO_2, H_2O, and heat through the boundary layer. These fluxes are shown as a network of conductances for the adaxial (upper) and abaxial (lower) leaf surfaces. For H_2O and CO_2, the conductance for each surface is obtained from stomatal and boundary layer conductances acting in series. The total conductance is defined by the adaxial and abaxial surfaces acting in parallel. (c) Stomata open to absorb CO_2 for photosynthesis, but, in doing so, water is lost as transpiration.

(b) Boundary layer processes

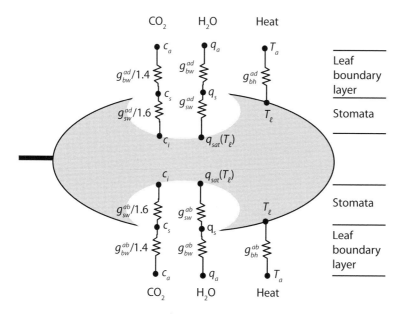

(c) Stomatal physiology

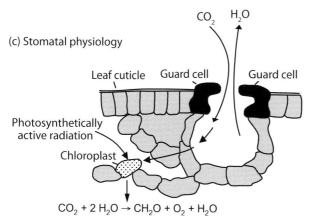

radiation and absorbed longwave radiation and is independent of leaf temperature. From this is subtracted emitted longwave radiation, which varies with leaf temperature. Then, the leaf energy balance is

$$c_L \frac{\partial T_\ell}{\partial t} = Q_a - 2\varepsilon_\ell \sigma T_\ell^4 - H(T_\ell) + \lambda E(T_\ell). \tag{10.2}$$

Longwave radiation is emitted from the upper and lower leaf surfaces with emissivity ε_ℓ so that the factor two appears in the longwave radiation flux. Sensible heat flux and latent heat flux are each written as a function of T_ℓ, denoted by $H(T_\ell)$ and $\lambda E(T_\ell)$, and (10.2) is solved for the temperature that balances the radiative forcing.

Solar and longwave radiation absorbed by the leaf provide the radiative forcing. Consider an individual leaf situated over a grass surface (Figure 10.2a). Solar radiation strikes the upper leaf surface, denoted as $S^\downarrow$. Solar radiation reflected from the underlying grass with albedo ρ_g strikes the lower leaf surface, as given by $\rho_g S^\downarrow$. The sum of these two fluxes is the total solar radiation incident on the leaf. Some of the incident radiation is transmitted through the leaf, some is reflected, and the rest is absorbed. This varies with wavelength, and the solar spectrum is commonly divided into visible and near-infrared wavebands. The leaf absorbs longwave radiation from the sky $L_{sky}^\downarrow$ and the grass $L_g^\uparrow$. The radiative forcing is the sum of these fluxes, equal to

$$Q_a = \sum_\Lambda S_\Lambda^\downarrow \left(1 + \rho_{g\Lambda}\right)\left(1 - \rho_{\ell\Lambda} - \tau_{\ell\Lambda}\right) + \varepsilon_\ell \left(L_{sky}^\downarrow + L_g^\uparrow\right), \tag{10.3}$$

with ρ_ℓ and τ_ℓ leaf reflectance and transmittance, respectively, and the subscript Λ denotes visible and near-infrared wavebands. The leaf receives longwave radiation from above and below, absorbed proportionally to leaf emissivity.

Sensible heat flux is described by Fick's law. Sensible heat exchange occurs from both sides of a leaf, and the total sensible heat flux is the sum of the adaxial (upper) and abaxial (lower) leaf surfaces so that

$$\begin{aligned} H &= c_p(T_\ell - T_a)g_{bh}^{ad} + c_p(T_\ell - T_a)g_{bh}^{ab} \\ &= c_p(T_\ell - T_a)\left(g_{bh}^{ad} + g_{bh}^{ab}\right). \end{aligned} \tag{10.4}$$

In this equation, $T_\ell - T_a$ is the leaf–air temperature difference (K), c_p is the specific heat of moist air at constant pressure ($c_p = 29.2$ J mol^{-1} K^{-1} is representative; equation [3.59]), and conductances have units mol m^{-2} s^{-1}. Sensible heat flux from the adaxial surface is proportional to the boundary layer conductance for heat g_{bh}^{ad}; the abaxial surface has the conductance g_{bh}^{ab}; and the total conductance is $g_{bh}^{ad} + g_{bh}^{ab}$. If these conductances are equal ($g_{bh}^{ad} = g_{bh}^{ab}$) and this one-sided conductance is denoted g_{bh}, the total leaf conductance is $2g_{bh}$. This gives the standard equation for sensible heat flux from a leaf in which

$$H = 2c_p(T_\ell - T_a)g_{bh}. \tag{10.5}$$

It is important to remember that here g_{bh} is defined for one side of the leaf, and $2g_{bh}$ is the total leaf conductance because convection occurs from both leaf surfaces acting in parallel.

Transpiration is also represented by Fick's law in which

$$E = \frac{q_{sat}(T_\ell) - q_a}{g_{bw}^{-1} + g_{sw}^{-1}}, \tag{10.6}$$

where $q = e/P$ is water vapor mole fraction (mol mol^{-1}) and $q_{sat}(T_\ell) - q_a$ is the water vapor difference between the leaf, which is assumed to be saturated with temperature T_ℓ, and the surrounding air. The term λ in (10.2) converts the transpiration water flux (mol H$_2$O m^{-2} s^{-1}) to an energy flux (W m^{-2}). A representative value is $\lambda = 44.44$ kJ mol^{-1} (Table 3.6). g_{bw} and g_{sw} are the boundary layer and stomatal conductance for water vapor (mol m^{-2} s^{-1}), respectively, acting in series. Stomatal conductance regulates water vapor flux from inside the stomatal cavity to the leaf surface; the boundary layer conductance regulates the flux from the leaf surface to air surrounding the leaf. Equation (10.6) is given for a leaf with stomata located on only one surface. For leaves with stomata located on both surfaces, the upper and lower conductances acting in parallel determine total conductance.

Care must be used in defining the total leaf conductance for water vapor and to correctly account for the adaxial (upper) and abaxial (lower) leaf surfaces (Figure 10.2b). The appropriate value for each surface is obtained from the boundary layer and stomatal conductances acting in series. The total leaf conductance is defined by the adaxial and abaxial surfaces acting in parallel, and the leaf conductance to water vapor is

$$g_{\ell w} = \frac{1}{1/g_{bw}^{ad} + 1/g_{sw}^{ad}} + \frac{1}{1/g_{bw}^{ab} + 1/g_{sw}^{ab}}. \qquad (10.7)$$

Leaves that have stomata only on the abaxial surface are termed hypostomatous. For this type of leaf, the abaxial surface represents the entire leaf, and the one-sided conductances are the conductances for the entire leaf ($g_{bw}^{ab} = g_{bw}$; $g_{sw}^{ab} = g_{sw}$). The total conductance is

$$g_{\ell w} = \frac{1}{g_{bw}^{-1} + g_{sw}^{-1}}. \qquad (10.8)$$

Some plants have stomata on both leaf surfaces; such leaves are termed amphistomatous. If the adaxial and abaxial surfaces have the same conductances ($g_{bw}^{ad} = g_{bw}^{ab} = g_{bw}$; $g_{sw}^{ad} = g_{sw}^{ab} = g_{sw}$), the leaf conductance is

$$g_{\ell w} = \frac{2}{g_{bw}^{-1} + g_{sw}^{-1}}, \qquad (10.9)$$

in which g_{bw} and g_{sw} are define as one-sided conductances. Most broadleaf tree species have hypostomatous leaves with stomata on the abaxial surface; grasses are amphistomatous with an equal distribution of stomata on both surfaces; and many herbaceous plants are amphistomatous but with stomatal primarily on one surface (Gates 1980; Bolhàr-Nordenkampf and Draxler 1993). It is common to approximate leaves as one-sided, and (10.8) is used to calculate leaf conductance in (10.6).

The storage flux is equal to the product of the time rate of change in temperature $\partial T_{\ell}/\partial t$ (K s^{-1}) and the heat capacity of leaves c_L (per unit leaf area; J m^{-2} K^{-1}). The fresh mass of a leaf is the sum of its dry mass m_{dry} and the mass of water m_{wat}. Heat capacity, then, is calculated from the specific heat of dry biomass and water (J kg^{-1} K^{-1}) and their respective mass per unit leaf area (kg m^{-2}), and the overall heat capacity is

$$c_L = m_{dry}c_{dry} + m_{wat}c_{wat}, \qquad (10.10)$$

with c_{dry} and c_{wat} the specific heat of dry biomass and water, respectively. In this equation, leaf heat capacity increases as the dry mass of the leaf or the water content increase. The dry mass of a leaf (per unit leaf area) is the leaf mass per area m_a (Chapter 2). The water content of a leaf is commonly expressed as a fraction f_w of total fresh mass (kg

H$_2$O kg^{-1}) so that the fresh weight is $m_a/(1-f_w)$ and the mass of water is $m_{wat} = f_w m_a/(1-f_w)$. Equation (10.10) can then be equivalently written as

$$c_L = \frac{m_a}{1-f_w}\left[(1-f_w)c_{dry} + f_w c_{wat}\right]. \qquad (10.11)$$

In this form, leaf heat capacity is calculated from the fresh mass (per unit leaf area) times the specific heat of dry biomass and water, each weighted by their respective mass fraction. This latter term, given in brackets in (10.11), is the specific heat of the leaf. Heat capacity is sometimes calculated from the specific heat of the leaf multiplied by biomass density (kg m^{-3}) times leaf thickness (Monteith 1981b; Jones 2014, p. 227). From the definition of m_a given by (2.18), it is evident that this is equivalent to (10.11). However, care must be given to distinguish projected leaf area from total leaf area if using leaf density to calculate c_L. In a cylinder, for example, the ratio of volume to area is $(\pi/4) \times$ diameter for projected area, but is diameter/4 for total area (Chapter 2).

A representative leaf water content for woody plants is 70% of fresh mass (Niinemets 1999), and the specific heat of water is about three times that of dry biomass (Ball et al. 1988; Blanken et al. 1997). With $c_{wat} = 4188$ J kg^{-1} K^{-1}, $c_{dry} = 1396$ J kg^{-1} K^{-1} so that the specific heat of leaves is 3350 J kg^{-1} K^{-1}, or 80% that of water. Using the leaf mass per area reported in Table 2.2, $c_L = 815$ J m^{-2} K^{-1} for deciduous broadleaf trees ($m_a = 73$ g m^{-2}) and 2937 J m^{-2} K^{-1} for evergreen needleleaf trees ($m_a = 263$ g m^{-2}). For comparison, Blanken et al. (1997) calculated a heat capacity of 1999 J m^{-2} K^{-1} for aspen (*Populus tremuloides*) leaves with $m_a = 111$ g m^{-2} and $f_w = 0.8$. Ball et al. (1988) obtained values of 1100–2200 J m^{-2} K^{-1} for mangrove leaves in which $f_w = 0.71$ and m_a varies from 93 g m^{-2} in thin leaves to 189 g m^{-2} in thick leaves. Leigh et al. (2012) reported values of 1040–1560 J m^{-2} K^{-1} in desert species.

Question 10.1 Leaf transpiration consists of diffusion through stomata and the boundary layer connected in series. At steady state, the fluxes of water vapor along each path are equal. Prove that the transpiration flux given by (10.6) can be obtained from this principle.

10.3 | Leaf Temperature

The full leaf energy balance is

$$c_L \frac{\partial T_\ell}{\partial t} = Q_a - 2\varepsilon_\ell \sigma T_\ell^4 - 2c_p(T_\ell - T_a)g_{bh}$$
$$- \lambda[q_{sat}(T_\ell) - q_a]g_{\ell w}, \quad (10.12)$$

with $g_{\ell w} = 1/(g_{bw}^{-1} + g_{sw}^{-1})$. T_ℓ is the leaf temperature that balances the energy budget. Equation (10.12) is, however, difficult to solve for T_ℓ. It is a nonlinear equation because of the dependence of longwave radiation on T_ℓ^4 and the dependence of transpiration on $q_{sat}(T_\ell)$. Various numerical methods are used to solve for temperature depending on whether the leaf is in steady state or has transient dynamics.

10.3.1 | Steady State Temperature

In the absence of heat storage, which is typically small due to the low heat capacity of leaves, $\partial T_\ell/\partial t = 0$. Leaf temperature is in steady state as determined by energy exchanges with the surrounding environment. One numerical solution for this condition solves the leaf energy balance using Newton–Raphson iteration to find T_ℓ (Appendix A5). In this method,

$$F(T_\ell) = Q_a - 2\varepsilon_\ell \sigma T_\ell^4 - 2c_p(T_\ell - T_a)g_{bh}$$
$$- \lambda[q_{sat}(T_\ell) - q_a]g_{\ell w} = 0 \quad (10.13)$$

is the function to solve. Leaf temperature is the root of this equation, which is obtained from

$$\Delta T_\ell = -\frac{F(T_\ell)}{\partial F/\partial T_\ell}. \quad (10.14)$$

The Newton–Raphson method is an iterative solution that calculates successive values of T_ℓ until the change in temperature is less than some small amount. Newton–Raphson iteration is possible because the temperature derivative is easily obtained as

$$\frac{\partial F}{\partial T_\ell} = -8\varepsilon_\ell \sigma T_\ell^3 - 2c_p g_{bh} - \lambda g_{\ell w} \frac{dq_{sat}(T_\ell)}{dT}. \quad (10.15)$$

The iteration may require many steps because the convergence criterion must be small (e.g., $\Delta T_\ell < 0.001$ K) to ensure that fluxes balance.

An alternative method that avoids iteration uses a Taylor series expansion (Appendix A1) to linearize the terms that are nonlinear in T_ℓ. In the simplest case, net radiation is specified so that the energy balance is

$$R_n = 2c_p(T_\ell - T_a)g_{bh} + \lambda[q_{sat}(T_\ell) - q_a]g_{\ell w}. \quad (10.16)$$

The only nonlinear term is $q_{sat}(T_\ell)$, and this is approximated by

$$q_{sat}(T_\ell) = q_{sat}(T_a) + s(T_\ell - T_a), \quad (10.17)$$

with $q_{sat}(T_a)$ saturation water vapor at air temperature and $s = dq_{sat}(T_a)/dT$ the temperature derivative evaluated at T_a. This expression for $q_{sat}(T_\ell)$ is substituted into (10.16), and after rearranging terms an equation for leaf temperature is found whereby

$$T_\ell - T_a = \frac{R_n - \lambda[q_{sat}(T_a) - q_a]g_{\ell w}}{2c_p g_{bh} + s\lambda g_{\ell w}}. \quad (10.18)$$

Equation (10.18) shows that leaf temperature increases above air temperature with greater net radiation and decreases as the water vapor deficit increases. In this solution, latent heat flux is linearized using

$$\lambda E = \lambda[q_{sat}(T_a) + s(T_\ell - T_a) - q_a]g_{\ell w}. \quad (10.19)$$

Substituting (10.18) for $T_\ell - T_a$ gives

$$\lambda E = \frac{sR_n + 2c_p g_b[q_{sat}(T_a) - q_a]}{s + 2(1 + g_b/g_{sw})c_p/\lambda}. \quad (10.20)$$

In this equation, $g_{bh} = g_{bw}$ and both are denoted by g_b. Equation (10.20) is the Penman–Monteith equation for a hypostomatous leaf (Monteith 1965; Jarvis and McNaughton 1986). It is similar to that given in Chapter 7 for a plant canopy. Latent heat flux increases with greater net radiation and greater water vapor deficit.

Equation (10.18) for leaf temperature and (10.20) for latent heat flux require that net radiation is known, but the emission of longwave radiation itself depends on temperature. A more general solution also linearizes longwave emission using a Taylor series expansion so that

$$\varepsilon_\ell \sigma T_\ell^4 = \varepsilon_\ell \sigma T_a^4 + 4\varepsilon_\ell \sigma T_a^3(T_\ell - T_a) \quad (10.21)$$

and

$$R_n = Q_a - 2\varepsilon_\ell \sigma T_\ell^4 = (Q_a - 2\varepsilon_\ell \sigma T_a^4) - 8\varepsilon_\ell \sigma T_a^3(T_\ell - T_a). \quad (10.22)$$

The first two terms in parentheses in the rightmost equation are the isothermal net radiation if the leaf was at air temperature (Monteith 1981b; Monteith and Unsworth 2013, p. 41; Jones 2014, p. 101). The second term represents the radiative heat loss due to non-isothermal conditions. This heat loss can be written in a form similar to sensible heat flux, whereby

$$2c_p(T_\ell - T_a)g_r = 8\varepsilon_\ell \sigma T_a^3 (T_\ell - T_a), \quad (10.23)$$

and $g_r = 4\varepsilon_\ell \sigma T_a^3 / c_p$ is the radiative conductance to heat transfer (mol m^{-2} s^{-1}). Replacing R_n in (10.18) with (10.22) gives

$$T_\ell - T_a = \frac{(Q_a - 2\varepsilon_\ell \sigma T_a^4) - \lambda[q_{sat}(T_a) - q_a]g_{\ell w}}{2c_p(g_r + g_{bh}) + s\lambda g_{\ell w}}.$$
$$(10.24)$$

This solution uses a Taylor series approximation for net radiation and latent heat flux. Other solutions are possible that retain higher-order terms in the Taylor series expansion – giving, for example, a quadratic equation for T_ℓ (Paw U 1987). An additional complexity in solving for leaf temperature is that stomatal conductance itself depends on leaf temperature (Chapter 12).

Question 10.2 Explain the difference between the Penman–Monteith equation for a leaf as given by (10.20) and that for a canopy of leaves given by (7.36).

Question 10.3 Equation (10.20) uses mole fraction to calculate latent heat flux. A more common expression instead uses vapor pressure with $\lambda E = (c_p/\gamma)[e_{sat}(T_\ell) - e_a]g_{\ell w}$. Derive the Penman–Monteith equation using this notation.

Question 10.4 Write the Penman–Monteith equation using the notation of radiative conductance.

10.3.2 Transient Behavior

Although heat storage is commonly neglected when solving the energy balance for leaf temperature, this term can, in fact, be non-negligible. Leaf temperature does not respond instantaneously to a change in its surrounding environment (radiation, air temperature, humidity, wind speed). Instead, the actual leaf temperature lags behind the temperature the leaf would attain if it was in equilibrium with its environment. If heat storage is considered, (10.12) must be stepped forward in time given some initial leaf temperature using numerical methods such as the fourth-order Runge–Kutta method (Appendix A3).

The leaf energy balance can also be linearized as in the preceding section to provide an analytical solution for transient leaf temperature (Monteith 1981b; Jones 2014, p. 355). The solution is obtained by first considering the energy balance in the absence of heat storage, where leaf temperature is described by

$$Q_a - 2\varepsilon_\ell \sigma T_a^4 - 2c_p(T_e - T_a)(g_r + g_{bh}) - \lambda[q_{sat}(T_a)$$
$$+ s(T_e - T_a) - q_a]g_{\ell w} = 0. \quad (10.25)$$

This equation uses linearized expressions for latent heat flux and net radiation; and T_e denotes steady state leaf temperature, which differs from the actual temperature T_ℓ. The latter is given by the full energy balance (with storage), whereby

$$Q_a - 2\varepsilon_\ell \sigma T_a^4 - 2c_p(T_\ell - T_a)(g_r + g_{bh}) - \lambda[q_{sat}(T_a)$$
$$+ s(T_\ell - T_a) - q_a]g_{\ell w} = c_L \frac{\partial T_\ell}{\partial t}. \quad (10.26)$$

Subtracting (10.25) from (10.26) gives

$$\frac{\partial T_\ell}{\partial t} = \frac{T_e - T_\ell}{\tau_L}, \quad (10.27)$$

in which

$$\tau_L = \frac{c_L}{2c_p(g_r + g_{bh}) + s\lambda g_{\ell w}} \quad (10.28)$$

is a time constant (s). Equation (10.27) is a first-order, linear differential equation. If a leaf has a steady state temperature T_{e1} and conditions are instantaneously altered to give a new steady state temperature T_{e2}, the solution to (10.27) is

$$T_\ell(t) = T_{e2} - (T_{e2} - T_{e1})e^{-t/\tau_L}. \quad (10.29)$$

For a leaf exposed to a step change in forcing, the actual leaf temperature approaches the equilibrium temperature over a time period that is related to τ_L. At time $t = \tau_L$, leaf temperature differs from T_{e2} by $0.37 \cdot \Delta T_e$. In the example shown in Figure 10.3a, T_ℓ approaches T_{e2} over a period of 60 s. The extent of the lag depends on heat capacity. For a leaf exposed to a periodic forcing, the transient temperature lags the equilibrium temperature and has a smaller amplitude (Figure 10.3b). The time constant for leaves is on the order of a few seconds to one minute, depending on leaf properties (Monteith 1981b; Michaletz et al. 2016). A smaller time constant

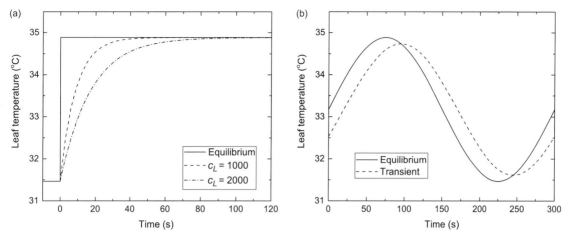

(a)

(b)

Figure 10.3 Leaf temperature in response to a change in radiative forcing. Shown are the equilibrium temperature (solid line) obtained from (10.25) and the transient temperature (dashed line) obtained by numerically integrating (10.26). (a) Step change in environmental forcing for two leaves with $c_L = 1000$ J m^{-2} K^{-1} ($\tau_L = 9$ s) and $c_L = 2000$ J m^{-2} K^{-1} ($\tau_L = 17$ s). (b) Sinusoidal change in forcing over 5 minutes for a leaf with $c_L = 2500$ J m^{-2} K^{-1} ($\tau_L = 21$ s). In these examples, $T_a = 35°$C, relative humidity is 50%, $P = 1013.25$ hPa, $\varepsilon_\ell = 0.98$, $g_{bh} = g_{bw} = 1$ mol m^{-2} s^{-1}, and $g_{sw} = 0.5$ mol m^{-2} s^{-1}. The step change instantaneously increases Q_a from 1000 to 1400 W m^{-2} to represent a change from cloudy to sunny. The sinusoidal forcing varies Q_a from 1000 to 1400 W m^{-2} over a period of 5 minutes.

means that leaves respond more rapidly to changes in environmental conditions (e.g., sunflecks).

Question 10.5 Calculate the time constant τ_L for deciduous broadleaf trees with leaf mass per area $m_a = 73$ g m^{-2} and evergreen needleleaf trees with $m_a = 263$ g m^{-2}. Use conditions for STP and with $\varepsilon_\ell = 0.98$, $g_{bh} = 1$ mol m^{-2} s^{-1}, and $g_{sw} = 0.5$ mol m^{-2} s^{-1}.

10.4 | Leaf Boundary Layer

The boundary layer conductance regulates heat and mass fluxes between the leaf surface and the surrounding air. This conductance varies with the size of a leaf and with wind speed. Equations for boundary layer conductance are derived from engineering studies of heat transfer for flat, rectangular plates as applied to leaves. Air moving across a leaf slows near the leaf surface and increases in speed with distance from the leaf. Full airflow occurs only at some distance beyond the leaf surface. The region below free air, in which wind speed decreases close to the surface, is termed the leaf boundary layer. It is typically 1–10 mm thick. The conductance

through this layer differs depending on fluid dynamics (free or forced convection; laminar or turbulent flow) and varies with the type of gas. Table 10.1 summarizes leaf boundary conductance equations derived in this chapter.

Convection is the transfer of heat or mass by a moving fluid such as air, and it is necessary to distinguish two types of convection in biometeorological studies. Free convection occurs due to temperature differences that affect the density, and therefore buoyancy, of air. Forced convection is flow caused by an external force that creates wind. Free convection occurs in still air while forced convection occurs with wind. Additionally, fluid motion is either laminar or turbulent. Smooth, constant motion characterizes laminar flow. Eddies, vortices, and instability characterize turbulent flow. For many biometeorological applications, forced convection and laminar boundary layer flow is valid (Gates 1980; Finnigan and Raupach 1987; Campbell and Norman 1998; Monteith and Unsworth 2013).

The conductance for heat transfer in forced convection is described in terms of the object itself; defined by a characteristic length d_ℓ; and the properties of air as defined by velocity u, viscosity v, thermal diffusivity D_h, and specific heat c_p. These properties are combined into the dimensionless Nusselt number Nu, which is defined in relation to

Table 10.1	Summary of boundary layer conductance equations for heat, H_2O, and CO_2		

	Coefficient a (for g_b in mol m^{-2} s^{-1})		
	Heat	H_2O	CO_2
Forced convection			
Laminar $g_b = au^{0.5}d_\ell^{-0.5}$	0.203	0.223	0.164
Turbulent $g_b = au^{0.8}d_\ell^{-0.2}$	0.311	0.343	0.252
Free convection			
Laminar $g_b = a(\Delta T/d_\ell)^{0.25}$	0.049	0.054	0.039

Note: For a standard atmosphere at sea level (1013.25 hPa and 15°C).

the molecular properties of the fluid as specified by the Prandtl number Pr and the flow regime as specified by the Reynolds number Re. A similar theory pertains to mass flux but replacing Nu with the nondimensional Sherwood number Sh and Pr with the Schmidt number Sc. Reviews of heat and mass transfer theory as applied to leaves are provided by Gates (1980), Finnigan and Raupach (1987), Schuepp (1993), Campbell and Norman (1998), and Monteith and Unsworth (2013).

The Nusselt number is a nondimensional heat transfer coefficient. It is the ratio of convective to conductive heat transfer, where conductive heat transfer is measured in still air. The convective heat flux is $c_p\Delta T g_{bh}$. The conductive heat flux in air with thermal diffusivity D_h (m^2 s^{-1}) and molar density ρ_m (mol m^{-3}) over the characteristic length d_ℓ (m) is equal to $c_p\rho_m D_h\Delta T/d_\ell$. The Nusselt number is, therefore, given as

$$Nu = \frac{g_{bh}d_\ell}{\rho_m D_h}, \quad (10.30)$$

and this equation relates g_{bh} to Nu. The Nusselt number also represents the ratio of the characteristic dimension d_ℓ to boundary layer thickness δ. The heat conducted across a boundary layer with thickness δ must equal the heat loss by convection. For this to be true, $g_{bh} = \rho_m D_h/\delta$ and, therefore, $Nu = d_\ell/\delta$. The thermal diffusivity of air at 15°C is $D_h = 20.8$ mm^2 s^{-1}, and the molar density is $\rho_m = 42.3$ mol m^{-3}. A 1 mm thick boundary layer has a conductance equal to 0.9 mol m^{-2} s^{-1}.

The Nusselt number is specified in terms of the Prandtl number Pr and the Reynolds number Re. The dimensionless Prandtl number describes the molecular diffusion properties of the fluid. It is the ratio between the molecular diffusivities for momentum and heat, and

$$Pr = v/D_h. \quad (10.31)$$

The molecular diffusivity for momentum v (m^2 s^{-1}) is commonly called the kinematic viscosity. For air, Pr = 0.70 (Table A.3). The Reynolds number is nondimensional and characterizes the flow regime. It represents the ratio of inertial forces to viscous forces and varies in relation to wind speed u so that

$$Re = ud_\ell/v. \quad (10.32)$$

Laminar flow occurs at low Re with large viscous forces. Turbulent flow occurs at high Re, where inertial forces dominate. Empirical studies of laminar flow with forced convection show that heat loss from one side of a flat rectangular plate is described by

$$Nu = 0.66Pr^{0.33}Re^{0.5}. \quad (10.33)$$

Equations (10.30) and (10.33) can be applied to a leaf to obtain g_{bh}. The characteristic length for a leaf is the leaf dimension, and a representative leaf dimension is $d_\ell = 5$ cm. The Nusselt number for leaves is generally larger than that for a flat rectangular plate, and Nu must be multiplied by an empirical correction factor b_1. A typical value is $b_1 = 1.5$, but this can vary (Finnigan and Raupach 1987; Schuepp 1993; Monteith and Unsworth 2013, p. 160). Combining equations gives

$$g_{bh} = a(u/\ell)^{0.5}, \quad (10.34)$$

with $a = b_1 0.66\rho_m D_h Pr^{0.33}/v^{0.5}$. The properties of air vary with temperature (Table A.3), and $a = 0.2$ mol m^{-2} s$^{-1/2}$ at 15°C. A representative value at a wind speed of 2 m s^{-1} is $g_{bh} = 1.3$ mol m^{-2} s^{-1}. Equation (10.34) is commonly used in models, with the assumption also that the boundary layer conductance for heat is the same as that for water vapor. In fact, however, these conductances differ.

A similar treatment describes mass transfer of H_2O and CO_2 from a leaf. Fick's law gives the flux of a gas F_j (mol m^{-2} s^{-1}) as the product of the concentration difference Δc_j (mol mol^{-1}) and a conductance g_{bj} (mol m^{-2} s^{-1}) so that

$$F_j = \Delta c_j g_{bj}. \tag{10.35}$$

The Sherwood number is a nondimensional mass transfer coefficient similar to Nu. It is the ratio of the flux F_j to the diffusive flux in still air with thickness d_ℓ. This latter flux is equal to $\rho_m D_j \Delta c_j / d_\ell$, and

$$Sh = \frac{g_{bj} d_\ell}{\rho_m D_j}, \tag{10.36}$$

with D_j the molecular diffusivity of mass (H_2O or CO_2) in air ($m^2\ s^{-1}$). The Schmidt number is analogous to the Prandtl number for convective heat exchange so that

$$Sc = v/D_j. \tag{10.37}$$

The Schmidt number equals 0.61 for H_2O and 0.96 for CO_2 (Table A.3). Similar to the Nusselt number,

$$Sh = 0.66 Sc^{0.33} Re^{0.5}, \tag{10.38}$$

and Sh for a leaf is also multiplied by the empirical factor b_1. The conductance relative to that for heat is, from (10.30) and (10.36), given by

$$\frac{g_{bj}}{g_{bh}} = \frac{Sh}{Nu} \frac{D_j}{D_h} = \left(\frac{D_j}{D_h}\right)^{0.67}. \tag{10.39}$$

The diffusivities for momentum, heat, H_2O, and CO_2 vary with temperature and pressure (Table A.3). At 15°C, the conductance for H_2O relative to heat is

$g_{bw} = 1.10 g_{bh}$ and relative to CO_2 is $g_{bw} = 1.36 g_{bc}$. More commonly, it is assumed that $g_{bw} = g_{bh}$ and $g_{bw} = 1.4 g_{bc}$.

Equation (10.33) for Nu and (10.38) for Sh apply to laminar flow. For turbulent flow,

$$Nu = 0.036 Pr^{0.33} Re^{0.8} \tag{10.40}$$

$$Sh = 0.036 Sc^{0.33} Re^{0.8}. \tag{10.41}$$

Both are again multiplied by the empirical factor b_1 to give values appropriate for leaves. The transition to turbulent flow occurs with Re > 20,000 for flat plates, but the distinction between laminar and turbulent flow is small over the range of Reynolds number typically found in biometeorological studies (Finnigan and Raupach 1987; Monteith and Unsworth 2013, p. 156).

Figure 10.4 shows g_{bh} with laminar and turbulent flow for leaves with $d_\ell = 5$ cm and $d_\ell = 10$ cm. Conductance increases with greater wind speed, illustrating the important influence of wind to transport heat away from a leaf. The turbulent conductance exceeds the laminar conductance for Re > 16,248. This corresponds to a wind speed of about 5 m s^{-1} for a 5 cm leaf. For this size leaf, typical of many leaves, the distinction between laminar and turbulent flow is insignificant at a wind speed of 5 m s^{-1}, but the turbulent conductance exceeds the laminar conductance by about 25% at a wind speed of 10 m s^{-1}. The larger leaf has a lower conductance

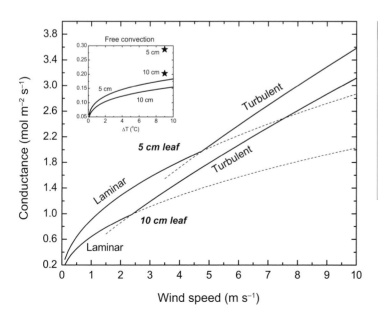

Figure 10.4 Leaf boundary layer conductance for heat g_{bh} in relation to wind speed for a 5 cm leaf and a 10 cm leaf. Shown are the laminar and turbulent conductance for each leaf. The turbulent conductance exceeds the laminar conductance for Re > 16,248. Dashed lines show laminar and turbulent conductance beyond this threshold. The inset panel shows conductance with laminar free convection. For comparison, the symbols show forced convection conductance with a wind speed of 0.1 m s^{-1}. Calculations are at STP (15°C and 1013.25 hPa).

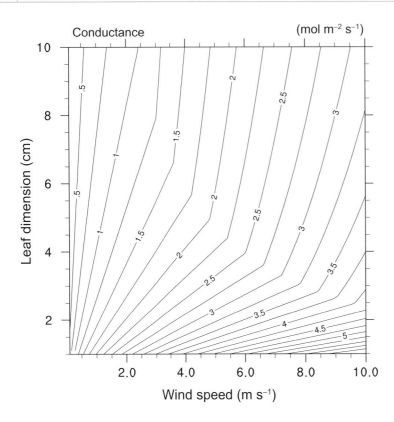

Figure 10.5 Leaf boundary layer conductance for heat g_{bh} with forced convection in relation to wind speed and leaf dimension. Calculations are at STP.

and a greater distinction between flow regimes. It is convenient to distinguish forced convection as laminar or turbulent based on the larger of the two conductances. Figure 10.5 illustrates the relationship of conductance to leaf dimension and wind speed. Conductance decreases with larger leaves and increases with higher wind speed. Note the transition from laminar to turbulent flow, which occurs at higher wind speed for smaller leaves.

In conditions of free convection, the Nusselt and Sherwood numbers are defined in terms of the Grashof number Gr. The Grashof number represents the product of buoyancy and inertial forces divided by the square of the viscous force, given by

$$Gr = \frac{g d_\ell^3 (T_\ell - T_a)}{v^2 T_a}, \tag{10.42}$$

with g gravitational acceleration (m s^{-2}). Free convection is strong with large Gr because the buoyancy and inertial forces that promote air movement exceed the viscous force that inhibits flow. For a horizontal flat plate or cylinder (Campbell and Norman 1998; Monteith and Unsworth 2013), laminar free convection occurs with Gr $< 10^5$, and

$$Nu = 0.54 Pr^{0.25} Gr^{0.25} \tag{10.43}$$

$$Sh = 0.54 Sc^{0.25} Gr^{0.25}. \tag{10.44}$$

For Gr $> 10^5$,

$$Nu = 0.15 Pr^{0.33} Gr^{0.33} \tag{10.45}$$

$$Sh = 0.15 Sc^{0.33} Gr^{0.33}. \tag{10.46}$$

Boundary layer conductances for free convection are generally smaller than those for forced convection (Figure 10.4). These equations apply only to still air and are frequently excluded from models.

The distinction between forced and free convection is given by the ratio of buoyancy to inertial forces, Gr/Re^2 (Gates 1980; Schuepp 1993; Campbell and Norman 1998; Monteith and Unsworth 2013). Forced convection dominates when Gr is much less than Re^2 (Gr/Re$^2 < 0.1$) and buoyancy forces are small. When Gr is much larger than Re^2 (Gr/Re$^2 > 10$), buoyancy forces are much larger than inertial forces and heat transfer is described by free convection. The transition between free and forced convection is not precise, and mixed regimes can prevail. A practical rule is that forced convection occurs with wind speed of 0.1 m s^{-1} or more, and free convections occurs with lower wind speed (Gates 1980). Consider, for example, a leaf with $d_\ell = 5$ cm and air at 15°C. At a wind speed equal to 0.1 m s^{-1}, $Gr > Re^2$ when leaf temperature exceeds air temperature by approximately 6°C, but the leaf must be almost 24°C warmer than air (an unrealistic condition) with wind speed equal to 0.2 m s^{-1}. At a leaf temperature 10°C warmer than air, such as may be found with still air on a warm sunny day, heat transfer is predominantly forced convection ($Gr/Re^2 < 0.1$) for wind greater than 0.4 m s^{-1} and is predominantly free convection ($Gr/Re^2 > 10$) for wind less than 0.04 m s^{-1}. Several techniques can be used to represent mixed convection regimes. For example, Gates (1980) and Schuepp (1993) suggested using the larger of the forced and free conductances. Leuning et al. (1995) added the conductances for forced and free convection, assuming that both forms of convection occur in parallel.

Question 10.6 Two estimates of the diffusivity of ozone (O_3) at sea level (1013.25 hPa) and 0°C are 13.0 and 14.4 mm^2 s^{-1} (Massman 1998). Using these values, compare the boundary layer conductance of O_3 to that of CO_2.

Question 10.7 Calculate the boundary layer conductance g_{bh} at STP for a small leaf (d_ℓ = 3 cm) and a large leaf (d_ℓ = 10 cm) with calm wind (u = 0.1 m s^{-1}) and strong wind (u = 5 m s^{-1}). Contrast the conductances for laminar and turbulent forced convection.

10.5 | Characteristic Leaf Dimension

For a rectangular plate, the characteristic dimension d_ℓ is the length of the plate in the direction of flow. A cylinder aligned parallel to the direction of flow similarly has a characteristic dimension equal to its length. A sphere or a cylinder oriented perpendicular to the flow has a characteristic dimension

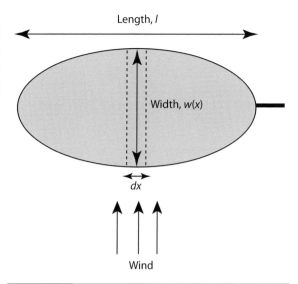

Figure 10.6 Plan view of a leaf with length l oriented perpendicular to the wind. The leaf has a width $w(x)$, which is the dimension of the leaf in the direction of flow at any point x.

equal to its diameter. Leaves have more complex shapes, but the characteristic dimension is related to leaf width (Gates and Papian 1971; Schuepp 1993; Campbell and Norman 1998; Monteith and Unsworth 2013). A leaf of length l oriented perpendicular to the wind has a width $w(x)$ at any position x along the leaf (Figure 10.6). The total leaf area is $\int_0^l w(x)\,dx$, and the leaf dimension is

$$d_\ell = \left[\int_0^l w(x)^n\,dx \bigg/ \int_0^l w(x)\,dx \right]^{\frac{1}{n-1}}. \qquad (10.47)$$

The coefficient n varies with flow regime; $n = 0.5$ for laminar forced convection; $n = 0.8$ for turbulent flow; and $n = 0.75$ for free convection (Monteith and Unsworth 2013, p. 158). For most applications, leaf dimension can be approximated as $d_\ell = 0.7w$, where w is the maximum leaf width in the direction of wind flow (Gates and Papian 1971, p. 14; Schuepp 1993; Campbell and Norman 1998, p. 107).

10.6 | Energy Budget Examples

Gates (1962, 1963) provided an example energy budget calculation for a leaf hanging over grass at local noon on a summer day with an air temperature of 34°C. The solar radiation incident on the

| Table 10.2 | Leaf temperature in relation to radiative forcing Q_a and wind speed for a leaf with closed stomata $(g_{sw} = 0 \text{ mol m}^{-2} \text{ s}^{-1})$ and open stomata $(g_{sw} = 0.4 \text{ mol m}^{-2} \text{ s}^{-1})$ |

		Leaf temperature (°C)							
		$g_{sw} = 0 \text{ mol m}^{-2} \text{ s}^{-1}$				$g_{sw} = 0.4 \text{ mol m}^{-2} \text{ s}^{-1}$			
		Wind speed (m s^{-1})				Wind speed (m s^{-1})			
Q_a (W m^{-2})		0.01	0.1	1	5	0.01	0.1	1	5
Clear sky	1444	49	46	41	38	40	38	36	35
Cloudy sky	1136	40	39	37	36	36	33	33	33

Note: Leaf temperature is calculated from (10.13). In this example, air temperature is 35°C, relative humidity is 50%, and air pressure is 1013.25 hPa. Solar radiation for clear sky is $S^{\downarrow} = 1100$ W m^{-2}, and $S^{\downarrow} = 550$ W m^{-2} for cloudy sky. Longwave fluxes are $L^{\downarrow}_{sky} = 300$ W m^{-2} and $L^{\uparrow}_{g} = 545$ W m^{-2} (ground temperature is 40°C). Leaf properties are $d_{\ell} = 0.05$m, $\varepsilon_{\ell} = 0.98$, $\rho_{\ell} = \tau_{\ell} = 0.1$ (visible), and $\rho_{\ell} = \tau_{\ell} = 0.4$ (near–infrared). Ground albedo is $\rho_g = 0.1$ (visible) and $\rho_g = 0.2$ (near–infrared).

upper leaf surface was 1117 W m^{-2}, of which 893 W m^{-2} was absorbed. The upper leaf surface absorbed an additional 279 W m^{-2} from atmospheric longwave radiation. The lower leaf surface absorbed 70 W m^{-2} of solar radiation reflected by the grass and 558 W m^{-2} radiated by the warm grass. The total radiative forcing on both sides of the leaf was therefore 1800 W m^{-2}. This radiation was dissipated through emission of longwave radiation, sensible heat flux, and latent heat flux. The leaf temperature was 46°C, and the upper and lower leaf surfaces each radiated 586 W m^{-2}. Sensible heat flux was calculated using heat transfer theory to be 13 W m^{-2}, and latent heat flux was inferred to be 615 W m^{-2} to balance the energy budget.

Table 10.2 gives examples of leaf temperature calculated using (10.13). Conditions are for a hot summer day with air temperature 35°C and relative humidity 50%. Solar radiation is given for clear sky and cloudy sky, and the corresponding radiative forcing is 1444 W m^{-2} and 1136 W m^{-2}, respectively. Leaf temperature with cloudy sky is several degrees cooler than that with clear sky. For both clear sky and cloudy sky, leaf temperature decreases with greater wind speed. This is because the boundary layer conductance increases with greater wind speed. With no transpiration ($g_{sw} = 0 \text{ mol m}^{-2} \text{ s}^{-1}$), leaf temperature in clear sky decreases from 49°C with still air (0.01 m s^{-1}) to 38°C with a gentle breeze (5 m s^{-1}). Convective heat loss is low with still air, radiative emission is the primary heat loss, and the leaf has near lethal temperature. A light breeze of 1 m s^{-1} decreases leaf temperature by

8°C, and a gentle breeze of 5 m s^{-1} decreases temperature by an additional 3°C. This illustrates the powerful effect of convection to carry heat away from the leaf surface and cool its temperature. The cooling effect of convection is less pronounced with cloudy sky. Similar behavior is seen with open stomata ($g_{sw} = 0.4 \text{ mol m}^{-2} \text{ s}^{-1}$). Here, transpiration additionally cools the leaf surface. With clear sky and still air, leaf temperature is 9°C cooler than without transpiration. With light and gentle breeze, leaf temperature is close to air temperature, and leaf temperature is below air temperature for cloudy sky. This illustrates the important effect of latent heat to cool a leaf.

The role of stomata in regulating transpiration can be seen in the two limiting cases of the Penman–Monteith equation (10.20) when leaf boundary layer conductance is very small or very large (Jarvis and McNaughton 1986). If the leaf boundary layer conductance is very small ($g_b \to 0$), the leaf is decoupled from free air by a thick boundary layer and

$$\lambda E = \frac{s R_n}{s + 2c_p / \lambda}. \tag{10.48}$$

This is known as the equilibrium rate. Latent heat flux is independent of stomatal conductance and depends on the net radiation available to evaporate water. If the boundary layer conductance is very large ($g_b \to \infty$), there is strong coupling between the leaf surface and free air, and latent heat flux is at a rate imposed by stomatal conductance, whereby

$$\lambda E = \lambda [q_{sat}(T_a) - q_a] g_{sw}. \tag{10.49}$$

In this case, latent heat flux depends on vapor pressure deficit, and an increase or decrease in stomatal conductance causes a proportional increase or decrease in transpiration.

Several leaf traits provide control of leaf temperature. The degree of coupling between a leaf and surrounding air depends on leaf size and wind speed. At moderate breezes, a small leaf, with high boundary layer conductance, is closely coupled to the air and has a temperature similar to that of air. A large leaf, with lower boundary layer conductance, is decoupled from the air, and leaf temperature can be several degrees above air temperature (see Figure 12.10). A high heat capacity, generally associated with thick leaves but more precisely related to leaf mass per area and water content, is another important leaf trait. Thick leaves with high water content have a larger heat capacity than thin leaves or leaves with low water content. The large heat capacity of thick leaves can dampen leaf temperature fluctuations compared with thin leaves and prevent heat stress from excessively high temperatures (Ball et al. 1988; Vogel 2009; Leigh et al. 2012; Michaletz et al. 2015, 2016). For this reason, Michaletz et al. (2015, 2016) argued that leaf mass per area is an important leaf trait that affects not only the carbon economics of leaves, but also their energetics and thermoregulation. Other analyses find minimal thermal protection in thin leaves with low heat capacity (Schymanski et al. 2013). A high leaf reflectance decreases absorption of solar radiation, and many desert plants have white coatings on their leaves to reduce temperature and prevent heat damage. An additional factor is leaf inclination angle. A larger inclination angle (relative to horizontal) decreases the solar radiation incident on the leaf, as discussed further in Chapter 14.

Evaporative cooling is necessary to avoid excessively high temperature in leaves exposed to full sunlight. Stomatal conductance, through its regulation of transpiration, is critical to leaf temperature. Current models of stomatal conductance link stomatal functioning with photosynthesis. As the rate of photosynthesis increases, stomatal conductance also increases, as modified by vapor pressure deficit and CO_2 concentration. This produces a tightly coupled system of equations governing leaf energy fluxes, leaf gas exchange, and leaf temperature. Subsequent chapters develop the equations for photosynthesis (Chapter 11) and stomatal conductance (Chapter 12).

Question 10.8 Use the Penman–Monteith equation for a hypostomatous leaf to calculate λE, H, and $T_\ell - T_a$ for each of the following leaf size, stomatal conductance, and net radiation: $d_\ell =$ 3 and 10 cm; $g_{sw} = 0.01$ and 0.4 mol m^{-2} s^{-1}; $R_n = 250$ and 750 W m^{-2}. Under what conditions is a small leaf favored? What conditions favor a large leaf? In a sunny environment, which leaf is favored (based on temperature)? How does leaf temperature differ between the small and large leaf in a shaded environment? How does soil moisture (mesic or dry environment) affect these conclusions? Use $u =$ 3 m s^{-1}, $T_a = 25°C$, and relative humidity is 75%.

Question 10.9 Boundary layer conductance increases with wind speed, but Grace (1981) noted that an increase in wind speed can, in many conditions, decrease transpiration rather than give the expected increase. Explain how this can occur.

10.7 | Supplemental Programs

10.1 Leaf Boundary Layer Conductance: This program calculates leaf boundary layer conductance for heat, water vapor, and CO_2 for forced convection (both laminar and turbulent flow) and free convection (laminar flow). It was used for Figure 10.4.

10.2 Leaf Temperature: This program calculates leaf temperature for different environmental forcings, leaf dimension, leaf absorptance, and stomatal conductance. It was used for Table 10.2.

10.8 | Modeling Projects

1. Many models only represent laminar forced convection when simulating boundary layer

conductance. (a) Use Supplemental Program 10.2 to repeat the calculation of leaf temperature and latent heat flux in Table 10.2. (b) Then, redo the calculations using only laminar forced convection. How do leaf temperature and latent heat flux differ? (c) Many models additionally assume that $g_{bw} = g_{bh}$. Repeat (b), but using this approximation.

2. Equation (10.24) is the linearized form of (10.13). Repeat the calculation of leaf temperature and latent heat flux in Table 10.2 using the linear form. How do these results compare with the solution of the nonlinear equation?

3. Key leaf parameters in the energy budget are ε_ℓ, d_ℓ, and b_1. For radiative forcing $Q_a = 1444$ W m^{-2} and 1136 W m^{-2}, repeat the calculations in Table 10.2 for $\pm 10\%$ variation in these three parameters. Which parameter has the greatest effect on leaf temperature? How do these changes in leaf temperature compare with the results of the previous project?

4. The size and reflectance of leaves are two leaf characteristics that control leaf temperature. Calculate leaf temperature in a hot sunny environment with and without stomata. Use a range of values for d_ℓ and absorptance.

Leaf Photosynthesis

Chapter Overview

Photosynthetic CO_2 assimilation provides the carbon input to ecosystems from the atmosphere and also regulates leaf temperature and transpiration through stomatal conductance. Current representations of these processes in terrestrial biosphere models link the dependencies of leaf biophysics with the biochemistry of photosynthesis. This chapter develops the physiological foundation and mathematical equations to describe photosynthesis for C_3 and C_4 plants. While the equations that describe C_3 photosynthesis are fairly well understood, key parameters such as $V_{c\,max}$ and J_{max} are less well known. Moreover, terrestrial biosphere models can differ greatly in how they implement the photosynthesis equations. C_4 photosynthesis is less well understood for global models.

11.1 C_3 Photosynthesis

Data from Dang et al. (1997a,b, 1998) illustrate environmental controls of photosynthesis for jack pine, a tree species utilizing the C_3 photosynthetic pathway (Figure 11.1). CO_2 uptake during photosynthesis increases with greater amounts of photosynthetically active radiation until light saturation. With more light beyond this amount, factors other than light availability limit photosynthesis. Photosynthetic rates increase with higher concentration of

CO_2 in the air up to a saturation point. Photosynthetic uptake of CO_2 decreases at temperature above or below an optimal value, decreases as the leaf becomes desiccated (as leaf water potential decreases), and declines with lower foliage nitrogen.

Farquhar et al. (1980) formulated a widely used mathematical model of leaf photosynthesis for plants utilizing the C_3 photosynthetic pathway (hereafter denoted FvCB). In C_3 plants, ribulose-1,5-bisphosphate (RuBP), a 5-carbon compound, and CO_2 combine to form two 3-carbon compounds called phosphoglycerate (PGA) as the initial products of photosynthesis. The enzyme RuBP carboxylase/oxygenase (Rubisco) catalyzes this reaction. Each PGA reduces to a 3-carbon compound in reactions fueled by NADPH (nicotinamide adenine dinucleotide phosphate) and ATP (adenosine triphosphate) produced during electron transport. Some of the resultant product is used in carbohydrate production, but most is used with ATP to regenerate RuBP. These reactions consume CO_2. However, Rubisco also fixes O_2 to RuBP, which releases CO_2 in a light-enhanced process known as photorespiration. This occurs because Rubisco has dual affinity for CO_2 and O_2; it catalyzes CO_2 fixation by RuBP, but also catalyzes oxidation of RuBP by O_2. The FvCB model (see also Farquhar et al. 1980; Farquhar and von Caemmerer 1982; von Caemmerer 2000, 2013; von Caemmerer et al. 2009; Diaz-Espejo et al. 2012; Bernacchi et al. 2013) summarizes the biochemistry of photosynthesis into a set of equations that describe the kinetic properties of Rubisco (its carboxylation and

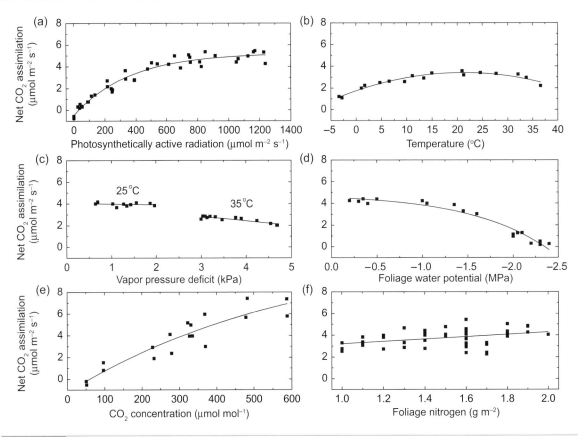

Figure 11.1 Photosynthesis in relation to (a) photosynthetically active radiation, (b) temperature, (c) vapor pressure deficit at 25°C and 35°C, (d) foliage water potential, (e) ambient CO_2 concentration, and (f) foliage nitrogen for jack pine (*Pinus banksiana*) trees. Data from Dang et al. (1997a,b, 1998). Reproduced from Bonan (2016)

oxygenation of RuBP), the ratio of CO_2 uptake during carboxylation to CO_2 loss during oxygenation (photorespiration), the regeneration of RuBP in response to the supply of NADPH and ATP produced during electron transport, the rate of carboxylation when RuBP is saturated, and the rate of carboxylation when RuBP is limited by regeneration via electron transport.

Oxygenation of 1 mol of RuBP releases 0.5 mol CO_2 so that the net rate of CO_2 assimilation per unit leaf area A_n is the balance between CO_2 uptake during carboxylation V_c, CO_2 loss during oxygenation (equal to $0.5V_o$), and CO_2 loss from mitochondrial respiration R_d (often called dark respiration). This is represented by the equation

$$A_n = V_c - 0.5V_o - R_d, \tag{11.1}$$

in which all fluxes commonly have the units μmol CO_2 m^{-2} s^{-1}. In this equation, V_c is the rate of

carboxylation by Rubisco, which follows the Michaelis–Menten response function given by (4.2) with

$$V_c = \frac{V_{c\,max}\,c_i}{c_i + K_c(1 + o_i/K_o)}. \tag{11.2}$$

The corresponding equation for oxygenation by Rubisco is

$$V_o = \frac{V_{o\,max}\,o_i}{o_i + K_o(1 + c_i/K_c)}. \tag{11.3}$$

In these equations, $V_{c\,max}$ and $V_{o\,max}$ (μmol m^{-2} s^{-1}) are the maximum rates of carboxylation and oxygenation, respectively; c_i (μmol mol^{-1}) and o_i (mmol mol^{-1}) are intercellular CO_2 and O_2 concentration, respectively; and K_c (μmol mol^{-1}) and K_o (mmol mol^{-1}) are the Michaelis–Menten constants for CO_2 and O_2, respectively. The ratio of oxygenation to carboxylation rates is

$$\phi = \frac{V_o}{V_c} = \left(\frac{V_{o\,max}}{V_{c\,max}}\frac{K_c}{K_o}\right)\frac{o_i}{c_i}. \qquad (11.4)$$

Substituting ϕ into (11.1) gives

$$A_n = (1 - 0.5\phi)V_c - R_d. \qquad (11.5)$$

For each carboxylation, ϕ oxygenations occur with a release of 0.5ϕ CO_2 during photorespiration.

The CO_2 compensation point is defined as the value of c_i at which no net CO_2 uptake occurs. In the absence of mitochondrial respiration, $A_n = 0$ when $\phi = 2$. Substituting $\phi = 2$ into (11.4), the corresponding value for c_i is the CO_2 compensation point in the absence of mitochondrial respiration. This value of c_i is denoted Γ_* (μmol mol^{-1}) and is equal to

$$\Gamma_* = 0.5\left(\frac{V_{o\,max}}{V_{c\,max}}\frac{K_c}{K_o}\right)o_i. \qquad (11.6)$$

Comparing (11.4) and (11.6) shows that $\phi = 2\Gamma_*/c_i$, and from (11.5),

$$A_n = \left(1 - \frac{\Gamma_*}{c_i}\right)V_c - R_d. \qquad (11.7)$$

The term $1 - \Gamma_*/c_i$ accounts for CO_2 release during photorespiration. It is common to define $S_{c/o}$ as the Rubisco specificity factor equal to

$$S_{c/o} = \frac{V_{c\,max}}{V_{o\,max}}\frac{K_o}{K_c} \qquad (11.8)$$

so that

$$\Gamma_* = 0.5\frac{o_i}{S_{c/o}}. \qquad (11.9)$$

The CO_2 compensation point in the presence of R_d is denoted Γ (μmol mol^{-1}) and is obtained by

$$\Gamma = \frac{\Gamma_* + K_c(1 + o_i/K_o)R_d/V_{c\,max}}{1 - R_d/V_{c\,max}}. \qquad (11.10)$$

In the simplest form of the model, carboxylation V_c is limited by either the activity of Rubisco, denoted by the rate w_c, or by the regeneration of RuBP, denoted by the rate w_j, so that

$$A_n = \left(1 - \frac{\Gamma_*}{c_i}\right)\min(w_c, w_j) - R_d. \qquad (11.11)$$

Equation (11.2) describes w_c, the Rubisco-limited carboxylation rate when RuBP is saturated, in which

$$w_c = \frac{V_{c\,max}c_i}{c_i + K_c(1 + o_i/K_o)}. \qquad (11.12)$$

The rate w_j is the carboxylation rate allowed by the capacity to regenerate RuBP via electron transport, commonly given by

$$w_j = \frac{Jc_i}{4c_i + 8\Gamma_*} \qquad (11.13)$$

or

$$w_j = \frac{Jc_i}{4.5c_i + 10.5\Gamma_*}, \qquad (11.14)$$

depending on NADPH (11.13) or ATP (11.14) requirements for photosynthesis. In these equations, J is the electron transport rate (μmol m^{-2} s^{-1}) for a given irradiance.

The derivation of w_j comes from consideration of requirements for regeneration of RuBP. With ϕ oxygenations occurring per carboxylation, the rate of RuBP regeneration required to satisfy RuBP consumption is

$$\text{RuBP regeneration rate} = (1 + \phi)V_c. \qquad (11.15)$$

Regeneration of RuBP depends on the supply of NADPH or ATP produced from electron transport. Equation (11.13) is derived using NADPH requirements. Carboxylation of 1 mol RuBP produces 2 mol PGA, which is used to regenerate RuBP with NADPH and ATP. Oxygenation of 1 mol RuBP produces an additional 1.5 mol PGA. Consequently,

$$\text{PGA production rate} = (2 + 1.5\phi)V_c. \qquad (11.16)$$

One mol NADPH is required for each mol PGA produced, with an additional requirement of 0.5 mol NADPH during the oxygenation of 1 mol RuBP, so that

$$\text{NADPH consumption rate} = (2 + 2\phi)V_c. \qquad (11.17)$$

Reduction of NADP$^+$ to NADPH requires two electrons so that the rate of electron transport needed to satisfy the NADPH consumption is

$$J = (4 + 4\phi)V_c = (4 + 8\Gamma_*/c_i)V_c. \qquad (11.18)$$

Equation (11.18) defines J, the rate of electron transport that satisfies RuBP regeneration. The corresponding value of V_c for that J is the carboxylation rate allowed by the capacity to regenerate RuBP via electron transport – i.e., $w_j = V_c = Jc_i/(4c_i + 8\Gamma_*)$, as given by (11.13). Implicit in this equation is the assumption that four electrons must pass through electron transport to regenerate each RuBP. Equation (11.14) is derived from ATP requirements in which

$$\text{ATP consumption rate} = (3 + 3.5\phi)V_c. \qquad (11.19)$$

If three H^+ are required to synthesize each ATP ($3H^+$/ ATP) and if their production is by electron transport alone with one electron yielding two H^+ ($1e^-/2H^+$), then the electron transport rate needed to meet the ATP production is

$$J = (4.5 + 5.25\phi)V_c = (4.5 + 10.5\Gamma_*/c_i)V_c. \qquad (11.20)$$

The value of V_c for this rate of electron transport is the w_j in (11.14).

The rate of electron transport is related to absorbed photosynthetically active radiation. A common expression is

$$\Theta_J J^2 - (I_{PSII} + J_{\max})J + I_{PSII}J_{\max} = 0, \qquad (11.21)$$

where $J_{\max}$ (μmol m^{-2} s^{-1}) is the maximum electron transport rate, I_{PSII} (μmol m^{-2} s^{-1}) represents the light utilized in electron transport by photosystem II, and Θ_J is a curvature parameter (von Caemmerer 2000, 2013; Bernacchi et al. 2003, 2013; von Caemmerer et al. 2009). The amount of light utilized by photosystem II is directly related to the photosynthetically active radiation incident on a leaf $I^\downarrow$ (μmol photon m^{-2} s^{-1}), but two different expressions are commonly given for I_{PSII}. In one,

$$I_{PSII} = \frac{(1-f)}{2}\alpha_\ell I^\downarrow, \qquad (11.22)$$

in which α_ℓ is leaf absorptance, $f = 0.15$ corrects for the spectral quality of light, and division by two arises because one photon must be absorbed by each of the two photosystems to move one electron (von Caemmerer 2000, 2013; von Caemmerer et al. 2009). An alternative expression is

$$I_{PSII} = \frac{\Phi_{PSII}}{2}\alpha_\ell I^\downarrow. \qquad (11.23)$$

The term Φ_{PSII} (mol mol^{-1}) is the quantum yield of photosystem II and only one-half of the absorbed light reaches photosystem II (Bernacchi et al. 2003, 2013). The quantum yield of photosystem II varies with temperature, but a representative value at 25°C is $\Phi_{PSII} = 0.7$. One-half is a representative value for the fraction of light that reaches photosystem II, but this can vary somewhat. Both equations are

functionally equivalent and differ only in that $1 - f = 0.85$ or $\Phi_{PSII} = 0.7$. Equation (11.21) provides co-limitation of electron transport between the light-limited rate I_{PSII} and the maximum rate $J_{\max}$ as described by (4.19). The electron transport rate is the smaller root of the quadratic equation so that

$$J = \frac{(I_{PSII} + J_{\max}) - \left[(I_{PSII} + J_{\max})^2 - 4\Theta_J I_{PSII}J_{\max}\right]^{1/2}}{2\Theta_J}. \qquad (11.24)$$

At low light, $J \approx 0.5\Phi_{PSII}\alpha_\ell I^\downarrow$ and the term $0.5\Phi_{PSII}\alpha_\ell$ is also the initial slope of the electron transport rate at low light.

Subsequent versions of the model introduced a third rate limited by the capacity to utilize the products of photosynthesis (triose phosphate) in synthesis of starch and sugar (Harley and Sharkey 1991; Harley et al. 1992; see also reviews by von Caemmerer 2000, 2013; von Caemmerer et al. 2009; Diaz-Espejo et al. 2012; Bernacchi et al. 2013). The rate of triose phosphate utilization T_p (μmol m^{-2} s^{-1}) ranges from $V_c/3 - V_o/6$ to $V_c/3 - V_o/6 - V_o/2$. A general expression is

$$T_p = \frac{V_c}{3} - \frac{V_o}{6} - B\frac{V_o}{2} = V_c\left(\frac{1}{3} - \frac{\phi}{6} - B\frac{\phi}{2}\right), \qquad (11.25)$$

where B is a coefficient that ranges from zero to one. The corresponding value for V_c defines w_p, the carboxylation rate limited by triose phosphate utilization. With $\phi = 2\Gamma_*/c_i$,

$$w_p = \frac{3T_p c_i}{c_i - \Gamma_* - 3B\Gamma_*}. \qquad (11.26)$$

In the simplest case, $B = 0$ and the product-limited assimilation rate is $A_p = (1 - \Gamma_*/c_i)w_p = 3T_p$. This rate of photosynthesis is insensitive to changes in CO_2 or O_2. Other formulations of triose phosphate utilization allow such interaction.

In summary, a common form of the FvCB model represents photosynthetic CO_2 assimilation for plants utilizing the C_3 photosynthetic pathway as limited by (i) A_c – the rate of Rubisco catalyzed carboxylation when RuBP is saturated (called Rubisco-limited photosynthesis because of its dependence on maximum Rubisco activity as set by $V_{c\max}$); and (ii) A_j – the rate of RuBP regeneration by light absorption and electron transport as determined by $J_{\max}$ (RuBP regeneration-limited, or light-limited, photosynthesis). Use of the product-limited rate A_p is less common, and there is less knowledge of T_p and its environmental dependencies (Lombardozzi et al. 2018). The net photosynthesis rate is

$$A_n = \left(1 - \frac{\Gamma_*}{c_i}\right) \min\left(w_c, w_j\right) - R_d. \qquad (11.27)$$

The gross photosynthesis rate when Rubisco is limiting is

$$A_c = \frac{V_{c\max}\left(c_i - \Gamma_*\right)}{c_i + K_c(1 + o_i/K_o)}, \qquad (11.28)$$

and the rate when RuBP regeneration is limiting is

$$A_j = \frac{J}{4}\left(\frac{c_i - \Gamma_*}{c_i + 2\Gamma_*}\right). \qquad (11.29)$$

In practice, (11.27) is commonly evaluated as

$$A_n = \min\left(A_c, A_j\right) - R_d. \qquad (11.30)$$

Use of the minimum rate in (11.30) produces an abrupt transition between the Rubisco-limited and RuBP regeneration-limited rates. This is particularly evident in photosynthetic light response curves or A_n-c_i curves, which pass through the transition point. In fact, leaf gas exchange data show a smooth transition, which relates to spatial heterogeneity in $V_{c\max}$ and $J_{\max}$ within a leaf (Chen et al. 2008). Furthermore, individual chloroplasts within a leaf may be exposed to different light and CO_2 regimes, and so both A_c and A_j can co-occur. Co-limitation of A_c and A_j smooths the transition (Collatz et al. 1990), and, as discussed in Chapter 4, can be achieved using the smaller root of the quadratic equation

$$\Theta_A A^2 - \left(A_c + A_j\right)A + A_c A_j = 0, \qquad (11.31)$$

where A is the co-limited assimilation rate and Θ_A is an empirical parameter that governs the transition between limitations. A representative value is $\Theta_A = 0.98$ (Collatz et al. 1990). A value of one gives results identical to the minimum rate. Here, co-limitation is applied to the gross assimilation rates, not the net rates (Collatz et al. 1990, 1991). A property of (11.31) is that $A < \min\left(A_c, A_j\right)$ for $\Theta_A < 1$, as discussed in Chapter 4. Co-limitation allows for a smooth, continuous photosynthetic response, which is useful when modeling stomatal conductance (Chapter 12).

Collatz et al. (1991) introduced a variant of the model, used by Sellers et al. (1996a,b) in SiB2. This model has an alternative equation for w_j in which the term $J/4$ in (11.13) is replaced with a quantum yield E (mol CO_2 mol^{-1} photon) that describes CO_2 uptake per absorbed photon so that

$$w_j = E\alpha_\ell I^\downarrow \left(\frac{c_i}{c_i + 2\Gamma_*}\right). \qquad (11.32)$$

In this variant, $J_{\max}$ never limits RuBP regeneration. Collatz et al. (1991) also represented the product-limited rate by $A_p = 0.5 V_{c\max}$, which implies that $T_p = 0.167 V_{c\max}$. Co-limitation among A_c, A_j, and A_p is given by the smaller roots of two quadratic equations,

$$\begin{aligned} 0.98A_i^2 - \left(A_c + A_j\right)A_i + A_c A_j = 0 \\ 0.95A^2 - \left(A_i + A_p\right)A + A_i A_p = 0 \end{aligned}, \qquad (11.33)$$

where A_i represents the co-limitation between A_c and A_j.

Figure 11.2 illustrates the behavior of the model in relation to CO_2 concentration and light. The C₃ photosynthetic response to c_i for saturating light shows that A_n increases with greater c_i up to some maximum rate (Figure 11.2a). At low c_i and high irradiance, photosynthesis is limited by the efficiency of carboxylation (i.e., the amount of Rubisco) because RuBP levels are saturating, and CO_2 assimilation follows the rate A_c. At high c_i, regeneration of RuBP limits photosynthesis, and CO_2 assimilation follows the rate A_j. Photosynthesis increases with greater c_i, but the increase is less than for the A_c-limited rate and dA_n/dc_i approaches zero at very high c_i. As c_i continues to increase, triose phosphate utilization (defined by A_p) limits photosynthesis.

In C₃ plants, the rate of photosynthesis increases with greater irradiance up to some saturated rate (Figure 11.2b). At low light, regeneration of RuBP is insufficient, and photosynthesis follows the rate A_j. Initially, CO_2 loss during respiration exceeds CO_2 uptake during photosynthesis until irradiance increases to the light compensation point. In Figure 11.2b, the light compensation point is about 20 μmol photon m^{-2} s^{-1}. Typical values for C₃ plants are 20–40 μmol photon m^{-2} s^{-1}. The rate of photosynthesis initially increases linearly with greater amounts of light. At higher light levels, the rate of electron transport with respect to irradiance becomes more curvilinear, and A_j begins to deviate from a linear light response. With even higher light, the supply of RuBP no longer limits photosynthesis. Instead, photosynthesis is limited by A_c.

Question 11.1 Describe the difference in A_j between the FvCB C_3 photosynthesis model as originally described by Farquhar et al. (1980) and its implementation in SiB2 (Sellers et al. 1996a,b).

Question 11.2 Most terrestrial biosphere models do not include triose phosphate utilization T_p when calculating C_3 photosynthesis. Discuss the pros and cons of including T_p-limitation in a model.

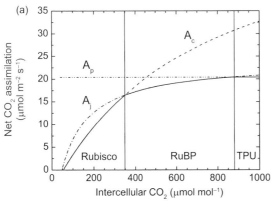

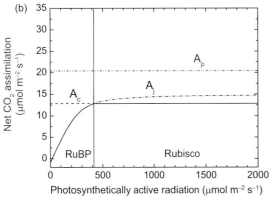

Figure 11.2 C_3 photosynthesis in relation to (a) intercellular CO_2 (at $I^\downarrow = 2000$ μmol m^{-2} s^{-1}) and (b) photosynthetically active radiation (at $c_i = 266$ μmol mol^{-1}). Shown are the Rubisco-, RuBP regeneration, and product-limited rates and the minimum rate (solid line). For this example, $V_{c\max} = 60$ μmol m^{-2} s^{-1}, $J_{\max} = 100.2$ μmol m^{-2} s^{-1}, $T_p = 7.1$ μmol m^{-2} s^{-1}, and $R_d = 0.9$ μmol m^{-2} s^{-1}. Rubisco parameters are from Bernacchi et al.(2001) (Table 11.1), and electron transport parameters are $\Theta_J = 0.9$, $\Phi_{PSII} = 0.85$, and $\alpha_\ell = 0.8$.

11.2 | Parameter Values and Temperature Dependencies

The FvCB model requires six physiological parameters: K_c, K_o, Γ_*, $V_{c\max}$, $J_{\max}$, and R_d. Additionally, the specification of electron transport in (11.21) requires Θ_J, Φ_{PSII}, and α_ℓ, and co-limitation requires the curvature factor Θ_A. Values for K_c, K_o, and Γ_* are defined by the biochemistry of Rubisco and are similar among species. Table 11.1 provides representative values at 25°C. Von Caemmerer (2000, 2013) and von Caemmerer et al. (2009) suggested values of $\Theta_J = 0.7$ and $\Phi_{PSII} = 0.85$ for electron transport (i.e., $1 - f = 0.85$), but some models use $\Theta_J = 0.9$ (Medlyn et al. 2002). The analyses of Bernacchi et al. (2003) suggest that $\Phi_{PSII} \approx 0.7$ and support the larger value for Θ_J. Both parameters vary with temperature (Bernacchi et al. 2003), though this dependency is generally omitted from models. Leaf absorptance for photosynthetically active radiation is commonly $\alpha_\ell = 0.8$.

The parameters K_c, K_o, Γ_*, $V_{c\max}$, $J_{\max}$, and R_d vary with temperature. The instantaneous responses of photosynthesis and respiration to temperature are driven by their underlying enzymatic

Table 11.1	Representative photosynthetic parameters for C_3 plants
Parameter	Value at 25°C
Collatz et al. (1991), Sellers et al. (1996a,b) [a]	
K_c	30 Pa
K_o	30,000 Pa
Γ_*	4.02 Pa
von Caemmerer et al. (1994), von Caemmerer (2000)	
K_c	40.4 Pa
K_o	24,800 Pa
Γ_*	3.69 Pa
Bernacchi et al. (2001) [b]	
K_c	404.9 μmol mol^{-1} (40.49 Pa)
K_o	278.4 mmol mol^{-1} (27,840 Pa)
Γ_*	42.75 μmol mol^{-1} (4.275 Pa)

Note: (a) Γ_* from (11.9) with $o_i = 20,900$ Pa and $S_{c/o} = 2600$. (b) Multiply by surface pressure (100 kPa) to convert from units of mol mol^{-1} to Pa.

responses. When warmed from low temperature, the enzymes involved in photosynthesis and respiration increase their activity as described by the Arrhenius function (4.6). Normalized to 25°C (298.15 K), this function is

$$f(T_\ell) = \exp\left[\frac{\Delta H_a}{298.15\Re}\left(1 - \frac{298.15}{T_\ell}\right)\right], \qquad (11.34)$$

where T_ℓ is leaf temperature (K), $\Re$ is the universal gas constant (8.314 J K^{-1} mol^{-1}), and ΔH_a is the activation energy (J mol^{-1}). Parameter values measured at 25°C are multiplied by (11.34) to obtain the temperature-adjusted value. Another common expression is the Q_{10} temperature function (4.7). Table 11.2 provides activation energy and Q_{10} values from various studies.

Table 11.2	Representative photosynthetic temperature parameters

Parameter	ΔH_a (J mol^{-1})	Q_{10} (25°C)
Collatz et al. (1991), Sellers et al. (1996a,b)		
K_c	–	2.1
K_o	–	1.2
Γ_*	–	1.75
$V_{c\,max}$	–	2.4 (2.0)
R_d	–	2.0
Farquhar et al. (1980), von Caemmerer (2000)		
K_c	59,356	2.23
K_o	35,948	1.63
Γ_*	23,400	1.37
$V_{c\,max}$	58,520	2.21
R_d	66,405	2.46
J_{max}	37,000	1.65
Bernacchi et al. (2001, 2003)		
K_c	79,430	2.93
K_o	36,380	1.64
Γ_*	37,830	1.67
$V_{c\,max}$	65,330	2.42
R_d	46,390	1.87
J_{max}	43,540	1.80
Leuning (2002)		
$V_{c\,max}$	73,637	2.71
J_{max}	50,300	1.97

Note: Sellers et al. (1996a,b) gave a Q_{10} of 2.0 for $V_{c\,max}$. Where the Arrhenius function is used, the equivalent Q_{10} at 25°C is calculated.

Bernacchi et al. (2001, 2003) estimated ΔH_a for transgenic tobacco plants, and Figure 11.3 shows the temperature responses. For $T_\ell < 35$°C, the Arrhenius and Q_{10} functions are very similar.

The parameters $V_{c\,max}$, J_{max}, and R_d vary with temperature following the Arrhenius function but have a peaked response in which enzymatic activity increases up to a temperature optimum beyond which the reaction rate decreases (Figure 11.3). This temperature response is represented by the peaked Arrhenius function, given for $V_{c\,max}$ by

$$V_{c\,max} = V_{c\,max\,25}f(T_\ell)f_H(T_\ell). \qquad (11.35)$$

The function $f_H(T_\ell)$ represents thermal breakdown of biochemical processes in which

$$f_H(T_\ell) = \frac{1 + \exp\left(\dfrac{298.15\Delta S - \Delta H_d}{298.15\Re}\right)}{1 + \exp\left(\dfrac{\Delta S T_\ell - \Delta H_d}{\Re T_\ell}\right)}, \qquad (11.36)$$

where ΔH_d (J mol^{-1}) is the energy of deactivation and ΔS (J K^{-1} mol^{-1}) is an entropy term (Farquhar et al. 1980; Leuning 2002). The term in the denominator is greater than one for high temperatures, and the term in the numerator scales (11.36) so that $f_H(T_\ell) = 1$ at 25°C (298.15 K). Equation (11.35) similarly applies to J_{max} and R_d. The temperature optimum of (11.35) is obtained from

$$T_{opt} = \frac{\Delta H_d}{\Delta S - \Re \ln\left(\dfrac{\Delta H_a}{\Delta H_d - \Delta H_a}\right)}. \qquad (11.37)$$

Bernacchi et al. (2001, 2003) estimated ΔH_a for $V_{c\,max}$, J_{max}, and R_d for transgenic tobacco plants (Table 11.2). Leuning (2002) derived values of ΔH_a for $V_{c\,max}$ and J_{max} from a synthesis of numerous published studies and demonstrated the generality of the Bernacchi et al. (2001, 2003) results. The temperature response of $V_{c\,max}$ and J_{max} shows little variation among different species for $T_\ell < 30$°C (Leuning 2002). Farquhar et al. (1980) used $\Delta H_d = 220$ kJ mol^{-1} and $\Delta S = 710$ J K^{-1} mol^{-1} for high temperature inhibition. In a synthesis of numerous published studies, Leuning (2002) obtained smaller values for these parameters (Table 11.3) and found that a common temperature response across species is appropriate for $T_\ell < 30$°C. These values provide a functionally similar temperature response as the Farquhar et al. (1980) values.

Values for $V_{c\,max}$, J_{max}, and R_d vary among plants and with environmental conditions, but covary and scale in relation to one another. For example, data from Wullschleger (1993) for 109 species of C_3 plants show that J_{max} increases in relation to $V_{c\,max}$. Correspondingly data show that T_p increases in relation to J_{max}, but estimates of T_p are less common. Leaf

respiration is typically 1%–2% of $V_{c\,max}$ (von Caemmerer 2000), and $R_{d25} = 0.015V_{c\,max\,25}$ at 25°C is frequently used for C_3 plants (Collatz et al. 1991). The parameters $V_{c\,max}$ and J_{max} are not directly observable but, rather, are estimated by fitting the FvCB model to leaf gas exchange measurements (Sharkey et al. 2007; Gu et al. 2010; Duursma 2015). The derived values depend on the particular form of the model used, the parameter values that describe Rubisco kinetics in the model, and the temperature response functions. Bernacchi et al. (2001) estimated K_{c25}, K_{o25}, and Γ_{*25} at 25°C and their temperature dependence based on A_c, as given by (11.28), and A_j, as given by (11.29). Medlyn et al. (2002) derived $J_{max\,25}/V_{c\,max\,25} = 1.67$ at 25°C using the Bernacchi et al. (2001) Rubisco parameters. This differs from the ratio 1.97 obtained from data in

Table 11.3	Representative photosynthetic temperature inhibition parameters	
Parameter	ΔH_d (J mol^{-1})	ΔS (J K^{-1} mol^{-1})
$V_{c\,max}$	149,250	485
J_{max}	152,040	495

Source: Leuning (2002).

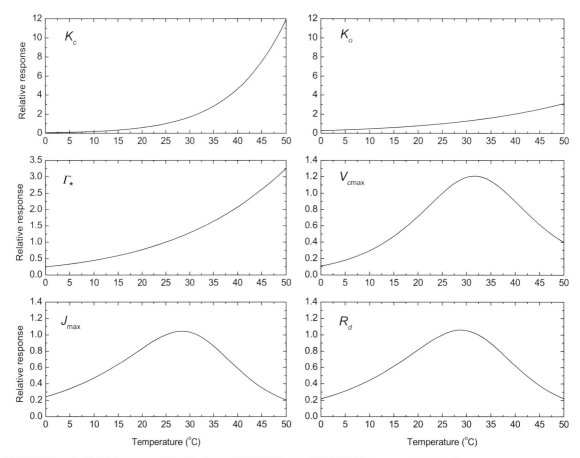

Figure 11.3 Relative temperature response functions using (11.34). Table 11.2 gives ΔH_a (Bernacchi et al. 2001, 2003). Response functions for $V_{c\,max}$, J_{max}, and R_d additionally include thermal breakdown at high temperatures using (11.36) with $\Delta H_d = 150,000$ J mol^{-1} and $\Delta S = 490$ J K^{-1} mol^{-1}.

Table 11.4 | $V_{c\,max}$ (at 25°C) estimated for C_3 plants from leaf trait databases

Plant type	i_v (μmol CO_2 m^{-2} s^{-1})	s_v (μmol CO_2 g^{-1} N s^{-1})	n_a (g N m^{-2})	$V_{c\,max\,25}$ (μmol m^{-2} s^{-1})
Tropical trees (oxisols)	8.90 ± 4.10	9.30 ± 1.62	2.17 ± 0.80	29.0 ± 7.7
Tropical trees (non-oxisols)	4.19 ± 4.89	26.19 ± 4.47	1.41 ± 0.56	41.0 ± 15.1
Temperate broadleaf trees				
Evergreen	5.73 ± 1.83	29.81 ± 1.41	1.87 ± 0.93	61.4 ± 27.7
Deciduous	5.73 ± 1.83	29.81 ± 1.41	1.74 ± 0.71	57.7 ± 21.2
Coniferous trees				
Evergreen	6.32 ± 2.78	18.15 ± 1.48	3.10 ± 1.35	62.5 ± 24.7
Deciduous	6.32 ± 2.78	18.15 ± 1.48	1.81 ± 0.64	39.1 ± 11.7
Shrubs				
Evergreen	14.71 ± 4.61	23.15 ± 3.16	2.03 ± 1.05	61.7 ± 24.6
Deciduous	14.71 ± 4.61	23.15 ± 3.16	1.69 ± 0.62	54.0 ± 14.5
C_3 herbaceous	6.42 ± 4.90	40.96 ± 3.21	1.75 ± 0.76	78.2 ± 31.1
C_3 crops	4.71 ± 5.35	59.23 ± 4.42	1.62 ± 0.61	100.7 ± 36.6

Note: $V_{c\,max\,25} = i_v + s_v n_a$. Shown are the mean ± one standard deviation.
Source: Kattge et al. (2009).

Wullschleger (1993) and 2.00 reported by Leuning (2002), but with different Rubisco kinetics. Kattge et al. (2009) derived $V_{c\,max\,25}$ at 25°C from leaf gas exchange measurements, also with the Bernacchi et al. (2001) parameters (Table 11.4). These studies present a consistent set of parameters for use with the model. However, the form of the model used does not distinguish intercellular CO_2 concentration c_i and chloroplastic CO_2 concentration c_c, essentially assuming infinite conductance between intercellular space and the site of carboxylation. This mesophyll conductance may, in fact, be significantly small, but its inclusion would necessitate new derivation of parameter values.

Question 11.3 Compare the temperature responses of K_c, K_o, and Γ_* to the corresponding Q_{10} function using the Bernacchi et al. (2001, 2003) values in Table 11.2. Then compare responses using SiB2 values (Sellers et al. 1996a,b).

Question 11.4 Calculate the temperature optimum for $V_{c\,max}$ and J_{max}. How does the temperature optimum vary with the different values for ΔH_a given in Table 11.2?

11.3 Photosynthetic Response to Environmental Factors

Table 11.5 summarizes the FvCB photosynthesis model. Environmental inputs for the model are c_i, o_i, $I^{\downarrow}$, and T_ℓ. The intercellular CO_2 concentration in C_3 plants is typically 65%–80% of ambient CO_2 concentration (Hetherington and Woodward 2003). The exact value depends on gas diffusion through stomata. Intercellular O_2 is the same as the atmosphere ($o_i = 209$ mmol mol^{-1}). Photosynthetically active radiation ranges up to 1500–2000 μmol photon m^{-2} s^{-1} for light saturation. Leaf temperature is required because the photosynthetic parameters vary with temperature. The model response to environmental factors (Figure 11.4) is similar to the observations shown in Figure 11.1. While the FvCB model is commonly used in terrestrial biosphere models, the exact implementation can vary greatly among models. The Rubisco kinetics constants, their temperature dependencies, whether A_j saturates with light or not, and the use of co-limitation are critical choices in the model implementation. These choices introduce significant differences among models in photosynthetic response to light,

Table 11.5 Summary equations for the C_3 photosynthesis model

Definition	Equation	Units
Rubisco-limited assimilation	$A_c = \dfrac{V_{c\max}(c_i - \Gamma_*)}{c_i + K_c(1 + o_i/K_o)}$	μmol CO_2 m^{-2} s^{-1}
Light-limited assimilation	$A_j = \dfrac{J}{4}\left(\dfrac{c_i - \Gamma_*}{c_i + 2\Gamma_*}\right)$	μmol CO_2 m^{-2} s^{-1}
Gross photosynthesis	$A = \min(A_c, A_j)$, or	μmol CO_2 m^{-2} s^{-1}
	$\Theta_A A^2 - (A_c + A_j)A + A_c A_j = 0$	
Net photosynthesis	$A_n = A - R_d$	μmol CO_2 m^{-2} s^{-1}
Electron transport rate	$\Theta_J J^2 - (I_{PSII} + J_{\max})J + I_{PSII}J_{\max} = 0$	μmol m^{-2} s^{-1}
PS II light utilization	$I_{PSII} = 0.5\Phi_{PSII}\alpha_\ell I^\downarrow$	μmol m^{-2} s^{-1}
Maximum carboxylation rate	$V_{c\max} = V_{c\max 25}\ f(T_\ell)f_H(T_\ell)$	μmol m^{-2} s^{-1}
Maximum electron transport rate	$J_{\max} = J_{\max 25}\ f(T_\ell)f_H(T_\ell)$	μmol m^{-2} s^{-1}
	$J_{\max 25} = 1.67 V_{c\max 25}$	
Leaf respiration	$R_d = R_{d25}\ f(T_\ell)f_H(T_\ell)$	μmol CO_2 m^{-2} s^{-1}
	$R_{d25} = 0.015 V_{c\max 25}$	
Michaelis–Menten constant, CO_2	$K_c = K_{c25}\ f(T_\ell)$	μmol mol^{-1}
Michaelis–Menten constant, O_2	$K_o = K_{o25}\ f(T_\ell)$	mmol mol^{-1}
CO_2 compensation point	$\Gamma_* = \Gamma_{*25}\ f(T_\ell)$	μmol mol^{-1}
Arrhenius function	$f(T_\ell) = \exp\left[\dfrac{\Delta H_a}{298.15\Re}\left(1 - \dfrac{298.15}{T_\ell}\right)\right]$	dimensionless
High temperature inhibition	$f_H(T_\ell) = \dfrac{1 + \exp\left(\frac{298.15\Delta S - \Delta H_d}{298.15\Re}\right)}{1 + \exp\left(\frac{\Delta S T_\ell - \Delta H_d}{\Re T_\ell}\right)}$	dimensionless

Note: With c_i (μmol mol^{-1}), o_i (mmol mol^{-1}), $I^\downarrow$ (μmol m^{-2} s^{-1}), and T_ℓ (K) specified.

temperature, CO_2 concentration, and other factors (Rogers et al. 2017).

A key outcome of the FvCB model is the photosynthetic response to CO_2 (Figure 11.2a). The amount of Rubisco, as quantified by $V_{c\max}$, and the Rubisco kinetic parameters K_c, K_o, and Γ_* determine the initial slope dA_n/dc_i at low c_i, where A_c (11.28) is the rate limiter. Differentiating $A_n = A_c - R_d$ with respect to c_i gives

$$\frac{dA_n}{dc_i} = \frac{V_{c\max}[\Gamma_* + K_c(1 + o_i/K_o)]}{[c_i + K_c(1 + o_i/K_o)]^2} \qquad (11.38)$$

and, at $c_i = \Gamma_*$,

$$\frac{dA_n}{dc_i} = \frac{V_{c\max}}{\Gamma_* + K_c(1 + o_i/K_o)}. \qquad (11.39)$$

Equation (11.39) provides an estimate of $V_{c\max}$ from experimental measurements of A_n in relation to c_i, but it is evident that this estimate depends on the values for K_c, K_o, and Γ_*. The transition from A_c to A_j as the limiting rate also determines the shape of the A_n-c_i curve. The value of c_i at which this transition occurs can be found when $A_c = A_j$ in (11.28) and (11.29). This value is

$$c_i^* = \left[K_c\left(1 + \frac{o_i}{K_o}\right)\frac{J}{4V_{c\max}} - 2\Gamma_*\right]\left[1 - \frac{J}{4V_{c\max}}\right]^{-1}. \qquad (11.40)$$

Photosynthesis is limited by A_c for $c_i < c_i^*$ and by A_j for $c_i > c_i^*$. For the example shown in Figure 11.2a, this transition occurs at $c_i \approx 350$ μmol mol^{-1}, but this value varies considerably among terrestrial biosphere models (Rogers et al. 2017).

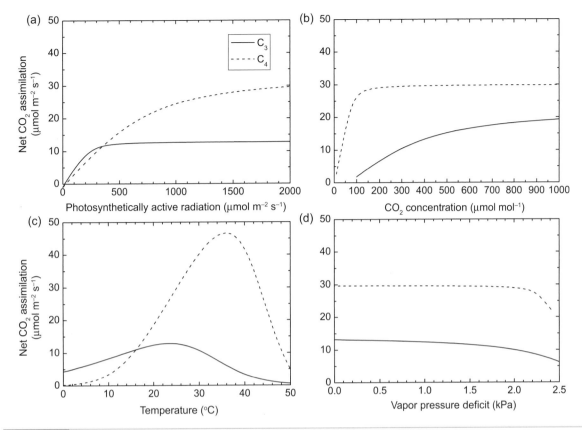

Figure 11.4 Comparison of C_3 and C_4 photosynthesis in response to (a) photosynthetically active radiation, (b) ambient CO_2 concentration, (c) leaf temperature, and (d) vapor pressure deficit. In this figure, stomatal conductance is calculated using the Ball–Berry model and c_i is obtained from the diffusion equation (Chapter 12).

The photosynthetic response to light (Figure 11.2b) is critical to scaling photosynthesis from an individual leaf to a canopy consisting of many leaves exposed to different light environments. The initial slope of the light response is the quantum yield (mol CO_2 fixed per mol quanta absorbed). The quantum yield can be theoretically derived at low irradiance, where A_j (11.29) is the rate limiter (Farquhar and von Caemmerer 1982; von Caemmerer 2000). At low light, the electron transport rate, given by (11.21) and (11.23), can be approximated as $J = 0.5\Phi_{PSII}\alpha_\ell I^\downarrow$ so that the net photosynthesis rate is

$$A_n = \left(\frac{c_i - \Gamma_*}{4c_i + 8\Gamma_*}\right)\frac{\Phi_{PSII}}{2}\alpha_\ell I^\downarrow - R_d. \qquad (11.41)$$

Differentiating A_n with respect to irradiance gives

$$E = \frac{dA_n}{dI^\downarrow} = \left(\frac{c_i - \Gamma_*}{8c_i + 16\Gamma_*}\right)\alpha_\ell \Phi_{PSII}. \qquad (11.42)$$

At high CO_2 (as $c_i \to \infty$) or low O_2 (as $\Gamma_* \to 0$), $c_i \gg \Gamma_*$ so that the effects of photorespiration are excluded and the quantum yield is $E = \alpha_\ell \Phi_{PSII}/8$. The maximum theoretical quantum yield (obtained with $\Phi_{PSII} = 1$ and $\alpha_\ell = 1$) is 0.125 mol mol^{-1}. The realized quantum yield is 0.090 mol mol^{-1} with $\Phi_{PSII} = 0.85$ and $\alpha_\ell = 0.85$, and is 0.074 mol mol^{-1} with $\Phi_{PSII} = 0.7$. Long et al. (1993) obtained a quantum yield equal to 0.093 mol mol^{-1} in experimental studies at ambient CO_2 and low O_2. A literature review finds an average quantum yield of 0.096 mol mol^{-1} for C_3 plants in non-photorespiratory conditions (Skillman 2008). Values used in terrestrial biosphere models have considerable variability (Rogers et al. 2017).

The transition from light-limited photosynthesis to light-saturated photosynthesis occurs over a range of light levels. Light saturation for C_3 plants typically occurs at 500–1000 μmol photon m^{-2} s^{-1}.

The transition is found when $A_c = A_j$ in (11.28) and (11.29). The rate of electron transport at which this occurs is

$$J^* = \left[\frac{4c_i + 8\Gamma_*}{c_i + K_c(1 + o_i/K_o)} \right] V_{c\,max} = B\ V_{c\,max}. \quad (11.43)$$

Photosynthesis is limited by A_j for $J < J^*$ and by A_c for $J > J^*$. From (11.21) and (11.23), the irradiance at which this occurs is

$$I^{\downarrow *} = \left(\frac{BV_{c\,max}}{0.5\Phi_{PSII}\alpha_\ell} \right) \left(\frac{\Theta_J B - J_{max}/V_{c\,max}}{B - J_{max}/V_{c\,max}} \right). \quad (11.44)$$

A_j limits photosynthesis for $I^\downarrow < I^{\downarrow *}$, and A_c limits photosynthesis for $I^\downarrow > I^{\downarrow *}$ (Wang 2000). For the example shown in Figure 11.2b, $I^{\downarrow *} = 416$ μmol photon m^{-2} s^{-1} (with $\Theta_J = 0.9$ and $\Phi_{PSII} = 0.85$). Light saturation is sensitive to parameters. A lower curvature parameter ($\Theta_J = 0.7$) gives a shallower increase in A_j with higher light, and light saturation occurs at $I^{\downarrow *} = 738$ μmol photon m^{-2} s^{-1} (Figure 11.5a). A higher ratio $J_{max}/V_{c\,max}$ decreases $I^{\downarrow *}$. Consequently, the parameterization of electron transport is an important determinant of photosynthesis both at the leaf scale and for plant canopies, where irradiance decreases with cumulative depth in the canopy and photosynthesis becomes light-limited deeper in the canopy. The irradiance at which transition from A_j limitation to A_c limitation occurs varies considerably among terrestrial biosphere models (Rogers et al. 2017).

The shape of the light response curve depends on whether the minimum or co-limited rate is used (Figure 11.5b). With co-limitation as in (11.31), the actual photosynthetic rate is less than the minimum rate for a curvature factor $\Theta_A < 1$ (Sellers et al. 1992; Bonan et al. 2011). A common value used in models is $\Theta_A = 0.98$ (Collatz et al. 1990), which smooths the transition between the A_j and A_c limitations but gives $A < \min(A_c, A_j)$. A value $\Theta_A = 0.995$ also smooths the transition and with less of a reduction in the actual rate.

Question 11.5 Calculate c_i^* for the A_n-c_i curve shown in Figure 11.2a. How do the different estimates for K_c, K_o, and Γ_* in Table 11.1 affect c_i^*? How do different estimates for $J_{max}/V_{c\,max}$ (1.67, 2.0) affect c_i^*?

Question 11.6 Calculate $I^{\downarrow *}$ for the light response curve shown in Figure 11.2b. How do the different estimates for K_c, K_o, and Γ_* in Table 11.1 affect $I^{\downarrow *}$? How do different estimates for $J_{max}/V_{c\,max}$ (1.67, 2.0) affect $I^{\downarrow *}$?

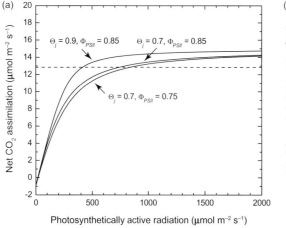

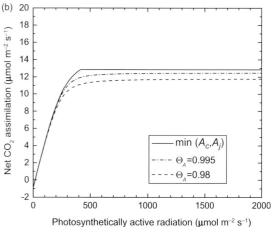

Figure 11.5 Sensitivity of photosynthetic light response to parameter values. Calculations are as in Figure 11.2b. (a) Shown is photosynthesis for different values of Θ_J and Φ_{PSII}. The dashed line shows the Rubisco-limited rate A_c, and the solid lines show three different RuBP regeneration-limited rates A_j. (b) Shown is the realized photosynthetic rate with different values of co-limitation. The three lines are the minimum of the A_c and A_j rates (equivalent to $\Theta_A = 1$) and the co-limited rate with $\Theta_A = 0.995$ and 0.98.

11.4 | Use of V_{cmax} in Models

$V_{c\max}$ is a key parameter in the FvCB model. It directly determines the Rubisco-limited rate A_c and, for C_3 plants, also influences the RuBP regeneration-limited rate A_j though its covariation with $J_{\max}$. The maximum rate of carboxylation has a wide range among plant species and environments; reported values for 109 species of C_3 plants vary from less than 10 to greater than 150 μmol m^{-2} s^{-1} (Wullschleger 1993). Maximum light-saturated photosynthesis increases with greater $V_{c\max}$ (Figure 11.6). The light-saturated rate ranges from less than 5 μmol CO_2 m^{-2} s^{-1} with $V_{c\max} < 20$ μmol m^{-2} s^{-1} to more than 30 μmol CO_2 m^{-2} s^{-1} with $V_{c\max} = 150$ μmol m^{-2} s^{-1}. This is comparable to the range in maximum photosynthetic rates observed in natural environments (Wright et al. 2004).

Despite its importance, $V_{c\max}$ is a poorly constrained parameter in terrestrial biosphere models (Rogers 2014). These models represent vegetation as physiologically distinct plant functional types with photosynthetic parameters assigned for each plant type. A comparison of 10 models shows more than a factor of four variation in $V_{c\max 25}$ within a given plant functional type (Figure 11.7). This arises from different methodologies to estimate $V_{c\max 25}$ from leaf trait databases, from empirical or theoretical relationships with leaf nitrogen, or from model calibration to achieve desired model output at the canopy or global scale. Seasonal changes in photosynthetic capacity have been found in trees (Niinemets et al. 1999a; Wilson et al. 2000; Xu and Baldocchi 2003; Bauerle et al. 2012; Rogers et al. 2017). Moreover, photosynthetic capacity decreases following long-term growth at elevated CO_2 concentration, seen in a reduction in $V_{c\max}$ (Ainsworth and Rogers 2007; Rogers et al. 2017). This photosynthetic acclimation to elevated CO_2 is poorly represented in terrestrial biosphere models. Rogers (2014) reviewed various methods to calculate $V_{c\max 25}$ in models. That review found considerable uncertainty associated with this parameter. The various methods to determine $V_{c\max 25}$ in terrestrial biosphere models greatly affect the global carbon cycle (Walker et al. 2017).

Measurements of leaf photosynthesis routinely show that plants with high amounts of nitrogen in their leaves have higher photosynthetic rates than plants with less leaf nitrogen (Field and Mooney 1986; Evans 1989; Wright et al. 2004). This is because nitrogen is an essential component of Rubisco and leaf chlorophyll. The prominent role of nitrogen in the biochemistry of photosynthesis is seen in the dependence of $V_{c\max}$ and $J_{\max}$ on leaf

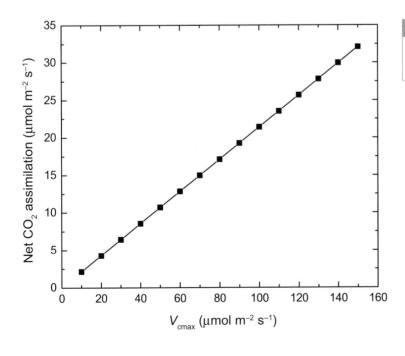

Figure 11.6 Light-saturated photosynthesis in relation to $V_{c\max}$. Simulations are as in Figure 11.2b with $J_{\max} = 1.67 V_{c\max}$ and $R_d = 0.015 V_{c\max}$.

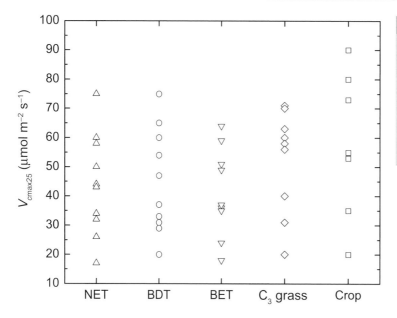

Figure 11.7 Values for $V_{c\max}$ at 25°C used in 10 different models for five common plant functional types. One anomalously high value (163 µmol m^{-2} s^{-1}) for BET was excluded. Not all models simulate crops. Abbreviations are: NET, temperate needleleaf evergreen tree; BDT, temperate broadleaf deciduous tree; BET, tropical broadleaf evergreen tree. Data from Rogers (2014).

nitrogen, and the relationship with leaf nitrogen provides an important constraint on these parameters (Rogers 2014). In a synthesis of leaf photosynthetic measurements among various plant functional types, Kattge et al. (2009) found that $V_{c\max}$ positively correlates with leaf nitrogen per unit leaf area n_a (g N m^{-2}) with the relationship:

$$V_{c\max 25} = i_v + s_v n_a. \tag{11.45}$$

The intercept i_v and slope s_v vary among plant types (Table 11.4). Other approaches employ a more mechanistic understanding of leaf physiology.

One general method calculates $V_{c\max}$ from leaf nitrogen using the relationship

$$V_{c\max} = 6.25 V_{cr} f_R n_m m_a, \tag{11.46}$$

where V_{cr} is the specific activity of Rubisco (i.e., the maximum rate of RuBP carboxylation per unit Rubisco protein; µmol CO$_2$ g^{-1} Rubisco s^{-1}); f_R is the fraction of leaf nitrogen in Rubisco (g N in Rubisco g^{-1} leaf N); n_m is the leaf nitrogen content per unit leaf mass (g N g^{-1} C); m_a is leaf mass per area (g C m^{-2}); and the nitrogen content of Rubisco is 6.25 g Rubisco per g N in Rubisco (Niinemets and Tenhunen 1997; Niinemets et al. 1998, 2004). Here, leaf nitrogen per unit leaf area is given by $n_a = n_m m_a$. Similarly,

$$J_{\max} = 8.06 J_{mc} f_B n_m m_a, \tag{11.47}$$

with J_{mc} the capacity of electron transport per unit cytochrome f (mol electron mol^{-1} cyt f s^{-1}) and f_B the fraction of leaf nitrogen in rate-limiting proteins of electron transport (g N in cyt f g^{-1} leaf N). The value 8.06 µmol cyt f per g N in cyt f converts nitrogen content to investment in bioenergetics. In these equations, $V_{c\max}$ and $J_{\max}$ vary with temperature through the temperature dependence of V_{cr} and J_{mc}. Representative values at 25°C are $V_{cr25} = 20.5$ µmol CO$_2$ g^{-1} Rubisco s^{-1} and $J_{mc25} = 156$ mol electron mol^{-1} cyt f s^{-1}.

Data from Niinemets et al. (2004) are illustrative. Representative values for mature leaves of aspen (*Populus tremula*) in the upper canopy are $f_R = 0.182$ and $f_B = 0.055$. Leaf nitrogen per unit leaf mass is fairly constant ($n_m = 0.02$), but m_a is highly variable among leaves and ranges from 20–110 g m^{-2} (both given here on a dry mass basis rather than carbon). Corresponding values are $V_{c\max 25} = 9$–51 µmol m^{-2} s^{-1} and $J_{\max 25} = 28$–152 µmol m^{-2} s^{-1}. In this case, $J_{\max 25}/V_{c\max 25} = 2.97$. Leaf mass per area varies with depth in the canopy and with leaf development, with concomitant changes in f_R and f_B. This reflects allocation of leaf nitrogen to photosynthetic capacity in response to the changing light environment. There is considerable uncertainty in the relationship with leaf nitrogen (Rogers 2014). Various models use different values for the specific activity of Rubisco V_{cr}, the fraction of leaf nitrogen in

Rubisco f_R, and the nitrogen content of Rubisco. Estimation of leaf nitrogen itself depends on n_m and m_a, both of which vary considerably among and within a species and can vary within a forest canopy depending on height.

Question 11.7 Contrast (11.45) and (11.46) as used to calculate $V_{c\max}$. Describe differences in parameters. Which equation is more easily fit to observations?

11.5 | The Kull and Kruijt (1998) Model

Most terrestrial biosphere models use the FvCB model of C_3 photosynthesis. Various forms of the model differ in their smoothing of the transition between the A_c and A_j limitations, with some implementations using the minimum of the two rates as in (11.30) and others using co-limitation as specified by an empirical curvature factor Θ_A (11.31). The FvCB model also requires an equation for electron transport. This is commonly specified as the co-limitation of a light-harvesting rate I_{PSII} and a maximum rate $J_{\max}$ using an empirical curvature factor Θ_J (11.21). Kull and Kruijt (1998) devised a leaf photosynthesis model that builds upon the FvCB model, but without the need for empirical curvature factors. Their model distinguishes two limitations to electron transport by separately considering restrictions imposed by light harvesting and by $J_{\max}$. The model explicitly considers these two processes by accounting for spatial heterogeneity inside an individual leaf in the light environment and also heterogeneity in photosynthetic properties. In particular, the portion of the leaf near the illuminated surface is light saturated, whereas other portions of the leaf are not. The model is less widely used than the FvCB model but has been implemented in the GISS (Friend and Kiang 2005) and O-CN (Zaehle and Friend 2010) models. It describes the photosynthetic response to light by accounting for separate light-saturated and light-limited chloroplasts. Photosynthesis is expressed per unit leaf nitrogen to scale over the entire leaf in relation to leaf nitrogen.

Photosynthesis by light-saturated chloroplasts follows the FvCB model with the rate constrained in relation to either $J_{\max}$ or $V_{c\max}$. The carboxylation rate from electron transport is w_j as in (11.13), but with $J = J_{\max}$ for light saturation. The Rubisco-limited carboxylation rate is w_c (11.12). For convenience, these rates are given here as

$$w_j = \frac{J_{\max}}{4} m_1, \qquad (11.48)$$

with

$$m_1 = \frac{c_i}{c_i + 2\Gamma_*}, \qquad (11.49)$$

and

$$w_c = V_{c\max} m_2, \qquad (11.50)$$

with

$$m_2 = \frac{c_i}{c_i + K_c(1 + o_i/K_o)}. \qquad (11.51)$$

$J_{\max}$ and $V_{c\max}$ relate to leaf nitrogen through a proportionality constant in which

$$\frac{J_{\max}}{4} = n_1 n_a \qquad (11.52)$$

and

$$V_{c\max} = n_2 n_a, \qquad (11.53)$$

where here n_a is expressed on a molar basis (mmol N m^{-2}). Substituting (11.52) into (11.48) shows that the RuBP regeneration-limited carboxylation rate per unit leaf nitrogen is $w_j/n_a = n_1 m_1$ (μmol CO_2 mmol^{-1} N s^{-1}) and, similarly, $w_c/n_a = n_2 m_2$ for Rubisco limitation. The actual rate of carboxylation at light saturation is

$$w_{sat} = \min(n_1 m_1, n_2 m_2). \qquad (11.54)$$

The Kull and Kruijt (1998) model explicitly represents light absorption through a leaf to separately calculate photosynthesis by light-limited chloroplasts. The carboxylation rate limited by light harvesting is

$$w_i = E\,\vec{I}\left(\frac{c_i}{c_i + 2\Gamma_*}\right) = E\vec{I} m_1. \qquad (11.55)$$

This is similar to (11.32), which was introduced previously and used in some models for electron transport, but with $\vec{I}$ the photosynthetically active radiation absorbed by the leaf. As light penetrates a leaf, the photons are absorbed by chlorophyll. The light absorption is represented by exponential attenuation in relation to chlorophyll content C_{chl} (µmol Chl m^{-2}) in which

$$\overrightarrow{I} = \alpha_\ell I^\downarrow \left(1 - e^{-K_a C_{chl}}\right) \quad (11.56)$$

and K_a is a light extinction coefficient (m^2 µmol^{-1} Chl). Chlorophyll concentration is specified in proportion to leaf nitrogen with

$$C_{chl} = n_3 n_a. \quad (11.57)$$

When the light-saturated and light-harvesting carboxylation rates are equal, the rate of change in light-harvesting carboxylation with nitrogen equals the light-saturated rate of carboxylation per unit nitrogen so that

$$\frac{dw_i}{dn_a} = w_{sat}. \quad (11.58)$$

The nitrogen concentration that satisfies this relationship, denoted as n_{sat}, is

$$n_{sat} = -\frac{\ln\left(\dfrac{w_{sat}}{\alpha_\ell I^\downarrow E m_1 K_a n_3}\right)}{K_a n_3}. \quad (11.59)$$

This is the cumulative leaf nitrogen concentration at which limitation by light harvesting exceeds limitation from light saturation. The calculated value is constrained by $0 \le n_{sat} \le n_a$. A value less than zero occurs when light is insufficient to saturate any portion of the leaf. Also, n_{sat} cannot exceed the

actual leaf nitrogen concentration, in which case the entire leaf is light saturated. The total photosynthesis for a leaf is found by integrating the light-saturated and light-harvesting carboxylation rates with respect to nitrogen concentration whereby

$$A = \left(1 - \frac{\Gamma_*}{c_i}\right)\left[\int_0^{n_{sat}} n_a w_{sat} dn_a + \int_{n_{sat}}^{n_a} w_i dn_a\right] \quad (11.60)$$

and

$$A = \left(1 - \frac{\Gamma_*}{c_i}\right)\left[w_{sat}n_{sat} + \alpha_\ell I^\downarrow E m_1 \left(e^{-K_a n_3 n_{sat}} - e^{-K_a n_3 n_a}\right)\right]. \quad (11.61)$$

Friend (2001) gave example simulations with $n_1 = 0.12$ and $n_2 = 0.23$ µmol CO$_2$ mmol^{-1} N s^{-1}, $n_3 = 2.5$ µmol Chl mmol^{-1} N, $K_a = 0.0055$ m^2 µmol^{-1} Chl, and E $= 0.08$ mol mol^{-1}. Comparison with the FvCB model for the same parameters shows markedly different photosynthetic responses to light (Figure 11.8). The FvCB model has a more sharply defined transition to light saturation and has higher photosynthetic rates at low light levels. This difference between models occurs at both high and low leaf nitrogen. Another distinction between models is that the FvCB model has similarly shaped light response curves with different leaf nitrogen while the Kull and Kruijt (1998) model differs between high and low leaf nitrogen, particularly with regard

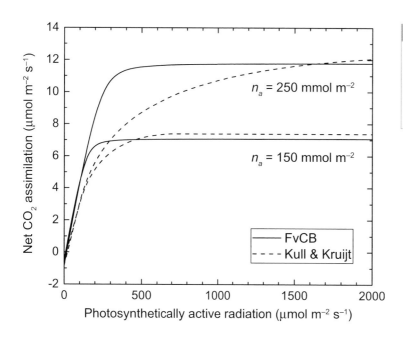

Figure 11.8 Photosynthetic response to light for the FvCB model and the Kull and Kruijt (1998) model. Model parameters are as in Figure 11.2b. $V_{c\,max}$, J_{max}, and R_d vary with leaf nitrogen and are the same for both models. The FvCB model has co-limitation with $\Theta_A = 0.98$.

to light saturation. The Kull and Kruijt (1998) model also does not need empirical smoothing between the A_c and A_j limitations. Friend (2001) examined model differences in more detail, though with a different implementation of electron transport and A_j.

Question 11.8 Discuss parameter differences between the FvCB model and the Kull and Kruijt (1998) model.

11.6 | Temperature Acclimation

The Arrhenius or Q_{10} temperature functions describe an instantaneous enzymatic temperature response that is constant even for long-term changes in temperature. In fact, many studies have found that plants can adjust their enzymatic photosynthetic or respiratory temperature response as a result of acclimation to a new growth temperature over periods of days to weeks (Smith and Dukes 2013). This acclimation changes the shape and/or basal rate of the temperature response. Type I acclimation describes changes in the shape of the instantaneous enzymatic response curve with little or no change in rates at low temperatures, but large changes in rates at high temperatures and shifts in the temperature optimum. Type II acclimation alters the basal rate but does not necessarily change the shape of the response.

Acclimation is seen in shifts in the instantaneous temperature response of photosynthesis, particularly the temperature optimum, depending on the temperature at which the plant is grown. This arises from changes in enzymatic processes such as the carboxylation rate of Rubisco and electron transport, as manifested in $V_{c\,max}$ and J_{max}. The parameterization of Kattge and Knorr (2007) illustrates temperature acclimation of $V_{c\,max}$ and J_{max}. Their analyses of leaf photosynthetic measurements found a linear relationship between plant growth temperature T_A (°C) and the parameter ΔS in the peaked Arrhenius function (11.36). This relationship varies for $V_{c\,max}$ and J_{max}, with

$$\Delta S = 668.39 - 1.07 T_A \qquad (11.62)$$

for $V_{c\,max}$ and

$$\Delta S = 659.70 - 0.75 T_A \qquad (11.63)$$

for J_{max}. In this parameterization, $\Delta H_d = 200$ kJ mol^{-1}, and ΔH_a is equal to 72 kJ mol^{-1} for $V_{c\,max}$ and 50 kJ mol^{-1} for J_{max}. The effect is to cause the temperature optimum of $V_{c\,max}$ and J_{max} to increase with warmer growth temperature. They also found that the ratio $J_{max\,25}/V_{c\,max\,25}$ decreases with warmer growth temperature, described by

$$J_{max\,25}/V_{c\,max\,25} = 2.59 - 0.035 T_A. \qquad (11.64)$$

These adjustments shift the temperature optimum of photosynthesis to higher values with warmer growth temperature (Figure 11.9a,b). Plants acclimated to warm temperatures have a higher photosynthetic rate at high temperatures compared with plants acclimated to cold temperatures.

Plants can also adjust their instantaneous respiration rate in response to a change in temperature (Smith and Dukes 2013). One aspect of this is a variable Q_{10} that decreases as temperature increases. For example, Tjoelker et al. (2001) used measurements for arctic, boreal, temperate, and tropical plants to develop a leaf respiration equation in which the Q_{10} parameter decreases with higher temperature. In this equation, the instantaneous respiration rate R_d at a given temperature T follows the standard Q_{10} relationship with

$$R_d = R_{d0} Q_{10}^{(T-T_0)/10}, \qquad (11.65)$$

where R_{d0} is the respiration rate at the reference temperature T_0, and the Q_{10} varies with temperature as

$$Q_{10} = 3.22 - 0.046 T \qquad (11.66)$$

for T given in °C. In this equation, the Q_{10} decreases from 3.0 at 5°C to 1.6 at 35°C. The decrease in Q_{10} with warmer temperature produces a diminished increase in respiration as temperature increases and an optimum temperature at which respiration peaks (Figure 11.9c).

Plants can also adjust the basal rate of respiration in relation to growth temperature. Plants acclimated to high temperatures typically have lower basal rates compared with plants acclimated to low temperatures. Atkin et al. (2008) accounted for this by adjusting the basal rate for growth temperature so that the rate of respiration for a leaf acclimated to temperature T_A is

$$R_d = 10^{-0.00794(T_A-T_0)} R_{d0} Q_{10}^{(T-T_0)/10}. \qquad (11.67)$$

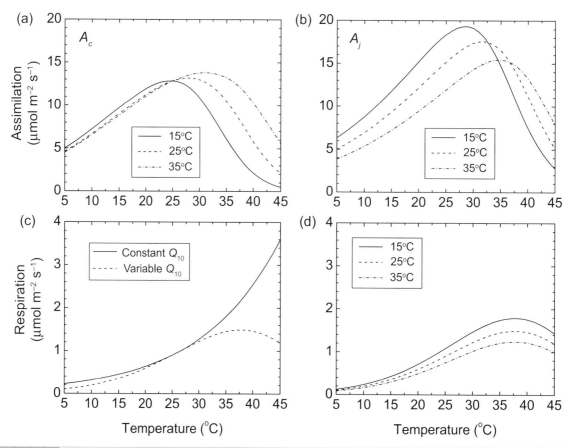

Figure 11.9 Photosynthesis and respiration with temperature acclimation. (a,b) Photosynthetic acclimation using the temperature functions of Kattge and Knorr (2007). Shown are the A_c- and A_j-limited rates at growth temperatures of 15°C, 25°C, and 35°C. (c) Instantaneous temperature response of respiration for constant $Q_{10} = 2$ and variable Q_{10} (Tjoelker et al. 2001). $R_{d0} = 0.9$ μmol m^{-2} s^{-1} at $T_0 = 25$°C. (d) Instantaneous temperature response of respiration with acclimation. Shown is the variable Q_{10} function of panel (c) with temperature acclimation (Atkin et al. 2008) at growth temperatures of 15°C, 25°C, and 35°C.

The first term adjusts the basal rate R_{d0} at the reference temperature T_0 for growth temperature T_A, and the second term is the standard Q_{10} relationship. The effect is that leaves acclimated to warm temperatures have a lower respiration rate at high temperatures compared with leaves acclimated to lower temperatures (Figure 11.9d).

Atkin et al. (2015) developed a more general equation that also accounts for the effects of foliage nitrogen on respiration. Similar to $V_{c\max}$ and $J_{\max}$, respiration rates increase with greater leaf nitrogen when the measurements are standardized to a common temperature. Observations for several plant functional types ranging from arctic to tropical show that R_d at a standard temperature (25°C) is

lower for tropical plants than for arctic plants. This is represented by the relationship:

$$R_{d25} = r_0 + 0.2061n_a - 0.0402T_A. \quad (11.68)$$

In this equation, R_{d25} increases with greater area-based leaf nitrogen n_a (g m^{-2}) and decreases with greater growth temperature T_A (see also Atkin 2016). The intercept parameter r_0 varies with plant functional type so that, for example, broadleaf trees have a larger basal respiration rate compared with needleleaf trees.

Terrestrial biosphere models that simulate the global carbon cycle and its response to climate change incorporate the instantaneous enzymatic response to temperature, but many do not account for photosynthetic or respiratory temperature acclimation (Smith and Dukes 2013). Some studies have incorporated photosynthetic and leaf respiration acclimation into global models (Atkin et al. 2008; Arneth et al. 2012; Lombardozzi et al. 2015a; Smith et al. 2016). These studies find that temperature acclimation can affect the terrestrial carbon cycle and its response to climate change. However, the exact way to represent temperature acclimation in models remains uncertain (Rogers et al. 2017).

Question 11.9 Discuss the effect of the peaked Arrhenius function on photosynthesis. How does this function affect the temperature response of photosynthesis? What is the consequence if the same parameter values are used globally for all plants? How does acclimation affect the model?

Question 11.10 Discuss how respiration temperature acclimation affects net CO_2 assimilation with warmer temperatures.

11.7 | C₄ Photosynthesis

The C₄ photosynthetic pathway differs from that of C₃ plants. In this pathway, CO_2 combines with phosphoenolpyruvate (PEP) in a reaction catalyzed by PEP carboxylase. The initial CO_2 fixation by PEP carboxylase into 4-carbon acids occurs in the leaf mesophyll. These acids are decarboxylated in the bundle sheath to provide CO_2 for Rubisco, whereupon the CO_2 combines with RuBP to form PGA as in C₃ plants. The separation of photosynthesis into mesophyll and bundle-sheath cells ensures high CO_2 concentration in the bundle sheath and reduces CO_2 loss during photorespiration. Von Caemmerer (2000, 2013) and Diaz-Espejo et al. (2012) reviewed C₄ photosynthesis models.

Collatz et al. (1992) devised a model for C₄ photosynthesis, which was first used in SiB2 (Sellers et al. 1996a,b) and has been adopted by other models (e.g., the Community Land Model). The C₄ model represents inorganic carbon fixation by PEP carboxylase, light-dependent generation of PEP and RuBP, the reaction kinetics of Rubisco, and diffusion of inorganic carbon and O_2 between the bundle-sheath cells and the leaf mesophyll. These processes are described by three rate limiting equations: (1) A_c – when light and CO_2 are not limiting, the capacity for CO_2 fixation by Rubisco controls the rate of photosynthesis; (2) – when light intensity is insufficient, the efficiency of CO_2 fixation with respect to absorbed light (quantum yield) determines the rate of photosynthesis; and (3)A_p – photosynthesis at low CO_2 concentration is limited by the initial rate of CO_2 fixation by PEP minus leakage of CO_2 from bundle-sheath cells.

For plants that utilize the C₄ photosynthetic pathway, the Rubisco-limited rate is

$$A_c = V_{c\max}.$$ (11.69)

The light-limited rate is

$$A_j = \alpha_\ell I^\downarrow E.$$ (11.70)

The PEP carboxylase-limited (or CO_2-limited) rate is

$$A_p = k_p c_i,$$ (11.71)

where k_p (mol m^{-2} s^{-1}) is the initial slope of the CO_2 response curve. In SiB2, Sellers et al. (1996a,b) used $k_{p25} = 0.02 V_{c\max 25}$, quantum yield E = 0.05 mol mol^{-1}, $R_{d25} = 0.025 V_{c\max 25}$, and co-limitation is given by (11.33) with the curvature factors 0.80 and 0.95 for A_i and A, respectively. The intercellular CO_2 concentration in C₄ plants is typically 40%–60% of ambient CO_2 concentration (Hetherington and Woodward 2003).

Temperature dependencies, as devised by Collatz et al. (1992) and implemented in SiB2 (Sellers et al. 1996a,b), are

$$V_{c\max} = V_{c\max 25} Q_{10}^{(T_\ell - 298.15)/10} \{1 + \exp[s_1(T_\ell - s_2)]\}^{-1}$$
$$\{1 + \exp[s_3(s_4 - T_\ell)]\}^{-1}.$$ (11.72)

The first term increases $V_{c\max}$ as temperature increases ($Q_{10} = 2$), and the two terms in brackets decrease $V_{c\max}$ for excessive high and low temperatures, respectively. The parameter s_1 controls the rate of decrease in $V_{c\max}$ with increasing temperature, and s_2 is the temperature at which $V_{c\max}$ is one-half its non-stressed value. The high temperature stress factors are $s_1 = 0.3$ K^{-1} and $s_2 = 313.15$ K (40°C). The corresponding low-temperature stress factors are $s_3 = 0.2$ K^{-1} and $s_4 = 288.15$ K (15°C). For respiration,

$$R_d = R_{d25} Q_{10}^{(T_\ell - 298.15)/10} \{1 + \exp[s_5(T_\ell - s_6)]\}^{-1},$$
(11.73)

where $Q_{10} = 2$ and the high-temperature stress factors are $s_5 = 1.3$ K^{-1} and $s_6 = 328.15$ K (55°C). The CO_2-limited rate A_p varies with temperature according to

$$k_p = k_{p25} Q_{10}^{(T_\ell - 298.15)/10},$$
(11.74)

with $Q_{10} = 2$.

The C_3 and C_4 photosynthetic pathways differ in their response to light, CO_2 concentration, and temperature (Figure 11.4). Net photosynthesis saturates at relatively low irradiance in the C_3 plant but shows little light saturation in the C_4 plant, except at high irradiance. Conversely, net photosynthesis saturates at low CO_2 concentration in the C_4 plant but increases with higher CO_2 in the C_3 plant. This difference arises because higher CO_2 concentration inhibits photorespiration in C_3 plants. For the C_4 plant, the initial slope parameter k_p determines the CO_2 concentration at which net photosynthesis saturates. The PEP carboxylase (CO_2-limited) rate A_p restricts photosynthesis at low CO_2, and the Rubisco-limited rate A_c operates at higher CO_2. Net photosynthesis decreases with temperature above or below some optimum in both plants, but the C_4 plant has higher rates at high temperature. Conversely, the C_3 plant performs better at low temperatures.

Question 11.11 Graph the rates A_c, A_j, and A_p in the C_4 photosynthetic response to light and c_i, and discuss how these determine the realized rate of photosynthesis.

11.8 | Diffusive Limitations on CO_2 Supply

Models of C_3 and C_4 photosynthesis represent the demand for CO_2 arising from the biochemistry of photosynthesis. The models require intercellular CO_2 concentration c_i. This depends on the diffusive supply of CO_2 through stomata. With reference to Figure 11.10, the net leaf CO_2 flux can also be described by a diffusion process in which

$$A_n = \frac{g_{bw}}{1.4}(c_a - c_s) = \frac{g_{sw}}{1.6}(c_s - c_i) = g_{\ell c}(c_a - c_i).$$
(11.75)

In these equations, c_a, c_s, and c_i are the CO_2 mole fraction (μmol mol^{-1}) for ambient air surrounding the leaf, at the leaf surface, and inside the stomatal cavity, respectively. The first equation describes diffusion of CO_2 from air with concentration c_a to the leaf surface with concentration c_s, which is proportional to leaf boundary layer conductance g_{bw}. The second equation describes diffusion of CO_2 from the leaf surface c_s to intercellular space c_i, which is proportional to stomatal conductance g_{sw}. The third equation is the net CO_2 flux from ambient air to intercellular space, which is proportional to the leaf conductance $g_{\ell c} = 1/(1.4g_{bw}^{-1} + 1.6g_{sw}^{-1})$. Conductance is given for water vapor (mol H$_2$O m^{-2} s^{-1}), and the factors 1.4 and 1.6 adjust for the lower diffusivity of CO_2 compared with H$_2$O. CO_2 has a larger molecular mass compared with H$_2$O, and $g_{bc} = g_{bw}/1.4$ and $g_{sc} = g_{sw}/1.6$ are conductances for CO_2 in contrast to H$_2$O. The factor 1.4 for boundary layer conductance is derived in (10.39). The factor 1.6 for stomatal conductance is the ratio of the diffusivity of H$_2$O to CO_2 (Table A.3).

If g_{sw} is known, c_i can be written in terms of c_a using the diffusion equation

$$c_i = c_a - A_n/g_{\ell c}.$$
(11.76)

This expression can be substituted into the photosynthesis model in place of c_i. The solution differs between the A_c- and A_j-limited rates. For C_3 plants, a general equation to describe the biochemical demand for CO_2 is

$$A_n = \frac{a(c_i - \Gamma_*)}{c_i + b} - R_d,$$
(11.77)

with $a = V_{c\,max}$ and $b = K_c(1 + o_i/K_o)$ for A_c-limited assimilation and $a = J/4$ and $b = 2\Gamma_*$ for A_j-limited assimilation. Substituting (11.76) for c_i in (11.77) gives a quadratic equation for A_n whereby

$$g_{\ell c}^{-1} A_n^2 - [c_a + b + (a - R_d)/g_{\ell c}]A_n$$
$$+ [a(c_a - \Gamma_*) - (c_a + b)R_d] = 0.$$
(11.78)

The correct solution is the smaller root. It is more common to solve for c_i by eliminating A_n. This gives a quadratic equation for c_i in which

$$g_{\ell c}c_i^2 + [a - R_d + (b - c_a)g_{\ell c}]c_i - [a\Gamma_* + (c_a g_{\ell c} + R_d)b] = 0.$$
(11.79)

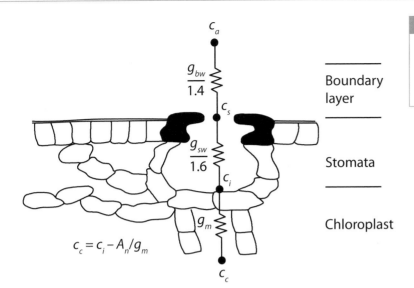

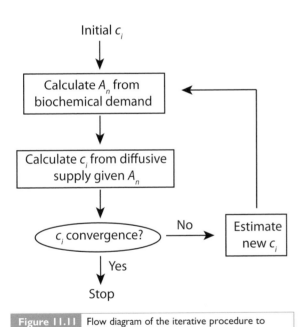

Figure 11.10 Diffusion of CO_2 from free air across the leaf boundary layer and through stomata to intercellular space. Diffusion of CO_2 to the chloroplast is additionally regulated by mesophyll conductance.

Figure 11.11 Flow diagram of the iterative procedure to numerically calculate c_i.

The correct expression is the larger root. The photosynthetic rate is obtained by using this value for c_i in (11.77). Equations (11.78) and (11.79) must be solved twice – once for A_c and again for A_j. The equations are often simplified by assuming that $g_{bw} \gg g_{sw}$ so that the term $1.4/g_{bw}$ is ignored and leaf conductance is approximated by $g_{\ell c} = g_{sw}/1.6$, and also by assuming that $R_d \ll A_n$ so that R_d is ignored.

A more general numerical solution is to iteratively find c_i that satisfies (11.76) and (11.77) using the secant method, Brent's method, or other numerical methods (Figure 11.11). These methods begin with an initial estimate for c_i to calculate A_n from the biochemical demand (11.77). With this value for A_n, a new estimate c_i' is obtained from the diffusive supply (11.76). The iteration is repeated until the change in c_i ($\Delta c_i = c_i' - c_i$) satisfies some convergence criteria (e.g., $\pm 10^{-3}$ µmol mol⁻¹). This numerical method is easily adapted for co-limitation and to represent C_4 photosynthesis.

Diffusion of CO_2 from free air through the boundary layer and stomata into intercellular air space requires a concentration gradient with $c_a > c_s > c_i$. Diffusion of CO_2 from intercellular space to the carboxylation sites in chloroplasts presents an additional limitation (Flexas et al. 2008, 2012; Diaz-Espejo et al. 2012; von Caemmerer 2013). As CO_2 diffuses from the intercellular air space through the mesophyll to the chloroplast, it encounters a series of structures (e.g., cell wall, cell membrane) that impose an additional resistance to CO_2 diffusion (Figure 11.10). An additional term is required to account for CO_2 diffusion from intercellular air space with concentration c_i through the mesophyll to the chloroplast with concentration c_c. This conductance, termed mesophyll conductance g_m, regulates the mesophyll diffusive flux such that $c_c = c_i - A_n/g_m$. Chloroplastic and intercellular CO_2 concentrations are equal only if mesophyll conductance is infinitely large, as assumed in this chapter and in many gas exchange studies. In fact, however, mesophyll conductance is finite and of similar magnitude as stomatal conductance so that $c_c < c_i$. There is considerable research into the magnitude of g_m and its environmental dependencies; e.g., g_m can vary with temperature and among species (von Caemmerer and Evans 2015). Some studies have implemented g_m in terrestrial biosphere models (Sun et al. 2014b), but greater process understanding is still required before robust parameterizations can be obtained (Rogers et al. 2017).

In the photosynthesis model as presented here, c_c is taken to be the same as c_i. Inclusion of g_m requires new estimation of Rubisco kinetics parameters. The parameters K_c and K_o are overestimated by the use of c_i (von Caemmerer et al. 1994; Bernacchi et al. 2002). Similarly, $V_{c\max}$ must be re-derived from leaf gas exchange measurements (Ethier and Livingston 2004; Niinemets et al. 2009; Diaz-Espejo et al. 2012; Sun et al. 2014a). Values obtained from A_n-c_c measurements are larger than those obtained from A_n-c_i measurements.

11.9 | Supplemental Programs

11.1 Farquhar et al. (1980) Photosynthesis: This program calculates the A_c- and A_j-limited photosynthetic rates for C$_3$ plants given a specified c_i as summarized in Table 11.5.

11.10 | Modeling Projects

1 Terrestrial biosphere models differ in values used for K_c, K_o, and Γ_*, which describe Rubisco kinetics. Figure 11.2 used values from Bernacchi et al. (2001) at 25°C. Compare this with results using the values in Table 11.1 for SiB2 (Sellers et al. 1996a,b).

2 Terrestrial biosphere models differ in the ratio $J_{\max}/V_{c\max}$. Calculate A_n-c_i curves and light response curves for C$_3$ photosynthesis as in Figure 11.2 with $J_{\max} = 1.67 V_{c\max}$ and $J_{\max} = 2.0 V_{c\max}$. Calculate the irradiance $I^{\downarrow *}$ and the CO$_2$ concentration c_i^* at which the transition between A_c- and A_j-limited photosynthesis occurs. How do these values compare with the A_n-c_i and light response curves? How does co-limitation affect the results?

3. Compare the FvCB and Kull and Kruijt (1998) models in their light response as in Figure 11.8. Compare model results using different values for K_c, K_o, and Γ_* (Table 11.1). Discuss the importance of model structural uncertainty versus parameter uncertainty.

4. Investigate the role of the product-limited rate in C$_3$ photosynthesis.

Stomatal Conductance

Chapter Overview

Of particular importance for leaf energy fluxes is that leaf temperature, transpiration, and photosynthesis are linked through stomatal conductance. Current representations of these processes in terrestrial biosphere models recognize that the biophysics of stomatal conductance is understood in relation to the biochemistry of photosynthesis. Stomata act to balance the need for photosynthetic CO_2 uptake while limiting water loss during transpiration. Consequently, an accurate depiction of photosynthesis is required to determine the stomatal conductance needed for transpiration and leaf temperature. This chapter develops the physiological theory and mathematical equations to describe stomatal conductance, photosynthesis, transpiration, and their interdependencies. Four main types of models are empirical multiplicative models, semiempirical models that relate stomatal conductance and photosynthesis, water-use efficiency optimization models, and plant hydraulic models. This chapter reviews the first three types of models. Plant hydraulic models are discussed in Chapter 13.

12.1 | Introduction

The photosynthesis models presented in the previous chapter provide CO_2 assimilation if intercellular CO_2 concentration c_i is known. Calculation of c_i requires stomatal conductance g_{sw}. This is evident from the diffusion equation,

$$A_n = \frac{g_{bw}}{1.4}(c_a - c_s) = \frac{g_{sw}}{1.6}(c_s - c_i) = g_{\ell c}(c_a - c_i), \quad (12.1)$$

which can be rewritten as

$$c_s = c_a - \frac{1.4}{g_{bw}}A_n \quad (12.2)$$

and

$$c_i = c_s - \frac{1.6}{g_{sw}}A_n. \quad (12.3)$$

More commonly, c_i is calculated from the overall diffusion equation,

$$c_i = c_a - \frac{A_n}{g_{\ell c}}, \quad (12.4)$$

with the leaf conductance to CO_2 given by,

$$g_{\ell c} = \frac{1}{1.4g_{bw}^{-1} + 1.6g_{sw}^{-1}}. \quad (12.5)$$

For a positive net photosynthesis rate, diffusion of CO_2 from ambient air through the boundary layer and stomata into intercellular air space requires a concentration gradient such that $c_a > c_s > c_i$. The magnitude of the difference relates to A_n and the conductances g_{bw} and g_{sw}.

Stomata open under favorable environmental conditions to allow CO_2 uptake during photosynthesis. At the same time, water diffuses out of the stomatal cavity to the ambient air during transpiration. Data from Dang et al. (1997a,b, 1998) illustrate

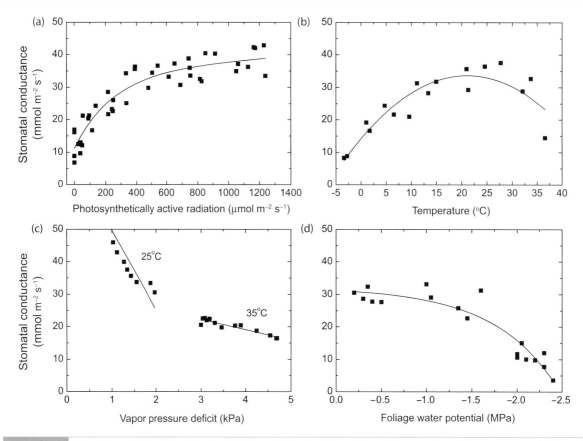

Figure 12.1 Stomatal conductance in relation to (a) photosynthetically active radiation, (b) temperature, (c) vapor pressure deficit at 25°C and 35°C, and (d) foliage water potential for jack pine (*Pinus banksiana*) trees. Data from Dang et al. (1997a,b, 1998). Reproduced from Bonan (2016).

the relationships among photosynthesis and stomatal conductance for jack pine trees. The rate of leaf photosynthesis responds to light, CO_2 concentration, temperature, and other factors (Figure 11.1). Stomatal conductance similarly responds to the same environmental conditions (Figure 12.1), and the rate of photosynthesis increases with higher stomatal conductance as shown in Figure 12.2 over a range of light from full illumination to darkness.

Equation (12.1) describes photosynthesis as a diffusion process dependent on c_i and leaf conductance, whereas the preceding chapter describes photosynthesis as a biochemical process dependent on c_i. The biochemical rate, given by (11.30) for C_3 plants and similarly for C_4 plants, represents the photosynthetic demand for CO_2, while the diffusive rate represents the supply of CO_2 to the leaf. These provide a system of equations that link the biophysics of stomatal conductance to the biochemistry of

photosynthesis. However, (11.30) and (12.1) are two equations with three unknowns (c_i, A_n, g_{sw}). A third equation – a stomatal constraint function – is needed to close the system of equations by providing an expression for g_{sw}. This gives a system of equations in which the solution is the c_i that satisfies all three equations.

Early models of stomatal conductance were empirical and did not link stomatal behavior with photosynthesis. A subsequent model of stomatal conductance recognized the empirical dependence between A_n and g_{sw}, such as is shown in Figure 12.2, in what is commonly referred to as the Ball–Berry model (Ball et al. 1987). Collatz et al. (1991) coupled photosynthesis and stomatal conductance models for C_3 plants, and Collatz et al. (1992) extended the work to C_4 plants. These equations, or some variant of the stomatal constraint equation, are widely used in terrestrial biosphere models. Alternatively, water-use

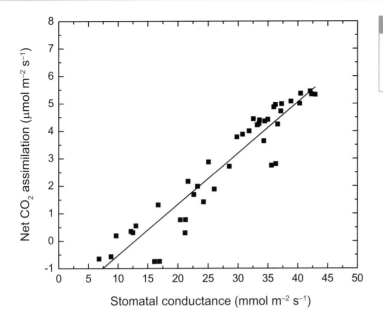

Figure 12.2 Relationship between photosynthesis and stomatal conductance for jack pine over a range of light from 0 to 1250 μmol photon m^{-2} s^{-1}. Data from Dang et al. (1997a,b, 1998)

efficiency optimization theory provides an expression for g_{sw} to close the system of equations derived from theoretical principles rather than empirical relationships. It is based on the principle that the physiology of stomata has evolved to maximize carbon uptake while minimizing water loss (Cowan 1977; Cowan and Farquhar 1977). A fourth class of models relates stomatal conductance to plant water uptake. Table 12.1 summarizes the various models.

12.2 | Empirical Multiplicative Models

Jarvis (1976) formulated an empirical model of stomatal conductance in which conductance increases with greater irradiance up to some light-saturated value scaled by the extent to which other environmental factors are suboptimal. In this model, stomatal conductance is

$$g_{sw} = g_{sw}(I^{\downarrow})f(T_{\ell})f(D_{\ell})f(\psi_{\ell})f(c_a), \quad (12.6)$$

where $g_{sw}(I^{\downarrow})$ is stomatal conductance in relation to photosynthetically active radiation, and $f(T_{\ell}), f(D_{\ell}), f(\psi_{\ell})$, and $f(c_a)$ are empirical functions scaled from zero to one that adjust conductance for leaf temperature, vapor pressure deficit, foliage water potential, and ambient CO_2 concentration, respectively. Stomatal response to light is described by

$$g_{sw}(I^{\downarrow}) = \frac{p_1 p_2 (I^{\downarrow} - p_3/p_1)}{p_1 + p_2 (I^{\downarrow} - p_3/p_1)}, \quad (12.7)$$

with p_1 the maximum value for g_{sw} at saturating light, p_2 the initial slope at $I^{\downarrow} = 0$, and p_3 a minimum value for g_{sw}. Change in stomatal conductance in relation to leaf temperature is represented by

$$f(T_{\ell}) = \left(\frac{T_{\ell} - T_L}{T_0 - T_L}\right)\left(\frac{T_H - T_{\ell}}{T_H - T_0}\right)^{p_4}, \quad (12.8)$$

in which $p_4 = (T_H - T_0)/(T_0 - T_L)$. In this function, $f(T_{\ell}) = 1$ at the optimal temperature T_0 and equals zero at the low and high temperatures (T_L and T_H, respectively). Stomatal conductance decreases linearly with greater vapor pressure deficit between the leaf – which is assumed to be saturated with vapor pressure $e_{sat}(T_{\ell})$ – and the air, with vapor pressure e_a so that

$$f(D_{\ell}) = 1 - p_5[e_{sat}(T_{\ell}) - e_a]. \quad (12.9)$$

Stomatal conductance also decreases as leaf water potential ψ_{ℓ} declines according to the equation

$$f(\psi_{\ell}) = 1 - \exp[-p_6(\psi_{\ell} - \psi_{\ell\,min})]. \quad (12.10)$$

The parameter p_6 determines the change in stomatal conductance with foliage water potential, and $\psi_{\ell\,min}$ is the foliage water potential at which $f(\psi_{\ell}) = 0$. The CO_2 function is a linear decline in

Table 12.1 Summary of stomatal models

Model	Equation
Empirical, multiplicative Jarvis (1976)	$g_{sw} = g_{sw}\left(I^{\downarrow}\right) f(T_{\ell}) f(D_{\ell}) f(\psi_{\ell}) f(c_a)$
Semiempirical based on A_n Ball et al. (1987)	$g_{sw} = g_0 + g_1 \dfrac{A_n}{c_s} h_s$
Leuning (1995)	$g_{sw} = g_0 + g_1 \dfrac{A_n}{c_s - \Gamma}\left(1 + D_s/D_0\right)^{-1}$
Medlyn et al. (2011)	$g_{sw} = g_0 + 1.6\left(1 + \dfrac{g_1}{\sqrt{D_s}}\right)\dfrac{A_n}{c_s}$
Water-use efficiency optimization Cowan (1977), Cowan and Farquhar (1977)	$\dfrac{\partial A_n/\partial g_{sw}}{\partial E/\partial g_{sw}} = \iota$
Arneth et al. (2002)	$g_{sw} = 1.6 \dfrac{\partial A_n}{\partial c_i}\left(\dfrac{c_a - c_i}{1.6 w_{\ell}\iota} - 1\right)$
Buckley et al. (2002)	$\dfrac{\partial A_n}{\partial E} = \left(\dfrac{c_a - c_i}{w_{\ell}}\right)\left(\dfrac{\partial A_n/\partial c_i}{\partial A_n/\partial c_i + g_{\ell c}}\right) 1.6 \dfrac{g_{\ell c}^2}{g_{\ell w}^2}$
Plant hydraulics Ewers et al. (2000)	$g_{sw} = \dfrac{K_L}{w_{\ell}}\left(\psi_s - \psi_{\ell} - \rho_w g h\right)$
Optimization and hydraulic safety Williams et al. (1996), Bonan et al. (2014)	g_{sw} such that $\Delta A_n \leq \iota w_s \Delta g_{sw}$ and $\psi_{\ell} > \psi_{\ell\,min}$

Table 12.2 Parameter values for Sitka spruce (*Picea sitchensis*) and Douglas fir (*Pseudotsuga menziesii*) for the Jarvis (1976) stomatal conductance model

Parameter	Sitka spruce	Douglas fir
p_1	0.43 cm s^{-1}	0.30 cm s^{-1}
p_2	0.0296 (cm s^{-1})/ (μmol photon m^{-2} s^{-1})	0.0135 (cm s^{-1})/ (μmol photon m^{-2} s^{-1})
p_3	0 cm s^{-1}	0.01 cm s^{-1}
T_L	−5°C	−5°C
T_O	9°C	20°C
T_H	35°C	45°C
p_5	0.26 kPa^{-1}	0.12 kPa^{-1}
p_6	40 MPa^{-1}	0.93 MPa^{-1}
$\psi_{\ell\,min}$	−2.4 MPa	−2.4 MPa

Note: Conductance with units m s^{-1} is converted to mol m^{-2} s^{-1} by multiplying by $\rho_m = P/\mathcal{R}T = 42.3$ mol m^{-3} (at STP).

stomatal conductance as CO_2 concentration increases in which:

$$f(c_a) = 1 - p_7 c_a. \tag{12.11}$$

Parameters vary among plant species and locations. Table 12.2 provides values for Sitka spruce and Douglas fir. Sitka spruce has a higher maximum conductance (p_1), a larger initial slope with respect to light (p_2), a lower temperature optimum (T_0) and maximum temperature (T_H), a sharper decline with increasing vapor pressure deficit (p_5), and an abrupt decline as foliage water potential approaches $\psi_{\ell\,min}$ (p_6).

Equation (12.6) was used in early generation models of Earth's vegetation for climate simulation such as BATS (Dickinson et al. 1986, 1993) and SiB (Sellers et al. 1986). However, the development of mathematical models of photosynthesis for C$_3$ and C$_4$ plants, combined with the formulation of empirical equations relating photosynthesis and stomatal conductance (Ball et al. 1987), led to a subsequent generation of models that explicitly linked leaf biophysics and biochemistry (Collatz et al. 1991, 1992). In particular, the Ball–Berry stomatal conductance model in combination with the Farquhar et al. (1980) model for C_3 photosynthesis was introduced into terrestrial biosphere models in the mid-1990s (Bonan 1995; Sellers et al. 1996a; Cox et al. 1998) and is now commonly used to simulate stomatal conductance.

12.3 | Semiempirical Photosynthesis-Based Models

Plants grown under a variety of irradiances, nutrient concentrations, CO_2 concentrations, and leaf water potentials show large variation in photosynthetic rate and stomatal conductance, but photosynthesis and stomatal conductance vary in near constant proportion for a given set of conditions (Wong et al. 1979). The measurements for jack pine given in Figure 12.2 show such a relationship, in which photosynthetic rates increase with higher stomatal conductance. Plants achieve this regulation by maintaining intercellular CO_2 at a nearly constant proportion of ambient CO_2 (i.e., c_i/c_a is constant) for particular environmental conditions. This is evident from the photosynthetic diffusion

equation. Because g_{bw} (~1–4 mol m^{-2} s^{-1}) is typically ten times greater than g_{sw} (~0.1–0.4 mol m^{-2} s^{-1}), the diffusion equation can be approximated as

$$A_n = \frac{g_{sw}}{1.6}(c_a - c_i) = \frac{g_{sw}}{1.6}c_a(1 - c_i/c_a). \quad (12.12)$$

Rearranging terms shows that g_{sw} varies with A_n according to

$$g_{sw} = \frac{1.6A_n}{c_a(1 - c_i/c_a)}. \quad (12.13)$$

A linear relationship between A_n and g_{sw} with a constant slope implies constant c_i/c_a for a given c_a. C_4 plants have a higher rate of photosynthesis for a given stomatal conductance, implying lower c_i/c_a. Indeed, c_i/c_a is about 0.65–0.80 for C_3 plants and 0.40–0.60 for C_4 plants (Hetherington and Woodward 2003). Equation (12.13) provides a simple equation for stomatal conductance if c_i is known, and it has been used in JSBACH with $c_i/c_a = 0.87$ for C_3 plants and $c_i/c_a = 0.67$ for C_4 plants (Knorr and Heimann 2001; Knauer et al. 2015).

The Ball et al. (1987) stomatal conductance model is an extension of (12.13). It is derived from empirical relationships, which demonstrate that stomatal conductance cannot achieve any arbitrary value but, rather, is constrained by the rate of photosynthesis (and the environmental factors that influence photosynthesis) and modulated by relative humidity and CO_2 concentration (Figure 12.3).

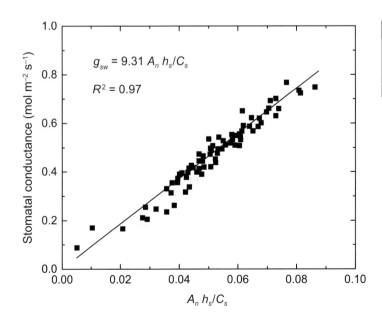

$$g_{sw} = 9.31\ A_n\ h_s/C_s$$

$$R^2 = 0.97$$

Stomatal conductance (mol m^{-2} s^{-1})

$A_n\ h_s/C_s$

Figure 12.3 Relationship between stomatal conductance and $A_n h_s/c_s$ for soybean (*Glycine max*) over a range of light, temperature, humidity, and CO_2 concentration. Adapted from Ball et al. (1987)

The equation is similar to (12.13) but is formulated with respect to the leaf surface rather than ambient air, and stomatal conductance varies directly with net photosynthesis scaled by relative humidity and CO_2 concentration whereby

$$g_{sw} = g_0 + g_1 \frac{A_n}{c_s} h_s. \tag{12.14}$$

Here, h_s is the fractional humidity at the leaf surface (dimensionless), c_s is the CO_2 concentration at the leaf surface (μmol mol^{-1}), g_1 is the slope of the relationship, and g_0 is the minimum conductance. A_n has the units μmol CO_2 m^{-2} s^{-1} while g_{sw} and g_0 have the units mol H_2O m^{-2} s^{-1}. Equation (12.14) specifies g_{sw} in relation to A_n and closes the system of equations for the supply and biochemical functions. A direct expression for c_i can be obtained by substituting (12.3) for A_n in (12.14) to get

$$\frac{c_i}{c_a} = 1 - \frac{1.6}{g_1 h_s}. \tag{12.15}$$

This derivation assumes that $g_0 \ll g_{sw}$ so that g_0 can be neglected. A further simplification needed to obtain (12.15) is $g_{bw} \gg g_{sw}$ so that $c_s \approx c_a$. In this equation, c_i/c_a decreases with lower h_s. Common parameter values used in global models are $g_1 = 9$ and $g_0 = 0.01$ mol H_2O m^{-2} s^{-1} for C_3 plants and $g_1 = 4$ and $g_0 = 0.04$ mol H_2O m^{-2} s^{-1} for C_4 plants (e.g., SiB2, Sellers et al. 1996a,b; CLM4.5, Oleson et al. 2013). However, g_1 varies considerably among species (Medlyn et al. 2011; Miner et al. 2017; Wolz et al. 2017); use of a single value for all plants ignores important variability in the biosphere.

Equation (12.14) is deceptively simple, but its use to simulate stomatal conductance requires knowledge of c_s and h_s. The CO_2 concentration at the leaf surface is given from the diffusion equation using (12.2). The complexity in the equation arises because h_s itself depends on g_{sw}. With reference to Figure 10.2b, $g_{bw}(e_a - e_s) = g_{sw}[e_s - e_{sat}(T_\ell)]$ so that

$$e_s = \frac{g_{bw}e_a + g_{sw}e_{sat}(T_\ell)}{g_{bw} + g_{sw}} \tag{12.16}$$

and $h_s = e_s/e_{sat}(T_\ell)$. Substituting this expression for h_s into (12.14), g_{sw} is the larger root of the quadratic equation:

$$g_{sw}^2 + \left[g_{bw} - g_0 - \frac{g_1 A_n}{c_s} \right] g_{sw} - g_{bw} \left[g_0 + \frac{g_1 A_n}{c_s} \frac{e_a}{e_{sat}(T_\ell)} \right] = 0 \tag{12.17}$$

In this form, h_s is not required. A simplification is to assume that g_{bw} is large so that c_s is replaced with c_a and e_s is replaced with e_a. This avoids the need for (12.17).

The Ball–Berry model is considered to be semi-empirical because although it is obtained from physiological theory it relies on empirical relationships between g_{sw} and A_n. A criticism of the model is that it uses relative humidity to represent the effect of vapor pressure on stomatal conductance. This is more evident when h_s is written as

$$h_s = 1 - D_s/e_{sat}(T_\ell), \tag{12.18}$$

with $D_s = e_{sat}(T_\ell) - e_s$ the vapor pressure deficit at the leaf surface. In this form, it can be seen that stomatal conductance declines linearly with larger vapor pressure deficit. Greater vapor pressure deficit causes stomata to close, which decreases c_i. Other variants of the model replace the linear dependence on D_s with more complex functions.

Leuning (1995) modified the model to adjust c_s for the CO_2 compensation point Γ (μmol mol^{-1}) and used D_s rather than h_s. In the modified equation,

$$g_{sw} = g_0 + g_1 \frac{A_n}{c_s - \Gamma} (1 + D_s/D_0)^{-1}. \tag{12.19}$$

The term $c_s - \Gamma$ improves model performance at low values of c_s. The function $(1 + D_s/D_0)^{-1}$, rather than h_s, better represents stomatal response to vapor pressure. In this equation:

$$\frac{c_i}{c_a} = 1 - \frac{1.6}{g_1} \left(1 - \frac{\Gamma}{c_a} \right) \left(1 + \frac{D_s}{D_0} \right). \tag{12.20}$$

The CO_2 compensation point is given by (11.10); a typical value is $\Gamma \sim 50$ μmol mol^{-1}. This stomatal model has been used in CABLE with $g_1 = 9$ for C_3 plants and $g_1 = 4$ for C_4 plants and $D_0 = 1.5$ kPa (De Kauwe et al. 2015).

Another variant of the model is derived from water-use efficiency optimization theory (Medlyn et al. 2011). In this form, stomatal conductance is given by

$$g_{sw} = g_0 + 1.6 \left(1 + \frac{g_1}{\sqrt{D_s}} \right) \frac{A_n}{c_s}. \tag{12.21}$$

As given here, D_s has the units kPa so that g_1 has the units kPa$^{0.5}$. A key feature of this model is that stomatal conductance varies inversely with $\sqrt{D_s}$ and

$$\frac{c_i}{c_a} = 1 - \frac{\sqrt{D_s}}{g_1 + \sqrt{D_s}}. \tag{12.22}$$

The difficulty in solving (12.21) for g_{sw} is that D_s varies with g_{sw}. With leaf surface vapor pressure given by (12.16), the g_{sw} that satisfies (12.21) is calculated as the larger root of the quadratic equation

$$g_{sw}^2 - \left[2\left(g_0 + \frac{1.6A_n}{c_s}\right) + \left(\frac{1.6A_n}{c_s}g_1\right)^2 \frac{1}{g_{bw}D_\ell}\right]g_{sw} + g_0^2$$

$$+ \left[2g_0 + \frac{1.6A_n}{c_s}\left(1 - \frac{g_1^2}{D_\ell}\right)\right]\frac{1.6A_n}{c_s} = 0, \tag{12.23}$$

with $D_\ell = e_{sat}(T_\ell) - e_a$ (kPa). The slope g_1 varies considerably among plant functional types (Lin et al. 2015). Table 12.3 gives values used in CABLE, for which $g_0 = 0$ (De Kauwe et al. 2015; Kala et al. 2015).

The minimum stomatal conductance g_0 is a parameter in the stomatal conductance models. Analysis of the Leuning (1995) model shows that g_0 is a key determinant of canopy transpiration rates but is poorly constrained (Barnard and Bauerle 2013; Bauerle et al. 2014). In recent studies, a common approach is to take g_0 as zero, both to estimate g_1

from observations (Lin et al. 2015; Wolz et al. 2017; Knauer et al. 2018) and in model implementation (De Kauwe et al. 2015; Kala et al. 2015). On the other hand, observations can show significant transpiration water loss at night, with important implications for the hydrologic cycle (Lombardozzi et al. 2017). A compromise might be to represent stomatal conductance as the maximum of g_{sw} or g_0 rather than as the sum of the two values (De Kauwe et al. 2015; Lin et al. 2015).

Question 12.1 Calculate e_s with $g_{bw} = 0$ and with $g_{bw} = \infty$.

Question 12.2 Derive an equation that relates the vapor pressure deficit D_s at the leaf surface to g_{sw} and g_{bw}. How does g_{bw} affect D_s?

Question 12.3 Equations (12.17) and (12.23) are expressions to calculate g_{sw} given A_n using the Ball et al. (1987) and Medlyn et al. (2011) stomatal conductance models. They are obtained by eliminating h_s and D_s, respectively, from the models. Derive the corresponding quadratic equation for g_{sw} using the Leuning (1995) model.

12.4 | Solution Techniques

Semiempirical A_n-g_{sw} models require photosynthesis to calculate stomatal conductance. The photosynthetic rate can be obtained from leaf gas exchange measurements in observational studies but must be calculated for terrestrial biosphere models, commonly using the Farquhar et al. (1980) model for C_3 plants. The environmental variables needed are c_a – CO_2 concentration of ambient air; o_i – intercellular O_2 concentration (209 mmol mol^{-1}); $I^\downarrow$ – incident photosynthetically active radiation; T_ℓ – leaf temperature; e_a – vapor pressure of ambient air; and g_{bw} – leaf boundary layer conductance for H_2O. Their use requires solving a set of equations simultaneously for A_n, g_{sw}, c_i, c_s, and D_s, as summarized in Table 12.4. This involves finding c_i that satisfies the equation set. Two mathematical techniques are available to do so. An analytical equation for c_i can be attained for C_3 photosynthesis under particular circumstances, as either a quadratic or cubic

	SiB2, CLM4.5		CABLE
	Ball et al.	Leuning	Medlyn
Plant type	(1987)	(1995)	et al. (2011)
Evergreen needleleaf tree	9	9	2.35
Evergreen broadleaf tree	9	9	4.12
Deciduous needleleaf tree	9	9	2.35
Deciduous broadleaf tree	9	9	4.45
Shrub	9	9	4.70
C_3 grass	9	9	5.25
C_4 grass	4	4	1.62
C_3 crop	9	9	5.79

Table 12.3 Values for g_1 in three stomatal conductance models as implemented in SiB2, CLM4.5, and CABLE

Source: SiB2 (Sellers et al. 1996a,b); CLM4.5 (Oleson et al. 2013); CABLE (De Kauwe et al. 2015; Kala et al. 2015).

Table 12.4 | Equations to solve C_3 photosynthesis, stomatal conductance, and leaf temperature

Variable	Equation	Units
Biochemical (demand-based) photosynthesis rate	$A_c = \dfrac{V_{c\max}(c_i - \Gamma_*)}{c_i + K_c(1 + o_i/K_o)}$	μmol CO_2 m^{-2} s^{-1}
	$A_j = \dfrac{J}{4}\left(\dfrac{c_i - \Gamma_*}{c_i + 2\Gamma_*}\right),$	
	$A = \min(A_c, A_j)$, or	
	$\Theta_A A^2 - (A_c + A_j)A + A_c A_j = 0$	
Net photosynthesis rate	$A_n = A - R_d$	μmol CO_2 m^{-2} s^{-1}
Leaf surface CO_2	$c_s = c_a - 1.4A_n/g_{bw}$	μmol mol^{-1}
Leaf surface vapor pressure	$D_s = \dfrac{e_{sat}(T_\ell) - e_a}{1 + g_{sw}/g_{bw}}$	Pa
Stomatal conductance		mol H_2O m^{-2} s^{-1}
Ball et al. (1987)	$g_{sw} = g_0 + g_1 \dfrac{A_n}{c_s}[1 - D_s/e_{sat}(T_\ell)]$	
Medlyn et al. (2011)	$g_{sw} = g_0 + 1.6\dfrac{A_n}{c_s}\left(1 + g_1 D_s^{-1/2}\right)$	
Intercellular CO_2 (diffusion, or supply-based, rate)	$c_i = c_a - A_n\left(1.4g_{bw}^{-1} + 1.6g_{sw}^{-1}\right)$	μmol mol^{-1}
Leaf temperature	$c_L \dfrac{\partial T_\ell}{\partial t} = Q_a - 2\varepsilon_\ell \sigma T_\ell^4 - 2c_p(T_\ell - T_a)g_{bh}$ $-\lambda g_{\ell w}[e_{sat}(T_\ell) - e_a]/P$	K

equation for c_i. Such a solution requires simplifying assumptions to make the equations more tractable to solve. Alternatively, numerical methods can be used to solve the equation set for c_i.

Leuning (1990) provided a commonly used analytical solution for C_3 photosynthesis when c_s and D_s are known, which simplifies the equation set to three equations with three unknowns (A_n, g_{sw}, c_i). In Chapter 11, an expression for c_i was obtained by combining the biochemical demand for CO_2 represented by the Farquhar et al. (1980) model with the diffusive supply of CO_2, resulting in a quadratic equation for c_i (11.79). That solution requires a value for g_{sw}. The stomatal constraint function, by relating g_{sw} to A_n (and therefore c_i), replaces g_{sw} and gives another quadratic equation for c_i, but now independent of g_{sw}. The solution requires solving for the c_i that satisfies the biochemical demand, the diffusive supply, and the stomatal constraint function. Here, the solution is given for the Ball–Berry conductance model. The equations for A_n differ between the Rubico-limited rate A_c and the RuBP

regeneration-limited rate A_j, and so the solution to g_{sw} also differs between these two rates. As in Chapter 11, a general equation for the biochemical demand is

$$A_n = \frac{a(c_i - \Gamma_*)}{c_i + b} - R_d, \tag{12.24}$$

with $a = V_{c\max}$ and $b = K_c(1 + o_i/K_o)$ when photosynthesis is limited by A_c or $a = J/4$ and $b = 2\Gamma_*$ when photosynthesis is limited by A_j. The supply function is $A_n = (c_s - c_i)g_{sw}/1.6$, and this is combined with the stomatal constraint function (12.14) to eliminate g_{sw}. The resulting supply-constraint function is

$$A_n = \frac{g_0}{1.6/(c_s - c_i) - g_1 h_s/c_s}. \tag{12.25}$$

The solution is the c_i that satisfies the demand function (12.24) and the supply-constraint function (12.25). Combining equations yields the quadratic equation,

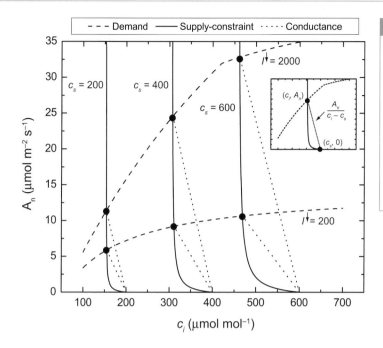

Figure 12.4 Photosynthesis A_n in relation to intercellular CO_2 concentration c_i for C_3 photosynthesis illustrating the graphical solution for the coupled photosynthesis–conductance model (Leuning 1990). Shown is the photosynthetic demand function (12.24) at $I^\downarrow =$ 200 and 2000 μmol photon m^{-2} s^{-1}. Also shown is the supply-constraint function (12.25) at, $c_s = 200, 400,$ and 600 μmol mol^{-1} and $h_s = 0.75$. The intersection of the two functions defines the solution for c_i and A_n. Stomatal conductance is given by the negative slope of the dotted line, as illustrated in the inset panel.

$$\left[\frac{g_0}{1.6} + v(a - R_d)\right]c_i^2$$

$$+ \left[(1 - vc_s)(a - R_d) + \frac{g_0}{1.6}(b - c_s) - v(a\Gamma_* + bR_d)\right]c_i$$

$$+ \left[(vc_s - 1)(a\Gamma_* + bR_d) - bc_s\frac{g_0}{1.6}\right] = 0$$

$$(12.26)$$

where $v = g_1 h_s/(1.6c_s)$. The correct solution is the larger root. Equation (12.26) must be solved twice, once to give c_i for A_c and again for A_j. Stomatal conductance is then obtained from (12.14).

Figure 12.4 illustrates the solution graphically. Demand curves for A_n are shown in relation to c_i at low light ($I^\downarrow = 200$ μmol photon m^{-2} s^{-1}) and high light (2000 μmol photon m^{-2} s^{-1}). Supply-constraint curves are shown for $c_s = 200, 400,$ and 600 μmol mol^{-1}. The solution is the value of c_i that satisfies both functions, found at the intersection of the demand and supply-constraint curves. Stomatal conductance is given by the negative value of the slope of the line drawn from the intersection point (which gives c_i and A_n) to the point on the x-axis defined by $c_i = c_s$ and $A_n = 0$. This slope is equal to $A_n/(c_i - c_s)$, and the supply function shows that $g_{sw}/1.6 = -A_n/(c_i - c_s)$. For a given c_s, the increase in light produces higher A_n and g_{sw}, but c_i is mostly constant.

Baldocchi (1994) and Su et al. (1996) obtained another analytical solution for C_3 photosynthesis coupled with the Ball–Berry stomatal conductance model. With h_s specified, a cubic equation for A_n eliminates c_i and g_{sw}. The cubic equation for A_n is obtained by solving four simultaneous equations for A_n, c_i, c_s, and g_{sw}. First, substitute (12.2) for c_s into (12.14) to eliminate c_s from g_{sw}. Then, substitute the resulting expression for g_{sw} into (12.4) to eliminate g_{sw} from c_i. Substitute the new expression for c_i into A_n as given by (12.24) to eliminate c_i. The resulting cubic equation for A_n is

$$c_3 A_n^3 + c_2 A_n^2 + c_1 A_n + c_0 = 0. \qquad (12.27)$$

The coefficients c_0–c_3 are given in the chapter appendix. The cubic equation is valid for $A_n > 0$, and the solution is the smaller of the positive roots.

Although (12.26) and (12.27) provide expressions for g_{sw} for C_3 plants, a more general iterative numerical calculation is required to account for the full complexity of stomatal models. Leaf surface vapor pressure itself depends on stomatal conductance, and stomatal conductance depends on leaf temperature as well (through A_n). Nor do the analytical solutions allow for co-limitation of A_c and A_j. Some stomatal models also account for hydraulic limits to plant water uptake (Chapter 13). A more general solution technique is to numerically calculate c_i, as illustrated in Figure 12.5. The inner loop solves for c_i given a known leaf temperature. Using an initial

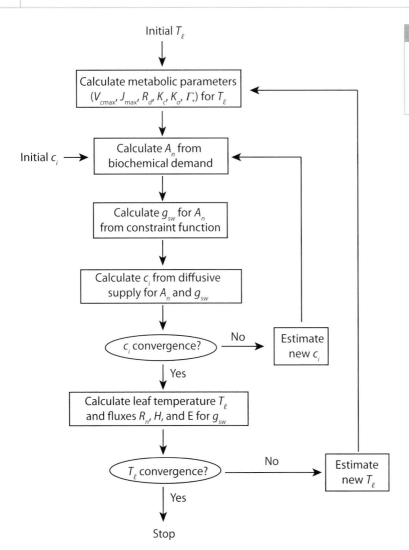

Initial T_ℓ

Calculate metabolic parameters
$(V_{cmax}, J_{max}, R_d, K_c, K_o, \Gamma_*)$ for T_ℓ

Initial c_i →

Calculate A_n from
biochemical demand

Calculate g_{sw} for A_n
from constraint function

Calculate c_i from diffusive
supply for A_n and g_{sw}

c_i convergence? — No → Estimate new c_i

Yes

Calculate leaf temperature T_ℓ
and fluxes R_n, H, and E for g_{sw}

T_ℓ convergence? — No → Estimate new T_ℓ

Yes

Stop

Figure 12.5 Flow diagram of leaf flux calculations. The inner iteration solves the coupled photosynthesis–stomatal conductance model for c_i. The outer iteration shows the leaf temperature calculation.

estimate for c_i, A_n is calculated from the biochemical (demand-based) rate. Stomatal conductance is next calculated depending on the particular stomatal constraint function – e.g., (12.17) or (12.23) – and with c_s given by (12.2). Once g_{sw} is known, a new estimate c_i' for intercellular CO_2 is obtained from the diffusion (supply-based) equation using (12.4). The iteration is repeated until the change in c_i ($\Delta c_i = c_i' - c_i$) satisfies some convergence criteria (e.g., $\pm 10^{-3}$ μmol mol^{-1}). The numerical solution is further complicated in that the parameters $V_{c\,max}$, J_{max}, R_d, K_c, K_o, and Γ_* that govern photosynthesis depend on leaf temperature, as does the vapor pressure deficit D_s used in the stomatal constraint function. Leaf temperature varies with transpiration rate and g_{sw} as specified through the leaf energy budget (Chapter 10), and the entire calculation is iterated until temperature converges within some specified tolerance. Numerical techniques such as the secant method and Brent's method (Appendix A5) can be used to minimize Δc_i and ΔT_ℓ. This is a general approach that works for both C$_3$ and C$_4$ photosynthesis. A critical difference among terrestrial biosphere models is the detail with which they represent the full dependencies among photosynthesis, stomatal conductance, and the leaf energy budget or simplify the equations for easier solution.

Question 12.4 Derive a quadratic equation similar to (12.26) for c_i but using the Medlyn et al. (2011) model for g_{sw}.

12.5 | Stomatal Response to Environmental Factors

Coupled A_n-g_{sw} models broadly capture the response of stomatal conductance to environmental factors, but there are considerable differences among model implementations, as seen in comparison of the Ball et al. (1987; denoted as BB) and Medlyn et al. (2011; denoted as MED) models for C_3 plants (Figure 12.6). These simulations use parameter values for deciduous broadleaf trees as implemented in two terrestrial biosphere models (Table 12.3). CLM4.5 uses BB with $g_0 = 0.01$ mol H_2O m^{-2} s^{-1} and $g_1 = 9$. CABLE uses MED with $g_0 = 0$ and $g_1 = 4.45$ kPa$^{0.5}$. Both

stomatal models show the same functional response to light, CO_2, and temperature, but MED has larger stomatal conductance compared with BB. The MED light-saturated conductance is more than 50% larger than BB. The shape of the CO_2 response is also similar between models, but offset with higher values for MED. The 16% decline in conductance from 370 to 570 μmol mol^{-1} in BB and the 19% decline in MED are consistent with the approximately 20% reduction seen in free-air CO_2 enrichment studies (Ainsworth and Rogers 2007). In both models, stomatal conductance increases with warmer temperature up to some optimum, beyond which conductance declines. The effect of temperature on stomatal conductance is felt from the

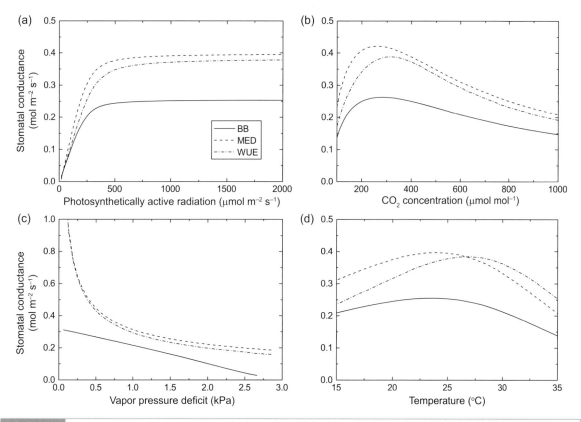

Figure 12.6 Stomatal conductance in response to (a) photosynthetically active radiation, (b) ambient CO_2 concentration, (c) leaf-to-air vapor pressure deficit, and (d) leaf temperature. Shown are simulations for a deciduous broadleaf tree ($V_{cmax25} = 57.7$ μmol m^{-2} s^{-1}, $J_{max25} = 96.4$ μmol m^{-2} s^{-1}) using Ball et al. (1987) with $g_0 = 0.01$ mol m^{-2} s^{-1} and $g_1 = 9$ (BB); Medlyn et al. (2011) with $g_0 = 0$ mol m^{-2} s^{-1} and $g_1 = 4.45$ kPa$^{0.5}$ (MED); and water-use efficiency optimization, as in Bonan et al. (2014) with $\iota = 750$ μmol CO_2 mol^{-1} H_2O (WUE). Standard conditions are $I^{\downarrow} = 2000$ μmol m^{-2} s^{-1}, $c_a = 380$ μmol mol^{-1}, $e_a = 2535$ Pa (80% relative humidity at 25°C), $T_{\ell} = 25$°C, and $g_{bw} = 2$ mol m^{-2} s^{-1}. Environmental factors were varied individually. For (c), e_a was evaluated at 25°C for a range of relative humidity (10%–100%). For (d), e_a was adjusted so that relative humidity remained constant (80%) or vapor pressure deficit remained constant (0.6 kPa) depending on stomatal model. See also Franks et al. (2017).

biochemical effects on photosynthesis and also from changes in the leaf-to-air vapor pressure deficit. It is important to distinguish these two effects (Lloyd and Farquhar 2008; Duursma et al. 2014). A difference between BB and MED is stomatal response to vapor pressure deficit at low and high extremes. One factor not included in the models is ozone. Increased concentrations of tropospheric O_3 damage stomata, but the appropriate way to parameterize this is uncertain (Amthor et al. 1994; Sitch et al. 2007; Wittig et al. 2007; Ainsworth et al. 2012; Lombardozzi et al. 2013, 2015b).

For C_3 plants, c_i/c_a is typically maintained at a value of approximately 0.65–0.80 (Hetherington and Woodward 2003). This ratio is relatively constant with variation in light or c_a, but varies with humidity. Figure 12.7 illustrates this for the light response curves shown in Figure 12.6a. The c_i/c_a simulated with BB is relatively constant (0.78) over a wide range of stomatal conductance, except at very low conductance. This value is the same as that given by (12.15). MED has a higher c_i/c_a (0.84), consistent with (12.22).

In addition to differences in the form of the equation, the two stomatal models differ in parameter values. When the slope of MED is decreased to $g_1 = 2.8$ kPa$^{0.5}$, the light, CO_2, and temperature responses are nearly identical to those for BB with its slope $g_1 = 9$ (Figure 12.8). The lower slope also decreases c_i/c_a (from 0.84 to 0.77). Stomatal response to vapor pressure deficit is similar over a moderate range but differs between models with extreme high or low vapor pressure deficit. In particular, whereas BB approaches zero at high vapor pressure deficit, MED asymptotes to near constant conductance. On the other hand, MED approaches infinite conductance with low vapor pressure deficit. The leaf simulations shown in Figure 12.6 and Figure 12.8 demonstrate that a primary difference between the two models relates to values of g_1, not to the form of the equation (Franks et al. 2017).

In Figure 12.6b and Figure 12.8b, stomatal conductance increases with c_a to about 300 µmol mol^{-1}, beyond which conductance decreases with elevated c_a. This behavior is seen in both BB and MED and relates to differences between the A_c- and A_j-limited rates, as has been discussed by others (Farquhar and Wong 1984; Arneth et al. 2002; Medlyn et al. 2013; Vico et al. 2013; Buckley and Schymanski 2014; Buckley 2017; Buckley et al. 2017; Franks et al. 2017). In these leaf simulations, photosynthesis is modeled as the co-limited of the A_c and A_j rates. Simulations without co-limitation show that stomatal response to c_a differs markedly between the A_c- and A_j-limited rates (Figure 12.9). Rubisco limits photosynthesis at high light or low c_a, when photosynthesis follows the A_c-limited rate (Figure 11.2). When g_{sw} is calculated from the A_c-limited rate, stomata open as c_a increases to 300–400 µmol mol^{-1} and thereafter close with elevated c_a. RuBP regeneration limits photosynthesis at low light or high c_a, when photosynthesis follows

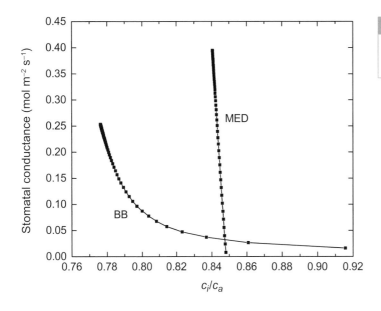

Figure 12.7 Intercellular CO_2 expressed in proportion to ambient CO_2 (c_i/c_a) in relation to stomatal conductance. Data are from the light response curves in Figure 12.6a

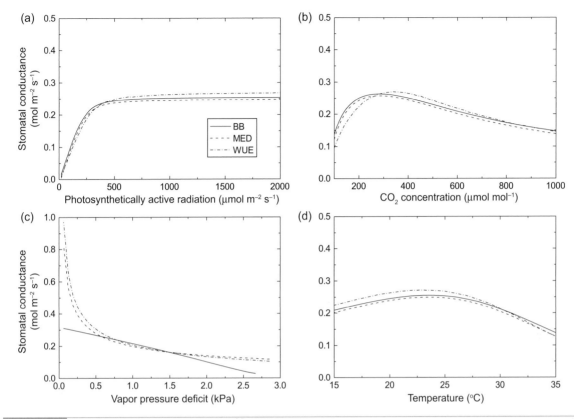

Figure 12.8 As in Figure 12.6, but with $g_1 = 2.8$ kPa$^{0.5}$ for MED and $\iota = 32.76\Gamma_*$ for WUE so that ι increases with temperature in proportion to Γ_* ($\iota = 1400$ μmol CO$_2$ mol^{-1} H$_2$O at 25°C). See also Franks et al. (2017)

the A_j-limited rate (Figure 11.2). When g_{sw} is calculated from the A_j-limited rate, stomata close with elevated c_a above about 150–200 μmol mol^{-1}. This difference in stomatal behavior arises from the different relationships of A_c and A_j with c_a (Figure 11.2a).

Stomatal response to c_a depends on light (Wong et al. 1978; Morison and Jarvis 1983; Morison 1987), as seen in Figure 12.9. With low light, A_j is the lesser of the two rates at low c_a (around 200 μmol mol^{-1}), and stomatal conductance declines as c_a increases above this value. At high light, A_c is limiting over a broader range of c_a (up to 400–500 μmol mol^{-1}). If stomata behavior followed the A_c-limited rate, conductance would not begin to decline until high c_a is attained. This is in contrast with numerous observations showing a decline in conductance with elevated c_a (Buckley 2017) and has led to the suggestion that stomata behave as if RuBP regeneration regulates their functioning (Medlyn et al. 2011, 2013). The distinction between the A_c- and A_j-limited rates

is, to some extent, an artifact of using the minimum rate. Co-limitation smooths the transition between these rates, and stomatal response to c_a more closely follows A_j, shown also by Vico et al. (2013). Buckley et al. (2017) and Franks et al. (2017) also noted the importance of co-limitation when modeling stomatal conductance and suggested that co-limitation is necessary to account for heterogeneity within the leaf.

The preceding examples calculated leaf gas exchange at a specified leaf temperature. However, stomatal conductance itself affects temperature. The interdependencies between leaf biophysics and leaf physiology are seen in the relationship between leaf fluxes and leaf dimension. Large leaves with a thick boundary layer are decoupled from the surrounding air and generally have a higher leaf temperature than do smaller leaves that have a thinner boundary layer and are closely coupled to the ambient air. Figure 12.10 illustrates this behavior for sunlit and shaded leaves on a warm, dry day with

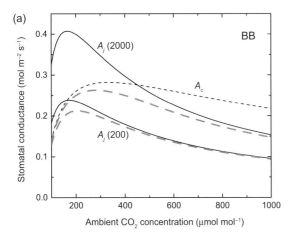

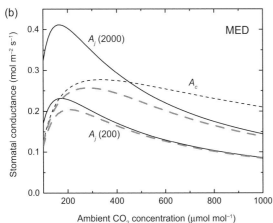

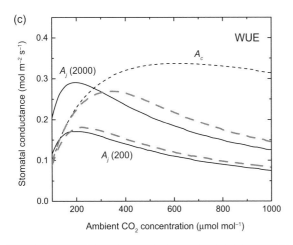

Figure 12.9 As in Figure 12.8b, but showing stomatal conductance separately for the A_c- and A_j-limited rates for (a) BB, (b) MED, and (c) WUE. The A_j conductance is given for $I^\downarrow$ = 200 and 2000 μmol m^{-2} s^{-1}. The dashed gray lines show the co-limited conductance for the two light regimes. See also Franks et al. (2017)

light wind. Boundary layer conductance decreases with larger leaf size, as discussed in Chapter 10. The temperature of the shaded leaf is insensitive to leaf size, but the sunlit leaf becomes warmer as leaf size increases. The small difference in boundary layer conductance between the sunlit and shaded leaves arises from the temperature difference between the leaves as it affects free convection. Stomatal conductance, photosynthesis, and transpiration decline with larger leaf size for the sunlit leaf but are relatively insensitive to leaf size for the shaded leaf. For the sunlit leaf, small leaves have a higher water-use efficiency than large leaves.

12.6 | Optimality Theory

Models that relate stomatal conductance to photosynthesis are empirical and provide the stomatal constraint function needed to close the system of equations that describes A_n, g_{sw}, and c_i. Water-use efficiency optimization theory provides an alternative stomatal constraint. Stomata open for carbon gain during photosynthesis (a benefit), but in doing so water is lost in transpiration (a cost). Cowan (1977) and Cowan and Farquhar (1977) formulated a theory of stomatal behavior based on the premise that stomata function in a manner to minimize the water cost of carbon gain over some time interval (see also Cowan 1982). Their derivation minimized the marginal water cost of carbon gain, which they expressed using the notation $\partial E / \partial A_n = \lambda$ with λ the Lagrange multiplier (Chapter 4). In the text that follows, the solution is specified in terms of the marginal carbon gain of water loss ($\partial A_n / \partial E = \iota$), which uses the same notation as water-use efficiency so that ι is marginal water-use efficiency and $\iota = 1/\lambda$.

Mathematically, the problem can be formulated as an optimization that maximizes carbon gain per unit of water loss (A_n/E). The instantaneous water-use efficiency is trivially large with $g_{sw} = 0$ (so that $E = 0$). Instead, the problem is posed so that optimal stomatal conductance is that which maximizes carbon uptake over some time interval (e.g., daily), which can be expressed as

$$\int A_n[g_{sw}(t)]dt, \tag{12.28}$$

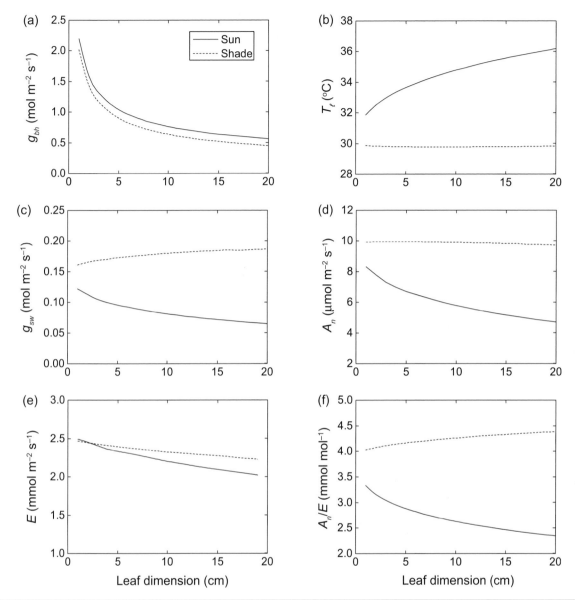

Figure 12.10 Leaf microclimate and boundary layer processes in relation to leaf dimension for sun and shade conditions. Shown are (a) boundary layer conductance for heat g_{bh}, (b) leaf temperature T_ℓ, (c) stomatal conductance g_{sw}, (d) photosynthesis A_n, (e) transpiration E, and (f) water-use efficiency A_n/E. Conditions are for a warm day ($T_a = 30°C$) with 60% relative humidity and calm wind ($u = 1$ m s^{-1}). Downwelling solar radiation from the sky is 800 W m^{-2} (sunlit leaf) and 300 W m^{-2} (shaded leaf). Simulations are for a C$_3$ plant with $V_{c\,max\,25} = 60$ μmol m^{-2} s^{-1}, $J_{max\,25} = 100$ μmol m^{-2} s^{-1}, and use the Ball–Berry stomatal model.

subject to a constraint imposed by water loss in which

$$\int E[g_{sw}(t)]dt = E_0. \tag{12.29}$$

The notation $g_{sw}(t)$ denotes that stomatal conductance varies with time t depending on environmental conditions, and both A_n and E depend on g_{sw}. The latter equation is a statement that transpiration equals some amount E_0 over the time interval. The optimal stomatal conductance satisfies the relationship:

$$\frac{\partial A_n/\partial g_{sw}}{\partial E/\partial g_{sw}} = \frac{\partial A_n}{\partial E} = \iota. \tag{12.30}$$

(See Buckley et al. (2002) and Konrad et al. (2008) for the mathematical proof.) The parameter ι is an arbitrary constant used for optimization but relates transpiration to a carbon cost (mol CO_2 mol^{-1} H_2O). A large value means that transpiration is costly, necessitating a conservative water-use strategy. Equation (12.30) shows that stomata behave optimally when g_{sw} varies to maintain the marginal carbon gain of water loss $\partial A_n / \partial E$ constant at a value equal to ι. The constraint E_0 on transpiration is unknown, but mathematical expressions for g_{sw} can be obtained if ι is specified. The challenge in using (12.30) to calculate g_{sw} is that the equation is merely a statement of the condition for optimal stomata behavior. It gives no insights to ι itself or what value it should be. A further difficulty pertains to the time span over which ι is constant – a necessary constraint for optimality. This is commonly taken as a day, but there is much debate concerning how ι varies over longer time periods and in response to environment.

An expression for $\partial A_n / \partial E$ in relation to g_{sw} is obtained from consideration of leaf gas exchange. Photosynthesis is described as a diffusion process by (12.1) in which:

$$A_n = (c_a - c_i)g_{\ell c}, \tag{12.31}$$

with $g_{\ell c} = 1/(1.4 g_{bw}^{-1} + 1.6 g_{sw}^{-1})$ the leaf conductance to CO_2. The comparable equation for transpiration is

$$E = w_\ell g_{\ell w}, \tag{12.32}$$

with $g_{\ell w} = 1/(g_{bw}^{-1} + g_{sw}^{-1})$ the leaf conductance to H_2O and $w_\ell = [e_{sat}(T_\ell) - e_a]/P$ the leaf-to-air water vapor deficit (mol mol^{-1}). With these equations,

$$\frac{\partial A_n}{\partial E} = \frac{\partial A_n / \partial g_{\ell w}}{\partial E / \partial g_{\ell w}} = \left(\frac{\partial A_n / \partial g_{\ell c}}{\partial E / \partial g_{\ell w}}\right)\frac{\partial g_{\ell c}}{\partial g_{\ell w}}, \tag{12.33}$$

as given by Buckley et al. (2002). Ignoring the dependence of leaf temperature on g_{sw} (i.e., for a constant leaf temperature):

$$\frac{\partial A_n}{\partial E} = \left(\frac{c_a - c_i}{w_\ell}\right)\left(\frac{\partial A_n / \partial c_i}{\partial A_n / \partial c_i + g_{\ell c}}\right)1.6\frac{g_{\ell c}^2}{g_{\ell w}^2}. \tag{12.34}$$

By setting $\partial A_n / \partial E = \iota$, (12.34) provides a mathematical expression for g_{sw}, but it is difficult to solve directly for g_{sw} because c_i depends on g_{sw} and $\partial A_n / \partial c_i$ depends on c_i. Instead, (12.34) must be solved using numerical methods unless it can be simplified. An even more complex equation for $\partial A_n / \partial E$ arises if the temperature dependencies of

A_n and E are included (Cowan 1977; Arneth et al. 2002; Buckley et al. 2002, 2017).

With simplifications, an analytical expression for g_{sw} can be obtained from (12.34) (Hari et al. 1986; Lloyd 1991; Arneth et al. 2002; Katul et al. 2010; Medlyn et al. 2011; Vico et al. 2013). Leaf gas exchange is greatly simplified by assuming that boundary layer conductance is much larger than stomatal conductance so that g_{bw} can be neglected. This is equivalent to assuming that the leaf is well-coupled to ambient air and that conditions at the leaf surface are equal to ambient air. Leaf conductance for H_2O is then g_{sw}, and leaf conductance for CO_2 is $g_{sw}/1.6$. Ignoring g_{bw}, (12.34) simplifies to

$$\frac{\partial A_n}{\partial E} = \left(\frac{c_a - c_i}{1.6 w_\ell}\right)\left(\frac{\partial A_n / \partial c_i}{\partial A_n / \partial c_i + g_{sw}/1.6}\right), \tag{12.35}$$

and with the requirement that $\partial A_n / \partial E = \iota$, (12.35) can be rearranged to give an expression for stomatal conductance whereby:

$$g_{sw} = 1.6\frac{\partial A_n}{\partial c_i}\left(\frac{c_a - c_i}{1.6 w_\ell \iota} - 1\right). \tag{12.36}$$

This provides the necessary stomatal constraint function akin to the Ball–Berry model or other such models. An analytical expression for g_{sw} can be obtained by finding c_i that satisfies the diffusive supply, the biochemical demand, and the stomatal constraint.

A difficulty in using (12.36) to calculate g_{sw} is that it requires an expression for $\partial A_n / \partial c_i$. That is obtained from the biochemical demand for CO_2, which for C$_3$ photosynthesis is a nonlinear function of c_i with the general form given by (12.24) so that $\partial A_n / \partial c_i$ depends on c_i. This dependence results in complex equations for g_{sw}, as described later. A common assumption that simplifies (12.36) is to assume that $\partial A_n / \partial c_i$ is independent of c_i. This can be achieved using a linear demand function with the form $A_n = k(c_i - \Gamma_*) - R_d$ so that $\partial A_n / \partial c_i = k$. The derivation is further simplified by the assumption that respiration is much smaller than photosynthesis so that R_d is ignored. Equation (12.36) can then be rewritten as

$$g_{sw} = 1.6k\left(\sqrt{\frac{c_a - \Gamma_*}{1.6 w_\ell \iota}} - 1\right). \tag{12.37}$$

Variants of this equation have been derived by Hari et al. (1986), Lloyd (1991), and Katul et al. (2010).

Arneth et al. (2002) obtained a more complex equation that retains the nonlinear form of the biochemical demand function. The derivation involves finding c_i for the A_c-limited and A_j-limited rates when stomatal behavior optimally. The solution requires expressions for A_n and $\partial A_n / \partial c_i$. These are obtained from the biochemical demand function, which has the general form given by (12.24) whereby

$$\frac{\partial A_n}{\partial c_i} = \frac{a(b + \Gamma_*)}{(c_i + b)^2}.$$ (12.38)

To obtain the solution, g_{sw} on the left-hand side of (12.36) is replaced using the diffusion equation $A_n = (c_a - c_i)g_{sw}/1.6$ (neglecting g_{bw}) so that

$$A_n = \frac{\partial A_n}{\partial c_i}(c_a - c_i)\left(\frac{c_a - c_i}{1.6w_\ell l} - 1\right).$$ (12.39)

Rearranging terms in (12.39) provides a quadratic equation for c_i in which

$$c_i^2 + (1.6w_\ell l - 2c_a)c_i + \left[(c_a - 1.6w_\ell l)c_a - \frac{1.6w_\ell l A_n}{\partial A_n / \partial c_i}\right] = 0.$$ (12.40)

A_n is replaced by the biochemical demand function given by (12.24), which is simplified by ignoring R_d, and (12.38) is substituted into (12.40) to obtain a new quadratic equation for c_i:

$$a_0 c_i^2 + b_0 c_i + c_0 = 0,$$ (12.41)

with

$$a_0 = 1 - \frac{1.6w_\ell l}{b + \Gamma_*}$$ (12.42)

$$b_0 = 1.6w_\ell l\left(\frac{2\Gamma_*}{b + \Gamma_*}\right) - 2c_a$$ (12.43)

$$c_0 = c_a^2 + 1.6w_\ell l\left(\frac{b\Gamma_*}{b + \Gamma_*} - c_a\right).$$ (12.44)

The correct solution is the smaller root of the quadratic equation. This c_i is used to calculate A_n from the biochemical demand equation, and g_{sw} is calculated from the diffusion equation.

The Medlyn et al. (2011) stomatal model is obtained from the Arneth et al. (2002) optimal stomatal conductance. Equation (12.41) is a complex expression for c_i that varies between the A_c- and A_j-limited rates. Medlyn et al. (2011) simplified the A_j-limited expression for c_i and used this to obtain

g_{sw}. With the assumption that $c_a \gg \Gamma_*$, the quadratic equation for c_i given by (12.41) simplifies so that c_i / c_a for A_j is approximated by

$$\frac{c_i}{c_a} = \frac{\sqrt{3\Gamma_*}}{\sqrt{3\Gamma_*} + \sqrt{1.6w_\ell l}}.$$ (12.45)

Substituting this equation into the diffusion equation $A_n = c_a(1 - c_i/c_a)g_{sw}/1.6$ gives an equation for stomatal conductance in which

$$g_{sw} = 1.6\left(1 + \frac{g_1}{\sqrt{w_\ell}}\right)\frac{A_n}{c_a},$$ (12.46)

with

$$g_1 = \sqrt{\frac{3\Gamma_*}{1.6l}}.$$ (12.47)

An equivalent expression for c_i / c_a calculated by substituting g_1 into (12.45) is

$$\frac{c_i}{c_a} = 1 - \frac{\sqrt{w_\ell}}{g_1 + \sqrt{w_\ell}}.$$ (12.48)

Equation (12.46) provides the theoretical basis for the Medlyn et al. (2011) model given previously by (12.21). It is given here with water vapor expressed as mole fraction (mol mol^{-1}) instead of pressure (kPa). However, a key difference is that (12.46) is derived using c_a and w_ℓ for ambient air rather than the leaf surface as in (12.21) and lacks a minimum conductance g_0. A defining feature of the model is that it is derived from A_j-limited photosynthesis, but this is thought to be appropriate (Medlyn et al. 2011, 2013). Equation (12.47) shows that g_1 is inversely related to the square root of marginal water-use efficiency. g_1, therefore, represents water-use efficiency; small values indicate high water-use efficiency and vice versa. Observations of stomatal conductance for different plant functional types and in different environments support this conclusion (Lin et al. 2015). g_1 also increases with growth temperature, in part because of its dependence on Γ_*.

Equation (12.48) for c_i/c_a is derived from water-use efficiency optimization. The same equation can be obtained from a different optimization. Prentice et al. (2014) proposed that plants minimize the combined costs for transpiration and carboxylation needed to achieve a given photosynthetic rate. For transpiration, the cost is to maintain tissues for water uptake and transport, while for carboxylation the cost is to maintain Rubisco. They posed the optimization such that c_i/c_a varies so as to minimize a weighted sum of E and $V_{c\max}$ expressed relative to photosynthetic rate, whereby the optimal c_i/c_a is defined by

$$\frac{\partial}{\partial(c_i/c_a)}\left[\frac{a_1E + b_1V_{c\max}}{A_n}\right] = 0. \tag{12.49}$$

With the assumptions that R_d is much smaller than photosynthetic CO_2 uptake and that $c_i \gg \Gamma_*$, $V_{c\max}/A_n$ can be expressed from the A_c-limited rate as

$$\frac{V_{c\max}}{A_n} = \frac{c_i + K_m}{c_i}, \tag{12.50}$$

with $K_m = K_c(1 + o_i/K_o)$. An expression for E/A_n can be obtained from the diffusion equations with the assumption that $g_{bw} \gg g_{sw}$ so that

$$\frac{E}{A_n} = \frac{1.6w_\ell}{c_a(1 - c_i/c_a)}. \tag{12.51}$$

Taking the partial derivatives of (12.50) and (12.51) with respect to c_i/c_a, (12.49) is equivalent to

$$\frac{1.6a_1w_\ell}{c_a(1 - c_i/c_a)^2} = \frac{b_1K_m/c_a}{(c_i/c_a)^2}. \tag{12.52}$$

Equation (12.52) can be rearranged in the same form as (12.48), but with

$$g_1 = \sqrt{\frac{b_1K_m}{1.6a_1}}. \tag{12.53}$$

This derivation obtains the optimal c_i/c_a for A_c-limited photosynthesis and so extends the generality of (12.48), but with a different interpretation of g_1. Equation (12.48) can also be obtained from other optimization criteria, with g_1 related to photosynthetic and plant hydraulic variables (Dewar et al. 2018).

The preceding derivations show that analytical expressions for g_{sw} can be obtained from optimality theory, but such equations are by necessity simplifications of complex biochemical and biophysical processes. The simplifications and approximations needed to obtain the analytical equations for optimal g_{sw} can affect the results compared with the full solution (Buckley et al. 2017). A further difficulty in using analytical solutions is that the form of the equations differ between the A_c- and A_j-limited rates of photosynthesis. Accounting for co-limitation

adds considerable complexity to the solution (Buckley et al. 2002, 2017; Vico et al. 2013; Buckley and Schymanski 2014). If the full complexity of the flux equations is retained, numerical modeling is required to solve the system of equations for optimization. One numerical approach is to maximize the objective function $A_n - \iota E$ (Duursma 2015). Alternatively, (12.34) can be solved numerically for g_{sw} such that $\partial A_n/\partial E = \iota$ (Buckley et al. 2002, 2017; Buckley and Schymanski 2014). Equation (12.38) provides $\partial A_n/\partial c_i$ for A_c and A_j and can be used in the absence of co-limitation. With co-limitation,

$$\frac{\partial A_n}{\partial c_i} = \frac{\dfrac{\partial A_c}{\partial c_i}\left(A_j - A\right) + \dfrac{\partial A_j}{\partial c_i}\left(A_c - A\right)}{A_c + A_j - 2\Theta_A A}, \tag{12.54}$$

as given by Buckley et al. (2002) with A the co-limited rate.

An alternative numerical approach is to solve the system of equations given by leaf temperature, transpiration, and photosynthesis such that further stomatal opening does not yield a sufficient carbon gain per unit water loss (Friend 1995; Williams et al. 1996). This is specified by the requirement that marginal water-use efficiency is $\Delta A_n/\Delta E = \iota$. Figure 12.11 illustrates one such implementation, adapted from Williams et al. (1996), who optimized intrinsic water-use efficiency $\Delta A_n/\Delta g_{sw}$, and modified by Bonan et al. (2014) to optimize $\Delta A_n/\Delta E$. Optimization is achieved when $\Delta A_n/\Delta E = \iota$, which is calculated as

$$\frac{\Delta A_n}{\Delta E} = \frac{\Delta A_n}{\Delta g_{sw}}\frac{1}{w_s}, \tag{12.55}$$

with $w_s = [e_{sat}(T_\ell) - e_s]/P$ the water vapor deficit (mol mol^{-1}) at the leaf surface. The solution is iterative and requires solving the system of equations twice at each iteration – once for g_{sw} and again for $g_{sw} - \Delta g_{sw}$, where Δg_{sw} is a small change in stomatal conductance (Δg_{sw}= 0.001 mol H_2O m^{-2} s^{-1}). The model solves for g_{sw} such that Δg_{sw} changes leaf photosynthesis by $\Delta A_n \leq \iota w_s\Delta g_{sw}$. Numerical techniques can efficiently solve for g_{sw}. The solution works with or without co-limitation and for both C_3 and C_4 plants.

Figure 12.12 compares simulated relationships between A_n and g_{sw} with observations from a leaf trait database. The database provides maximum A_n and g_{sw} measured at high light, moist soil, and ambient CO_2. For C_3 plants, A_n ranged from 0.1 to

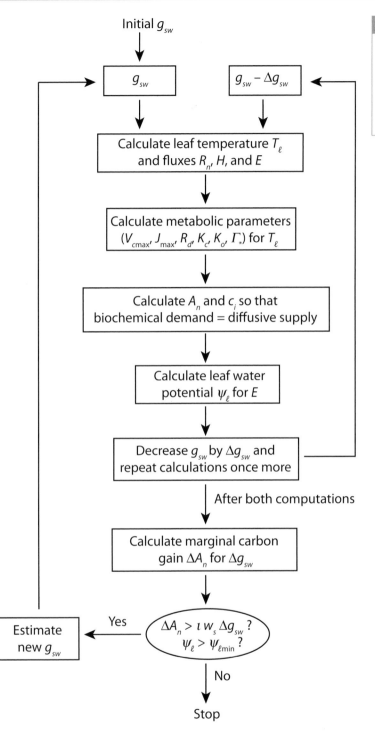

Initial g_{sw}

g_{sw}

$g_{sw} - \Delta g_{sw}$

Calculate leaf temperature T_ℓ
and fluxes R_n, H, and E

Calculate metabolic parameters
(V_{cmax}, J_{max}, R_d, K_c, K_o, Γ_*) for T_ℓ

Calculate A_n and c_i so that
biochemical demand = diffusive supply

Calculate leaf water
potential ψ_ℓ for E

Decrease g_{sw} by Δg_{sw} and
repeat calculations once more

After both computations

Calculate marginal carbon
gain ΔA_n for Δg_{sw}

$\Delta A_n > \iota\, w_s\, \Delta g_{sw}$?
$\psi_\ell > \psi_{\ell min}$?

Yes

Estimate
new g_{sw}

No

Stop

Figure 12.11 Flow diagram of leaf flux calculations to numerically solve for stomatal conductance that optimizes water-use efficiency. The calculation shows an additional constraint that leaf water potential ψ_ℓ must be greater than some critical value $\psi_{\ell min}$ (see Chapter 13 for details). Adapted from Bonan et al. (2014)

35 μmol CO_2 m^{-2} s^{-1} and g_{sw} varied from less than 0.05 to greater than 1 mol H_2O m^{-2} s^{-1}. This reflects a range in photosynthetic capacity, seen in leaf nitrogen concentration that varied from 0.5% to

more than 4% (by mass). Similar relationships can be simulated for 100 theoretical leaves that differ in photosynthetic capacity, specified by varying $V_{c\,max}$ from 1.5 to 150 μmol m^{-2} s^{-1}. The optimal A_n and

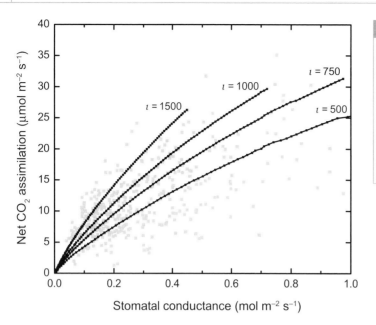

Figure 12.12 Observed and simulated relationships between A_n and g_{sw}. Observations (light gray symbols) are from a leaf trait database (Wright et al. 2004) for C_3 plants ($n = 421$). Simulations show optimal A_n and g_{sw} calculated for 100 theoretical leaves that differ in photosynthetic capacity, specified by varying V_{cmax} from 1.5 to 150 μmol m^{-2} s^{-1}. Results are shown for $\iota = 500$, 750, 1000, and 1500 μmol CO_2 mol^{-1} H_2O. Adapted from Bonan et al. (2014)

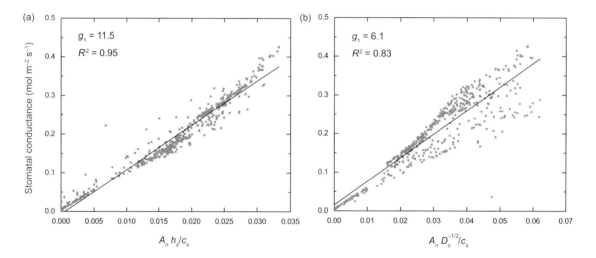

Figure 12.13 Water-use efficiency optimization stomatal conductance in relation to (a) $A_n h_s / c_s$ and (b) $A_n D_s^{-1/2} / c_s$. The linear regression line (solid line) is shown, with the slope g_1 and R^2. For these simulations $\iota = 750$ μmol CO_2 mol^{-1} H_2O. Environmental conditions varied over the range: absorbed photosynthetically active radiation, 7–1288 μmol m^{-2} s^{-1}; T_ℓ, 12°C–33°C; h_s, 0.42–1.0; D_s, 0–2.6 kPa; and A_n, 0–13 μmol CO_2 m^{-2} s^{-1}. Unpublished data from Bonan et al. (2014)

g_{sw} increase in relation to each other, consistent with the range of observations of maximum A_n and g_{sw}. The slope of the simulated A_n-g_{sw} relationship increases with larger values of ι (i.e., larger ι produces higher A_n for the same g_{sw}).

The model produces leaf gas exchange results consistent with theory and observations. Over a range of light, temperature, and vapor pressure deficit, optimized g_{sw} linearly relates with $A_n h_s / c_s$ and $A_n D_s^{-1/2} / c_s$ (Figure 12.13). The slope g_1 of these relationships decreases with higher ι in relation to $\iota^{-1/2}$ as expected from theory (Bonan et al. 2014). An outcome of water-use efficiency optimization models is that the dependence of g_{sw} on vapor

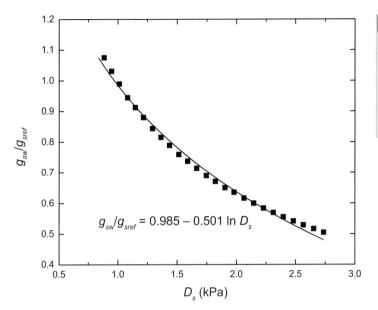

Figure 12.14 Relationship between g_{sw} and D_s obtained from water-use efficiency optimization. g_{sref} is the stomatal conductance at $D_s = 1$ kPa. The solid line shows the best-fit regression equation using the relationship $g_s/g_{sref} = y_0 + m \ln D_s$ from Katul et al. (2009). Calculations used $\iota = 750$ μmol CO_2 mol^{-1} H_2O. Adapted from Bonan et al. (2014)

pressure deficit emerges from the optimization and does not require a priori specification (Figure 12.8c). Water-use efficiency optimization predicts a relationship in which g_{sw} varies in relation to $1 - 0.5 \ln D_s$ (Figure 12.14), consistent with observations (Oren et al. 1999; Katul et al. 2009). Analytical stomatal conductance models obtained from water-use efficiency optimization show that g_{sw} varies inversely with the square root of vapor pressure deficit, e.g., (12.37) and (12.46), which approximates $1 - 0.5 \ln D_s$.

Water-use efficiency optimization provides stomatal conductance that is comparable to Ball et al. (1987) and Medlyn et al. (2011). The third simulation in Figure 12.6 is the Bonan et al. (2014) water-use efficiency optimization model (denoted WUE) with $\iota = 750$ μmol CO_2 mol^{-1} H_2O. WUE is comparable to MED in terms of its response to light, CO_2, and vapor pressure deficit, though with slightly lower conductance. Increasing ι from 750 to 1400 μmol CO_2 mol^{-1} H_2O more closely matches WUE with BB and MED (Figure 12.8). The light, CO_2, and vapor pressure deficit simulations shown in Figure 12.8 are at a constant leaf temperature (25°C). When leaf temperature varies, the WUE temperature response also closely matches BB and MED, but only if $\iota \propto \Gamma_*$. This scaling of ι is obtained from (12.47), in which ι increases with temperature according to the temperature dependence of Γ_*.

These leaf simulations show that both the Ball et al. (1987) and Medlyn et al. (2011) stomatal models are consistent with water-use efficiency optimization, as noted also by Buckley et al. (2017) and Franks et al. (2017). In particular, with careful choice of g_1 and ι, the three models are similar to one another (Figure 12.8). The key difference among models is in stomatal conductance at low and high vapor pressure deficit, but over the range of about 0.5–2 kPa, the three models are quite similar (Figure 12.8c). The three stomatal conductance models have markedly different underpinnings: Ball et al. (1987) is empirical; Medlyn et al. (2011) is derived from water-use efficiency optimization (using the A_j-limited rate and after several simplifications); and the numerical water-use efficiency model retains the full complexity of stomatal optimization. Yet all three models give similar stomatal conductances for the same conditions (with the exception of low and high vapor pressure deficit). All three models have a critical parameter (g_1 or ι) that determines the relationship between conductance and photosynthesis.

Optimality models have the appeal that stomatal behavior emerges from fundamental physiological principles (e.g., water-use efficiency optimization) rather than being prescribed from empirical relationships between g_{sw} and A_n. A challenge in using these models, however, is that the target value for $\partial A_n / \partial E$ (namely ι) must be specified. In practice, this can be estimated as a fitting parameter to match g_{sw} and A_n with observations (Figure 12.12), similar to the empirical estimation of g_1. A related challenge pertains to the timescale over which ι remains constant. Does ι need to be treated as a variable that changes in response to prevailing environmental conditions such as elevated c_a or drought? In principle it should because ι relates to how much water loss can be tolerated, but how (or if) ι varies with elevated c_a or soil moisture stress is the subject of considerable debate (Katul et al. 2010, 2012; Manzoni et al. 2011; Medlyn et al. 2013; Vico et al. 2013; Buckley and Schymanski 2014; Buckley 2017; Buckley et al. 2017). A critical point of contention is whether ι increases with elevated c_a, as has been argued by some (Katul et al. 2010, 2012; Manzoni et al. 2011). Regardless, it is evident that g_1 relates to ι so that any debate concerning how ι varies on long timescales should also pertain to g_1. Prescribing ι for water-use efficiency optimization is analogous to prescribing g_1 for semiempirical models. Some observations find that g_1 does not change with elevated c_a, except, perhaps, with water stress (Medlyn et al. 2001), though a robust synthesis of the effect of elevated CO_2 on g_1 is lacking (Miner et al. 2017). A further challenge in using optimality theory is to define the cost associated with stomatal opening. Much research has focused on transpiration water loss per se. Alternative hypotheses have been posed that relate the cost directly to hydraulic transport (Wolf et al. 2016; Sperry et al. 2017; Dewar et al. 2018).

Question 12.5 Derive $\partial A_n / \partial E$ as given by (12.34).

Question 12.6 Show that (12.35) is equivalent to (12.34) if g_{bw} is neglected.

Question 12.7 Derive (12.37) for optimal g_{sw}. What assumptions have been made to obtain (12.37)?

Question 12.8 What are the assumptions and simplifications needed to derive the Medlyn et al. (2011) equation for g_{sw}?

Question 12.9 Show that (12.22) and (12.48) are equivalent expressions for c_i / c_a obtained from the Medlyn et al. (2011) optimal g_{sw}.

Question 12.10 Prentice et al. (2014) obtained the expression $c_i / c_a = g_1 / (g_1 + \sqrt{w_\ell})$. Show that this gives the same stomatal conductance model as (12.46).

12.7 | Soil Moisture Stress

The Ball–Berry stomatal model and its variants are empirical, based on correlations between stomatal conductance and photosynthesis from leaf gas exchange measurements. Such relationships are typically obtained for well-watered soils and do not account for soil moisture stress. How to represent stomatal closure with soil moisture stress is problematic in these empirical models. Most terrestrial biosphere models introduce a soil moisture stress factor β_w that reduces stomatal conductance as root zone soil water θ decreases below some critical value. This stress factor has the general form:

$$\beta_w = \begin{cases} 1 & \theta \geq \theta_c \\ \dfrac{\theta - \theta_w}{\theta_c - \theta_w} & \theta_w < \theta < \theta_c \\ 0 & \theta \leq \theta_w \end{cases} \tag{12.56}$$

The parameters θ_c and θ_w define the water content at which β_w has a value of one and zero, respectively. These are typically taken as field capacity and wilting point. Equation (12.56) is a linear function of soil moisture, but many plants show a curvilinear response to drying in which stomatal closure is initially gradual and increases sharply as soil becomes very dry. The equation can be generalized to represent curvilinear relationships using $[(\theta - \theta_w)/(\theta_c - \theta_w)]^p$ (Egea et al. 2011). This function gives a family of relationships from linear to a step response; values for p that are less than one give increasingly nonlinear declines in β_w as soil moisture decreases (Figure 12.15).

Although all models reduce stomatal conductance and photosynthesis for soil moisture stress, they differ in how this is implemented. Some models directly impose diffusive limitations in stomatal conductance and either reduce the slope g_1 (multiplying g_1 by β_w) or calculate a maximum conductance for unstressed conditions that is then multiplied by β_w. Other models impose biochemical limitations and indirectly reduce stomatal conductance by reducing A_n as soil moisture stress increases (multiplying $V_{c\max}$ and $J_{\max}$ by β_w). Neither approach completely replicates observed stomatal response to soil moisture stress (Damour et al. 2010; Egea et al. 2011; De Kauwe et al. 2013). Some evidence suggests that both diffusive and biochemical limitations must be considered (Zhou et al.

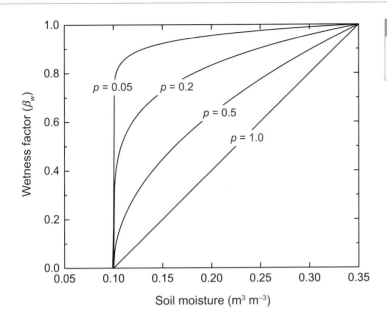

2013). Other modeling studies find a role for mesophyll conductance to limit photosynthesis during drought (Keenan et al. 2010). There is also uncertainty about the form of the soil moisture stress function (Verhoef and Egea 2014).

Some models replace volumetric water content θ with matric potential ψ because plants respond more directly to matric potential than to water content. An alternative formulation recognizes that stomata respond to leaf water potential rather than soil water potential. As soil moisture decreases, plant water uptake declines, leaf water potential decreases, and stomata close to prevent desiccation. Tuzet et al. (2003) provided a soil moisture stress function to represent this in which

$$\beta_w = \frac{1 + \exp\left(s_c \psi_c\right)}{1 + \exp\left[s_c(\psi_c - \psi_\ell)\right]}, \tag{12.57}$$

and Duursma and Medlyn (2012) used this in their plant canopy model. In this equation, ψ_ℓ is leaf water potential (MPa), ψ_c is a reference water potential, and s_c describes the sensitivity to low water potential. These parameters vary among species depending on adaptation to water stress. Representative values used in Tuzet et al. (2003) are $s_c = 2$–5 MPa^{-1}, and ψ_c has a value of about –2 MPa. With these parameters, stomata are not very sensitive to leaf water potential at values close to zero but close sharply with decreasing ψ_ℓ. Equation (12.57) requires leaf water potential, which is calculated

from plant water uptake, transpiration loss, and hydraulic transport from the root to leaf. The next chapter describes models of plant hydraulics and stomatal conductance.

12.8 Chapter Appendix – Cubic Equation for A_n

The cubic equation for A_n given by (12.27) has the coefficients

$$c_3 = f_1 \tag{12.58}$$

$$c_2 = bf_3 + f_2 + (R_d - a)f_1 \tag{12.59}$$

$$c_1 = (c_a + b)g_0 c_a + R_d(bf_3 + f_2) + a(\Gamma_* f_3 - f_2) \tag{12.60}$$

$$c_0 = [R_d(c_a + b) - a(c_a - \Gamma_*)]g_0 c_a, \tag{12.61}$$

with

$$f_1 = \frac{1.4}{g_{bw}}\left(1.6 + g_0\frac{1.4}{g_{bw}} - g_1 h_s\right) \tag{12.62}$$

$$f_2 = c_a\left(g_1 h_s - 1.6 - 2g_0\frac{1.4}{g_{bw}}\right) \tag{12.63}$$

$$f_3 = g_1 h_s - g_0\frac{1.4}{g_{bw}}. \tag{12.64}$$

12.9 | Supplemental Programs

12.1 Leaf Gas Exchange: This program calculates leaf gas exchange for C_3 and C_4 plants at a specified leaf temperature. Photosynthesis is calculated as in Chapter 11. Numerical iteration is used to solve for c_i that satisfies the biochemical demand, stomatal constraint, and diffusive supply functions as in the inner loop shown in Figure 12.5. Leaf boundary layer conductance is from Chapter 10. Stomatal conductance is obtained from the Ball et al. (1987) or Medlyn et al. (2011) models. An additional option calculates stomatal conductance from water-use efficiency optimization as in Bonan et al. (2014). The program was used to calculate photosynthesis in Figure 11.4 and stomatal conductance in Figure 12.6 and Figure 12.8.

12.2 Leaf Gas Exchange Coupled with the Leaf Energy Budget: The previous program simulates leaf gas exchange for a specified leaf temperature. The code here illustrates a coupled photosynthesis–stomatal conductance–energy balance model to calculate A_n and g_{sw} when the leaf energy balance is included to calculate leaf temperature. It is the same as the previous program but with steady state leaf temperature calculated, as in Chapter 10. For the current estimate of leaf temperature, the inner loop in Figure 12.5 is executed to solve for c_i that satisfies the biochemical demand, stomatal constraint, and diffusive supply. The outer loop iterates on leaf temperature, and the calculations are repeated until leaf temperature converges.

Figure 12.11 shows the complete water-use efficiency optimization iteration. The program was used for Figure 12.10.

12.10 | Modeling Projects

1 Photosynthetic capacity, as represented by $V_{c\,\mathrm{max}}$ at 25°C, is an important parameter that affects leaf gas exchange. Use Supplemental Program 12.2 to investigate the effect of $V_{c\,\mathrm{max}}$ on leaf temperature and gas exchange. How does wind speed affect the results?

2 Write a program to calculate the Arneth et al. (2002) optimal g_{sw} for a specified leaf temperature. Compare stomatal response to c_a with that shown in Figure 12.9.

3 Write a program that calculates optimal g_{sw} following Buckley et al. (2002) using (12.34) and with co-limitation specified by (12.54). Compare results with the numerical method of Bonan et al. (2014) used in Supplemental Program 12.1 and shown in Figure 12.11.

4 Investigate the effects of leaf boundary layer conductance on g_{sw}. Use your own model or Supplemental Program 12.2 with different values of g_{bw} and also with $g_{bw} = \infty$.

5 The photosynthesis model in Supplemental Program 12.2 uses the peaked Arrhenius function to account for high temperature on $V_{c\,\mathrm{max}}, J_{\mathrm{max}}$, and R_d. Investigate the effect of this on stomatal conductance and leaf temperature.

Plant Hydraulics

Chapter Overview

Additional understanding of stomatal behavior comes from transport of water through the soil–plant–atmosphere continuum based on the principle that plants reduce stomatal conductance as needed to regulate transpiration and prevent hydraulic failure. As xylem water potential decreases, the supply of water to foliage declines and leaves may become desiccated in the absence of stomatal control. Stomata close as needed to prevent desiccation within the constraints imposed by soil water availability and plant hydraulic architecture. This chapter develops the physiological theory and mathematical equations to model plant water relations.

13.1 | Introduction

Water loss during transpiration drives plant water uptake from soil by roots and transport through the stem to leaves. According to cohesion–tension theory, leaves draw water from soil along a water potential gradient generated by transpiration. Transpiration provides the force that pulls water from soil. Cohesion binds water molecules together so that the pull exerted by transpiration extends down the stem to roots. As transpiration increases during the day, water is initially drawn from internal plant storage and then from soil. Movement of water from soil through plants into the atmosphere occurs along a

continuum of decreasing water potential (Figure 13.1). When wet, soil particles exert minimal suction and have high water potential. The atmosphere exerts the most suction and has the lowest water potential. Water flows from soil into roots through the stem and out through foliage into the surrounding air, moving from high to low water potential. This gradient of water potential is commonly referred to as the soil–plant–atmosphere continuum. A water deficit builds up during the day when the rate at which internally stored plant water is drawn down by transpiration exceeds the rate of root uptake. At night, plant water uptake replenishes water depleted during the day, provided the soil is sufficiently moist. By morning, before transpiration begins, water in soil near the roots, water in the stem, and water in foliage are again nearly equal in potential. The gradient in water potential re-establishes during the day as transpiration increases and leaf water potential decreases (Figure 13.2).

The water column in a plant is connected and under tension, and trees, in particular, must maintain water potentials above a minimum threshold to avoid cavitation. Hydraulic failure occurs when the water column in xylem conduits breaks (i.e., becomes air-filled, also referred to as cavitation) because of high transpiration demand and low soil water supply. Cavitation decreases the conductance of water through the stem, impairs water flow, and leads to tissue dehydration. One means to maintain water potential sufficient to prevent cavitation is by stomatal regulation of transpiration. A reduction in stomatal conductance during periods of water

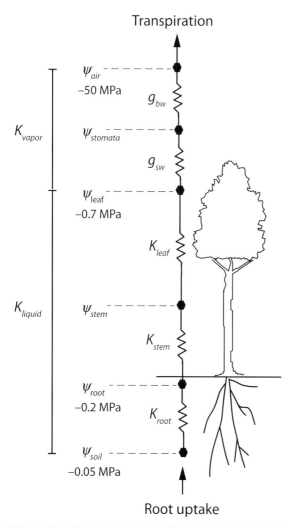

Transpiration

ψ_{air}
–50 MPa

g_{bw}

K_{vapor}

$\psi_{stomata}$

g_{sw}

ψ_{leaf}
–0.7 MPa

K_{leaf}

K_{liquid}

ψ_{stem}

K_{stem}

ψ_{root}
–0.2 MPa

K_{root}

ψ_{soil}
–0.05 MPa

Root uptake

Figure 13.1 Flow of water and representative water potentials along the soil–plant–atmosphere continuum. Also shown are conductances along the hydraulic pathway. Adapted from Tyree and Ewers (1991)

stress, such as midday with high vapor pressure deficit and high transpiration demand or during soil drydown, can prevent large decreases in leaf water potential and loss of hydraulic conductance in xylem. This behavior is characteristic of isohydric plants, which maintain relatively constant leaf water potential by reducing stomatal conductance so as to decrease transpiration loss during periods of water stress. This reduces the risk of xylem cavitation but decreases carbon uptake during photosynthesis. In contrast, anisohydric plants have weak stomatal regulation of leaf water potential. These

plants maintain relatively high stomatal conductance so that leaf water potential decreases with drier soil or greater vapor pressure deficit. This results in a greater risk of cavitation. Differences in xylem characteristics (diffuse-porous or ring-porous) and rooting depth also contribute to the whole-plant hydraulic strategy.

Red maple and red oak trees growing in Michigan illustrate these two different hydraulic strategies (Thomsen et al. 2013; Matheny et al. 2017). The diffuse-porous and shallow-rooted red maple follows an isohydric hydraulic strategy while the ring-porous and deep-rooted red oak exhibits anisohydric behavior. Red maple maintains higher leaf water potential compared with red oak, and leaf water potential is relatively constant as the soil dries (lower soil water potential) or as vapor pressure deficit increases (Figure 13.3). In contrast, leaf water potential in red oak decreases sharply with drier soil and as vapor pressure deficit increases. Similar differences in isohydric and anisohydric strategies can be seen in gas exchange measurements for many plants.

Question 13.1 Discuss the stomatal response of red maple and red oak to vapor pressure deficit. Characterize the general stomatal response of isohydric and anisohydric plants to vapor pressure deficit.

13.2 | Plant Water Uptake and Leaf Water Potential

Models of the soil–plant–atmosphere continuum explicitly simulate water movement through plants from soil to the atmosphere. Uptake of water by plants and hydraulic transport to leaves can be described mathematically by Darcy's law, similar to that for soil moisture. The flow of water through a unit area of a medium is given by:

$$Q = -K' \frac{\partial \psi}{\partial z} \tag{13.1}$$

where Q is water flux density (mol H_2O m^{-2} s^{-1}), K' is hydraulic conductivity (here with units mol H_2O m^{-1} s^{-1} Pa^{-1}), and $\partial \psi / \partial z$ is the potential gradient (Pa m^{-1}). Integrated over the flow path length Δz over which the potential differs by $\Delta \psi$:

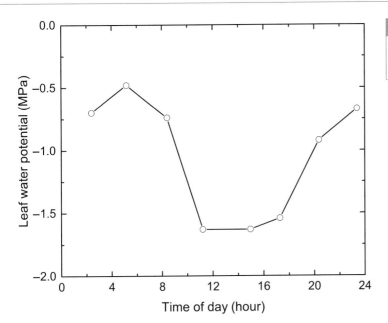

$$Q = \frac{K'}{\Delta z}\Delta\psi. \tag{13.2}$$

The term $K'/\Delta z$ is the hydraulic conductance (mol H_2O m^{-2} s^{-1} Pa^{-1}) over the length Δz. The distinction between conductivity K' and conductance $K'/\Delta z$ is important for plant water relationships, and one must pay careful attention to usage and units. Conductivity is a property of the plant material and is independent of the length of flow, whereas conductance additionally depends on length.

Models of the soil–plant–atmosphere continuum use Darcy's law to simulate water flow based on water potentials and conductances, such as in Figure 13.1. Van den Honert (1948) outlined the basic principles of this approach (see also Campbell (1985) and Jones (2014) for the governing equations). In the absence of storage, steady state water transport in plants requires that the transpiration flux E is balanced by a series of fluxes described by

$$E = K_\ell(\psi_{x\ell} - \psi_\ell) = K_x(\psi_{xr} - \psi_{x\ell}) = K_r(\psi_r - \psi_{xr})$$
$$= K_s(\psi_s - \psi_r), \tag{13.3}$$

in which ψ_s, ψ_r, ψ_{xr}, $\psi_{x\ell}$, and ψ_ℓ are the potentials of the bulk soil, at the surface of the roots, at the base of the stem, at the top of the stem, and at the leaf,

respectively; and K_s, K_r, K_x, and K_ℓ are the corresponding conductances.

Early examples of such models include Running et al. (1975), Waring and Running (1976), and Federer (1979). These models have two compartments for root zone water and plant water. A series of equations that describe water uptake by roots, water storage within the plant (in sapwood for trees), and water loss during transpiration connect the two compartments. The plant is represented by a small number of conductances (root, stem, leaf) connected in series. More complex models represent branching structure by dividing the plant into numerous segments and calculate water flow along each segment based on the hydraulic conductivity for that segment using principles outlined by Tyree (1988) and Sperry et al. (1998). Such methods were first considered too complex for implementation in global terrestrial biosphere models, and, instead, the effects of plant water relations on stomatal conductance and transpiration were represented through an ad hoc soil wetness factor (Chapter 12). More recently, however, there has been renewed interested in plant hydraulics as a physiologically realistic scheme to overcome the limitation of soil wetness factors (Bonan et al. 2014; Christoffersen et al. 2016; Xu et al. 2016).

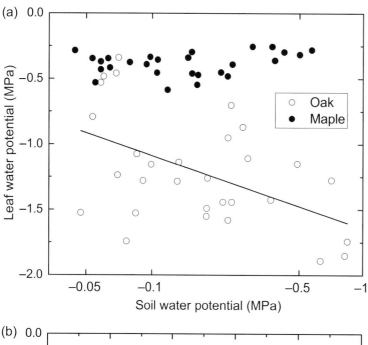

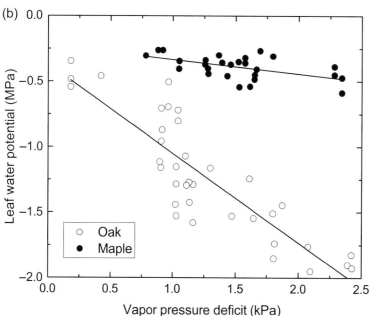

Figure 13.3 Leaf water potential in relation to (a) soil water potential and (b) vapor pressure deficit for red oak (*Quercus rubra*) and red maple (*Acer rubrum*) growing in Michigan. Leaf water potential is at midday. Soil water potential is for the top 30 cm of soil. Redrawn from Thomsen et al. (2013)

One application of (13.2) is to calculate plant water uptake and transpiration. The rate of plant water uptake Q_s can be represented as the flow along a pathway between soil and leaves, where the driving forcing is the gradient in water potential. In this formulation,

$$Q_s = K_L(\psi_s - \psi_\ell - \rho_{wat}gh). \quad (13.4)$$

Here, K_L is the whole-plant hydraulic conductance from soil to leaves (mol H_2O m^{-2} s^{-1} Pa^{-1}), and ψ_s and ψ_ℓ are soil and leaf water potential (Pa), respectively. The term $\rho_{wat}gh$ is the gravitational potential with h the height (m) of the water column, ρ_{wat} the density of water (kg m^{-3}), and g gravitational acceleration (m s^{-2}), equivalent to ~0.01 MPa m^{-1}. If water uptake (mol H$_2$O s^{-1}) is normalized by the leaf area supplied by the flow, K_L is also expressed per unit leaf area, in which case it is termed leaf-specific plant hydraulic conductance. It is the flux of water per m^2 of leaf area per unit change in potential from soil to leaf. If Q_s is normalized by the stem cross-sectional area through which the flow occurs (sapwood area in trees), conductance is similarly normalized by sapwood area and is called sapwood-specific hydraulic conductance K_S.

A second equation describes the change in plant water storage. Continuity requires that the rate of change of plant water W_s (mol H$_2$O m^{-2}) is the difference between plant water uptake and transpiration loss so that

$$\frac{dW_s}{dt} = Q_s - E. \tag{13.5}$$

Combining (13.4) and (13.5),

$$\frac{dW_s}{dt} = K_L(\psi_s - \psi_\ell - \rho_{wat}gh) - E. \tag{13.6}$$

At steady state, transpiration loss equals water uptake so that $dW_s/dt = 0$ and (13.6) can be rearranged to give an expression for leaf water potential whereby

$$\psi_\ell = \psi_s - \rho_{wat}gh - \frac{E}{K_L}. \tag{13.7}$$

This equation shows that leaf water potential decreases with greater transpiration, all other factors being equal. The decline in ψ_ℓ is inversely proportional to hydraulic conductance, and K_L must be sufficiently large to prevent low ψ_ℓ. Leaf water potential becomes increasingly more negative (for a given transpiration rate) with smaller K_L. Leaf water potential also declines with height above the ground. The difference between soil and leaf water potential is determined by transpiration loss and the conductance to water flow within the plant.

Leaf-specific plant hydraulic conductance is the conductance of the soil-to-leaf pathway and integrates the hydraulic conductance of roots, stems, and branches. It is the ratio of water flow to the driving force for the flow and can be estimated from sap flux measurements as $K_L = Q_s/\Delta\psi$ and

normalized per unit leaf area. Many plant hydraulics model use a sapwood-specific conductivity K_S' rather than K_L. These are related by

$$K_L = \frac{1}{h}K_S'\frac{A_S}{A_L}, \tag{13.8}$$

with A_L leaf area and A_S sapwood cross-sectional area, respectively. Sapwood-specific hydraulic conductivity K_S' has units mol H$_2$O m^{-1} s^{-1} Pa^{-1}. K_S' is the sapwood-specific hydraulic conductivity (per m of sapwood), K_S'/h is the sapwood-specific hydraulic conductance (per unit sapwood area), and K_L is the leaf-specific hydraulic conductance (per leaf area). Equation (13.8) shows that K_L increases with greater sapwood area and decreases with greater soil-to-leaf water flow path length (i.e., tree height h). Some values for K_L reported in the literature are 1.1 mmol H$_2$O m^{-2} s^{-1} MPa^{-1} for loblolly pine (*Pinus taeda*) (Ewers et al. 2000); 0.5–1 for aspen (*Populus tremuloides*) and black spruce (*Picea mariana*) and 6–11 for jack pine (*Pinus banksiana*) boreal forest (Ewers et al. 2005); and 1–10 for tropical trees (Meinzer et al. 1995).

A more complex equation for ψ_ℓ arises in the absence of steady state. Equation (13.6) can be converted to the change in ψ_ℓ over some time period by defining plant capacitance C_p (mol H$_2$O m^{-2} Pa^{-1}) as the ratio of the change in plant water to the change in water potential so that

$$C_p = \frac{dW_s}{d\psi_\ell}. \tag{13.9}$$

Capacitance can be thought of as the amount of water that can be extracted per unit change in water potential. Dividing both sides of (13.6) by C_p gives the change in leaf water potential in which

$$\frac{d\psi_\ell}{dt} = \frac{K_L(\psi_s - \psi_\ell - \rho_{wat}gh) - E}{C_p}. \tag{13.10}$$

Plant conductance, capacitance, transpiration, and water storage are expressed per unit area and must have the same areal basis (i.e., per unit leaf area or stem cross-sectional area). Each of the terms in (13.10) provides a critical constraint on plant water relations. A high transpiration rate requires large water uptake from soil to maintain high leaf water potential. The rate of uptake depends directly on hydraulic conductance. Plant capacitance controls the timing of water use throughout the day. Plants can lose water from internal storage due to a faster rate of water loss from transpiration than gain from root uptake. A high capacitance means there is a large buffer (storage) at the beginning of the day, before water use is limited by the rate of supply from soil. Equation (13.10) is a first-order, linear differential equation (Appendix A3). Integration with respect to time gives the change in leaf water potential over the time interval Δt from an initial value ψ_0 so that

$$\Delta\psi_\ell = \left[\psi_s - \psi_0 - \rho_{wat}gh - \frac{E}{K_L}\right]\left(1 - e^{-K_L\Delta t/C_p}\right).$$

(13.11)

The term C_p/K_L in the exponential is a time constant (s) and is the time at which 63% of the total change in ψ_ℓ occurs.

While these equations provide a simple model of leaf water potential, they are, in fact, complex equations to solve. One difficulty is that transpiration depends on stomatal conductance, which varies with leaf water potential. A further complication is that K_L depends on plant water potential (similar to soil hydraulic conductivity). Furthermore, K_L represents the whole-plant conductance and must also include a soil-to-root conductance connected in series to the stem conductance. An additional complexity is that C_P depends on plant water potential, similar to the specific moisture capacity for soil.

Question 13.2 Darcy's law is commonly expressed as a flux of water (mm H_2O s^{-1}) with conductivity in the same units and potential with the units mm. Equation (13.1) is a form of Darcy's law, but the flux of water is mol H_2O m^{-2} s^{-1}, conductivity is mmol H_2O m^{-1} s^{-1} MPa^{-1}, and potential is MPa. Derive the conversion factor between these units of conductivity.

Question 13.3 The latent heat flux measured above a 20 m forest canopy with a leaf area index of 5 m^2 m^{-2} is 400 W m^{-2}. The soil is moist with a water potential of –0.01 MPa, and leaf water potential is –1 MPa. Calculate the leaf-specific plant hydraulic conductance (assuming all the latent heat flux is from transpiration).

13.3 | Plant Hydraulics and Stomatal Conductance

Plant hydraulics plays a critical role in leaf gas exchange by regulating stomatal conductance. A class of stomatal models calculates soil water control of stomatal conductance directly from plant water relations (Whitehead 1998; Oren et al. 1999; Ewers et al. 2000, 2005; Kumagai 2011). These models relate stomatal conductance to water flow along the soil–plant–atmosphere continuum using the water potential difference between the leaf and soil and the hydraulic conductance along the flow path. The model is most commonly applied to trees, which conduct water from soil to leaves through the xylem in sapwood.

Transpiration can be represented as a diffusion process in which the water loss per unit leaf area is approximated as

$$E = w_\ell g_{sw},$$

(13.12)

with $w_\ell = [e_{sat}(T_\ell) - e_a]/P$ the leaf-to-air water vapor gradient (mol mol^{-1}) and g_{sw} stomatal conductance (mol H_2O m^{-2} s^{-1}). The equation describes transpiration in terms of the atmospheric demand w_ℓ. Transpiration can be equivalently represented based on supply of water to leaves. With the assumption of steady state (i.e., no storage of water in the plant), transpiration is equivalent to plant water uptake, and

$$E = K_L(\psi_s - \psi_\ell - \rho_{wat}gh).$$

(13.13)

Combining (13.12) and (13.13) gives an equation for stomatal conductance in which

$$g_{sw} = \frac{K_L}{w_\ell}(\psi_s - \psi_\ell - \rho_{wat}gh).$$

(13.14)

Equation (13.14) highlights the role of transpiration in stomatal control. Stomata conductance decreases so as to prevent excessive leaf desiccation as vapor pressure deficit increases. Furthermore, stomata close as leaf water potential decreases. Stomatal conductance determines the transpiration rate for a given vapor pressure deficit, and plant hydraulic conductivity determines the leaf water potential for that transpiration rate. In particular, K_L determines how high stomatal conductance can be without desiccating the leaf. A reduction in K_L necessitates a decrease in stomatal conductance to maintain the same leaf water potential. This can arise, for example, because of an increase in leaf area relative to sapwood area or lower sapwood conductivity. Plants, especially trees, can experience water stress at high transpiration rates due to hydraulic limitation in the flow of water from soil to leaves.

Question 13.4 A representative leaf-specific plant hydraulic conductance is $K_L = 2$ mmol H_2O m^{-2} s^{-1} MPa^{-1}. Calculate stomatal conductance for a leaf located 20 m above the ground with a water potential of -1 MPa. The soil is moist with a water potential of -0.01 MPa, and the vapor pressure deficit is 1 kPa. What is the transpiration rate? Calculate the change in leaf water potential over 30 minutes if $C_p = 2000$ mmol H_2O m^{-2} MPa^{-1}. How does a lower plant conductance (1 mmol H_2O m^{-2} s^{-1} MPa^{-1}) affect the results?

13.4 | The Soil–Plant–Atmosphere (SPA) Model

Although (13.14) links plant hydraulics and stomatal conductance, it neglects the relationship between stomatal conductance and photosynthesis. The Soil–Plant–Atmosphere (SPA) model (Williams et al. 1996) is a multilayer canopy model of leaf gas exchange that extends the water-use efficiency hypothesis of stomatal behavior to also consider whether the rate of water loss is physiologically plausible. The model uses an iterative solution to calculate stomatal conductance for each canopy layer so as to maximize photosynthesis within the limitations imposed by water-use efficiency, plant water storage, and soil-to-leaf water transport (Figure 13.4). Stomatal conductance is calculated such that further opening does not yield a sufficient carbon gain per unit water loss (defined by marginal water-use efficiency ι) or further opening causes leaf water potential to decrease below a minimum sustainable threshold (defined by $\psi_{\ell\,min}$). The model is therefore an optimality model with two criteria of water-use efficiency and hydraulic safety. The

model has been successfully applied to arctic ecosystems and black spruce (*Picea mariana*) boreal forest (Williams et al. 2000; Hill et al. 2011), ponderosa pine (*Pinus ponderosa*) forest (Williams et al. 2001a,b; Schwarz et al. 2004), temperate forests (Williams et al. 1996; Bonan et al. 2014), tropical rainforest (Williams et al. 1998; Fisher et al. 2007), and Australian woodland (Zeppel et al. 2008). The description that follows is based on the implementation described by Bonan et al. (2014).

Leaf water potential is calculated with (13.10) applied to each leaf layer. The leaf-specific hydraulic conductance of the soil-to-leaf pathway integrates the hydraulic conductance of plant parts (roots, stems, branches) and is given by a belowground resistance $R_{b,i}$ and aboveground plant resistance $R_{a,i}$ (Pa s m^2 leaf area mol^{-1} H_2O) in series so that

$$\frac{1}{K_{L,i}} = R_{b,i} + R_{a,i}, \tag{13.15}$$

where here the subscript i denotes canopy layer. These resistances influence leaf water potential such that stomata close. In wet soils, for example, belowground resistance is small and K_L is large. In dry soils, belowground resistance is large and K_L imposes hydraulic constraints on leaf water potential and stomatal conductance. The aboveground plant resistance governing flow through stems to leaves is

$$R_{a,i} = \frac{1}{K_p}, \tag{13.16}$$

where K_p (mol H_2O m^{-2} leaf area s^{-1} Pa^{-1}) is the leaf-specific stem hydraulic conductance (i.e., the stem-to-leaf path). An alternative expression uses stem hydraulic conductivity (mol H_2O m^{-1} s^{-1} Pa^{-1}) in which the resistance is height divided by conductivity. In the latter formulation, hydraulic resistance increases with greater height in the canopy, as often seen in the lower conductance (higher resistance) of branches high in the canopy.

Belowground resistance is the resistance to water uptake imposed by water movement in soil and by fine roots. It is represented by a soil-to-root system in which each soil layer consists of a soil-to-root conductance K_s and a root-to-stem conductance K_r connected in series and with multiple soil layers arranged in parallel (Figure 13.4a). For soil layer j, it depends on soil hydraulic conductivity K_j (here with units mol H_2O m^{-1} s^{-1} Pa^{-1}; for hydraulic

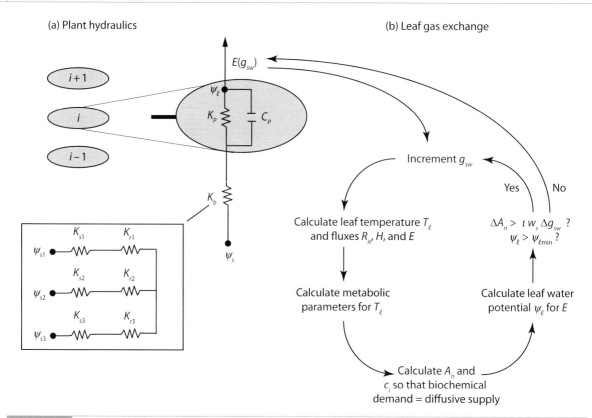

(a) Plant hydraulics

(b) Leaf gas exchange

Figure 13.4 Depiction of (a) plant hydraulics and (b) leaf gas exchange in the Soil–Plant–Atmosphere (SPA) model. SPA is a multilayer canopy model. Shown are three leaf layers and the water balance for layer i. Soil and root conductances for each soil layer are aggregated into a bulk belowground conductance K_b that contributes to total plant conductance. Transpiration of layer i is calculated from leaf gas exchange using principles of water-use efficiency optimization as described in Figure 12.11. Adapted from Williams et al. (1996)

conductivity in mm s^{-1} the conversion is $10^{-3}/(M_w g)$; see Question 13.2) and the characteristics of the rooting system so that

$$K_{s,j} = \frac{2\pi L_{r,j}\Delta z_j K_j}{\ln\left(r_{s,j}/r_r\right)}, \qquad (13.17)$$

where $L_{r,j}$ is root length per unit volume of soil (root length density, m m^{-3}); $L_{r,j}\Delta z_j$ is root length per unit area of soil (root length index, m m^{-2}) in a layer with thickness Δz_j (m); r_r is mean fine root radius (m); and $r_{s,j} = \left(\pi L_{r,j}\right)^{-1/2}$. In the absence of direct measurements, root length density can be calculated from fine root biomass M_r, average fine root radius r_r, and root tissue density ρ_r using (2.24) and (2.25). The conductance of the root-to-stem path is calculated from root resistivity R_r^* (Pa s kg mol^{-1} H$_2$O) and root biomass $M_{r,j}$ (kg m^{-2}) as

$$K_{r,j} = \frac{M_{r,j}}{R_r^*}. \qquad (13.18)$$

The total belowground resistance is obtained assuming the layers are arranged in parallel whereby

$$R_{b,i} = \left(\sum_j \frac{1}{K_{s,j}^{-1} + K_{r,j}^{-1}}\right)^{-1} L. \qquad (13.19)$$

Multiplication by the canopy leaf area index L arises because the belowground resistance is calculated on a ground area basis; multiplying by L converts to leaf area. This formulation assumes that each canopy layer is connected independently to each soil layer so that the roots in each soil layer supply water to each canopy layer and the fraction of roots supplying each canopy layer is the same as the leaf area in that layer. Soil hydraulic conductivity is large in a wet soil, and most of the belowground resistance is from roots. As the soil becomes drier, hydraulic conductivity decreases and the soil contributes more to the total resistance.

SPA simplifies the treatment of belowground hydraulics into a bulk representation of belowground conductance and soil water. The soil water potential for use with (13.10) is weighted over the soil profile according to the fraction of total transpiration obtained from each soil layer. The maximum water uptake rate for a soil layer is determined by the difference between soil water potential and the minimum leaf water potential so that

$$E_{\max,j} = \frac{\psi_{s,j} - \psi_{\ell\min}}{K_{s,j}^{-1} + K_{r,j}^{-1}}, \tag{13.20}$$

and the fraction of transpiration supplied by an individual soil layer is

$$f_j = E_{\max,j} \Big/ \sum_j E_{\max,j} \tag{13.21}$$

so that

$$\psi_s = \sum_j \psi_{s,j} f_j. \tag{13.22}$$

The SPA stomatal optimization requires four physiological parameters that describe plant water relations ($\psi_{\ell\min}$, C_p, K_p, ι) and four parameters for fine roots that are needed to calculate belowground conductance (M_r, r_r, ρ_r, R_r^*). Values of $\psi_{\ell\min}$ vary greatly among species. A representative value is $\psi_{\ell\min} = -2$ to -3 MPa but can be much lower for drought-tolerant species. Representative values used in SPA simulations are $C_p = 2000$–8000 mmol H_2O m^{-2} leaf area MPa^{-1}. SPA simulations commonly use stem hydraulic conductivity (not conductance) with values of 3.5–100 mmol H_2O m^{-1} s^{-1} MPa^{-1}. Bonan et al. (2014) used a leaf-specific stem hydraulic conductance $K_p = 4$ mmol H_2O m^{-2} leaf area s^{-1} MPa^{-1}, which gives a leaf-specific whole-plant conductance $K_L = 2$ mmol H_2O m^{-2} s^{-1} MPa^{-1} for moist soil with negligible soil resistance and if root and stem conductances are equal. The marginal water-use efficiency parameter defines the water-use strategy. Low values, with a low marginal carbon gain, optimize at high photosynthetic rates, high stomatal conductance, and high transpiration rates; plant water storage

can become depleted, causing stomata to close in early-afternoon. Higher values, with a larger marginal return, describe a more conservative strategy. Optimization is achieved at lower stomatal conductance, with correspondingly lower photosynthesis and transpiration. This reduces afternoon water stress, but restricts daily photosynthetic carbon gain. The parameters M_r, r_r, and ρ_r are discussed in Chapter 2. Bonan et al. (2014) used $R_r^* = 25$ MPa s g mmol^{-1} H_2O, but this number is highly uncertain. With fine root biomass $M_r = 500$ g m^{-2}, this value for R_r^* gives a total root conductance of 20 mmol m^{-2} ground area s^{-1} MPa^{-1}, or 4 mmol m^{-2} leaf area s^{-1} MPa^{-1} in a forest with a leaf area index of 5 m^2 m^{-2}.

The SPA plant hydraulics model uses a constant plant conductance (or conductivity). This is a simplification compared with more complex models that assess changes in conductance caused by xylem embolism under tension. It is based on the assumption that the majority of soil-to-leaf resistance is belowground and that the soil-to-root resistance provides an adequate explanation of variability in observed soil-to-leaf resistance. The model also treats capacitance as a constant. This buffers transpiration but does not simulate plant water content and its effect on capacitance as in more complex plant hydraulics models. A further assumption is that each leaf layer is independently connected by roots to soil water, and there is no intermediate xylem water potential between leaves and roots. These assumptions greatly simplify the numerical solution and allow each leaf layer to be simulated separately rather than as a system of equations. The advantage of the model is that it resolves vertical profiles of leaf water potential in the canopy. In Figure 13.5, for example, a sharp gradient in leaf water potential develops by midday in the upper canopy. Whereas the lower canopy has low transpiration rates and has leaf water potential comparable to predawn values, the upper canopy develops low leaf water potential by midday as a result of high transpiration rates.

13.5 | Soil-to-Root Conductance

Equation (13.17) for soil-to-root conductance is derived from the Richards equation as applied to water flow for a single root (Gardner 1960). In this

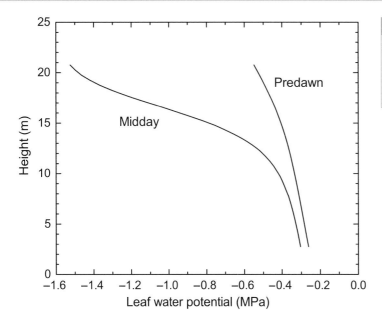

Figure 13.5 Vertical profile of leaf water potential at predawn and midday for a temperate deciduous forest on a typical summer day as simulated with a multilayer canopy model using the SPA plant hydraulics. Unpublished data from Bonan et al. (2014)

framework, the root is treated as a cylinder with radius r_r and has access to water in a surrounding cylinder of soil with radius r_s (Figure 13.6a). Water flow to the root is in the radial direction (perpendicular to the root length). In radial coordinates, the θ-based form of the Richards equation is

$$\frac{\partial \theta}{\partial t} = \frac{1}{r}\frac{\partial}{\partial r}\left[rD(\theta)\frac{\partial \theta}{\partial r}\right], \qquad (13.23)$$

where r is the radial coordinate. This equation can be solved to give the flux of water to the root in which

$$Q = \left[\frac{2\pi}{\ln{(r_s/r_r)}}\right]K\Delta\psi, \qquad (13.24)$$

with Q the rate of water uptake by the root given here as the volume of water per unit length of root per unit time (m^3 m^{-1} s^{-1}) and $\Delta\psi$ the difference in water potential between the soil and root. The term in brackets adjusts soil hydraulic conductivity K for the soil cylinder. A similar equation for conductivity can be derived from radial heat flow in a cylinder or current flow in a cable (Figure 13.6b,c). For a soil layer with thickness Δz_j, multiplying both sides of (13.24) by $L_{r,j}\Delta z_j$, the root length per unit area of soil (m m^{-2}), converts from the volume flux per unit length of root (m^3 m^{-1} s^{-1}) to the volume flux per unit area of soil (m^3 m^{-2} s^{-1}) and gives the effective soil-to-root conductance in (13.17). This formulation of soil-to-root conductance is commonly used in plant hydraulics models (e.g., Williams et al. 1996; Sperry et al. 1998).

In calculating soil-to-root conductance, the root system is described by the root radius r_r, the radius r_s of the soil cylinder occupied by the root, and the root length density L_r. Feddes and Raats (2004) and Raats (2007) summarized the geometry of roots as commonly used in root water uptake models. The volume of a root with length ℓ is $\pi r_r^2\ell$, and the volume of the soil cylinder is $\pi r_s^2\ell$. For uniformly distributed roots, root length density is equal to the inverse of the volume of soil associated with a unit length of root so that

$$L_r = \left(\pi r_s^2\right)^{-1} \qquad (13.25)$$

and

$$r_s = \left(\pi L_r\right)^{-1/2}. \qquad (13.26)$$

In this connotation, r_s is one-half the distance between uniformly spaced roots. Root length density is itself related to the mass of roots per unit volume of soil by the length of root per unit mass of root. This latter term is the specific root length r_l (m kg^{-1}) and is

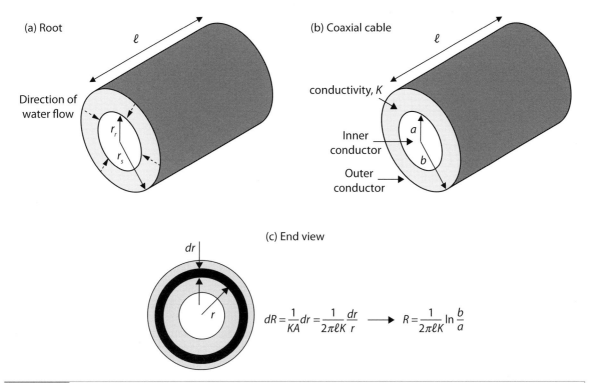

Figure 13.6 Derivation of soil-to-root conductance. (a) Radial water flow to a root is obtained by representing the root as a cylinder with radius r_r and length ℓ surrounded by a cylinder of soil with radius r_s. (b) Resistance to current flow in the radial direction in coaxial cable. Shown is coaxial cable with length ℓ consisting of two concentric cylindrical conductors. The inner conductor has radius a, and the outer conductor has radius b. The medium between the two conductors has conductivity K. (c) End view showing current leakage from the inner to outer conductor through the medium. The resistance of the medium with thickness dr and surface area $A = 2\pi r \ell$ is $dR = dr/(KA)$. The total radial resistance is found by integrating dR from a to b so that $R = \ln(b/a)/2\pi\ell K$.

$$r_l = \frac{\text{Root length}}{\text{Root mass}} = \frac{\ell}{\rho_r \pi r_r^2 \ell} = \frac{1}{\rho_r \pi r_r^2}. \quad (13.27)$$

The terms in the denominator are root tissue density ρ_r (kg m^{-3}) and root volume $\pi r_r^2 \ell$ (m^3). The mass of roots in a unit volume of soil is related to the length of roots in the soil volume by

$$M_{root(vol)} = L_r/r_l. \quad (13.28)$$

This is the same expression given by (2.24) to describe root systems in Chapter 2 and provides a means to calculate root length density from the more easily measured mass of roots. The root volume fraction is

$$V_r = \frac{\text{Root volume}}{\text{Soil volume}} = (r_r/r_s)^2 \quad (13.29)$$

so that an equivalent expression is

$$M_{root(vol)} = \frac{\text{Root mass}}{\text{Root volume}} \times \frac{\text{Root volume}}{\text{Soil volume}} = \rho_r V_r$$

$$= \rho_r (r_r/r_s)^2. \quad (13.30)$$

These equations show that a uniform root system can be described by three parameters: root radius, root tissue density, and root mass per unit volume of soil. In practice, these are combined with a vertical profile of roots so that the root system is layered as in (2.24).

Question 13.5 Derive an equation that relates the root volume fraction V_r to the root length density L_r.

13.6 | Stem Conductivity

The hydraulic conductance of plant stems is not constant but, rather, varies with xylem water potential. Xylem pressure decreases with drought stress, and hydraulic conductivity similarly declines as air enters the xylem and the air-filled conduits cannot transport water. The decline in hydraulic conductance in relation to water potential for plant elements can be modeled based on a two-parameter Weibull function in which

$$K_i = K_{\max} \exp\left[-\left(\frac{\psi_i}{d}\right)^c\right], \tag{13.31}$$

with c and d empirical parameters obtained by fitting (13.31) to data (Sperry et al. 1998; Bohrer et al. 2005; Mirfenderesgi et al. 2016). Another common equation is

$$K_i = K_{\max}\left[1 + \left(\frac{\psi_i}{\psi_{50}}\right)^a\right]^{-1}, \tag{13.32}$$

where ψ_{50} is the water potential at 50% loss of conductivity and a is a shape parameter (Manzoni et al. 2013; Christoffersen et al. 2016; Xu et al. 2016). ψ_{50} is a commonly used measure of resistance to cavitation; when xylem water potential decreases below this value, water transport is severely restricted and cavitation can occur. Both functions are empirical and conceptually similar (Figure 13.7).

Plant water uptake increases with a larger soil-to-leaf driving force in water potential. However, stem hydraulic conductivity decreases with lower water potential as described earlier. Consequently, plant water uptake is the balance of two opposing factors: a large decrease in water potential from soil to leaf is needed to provide the driving force for water uptake to supply transpiration loss, but low xylem water potential decreases stem conductivity. The loss in stem conductivity in relation to water potential is known as the vulnerability curve and determines a minimum leaf water potential that causes hydraulic failure. Plants with high vulnerability to cavitation tend to have high $K_{\max}$ and high (less negative) ψ_{50}. Species with low vulnerability to cavitation tend to have low $K_{\max}$ and low (more negative) ψ_{50}. A study of tropical dry forest species, for example, found that early successional species have high stem conductivity to supply the water needed to support their high growth rates, but the cost is reduced resistance to cavitation (Markesteijn et al. 2011). These species maintain high leaf water potential, but close to the point of hydraulic failure. In contrast, slower-growing, shade-tolerant species have low stem conductivity and operate with more hydraulic safety. Conifers are less vulnerable, on average, than are

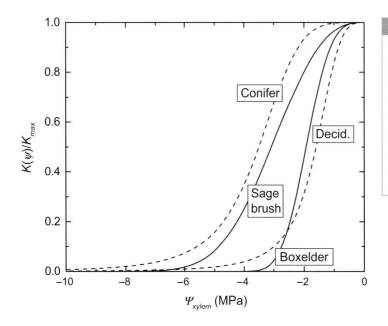

Figure 13.7 Stem conductance in relation to xylem water potential. Shown is the fraction of maximum conductance. The solid lines are from Sperry et al. (1998) using (13.31) with $d = -2.15$ MPa and $c = 3.43$ (boxelder; *Acer negundo*) and $d = -3.54$ MPa and $c = 2.64$ (mountain sagebrush; *Artemisia tridentata* ssp. *vaseyana*). The dashed lines are from Manzoni et al. (2013) using (13.32) with $\psi_{50} = -1.6$ MPa and $a = 3.5$ (temperate deciduous angiosperm) and $\psi_{50} = -3.5$ MPa and $a = 4.7$ (temperate conifer).

angiosperms; they have lower ψ_{50}, lower $K_{\max}$, and operate at lower xylem water potential than angiosperms (Choat et al. 2012; Manzoni et al. 2013).

Question 13.6 Characterize the species in Figure 13.7 in relation to sensitivity to cavitation.

13.7 | Multi-Node Models

The SPA model is a simplification of plant hydraulics. Its advantage is that it is used in a multilayer canopy and provides profiles of leaf water potential in the canopy. More complex plant hydrodynamics models use the Richards equation to describe water movement and storage, similar to soil moisture, by dividing the soil-to-leaf continuum into multiple nodes and conducting elements to represent water uptake by roots, flow through the stem to leaves, and the water potential gradient from soil to leaf. When applied to plant hydrodynamics, the governing equation is

$$C(\psi)\frac{\partial \psi}{\partial t} = \frac{\partial}{\partial z}\left\{K(\psi)\left[\frac{\partial \psi}{\partial z} - \rho_{wat}g\right]\right\} - \frac{E}{\Delta z} \quad (13.33)$$

as used in FETCH2 (Mirfenderesgi et al. 2016; note the different units in that model). $K(\psi)$ and $C(\psi)$ denote that conductivity and capacitance depend on water potential, and these relationships vary among species. The original model of Sperry et al. (1998) abstracted the system into leaf, stem, transporting root, absorbing root, and multiple soil nodes (Figure 13.8a). Similar models have since been implemented in terrestrial biosphere models (Christoffersen et al. 2016; Xu et al. 2016), some with or without capacitance (Figure 13.8b,c). These models generally abstract soil-to-leaf transport as a single beam system (similar to Figure 13.1) and use a single layer of transpiring leaves or separate sunlit and shaded leaves. The FETCH model (Bohrer et al. 2005; Mirfenderesgi et al. 2016) represents more complex plant architectures and branching patterns or simplifies to a single beam with multiple leaf layers (Figure 13.8d). Boundary conditions in solving the Richards equations are soil water potential at the soil–root interface and transpiration at the leaf node. These plant hydraulics models are solved for multiple soil layers each with separate roots. The models can have redistribution of soil water from one layer to another along the root

pathway, in contrast with the use of a single bulk soil water potential and soil-to-root conductance in SPA, and do not require a specific hydraulic redistribution parameterization as described in Chapter 8.

A numerical challenge in plant hydrodynamics models is that transpiration is a required boundary condition at the leaf node. This flux depends on stomatal conductance, which itself depends on leaf water potential. Multi-node models specify transpiration directly (Sperry et al. 1998), calculate it by reducing a potential transpiration rate for plant water potential so as to mimic stomatal regulation (Mirfenderesgi et al. 2016), or downregulate a non–water stressed stomatal conductance for water limitation (Christoffersen et al. 2016). The model of Christoffersen et al. (2016), for example, reduces stomatal conductance using a function similar to that for stem conductivity whereby

$$g_{sw} = g_{s\max}\left[1 + \left(\frac{\psi_\ell}{\psi_{50}}\right)^a\right]^{-1}. \quad (13.34)$$

This is a general method that works independent of the specific stomatal conductance model. When used with (13.33), $K(\psi)$, $C(\psi)$, and $g_{sw}(\psi)$ define the hydraulic strategy.

Question 13.7 Graph the relative stomatal closure given by (13.34) for $a = 6$ with $\psi_{50} = -1.5$ MPa and again for $\psi_{50} = -3.0$ MPa. How do the curves differ? Which parameter set describes an isohydric plant and an anisohydric plant?

Question 13.8 SPA calculates stomatal conductance from principles of water-use efficiency optimization provided that leaf water potential does not drop below some minimum threshold. Discuss the conceptual difference between this method and the use of (13.34) in other plant hydraulics models. How do these two approaches represent isohydric and anisohydric behavior?

Question 13.9 In the FETCH2 implementation of the Richards equation, ψ is water potential with units Pa, K is conductivity in m^2 s, C is capacitance in kg H_2O m^{-1} Pa^{-1}, and E is transpiration in kg H_2O s^{-1}

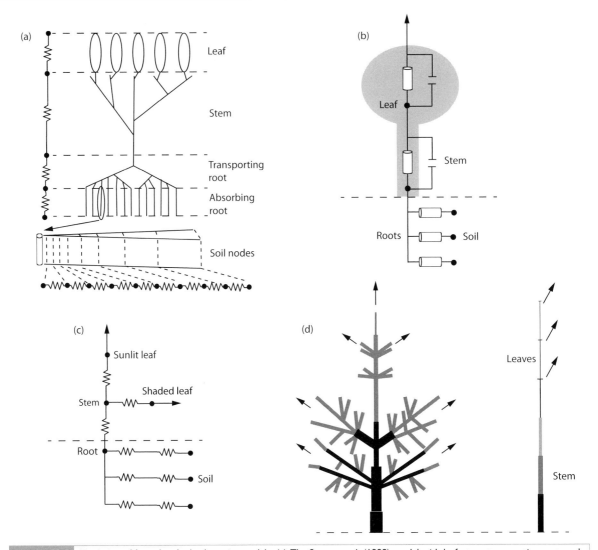

Figure 13.8 Depiction of four plant hydrodynamics models. (a) The Sperry et al. (1998) model with leaf, stem, transporting root, and absorbing root components. The root system is depicted as a cylinder connected to surrounding soil with multiple nodes. (b) The two-node model of Xu et al. (2016) for xylem and leaf water potential with capacitance. The root system is modeled by individual layers. (c) The four-node model used in version 5 of the Community Land Model (Lawrence et al. 2018) consisting of root, stem, sunlit leaf, and shaded leaf water potentials. The root system is modeled by individual layers. Capacitance is neglected. (d) The FETCH model (Bohrer et al. 2005) depicting a tree as multiple branching patterns with leaves that transpire or as a single beam. FETCH2 uses the single-beam representation (Mirfenderesgi et al. 2016)

(Mirfenderesgi et al. 2016). (a) Show that these units are dimensionally correct. (b) Relate the conductivity in m^2 s to sapwood-specific conductivity with units mol H_2O m^{-1} s^{-1} Pa^{-1}. Hint: Water flow is expressed as a mass flux (kg H_2O s^{-1}) in (13.33), not as a mass flux density (kg H_2O m^{-2} s^{-1}).

13.8 | Supplemental Programs

13.1 Plant Hydraulics: This program uses the numerical method shown in Figure 13.4 applied to a single leaf layer to calculate leaf gas exchange. It is similar to Supplemental Program 12.2 for leaf gas exchange and leaf temperature but additionally includes the SPA hydraulic constraint on stomatal conductance.

13.9 | Modeling Projects

1. Supplemental Program 13.1 describes SPA plant hydraulics as applied to a single leaf. Use this to investigate the sensitivity of leaf water potential and stomatal conductance to plant physiological parameters ($\psi_{\ell \min}$, C_p, K_p, ι) and fine root parameters (M_r, r_r, ρ_r, R_r^*).

2. Use the same code to investigate the contributions of aboveground and belowground resistances to leaf water potential and stomatal conductance. Compare results for a loam soil ranging from wet to dry conditions.

14

Radiative Transfer

Chapter Overview

Absorption and reflection of solar radiation by plant canopies are related to the amount of leaves, stems, and other phytoelements, their optical properties, and their orientation. This chapter develops the biophysical theory and mathematical equations to describe radiative transfer for plant canopies, incorporating these concepts as well as accounting for the spectral composition of light and distinguishing between direct beam and diffuse radiation. Key derivations are K_b and K_d, which describe the extinction of direct beam and diffuse radiation, respectively, in relation to leaf angle, solar zenith angle, and leaf area index. Equations for radiative transfer describe horizontally homogeneous, plane-parallel canopies in which variation in radiation is in the vertical direction. The distinction between sunlit and shaded leaves is also important. Similar equations pertain to longwave fluxes.

14.1 | Introduction

Plant canopies have a prominent vertical profile of foliage. Some canopies have a uniform distribution; others more commonly have most foliage concentrated at the top of the canopy. Absorption of solar radiation by vegetation follows a vertical profile within the canopy related to the distribution of foliage. In the herbaceous vegetation shown in Figure 14.1, only about 1% of the incident sunlight penetrates through the bottom of the canopy. About 50% of the sunlight is absorbed in the upper 30 cm of the canopy. The sunlight that strikes an individual leaf is reflected upward, transmitted downward deeper into the canopy, or absorbed by the leaf. As the canopy becomes denser with leaves, more radiation is absorbed and less is reflected upward or transmitted through the canopy. As a result, little of the sunlight at the top of the canopy penetrates through the bottom of a dense canopy of leaves. Moreover, a plant canopy generally has a lower albedo than an individual leaf because of multiple scattering among leaves and subsequent trapping of radiation in the canopy.

A theory of radiative transfer must integrate the light environment of an individual leaf over all leaves in the canopy, as well as account for other absorbing and scattering surfaces such as stems and branches. The theory must distinguish direct beam radiation, which originates from the Sun's position in the sky, from diffuse radiation, which is scattered light that emanates from the entire sky hemisphere. This distinction is important because absorption of sunlight by a leaf depends on the angle of the incident radiation relative to the leaf surface. It is also necessary to account for wavelength because the spectral properties of leaves and other materials vary with wavelength.

This chapter develops the theory of radiative transfer through plant canopies. The plant canopy consists of individual leaves with soil underneath. The optical properties of leaves and their orientation

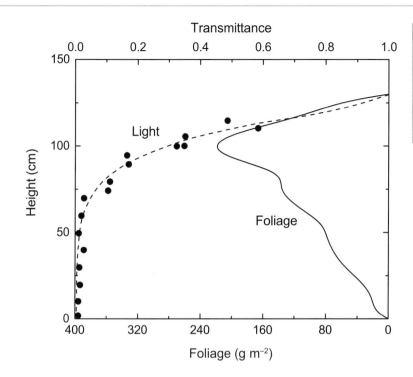

Figure 14.1 Profile of light and foliage in a stand of herbaceous plants approximately 130 cm tall. The horizontal axis shows transmittance as a fraction of incident radiation (top axis) and foliage mass (bottom axis) at various heights in the canopy. Redrawn from Monsi and Saeki (1953; reprinted as Monsi and Saeki 2005)

are described first. Radiative transfer theory is then developed for a canopy of non-scattering (black) leaves, for both direct beam and diffuse radiation. This theory is extended to describe multilayer radiative transfer with scattering using the numerical methods of Norman (1979) and the analytical solutions of Goudriaan (1977), both commonly used in plant canopy models. Sunlit and shaded leaves receive different light environments, and equations for both are presented. The two-stream approximation, a whole-canopy analytical model widely used in terrestrial biosphere models, is also described. Each of these three theories calculates radiative transfer from knowledge of leaf orientation, leaf reflectance and transmittance, leaf area index, soil albedo, the angle of the incident beam of radiation, and the partitioning of solar radiation into direct beam and diffuse components and visible and near-infrared wavebands. Longwave radiative transfer is described similarly as diffuse radiation, but with the addition of a leaf source term related to leaf temperature.

The theory, as developed here, treats the canopy as an infinitely long, plane-parallel, and horizontally homogeneous layer of leaves randomly distributed in space, or as multiple such layers distributed vertically through the canopy (Figure 14.2). This greatly simplifies radiative transfer and describes variation in radiation solely in the vertical direction, but it ignores the complexity of spatial structure – especially crown geometry, gaps, and the distribution of phytoelements in three dimensions throughout the canopy. More complex three-dimensional models of radiative transfer are needed to account for this canopy structure, as well as the amount and optical properties of green leaves, brown leaves, and wood (Wang and Jarvis 1990; Gastellu-Etchegorry et al. 1996; Cescatti 1997; Cescatti and Niinemets 2004; Kobayashi et al. 2012). Some terrestrial biosphere models have incorporated patchiness and canopy gaps (Yuan et al. 2014), but most continue to use the plane-parallel, homogenous canopy simplification. The nonrandom distribution of foliage can be accounted for with a clumping index such as used in CANOAK (Baldocchi and Wilson 2001) and other canopy models (Walcroft et al. 2005; Ryu et al. 2011; Chen et al. 2012, 2016a), but the effect of branches and stems on radiative transfer is problematic.

14.2 Leaf Optical Properties

When solar radiation strikes a leaf, a fraction of the radiation is absorbed, and the remainder is

(a) 1-D

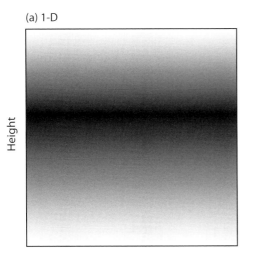

Height

(b) 3-D

Figure 14.2 Representation of a canopy as (a) one-dimensional with a vertical profile of leaf area (shown by grayscale gradation in which darker shading denotes more leaves) that is horizontally homogenous and (b) three-dimensional with vertical and spatial structure determined by crown geometry and spacing. See also Cescatti and Niinemets (2004)

scattered in all directions from both sides of the leaf. Radiation scattered into the backward hemisphere is referred to as reflected radiation, and that scattered into the forward hemisphere is the radiation transmitted through the leaf. Mathematically, this is described as

$$\alpha_\ell = 1 - \rho_\ell - \tau_\ell = 1 - \omega_\ell. \qquad (14.1)$$

In this equation, α_ℓ is the fraction of the radiation that is absorbed, ρ_ℓ is the fraction that is reflected, τ_ℓ is the fraction that is transmitted, and $\omega_\ell = \rho_\ell + \tau_\ell$ is the leaf scattering coefficient. Absorption, reflection, and transmission of solar radiation by foliage have strong wavelength dependence. Solar radiation is divided into two broad spectral bands: the visible waveband at wavelengths less than 0.7 μm and the near-infrared waveband at wavelengths greater than 0.7 μm. Plants utilize light in these two wavebands differently (Table 14.1). Green leaves typically absorb more than 80% of the solar radiation in the visible waveband that strikes the leaf. This light is used during photosynthesis. Light in the near-infrared waveband is not utilized during photosynthesis, and rather than absorbing this radiation, and thus possibly overheating, leaves typically absorb less than 50% of the radiation in the near-infrared waveband.

An additional distinction is between the optical properties of foliage and stems. The latter is generally represented in models by stem or wood area index, in contrast with leaf area index. In the absence of three-dimensional models – which explicitly distinguish green leaves, brown leaves, wood, and other phytoelements – homogeneous canopy models use optical properties weighted for the amount of foliage and stem material. When snow is on the canopy, the optical properties must further be adjusted for the amount of intercepted snow.

14.3 Light Transmission without Scattering

The Beer–Bouguer–Lambert law, which describes attenuation of light through a homogeneous medium in the absence of scattering, also describes radiative transfer through a canopy of leaves. In a simple derivation of this law, an amount of light denoted by $I^\downarrow$ strikes a medium of thickness dz that contains absorbing particles. Absorption is assumed to be proportional to $I^\downarrow$ and dz so that $dI^\downarrow = -KI^\downarrow dz$, where K is a proportionality constant. The negative sign denotes loss of light as it passes through the medium. Integration of this equation shows that the light at some distance z in the medium is $I^\downarrow = I_0^\downarrow \exp(-Kz)$, where $I_0^\downarrow$ is the radiation incident

Table 14.1	Leaf orientation χ_ℓ and reflectance ρ_ℓ, transmittance τ_ℓ, and absorptance α_ℓ of solar radiation by a leaf for visible and near-infrared wavebands						
Plant type	χ_ℓ	Visible			Near-infrared		
		ρ_ℓ	τ_ℓ	α_ℓ	ρ_ℓ	τ_ℓ	α_ℓ
Needleleaf tree	0.01	0.07	0.05	0.88	0.35	0.10	0.55
Broadleaf tree	0.25	0.10	0.05	0.85	0.45	0.25	0.30
Grass, crop	−0.30	0.11	0.05	0.84	0.35	0.34	0.31

Source: From the Community Land Model.

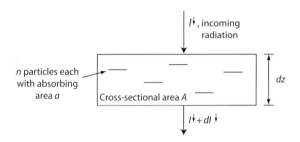

Figure 14.3 Transmission of solar radiation through a homogeneous medium in the absence of scattering. In this example, n non-overlapping opaque particles each with cross-sectional area a oriented perpendicular to the path of light are placed in a medium with cross-sectional area A and thickness dz. The radiation absorbed in the medium is $dI^\downarrow$.

on the medium. Two alternative derivations of the law provide a more physical interpretation with extension to plant canopies.

With reference to Figure 14.3, consider a medium of cross-sectional area A and thickness dz oriented perpendicular to the direction of light penetration. The medium contains random, non-overlapping opaque (i.e., black) particles. A photon of light is absorbed if it strikes the particle and is otherwise transmitted through the medium. The number of absorbing particles in the medium is n, and the density of particles (per unit volume) is $N = n/(A\,dz)$. Each particle has a cross-sectional area denoted as a that is oriented perpendicular to the path of light. The total absorbing area is $a \cdot n$, and the fraction of light absorbed is $a \cdot n/A = aN\,dz$. The amount of radiation absorbed in the medium is $dI^\downarrow = -I^\downarrow aN\,dz$, and the radiation transmitted through the medium is $I^\downarrow = I_0^\downarrow \exp(-aNz)$. For a canopy of horizontal leaves, aN is the leaf area density, and $L = aNz$ is the leaf area index so that

$$I^\downarrow = I_0^\downarrow e^{-L}. \tag{14.2}$$

The ratio $\tau_b = I^\downarrow/I_0^\downarrow$ is the fraction of incoming radiation transmitted through the leaf area index L, and in the absence of scattering, the fraction $1 - \tau_b$ is absorbed.

Equation (14.2) can also be derived from probability theory (Monsi and Saeki 1953; reprinted as Monsi and Saeki 2005). Consider a horizontal ground surface with area A. The Sun is directly overhead so that the solar beam is perpendicular to the surface. If one black horizontal leaf with area a is randomly placed over the area A, the probability that the leaf intercepts radiation is a/A, which is also the fraction of the area A that is in shadow. The probability that the leaf does not intercept the beam (i.e., light reaches the ground) is $1 - a/A$. If a second leaf is placed at random and the two leaves do not overlap, the probability that neither leaf intercepts the beam is $(1 - a/A)(1 - a/A)$. More generally, $\Delta L = a/A$ is the area of a leaf per unit ground area, and the probability that no leaves intercept radiation in a canopy of n randomly placed, non-overlapping black horizontal leaves is

$$P_0 = (1 - \Delta L)^n. \tag{14.3}$$

In this equation, $P_0 = I^\downarrow/I_0^\downarrow$ is the fraction of incoming radiation that is not intercepted and describes light transmittance through the canopy. It is also the fraction of the ground illuminated by the incident beam. The total leaf area index is $L = n\Delta L$, and (14.3) can be rewritten in the form:

$$P_0 = (1 - \Delta L)^{L/\Delta L}. \tag{14.4}$$

Taking the limit $\Delta L \to 0$ gives the exponential transmittance $P_0 = \exp(-L)$ as in (14.2).

Equation (14.2) applies to horizontal leaf surfaces and for sunlight received from directly overhead. For leaves of any orientation or for sunlight at directions other than vertical, leaf area L must be adjusted for the projected leaf area onto a horizontal plane L_H. This is achieved by introducing the extinction coefficient for direct beam K_b (also known as the optical depth of direct beam per unit leaf area), which is defined as L_H/L. A general expression that describes transmission of direct beam radiation through a homogenous canopy of randomly placed leaves in the absence of scattering is

$$\frac{dI^{\downarrow}}{dL} = -K_b I^{\downarrow} \tag{14.5}$$

and

$$I^{\downarrow} = I^{\downarrow}_{sky,b} e^{-K_b L}, \tag{14.6}$$

with $I^{\downarrow}_{sky,b}$ the radiation incident at the top of the canopy. A small increment of leaf area intercepts an amount of radiation equal to $dI^{\downarrow} = -K_b\ I^{\downarrow} dL$, and the mean interception (absorption) per leaf area is $dI^{\downarrow}/dL = -K_b I^{\downarrow}$. The value of K_b varies with leaf orientation and solar angle. Figure 14.4 illustrates (14.6) for representative values of K_b. For $K_b > 0.5$, less than 10% of the incident radiation is transmitted through a canopy with leaf area index $L = 5$ m^2 m^{-2}.

With $K_b > 1$, less than 1% penetrates through the same canopy.

The direct beam transmittance $\tau_b = P_0 = \exp(-K_b x)$ describes the mean irradiance on a horizontal surface at a depth in the canopy with cumulative leaf area index x. This also describes f_{sun}, the fractional area on a horizontal plane at x that is illuminated by beam radiation – more commonly referred to as the sunlit fraction. The sunlit fraction of leaf area at x is

$$f_{sun}(x) = e^{-K_b x}, \tag{14.7}$$

and $1 - f_{sun}(x)$ is shaded. This equation can be derived by finding the sunlit leaf area in a thickness ΔL at a depth x in the canopy (Campbell and Norman 1998, p. 259). At x, the sunlit leaf area in an increment of leaf area ΔL is

$$\Delta L_{sun} = \frac{e^{-K_b x}\left(1 - e^{-K_b \Delta L}\right)}{K_b}. \tag{14.8}$$

The first term in the numerator is the probability that a beam penetrates to depth x, and the second term is the probability that the beam is intercepted in the layer ΔL. Division by K_b adjusts the projection of leaf area on a horizontal surface to actual leaf area. As $\Delta L \to 0$,

$$\Delta L_{sun} = e^{-K_b x} \Delta L, \tag{14.9}$$

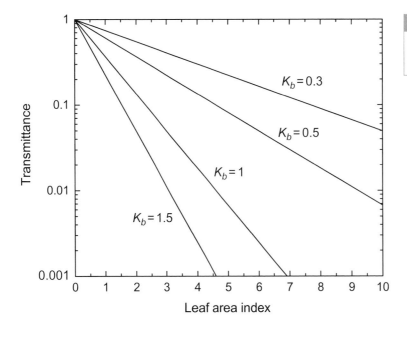

Figure 14.4 Transmission of direct beam radiation τ_b in relation to leaf area index for typical values of the extinction coefficient K_b.

and the sunlit fraction is $\Delta L_{sun}/\Delta L$. An equivalent derivation of (14.7) from Gutschick (1991) is to define

$$f_{sun}(x) = -\frac{1}{K_b}\frac{dP_0}{dL}, \tag{14.10}$$

which is equal to (14.7). The same result can be obtained using the numerical approximation to (14.10), which is

$$f_{sun}(x) = \frac{e^{-K_b x} - e^{-K_b(x+\Delta L)}}{K_b \Delta L}. \tag{14.11}$$

As $\Delta L \to 0$, this equation reduces to (14.7). For a canopy with leaf area index L, the sunlit leaf area index is

$$L_{sun} = \int_0^L f_{sun}(x)\,dx = \frac{1 - e^{-K_b L}}{K_b}, \tag{14.12}$$

and the shaded leaf area index is $L - L_{sun}$. For very large values of L, $L_{sun} \approx 1/K_b$.

In a canopy of horizontal foliage, much of the direct beam radiation is intercepted by leaves, and little of the canopy is sunlit (Figure 14.5). In contrast, vertical leaves intercept little radiation when the Sun is high in the sky (the solar zenith angle is $10°$ in Figure 14.5), and a large fraction of the canopy is sunlit. In general, when the solar zenith angle is small, horizontal leaves attenuate the most radiation and have the lowest sunlit leaf area index (Figure 14.6). Vertical leaves have the lowest

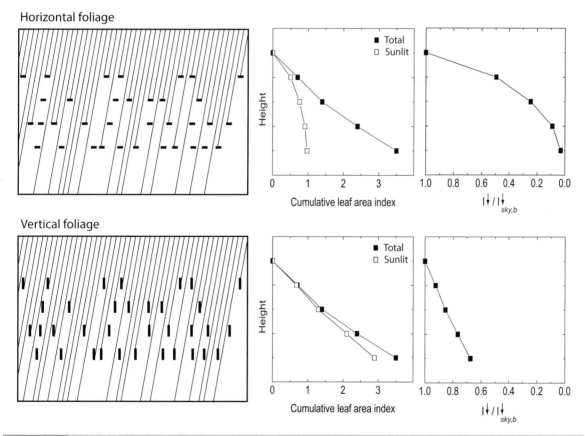

Figure 14.5 Radiative transfer and sunlit leaf area index for a canopy of horizontal leaves (top panels) with $K_b = 1$ and vertical leaves (bottom panels) with $K_b = 0.112$. The left-hand panels show a canopy consisting of four layers of leaves. Each thick black line represents a leaf area index of 0.1 $m^2 \ m^{-2}$. The thin lines depict interception or transmission of beam radiation with a zenith angle of $10°$. The middle panels show cumulative leaf area index and sunlit leaf area index with depth in the canopy. The right-hand panels show direct beam transmittance with depth in the canopy.

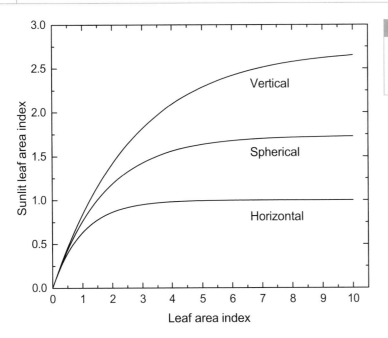

extinction coefficient and have greater sunlit leaf area. For a zenith angle of 30°, as in the example shown in Figure 14.6, maximum sunlit leaf area index is 1 m² m⁻² for horizontal leaves, 1.7 m² m⁻² for spherical leaves, and 2.7 m² m⁻² for vertical leaves.

In the absence of scattering, $dI^\downarrow/dL = -K_b I^\downarrow$ is the mean beam irradiance absorbed per leaf area. At a depth in the canopy with cumulative leaf area index x, the mean beam irradiance of foliage $\overrightarrow{I}_{\ell b}(x)$ (per unit leaf area) is, therefore, given by

$$\overrightarrow{I}_{\ell b}(x) = -\frac{dI^\downarrow}{dx} = K_b I^\downarrow_{sky,b} e^{-K_b x}. \tag{14.13}$$

As illustrated in Figure 14.5, this is the average of the sunlit portion of the canopy in which light is unattenuated (i.e., sunflecks with $I^\downarrow_{sky,b}$) and the shaded portion with zero irradiance. Equation (14.13), therefore, can be equivalently written as the average absorption of sunlit leaves with mean irradiance per sunlit leaf area $\overrightarrow{I}_{\ell sun}(x)$ and shaded leaves with $\overrightarrow{I}_{\ell sha}(x) = 0$, each weighted by the sunlit and shaded fraction, respectively. This means that

$$\overrightarrow{I}_{\ell b}(x) = \overrightarrow{I}_{\ell sun}(x)f_{sun}(x) + \overrightarrow{I}_{\ell sha}(x)[1 - f_{sun}(x)]. \tag{14.14}$$

Combining (14.13) and (14.14), and with f_{sun} given by (14.7), it is evident that

$$\overrightarrow{I}_{\ell sun}(x) = K_b I^\downarrow_{sky,b}. \tag{14.15}$$

This expression shows that the mean irradiance absorbed by a sunlit leaf is invariant with depth in the canopy. The mean absorption of sunlit leaves can be similarly derived by considering the canopy as a whole. For a canopy with leaf area index L, the mean irradiance of sunlit leaves is equal to the radiation absorbed by the canopy divided by the sunlit leaf area so that

$$\overrightarrow{I}_{\ell sun} = \frac{\left(1 - e^{-K_b L}\right)I^\downarrow_{sky,b}}{L_{sun}} = K_b I^\downarrow_{sky,b}. \tag{14.16}$$

In this context, K_b can be seen to represent the mean flux absorbed by the sunlit canopy divided by the flux on the horizontal plane above the canopy.

These equations describe light transmission through a canopy in which leaves are randomly spaced. In fact, however, foliage is clumped along branches and shoots in plant crowns, and crowns themselves can be nonrandomly grouped. When foliage is clumped, light transmission still follows an exponential function but multiplied by a clumping factor Ω (Nilson 1971) so that irradiance is attenuated in relation to leaf area as

$$I^\downarrow = I^\downarrow_{sky,b} e^{-K_b \Omega L}, \tag{14.17}$$

and the sunlit fraction is

$$f_{sun}(x) = \Omega e^{-K_b \Omega x}. \qquad (14.18)$$

The clumping factor Ω characterizes the extent to which foliage has a nonrandom spatial distribution. It has a value of one for randomly spaced leaves and less than one for clumped foliage. Clumping increases the probability of overlapping leaves compared with a random spatial distribution so that the probability of light transmission increases and the sunlit fraction decreases. In a canopy of clumped leaves ($\Omega < 1$), the effective leaf area is reduced by the factor Ω and intercepts less radiation than a canopy of randomly spaced leaves ($\Omega = 1$). Values for Ω vary geographically (Chen et al. 2005, 2012), but representative values are broadleaf evergreen, 0.66; broadleaf deciduous, 0.70; needleleaf evergreen, 0.74; needleleaf deciduous, 0.78; shrub, 0.75; herbaceous plants such as grasses and crops, 0.75 (Chen et al. 2012).

Question 14.1 The leaf area index of a forest can be directly sampled in the field by harvesting leaves and measuring the one-sided area of the leaves or by allometric relationships between leaf area and tree diameter. Alternatively, optical techniques indirectly estimate leaf area index by inverting measurements of sunlight at the ground using simple radiative transfer theory. Contrast the estimation of leaf area index by the direct and indirect methods. What are the uncertainties of the indirect techniques?

Question 14.2 Derive an equation for sunlit leaf area index L_{sun} that accounts for nonrandom foliage using the clumping factor Ω. What is the corresponding sunlit leaf irradiance $\vec{I}_{\ell sun}(x)$? Describe how clumping affects light absorption in a plant canopy.

14.4 | Direct Beam Extinction Coefficient

The extinction coefficient K_b represents the ratio between the leaf area projected onto a horizontal

surface and the actual leaf area. This is equivalent to the ratio between the area of shadow cast onto a horizontal surface by a leaf and the area of the leaf. For a canopy of leaves, K_b can be derived by approximating the distribution of foliage as similar to the distribution of surface area on cylinders, spheres, and cones. The extinction coefficient is the ratio of the shadow area to surface area for these geometric shapes. In the text that follows, K_b is derived first for an individual leaf with a specific orientation and then integrated over all leaf orientations in the canopy.

As a simple example, a thin, flat leaf intercepts no sunlight when it is oriented vertically and the Sun is directly overhead. Figure 14.7 illustrates more complex geometries. The solar zenith angle Z is the angle of the Sun's beam from vertical ($0°$ for vertical to $90°$ for horizontal); the solar elevation angle B is the angle above the horizon ($B = 90° - Z$). In this

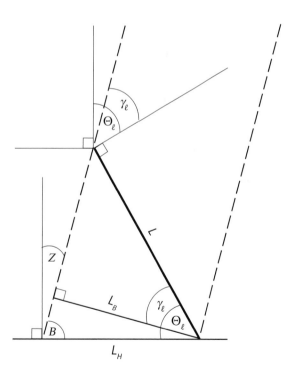

Figure 14.7 Derivation of the extinction coefficient K_b from leaf inclination angle Θ_ℓ, solar elevation angle B or zenith angle Z, and the incidence angle of the solar beam on the leaf γ_ℓ. In this diagram, the thick line represents a leaf with dimension L and the thin lines are the projection normal to the solar beam (denoted L_B) and onto horizontal (denoted L_H). The dashed lines denote the Sun's beam. Additional lines are drawn for reference.

diagram, the leaf is oriented facing the Sun's rays with an inclination angle Θ_ℓ above horizontal, which is also the angle between vertical and an imaginary line normal (perpendicular) to the upper surface of the leaf. The Sun's rays strike the leaf at an angle γ_ℓ formed by the leaf normal and the Sun's beam. For a leaf area L, the area L_B that is projected onto a plane normal (i.e., perpendicular) to the direction of the solar beam is obtained from the trigonometric relationship

$$\cos\gamma_\ell = \frac{L_B}{L}, \tag{14.19}$$

and the area L_H that is projected onto a horizontal plane is

$$\cos Z = \sin B = \frac{L_B}{L_H}. \tag{14.20}$$

The extinction coefficient is defined from these two relationships by

$$K_b = \frac{L_H}{L} = \frac{\cos\gamma_\ell}{\cos Z} = \frac{G(Z)}{\cos Z}. \tag{14.21}$$

The term $\cos\gamma_\ell$ is the projection of leaf area in the direction of the solar beam and is referred to as the function $O(B)$ (Goudriaan 1977) or $G(Z)$ (Ross 1975, 1981). The term $\cos Z$ adjusts this projection to a horizontal plane.

In (14.21), γ_ℓ is the incidence angle of the solar beam on the leaf, and its cosine gives the solar radiation incident on a sloped surface. γ_ℓ is determined by solar zenith angle Z, leaf inclination angle Θ_ℓ, and the difference between the solar azimuth angle A and the leaf azimuth angle A_ℓ according to

$$\cos\gamma_\ell = \cos Z\cos\Theta_\ell + \sin Z\sin\Theta_\ell\cos(A_\ell - A). \tag{14.22}$$

Figure 14.8 illustrates K_b for various leaf and solar geometries. The extinction coefficient has a value of one for a horizontal leaf regardless of zenith angle. As the leaf is inclined at a greater angle from horizontal, K_b declines because the projected leaf area decreases. A leaf facing the solar beam has a larger extinction coefficient than does a leaf oriented away from the Sun's azimuth.

Leaves in a plant canopy do not have a single inclination angle or aspect, but rather have numerous orientations defined by a leaf angle distribution, denoted by the probability density function $f(\Theta_\ell)$, and an azimuth angle distribution, denoted by the probability density function $g(A_\ell)$, as given by (2.5) and (2.7). The function $G(Z)$ is equal to the mean value of $|\cos\gamma_\ell|$, which is found by integrating over all azimuth differences (0 to 2π) and over all leaf inclination angles (0 to $\pi/2$) so that

$$G(Z) = \int_0^{\pi/2}\int_0^{2\pi}|\cos\gamma_\ell|f(\Theta_\ell)g(A_\ell)d\Theta_\ell\ dA_\ell. \tag{14.23}$$

Lemeur (1973) provided a general numerical solution for any leaf angle distribution and azimuth angle distribution that involves summing $G(Z)$ over 16 discrete values of azimuth angles (22.5° increments) and 12 discrete values of inclination angles (7.5° increments) weighted by the relative frequency of each azimuth and inclination.

It is more common to assume an absence of azimuth preference, and Goudriaan (1977, 1988) provided a specific solution for this case. If leaf azimuth is distributed at random, the azimuth density function is uniform and given by $g(A_\ell) = 1/(2\pi)$. In this case, $G(Z)$ for a particular leaf inclination angle Θ_ℓ is

$$G(Z, \Theta_\ell) = \frac{1}{2\pi}\int_0^{2\pi}|\cos\gamma_\ell|\ dA_\ell. \tag{14.24}$$

Substitution of (14.22) for $\cos\gamma_\ell$ gives

$$G(Z, \Theta_\ell) = \begin{cases} \cos Z\ \cos\Theta_\ell & \Theta_\ell \leq 90° - Z \\ \frac{2}{\pi}\left[c + a\sin^{-1}(a/b)\right] & \Theta_\ell > 90° - Z \end{cases}, \tag{14.25}$$

with $a = \cos Z\cos\Theta_\ell$, $b = \sin Z\sin\Theta_\ell$, and $c = (\sin^2\Theta_\ell - \cos^2 Z)^{1/2}$. The $G(Z)$ function requires integrating over all leaf angles. This is obtained numerically by evaluating $G(Z, \Theta_\ell)$ for each of nine leaf angle classes in 10° increments ($\Theta_{\ell,i} = 5°, \ldots, 85°$) and weighting by the fractional abundance of leaves F_i in each angle class i, given by (2.15), so that

$$G(Z) = \sum_{i=1}^9 G(Z, \Theta_{\ell,i})F_i. \tag{14.26}$$

Goudriaan (1988) showed that three leaf angle classes of 30° is sufficient.

Analytical solutions are available for specific leaf orientations. For horizontal leaves,

$$G(Z) = \cos Z, \qquad K_b = 1, \tag{14.27}$$

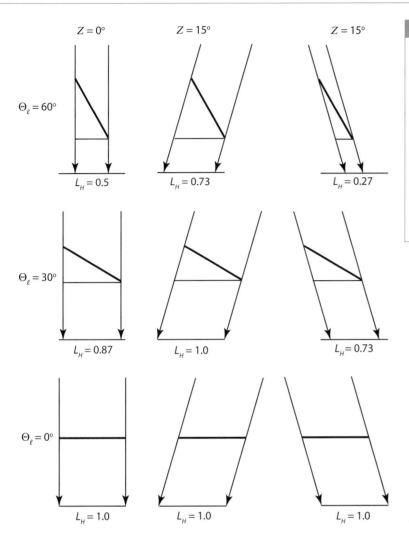

Figure 14.8 Extinction coefficient in relation to solar zenith angle Z and leaf inclination angle Θ_ℓ. In each panel, a unit leaf area ($L = 1$), shown with a thick line, is projected onto a horizontal surface L_H so that $K_b = L_H$. The leaf inclination angle is $0°$ (bottom panels), $30°$ (middle panels), and $60°$ (top panels). In the left and middle columns, the leaf is oriented towards the Sun ($A_\ell - A = 0°$) and the solar zenith angle is $0°$ (left column) and $15°$ (middle column). In the right column, Z = $15°$, but the leaf is oriented away from the Sun ($A_\ell - A = 180°$). In each panel, the arrows indicate the solar beam.

and direct beam transmission is independent of solar angle. For vertical leaves,

$$G(Z) = \frac{2}{\pi} \sin Z, \qquad K_b = \frac{2}{\pi} \tan Z. \qquad (14.28)$$

If all the leaves in a canopy are vertical but have a random azimuth, they can be thought of as being arranged along the surface of a vertical cylinder. The extinction coefficient for vertical leaves is two times the ratio of the projection of a cylinder with radius r and length z onto a horizontal plane ($2rz \tan Z$) to its surface area ($2\pi rz$). The factor two in (14.28) arises because the leaf intercepts radiation on two sides. For a spherical leaf distribution:

$$G(Z) = 0.5, \qquad K_b = 0.5/\cos Z. \qquad (14.29)$$

For this distribution, the leaves can be thought of as being arranged on the surface of a sphere, and the extinction coefficient is twice the ratio of the projection of the sphere with radius r onto a horizontal plane ($\pi r^2/\cos Z$) to its surface area ($4\pi r^2$). Campbell (1986, 1990) provided the solution for the ellipsoidal distribution described by (2.11) whereby

$$K_b = \frac{1}{l}\sqrt{x^2 + \tan^2 Z}, \qquad (14.30)$$

with x defined as in (2.11) and l provided specifically by (2.12) and (2.13) or more generally by (2.14). Equation (14.30) is a generalization for a variety of leaf orientations. The solution for horizontal leaves ($K_b = 1$) is obtained for large x (as $x \to \infty$). Vertical leaves are represented with $x = 0$. The spherical leaf angle distribution is obtained with $x = 1$ and $l = 2$.

Goudriaan (1977) simplified much of the complexity of these equations into a semiempirical relationship using the Ross index χ_ℓ, which quantifies the departure of leaf angles from a spherical distribution using (2.16). In the Ross–Goudriaan function, the projection of leaf area in the direction of the solar beam is

$$G(Z) = \phi_1 + \phi_2 \cos Z, \qquad (14.31)$$

with $\phi_1 = 0.5 - 0.633\chi_\ell - 0.33\chi_\ell^2$ and $\phi_2 = 0.877(1 - 2\phi_1)$. This equation is restricted to $-0.4 < \chi_\ell < 0.6$.

Figure 14.9a illustrates the direct beam extinction coefficient in relation to solar zenith angle for various leaf angle distributions. When the Sun is high in the sky, K_b is smaller for spherical and vertical leaves than for horizontal leaves. Light penetrates deeper into these canopies than into a canopy of horizontal leaves. This is why spherical and vertical canopies have a greater sunlit fraction than does a horizontal canopy (Figure 14.6). The opposite is true when the Sun is low in the sky. For zenith angles greater than about 60°, $K_b > 1$ for spherical and vertical leaves and light is attenuated more rapidly than in a canopy of horizontal leaves. Transmittance declines markedly as leaf area index increases (Figure 14.10). Canopies with a spherical leaf angle distribution and $L = 4$ m^2 m^{-2} transmit at most 14% of incident sunlight, and less than 5% passes through the canopy for $Z > 50°$.

A canopy transmits less than 5% of sunlight with $L = 6$ m^2 m^{-2} and less than 1% with $L = 9$ m^2 m^{-2}, regardless of zenith angle.

Question 14.3 Calculate the shadow cast by a unit area of leaf inclined at an angle $\Theta_\ell = 30°$ when the Sun has a zenith angle $Z = 30°$ and has an azimuth angle A that is 90° east of the direction of the leaf ($A_\ell - A = -90°$). Then repeat the calculation for all azimuth directions from 0°–360° using 30° increments. In which directions does the Sun cast the largest and smallest shadow?

Question 14.4 Use Supplemental Program 14.1 to calculate the direct beam extinction coefficient K_b for planophile, erectophile, plagiophile, uniform, and spherical leaf distributions (Table 2.1) with a solar zenith angle $Z = 30°$. (a) Use Goudriaan's numerical method with nine angle classes. (b) Compare answers using the Ross index χ_ℓ to calculate $G(Z)$. How different are the results between these two methods?

Question 14.5 Figure 14.6 shows sunlit leaf area index for horizontal, vertical, and spherical leaf angle distributions with $Z = 30°$. Using the preceding results, redraw this figure for the planophile, erectophile, and

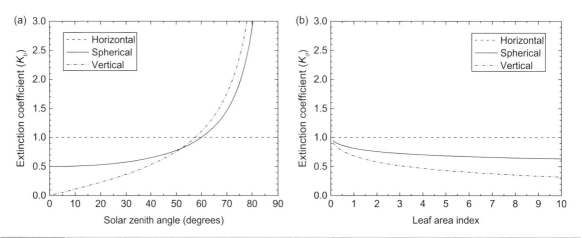

Figure 14.9 Extinction coefficients for horizontal, spherical, and vertical leaf angle distributions. (a) Direct beam radiation K_b in relation to solar zenith angle. (b) Diffuse radiation K_d in relation to leaf area index.

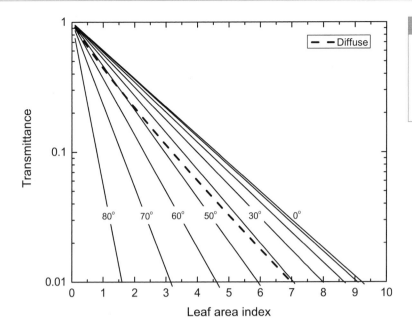

Figure 14.10 Transmission of solar radiation through a canopy with spherical leaf distribution in relation to leaf area index. The solid lines show direct beam transmittance τ_b for solar zenith angles of $0°–80°$ (in $10°$ increments). The dashed line shows the diffuse transmittance τ_d.

plagiophile leaf angle distributions. Discuss uncertainty in leaf distribution and K_b as they affect sunlit leaf area.

Question 14.6 Figure 14.9a shows the direct beam extinction coefficient in relation to solar zenith angle for horizontal, vertical, and spherical leaf angle distributions. Redraw this figure for the planophile, erectophile, and plagiophile leaf angle distributions using Goudriaan's numerical method and the Ross index χ_ℓ. Discuss how different leaf angle distributions and the two numerical methods affect K_b.

Question 14.7 γ_ℓ is the angle at which the solar beam strikes a leaf. Discuss how leaf inclination angle affects leaf temperature.

14.5 | Diffuse Transmittance

Diffuse radiation is attenuated similarly as direct beam but is received from all parts of the sky. It can be treated as independent beams of radiation from all directions (Figure 14.11). Radiation received from regions of the sky with different zenith angles will vary in extinction coefficient. The extinction coefficient for diffuse radiation is obtained by substituting the sky zenith angle for the solar zenith angle and integrating over the sky hemisphere. The transmission of diffuse radiation through a canopy with leaf area index L is

$$\tau_d = \frac{I^\downarrow}{I^\downarrow_{sky,d}} = 2 \int_0^{\pi/2} \exp\left[-\frac{G(Z)}{\cos Z}L\right] \sin Z \cos Z \, dZ, \quad (14.32)$$

where $G(Z)/\cos Z$ is the direct beam extinction coefficient and varies with sky zenith angle, denoted here by Z. The term $2 \sin Z \cos Z \, dZ$ is the contribution of Z to total sky irradiance. This is derived assuming a uniform sky source of diffuse radiation with equal radiance in all directions. In practice, (14.32) can be evaluated using nine sky zones with $10°$ increments (Norman 1979; Goudriaan 1977) or three $30°$ zones (Goudriaan 1988). The numerical approximation for nine sky zones ($Z_i = 5°, \ldots, 85°$) is

$$\tau_d = 2 \sum_{i=1}^{9} \exp\left[-\frac{G(Z_i)}{\cos Z_i}L\right] \sin Z_i \cos Z_i \, \Delta Z_i, \quad (14.33)$$

with $\Delta Z_i = 0.1745$ radians. The effective extinction coefficient is

$$K_d = -\frac{\ln \tau_d}{L}. \quad (14.34)$$

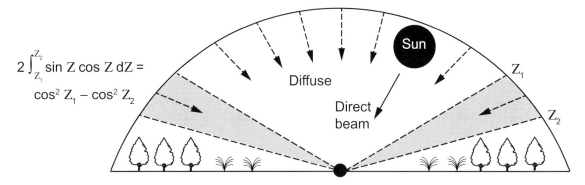

$$2\int_{Z_1}^{Z_2} \sin Z \cos Z \, dZ = \cos^2 Z_1 - \cos^2 Z_2$$

Figure 14.11 Illustration of direct beam and diffuse radiation. The sky forms a bowl, or inverted hemisphere, over a horizontal surface. Shown is a cross section of the sky hemisphere. Direct beam (solid line) originates from the direction of the Sun with zenith angle Z. Diffuse radiation (dashed lines) can be treated as independent beams of radiation each with an angle Z. The shaded region is the relative contribution between sky angles Z_1 and Z_2 to total sky irradiance.

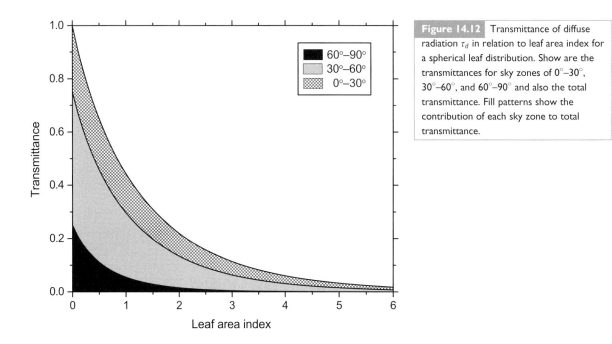

Figure 14.12 Transmittance of diffuse radiation τ_d in relation to leaf area index for a spherical leaf distribution. Show are the transmittances for sky zones of $0°–30°$, $30°–60°$, and $60°–90°$ and also the total transmittance. Fill patterns show the contribution of each sky zone to total transmittance.

For a canopy of horizontal leaves, $K_b = 1$ and, therefore, $K_d = 1$. However, K_d decreases with greater leaf area index in other canopies (Figure 14.9b). This is because the extinction coefficient $G(Z)/\cos Z$ varies with zenith angle, and, therefore, diffuse transmittance varies with sky angle. In canopies of spherical leaves, for example, diffuse radiation originating from low in the sky ($Z > 60°$) contributes little to total transmittance (Figure 14.12). In closed canopies with high leaf area index, virtually all diffuse radiation transmitted through the canopy originates high in the sky at zenith angles less than $60°$. The extinction coefficient for diffuse radiation depends on leaf area index because the exponential attenuation of radiation is weighted for the various sky angles.

It is evident that diffuse radiation penetrates deeper into the canopy than does direct beam radiation. Consider, for example, a canopy with a leaf area index $L = 4\ \mathrm{m^2\ m^{-2}}$ and spherical leaf angle distribution (Figure 14.9). The extinction coefficient for diffuse radiation is $K_d = 0.70$. This is smaller than the direct beam extinction coefficient when the Sun is not high in the sky ($K_b > 0.70$ for $Z \geq 45°$). Indeed, for a spherical leaf angle distribution the diffuse transmittance is greater than the direct beam transmittance for solar zenith angles greater than 40°–50° (Figure 14.10).

Question 14.8 Divide the sky into nine zones with 10° increments. Calculate the relative contribution of each zone to total irradiance assuming a uniform sky source of diffuse radiation with equal radiance in all directions. Does most of the diffuse radiation originate at angles low or high in the sky?

14.6 | The Norman (1979) Model

Leaves in real canopies are not black, and intercepted solar radiation is either absorbed or scattered as diffuse radiation. This scattering occurs in all directions but can be simplified into two fluxes of upward and downward diffuse radiation. Then, the total radiation at any depth in the canopy is the sum of the following: direct beam radiation that is not intercepted; diffuse radiation that similarly is not intercepted; and direct beam and diffuse radiation that is intercepted by leaves and scattered upward and downward within the canopy. Goudriaan (1977) and Goudriaan and van Laar (1994) described a framework to model the downward and upward fluxes for a single layer of horizontal leaves with leaf area index ΔL. In the notation that follows, leaf layer $i + 1$ is above layer i and the downward and upward fluxes (per ground area) are

$$I_i^{\downarrow} = I_{i+1}^{\downarrow}[(1 - \Delta L) + \tau_\ell \Delta L] + I_i^{\uparrow} \rho_\ell \Delta L \qquad (14.35)$$

$$I_{i+1}^{\uparrow} = I_i^{\uparrow}[(1 - \Delta L) + \tau_\ell \Delta L] + I_{i+1}^{\downarrow} \rho_\ell \Delta L. \qquad (14.36)$$

The flux $I_i^{\downarrow}$ is the downward flux below layer $i + 1$ onto layer i and $I_{i+1}^{\uparrow}$ is the upward flux above layer $i + 1$ (Figure 14.13). These equations are a

multilayer extension of (14.3) but with scattering. The term ΔL represents the portion of the flux that is intercepted, of which $\tau_\ell \Delta L$ is transmitted through the leaf and $\rho_\ell \Delta L$ is reflected. The term $1 - \Delta L$ is the non-intercepted portion that passes through the layer.

In (14.35), the downward flux $I_i^{\downarrow}$ onto leaf layer i consists of three component fluxes. A portion of the downward flux $I_{i+1}^{\downarrow}$ incident on leaf layer $i + 1$ is not intercepted and passes through the layer, given by $1 - \Delta L$; another portion is intercepted and scattered in the forward direction (downward), given by $\tau_\ell \Delta L$. Additionally, some of the upward flux $I_i^{\uparrow}$ above leaf layer i is intercepted by layer $i + 1$ and scattered in the backward direction (downward), given by $\rho_\ell \Delta L$. Similarly, the upward flux $I_{i+1}^{\uparrow}$ above layer $i + 1$ in (14.36) consists of: upward radiation from leaf layer i that passes unimpeded through layer $i + 1$ or that is intercepted and scattered in the forward direction; and downward radiation onto layer $i + 1$ that is intercepted and reflected upward.

Equations (14.35) and (14.36) can be extended to include direct beam radiation that is intercepted and scattered, and Norman (1979) gave a numerical solution to these equations, which is used in some canopy models (e.g., CANOAK, SPA). The equations are solved separately for visible and near-infrared radiation to account for differences in leaf reflectance and transmittance across the solar spectrum. In the equations that follow, the canopy is represented as N layers of leaves each with leaf area index ΔL and oriented such that $i = 1$ is the leaf layer at the bottom of the canopy, leaf layer $i + 1$ is above layer i, and $i = N$ is the leaf layer at the top of the canopy (Figure 14.13).

With reference to Figure 14.14, the downward and upward scattered fluxes are

$$I_i^{\downarrow} = I_{i+1}^{\downarrow}[\tau_{d,i+1} + (1 - \tau_{d,i+1})\tau_{\ell,i+1}]$$
$$+ I_i^{\uparrow}\left[(1 - \tau_{d,i+1})\rho_{\ell,i+1}\right]$$
$$+ I_{sky,b}^{\downarrow}[T_{b,i+1}(1 - \tau_{b,i+1})]\tau_{\ell,i+1} \qquad (14.37)$$

$$I_{i+1}^{\uparrow} = I_i^{\uparrow}[\tau_{d,i+1} + (1 - \tau_{d,i+1})\tau_{\ell,i+1}]$$
$$+ I_{i+1}^{\downarrow}\left[(1 - \tau_{d,i+1})\rho_{\ell,i+1}\right]$$
$$+ I_{sky,b}^{\downarrow}[T_{b,i+1}(1 - \tau_{b,i+1})]\rho_{\ell,i+1}. \qquad (14.38)$$

These equations are similar to (14.35) and (14.36), but with the addition of a direct beam source term. Some direct beam radiation is intercepted by the

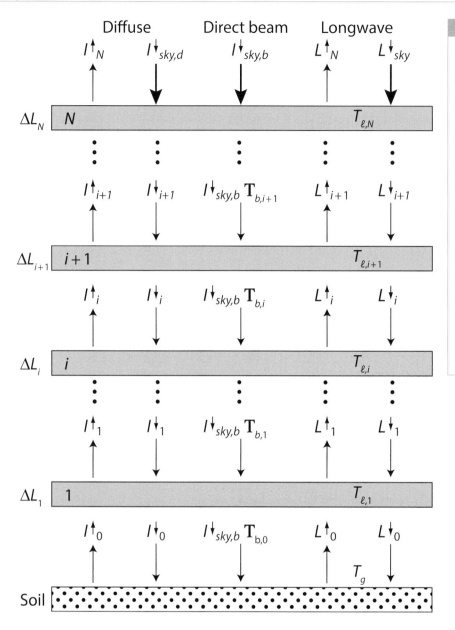

Figure 14.13 Radiative fluxes in a canopy of N leaf layers. The vertical profile is oriented with $i = 1$ the leaf layer at the bottom of the canopy, leaf layer $i + 1$ above layer i, and $i = N$ the leaf layer at the top of the canopy. Each layer has a leaf area index ΔL. $I_i^\downarrow$ is the downward diffuse shortwave flux onto layer i, $I_i^\uparrow$ is the upward diffuse shortwave flux above layer i, and $I_{sky,b}^\downarrow T_{b,i}$ is the unscattered direct beam flux onto layer i. $L_i^\downarrow$ and $L_i^\uparrow$ are the corresponding downward and upward fluxes of longwave radiation. These depend on leaf T_ℓ and ground T_g temperatures. Thick arrows denote boundary conditions of diffuse solar radiation $I_{sky,d}^\downarrow$, direct beam solar radiation $I_{sky,b}^\downarrow$, and atmospheric longwave radiation $L_{sky}^\downarrow$ at the top of the canopy.

leaf layer and is scattered downward (transmitted) or upward (reflected), as represented by the last term in (14.37) and (14.38). In these equations, leaf optical properties vary with canopy layer. $\tau_{d,i+1}$ is the diffuse transmittance through layer $i + 1$ with leaf area index ΔL_{i+1} obtained by summing $\exp\left[-G(Z)\Omega\Delta L_{i+1}/\cos Z\right]$ over all sky angles similar to (14.33) and $\tau_{b,i+1} = \exp\left(-K_{b,i+1}\Omega\Delta L_{i+1}\right)$ is the direct beam transmittance through ΔL_{i+1} with the extinction coefficient $K_{b,i+1}$. The clumping index Ω is included to account for the non-random distribution of foliage. A portion $1 - \tau_{d,i+1}$ of the diffuse radiation and $1 - \tau_{b,i+1}$ of the direct beam radiation is intercepted and is scattered forward and backward. $I_{sky,b}^\downarrow$ is the direct beam radiation incident on the top of the canopy, and $T_{b,i+1}$ is the fraction of this radiation that is not intercepted through the cumulative leaf area above layer $i + 1$. For a uniform canopy, the cumulative leaf area index is $x_i = (N - i)\Delta L$ and $T_{b,i} = \exp\left(-K_b\Omega x_i\right)$. For a

(a) Diffuse

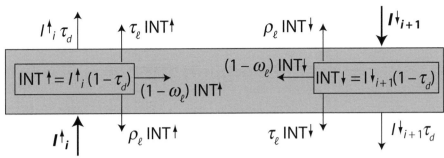

(b) Direct beam

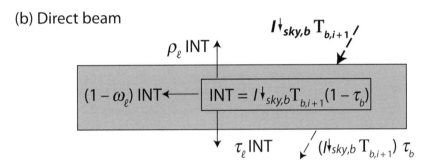

Figure 14.14 Radiative fluxes for a single leaf layer consisting of (a) diffuse and (b) direct beam (Norman 1979). Shown is the radiative balance for leaf layer $i+1$ located above leaf layer i. Dashed lines show direct beam fluxes and solid lines show diffuse fluxes. Thick lines show fluxes incident on leaf layer $i+1$. The top leaf surface depicts the flux $I^{\downarrow}_{i+1}$ above layer $i+1$. The bottom leaf surface depicts the flux $I^{\uparrow}_i$ below layer $i+1$ onto layer i. Leaf optical properties τ_ℓ, ρ_ℓ, ω_ℓ, τ_d, and τ_b are for layer $i+1$.

nonuniform canopy, the cumulative transmittance of direct beam radiation onto a layer is

$$T_{b,i} = \begin{cases} 1 & i = N \\ \prod_{j=i+1}^{N} e^{-K_{b,j}\Omega_j \Delta L_j} & i \leq N-1 \end{cases}. \quad (14.39)$$

A layer thickness less than 0.5 m² m⁻² is recommended (Norman 1979; Baldocchi et al. 1985). Boundary conditions are $I^{\downarrow}_N = I^{\downarrow}_{sky,d}$ the incident diffuse radiation at the top of the canopy and $I^{\uparrow}_0 = \rho_{gd} I^{\downarrow}_0 + \rho_{gb} I^{\downarrow}_{sky,b} T_{b,0}$ the upward flux at the soil surface, with ρ_{gd} and ρ_{gb} the ground albedos for diffuse and direct beam radiation, respectively.

Norman (1979) described an iterative numerical solution for (14.37) and (14.38) in which the downward and upward diffuse fluxes are first solved without the direct beam term. This provides a first approximation to the downward and upward fluxes. These values for $I^{\downarrow}_{i+1}$ and $I^{\uparrow}_i$ are substituted into the right-hand side of (14.37) and (14.38) to yield better estimates of the downward and upward fluxes on the left-hand side, now with the direct beam term. Several iterations provide the final flux estimates. An alternative solution is to directly solve (14.37) and (14.38) without iteration. The equations can be rewritten as a tridiagonal system of linear equations that is solved for $I^{\downarrow}_i$ and $I^{\uparrow}_i$ at each level in the canopy. Equation (14.37) is rewritten by substituting an expression for $I^{\downarrow}_{i+1}$ from (14.38), and, similarly, (14.38) is rewritten by substituting $I^{\uparrow}_i$ from (14.37). This gives the paired equations:

$$-a_i I^{\uparrow}_i + I^{\downarrow}_i - b_i I^{\downarrow}_{i+1} = d_i \quad (14.40)$$

$$-e_{i+1} I^{\downarrow}_i + I^{\uparrow}_{i+1} - f_{i+1} I^{\uparrow}_{i+1} = c_{i+1}. \quad (14.41)$$

For a canopy with N layers, there are $2(N+1)$ equations to solve, in which the matrix form of the equations (Appendix A6) is

$$
\begin{bmatrix}
1 & -\rho_{gd} & & & & & & & \\
-a_0 & 1 & -b_0 & & & & & & \\
& -e_1 & 1 & -f_1 & & & & & \\
& & -a_1 & 1 & -b_1 & & & & \\
& & & -e_2 & 1 & -f_2 & & & \\
& & & & -a_2 & 1 & -b_2 & & \\
& & & & & \ddots & \ddots & \ddots & \\
& & & & & & \ddots & \ddots & \ddots \\
& & & & & & & \ddots & \ddots & \ddots \\
& & & & & & & & -e_N & 1 & -f_N \\
& & & & & & & & & 0 & 1
\end{bmatrix}
\times
\begin{bmatrix}
I{\uparrow}_0 \\
I{\downarrow}_0 \\
I{\uparrow}_1 \\
I{\downarrow}_1 \\
I{\uparrow}_2 \\
I{\downarrow}_2 \\
\vdots \\
\vdots \\
\vdots \\
I{\uparrow}_N \\
I{\downarrow}_N
\end{bmatrix}
=
\begin{bmatrix}
c_0 \\
d_0 \\
c_1 \\
d_1 \\
c_2 \\
d_2 \\
\vdots \\
\vdots \\
\vdots \\
c_N \\
d_N
\end{bmatrix}.
\tag{14.42}
$$

The non-diagonal elements are zero and other terms are

$$
a_i = f_{i+1} = (1 - \tau_{d,i+1})\rho_{\ell,i+1} - \frac{[\tau_{d,i+1} + (1-\tau_{d,i+1})\tau_{\ell,i+1}]^2}{(1-\tau_{d,i+1})\rho_{\ell,i+1}}
\tag{14.43}
$$

$$
b_i = e_{i+1} = \frac{\tau_{d,i+1} + (1-\tau_{d,i+1})\tau_{\ell,i+1}}{(1-\tau_{d,i+1})\rho_{\ell,i+1}}
\tag{14.44}
$$

$$
c_i = I^{\downarrow}_{sky,b}T_{b,i}(1-\tau_{b,i})(\rho_{\ell,i} - \tau_{\ell,i}e_i); \qquad c_0 = \rho_{gb}I^{\downarrow}_{sky,b}T_{b,0}
\tag{14.45}
$$

$$
d_i = I^{\downarrow}_{sky,b}T_{b,i+1}(1-\tau_{b,i+1})(\tau_{\ell,i+1} - \rho_{\ell,i+1}b_i); \qquad d_N = I^{\downarrow}_{sky,d}.
\tag{14.46}
$$

A tridiagonal system of equations is easily solved with numerical methods (Appendix A8).

With reference to Figure 14.14, the diffuse and direct beam fluxes absorbed by leaf layer i (per ground area) are

$$
\overrightarrow{I}_{cd,i} = \left(I^{\downarrow}_i + I^{\uparrow}_{i-1}\right)(1-\tau_{d,i})(1-\omega_{\ell,i})
\tag{14.47}
$$

$$
\overrightarrow{I}_{cb,i} = I^{\downarrow}_{sky,b}T_{b,i}(1-\tau_{b,i})(1-\omega_{\ell,i}).
\tag{14.48}
$$

These equations describe the intercepted flux, of which the non-scattered portion $1-\omega_{\ell,i}$ is absorbed. Division by ΔL_i yields the corresponding fluxes per leaf area, $\overrightarrow{I}_{\ell d,i}$ and $\overrightarrow{I}_{\ell b,i}$. The radiation absorbed by the canopy is summed over all layers so that

$$
\overrightarrow{I}_c = \sum_{i=1}^{N} \overrightarrow{I}_{cd,i} + \overrightarrow{I}_{cb,i},
\tag{14.49}
$$

and the radiation absorbed at the ground is

$$
\overrightarrow{I}_g = \left(1-\rho_{gd}\right)I^{\downarrow}_0 + \left(1-\rho_{gb}\right)I^{\downarrow}_{sky,b}T_{b,0}.
\tag{14.50}
$$

The first term on the right-hand side in (14.50) is the diffuse absorption $\overrightarrow{I}_{gd}$, and the second term is the direct beam absorption $\overrightarrow{I}_{gb}$. Radiation absorbed by the canopy and ground balances the net flux at the top of the canopy so that the overall radiative balance is

$$
\overrightarrow{I}_c + \overrightarrow{I}_g = I^{\downarrow}_{sky,d} + I^{\downarrow}_{sky,b} - I^{\uparrow}_N.
\tag{14.51}
$$

The first two terms on the right-hand side are the incoming diffuse and direct beam fluxes, respectively, at the top of the canopy, and $I^{\uparrow}_N$ is the upward scattered flux above the canopy.

The radiative transfer model simulates the absorption by the average leaf in a layer but can be extended to separately consider sunlit and shaded leaves. A common assumption is that shaded leaves receive only diffuse radiation while sunlit leaves receive diffuse and direct beam radiation. The absorbed diffuse radiation (per leaf area) for layer i is $\overrightarrow{I}_{cd,i}/\Delta L_i$, and this is also the absorption by shaded leaves (per shaded leaf area) so that

$$
\overrightarrow{I}_{\ell sha,i} = \frac{\overrightarrow{I}_{cd,i}}{\Delta L_i}.
\tag{14.52}
$$

The sunlit leaf absorption (per sunlit leaf area) is the diffuse absorption plus the direct beam absorption so that

$$
\overrightarrow{I}_{\ell sun,i} = \overrightarrow{I}_{\ell sha,i} + \frac{\overrightarrow{I}_{cb,i}}{f_{sun,i}\Delta L_i}.
\tag{14.53}
$$

The corresponding canopy fluxes (per ground area) are

$$
\overrightarrow{I}_{cSha} = \sum_{i=1}^{N} \overrightarrow{I}_{\ell sha,i}\left(1-f_{sun,i}\right)\Delta L_i = \sum_{i=1}^{N} \overrightarrow{I}_{cd,i}\left(1-f_{sun,i}\right)
\tag{14.54}
$$

for the shaded portion of the canopy, and

$$\vec{I}_{cSun} = \sum_{i=1}^{N} \vec{I}_{\ell sun,i} f_{sun,i} \Delta L_i = \sum_{i=1}^{N} \vec{I}_{cd,i} f_{sun,i} + \vec{I}_{cb,i}$$

(14.55)

for the sunlit canopy. It can be seen, then, that the diffuse flux $\vec{I}_{cd,i}$ is partitioned to the sunlit and shaded fractions of leaf layer i in proportion to $f_{sun,i}$ and $1 - f_{sun,i}$, respectively. The sunlit leaves additionally receive the direct beam flux $\vec{I}_{cb,i}$. Figure 14.15 shows vertical profiles of absorbed radiation for sunlit and shaded leaves in sparse and dense canopies.

Question 14.9 The Norman (1979) model gives fluxes integrated for a layer of leaves with leaf area index ΔL. Explain how this integration is achieved. How may results be affected by using smaller or larger ΔL?

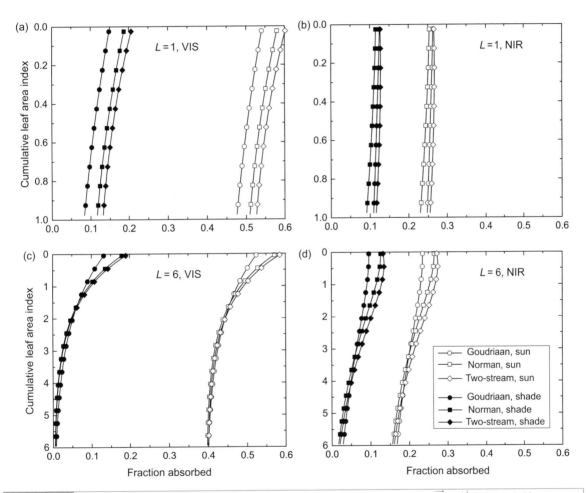

Figure 14.15 Vertical profiles of the fraction of radiation absorbed (per leaf area) for sunlit leaves ($\vec{I}_{\ell sun}/I_{sky}^{\downarrow}$) and shaded leaves ($\vec{I}_{\ell sha}/I_{sky}^{\downarrow}$) calculated using the Norman, Goudriaan, and two-stream models. (a) Visible waveband (VIS) fluxes for a sparse canopy with $L = 1$ m² m⁻². (b) Same, but for near-infrared (NIR). (c) Visible waveband (VIS) fluxes for a dense canopy with $L = 6$ m² m⁻². (d) Same, but for near-infrared (NIR). In this example, solar zenith angle is $Z = 30°$, 80% of the incident radiation is direct beam, leaf orientation is $\chi_\ell = 0$ (spherical), leaf reflectance is $\rho_\ell = 0.10$ (visible) and 0.45 (near-infrared), leaf transmittance is $\tau_\ell = 0.05$ (visible) and 0.25 (near-infrared), and soil albedo is $\rho_{gb} = \rho_{gd} = 0.10$ (visible) and 0.20 (near-infrared). See also Wang (2003) for a comparison of the Goudriaan and two-stream models.

14.7 | The Goudriaan and van Laar (1994) Model

Goudriaan and van Laar (1994) described an alternative radiative transfer model utilizing the theory of Goudriaan (1977). In contrast with Norman's model, which requires a multilayer canopy and must be solved numerically, this approach can be analytically integrated to give canopy-scale radiative fluxes. The basis for the model is that light transmission with scattering follows an exponential profile, but with an effective extinction coefficient that is adjusted for scattering. Because of downward scattering, transmission of radiation through the canopy is greater, and the effective extinction coefficient is smaller, compared with black leaves. Goudriaan (1977) showed that exponential profiles also describe light transmission through non-black leaves, but with an effective extinction coefficient given by

$$K'_b = K_b \sqrt{1 - \omega_\ell} \qquad (14.56)$$

for direct beam and

$$K'_d = K_d \sqrt{1 - \omega_\ell} \qquad (14.57)$$

for diffuse radiation, with $\alpha_\ell = 1 - \omega_\ell$ the leaf absorptance. A leaf layer that does not absorb all the intercepted radiation ($\alpha_\ell < 1$) has a smaller extinction coefficient and transmits more light than does a leaf layer that absorbs all the intercepted radiation ($\alpha_\ell = 1$).

Goudriaan's theory describes exponential profiles of light within the canopy and accounts for scattering using the direct beam extinction coefficient for black leaves K_b, the diffuse extinction coefficient K_d, and the extinction coefficients K'_b and K'_d adjusted for scattering. The equations are solved separately for visible and near-infrared radiation and have been used in many plant canopy models (Spitters 1986; de Pury and Farquhar 1997; Wang and Leuning 1998; Wang 2003) and in CABLE (Kowalczyk et al. 2006; Wang et al. 2011). Figure 14.16 illustrates the derivation of absorbed radiation, given here with the clumping index Ω to account for the nonrandom distribution of foliage. The direct beam radiation I_1 (per ground area) incident upon a layer of leaves with leaf area index ΔL is transmitted through the layer according to $I_2 = I_1 \exp\left(-K'_b \Omega \Delta L\right)$. This is the total direct beam flux, including scattering. Some of the incident radiation is reflected, equal to $I_4 = \rho_c I_1$, and some of the transmitted radiation is reflected from underlying leaves, equal to $I_3 = \rho_c I_2$. The radiation absorbed by the leaf layer (per ground area) is the balance of these fluxes so that

$$\vec{I} = I_1 + I_3 - I_2 - I_4 = (1 - \rho_c)I_1\left(1 - e^{-K'_b \Omega \Delta L}\right) \qquad (14.58)$$

and the flux absorbed per leaf area, found by differentiation with respect to ΔL, is

$$\frac{d\vec{I}}{d\Delta L} = (1 - \rho_c)K'_b \Omega I_1 e^{-K'_b \Omega \Delta L}. \qquad (14.59)$$

A similar equation describes diffuse radiation absorbed per leaf area using the extinction coefficient K'_d.

The total direct beam flux passing through the leaf layer is the sum of non-intercepted direct beam

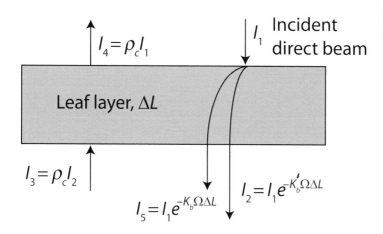

Figure 14.16 Derivation of absorbed direct beam solar radiation for a leaf layer with leaf area index ΔL (Goudriaan 1982). ρ_c is the reflectance of the leaf layer.

plus intercepted direct beam that is scattered downward. In Figure 14.16, the total flux I_2 obtained with K_b' decreases less than the flux I_5 obtained with K_b because of the additional scattered flux in I_2. The difference between these two fluxes is the scattered direct beam radiation and is obtained as follows. The intercepted direct beam flux without scattering is $I_1 - I_5$, and the flux (per ground area) absorbed by the leaf layer is

$$\overrightarrow{I} = (1 - \omega_\ell)(I_1 - I_5) = (1 - \omega_\ell)I_1\left(1 - e^{-K_b\Omega\Delta L}\right).$$

$$(14.60)$$

The absorption per leaf area is

$$\frac{d\overrightarrow{I}}{d\Delta L} = (1 - \omega_\ell)K_b\Omega I_1 e^{-K_b\Omega\Delta L}.$$

$$(14.61)$$

This is less than the total direct beam absorbed in (14.59), and the difference is the scattered beam irradiance that is absorbed.

In a canopy of leaves, the diffuse radiation absorbed per leaf area (average for all leaves) at a canopy depth defined by the cumulative leaf area index x is

$$\overrightarrow{I}_{\ell d}(x) = (1 - \rho_{cd})K_d'\Omega I_{sky,d}^{\downarrow} e^{-K_d'\Omega x},$$

$$(14.62)$$

and the total direct beam flux (unscattered and scattered) absorbed per leaf area (average for all leaves) is

$$\overrightarrow{I}_{\ell b}(x) = (1 - \rho_{cb})K_b'\Omega I_{sky,b}^{\downarrow} e^{-K_b'\Omega x}.$$

$$(14.63)$$

These are the same as (14.59), but with $I_{sky,b}^{\downarrow}$ and $I_{sky,d}^{\downarrow}$ the incident direct beam and diffuse radiation at the top of the canopy, respectively, and ρ_{cb} and ρ_{cd} the effective canopy albedo for direct beam and diffuse radiation, respectively. The total absorbed direct beam flux consists of unscattered direct beam $\overrightarrow{I}_{\ell bb}(x)$ and scattered beam $\overrightarrow{I}_{\ell bs}(x)$. Using (14.61), the unscattered direct beam radiation absorbed per leaf area (average for all leaves) is

$$\overrightarrow{I}_{\ell bb}(x) = (1 - \omega_\ell)K_b\Omega I_{sky,b}^{\downarrow} e^{-K_b\Omega x},$$

$$(14.64)$$

and the scattered beam radiation absorbed per leaf area (average for all leaves) is

$$\overrightarrow{I}_{\ell bs}(x) = \overrightarrow{I}_{\ell b}(x) - \overrightarrow{I}_{\ell bb}(x).$$

$$(14.65)$$

These equations can be extended to distinguish sunlit and shaded leaves. All leaves absorb diffuse radiation and scattered beam radiation. Sunlit leaves additionally absorb the unscattered direct beam radiation. The radiation absorbed by shaded leaves (per shaded leaf area) is the sum of diffuse and scattered beam radiation,

$$\overrightarrow{I}_{\ell sha}(x) = \overrightarrow{I}_{\ell d}(x) + \overrightarrow{I}_{\ell bs}(x),$$

$$(14.66)$$

and the radiation absorbed by sunlit leaves (per sunlit leaf area) is

$$\overrightarrow{I}_{\ell sun}(x) = \overrightarrow{I}_{\ell sha}(x) + (1 - \omega_\ell)K_b I_{sky,b}^{\downarrow}.$$

$$(14.67)$$

The latter term in (14.67) is the unscattered direct beam radiation absorbed by sunlit leaves only, given by $\overrightarrow{I}_{\ell bb}(x)/f_{sun}(x)$. It is comparable to that for a canopy of leaves given by (14.16), but for non-black leaves ($\omega_\ell > 0$), and can be similarly derived as $\int_0^L \overrightarrow{I}_{\ell bb}(x)dx$ divided by L_{sun}.

Equations (14.62)–(14.67) can be applied in a multilayer model but can also be integrated over the leaf area profile to provide canopy fluxes. The total radiation absorbed (per ground area) by a canopy with leaf area index L is

$$\overrightarrow{I}_c = \int_0^L [\overrightarrow{I}_{\ell b}(x) + \overrightarrow{I}_{\ell d}(x)]dx$$
$$= (1 - \rho_{cb})I_{sky,b}^{\downarrow}(1 - e^{-K_b'\Omega L}) + (1 - \rho_{cd})I_{sky,d}^{\downarrow}(1 - e^{-K_d'\Omega L})$$

$$(14.68)$$

The first term on the right-hand side is the direct beam radiation $\overrightarrow{I}_{cb}$ absorbed by the canopy, and the second term is the diffuse absorption $\overrightarrow{I}_{cd}$. The total radiation absorbed equals the sum of the canopy $\overrightarrow{I}_c$ and ground $\overrightarrow{I}_g$ absorption so that

$$(1 - \rho_{cb})I_{sky,b}^{\downarrow} + (1 - \rho_{cd})I_{sky,d}^{\downarrow} = \overrightarrow{I}_c + \overrightarrow{I}_g,$$

$$(14.69)$$

and the solar radiation absorbed at the ground is

$$\overrightarrow{I}_g = (1 - \rho_{cb})I_{sky,b}^{\downarrow}e^{-K_b'\Omega L} + (1 - \rho_{cd})I_{sky,d}^{\downarrow}e^{-K_d'\Omega L}.$$

$$(14.70)$$

The first term on the right-hand side is the direct beam radiation $\overrightarrow{I}_{gb}$ absorbed by the ground, and the second term is the diffuse absorption $\overrightarrow{I}_{gd}$.

Canopy radiation is partitioned between sunlit and shaded fractions of the canopy so that $\overrightarrow{I}_c = \overrightarrow{I}_{cSun} + \overrightarrow{I}_{cSha}$. The radiation absorbed by the sunlit canopy (per ground area) is the integral of the sunlit leaf flux $\overrightarrow{I}_{\ell sun}$, obtained as

$$\vec{I}_{cSun} = \int_0^L \vec{I}_{\ell sun}(x) f_{sun}(x)\, dx$$

$$= I^{\downarrow}_{sky,b}(1 - \omega_\ell)(1 - e^{-K_b\Omega L}) + I^{\downarrow}_{sky,d}(1 - \rho_{cd})\Omega[1 - e^{-(K'_d + K_b)\Omega L}]\frac{K'_d}{K'_d + K_b} \tag{14.71}$$

$$+ I^{\downarrow}_{sky,b}\left\{(1 - \rho_{cb})\Omega[1 - e^{-(K'_b + K_b)\Omega L}]\frac{K'_b}{K'_b + K_b} - \frac{1 - \omega_\ell}{2}\Omega(1 - e^{-2K_b\Omega L})\right\}.$$

The first term on the right-hand side is the direct beam radiation absorbed by sunlit leaves, the second term is the absorbed diffuse radiation, and the third term is the absorbed scattered beam radiation. For the shaded canopy,

$$\vec{I}_{cSha} = \int_0^L \vec{I}_{\ell sha}(x)[1 - f_{sun}(x)]\, dx$$

$$= I^{\downarrow}_{sky,d}(1 - \rho_{cd})\left\{1 - e^{-K'_d\Omega L} - \Omega[1 - e^{-(K'_d + K_b)\Omega L}]\frac{K'_d}{K'_d + K_b}\right\}$$

$$+ I^{\downarrow}_{sky,b}(1 - \rho_{cb})\left\{1 - e^{-K'_b\Omega L} - \Omega[1 - e^{-(K'_b + K_b)\Omega L}]\frac{K'_b}{K'_b + K_b}\right\} \tag{14.72}$$

$$- I^{\downarrow}_{sky,b}(1 - \omega_\ell)\left[1 - e^{-K_b\Omega L} - \frac{\Omega}{2}(1 - e^{-2K_b\Omega L})\right]$$

The first term is the diffuse radiation absorbed by shaded leaves. The latter two terms represent absorbed scattered beam radiation.

The canopy albedo for direct beam ρ_{cb} and diffuse ρ_{cd} radiation are described by Goudriaan and van Laar (1994). For a thick canopy of horizontal leaves, the albedo is

$$\rho'_h = \frac{1 - \sqrt{1 - \omega_\ell}}{1 + \sqrt{1 - \omega_\ell}}. \tag{14.73}$$

Typical values are $\omega_\ell = 0.2$ for the visible waveband ($\rho_\ell = \tau_\ell = 0.1$) and $\omega_\ell = 0.8$ for near-infrared ($\rho_\ell = \tau_\ell = 0.4$) so that ρ'_h equals 0.06 and 0.38, respectively, for visible and near-infrared radiation. The canopy albedo ρ'_h is less than the leaf reflectance ρ_ℓ because of multiple scattering and resultant light trapping in the canopy. This is especially evident for

visible radiation, where leaf absorptance is large. The albedo of a canopy of horizontal leaves is adjusted for other leaf orientations to obtain the direct beam and diffuse albedos. The albedo for direct beam is obtained using the relationship:

$$\rho'_b = \frac{2K_b\rho'_h}{K_b + K_d}. \tag{14.74}$$

Radiation impinging on the canopy from a vertical orientation penetrates deeply into the canopy, little scattered radiation emerges above the canopy, and the albedo is small. At high zenith angles, however, albedo increases. The albedo for diffuse radiation is found by integrating the direct beam albedo over the sky hemisphere, similar to K_d in (14.33), so that

$$\rho'_d = 2\sum_{i=1}^9 \frac{2K_{b,i}\rho'_h}{K_{b,i} + K_d} \sin Z_i \cos Z_i \, \Delta Z_i, \tag{14.75}$$

where $K_{b,i} = G(Z_i)/\cos Z_i$ is the direct beam extinction coefficient evaluated for each sky zenith angle Z_i. The preceding equations describe albedo above an infinitely thick canopy of leaves in which the albedo of soil is neglected. Most plant canopies have an effective albedo that is influenced by soil albedo. The effective canopy albedos are

$$\rho_{cb} = \rho_b' + \left(\rho_{gb} - \rho_b'\right)e^{-2K_b'\Omega L} \qquad (14.76)$$

$$\rho_{cd} = \rho_d' + \left(\rho_{gd} - \rho_d'\right)e^{-2K_d'\Omega L}, \qquad (14.77)$$

with ρ_{gb} and ρ_{gd} the ground albedos for direct beam and diffuse radiation, respectively.

Figure 14.15 compares vertical profiles of light absorption for sunlit and shaded leaves calculated using the Norman and Goudriaan models. With both models, differences in absorbed radiation between sunlit and shaded leaves are larger for visible radiation than for near-infrared radiation. In

the sparse canopy, the fraction of incident radiation absorbed is similar in both models for near-infrared, but the Goudriaan model underestimates absorption for visible radiation. In the dense canopy, the light profiles differ in the upper canopy, where the Goudriaan model underestimates absorption for both visible and near-infrared radiation.

Figure 14.17 compares the two models in terms of absorption of visible radiation by the entire canopy and by the sunlit and shaded fractions of the canopy. Absorbed radiation increases with greater leaf area index in both models, saturating at values greater than 95% of incident radiation at high leaf area. A dense canopy with $L = 6$ m^2 m^{-2} absorbs approximately 93% of the incident radiation. There is little difference in direct beam

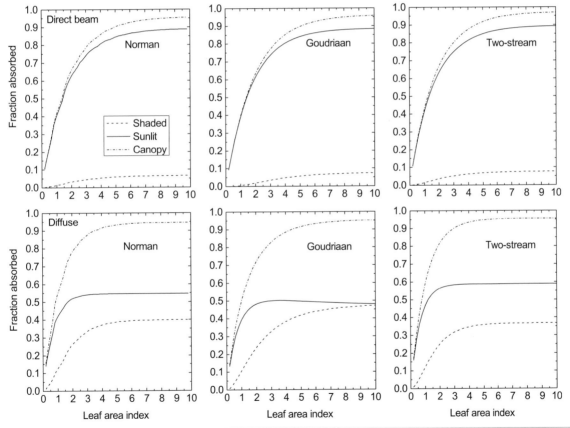

Figure 14.17 Fraction of photosynthetically active radiation (visible waveband) absorbed by sunlit leaves ($\vec{I}_{cSun}/I_{sky}^{\downarrow}$), shaded leaves ($\vec{I}_{cSha}/I_{sky}^{\downarrow}$), and the entire canopy ($\vec{I}_c/I_{sky}^{\downarrow}$) in relation to leaf area index for direct beam radiation (top panels) and diffuse radiation (bottom panels). Shown are the Norman, Goudriaan, and the two-stream models. In this example, solar zenith angle is $Z = 30°$, leaf orientation is $\chi_\ell = 0$ (spherical), leaf reflectance is $\rho_\ell = 0.10$ (visible), leaf transmittance is $\tau_\ell = 0.05$ (visible), and soil albedo is $\rho_{gb} = \rho_{gd} = 0.10$ (visible).

radiation between models. However, the Norman model allocates a greater amount of diffuse radiation to sunlit leaves than to shaded leaves. The Goudriaan model has a more equal distribution of diffuse radiation between sunlit and shaded leaves at high leaf area index.

Question 14.10 Explain how the Goudriaan model can be used in a multilayer canopy. Can the model accommodate inhomogeneous canopies in which optical properties vary with depth in the canopy?

14.8 | The Two-Stream Approximation

The two-stream approximation, adapted by Dickinson (1983) and Sellers (1985) for plant canopies from atmospheric radiative transfer theory (Meador and Weaver 1980; Toon et al. 1989; Liou 2002), provides a framework to calculate surface albedo and light absorption in a canopy while allowing for multiple reflections by leaves. The two-stream approximation is a solution to the basic equation for radiative transfer in a plane-parallel atmosphere. It reduces the multidirectional scattering into two discrete fluxes of radiation in the upward and downward directions. It is analytically integrated over the canopy and was first included in terrestrial biosphere models with SiB (Sellers et al. 1986, 1996a). Its derivation and solution is given here with the clumping index Ω.

The change in scattered (diffuse) radiation through some optical depth is the balance of reduction from attenuation, increase from scattering of the diffuse flux, and increase from scattering of the direct beam flux. Figure 14.18 illustrates these fluxes (per leaf area) in a plant canopy. $\omega_\ell = \rho_\ell + \tau_\ell$ is the leaf scattering coefficient, and β is an upscatter parameter whereby β is the fraction scattered in the backward direction and $1 - \beta$ is the fraction scattered in the forward direction. K_d is the extinction coefficient for diffuse radiation, more formally referred to as the optical depth of diffuse radiation per unit leaf area. (The original formulation of Dickinson [1983] and Sellers [1985] used the notation $\bar{\mu} = 1/K_d$ to be consistent with atmospheric

radiative transfer.) The downward flux through an increment of leaf area dx is attenuated by the amount $dI^\downarrow/dx = K_d \Omega I^\downarrow$, but a fraction ω_ℓ of this intercepted radiation is scattered (Figure 14.18a). The portion $\beta\omega_\ell$ is scattered in the backward (upward direction) and $(1 - \beta)\omega_\ell$ is scattered in the forward (downward) direction. The downward scattering lessens the attenuation so that the flux below

(a) Downward diffuse

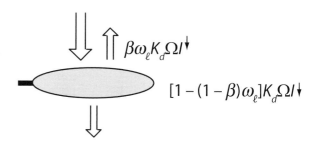

(b) Upward diffuse

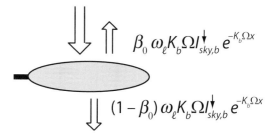

(c) Direct beam

Figure 14.18 Upward and downward fluxes in the two-stream approximation. The downward diffuse flux (a), upward diffuse flux (b), and direct beam flux (c) are scattered in the forward and backward directions.

dx is $dI^\downarrow/dx = [1 - (1-\beta)\omega_\ell]K_d\Omega I^\downarrow$. The downward flux is enhanced by scattering of the upward flux from below, given by $\beta\omega_\ell K_d\Omega I^\uparrow$ (Figure 14.18b), and also by the scattered direct beam (Figure 14.18c). The term $K_b\Omega I^\downarrow_{sky,b}\exp(-K_b\Omega x)$ is the direct beam radiation at the cumulative leaf area index x, of which ω_ℓ is the total portion that is scattered and $(1-\beta_0)\omega_\ell$ is scattered downward. Similar fluxes pertain to $I^\uparrow$. These fluxes are similar to those in (14.37) and (14.38) presented earlier, but are written here as differential equations.

In the two-stream approximation, the upward and downward scattered (diffuse) radiation (per leaf area) at a cumulative leaf area index x in the canopy are described by

$$\frac{dI^\uparrow}{dx} = [1 - (1-\beta)\omega_\ell]K_d\Omega I^\uparrow - \beta\omega_\ell K_d\Omega I^\downarrow$$
$$- \beta_0\omega_\ell K_b\Omega I^\downarrow_{sky,b}e^{-K_b\Omega x} \quad (14.78)$$

$$\frac{dI^\downarrow}{dx} = -[1 - (1-\beta)\omega_\ell]K_d\Omega I^\downarrow + \beta\omega_\ell K_d\Omega I^\uparrow$$
$$+ (1-\beta_0)\omega_\ell K_b\Omega I^\downarrow_{sky,b}e^{-K_b\Omega x}. \quad (14.79)$$

Equation (14.79) defines the vertical profile of the downward scattered flux $I^\downarrow$ in the canopy. The first term on the right-hand side of the equation is the portion of $I^\downarrow$ that passes through dx. The second term is the portion of the upward flux $I^\uparrow$ that is scattered in the downward direction. The last term is the contribution to $I^\downarrow$ from interception and scattering of direct beam radiation penetrating to depth x in the canopy. Equation (14.78) is the analogous equation for $I^\uparrow$. These two equations describe the upward and downward diffuse fluxes resulting from scattering of direct beam radiation incident on the canopy. For clarity, these fluxes are hereafter denoted as $I^\uparrow_b$ and $I^\downarrow_b$. The equations can also be used to calculate scattering of diffuse radiation incident on the canopy, in which case the direct beam term on the right-hand side of the equations is dropped. Scattered fluxes originating from diffuse radiation are denoted as $I^\uparrow_d$ and $I^\downarrow_d$. The equations are solved separately for visible and near-infrared radiation.

The two-stream parameters $K_d = 1/\bar{\mu}$, β, and β_0 vary with leaf orientation and were first provided by Dickinson (1983) for specific orientations. Sellers (1985) generalized these expressions using the Ross index χ_ℓ and the Ross–Goudriaan function (14.31) for $G(\mu)$, given here using the notation $\mu = \cos Z$.

The parameter $\bar{\mu}$ is integrated over all angles of the incident beam, similar to that given previously for K_d (14.32), so that

$$\frac{1}{K_d} = \bar{\mu} = \int_0^1 \frac{\mu'}{G(\mu')}d\mu' = \frac{1}{\phi_2} - \frac{\phi_1}{\phi_2^2}\ln\left(\frac{\phi_1+\phi_2}{\phi_1}\right), \quad (14.80)$$

where μ' is the direction of the scattered flux. The upscatter parameter β for diffuse radiation is

$$\omega_\ell\beta = \frac{1}{2}\left[\rho_\ell + \tau_\ell + (\rho_\ell - \tau_\ell)\cos^2\bar{\Theta}_\ell\right], \quad (14.81)$$

in which $\bar{\Theta}_\ell$ is the mean leaf inclination angle and $\cos^2\bar{\Theta}_\ell$ is approximated by

$$\cos^2\bar{\Theta}_\ell = \left(\frac{1+\chi_\ell}{2}\right)^2. \quad (14.82)$$

In this approximation, $\omega_\ell\beta = \rho_\ell$ for horizontal leaves with $\chi_\ell = 1$ ($\bar{\Theta}_\ell = 0°$) and $\omega_\ell\beta = 0.5(\rho_\ell + \tau_\ell)$ for vertical leaves with $\chi_\ell = -1$ ($\bar{\Theta}_\ell = 90°$), which agree with the solutions of Dickinson (1983). For a spherical leaf angle distribution with $\chi_\ell = 0$, (14.82) gives $\bar{\Theta}_\ell = 60°$ and $\omega_\ell\beta = 0.625\rho_\ell + 0.375\tau_\ell$, which approximates the exact solution. The discrepancy arises because a spherical leaf angle distribution has a true mean leaf inclination $\bar{\Theta}_\ell = 57.3°$. The upscatter parameter β_0 for direct beam is

$$\omega_\ell\beta_0 = \left[\frac{K_b + K_d}{K_b}\right]a_s(\mu), \quad (14.83)$$

in which $a_s(\mu)$ is the single scattering albedo and is obtained by integration over all angles so that

$$a_s(\mu) = \frac{\omega_\ell}{2}\int_0^1 \frac{\mu'G(\mu)}{\mu G(\mu') + \mu'G(\mu)}d\mu'$$
$$= (\omega_\ell/2)\frac{G(\mu)}{G(\mu)+\mu\phi_2}\left\{1 - \frac{\mu\phi_1}{G(\mu)+\mu\phi_2}\ln\left[\frac{G(\mu)+\mu\phi_1+\mu\phi_2}{\mu\phi_1}\right]\right\}. \quad (14.84)$$

In this implementation of the two-stream approximation, the fundamental leaf optical parameters are ρ_ℓ, τ_ℓ, and χ_ℓ; β, β_0, ω_ℓ, K_d, and K_b are calculated from these parameters.

Liou (2002, p. 310) provided the solution to (14.78) and (14.79). Integrated over an amount of leaf area equal to ΔL, the upward and downward fluxes (per ground area) are

$$I^{\uparrow} = -\gamma_1 e^{-K_b \Omega \Delta L} + \eta_1 u e^{-h\Omega \Delta L} + \eta_2 v e^{h\Omega \Delta L} \qquad (14.85)$$

$$I^{\downarrow} = \gamma_2 e^{-K_b \Omega \Delta L} - \eta_1 v e^{-h\Omega \Delta L} - \eta_2 u e^{h\Omega \Delta L}. \qquad (14.86)$$

Equation (14.85) solved with $\Delta L = 0$ gives the scattered flux $I^{\uparrow}(0)$ above the leaves. Equation (14.86) solved with $\Delta L = L$ gives the scattered flux $I^{\downarrow}(L)$ below the leaves. The parameters h, u, v, γ_1, and γ_2 are

$$b = [1 - (1 - \beta)\omega_\ell]K_d, \quad c = \beta \omega_\ell K_d \qquad (14.87)$$

$$h = \sqrt{b^2 - c^2} \qquad (14.88)$$

$$u = \frac{h - b - c}{2h}, \qquad v = \frac{h + b + c}{2h} \qquad (14.89)$$

$$\gamma_1 = [\beta_0 K_b - b\beta_0 - c(1 - \beta_0)]\frac{\omega_\ell K_b I_{sky,b}^{\downarrow}}{h^2 - K_b^2} \qquad (14.90)$$

$$\gamma_2 = [(1 - \beta_0)K_b + c\beta_0 + b(1 - \beta_0)]\frac{\omega_\ell K_b I_{sky,b}^{\downarrow}}{h^2 - K_b^2}. \qquad (14.91)$$

Equations (14.85) and (14.86) have two unknowns – the constants η_1 and η_2. These are determined from boundary conditions. The boundary conditions specify continuity of fluxes above and below the leaves. The downward flux $I^{\downarrow}(0)$ is equal to the flux incident on the leaves from above, and, similarly, the upward flux $I^{\uparrow}(L)$ is equal to the upward flux below the leaves. The solution for η_1 and η_2 is found by considering the radiative balance of a plant canopy.

Figure 14.19a illustrates the radiative balance for direct beam in a canopy with leaf area index L. The direct beam radiation $I_{sky,b}^{\downarrow}$ at the top of the canopy is transmitted through the canopy following the exponential profile $I_{sky,b}^{\downarrow} \exp(-K_b \Omega L)$. The direct beam radiation intercepted by the leaves is scattered upward and downward. The flux $I_b^{\uparrow}(0)$ above the canopy calculated from (14.85) with $\Delta L = 0$ is the reflected direct beam radiation so that $\rho_{cb} = I_b^{\uparrow}(0)/I_{sky,b}^{\downarrow}$ is the albedo of the canopy for direct beam radiation. The downward scattered flux $I_b^{\downarrow}(L)$ calculated from (14.86) with $\Delta L = L$ is the intercepted direct beam radiation scattered downward below the canopy. The total radiation below the canopy is the downward scattered flux plus the unscattered flux.

For direct beam radiation, the boundary condition is $I_b^{\downarrow}(0) = 0$ at the top of the canopy. Applying this constraint to (14.86) gives

$$\eta_1 = \frac{\gamma_2 - \eta_2 u}{v}. \qquad (14.92)$$

Below the canopy, the ground reflection is $\rho_{gd}I_b^{\downarrow}(L) + \rho_{gb}I_{sky,b}^{\downarrow}\exp(-K_b\Omega L)$. This is the lower boundary condition and must equal $I_b^{\uparrow}(L)$ from (14.85). Equating these two fluxes results in the expression:

$$\eta_2 = \frac{v\left(\gamma_1 + \gamma_2\rho_{gd} + \rho_{gb}I_{sky,b}^{\downarrow}\right)e^{-K_b\Omega L} - \gamma_2\left(u + v\rho_{gd}\right)e^{-h\Omega L}}{v\left(v + u\rho_{gd}\right)e^{h\Omega L} - u\left(u + v\rho_{gd}\right)e^{-h\Omega L}}. \qquad (14.93)$$

The direct beam flux scattered upward in the canopy at a depth equal to the cumulative leaf area index x is

$$I_b^{\uparrow}(x) = -\gamma_1 e^{-K_b\Omega x} + \eta_1 u e^{-h\Omega x} + \eta_2 v e^{h\Omega x}, \qquad (14.94)$$

and the downward scattered flux at x is

$$I_b^{\downarrow}(x) = \gamma_2 e^{-K_b\Omega x} - \eta_1 v e^{-h\Omega x} - \eta_2 u e^{h\Omega x}. \qquad (14.95)$$

The overall radiative balance for direct beam is

$$\vec{I}_{cb} + \vec{I}_{gb} = I_{sky,b}^{\downarrow} - I_b^{\uparrow}(0). \qquad (14.96)$$

The left-hand side of this equation is the direct beam radiation absorbed by the vegetation and ground, respectively. The right-hand side is the difference between the incoming and reflected radiation above the canopy. The direct beam flux absorbed by the vegetation is obtained from the radiative balance by considering fluxes above and below the canopy whereby

$$\vec{I}_{cb} = \left(1 - e^{-K_b\Omega L}\right)I_{sky,b}^{\downarrow} - I_b^{\uparrow}(0) + I_b^{\uparrow}(L) - I_b^{\downarrow}(L). \qquad (14.97)$$

The first term on the right-hand side is the difference between the direct beam radiation incident at the top of the canopy and the unscattered radiation transmitted through the canopy. The remaining terms are the scattered radiation above and below the canopy. The ground flux is

$$\vec{I}_{gb} = \left(1 - \rho_{gd}\right)I_b^{\downarrow}(L) + \left(1 - \rho_{gb}\right)I_{sky,b}^{\downarrow}e^{-K_b\Omega L}. \qquad (14.98)$$

Figure 14.19b illustrates the diffuse radiative balance. In solving (14.85) and (14.86) for diffuse radiation, the direct beam term is excluded. The boundary conditions are $I_d^{\downarrow}(0) = I_{sky,d}^{\downarrow}$ at the top of the canopy and $I_d^{\uparrow}(L) = \rho_{gd}I_d^{\downarrow}(L)$ below the canopy. The constants η_1 and η_2 are

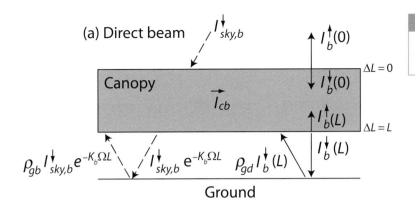

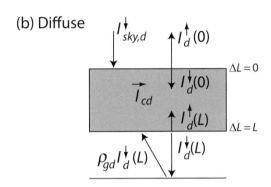

Figure 14.19 Fluxes for (a) direct beam and (b) diffuse radiation in the two-stream approximation for a canopy with leaf area index L.

$$\eta_1 = \frac{-I^\downarrow_{sky,d} - \eta_2 u}{v} \tag{14.99}$$

$$\eta_2 = \frac{I^\downarrow_{sky,d}\left(u + v\rho_{gd}\right)e^{-h\Omega L}}{v\left(v + u\rho_{gd}\right)e^{h\Omega L} - u\left(u + v\rho_{gd}\right)e^{-h\Omega L}}. \tag{14.100}$$

The fluxes for diffuse radiation are

$$I^\uparrow_d(x) = \eta_1 u e^{-h\Omega x} + \eta_2 v e^{h\Omega x} \tag{14.101}$$

$$I^\downarrow_d(x) = -\eta_1 v e^{-h\Omega x} - \eta_2 u e^{h\Omega x}. \tag{14.102}$$

The overall diffuse radiative balance is

$$\vec{I}_{cd} + \vec{I}_{gd} = I^\downarrow_{sky,d} - I^\uparrow_d(0), \tag{14.103}$$

with the vegetative flux,

$$\vec{I}_{cd} = I^\downarrow_{sky,d} - I^\uparrow_d(0) + I^\uparrow_d(L) - I^\downarrow_d(L), \tag{14.104}$$

and the ground flux,

$$\vec{I}_{gd} = \left(1 - \rho_{gd}\right)I^\downarrow_d(L). \tag{14.105}$$

The solution to the two-stream approximation gives the total radiation absorbed by the canopy ($\vec{I}_c = \vec{I}_{cb} + \vec{I}_{cd}$). It can be extended to include sunlit and shaded leaves, as first described by Dai et al. (2004) and later implemented in the Community Land Model (Bonan et al. 2011). As given previously, the radiation absorbed by shaded leaves (per shaded leaf area) at the cumulative leaf area index x is the sum of diffuse and scattered beam radiation:

$$\vec{I}_{\ell sha}(x) = \vec{I}_{\ell d}(x) + \vec{I}_{\ell bs}(x). \tag{14.106}$$

The diffuse radiation absorbed (per leaf area) is the change in the net scattered flux,

$$\vec{I}_{\ell d}(x) = \frac{d}{dx}\left(I^\uparrow_d - I^\downarrow_d\right), \tag{14.107}$$

and the scattered direct beam radiation (per leaf area) is

$$\vec{I}_{\ell bs}(x) = \omega_\ell K_b \Omega I^\downarrow_{sky,b} e^{-K_b\Omega x} + \frac{d}{dx}\left(I^\uparrow_b - I^\downarrow_b\right). \tag{14.108}$$

The scattered beam flux consists of the unscattered direct beam penetrating to depth x, of which the fraction ω_ℓ is scattered, given by the first term on the right-hand side of the equation. The portion $1 - \omega_\ell$ of the direct beam appears in the sunlit leaf flux. An additional term is the change in the net scattered flux, given by the second term in (14.108). The radiation absorbed by sunlit leaves (per sunlit leaf area) is

$$\vec{I}_{\ell sun}(x) = \vec{I}_{\ell sha}(x) + (1 - \omega_\ell)K_b I^\downarrow_{sky,b}. \qquad (14.109)$$

The second term on the right-hand side is the portion $1 - \omega_\ell$ of the unscattered direct beam flux, but expressed on a per sunlit leaf area basis.

Integrated over a canopy with leaf area index L, the solar radiation $\vec{I}_{cSun}$ absorbed by the sunlit canopy (per ground area) is

$$\vec{I}_{cSun} = \int_0^L \vec{I}_{\ell sun}(x) f_{sun}(x) dx, \qquad (14.110)$$

and the comparable absorption by the shaded canopy is

$$\vec{I}_{cSha} = \int_0^L \vec{I}_{\ell sha}(x)[1 - f_{sun}(x)] dx, \qquad (14.111)$$

with $\vec{I}_c = \vec{I}_{cSun} + \vec{I}_{cSha}$.

The solution to these equations is from Dai et al. (2004), modified here to include foliage clumping. For the sunlit canopy,

$$\vec{I}_{cSun} = \vec{I}_{cSun,b} + \vec{I}_{cSun,d}, \qquad (14.112)$$

and for the shaded canopy,

$$\vec{I}_{cSha} = \vec{I}_{cSha,b} + \vec{I}_{cSha,d}. \qquad (14.113)$$

The absorption of direct beam radiation by the sunlit canopy (per ground area) is

$$\vec{I}_{cSun,b} = (1 - \omega_\ell)\left[(1 - e^{-K_b \Omega L})I^\downarrow_{sky,b} + K_d \Omega(a_1 + a_2)\right], \qquad (14.114)$$

and for the shaded canopy is

$$\vec{I}_{cSha,b} = \vec{I}_{cb} - \vec{I}_{cSun,b}, \qquad (14.115)$$

with

$$a_1 = -\gamma_1\left[\frac{1 - e^{-2K_b \Omega L}}{2K_b}\right] + \eta_1 u\left[\frac{1 - e^{-(K_b + h)\Omega L}}{K_b + h}\right]$$
$$+ \eta_2 v\left[\frac{1 - e^{(-K_b + h)\Omega L}}{K_b - h}\right] \qquad (14.116)$$

$$a_2 = \gamma_2\left[\frac{1 - e^{-2K_b \Omega L}}{2K_b}\right] - \eta_1 v\left[\frac{1 - e^{-(K_b + h)\Omega L}}{K_b + h}\right]$$
$$- \eta_2 u\left[\frac{1 - e^{(-K_b + h)\Omega L}}{K_b - h}\right]. \qquad (14.117)$$

For diffuse radiation, the absorbed radiation (per ground area) is

$$\vec{I}_{cSun,d} = (1 - \omega_\ell)K_d \Omega(a_1 + a_2) \qquad (14.118)$$

$$\vec{I}_{cSha,d} = \vec{I}_{cd} - \vec{I}_{cSun,d}, \qquad (14.119)$$

with

$$a_1 = \eta_1 u\left[\frac{1 - e^{-(K_b + h)\Omega L}}{K_b + h}\right] + \eta_2 v\left[\frac{1 - e^{(-K_b + h)\Omega L}}{K_b - h}\right] \qquad (14.120)$$

$$a_2 = -\eta_1 v\left[\frac{1 - e^{-(K_b + h)\Omega L}}{K_b + h}\right] - \eta_2 u\left[\frac{1 - e^{(-K_b + h)\Omega L}}{K_b - h}\right]. \qquad (14.121)$$

The two-stream approximation is more similar to the Norman model than to the Goudriaan model in the vertical profile of light absorption (Figure 14.15) and absorption of radiation by the sunlit and shaded canopy (Figure 14.17).

The two-stream approximation is used in many models because it can be analytically integrated over the plant canopy, as needed for a big-leaf canopy (Chapter 15). Some models, however, explicitly resolve light, photosynthesis, and stomatal conductance at multiple levels in the canopy. The two-stream approximation can be used in a multilayer model by applying (14.106)–(14.109) for shaded and sunlit leaves at depth x, but this requires a homogenous canopy in which leaf properties do not vary with depth. The solution can be extended to a vertically inhomogeneous canopy by dividing the canopy into discrete homogenous layers and by recognizing that the upward and downward radiative fluxes are continuous at the interface of each layer (i.e., the upward flux above layer i is also the upward flux below layer $i + 1$, and similarly for the downward flux). This produces a system of equations that can be written in matrix form. Atmospheric models use this procedure to calculate radiative transfer in multiple layers of the atmosphere (Toon et al. 1989), and a similar solution has been used for snowpack (Flanner et al. 2007) and plant canopies (Knox et al. 2015). Another approach is to apply the integrated solution to successive canopy layers in an upward and downward sweep through the canopy (Bonan 1991). Beginning with the bottom canopy layer (in which ground albedos are the lower boundary condition), the equations are solved for the layer leaf area index ΔL with a unit of incoming radiation above the layer to obtain fluxes absorbed, transmitted, and scattered for that layer (normalized by incident radiation). The upward fluxes above the layer are the albedos, and these are used as the lower boundary conditions to obtain the normalized fluxes for layer $i + 1$, and so forth up through the canopy. The actual fluxes are obtained by working successively downward through the canopy from the top layer to obtain the direct beam and diffuse fluxes incident on each layer, with the boundary conditions $I^{\downarrow}_{sky,b}$ and $I^{\downarrow}_{sky,d}$ at the top layer.

Question 14.11 In the two-stream approximation, the portion $\beta \omega_\ell$ of the intercepted radiation is scattered in the backward direction, and $(1 - \beta)\omega_\ell$ is scattered in the forward direction. β_0 is the comparable parameter for diffuse radiation. Contrast these parameters with the forward and backward scattering in Norman's model.

Question 14.12 Use Supplemental Program 14.2 to verify that the integration of canopy fluxes from leaf fluxes is correct for the two-stream approximation.

Question 14.13 Contrast the Goudriaan and two-stream models with the Norman model. What is an important conceptual difference between the first two models and the Norman model?

Question 14.14 Explain how the two-stream model can be used in a multilayer canopy that is homogenous with depth. What changes are needed to accommodate a canopy in which optical properties vary with depth?

14.9 | Surface Albedo

The overall surface albedo is determined by the optical properties and amount of plant material, as well as the albedo of the ground. Figure 14.20 illustrates canopy albedo for direct beam radiation obtained from the two-stream approximation. With low leaf area index, leaves absorb little solar radiation, and the overall albedo of the plant canopy is largely that of the underlying ground. As leaf area index increases, canopy albedo responds more to the optical properties of the leaves. In this example, albedo is relatively constant for leaf area index greater than about 3 m^2 m^{-2}. Even when the ground has a high albedo, such as with snowy surfaces, leaves effectively mask the underlying bright surface.

Ground albedo is calculated from soil albedo and snow albedo, weighted by the fraction of the ground covered with snow. Soil albedo is specified for various soil color classes, with higher albedo for the near-infrared waveband compared with the visible waveband and higher albedo for dry soil compared with wet soil (see, for example, the Community Land Model; Oleson et al. 2013). Snow albedo is a complex function of solar zenith angle, the mass concentration of aerosols such as black carbon and mineral dust, snow grain size, snow age, and wavelength. A key factor in determining ground albedo is the fraction of the ground covered by snow (Figure 9.14).

14.10 | Longwave Radiation

Longwave radiation fluxes can be described similarly to diffuse solar radiation, but dropping the direct beam radiation scattering term and with the addition of a thermal radiation source term emitted by foliage. Figure 14.21a extends the multilayer theory of Norman (1979) for infrared wavelengths. The longwave fluxes are

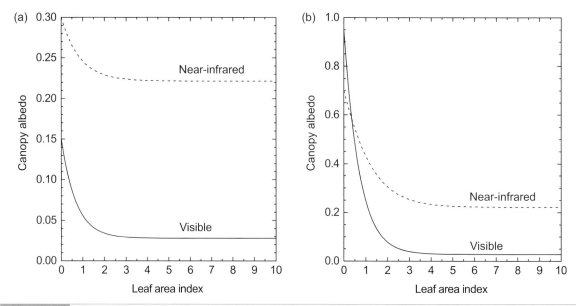

Figure 14.20 Canopy albedo for direct beam in the visible and near-infrared wavebands in relation to leaf area index. (a) Canopy albedo with ground albedo $\rho_{gb} = 0.15$ (visible) and 0.30 (near-infrared). (b) Canopy albedo for snow-covered ground with $\rho_{gb} = 0.95$ (visible) and 0.70 (near-infrared). Simulations use the two-stream approximation with solar zenith angle $Z = 30°$, leaf orientation $\chi_\ell = 0$ (spherical), leaf reflectance $\rho_\ell = 0.10$ (visible) and 0.45 (near-infrared), and leaf transmittance $\tau_\ell = 0.05$ (visible) and 0.25 (near-infrared).

$$L_i^\downarrow = L_{i+1}^\downarrow [\tau_{d,i+1} + (1 - \tau_{d,i+1})\tau_{\ell,i+1}]$$
$$+ L_i^\uparrow [(1 - \tau_{d,i+1})\rho_{\ell,i+1}] + \varepsilon_\ell \sigma T_{\ell,i+1}^4 (1 - \tau_{d,i+1})$$
$$(14.122)$$

$$L_{i+1}^\uparrow = L_i^\uparrow [\tau_{d,i+1} + (1 - \tau_{d,i+1})\tau_{\ell,i+1}]$$
$$+ L_{i+1}^\downarrow [(1 - \tau_{d,i+1})\rho_{\ell,i+1}]$$
$$+ \varepsilon_\ell \sigma T_{\ell,i+1}^4 (1 - \tau_{d,i+1}).$$
$$(14.123)$$

These equations are the same as (14.37) and (14.38) for diffuse radiation, but replace the direct beam scattering with emitted thermal radiation. The emitted thermal radiation is in both the downward and upward directions with ε_ℓ leaf emissivity and T_ℓ leaf temperature (K).

A leaf layer emits longwave radiation in proportion to T_ℓ^4. The factor $1 - \tau_d$ describes the emittance from the leaf layer. Longwave radiation emitted from a leaf layer is similar to interception of radiation by the layer, but in reverse; the fraction of the leaf layer that intercepts radiation is also the fraction occupied by leaves that emit radiation. This is evident by considering the derivation of transmittance for a canopy of leaves given by (14.3). One black horizontal leaf with area a randomly placed over the surface area A occupies the fractional area a/A. This is the fraction of the surface area that intercepts radiation, and the transmittance is

$1 - a/A$. For a canopy of n randomly placed, nonoverlapping black horizontal leaves, transmittance is $P_0 = (1 - a/A)^n$ and $1 - P_0$ is intercepted. The term a/A is also the fraction of the surface that has leaves that emit radiation, and the fraction that has no leaves to emit radiation is $1 - a/A$. For a canopy of n leaves, P_0 is the fraction of the canopy that does not emit radiation, and $1 - P_0$ is the effective fraction of the canopy that emits radiation.

For longwave radiation, leaf absorptance is equal to emittance so that $1 - \varepsilon_\ell = \rho_\ell + \tau_\ell = \omega_\ell$ is the leaf scattering coefficient. A typical value is $\varepsilon_\ell = 0.96$–0.98. Equations (14.122) and (14.123) can be solved with $\varepsilon_\ell = 1$ and $\rho_\ell = \tau_\ell = 0$ (Norman 1979). Other possibilities are to allow only backward scattering ($\tau_\ell = 0$), forward scattering ($\rho_\ell = 0$), or equally in both directions so that $\rho_\ell = \tau_\ell = (1 - \varepsilon_\ell)/2$. Longwave radiative transfer theory can be applied to sunlit and shaded leaves by defining an effective leaf temperature T_ℓ^4 as the contribution from sunlit and shaded leaves, $T_{\ell sun}^4$ and $T_{\ell sha}^4$, weighted by the sunlit and shaded fraction, respectively.

The numerical solution utilizes a tridiagonal system of linear equations similar to that given previously for $I_i^\uparrow$ and $I_i^\downarrow$, but the boundary conditions are as follows: $L_N^\downarrow = L_{sky}^\downarrow$ is the atmosphere longwave radiation above the canopy; and $L_0^\uparrow = (1 - \varepsilon_g)L_0^\downarrow + \varepsilon_g\sigma T_g^4$ is the upward flux at the soil surface, with ε_g the emissivity of the ground surface and T_g the temperature of the ground (K). In matrix form:

and the radiation absorbed at the ground is

$$\vec{L}_g = \varepsilon_g L_0^\downarrow - \varepsilon_g\sigma T_g^4 = L_0^\downarrow - L_0^\uparrow. \qquad (14.132)$$

$$\begin{bmatrix} 1 & -(1-e_g) & & & & & & & \\ -a_0 & 1 & -b_0 & & & & & & \\ & -e_1 & 1 & -f_1 & & & & & \\ & & -a_1 & 1 & -b_1 & & & & \\ & & & -e_2 & 1 & -f_2 & & & \\ & & & & -a_2 & 1 & -b_2 & & \\ & & & & & \ddots & \ddots & \ddots & \\ & & & & & & \ddots & \ddots & \ddots \\ & & & & & & -e_N & 1 & -f_N \\ & & & & & & & 0 & 1 \end{bmatrix} \times \begin{bmatrix} L\uparrow_0 \\ L\downarrow_0 \\ L\uparrow_1 \\ L\downarrow_1 \\ L\uparrow_2 \\ L\downarrow_2 \\ \vdots \\ \vdots \\ L\uparrow_N \\ L\downarrow_N \end{bmatrix} = \begin{bmatrix} c_0 \\ d_0 \\ c_1 \\ d_1 \\ c_2 \\ d_2 \\ \vdots \\ \vdots \\ c_N \\ d_N \end{bmatrix}. \qquad (14.124)$$

The non-diagonal elements are zero and other terms are

$$a_i = f_{i+1} = (1 - \tau_{d,i+1})\rho_{\ell,i+1} - \frac{[\tau_{d,i+1} + (1 - \tau_{d,i+1})\tau_{\ell,i+1}]^2}{(1 - \tau_{d,i+1})\rho_{\ell,i+1}} \qquad (14.125)$$

$$b_i = e_{i+1} = \frac{\tau_{d,i+1} + (1 - \tau_{d,i+1})\tau_{\ell,i+1}}{(1 - \tau_{d,i+1})\rho_{\ell,i+1}} \qquad (14.126)$$

$$c_i = (1 - e_i)(1 - \tau_{d,i})\varepsilon_\ell\sigma T_{\ell,i}^4; \qquad c_0 = \varepsilon_g\sigma T_g^4 \qquad (14.127)$$

$$d_i = (1 - b_i)(1 - \tau_{d,i+1})\varepsilon_\ell\sigma T_{\ell,i+1}^4; \qquad d_N = L_{sky}^\downarrow. \qquad (14.128)$$

With reference to Figure 14.21a, the net longwave flux (per ground area) for layer i is

$$\vec{L}_i = \varepsilon_\ell\left(L_i^\downarrow + L_{i-1}^\uparrow\right)(1 - \tau_{d,i}) - 2\varepsilon_\ell\sigma T_{\ell,i}^4(1 - \tau_{d,i}). \qquad (14.129)$$

The first term on the right-hand side is the intercepted flux, of which the portion ε_ℓ is absorbed. The second term is the thermal flux emitted upward and downward. The net flux per leaf area is

$$\vec{L}_{\ell,i} = \frac{\vec{L}_i}{\Delta L_i}. \qquad (14.130)$$

The radiation absorbed by the canopy is

$$\vec{L}_c = \sum_{i=1}^N \vec{L}_i, \qquad (14.131)$$

The radiation absorbed by the canopy and ground balance the net flux at the top of the canopy so that

$$\vec{L}_c + \vec{L}_g = L_{sky}^\downarrow - L_N^\uparrow. \qquad (14.133)$$

A more simple analytical solution can be obtained by reducing the complexities of scattering. Figure 14.21b simplifies the radiative balance of a leaf layer so that scattering occurs in the forward direction only ($\rho_\ell = 0$ and $\tau_\ell = \omega_\ell = 1 - \varepsilon_\ell$). Excluding leaf emission, the downward flux below the leaf layer is $L^\downarrow[1 - \varepsilon_\ell(1 - \tau_d)]$. Extended to a canopy, the downward longwave flux at a canopy depth defined by the cumulative leaf area index x is

$$L^\downarrow(x) = L_{sky}^\downarrow\left[1 - \varepsilon_\ell\left(1 - e^{-K_dx}\right)\right] + \varepsilon_\ell\sigma T_\ell^4\left(1 - e^{-K_dx}\right). \qquad (14.134)$$

The first term on the right-hand side is the transmittance of atmospheric longwave radiation through the canopy, and the second term is the longwave radiation emitted by the canopy with temperature T_ℓ. Here, K_d is the extinction coefficient for diffuse radiation evaluated for a canopy with leaf area index L using (14.34). Similarly, the upward flux is

$$L^\uparrow(x) = L_g^\uparrow\left\{1 - \varepsilon_\ell\left[1 - e^{-K_d(L-x)}\right]\right\} + \varepsilon_\ell\sigma T_\ell^4\left[1 - e^{-K_d(L-x)}\right], \qquad (14.135)$$

where $L_g^\uparrow = \varepsilon_g\sigma T_g^4$ is the longwave radiation emitted by the ground. This ground source ignores the portion of the downward flux below the canopy that is

(a) Numerical

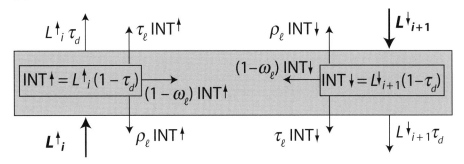

(b) Analytical

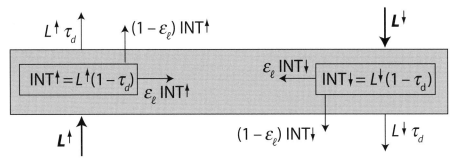

Figure 14.21 Longwave radiation fluxes represented for a single leaf layer. (a) Norman's (1979) numerical model, as in Figure 14.14a. Shown is the radiative balance for leaf layer $i + 1$ located above leaf layer i. (b) A simplified model to allow only forward scattering ($\rho_\ell = 0$ and $\tau_\ell = \omega_\ell = 1 - \varepsilon_\ell$) and to permit an analytical solution integrated over a canopy. In both panels, emitted radiation is excluded. Thick lines denote fluxes incident onto the layer.

scattered upward by the ground (unless $\varepsilon_g = 1$). The radiation absorbed per leaf area is

$$\overrightarrow{L}_\ell(x) = \frac{d}{dx}\left[L^\uparrow(x) - L^\downarrow(x)\right]$$
$$= \varepsilon_\ell\left(L_g^\uparrow - \sigma T_\ell^4\right)K_d e^{-K_d(L-x)} + \varepsilon_\ell\left(L_{sky}^\downarrow - \sigma T_\ell^4\right)K_d e^{-K_d x}. \tag{14.136}$$

For a canopy with leaf area index L, the radiation absorbed by the canopy (per ground area) is

$$\overrightarrow{L}_c = \int_0^L \overrightarrow{L}_\ell(x)dx$$
$$= \varepsilon_\ell\left(L_{sky}^\downarrow + L_g^\uparrow\right)\left(1 - e^{-K_d L}\right) - 2\varepsilon_\ell\sigma T_\ell^4\left(1 - e^{-K_d L}\right). \tag{14.137}$$

The first term in the second equation is the longwave radiation from the sky and ground absorbed

by the canopy, and the second term is the longwave radiation emitted by the canopy. Equation (14.135) solved with $x = 0$ is the longwave radiation emitted at the top of the canopy, and (14.134) solved with $x = L$ is the longwave radiation below the canopy onto the ground. The longwave radiation absorbed at the ground is

$$\overrightarrow{L}_g = L^\downarrow(L) - L_g^\uparrow, \tag{14.138}$$

and the overall energy balance is

$$\overrightarrow{L}_c + \overrightarrow{L}_g = L_{sky}^\downarrow - L^\uparrow(0). \tag{14.139}$$

Amthor et al. (1994) used these equations with $\varepsilon_\ell = \varepsilon_g = 0.96$. Sellers et al. (1996a) and Dai et al. (2004) used $\varepsilon_\ell = \varepsilon_g = 1$. Equation (14.137) can be extended to sunlit and shaded canopies. For the sunlit canopy,

$$\overrightarrow{L}_{cSun} = \int_0^L \overrightarrow{L}_\ell(x) f_{sun}(x) dx$$

$$= \frac{\varepsilon_\ell (L_{sky}^\downarrow - \sigma T_\ell^4) K_d}{K_d + K_b} [1 - e^{-(K_d + K_b)L}]$$

$$+ \frac{\varepsilon_\ell (L_g^\uparrow - \sigma T_\ell^4) K_d}{K_d - K_b} (e^{-K_b L} - e^{-K_d L}), \quad (14.140)$$

and for the shaded canopy,

$$\overrightarrow{L}_{cSha} = \int_0^L \overrightarrow{L}_\ell(x) [1 - f_{sun}(x)] dx = \overrightarrow{L}_c - \overrightarrow{L}_{cSun}. \quad (14.141)$$

Question 14.15 Discuss different assumptions between the Norman (1979) model given by (14.122) and (14.123) and the analytical model given by (14.134) and (14.135).

14.11 | Supplemental Programs

14.1 Direct Beam Extinction Coefficient: This program calculates K_b for the leaf angle distributions in Table 2.1. It obtains $G(Z)$ using a numerical integration of (14.25) with nine leaf inclination angles.

14.2 Two-Stream Approximation: This program describes the two-stream approximation at both the leaf and canopy scales. It gives the analytical solution to canopy absorption and compares that with the numerical solution obtained by integrating leaf fluxes over the canopy.

14.3 Radiative Transfer: This program calculates radiative transfer and light profiles in a canopy using the Norman, Goudriaan, and two-stream models. It was used for Figure 14.15.

14.4 Longwave Radiation: This program calculates longwave radiation using the Norman model and the analytical canopy-integrated model.

14.12 | Modeling Projects

1 Using the radiative transfer models (Supplemental Program 14.3), calculate the photosynthetically active radiation absorbed by a small increment in leaf area ($\Delta L = 0.1$ m^2 m^{-2}) at the bottom of canopies with total leaf area index 1, 2, 3, 4, 5, and 6 m^2 m^{-2}. For which canopies is it photosynthetically advantageous to add extra leaf area?

2 In Figure 14.15, 80% of the incident radiation is direct beam. Repeat the calculations using a higher fraction of diffuse radiation (80% diffuse, 20% direct beam). How does more diffuse radiation affect the light profile?

3 Examine the effect of foliage clumping for radiative transfer. Calculate light profiles as in Figure 14.15, but for different values of the clumping index Ω.

4 Figure 14.20a shows canopy albedo for direct beam radiation from the two-stream approximation. Compare this to the Norman and Goudriaan models. Repeat the calculations for a Sun that is lower on the horizon (Z = 60°) and for semihorizontal ($\chi_\ell = 0.25$) and semivertical ($\chi_\ell = -0.30$) leaf distributions. Compare direct beam and diffuse albedos.

5 Use Supplemental Program 14.4 to compare the two models of longwave radiative transfer. Compare results for the analytical model and the Norman model when $\varepsilon_g = 1$ and scattering is in the forward direction only ($\rho_\ell = 0$; $\tau_\ell = \omega_\ell = 1 - \varepsilon_\ell$). Compare models with $\varepsilon_g < 1$ and again for different assumptions about scattering (forward and backward directions). How important are the simplifications needed to attain the analytical solution?

Plant Canopies

Chapter Overview

Terrestrial biosphere models scale physiological and biophysical processes such as photosynthesis, stomatal conductance, and energy fluxes from individual leaves to the entire plant canopy. Critical to this is an understanding of leaf gas exchange, plant hydraulics, and radiative transfer presented in Chapters 10–14, and a theory and numerical parameterization to scale over all leaves in the canopy. This chapter focuses on the latter requirement and considers how to scale leaf fluxes to the canopy. Three general approaches to do so treat the canopy as: analogous to a big leaf with a single flux exchange surface without vertical structure; a dual source with separate fluxes for vegetation and soil; and in a multilayer framework in which the canopy is vertically structured and fluxes are explicitly resolved at multiple levels in the canopy.

15.1 | Introduction

Scaling fluxes from individual leaves to the entire plant canopy is an essential requirement of terrestrial biosphere models. Because of differences in leaf physiology (e.g., photosynthesis, stomatal conductance) and canopy structure (e.g., leaf area index, height), surface fluxes can be quite dissimilar among various types of vegetation, even when exposed to the same prevailing atmospheric conditions. Flux measurements at the Duke Forest (North Carolina) for an old-field grassland, an 18 m tall pine plantation, and a 25 m tall late successional oak–hickory forest separated by less than 1 km illustrate such differences (Figure 15.1). The grassland has the lowest midday net radiation, followed by the deciduous forest; the pine plantation has the highest net radiation. Latent heat flux is comparable among all three sites, though the forests have somewhat larger midday fluxes than the grassland. Midday sensible heat flux is markedly lower for the deciduous forest compared with the grassland and pine plantation. Another prominent difference is the lower friction velocity for the short grassland compared with the taller forests. These flux differences arise from differences in surface albedo, ecophysiology, and aerodynamics and manifest in cooler surface temperatures for the forests compared with the grassland (Stoy et al. 2006; Juang et al. 2007).

There are three general classes of plant canopy models (Figure 7.3b–d). One class of models simulates energy and mass exchanges between the biosphere and atmosphere by abstracting the canopy as a single big leaf. These models do not represent the vertical structure of a canopy, such as an overstory and understory, and instead characterize the canopy as a single bulk exchange surface. The big-leaf approximation can be extended to represent the canopy as two leaves – a sunlit and a shaded leaf – to better account for light attenuation within the canopy. A second related approach, commonly used in terrestrial biosphere models, is to represent

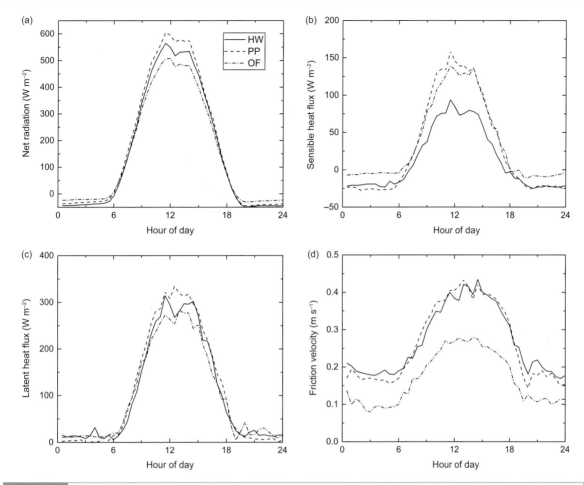

Figure 15.1 Above-canopy measurements of (a) net radiation, (b) sensible heat flux, (c), latent heat flux, and (d) friction velocity for an old-field grassland (OF), loblolly pine (*Pinus taeda*) plantation (PP), and oak–hickory (*Quercus–Carya*) hardwood forest (HW). Shown are the average diurnal cycles for July 2004. Stoy et al. (2006) described the sites.

the canopy as a big leaf, but with dual sources of vegetation and soil fluxes. In contrast, multilayer models recognize the nonuniform vertical structure of the canopy with respect to radiation, leaf temperature and energy fluxes, carbon assimilation, and scalar profiles. In developing canopy parameterizations for terrestrial biosphere models, it is important to have consistency among parameters, theory, and observations across scales from leaf to canopy to global (Figure 15.2). In particular, one desires to have observations of leaf traits and gas exchange that when combined with appropriate theory and numerical parameterizations give simulated canopy fluxes consistent with eddy covariance flux measurements and in which the generality is confirmed through comparison with global scale fluxes.

15.2 Big-Leaf Models

Canopy fluxes represent an integration of leaf-scale processes over all leaves in the canopy. For example, canopy photosynthesis (gross primary production, GPP) is leaf CO_2 assimilation integrated over the leaf area index L such that

$$\text{GPP} = \int_0^L A(x)\,dx. \tag{15.1}$$

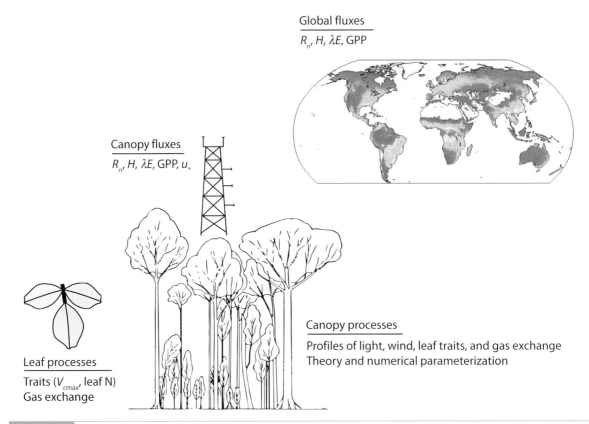

Global fluxes

R_n, H, λE, GPP

Canopy fluxes

R_n, H, λE, GPP, u_*

Canopy processes

Profiles of light, wind, leaf traits, and gas exchange
Theory and numerical parameterization

Leaf processes

Traits (V_{cmax}, leaf N)
Gas exchange

Figure 15.2 Scaling from the leaf to the canopy requires datasets of leaf traits and models of leaf gas exchange, methods to scale over vertical profiles of light and the microclimate in canopies, measurements of fluxes above canopies, and global flux datasets to test the model over a broad range of ecological and climatic conditions.

Here, $A(x)$ denotes leaf photosynthesis at a depth in the canopy given by the cumulative leaf area index x. Direct integration of leaf processes is not possible if one accounts for the full complexities of radiative transfer and the nonlinearity of photosynthesis, stomatal conductance, transpiration, and leaf temperature. Instead, a class of models simulates the plant canopy as a single big-leaf in which the leaf-level equations are solved with parameters that are scaled to the canopy (Figure 15.3). The Penman–Monteith equation for evapotranspiration is an example of a big-leaf model (Chapter 7). In the Penman–Monteith equation, canopy conductance is the sum of the stomatal conductances of individual leaves acting in parallel. For a canopy with multiple layers of leaves, each with conductance $g_{sw,i}$ and leaf area index ΔL_i, canopy conductance is

$$g_c = \sum_i g_{sw,i} \Delta L_i, \tag{15.2}$$

which is approximated as the product of the mean leaf conductance and the leaf area index of the canopy ($g_c = \bar{g}_{sw} L$). The Penman–Monteith representation of the canopy as a big leaf can be formally related to multilayer canopy models (Shuttleworth 1976).

The theoretical justification for big-leaf models can be shown by considering leaf photosynthesis. One of the primary drivers of vertical variation in photosynthesis is the light profile in the canopy. Analytical solutions to (15.1) can be obtained using simple radiative transfer models such as an exponential light profile obtained from Beer's law in combination with simple models of leaf photosynthesis (Johnson and Thornley 1984; Sands 1995). Leaf photosynthesis can be represented using the non-rectangular hyperbola (4.21) in which

$$\Theta A^2(x) - \left[\mathrm{E}\vec{I_\ell}(x) + A_{max} \right] A(x) + \mathrm{E}\vec{I_\ell}(x) A_{max} = 0, \tag{15.3}$$

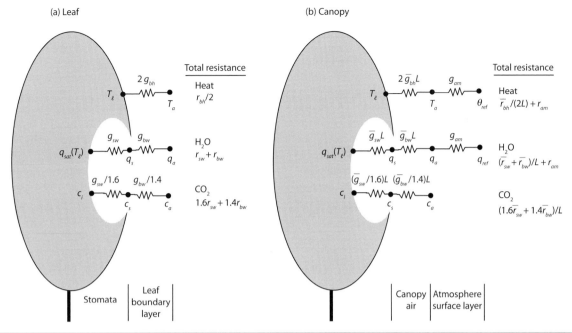

Figure 15.3 Scaling of leaf fluxes to the canopy using a big-leaf model. (a) Shown are leaf sensible heat, transpiration, and CO_2 fluxes in relation to various conductances. Fluxes are exchanged between the leaf and air around the leaf. Also shown is the total resistance. (b) Shown are big-leaf canopy fluxes in which leaf fluxes are scaled by the average conductance and leaf area index and are further modified by turbulent transport in the atmospheric surface layer. Surface layer processes are commonly omitted for CO_2 exchange. Only a single big leaf is shown, but separate sunlit and shaded big leaves can be similarly depicted.

with $\overrightarrow{I}_\ell(x)$ leaf irradiance at the cumulative leaf area index x. As described in Chapter 14, a simple expression for the mean irradiance absorbed by a leaf is

$$\overrightarrow{I}_\ell(x) = K_b I_0^\downarrow e^{-K_b x}. \tag{15.4}$$

The integral of (15.3) over the canopy is complex if the maximum photosynthetic rate A_{max} is constant throughout the canopy, but the solution simplifies if A_{max} varies in relation to the light profile such that

$$A_{max}(x) = A_{max\,0} e^{-K_b x}.$$

In this case, the total carbon uptake integrated over the canopy is

$$GPP = \int_0^L A(x)dx = A(0)\left[\frac{1 - e^{-K_b L}}{K_b}\right]. \tag{15.5}$$

This equation shows that canopy-integrated photosynthesis is the leaf rate evaluated at the top of the canopy $A(0)$ times a canopy scaling factor given in brackets. Sellers et al. (1992) generalized the canopy scaling to include advanced models of photosynthesis and stomatal conductance, and Sellers et al. (1996a) implemented the parameterization in SiB2.

Big-leaf models are simple representations of canopy gas exchange that are scaled versions of leaf models. However, representing the canopy as a single bulk leaf causes errors because of the non-linear dependence of leaf gas exchange on irradiance. A better method to integrate leaf processes over the canopy is to consider the portion of the canopy that is sunlit and shaded so as to treat the canopy as two separate big leaves (Sinclair et al. 1976; Norman 1993). Different photosynthetic rates and stomatal conductance are calculated for sunlit and shaded leaves based on the amount of absorbed photosynthetically active radiation. As explained in Chapter 14, shaded leaves receive only diffuse radiation, from both the sky and direct beam radiation scattered within the canopy, and are typically in the linear portion of the photosynthetic light response curve. Sunlit leaves receive the same diffuse radiation and additionally direct beam radiation intercepted in the canopy. Because they receive much more radiation than shaded leaves, sunlit leaves are typically near light saturation.

Analytical solutions to canopy photosynthesis such as (15.5) can be extended to include sunlit and shaded leaves, though the equations are complex (Medlyn et al. 2000; Thornley 2002). A simple numerical method is to calculate stomatal conductance and photosynthesis at the leaf scale using the leaf-average absorbed irradiance (per unit leaf area) and the leaf-average photosynthetic parameters (per unit leaf area). Canopy photosynthesis is the sum of the sunlit and shaded leaf rates multiplied by their respective leaf area index so that

$$\text{GPP} = [A_{sun} f_{sun} + A_{shade}(1 - f_{sun})]L. \qquad (15.6)$$

Similar calculations pertain to canopy conductance. The governing equations to calculate the canopy energy budget, photosynthesis, and stomatal conductance are the same as those for a leaf, but the radiative fluxes are integrated over the canopy and the conductances are scaled for the canopy. Terms in these equations also represent bulk canopy values for sunlit or shaded leaves.

The canopy scaling factor in (15.5) is obtained if photosynthetic capacity has the same exponential profile in the canopy as does light. Because photosynthetic capacity varies linearly with leaf nitrogen content (area-based), this is equivalent to assuming that leaf nitrogen content varies in relation to photosynthetically active radiation. It is now recognized that the leaf nitrogen gradient is not as steep as the light gradient but can still be described by an exponential profile (Chapter 2). The current generation of big-leaf models uses the concept of sunlit and shaded leaves to extend the single big-leaf approach to a two-leaf model and additionally uses exponential profiles to account for vertical gradients in leaf nitrogen and photosynthetic capacity in the canopy (de Pury and Farquhar 1997; Wang and Leuning 1998; Dai et al. 2004; Bonan et al. 2011). The photosynthetic parameters $V_{c\max}$ and $J_{\max}$ vary linearly with area-based leaf nitrogen (Chapter 11). A simple scaling approach describes gradients in these parameters using an exponential decay coefficient and given in relation to canopy depth, expressed as the cumulative leaf area index from the canopy top (Figure 15.4). For example, $V_{c\max}$ at cumulative leaf area index x from the canopy top decreases exponentially such that

$$V_{c\max}(x) = V_{c\max 0} e^{-K_n x}, \qquad (15.7)$$

in which $V_{c\max 0}$ is defined at the top of the canopy and K_n describes the steepness of the vertical profile. This equation is similar to the vertical profile of area-based leaf nitrogen given by (2.21).

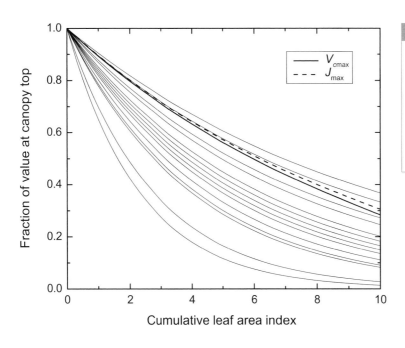

Figure 15.4 Canopy profiles of relative photosynthetic capacity in relation to cumulative leaf area index. Thin lines show exponential profiles using values of K_n for 16 temperate broadleaf forests and two tropical forests ranging from 0.10 to 0.43 (Lloyd et al. 2010). The two thick lines show observed profiles of $V_{c\max}$ and $J_{\max}$ from Niinemets and Tenhunen (1997) obtained for sugar maple (*Acer saccharum*).

Canopy values for $V_{c\max}$ are found by integrating (15.7) over the sunlit and shaded fractions of the canopy. For the sunlit canopy,

$$V_{c\max}(\text{sun}) = \int_0^L V_{c\max}(x)f_{sun}(x)dx$$

$$= V_{c\max 0}\left\{[1 - e^{-(K_n+K_b)L}]\frac{1}{K_n + K_b}\right\},$$

$$(15.8)$$

and for the shaded canopy,

$$V_{c\max}(\text{sha}) = \int_0^L V_{c\max}(x)[1 - f_{sun}(x)]dx$$

$$= V_{c\max 0}\left\{[1 - e^{-K_n L}]\frac{1}{K_n} - [1 - e^{-(K_n+K_b)L}]\frac{1}{K_n + K_b}\right\}.$$

$$(15.9)$$

In (15.8) and (15.9), $V_{c\max}$ has been directly scaled to represent the sunlit and shaded portions of the canopy. The terms in braces are scaling factors, analogous to (15.5), but now distinguished by sunlit and shaded leaves and additionally recognizing that the profile of nitrogen is separate from that of light (Figure 15.5). Dividing by sunlit and shaded leaf area index gives the corresponding leaf values of $V_{c\max}$

for the average sunlit and shaded leaf. Other photosynthetic parameters (e.g., $J_{\max}$, R_d) scale similarly.

The nitrogen decay parameter K_n is critical to the canopy scaling. The original theory postulated that plants optimally allocate resources to maximize carbon gain such that area-based leaf nitrogen is distributed through the canopy in relation to the time-mean profile of photosynthetically active radiation, but it is now recognized that the nitrogen gradient is shallower than the light gradient (Chapter 2). Observational estimates of K_n in forests range from 0.10 to 0.43 (mean, 0.19; median, 0.18) (Lloyd et al. 2010), as shown in Figure 15.4. Moreover, theory suggests that trees with low overall photosynthetic capacity should have a shallow gradient in photosynthetic capacity with canopy depth, while trees with higher photosynthetic capacity should have a stronger gradient (Lloyd et al. 2010). Observations support this across a range of broadleaf trees from various biomes. Trees with high photosynthetic capacity in the upper canopy have a sharper gradient than trees with low photosynthetic capacity so that K_n scales with $V_{c\max}$ at the canopy top following the relationship:

$$K_n = \exp(0.00963V_{c\max 0} - 2.43). \qquad (15.10)$$

In this equation, larger values of $V_{c\max 0}$ imply steeper declines in photosynthetic capacity through

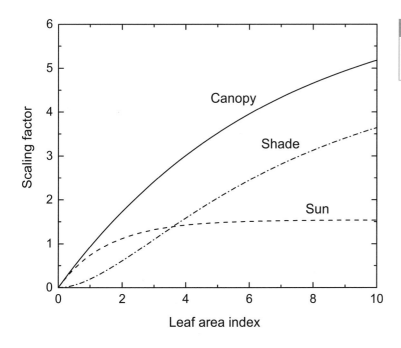

Figure 15.5 Canopy scaling factors for $V_{c\max}$. Shown are scaling factors for the sunlit and shaded portions of the canopy and the entire canopy with $K_n = 0.15$ and $K_b = 0.5$.

the canopy with respect to cumulative leaf area index.

The partitioning of the canopy into sunlit and shaded fractions changes over the day with the Sun's position in the sky, and consequently the photosynthetic capacity of the canopy varies over the day in relation to the nitrogen profile and sunlit/shaded fractions. An additional factor relates to foliage clumping. Big-leaf canopy models commonly calculate radiative transfer based on a plane-parallel, homogenous canopy in which leaves are randomly spaced. The nonrandom distribution of foliage can be accounted for with the clumping index Ω (Chapter 14). Foliage clumping, through its influence on radiative transfer and sunlit and shaded fractions, affects canopy fluxes of evapotranspiration and gross primary production (Baldocchi and Wilson 2001; Walcroft et al. 2005; Ryu et al. 2011; Chen et al. 2012, 2016a).

While many models utilize sunlit and shaded leaves to calculate canopy fluxes, the models can differ in the details of how this is implemented. Some models apply canopy-integrated parameters to the leaf equations so that fluxes are directly scaled to the canopy, while others use leaf-average values so that fluxes must be scaled by sunlit or shaded leaf area index. Models that include the leaf energy budget may calculate separate sunlit and shaded leaf conductances but use the total canopy conductance in a single big-leaf calculation of the leaf energy balance (e.g., the Community Land Model; Oleson et al. 2013). More advanced models have separate energy balance calculations for sunlit and shaded leaves (e.g., Wang and Leuning 1998; Dai et al. 2004). Models also differ in whether or not co-limitation of photosynthesis, as described in Chapter 11, is implemented. Co-limitation is necessary, especially in a single big leaf but even in sunlit/shaded big-leaf models, to account for leaf variability in photosynthetic machinery and to allow for Rubisco- and light-limited photosynthesis to co-occur (de Pury and Farquhar 1997; Wang 2000).

Sunlit/shaded big-leaf models are useful approximations to study plant canopies, but there are two contrasting approaches to estimate critical model parameters. In one approach, a parameter such as $V_{c\,max}$ is treated as a free parameter that is found by fitting the model to eddy covariance flux tower measurements. Such parameter estimation ensures good fit between the model and observations but

can compensate for or obscure structural deficiencies in the model. An alternative approach is to use leaf-scale measurements and devise an appropriate procedure to scale leaf fluxes to the canopy. The canopy scaling should be consistent with observations of leaf gas exchange, leaf traits such as nitrogen concentration and $V_{c\,max}$, their profiles in the canopy, and should be theoretically justified (Figure 15.2). Such an approach is more difficult than simply fitting the model to canopy flux measurements and must satisfy multiple constraints but is likely to be more robust under changing environmental conditions.

Question 15.1 Equation (15.1) describes photosynthesis in relation to cumulative leaf area index x. (a) Give an equivalent equation that calculates canopy conductance from stomatal conductance. (b) Rewrite this equation so that stomatal conductance is specified in relation to height z.

Question 15.2 The big-leaf canopy scaling shown in Figure 15.3b uses the aerodynamic conductance for momentum g_{am}. (a) Why is this conductance used rather than the aerodynamic conductance for scalars g_{ac}? (b) The Penman–Monteith equation (7.36) uses the aerodynamic conductance for scalars g_{ac} to calculate latent heat flux. Is this correct?

Question 15.3 Equations (15.8) and (15.9) give $V_{c\,max}$ integrated over the sunlit and shaded portions of the canopy, respectively. Derive a comparable equation for the total canopy-integrated $V_{c\,max}$. Show that this is equal to the sum of the sunlit and shaded $V_{c\,max}$.

Question 15.4 Show that the average $V_{c\,max}$ for sunlit leaves is $V_{c\,max}(\mathrm{sun})/L_{sun}$.

15.3 | Dual-Source Models

Big-leaf models represent fluxes from a plant canopy without accounting for soil fluxes. As an alternative, two-layer dual-source canopy models represent the flux exchange between land and atmosphere as separate fluxes from vegetation and soil (Figure 7.3c). Sensible heat is partitioned into that from foliage and that from soil, each regulated by different processes. Leaf boundary layer conductance integrated over all leaves in the canopy governs sensible heat flux from foliage. Turbulent processes within the canopy govern sensible heat flux from soil. Latent heat is partitioned into soil evaporation and transpiration. Transpiration is regulated by a canopy conductance that is an integration of leaf stomatal and boundary layer conductances over all leaves. Soil evaporation is regulated by aerodynamic processes within the canopy and by soil moisture. Deardorff (1978) described the fundamental principles of the dual-source canopy model, which were introduced into land surface models with BATS (Dickinson et al. 1986) and SiB (Sellers et al. 1986), though Shuttleworth (1976) proposed a similar concept.

The leaf and ground energy budgets represent a system of equations that can be solved separately for vegetation T_ℓ and ground T_0 temperatures. With reference to Figure 15.6a, three sensible heat fluxes are represented in the soil–plant–atmosphere system: a flux from vegetation to canopy air; a flux from the ground to canopy air; and a flux from canopy air to the atmosphere. Assuming canopy air has negligible capacity to store heat, the total sensible heat flux to the atmosphere is the sum of fluxes from vegetation and the ground so that

$$H = c_p(T_a - \theta_{ref})g_{am}$$
$$= 2c_p(T_\ell - T_a)\bar{g}_{bh}L + c_p(T_0 - T_a)g_{a,0}. \quad (15.11)$$

Here, g_{am} is the aerodynamic conductance, $2\bar{g}_{bh}L$ is leaf boundary layer conductance scaled to the canopy, and $g_{a,0}$ is an aerodynamic conductance within the canopy. Rearranging terms in these equations, canopy air temperature T_a, which is common to all three fluxes, is diagnosed as a weighted average of the atmospheric, vegetation, and ground temperatures in which

$$T_a = \frac{g_{am}\theta_{ref} + (2\bar{g}_{bh}L)T_\ell + g_{a,0}T_0}{g_{am} + 2\bar{g}_{bh}L + g_{a,0}}. \quad (15.12)$$

A similar expression can be derived for canopy air water vapor from equations for latent heat flux. The energy balance of the canopy is solved for the temperature T_ℓ that balances net radiation, sensible heat flux, and latent heat flux. Then, with canopy fluxes known, the ground fluxes and temperature T_0 are updated. The conductance network in this formulation is arranged so that the soil and canopy are treated in a two-layer framework, and fluxes interact through the canopy airspace (Figure 15.6a). An aerodynamic conductance describes vertical turbulent transport through the atmosphere to the canopy airspace (Chapter 6); leaf boundary layer conductance (Chapter 10) accounts for sensible heat transport from leaves to the canopy air and stomatal conductance (Chapter 12) additionally accounts for transpiration; and an additional aerodynamic conductance is needed for within-canopy transport from the soil to the canopy air. This approach can be extended to chemical fluxes in the canopy (Nemitz et al. 2001).

Two-layer models are an advancement over big-leaf models because they account for the different meteorological and physiological controls of leaf and soil fluxes, particularly for transpiration and soil evaporation. Two-layer models also distinguish multiple temperatures at the land surface. Vegetation temperature and a separate ground temperature are obtained by solving the respective energy budgets. Canopy air temperature is diagnosed from these temperatures. These are different from radiometric surface temperature, which is calculated from the upward longwave radiation above

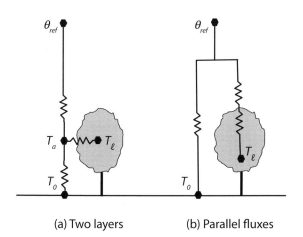

(a) Two layers   (b) Parallel fluxes

Figure 15.6 Conductance networks for dual-source models showing sensible heat fluxes in (a) the two-layer approach in which soil and vegetation fluxes interact with a common canopy air temperature and (b) the parallel flux approach. T_ℓ is canopy temperature, and T_0 is ground temperature.

the canopy. The methodology can accommodate sparse canopies (Shuttleworth and Wallace 1985; Shuttleworth and Gurney 1990) and can be modified so that canopy air has some capacity to store heat (Vidale and Stöckli 2005). A difficulty, however, is to parameterize turbulent exchange within the canopy and the soil-to-canopy aerodynamic conductance. Chapter 16 presents some approaches to do this.

Alternatively, the soil–canopy system can be configured as a parallel flux network in which vegetation and soil are treated as separate patches without interactions; instead, both surfaces exchange fluxes directly with the atmosphere (Figure 15.6b). This approach treats vegetation as a big leaf to calculate fluxes and separately calculates soil fluxes for non-vegetated bare ground. It is best for patchy canopies in which plants are widely spaced among patches of bare soil. An example is the model of Norman et al. (1995), subsequently modified by Kustas and Norman (1999), which has been used to obtain remote sensing estimates of evapotranspiration from radiometric surface temperature (Anderson et al. 2007, 2011).

Question 15.5 In the dual-source canopy shown in Figure 15.6a, is the aerodynamic conductance to the atmosphere calculated using z_{0m} or z_{0c}?

Question 15.6 Explain the difference between radiometric surface temperature and aerodynamic surface temperature. Which surface energy balance model (bulk surface or dual-source) better distinguishes these two temperatures in a sparse canopy?

15.4 | Multilayer Models

Many plant traits vary in relation to the light environment, and the strong decrease in light from the top to the bottom of the canopy results in pronounced within-canopy gradients of leaf physiological, structural, and chemical traits. Light-driven variation in area-based traits such as leaf mass per area, photosynthetic capacity, and leaf nitrogen content and also the partitioning of nitrogen among photosynthetic machinery (Rubisco,

electron transport) is seen in many plant functional groups (Niinemets et al. 2015; Keenan and Niinemets 2016). Within-canopy variation in photosynthetic capacity can be accounted for, in part, by using simple scaling rules as described previously. However, some measurements in forests also show that leaf angle distribution varies with height. Leaves in the understory tend to be more horizontal while those in the upper canopy are more vertically inclined (Ford and Newbould 1971; Hutchison et al. 1986; Hollinger 1989; Niinemets 1998; Kull et al. 1999). Additional vertical gradients can arise from plant hydraulics. One of the predictions of the SPA plant hydraulics model is that leaves in the upper canopy, with high solar radiation and high transpiration rates, close their stomata at midday in response to low leaf water potential to avoid desiccation (Figure 13.5).

Vertical gradients can also be seen in meteorological measurements of wind speed, air temperature, and vapor pressure. These follow logarithmic profiles above the canopy (Chapter 6), but within-canopy profiles deviate from above-canopy profiles, as illustrated in Figure 15.7 for a Norway spruce forest. In this particular forest, wind speed decreases sharply in the upper canopy where most of the plant material is located. A local minimum occurs at about $0.5h_c$, with a secondary maximum under the tree crowns. Daytime air temperature and vapor pressure deficit increase with height within the canopy but decrease with height above the canopy. Both have a local maximum at a height of about $0.8h_c$. Geiger (1927) first described such daytime profiles with a mid-canopy maximum, seen also in other studies (Jarvis and McNaughton 1986; Pyles et al. 2000). Measurements in a walnut orchard show that complex temperature patterns can emerge at night (Figure 15.8). During the particular night studied, the air initially cooled near the ground, but as night progressed a temperature minimum developed in mid-canopy. Heights greater than $0.5h_c$ were stably stratified as expected, though the temperature gradient was much greater in the upper canopy compared with air above the canopy. The lower canopy was unstably stratified. This profile develops because leaves in the upper canopy cool from exposure to the overlying cold air and loss of longwave radiation to the sky. At the same time, the leaves absorb and reemit longwave radiation back to the soil surface, keeping the lower canopy

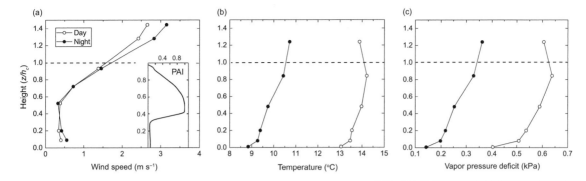

Figure 15.7 Mean daytime and nighttime profiles of (a) wind speed, (b) air temperature, and (c) vapor pressure deficit measured within and above a Norway spruce (*Picea abies*) forest with a plant area index of about 5 m^2 m^{-2} (20–24 September 2007). Height is given as a fraction of canopy height (z/h_c), and the dashed line denotes the canopy height (25 m). The inset panel shows the plant area index profile (normalized to the maximum value). Redrawn from Staudt et al. (2011)

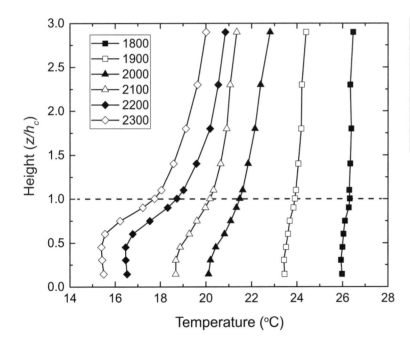

Figure 15.8 Hourly temperature profiles from 1800 to 2300 local time measured in a 25-year-old walnut (*Juglans regia*) orchard in California during the evening of 21 May 2007. At the time of measurement, the canopy height averaged 10 m, and the plant area index was 2.5 m^2 m^{-2}. The dashed line denotes the canopy height. Adapted from Patton et al. (2011)

air warm. Enhanced diffusivity because of convective instability in the canopy makes the temperature profile in the lower canopy uniform.

Big-leaf models ignore vertical structure within canopies, but nonlinearity in the canopy microenvironment (especially light, but also temperature and vapor pressure deficit), leaf biochemistry, and leaf physiological responses preclude simple aggregation of leaf processes over the canopy (Niinemets and Anten 2009). Instead, multilayer models are vertically resolved and partition the canopy into multiple layers of leaves and stems to explicitly calculate gradients within the canopy. Each layer can have a unique amount of plant material, leaf orientation, optical properties, photosynthetic capacity, and other physiological parameters. Some multilayer models are radiation-driven and neglect turbulent transport within the canopy, e.g., MAESTRA (Wang and Jarvis 1990; Medlyn 2004), SPA (Williams et al. 1996), and other models (Leuning et al. 1995; de Pury and Farquhar 1997). These models assume that wind speed, temperature, and water vapor in the canopy airspace are the same as above the canopy (known as the well-mixed assumption as discussed in Chapter 16).

The more complete multilayer models additionally simulate turbulent transport and scalar profiles above and within the canopy by dividing the canopy airspace into discrete layers. In these models, temperature, water vapor, CO_2, and other scalars in the canopy depend on sources and sinks from leaves and soil and also on turbulent transport in the canopy. Leaf fluxes are calculated for each layer in the canopy, and these are linked to conservation equations for energy and mass exchanges vertically between layers and horizontally between leaves and the surrounding air. The canopy is represented using a network of conductances that regulate fluxes from leaves to the canopy air and between layers of air in the canopy (Figure 7.3d), as initially described by Waggoner and Reifsnyder (1968) and Waggoner et al. (1969) and which can be expressed as a system of equations to solve using matrix methods (Waggoner 1975).

Examples of such models include: CUPID (Norman 1979, 1982, 1989; Norman and Campbell 1983; Wilson et al. 2003; Kustas et al. 2007); CANOAK (Baldocchi and Harley 1995; Baldocchi and Meyers 1998; Baldocchi and Wilson 2001; Baldocchi et al. 2002) and variants (Wohlfahrt et al. 2001); and other models (Goudriaan 1977; Choudhury and Monteith 1988; Gu et al. 1999; Ogée et al. 2003; Drewry et al. 2010; Launiainen et al. 2011, 2015). Some multilayer models simulate dry deposition (Baldocchi 1988; Meyers et al. 1998; Wu et al. 2003; Launiainen et al. 2013) and also chemical species in the canopy airspace (Gao et al. 1993; Makar et al. 1999; Stroud et al. 2005; Boy et al. 2011; Wolfe and Thornton 2011; Bryan et al. 2012; Saylor 2013; Ashworth et al. 2015). The effects of leaves on light and vertical diffusion within canopies are particularly important for modeling ozone chemistry (Makar et al. 2017). Other models have treated interception and evaporation in a multilayer framework (Calder 1977; Sellers and Lockwood 1981; Bouten et al. 1996; Jetten 1996). Multilayer canopies have not, generally, been used in the land surface models for climate simulation because they are computationally expensive, though some centers are developing such models (Community Land Model, Bonan et al. 2014, 2018; ORCHIDEE, Ryder et al. 2016; Chen et al. 2016b).

Multilayer models require a mathematical description of temperature, humidity, and wind speed profiles within the canopy, which are needed to calculate leaf fluxes at each level. These are obtained from turbulence theory. The most common turbulence parameterization is the first-order flux–gradient closure, or K-theory, in which scalar fluxes are calculated as the product of the mean concentration gradient and a diffusivity (similar to fluxes in the atmospheric surface layer as described in Chapter 6). Higher-order closure schemes allow for more complex turbulent flux parameterizations but are computationally difficult to solve. Accounting for turbulent transport greatly improves model performance compared with simulations in which above-canopy wind speed, temperature, and humidity are used within the canopy airspace, even if the turbulence parameterization is rudimental. The theory and equations to calculate scalar profiles within the canopy are given in the next chapter.

Multilayer canopy models explicitly resolve the vertical profile of leaf and stem area within a canopy. This profile drives gradients in light, temperature, humidity, CO_2 concentration, wind speed, turbulent transport, and leaf gas exchange. In a multilayer model, the canopy is divided into several discrete layers, similar to that used for soil temperature (Chapter 5) and soil moisture (Chapter 8). Each layer has a leaf area density. Many models divide the layer into sunlit and shaded fractions, and leaf processes are calculated separately for these. Component submodels include radiative transfer, the leaf energy balance, photosynthesis and stomatal conductance, plant hydraulics, turbulent diffusion to calculate scalar profiles, the soil energy balance and soil temperature, and interception of precipitation by the canopy (Figure 15.9).

Multilayer models are computationally expensive and can require iterative calculations to compute scalar fluxes and profiles within the canopy. The complexity is that longwave radiation depends on leaf temperature; stomatal conductance depends on leaf temperature and plant hydraulics; leaf fluxes depend on scalar profiles; evaporation depends on the amount of intercepted water; and canopy turbulence depends on surface fluxes. Figure 15.10 illustrates the flow diagram of a typical such model. The primary loop is the iterative calculation of leaf temperature, energy fluxes, and scalar

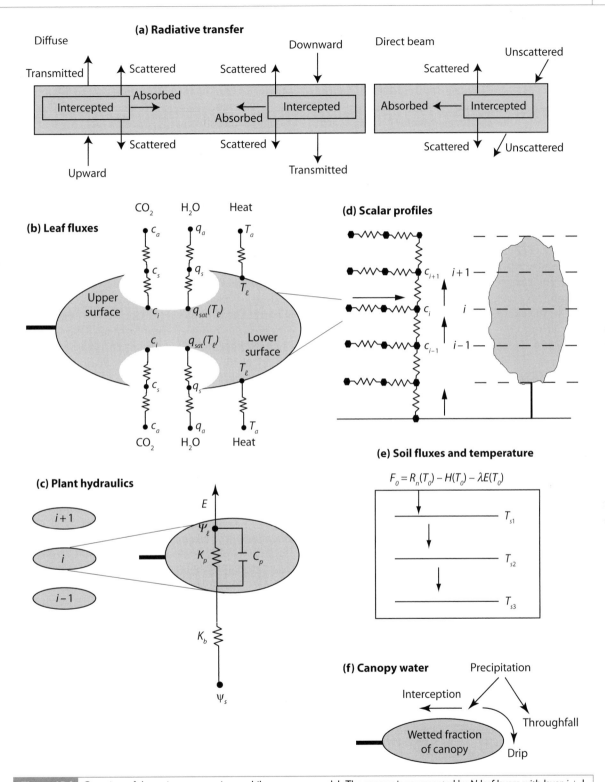

Figure 15.9 Overview of the main processes in a multilayer canopy model. The canopy is represented by N leaf layers with layer $i + 1$ above layer i. (a) Diffuse and direct beam solar radiation is transmitted or intercepted. The intercepted portion is absorbed or scattered in

profiles. This begins with calculation of longwave radiation given an initial estimate of leaf temperature. With net radiation known, the leaf temperature that balances the leaf energy budget is obtained. This can involve an iterative calculation of stomatal conductance and photosynthesis, and because the metabolic parameters that govern photosynthesis depend on leaf temperature, the entire calculation is iterated until leaf temperature converges within some specified tolerance. Once leaf temperature is known for each canopy layer, longwave radiation is recalculated with the updated temperatures, and the leaf calculations are repeated until the change in fluxes is less than some specified tolerance. Then, with solar and longwave radiation at the soil surface known, the soil energy budget is solved for surface temperature. Calculation of leaf and soil fluxes requires specification of air temperature, water vapor, wind speed, and CO_2 concentration within the canopy. These scalar profiles themselves depend on the leaf and soil fluxes. The entire set of calculations is repeated for updated scalar profiles, and the iteration continues until some convergence criteria is satisfied.

Multilayer canopy models can be criticized because many of the processes are problematic to resolve and cannot be described with simple, computationally efficient mathematics (especially within-canopy turbulence) and because vertical variation in leaf morphology and physiology are challenging to characterize (Raupach and Finnigan 1988). However, the prevailing class of big-leaf or dual-source canopy models still must represent these processes in some manner. For example, canopy radiative transfer requires leaf optical properties, leaf angle distribution, and foliage clumping. Big-leaf and dual-source models simply assume that vertical variation within the canopy is unimportant. These models must also make assumptions about

canopy scaling of photosynthetic capacity in relation to the vertical profile of leaf nitrogen. An advantage of multilayer models is that they explicitly represent leaf processes at the same scale at which they can be measured, and the calculated leaf fluxes can be directly compared with leaf gas exchange measurements. In contrast, big-leaf and dual-source models aggregate leaf-scale processes to the canopy scale. The effective model parameters integrated over the canopy cannot be compared with leaf measurements; instead required parameters are often obtained by calibrating the model to canopy flux measurements. It is generally thought that the computational efficiency and generality of big-leaf or dual-source models make them appropriate for calculating surface fluxes to the atmosphere in global models, whereas multilayer models are useful to study the microenvironment of particular canopies (Raupach and Finnigan 1988). However, the renewed interest in multilayer models suggests important benefits of this class of models (Bonan et al. 2014, 2018; Ryder et al. 2016; Chen et al. 2016b).

Bonan et al. (2014, 2018) developed a multilayer model that uses principles of water-use efficiency optimization (Chapter 12) and plant hydraulics (Chapter 13) to calculate stomatal conductance. A first-order turbulence closure combined with the Harman and Finnigan (2007, 2008) roughness sublayer parameterization (Chapter 6) is used to calculate scalar profiles (described in Chapter 16). Figure 15.11 illustrates the general behavior of the model over the course of a day at a forest site compared with observations and with the Community Land Model (CLM4.5), which uses a dual-source canopy. CLM4.5 overestimates midday sensible heat flux, underestimates peak latent heat flux, and underestimates gross primary production. Friction velocity is underestimated at night. Radiative

the forward and backward direction. Longwave radiation is similar to diffuse radiation. See Figure 14.14 for more details. (b) Leaf sensible heat, transpiration, and CO_2 fluxes depend on absorbed radiation and leaf boundary layer and stomatal conductances. Sensible heat is exchanged from both sides of the leaf. Water vapor and CO_2 can be exchanged from one or both sides of the leaf depending on stomata. Leaf temperature is the temperature that balances the energy budget. See Figure 10.2b for more details. (c) Stomatal conductance depends on leaf water potential. Plant water uptake for a canopy layer is in relation to belowground soil and root conductance and aboveground stem conductance acting in series and also a capacitance term. See Figure 13.4a for more details. (d) Scalar profiles are calculated from a conductance network. Leaf fluxes provide the source or sink of heat, water vapor, and CO_2, along with soil fluxes. See Figure 16.6 for more details. (e) Sensible heat, latent heat, and heat storage in soil depend on the ground temperature that balances the soil energy budget. (f) The wetted fraction of the canopy layer depends on the portion of precipitation that is intercepted.

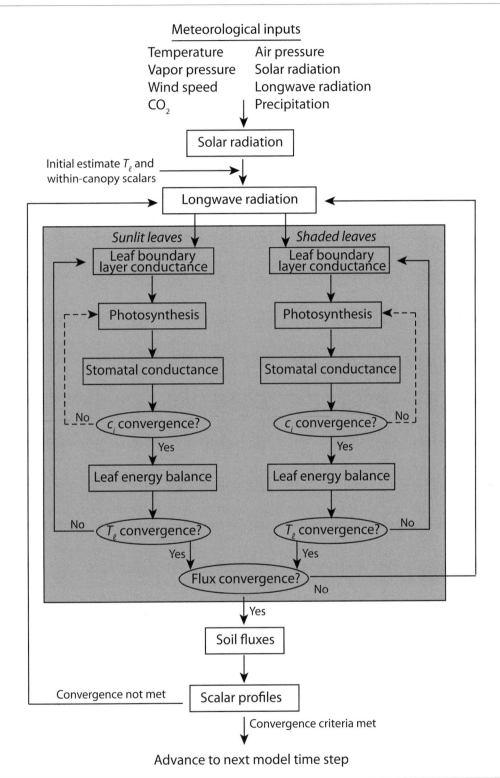

Figure 15.10 Flow diagram of processes in a multilayer canopy model. The shaded area denotes leaf processes resolved at each layer in the canopy. This is a generalized diagram of the required calculations for a dry leaf. Specific models differ in how the equation set is solved and the iterative calculations. Evaporation of intercepted water requires additional complexity.

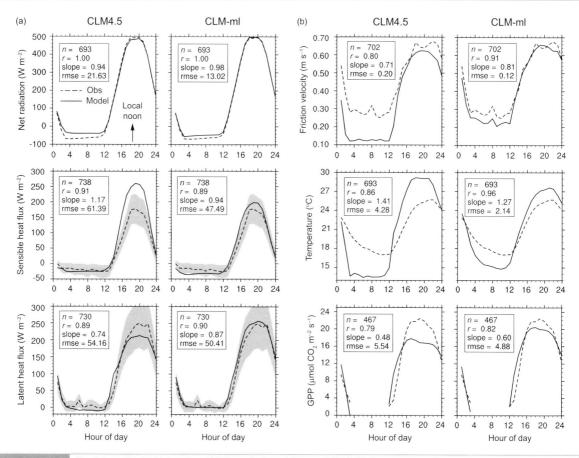

Figure 15.11 Simulations for the University of Michigan Biological Station (US-UMB) flux tower site (July 2006) using the Community Land Model (CLM4.5) and a multilayer canopy (CLM-ml). Shown are the average diurnal cycle (GMT) of (a) net radiation, sensible heat flux, and latent heat flux and (b) friction velocity, radiative temperature, and gross primary production (GPP) for the observations (dashed line) and model (solid line). Local noon is at approximately 1800. Shading denotes ± 1 standard deviation of the random flux error for the observations (sensible and latent heat flux only). Statistics show sample size (*n*), correlation coefficient (*r*), slope of the regression line, and root mean square error (rmse) between the model and observations. Adapted from Bonan et al. (2018)

temperature (i.e., the temperature inverted from the upward longwave flux) has a larger than observed diurnal range with colder temperatures at night and warmer temperatures during the day. The multilayer canopy improves the simulation, decreasing midday sensible heat flux, increasing midday latent heat flux and gross primary production, increasing nighttime friction velocity, and reducing the diurnal temperature range. These improvements arise from advances in stomatal conductance and canopy physiology beyond that in CLM4.5, as well as the treatment of within- and above-canopy turbulent transport (Bonan et al. 2018). The larger friction velocity at night results from the influence of the forest canopy on

turbulence. The improvement in radiative temperature arises from cooler leaf temperatures during the day and warmer leaf temperatures at night.

Wind speed and temperature profiles simulated with the multilayer canopy are noticeably different compared with CLM4.5 and that predicted from Monin–Obukhov similarity theory (MOST; Chapter 6), as shown in Figure 15.12. Wind speed decreases to zero at the surface with MOST. CLM4.5, which uses MOST for its surface fluxes, instead takes friction velocity as representative of the within-canopy wind speed. In a multilayer canopy, wind speed remains finite within the canopy. During the day, CLM4.5 simulates a warmer canopy air space than the multilayer canopy and is

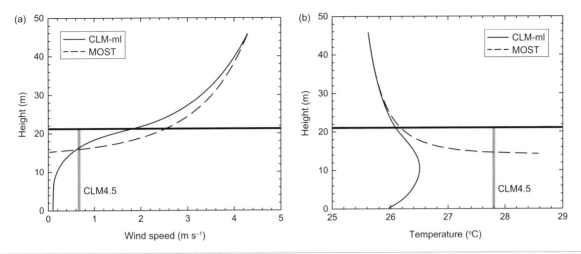

Figure 15.12 Profiles of (a) wind speed and (b) air temperature for US-UMB (July 2006) at 1400 local time. Shown are simulations for CLM4.5 and CLM-ml as in Figure 15.11. The thick horizontal line denotes canopy height. The CLM4.5 canopy wind speed and air temperature are shown as a thick gray vertical line but are not vertically resolved. Also shown are the profiles obtained using Monin–Obukhov similarity theory (MOST) extrapolated to the surface. Adapted from Bonan et al. (2018).

consistent with MOST, in which temperature increases monotonically to a maximum in the canopy. The multilayer canopy has a more complex profile with a local temperature maximum in mid-canopy and lower temperature near the ground. This is similar to the daytime temperature profile shown in Figure 15.7.

15.5 | Canopy Hydrology

A portion of the precipitation that falls onto a plant canopy is intercepted by foliage, stems, and woody structural elements and evaporates directly back to the atmosphere rather than replenishing soil water. The remainder reaches the ground as throughfall (directly through canopy gaps or by dripping off leaves, twigs, and branches) or as stemflow (flowing down stems and trunks). The amount of interception loss in forests is about 10%–30% of annual precipitation but can be larger (Miralles et al. 2010; Carlyle-Moses and Gash 2011). More water can be held as snow than as rain. Evaporation of intercepted water can greatly exceed transpiration under the same meteorological conditions so that total evapotranspiration must consider these two processes separately (Rutter 1967; Stewart 1977).

The thermodynamic effects of evaporation of intercepted water can be seen in the cooler temperature of wet leaves compared with dry leaves (Rutter 1967; Pereira et al. 2009) and needs to be treated as part of the canopy energy balance. Whereas transpiration is controlled by stomatal and aerodynamic conductances acting in series, evaporation of water held externally on plant surfaces depends on turbulent transport only, which imposes a much smaller resistance to water transport compared with stomata. Rutter (1967) pioneered the application of the Penman–Monteith equation, given by (7.36), to rainfall interception. For a wet canopy, stomatal conductance is neglected and evaporation is

$$E_p = \frac{s(R_n - G) + c_p[q_{sat}(T_a) - q_a]g_{ac}}{\lambda s + c_p}. \tag{15.13}$$

Equation (15.13) is used in many canopy interception models, but it may not account for all the processes that regulate evaporation of intercepted water (Carlyle-Moses and Gash 2011; van Dijk et al. 2015). High evaporation rates can occur during rainfall because evaporative cooling produces a downward sensible heat flux toward the surface, and heat stored in biomass and soil can be an additional source of energy. An alternative method is to calculate evaporation directly from the diffusion equation:

$$E_p = [q_{sat}(T_s) - q_a]g_{ac}. \tag{15.14}$$

This is the same as presented in (7.6) for surface evapotranspiration, but neglecting stomatal conductance. In either formulation, evaporation should be considered a potential rate in that more water cannot evaporate than is intercepted by the canopy.

The total water loss from a canopy combines evaporation from the wet portion of the canopy that holds intercepted water and transpiration from the dry portion. This can be represented using the big-leaf concept but dividing the canopy into separate wet and dry fractions and modeling each flux independently (Shuttleworth 1978). Evaporation is

$$E_{wet} = [q_{sat}(T_\ell) - q_a]\bar{g}_{bw}f_{wet}L \tag{15.15}$$

for the wet fraction of the canopy f_{wet}, and transpiration is

$$E_{dry} = [q_{sat}(T_\ell) - q_a]\frac{(1 - f_{wet})L}{\bar{g}_{bw}^{-1} + \bar{g}_{sw}^{-1}} \tag{15.16}$$

for the dry fraction of the canopy. This formulation linearly weights evaporation and transpiration by the wet and dry fractions of leaf area, respectively, and has the effect of suppressing transpiration over wet leaves. The actual rate is constrained by the amount of water intercepted by the foliage. The concept originates with Deardorff (1978) and was included in early models such as BATS (Dickinson et al. 1986, 1993) and SiB (Sellers et al. 1986). Deardorff (1978) gave the wetted fraction of the canopy as

$$f_{wet} = (W_c/W_{c\max})^{2/3} \tag{15.17}$$

for canopy water W_c less than some maximum $W_{c\max}$, and this has been used in many models.

Interception loss must consider the full water balance of the canopy. The model of Rutter et al. (1971, 1975) illustrates the calculations needed to estimate evaporation of intercepted rainfall and is a common method for numerical interception modeling (Muzylo et al. 2009). The model calculates throughfall, stemflow, and interception loss in relation to rainfall and meteorological data (as needed to calculate evaporation) and with separate water balance calculations for canopy foliage and tree trunks (Figure 15.13). The canopy is represented by a bulk pool of stored water on leaves that is the balance between intercepted rainfall, evaporation loss, and drip from leaves. A second pool of water

describes rainfall that is stored on trunks and becomes stemflow. Several parameters define the structure of the canopy. Each pool of water has a saturation storage capacity for the canopy and trunks, respectively. A portion f of the rain falls through gaps in the canopy as throughfall, and another portion f_t is intercepted by tree trunks to become stemflow. The remainder $(1 - f - f_t)$ is input to canopy storage.

The water balance of the canopy is governed by the equation

$$\frac{dW_c}{dt} = (1 - f - f_t)P_R - E_c - D_c, \tag{15.18}$$

in which W_c is the amount of water held in the canopy and has the units depth (mm) or equivalently mass per unit area (kg m^{-2}). The first term on the right-hand side is the amount of rainfall P_R that is intercepted by the canopy; E_c is the intercepted water that evaporates; and D_c is the rate at which intercepted water drips from the canopy. Evaporation is given by a potential rate E_p obtained from the Penman–Monteith equation applied to a wet canopy and modified by the relative canopy wetness whereby

$$E_c = \begin{cases} E_p & W_c \geq W_{c\max} \\ E_p(W_c/W_{c\max}) & W_c < W_{c\max} \end{cases}, \tag{15.19}$$

with $W_{c\max}$ the saturation storage capacity equal to the minimum amount of water at which all canopy surfaces are wetted (but W_c can exceed $W_{c\max}$). The drip, or drainage, rate is

$$D_c = \begin{cases} D_0 \exp[b(W_c - W_{c\max})] & W_c \geq W_{c\max} \\ 0 & W_c < W_{c\max} \end{cases}, \tag{15.20}$$

with D_0 the drainage rate when $W_c = W_{c\max}$. While $W_c < W_{c\max}$, evaporation varies with the amount of water held on the canopy. This water, in turn, is the balance between rainfall, evaporation, and drip. Drip occurs when $W_c \geq W_{c\max}$. The water balance of trunks is

$$\frac{dW_t}{dt} = f_t P_R - E_t - D_t. \tag{15.21}$$

Evaporation of water held on stems is similar to that for the canopy so that

$$E_t = \begin{cases} eE_p & W_t \geq W_{t\max} \\ eE_p(W_t/W_{t\max}) & W_t < W_{t\max} \end{cases}, \tag{15.22}$$

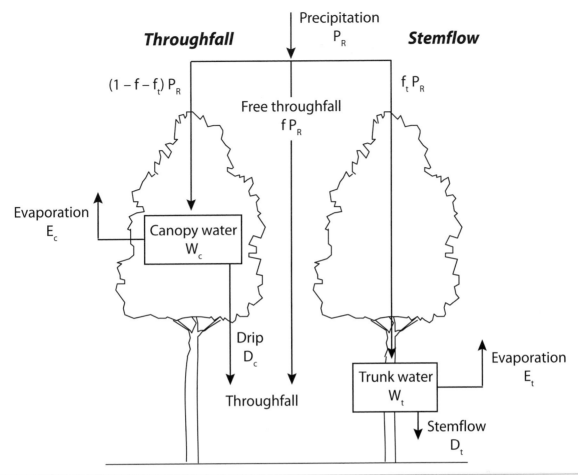

Precipitation
P_R

Throughfall

Stemflow

$(1 - f - f_t) P_R$

$f_t P_R$

Free throughfall
$f P_R$

Evaporation
E_c

Canopy water
W_c

Drip
D_c

Throughfall

Trunk water
W_t

Evaporation
E_t

Stemflow
D_t

Figure 15.13 Processes represented in the Rutter et al. (1971, 1975) interception model. Canopy hydrology for snow is more complex and must include energy used to melt intercepted snow. Adapted from Valente et al. (1997)

with e a constant that relates evaporation from stems to that from the canopy. Stemflow is

$$D_t = \begin{cases} (W_t - W_{t\max})/\Delta t & W_t \geq W_{t\max} \\ 0 & W_t < W_{t\max} \end{cases}. \quad (15.23)$$

The total interception loss is the time integral of evaporation from the canopy and trunk. Throughfall is the sum of free throughfall and canopy drip. This plus stemflow is the total water reaching the ground.

Implementation of the model requires several parameters, which are empirical, vary with canopy structure, and are estimated from measured fluxes. Representative parameter values are $D_0 = 0.12$ mm h^{-1} and $b = 3.7$ mm^{-1} for canopy drip, and specific parameters for a Douglas fir (*Pseudotsuga menziesii*)

stand are $f = 0.09$, $f_t = 0.15$, $W_{c\max} = 1.2$ mm, $W_{t\max} = 0.9$ mm, and $e = 0.04$ (Rutter et al. 1975). The model is typically run with relatively short time steps of one hour for precipitation and other meteorological input data, but the equations themselves are numerically integrated with shorter time steps because of the dependencies among pools and fluxes (e.g., two minutes; Lloyd et al. 1988). The model has been applied to many types of vegetation and performs well (Muzylo et al. 2009; Carlyle-Moses and Gash 2011). Valente et al. (1997) modified the model to account for sparse canopies by separating the calculations into a vegetated fraction with interception and a bare fraction of the surface without interception. Comparable multilayer interception models have been developed based on a cascade of precipitation and drip from higher layers

(Calder 1977; Sellers and Lockwood 1981; Bouten et al. 1996; Jetten 1996).

Question 15.7 Write an equation for the total leaf conductance to evaporation from the wetted fraction and transpiration from the dry fraction.

Question 15.8 Terrestrial biosphere models typically have a time step of 30 minutes when coupled to a climate model. What is the consequence of this when calculating canopy evaporation using (15.15)–(15.17)? How does the Rutter et al. (1971, 1975) model avoid this problem?

Question 15.9 Many terrestrial biosphere models have a single water pool in the canopy. The Rutter et al. (1971, 1975) model has separate pools for leaves and stems. What is a consequence of this compared with one pool?

15.6 | Optimality Theory

Optimality theory provides a framework to understand and model plant canopies. The fundamental premise is that canopies are organized so as to maximize photosynthesis, though this basic principle is much more complex than it seems (Niinemets 2012). Optimization helps to explain maximum leaf area index in different environments. Leaves provide the photosynthetic machinery for carbon gain, but too many leaves block sunlight and cast shade over lower leaves. Canopy photosynthesis is maximized when the total amount of leaves is such that those at the bottom of the canopy receive enough light to maintain a positive photosynthetic rate after accounting for respiration (Monsi and Saeki 1953; reprinted as Monsi and Saeki 2005). Horn (1971, 1975) extended this to show that productivity is maximized if trees growing in full sunlight have leaves distributed throughout the crown, whereas those growing in shade have leaves arranged in a shell along the tips of branches. A further optimization analysis explained the large leaf mass per area commonly found in leaves at the top of the canopy growing in high light conditions compared with leaves at the bottom of the canopy growing in shade (Gutschick and Wiegel 1988).

In the work of Monsi and Saeki (1953; reprinted as Monsi and Saeki 2005), leaf net photosynthetic rate varies with photosynthetically active radiation following the rectangular hyperbola, given by (15.3) with $\Theta = 0$, and photosynthetic capacity A_{max} is invariant with canopy height. At the cumulative leaf area index x, the net photosynthetic rate is

$$A_n(x) = \frac{A_{max}E\vec{I}_\ell(x)}{A_{max} + E\vec{I}_\ell(x)} - R_d, \tag{15.24}$$

and (15.4) can be used to describe leaf irradiance. Combining equations,

$$A_n(x) = \frac{A_{max}EK_bI_0^\downarrow e^{-K_bx}}{A_{max} + EK_bI_0^\downarrow e^{-K_bx}} - R_d \tag{15.25}$$

so that the total net carbon uptake integrated over the canopy is

$$A_n = \int_0^L A_n(x)dx$$
$$= \frac{A_{max}}{K_b} \ln\left(\frac{A_{max} + EK_bI_0^\downarrow}{A_{max} + EK_bI_0^\downarrow e^{-K_bL}}\right) - R_dL. \tag{15.26}$$

The optimal leaf area index that maximizes net carbon uptake is found when $dA_n/dL = 0$ (see also Figure 4.6). The solution is easily obtained because it is evident from (15.26) that $A_n(x)$ is dA_n/dL. The optimal leaf area index is that for which $A_n(x) = 0$, whereby

$$L_{max} = \frac{1}{K_b} \ln\left[\frac{(A_{max} - R_d)EK_bI_0^\downarrow}{A_{max}R_d}\right]. \tag{15.27}$$

This analysis shows that with low irradiance $I_0^\downarrow$, canopies have small leaf area index, but the exact amount varies also with leaf inclination angle (through K_b) and photosynthetic capacity (through A_{max} and R_d).

The vertical distribution of nitrogen in plant canopies (Chapter 2) can be understood in the context of optimality theory. Models of plant canopies show that daily canopy photosynthesis is maximized when nitrogen is allocated such that leaves that receive high levels of light also have high nitrogen concentration so as to have high photosynthetic capacity to utilize the light (Field 1983; Hirose and Werger 1987; Sellers et al. 1992; Anten et al. 1995; Buckley et al. 2002). In these models, plants maximize photosynthetic nitrogen-use efficiency (i.e., photosynthetic carbon gain per unit of nitrogen invested in photosynthetic capacity). Optimization is achieved when the marginal increase in leaf photosynthesis with an increase in leaf nitrogen is constant throughout the canopy. Optimal allocation of nitrogen manifests in higher light-saturated photosynthesis rates for leaves growing in full sunlight compared with leaves growing in shade, arising from different investments in light utilization and Rubisco. These investments require nitrogen, but plants that invest too much in photosynthetic capacity are at a competitive disadvantage if the machinery is underutilized, such as occurs in a shaded understory. Optimization can be achieved if plants allocate nitrogen to leaves in such a way that the Rubisco- and RuBP-limited rates are balanced (Chen et al. 1993). A broader principle is that nitrogen is optimally allocated among carboxylation, electron transport, and light capture. However, the optimal allocation of nitrogen for photosynthesis cannot be considered independent of stomatal conductance (Farquhar et al. 2002; Buckley et al. 2002, 2013). The model of Buckley et al. (2013), for example, simultaneously adjusts $V_{c\,max}$, J_{max}, leaf absorptance, and stomatal conductance to attain target marginal costs of photosynthetic carbon gain related to nitrogen and transpiration. Hydraulic limits on transpiration placed by stomatal conductance can affect the optimal distribution of leaf nitrogen in a canopy (Peltoniemi et al. 2012; Buckley et al. 2013).

Xu et al. (2012) and Ali et al. (2016) developed a model that calculates the photosynthetic parameters $V_{c\,max}$ and J_{max} based on optimal allocation of leaf nitrogen for light capture, electron transport, carboxylation, and respiration. Leaf nitrogen is divided into separate pools of structural nitrogen, storage nitrogen, respiratory nitrogen, and photosynthetic nitrogen. The photosynthetic pool is further divided into nitrogen for light capture, nitrogen for electron transport, and nitrogen for carboxylation. Nitrogen is allocated among storage, respiration, and photosynthesis to maximize daily net photosynthesis with the constraints that storage nitrogen is allocated to meet the requirements of new tissue production; nitrogen is allocated to meet the respiration demand; and nitrogen allocation for light capture, electron transport, and carboxylation

is co-limiting. An outcome of the model is that $V_{c\,max}$ and J_{max} are predicted rather than prescribed inputs to the model, and they vary spatially and temporally in response to environmental conditions such as temperature, radiation, and CO_2 concentration.

Interest in optimality theory is widespread, and optimality theory is being used to model not just plant canopies, but also stomatal conductance (Chapter 12) and other biological processes. The premise is that natural selection acts to optimize behavior and that optimization can be seen in plant processes or traits. An optimality model is constructed in terms of costs and benefits and seeks a solution that maximizes benefits relative to costs. A system that performs in such a manner is said to be optimized or acting optimally. This differs from another usage of the term optimization, in which an optimal parameter set is one that minimizes errors in model output compared with observations. Optimality theory provides a framework in which plant functioning emerges from theory rather than being parameterized with empirical or semiempirical relationships. The predictive power of optimality theory is seen, for example, in new models of gross primary production (Wang et al. 2014, 2017). Yet not all systems may be acting optimally; some may be better considered as optimizing (i.e., approaching optimality) rather than being optimal per se.

15.7 | Modeling Projects

1 Calculate the canopy scaling factors for $V_{c\,max}$ as in Figure 15.5. Include the calculation of K_b in relation to leaf angle distribution and zenith angle. Examine how the scaling factors vary with zenith angle (compare small and large zenith angles) and leaf angle distribution (compare vertical, horizontal, and spherical leaves). How does foliage clumping affect the scaling factors?

2 Write a program that uses the Rutter et al. (1971, 1975) interception model. Calculate interception with and without storage on trunks. How does the presence of stems affect interception?

3 Calculate optimal leaf area index using (15.27) for various values of $I_0^{\downarrow}$, K_b, A_{max}, and R_d. What is the maximum leaf area index possible?

Scalar Canopy Profiles

Chapter Overview

Vertical profiles of temperature, water vapor, CO_2, and other scalars within plant canopies reflect a balance between turbulent transport and the distribution of scalar sources and sinks. Scalar sources and sinks depend on leaf biophysical and physiological processes, as well as processes at the soil surface. Vertical transport depends on turbulence within the canopy. This chapter develops the theory and mathematical equations to calculate scalar profiles in plant canopies. A common approach calculates scalar concentrations from one-dimensional conservation equations with a first-order turbulence closure in an Eulerian framework. Analytical solutions are possible with some assumptions. A more general numerical method requires an iterative solution because of the coupling between fluxes and concentrations. An implicit solution is possible for temperature and water vapor using the leaf energy budget as an additional constraint. An alternative approach represents fluid transport with a Lagrangian framework, either directly through use of a dispersion matrix to characterize turbulent motion or through use of localized near-field theory.

16.1 | Introduction

Multilayer canopy models require profiles of wind speed, air temperature, water vapor, and CO_2 within the canopy. However, the conservation equation that describes the time rate of change in a scalar contains the mean scalar concentration c (known as the first moment in statistics), the turbulent scalar flux $\overline{w'c'}$ (i.e., the covariance of the scalar concentration and the vertical velocity; a second moment term), and, depending on complexity, other higher-order terms. The closure problem in turbulence is to define the high-order terms in the scalar conservation equation in relation to the lower-order moments. K-theory, which is used to calculate fluxes in the atmospheric surface layer (Chapter 6), is one such turbulence closure. The theory describes turbulent transfer as analogous to molecular diffusion, in which a scalar is transported down concentration gradients in proportion to a turbulent diffusivity. K-theory is a first-order closure; the unknown second-order term (i.e., the turbulent flux $\overline{w'c'}$) is parameterized in terms of the vertical gradient of the first-order moment (i.e., $\partial c/\partial z$). Second-order closure models replace the gradient–diffusion assumption with equations for the second moments, but the equations for the second moments themselves contain unknown third moments. This is the essence of the closure problem.

Canopy models span a large range of complexity in their parameterization of turbulence. The simplest models assume that within-canopy values are the same as at some reference height above the canopy. This is called the well-mixed assumption because air above and within the canopy is assumed to be well mixed. It neglects turbulent transport and can produce large errors in modeled fluxes (Juang et al. 2008; Bonan et al. 2018). Many models use K-theory for its simplicity, but others use higher-order turbulence closure. The Advanced Canopy–Atmosphere–Soil Algorithm (ACASA) multilayer canopy model, for example, uses a third-order turbulence closure, which has been tested for multiple vegetation types (Pyles et al. 2000; Staudt et al. 2010) and has been coupled to an atmospheric model (Xu et al. 2014). Some canopy–chemistry models also use higher-order turbulence closure (Gao and Wesely 1994; Boy et al. 2011). Other multilayer models utilize concepts of Lagrangian theory and track the diffusion of fluid parcels in a canopy (e.g., CANOAK) or use large-eddy simulation (e.g., Patton et al. 2016), but such models are computationally intensive and are not practical for global implementation.

The first multilayer canopy models developed in the 1960s and 1970s used K-theory with a mixing length diffusivity, and that use continues today. Such first-order closure models do not fully represent the complexity of turbulence in plant canopies, but they make the mathematics tractable and even allow for analytical solutions. The resulting wind speed, for example, has a smooth exponential vertical profile, in contrast with observed profiles that show a secondary wind maximum (Figure 15.7, Figure 16.1). K-theory assumes that the length scale of the eddies that produce vertical transport is small compared with the height over which the concentration substantially changes. This requirement that vertical mixing is fine-grained with respect to the concentration gradient is violated in plant canopies because the turbulent length scale is comparable to the canopy height (Finnigan and Raupach 1987; Finnigan et al. 2009). Further evidence for failure of K-theory is seen in counter-gradient fluxes (e.g., Denmead and Bradley 1985). As such, application of K-theory in plant canopies is problematic compared with higher-order closure or Lagrangian models. However, a sharp decrease in wind speed in the upper canopy has been observed in many plant canopies, and exponential profiles approximate this fairly well (Raupach 1988; Raupach et al. 1996). Despite its limitation, first-order closure remains a useful approach to model flow in canopies (Finnigan et al. 2015), and the parameterizations described in the next sections have a long history of use. Comparison of first-order (K-theory) and higher-order models shows comparable results in terms of scalar fluxes above the canopy (heat, water vapor, CO_2), but scalar profiles within the canopy can differ markedly depending on the closure scheme (Juang et al. 2008).

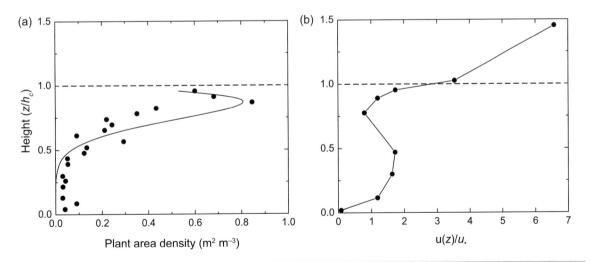

Figure 16.1 Vertical profiles of (a) plant area density and (b) wind speed above and within the canopy of an oak–hickory (*Quercus–Carya*) forest. Height is given as a fraction of canopy height (z/h_c), and the dashed line shows the top of the canopy. Wind speed is normalized by friction velocity measured above the canopy. The solid line shown in (a) is the beta distribution using (2.4). Plant area index is 5.5 m² m⁻², height is 23 m, $p = 8$, and $q = 2$. Adapted from Baldocchi (1989)

16.2 | Wind Profile

Plant canopies exert considerable drag on the flow of air in the surface layer. This is especially evident in forests, where leaves, branches, and trunks impede airflow. The effect of drag is apparent in the reduction of wind speed above the canopy (Chapter 6), and it is also seen in a decrease in wind speed within the canopy (Figure 16.1). The vertical profile of wind speed is derived from the momentum balance equation, which, with some assumptions, has a simple analytical solution in which wind speed decreases exponentially with depth in the canopy (Inoue 1963; Cionco 1965; Brutsaert 1982; Harman and Finnigan 2007).

The equation of motion gives the momentum balance in which for steady state conditions, and in the absence of a pressure gradient, conservation of momentum τ (mol m^{-1} s^{-2}) requires that the vertical gradient $\partial\tau/\partial z$ is balanced by the drag force from vegetation D_f (i.e., the momentum sink with units mol m^{-2} s^{-2}). In this form, the conservation equation for momentum is

$$\frac{\partial \tau}{\partial z} = D_f(z). \tag{16.1}$$

The drag force for vegetation is related to wind speed u by

$$D_f(z) = \rho_m c_d(z) a(z) u^2(z), \tag{16.2}$$

in which a is leaf area density (m^2 m^{-3}) and c_d is the dimensionless leaf aerodynamic drag coefficient (a common value is $c_d = 0.2$) at height z in the canopy. In this equation for D_f, leaf area is given as the planform area (the projected area viewed from above), which can be taken as the one-sided leaf area. Equations (16.1) and (16.2) can be solved for u, provided there is an equation for τ. A convenient expression is to relate τ to the vertical gradient in wind speed using K-theory, similar to that given by (6.11) in Chapter 6 for the momentum flux above the canopy, so that

$$\tau(z) = \rho_m K_m(z) \frac{\partial u}{\partial z}. \tag{16.3}$$

Combining (16.1), (16.2), and (16.3) gives

$$\frac{\partial}{\partial z}\left[K_m(z)\frac{\partial u}{\partial z}\right] = c_d(z) a(z) u^2(z), \tag{16.4}$$

and expanding the $\partial/\partial z$ term:

$$K_m(z)\frac{\partial^2 u}{\partial z^2} + \frac{\partial K_m}{\partial z}\frac{\partial u}{\partial z} = c_d(z) a(z) u^2(z). \tag{16.5}$$

Equation (16.5) is a second-order ordinary differential equation and can be solved with numerical methods. Boundary conditions are a specified wind speed at the top of the canopy and zero wind speed at the ground.

In this derivation, (16.3) closes the conservation equation by relating τ to u. Mixing length theory provides a convenient expression for K_m, similar to that above the canopy (Chapter 6), in which

$$K_m(z) = l_m^2 \frac{\partial u}{\partial z}, \tag{16.6}$$

with l_m (m) the mixing length in the canopy. Observations suggest that l_m is constant in plant canopies, and one simple parameterization is to assume l_m is proportional to canopy height. A more complex expression links l_m to canopy structure and the flow regime using an analytical solution for (16.4), which is described as follows.

If leaf area density and the drag coefficient are constant with height in the canopy (i.e., are uniform through the canopy), (16.4) simplifies to

$$\frac{\partial}{\partial z}\left[l_m^2\left(\frac{\partial u}{\partial z}\right)^2\right] = c_d a u^2(z). \tag{16.7}$$

The solution is the exponential equation,

$$u(z) = u(h_c)\exp\left[-\eta(1 - z/h_c)\right], \tag{16.8}$$

where $u(h_c)$ is wind speed at the canopy height h_c, and

$$\eta = h_c\left(\frac{c_d a}{2 l_m^2}\right)^{1/3}. \tag{16.9}$$

Equation (16.8) relates wind speed to height in the canopy such that $u \to 0$ as $z \to -\infty$. The parameter η is an attenuation factor. In practice, it is estimated empirically by fitting (16.8) to measurements of wind speed and has values of about 2–4 (Thom 1975; Cionco 1978; Brutsaert 1982). Figure 16.2 illustrates this relationship for different values of η. In a dense canopy with leaf area index 5 m^2 m^{-2}, there is a sharp decline in wind speed in the upper one-third of the canopy.

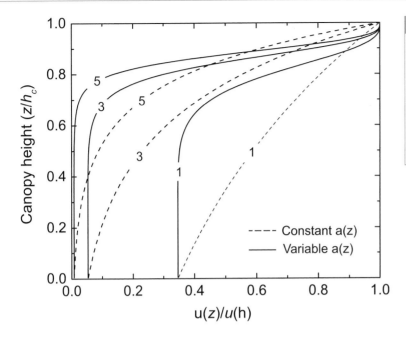

Figure 16.2 Wind-speed profiles in plant canopies for constant foliage distribution (dashed lines) and variable foliage distribution with maximum foliage density at approximately $z/h_c = 0.8$ (solid lines). $c_d = 0.2$ and $u_*/u(h_c)$ is approximately equal to 0.3 using (16.23). Data are shown for a 20 m canopy with leaf area index of 1, 3, and 5 m^2 m^{-2} ($\eta = 1.1$, 2.9, and 4.9, respectively). See also Massman (1997).

An expression for l_m can be obtained by considering continuity of momentum at the top of the canopy. The momentum flux in the surface layer above the canopy is constant with height as given by $\tau = \rho_m u_*^2$ (Chapter 6), and so $\rho_m u_*^2$ is also the momentum flux at the top of the canopy. An alternative expression can be obtained from (16.1) and (16.2), in which the momentum flux at height z in the canopy is

$$\tau(z) = \rho_m \int_{-\infty}^{z} c_d(z')a(z')u^2(z')\,dz'. \tag{16.10}$$

If a and c_d are constant, integration gives

$$\tau(z) = \rho_m \frac{h_c c_d a}{2\eta} u^2(h_c) \exp\left[-2\eta(1 - z/h_c)\right]. \tag{16.11}$$

The momentum flux at the canopy top (i.e., at $z = h_c$) is

$$\tau(h_c) = \rho_m \frac{h_c c_d a}{2\eta} u^2(h_c). \tag{16.12}$$

Continuity at the top of the canopy requires that $\tau(h_c) = \rho_m u_*^2$. Relating the two equations gives

$$\eta = \frac{h_c c_d a}{2u_*^2/u^2(h_c)}. \tag{16.13}$$

Mixing length is obtained by equating (16.9) and (16.13) whereby

$$l_m = \frac{h_c}{\eta}\frac{u_*}{u(h_c)} = 2\beta^3 L_c. \tag{16.14}$$

The right-most form of (16.14) uses the notation of Harman and Finnigan (2007, 2008) with

$$L_c = \frac{1}{c_d a} \tag{16.15}$$

the canopy length scale and $\beta = u_*/u(h_c)$ as in Chapter 6. The lower limit of $-\infty$ rather than zero for the integration in (16.10) requires that all of the momentum is absorbed by the canopy rather than as stress on the ground. This is reasonably valid for a uniform, dense, homogenous canopy.

The coupling of wind speed and momentum above and within the canopy provides equations for displacement height d and roughness length z_{0m} needed to calculate surface fluxes (Chapter 6). Displacement height can be estimated from the within-canopy momentum profile as the effective level of the mean drag on the canopy (Massman 1997; Harman and Finnigan 2007). This is given by

$$h_c - d = \int_{-\infty}^{h_c} \tau(z)/\tau(h_c)\, dz = \int_{-\infty}^{h_c} \exp\left[-2\eta(1 - z/h_c)\right] dz,$$

$$(16.16)$$

from which

$$h_c - d = \frac{h_c}{2\eta} = \frac{l_m}{2\beta} = \beta^2 L_c. \qquad (16.17)$$

Displacement height depends on canopy structure through canopy height h_c and the canopy length scale L_c as defined by leaf area density a and the leaf drag coefficient c_d. Roughness length is calculated as in (6.79) from the logarithmic wind profile evaluated at the canopy height (i.e., $z = h_c$), where the wind speed is equal to $u(h_c)$. Both d and z_{0m} additionally depend on $\beta = u_*/u(h_c)$. A representative value is 0.3–0.5 in neutral conditions, but its value varies with canopy structure and atmospheric stability (Harman and Finnigan 2007; Belcher et al. 2012). Under neutral conditions and with $u_*/u(h_c) = 0.5$, then $d/h_c = 0.79$ and $z_{0m}/h_c = 0.09$ for a 20 m canopy with $c_d = 0.2$ and $a = 0.3$ m^2 m^{-3}.

The preceding derivation of wind speed requires uniform vertical profiles of leaf area density and the leaf drag coefficient in the canopy. In fact, however, leaf area density is not constant in the canopy (Figure 2.2), and wind speed, displacement height, and roughness length vary with foliage distribution. Massman (1997) extended the model to allow for variation with depth using the cumulative leaf drag area $\xi(z)$ instead of the normalized height z/h_c. In this parameterization of wind speed,

$$u(z) = u(h_c) \exp\left\{-\eta\left[1 - \frac{\xi(z)}{\xi(h_c)}\right]\right\}. \qquad (16.18)$$

The momentum profile in the canopy is

$$\frac{\tau(z)}{\tau(h_c)} = \exp\left\{-2\eta\left[1 - \frac{\xi(z)}{\xi(h_c)}\right]\right\}, \qquad (16.19)$$

and displacement height is

$$h_c - d = \int_0^{h_c} \exp\left\{-2\eta\left[1 - \frac{\xi(z)}{\xi(h_c)}\right]\right\} dz. \qquad (16.20)$$

The cumulative leaf drag area at height z is

$$\xi(z) = \int_0^z \frac{c_d(z')a(z')}{P_m(z')}\, dz'. \qquad (16.21)$$

Here, leaf area density a, leaf drag coefficient c_d, and the foliage shelter factor P_m are each evaluated at height z. The shelter factor represents the effects from aerodynamic interference of neighboring leaves in the canopy, in which leaves block or reduce the exposure of nearby leaves to wind. P_m equals one if there are no sheltering effects and is greater than one when neighboring leaves reduce exposure to wind. These three variables define the aerodynamic structure of the canopy. If they are constant with height, $\xi(z) = z c_d a/P_m$ and $\xi(z)/\xi(h_c) = z/h_c$, similar to the wind profile given previously in (16.8). If c_d is constant and ignoring shelter effects (so that $P_m = 1$), then $\xi(h_c)$ is equal to c_d multiplied by leaf area index L. The attenuation coefficient η is

$$\eta = \frac{\xi(h_c)}{2\beta^2}. \qquad (16.22)$$

Equation (16.22) is equivalent to (16.13) for constant c_d and $a = L/h_c$. Massman (1997) parameterized β using the relationship

$$\frac{u_*}{u(h_c)} = c_1 - c_2 \exp\left[-c_3\xi(h_c)\right], \qquad (16.23)$$

with $c_1 = 0.320$, $c_2 = 0.264$, and $c_3 = 15.1$ constants. These equations have been extended for meadow vegetation (Wohlfahrt and Cernusca 2002).

The Massman (1997) model produces, similar to uniform foliage distribution, a monotonic decline in wind speed (Figure 16.2). In this example, maximum leaf area density occurs at a height of about $0.8 h_c$, and there is consequently a sharper decline in wind speed in the upper canopy compared with uniform foliage distribution. Figure 16.3 illustrates the dependency of displacement height and roughness length on leaf area and its distribution in the canopy. The ratio d/h_c is about 0.4–0.9, with highest values in canopies with large amounts of leaf area. The leaf area density profile affects displacement height, particularly when leaf area is largest in mid-canopy. Canopies with foliage concentrated in the upper canopy have larger displacement height. Similarly, z_{0m}/h_c varies with leaf area index. Maximum values range from about 0.08 to 0.13 in sparse canopies depending on foliage distribution, decreasing to less than 0.05 in dense canopies.

Equation (16.8) for wind speed is easy to implement in a model with η specified as a parameter.

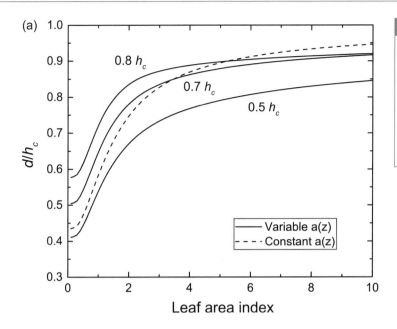

(a)

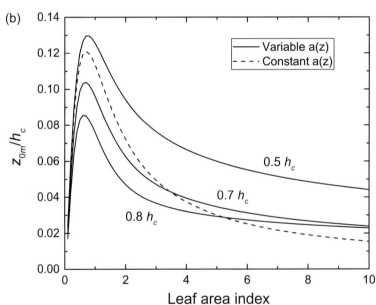

(b)

Figure 16.3 Effect of canopy structure on (a) displacement height d/h_c and (b) roughness length z_{0m}/h_c. Calculations use the parameterization of Massman (1997) for constant and variable leaf area density. Variable leaf area density is shown for the three canopies depicted in Figure 2.3 with maximum leaf area density at $0.5h_c$, $0.7h_c$, and $0.8h_c$. The canopy height is 20 m and $c_d = 0.2$.

Equation (16.18) extends the derivation to include variable foliage distribution but relies on an empirical parameterization of β to obtain η. Harman and Finnigan (2007, 2008) used an alternative form of the equation with a parameterization of β that accounts for the roughness sublayer (Chapter 6). In their notation,

$$u(z) = u(h_c) \exp\left(\frac{z - h_c}{l_m/\beta}\right), \qquad (16.24)$$

in which $z - h_c$ is the deviation from the canopy height (positive above the canopy; negative below the canopy). In this form, the wind profile is given by the height $z - h_c$ normalized by the length scale

l_m/β. Equation (16.24) relates to (16.8) with $l_m/\beta = h_c/\eta$. Mixing length is obtained from $l_m = 2\beta^3 L_c$, as in (16.14), and relates to canopy aerodynamic structure as defined by $L_c = (c_d a)^{-1}$. The length scale l_m/β determines the attenuation of wind speed in the canopy, provides displacement height through the expression $l_m/\beta = 2(h_c - d)$ as in (16.17), and is a key length scale in the roughness sublayer theory. l_m/β is the dominant scale of shear-driven turbulence generated at or near the top of the canopy and is equal to $u/(\partial u/\partial z)$ at $z = h_c$. Equation (16.24) requires no assumption that leaf area density is uniform, only that L_c is constant; indeed, this is thought to be more conservative than either leaf area density or leaf drag coefficient separately (Harman and Finnigan 2007).

Equation (16.6) gives the vertical profile of K_m within the canopy. With (16.24) for $u(z)$, so that $\partial u/\partial z = u(z)\beta/l_m$, K_m is obtained from

$$K_m(z) = 2\beta^3 L_c u_* \exp\left(\frac{z - h_c}{l_m/\beta}\right). \tag{16.25}$$

In this expression, $K_m(h_c) = 2\beta^3 L_c u_*$ is the eddy diffusivity at the top of the canopy. This equation for K_m is consistent with the exponential wind profile. If K_m is constant in the canopy or varies differently, other expressions for the wind profile are obtained (Cowan 1968; Thom 1971, 1975; Brutsaert 1982).

Question 16.1 The derivation of wind profiles above the canopy in Chapter 6 and within the canopy as given in this chapter both use K-theory with eddy diffusivity K_m parameterized by the mixing length l_m. Give the formulas for l_m above and within the canopy. What is a key difference in these formulas?

Question 16.2 A canopy model uses (16.5) to calculate wind speed. Another model uses (16.24) with (16.14) for l_m. Explain the differences between these two approaches.

Question 16.3 Calculate η for a canopy with height $h_c = 20$ m, leaf area index $L = 5$ m^2 m^{-2}, and drag coefficient $c_d = 0.2$. Compare results for $u_*/u(h_c) = 0.3$ and 0.5. How do these compare with estimates of 2–4 for η?

Question 16.4 Eddy diffusivity at the top of the canopy can be obtained from $K_m(h_c) = ku_*(h_c - d)\phi_m^{-1}[(h_c - d)/L_{MO}]$ using surface flux theory (6.51). Prove that this is the same expression as $K_m(h_c) = 2\beta^3 L_c u_*$ in (16.25). Hint: Review (6.83) and (6.84) in Chapter 6.

Question 16.5 Equation (16.20) provides displacement height in a canopy with variable foliage distribution. In this equation, $\xi(z)/\xi(h_c)$ simplifies to z/h_c for a uniform canopy. Equation (16.17) is also an expression for displacement height obtained from (16.16) for a uniform canopy. Are (16.16) and (16.20) equivalent for a uniform canopy?

16.3 | Scalar Continuity Equation

Profiles of temperature, water vapor, CO_2, and other scalars in a plant canopy can be calculated from the one-dimensional continuity equation for a scalar. Conservation requires that the change in amount of material $\rho_m c$ (mol m^{-3}) or heat $\rho_m c_p \theta$ (J m^{-3}; here θ denotes potential temperature) in a unit volume of canopy airspace over a unit of time is equal to the difference between the scalar flux into the volume and the flux out of the volume plus the source flux (or minus the sink flux) from vegetation within the canopy volume. With reference to Figure 16.4, the amount of scalar c (mol mol^{-1}) in a volume of air with dimensions $\Delta x \Delta y \Delta z$ (m^3) and molar density ρ_m (mol m^{-3}) is $\rho_m c \Delta x \Delta y \Delta z$. The flux $F_{c,in}$ (mol m^{-2} s^{-1}) flows into the volume across a cross-sectional area $\Delta x \Delta y$. Similarly, an amount of material equal to $F_{c,out} \Delta x \Delta y$ flows out. The source (or sink) term within the canopy volume is S_c (mol m^{-3} s^{-1}), with a positive flux denoting a source and a negative flux denoting a sink. Conservation requires that

$$\rho_m \frac{\Delta c}{\Delta t} \Delta x \Delta y \Delta z = F_{c,in} \Delta x \Delta y + S_c \Delta x \Delta y \Delta z - F_{c,out} \Delta x \Delta y,$$

$$\tag{16.26}$$

which simplifies to

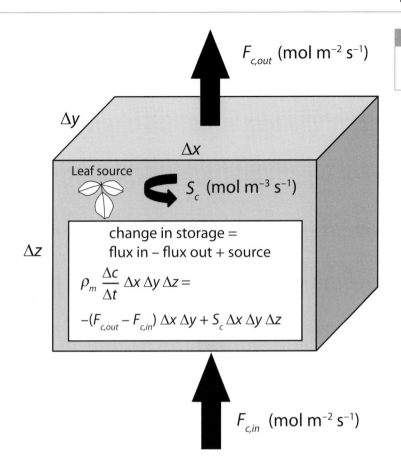

$F_{c,out}$ (mol m^{-2} s^{-1})

Δy

Δx

Leaf source

S_c (mol m^{-3} s^{-1})

Δz

change in storage =
flux in − flux out + source

$$\rho_m \frac{\Delta c}{\Delta t} \Delta x \, \Delta y \, \Delta z =$$

$$-(F_{c,out} - F_{c,in}) \, \Delta x \, \Delta y + S_c \, \Delta x \, \Delta y \, \Delta z$$

$F_{c,in}$ (mol m^{-2} s^{-1})

Figure 16.4 Conservation for a scalar c (mol mol^{-1}) in a canopy volume with dimensions $\Delta x \Delta y \Delta z$ given the vertical fluxes $F_{c,in}$ and $F_{c,out}$ and the vegetation source flux S_c.

$$\rho_m \frac{\Delta c}{\Delta t} = -\frac{\Delta F_c}{\Delta z} + S_c \tag{16.27}$$

or, in the notation of calculus,

$$\rho_m \frac{\partial c(z)}{\partial t} + \frac{\partial F_c}{\partial z} = S_c(z). \tag{16.28}$$

With the assumption that $\partial c / \partial t = 0$, which occurs over sufficient time averaging so that conditions are steady, the one-dimensional scalar continuity equation simplifies to

$$\frac{\partial F_c}{\partial z} = S_c(z). \tag{16.29}$$

Equation (16.29) shows that the change in scalar flux across some height (i.e., the flux divergence) is equal to the source strength at that height. The soil flux must also be considered so that the turbulent scalar flux at height z in the canopy is found by integrating S_c and adding the soil flux $F_{c,0}$ whereby

$$F_c(z) = F_{c,0} + \int_0^{h_c} S_c(z) dz. \tag{16.30}$$

Similar equations pertain to sensible heat but replace c with $c_p \theta$.

Equation (16.29) can be used to calculate the scalar concentration profile, provided there exists an expression for F_c. One simple approach is first-order closure (K-theory), as described previously for momentum, so that

$$F_c(z) = -\rho_m K_c(z) \frac{\partial c}{\partial z}, \tag{16.31}$$

with K_c the scalar diffusivity (m^2 s^{-1}). Combining (16.29) and (16.31) provides the scalar concentration balance in the form

$$\rho_m \frac{\partial}{\partial z} \left[K_c(z) \frac{\partial c}{\partial z} \right] = -S_c(z), \tag{16.32}$$

and expanding the $\partial/\partial z$ term gives the second-order ordinary differential equation,

$$K_c(z)\frac{\partial^2 c}{\partial z^2} + \frac{\partial K_c}{\partial z}\frac{\partial c}{\partial z} = -\frac{S_c(z)}{\rho_m}. \quad (16.33)$$

This equation can be used to calculate scalar concentration profiles in a canopy given the source flux S_c. It applies to any scalar, such as CO_2 with concentration c (mol mol^{-1}) and CO_2 flux (mol m^{-3} s^{-1}). For water vapor, the concentration is the mole fraction e/P (mol mol^{-1}), and S_c is the flux of evapotranspiration (mol m^{-3} s^{-1}). For temperature, c is replaced with $c_p\theta$ (J mol^{-1}), and S_c is the sensible heat flux (J m^{-3} s^{-1}).

Equation (16.33) is written numerically using a central difference approximation for the spatial derivatives (Appendix A2). For a vertical grid with discrete layers denoted by the subscript i each spaced at an equal distance Δz, the numerical form of (16.33) is

$$K_{c,i}\left(\frac{c_{i-1} - 2c_i + c_{i+1}}{\Delta z^2}\right) + \left(\frac{K_{c,i+1} - K_{c,i-1}}{2\Delta z}\right)\left(\frac{c_{i+1} - c_{i-1}}{2\Delta z}\right)$$
$$= -\frac{S_{c,i}}{\rho_m}. \quad (16.34)$$

Rearranging terms gives an equivalent form in which

$$\rho_m\frac{K_{c,i-1/2}}{\Delta z}(c_{i-1} - c_i) - \rho_m\frac{K_{c,i+1/2}}{\Delta z}(c_i - c_{i+1}) = -S_{c,i}\Delta z \quad (16.35)$$

with

$$K_{c,i\pm1/2} = K_{c,i} \pm \frac{K_{c,i+1} - K_{c,i-1}}{4}. \quad (16.36)$$

Equation (16.35) is similar to expressions for soil temperature (Chapter 5) and soil moisture (Chapter 8) and describes a tridiagonal system of equations. It can be derived directly from principles of diffusion and conservation in a cell-centered grid (Figure 16.5). The first term on the left-hand side of (16.35) represents the flux $F_{c,i-1}$, and the second term is the flux $F_{c,i}$.

A more common form of the scalar conservation equation describes the canopy by a network of conductances and source fluxes at multiple levels in the canopy. The terms $K_c/\Delta z$ in (16.35) are conductances with units m s^{-1}; multiplication by ρ_m converts to mol m^{-2} s^{-1}. An equivalent form is, therefore,

$$g_{a,i-1}(c_{i-1} - c_i) - g_{a,i}(c_i - c_{i+1}) = -S_{c,i}\Delta z, \quad (16.37)$$

in which the conductances are evident by comparison with (16.35). In this notation, the turbulent flux is given by the diffusion equation,

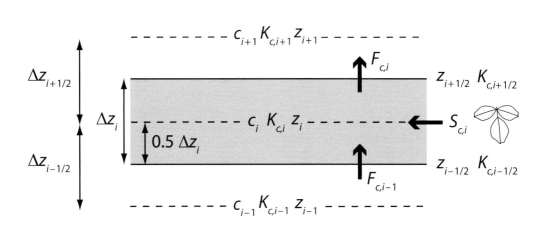

Figure 16.5 Depiction of scalar transport in a plant canopy. Shown are three canopy layers denoted $i-1$, i, and $i+1$. The scalar concentration c_i, diffusivity $K_{c,i}$, and source flux $S_{c,i}$ are defined at the center of layer i at height z_i and are uniform over the layer with thickness Δz_i. Layers $i-1$ and $i+1$ have similar definitions. An effective diffusivity $K_{c,i+1/2}$ is defined at the interface between layers at height $z_{i\pm1/2}$.

$$F_{c,i} = -(c_{i+1} - c_i)g_{a,i}, \tag{16.38}$$

where $g_{a,i}$ is the aerodynamic conductance (mol m^{-2} s^{-1}) between layers i and $i + 1$. A similar equation pertains to the flux $F_{c,i-1}$ between i and $i - 1$ with conductance $g_{a,i-1}$. Dual-source canopy models use (16.37) to calculate surface fluxes, albeit with one leaf layer as seen, for example, in (15.11) for sensible heat flux. The next section provides an expression for K_c and derives a precise definition of the corresponding conductances.

Question 16.6 Equations (16.35) for scalar concentration and (16.36) for scalar diffusivity are for uniform spacing between layers. With reference to Figure 16.5, derive the equivalent equations for nonuniform grid spacing. How does K_c compare with that in (16.36)? Are the two expressions the same for uniform Δz? Why do they differ?

Question 16.7 Equation (16.37) is a tridiagonal system of equations that can be solved for scalar concentration. Write the equation in tridiagonal form. What are the boundary conditions at the top and bottom of the canopy?

16.4 | Aerodynamic Conductance

A convenient expression for scalar diffusivity utilizes mixing length theory so that

$$K_c(z) = l_c l_m \frac{\partial u}{\partial z}. \tag{16.39}$$

This is the same as that given by (16.6) for momentum but with l_c the scalar mixing length. That term can be related to the mixing length for momentum by the Schmidt number Sc. The Schmidt number is a dimensionless quantity used in fluid dynamics to describe the molecular diffusion properties of a fluid and is equal to the ratio of molecular diffusivities for momentum and scalars. In boundary layer meteorology, it is defined as the ratio of the eddy diffusivities whereby

$$\mathrm{Sc} = \frac{K_m(z)}{K_c(z)} = \frac{l_m}{l_c}. \tag{16.40}$$

Equation (16.25) for K_m can then be extended to scalars using

$$K_c(z) = \frac{2\beta^3 L_c u_*}{\mathrm{Sc}} \exp\left(\frac{z - h_c}{l_m/\beta}\right). \tag{16.41}$$

Equation (16.38) is the integrated form of (16.31) and expresses the turbulent flux in proportion to an aerodynamic conductance. This conductance is obtained from the relationship

$$\frac{1}{g_{a,i}} = \frac{1}{\rho_m} \int_{z_i}^{z_{i+1}} \frac{dz}{K_c(z)} \tag{16.42}$$

and

$$\frac{1}{g_{a,i}} = \frac{1}{\rho_m \beta u_*} \mathrm{Sc} \left\{ \exp\left[\frac{-(z_i - h_c)}{l_m/\beta}\right] - \exp\left[\frac{-(z_{i+1} - h_c)}{l_m/\beta}\right] \right\}. \tag{16.43}$$

For thin layers, the integrated conductance $g_{a,i}$ between layers i and $i + 1$ is approximately equal to the finite difference conductance $K_{c,i+1/2}/\Delta z_{i+1/2}$ (Figure 16.5).

K-theory is routinely used to model plant canopies, though models differ in their implementation. Exponential profiles for u (16.24), K_m (16.25), and K_c (16.41) are derived from first-order turbulence closure with constant mixing length in the canopy. Similar equations have been used to parameterize within-canopy wind and aerodynamic conductance in dual-source canopy models (Shuttleworth and Wallace 1985; Shuttleworth and Gurney 1990; Nemitz et al. 2001), land surface models (Dolman 1993; Bonan 1996; Niu and Yang 2004), and hydrologic models (Mahat et al. 2013; Clark et al. 2015c), but with η specified as a model parameter rather than using the notation l_m/β. Some models extend the exponential profile to the ground while others assume a logarithmic profile between the bottom of the canopy and the ground. Exponential profiles have also been used in multilayer canopy models (Choudhury and Monteith 1988; Bonan 1991). Another variant is to assume that mixing length is proportional to canopy height (Katul et al. 2004; Drewry and Albertson 2006; Juang et al. 2008; Launiainen et al. 2011, 2015; Mirfenderesgi et al. 2016). A challenge has been to obtain a unified parameterization of within-canopy transport that is consistent with transport processes above the canopy (Chapter 6). The form of the equations given in this chapter uses the roughness sublayer theory of Harman and Finnigan (2007, 2008), as described in Chapter 6. That theory provides a unified treatment from the ground through the canopy and the roughness sublayer and has been implemented in a multilayer canopy model (Bonan et al. 2018).

Question 16.8 The expressions for K_c and g_a given by (16.41) and (16.43) use the length scale l_m/β from the Harman and Finnigan (2007, 2008) roughness sublayer parameterization. Many models instead use the empirical parameter η. Write the comparable equations in terms of η.

Question 16.9 Describe how (16.42) is used to calculate the two aerodynamic conductances needed for dual-source canopy models, as shown in Figure 15.6a.

Question 16.10 One model solves for scalar concentrations using (16.35). Another model uses (16.37). Describe the differences between these two implementations.

16.5 | Scalar Concentration Profiles

Further understanding of scalar profiles is gained by directly solving for c_i. A solution is obtained by solving the conservation equation $F_{c,i} = F_{c,i-1} + S_{c,i}\Delta z_i$ sequentially through the canopy. This equation states that without storage in the canopy air, the turbulent flux $F_{c,i}$ at layer i is equal to the turbulent flux $F_{c,i-1}$ from below plus the vegetation flux $S_{c,i}\Delta z_i$ at that level. Substituting (16.38) for $F_{c,i}$ relates the scalar concentration at layer i to that of the layer above and additionally to the fluxes from below and within the layer so that

$$c_i = c_{i+1} + \frac{F_{c,i-1} + S_{c,i}\Delta z_i}{g_{a,i}}. \tag{16.44}$$

The flux $F_{c,i-1}$ equals the sum of the source fluxes below layer i so that a general equation for the concentration at i is

$$c_i = c_{i+1} + \left(F_{c,0} + \sum_{j=1}^{i} S_{c,j}\Delta z_j\right)\frac{1}{g_{a,i}}. \tag{16.45}$$

This equation can be solved sequentially downward through the canopy starting at the top ($i = N$) with the boundary condition c_{ref} specified above the canopy.

It is convenient to include the ground flux with the source flux for the first layer so that the total source is $F_{c,0} + S_{c,1}\Delta z_1$. Then, (16.45) can be rewritten as

$$c_i - c_{ref} = \sum_{j=1}^{i-1}\left(S_{c,j}\Delta z_j \sum_{k=i}^{N} g_{a,k}^{-1}\right) + \sum_{j=i}^{N}\left(S_{c,j}\Delta z_j \sum_{k=j}^{N} g_{a,k}^{-1}\right). \tag{16.46}$$

At any level i, the deviation from the reference concentration above the canopy is the weighted sum of the source flux $S_{c,j}\Delta z_j$ for each of the N canopy layers. The weights are determined from the appropriate aerodynamic conductances and represent the contribution of the source flux for each layer to the concentration at layer i. Equation (16.46) can be written in an equivalent form as

$$c_i - c_{ref} = \sum_{j=1}^{N} S_{c,j}\Delta z_j D_{ij}, \tag{16.47}$$

where D_{ij} has the units of resistance (s m^2 mol^{-1}). For a canopy with N layers, there are $N \times N$ weights in which

$$D_{ij} = \begin{cases} \sum_{k=i}^{N} g_{a,k}^{-1} & j = 1 \text{ to } i-1 \\ \sum_{k=j}^{N} g_{a,k}^{-1} & j = i \text{ to } N \end{cases} \tag{16.48}$$

The resulting matrix is known as a dispersion matrix and, as given here, represents only the diffusive contribution to scalar concentration. Lagrangian theory extends the dispersion matrix to include a nondiffusive component. In that usage, D_{ij} more commonly has the units s m^{-1}. This is obtained when S_c is divided by ρ_m for evapotranspiration and CO$_2$ (to have the units mol mol^{-1} m s^{-1}) and by $\rho_m c_p$ for sensible heat flux (to have the units K m s^{-1}).

The equations for scalar concentration require the source flux. They are used most commonly not to calculate scalar concentrations but, rather, in the inverse problem to estimate the vertical profile of sources given an observed concentration profile. The difficulty in solving for scalar concentration is that the source flux depends on the scalar concentration. A general expression for the vegetation source is

$$S_c(z) = \left[\frac{c_\ell(z) - c(z)}{g_{bc}^{-1}(z) + g_{sc}^{-1}(z)} \right] a(z), \tag{16.49}$$

in which g_{bc} and g_{sc} are leaf boundary layer and stomatal conductances (mol m^{-2} s^{-1}), respectively, for scalar c, and c_ℓ is the leaf concentration. This equation is similar to that presented in Chapter 10 for leaf fluxes, but here leaf area density (i.e., the surface area of leaves per unit volume of air; m^2 m^{-3}) scales the leaf flux to a unit volume of canopy air. The leaf area density in a layer of the canopy with thickness Δz that contains ΔL m^2 of leaves per m^2 of ground is $a = \Delta L / \Delta z$. The interdependence of c and S_c necessitates an iterative solution in which an initial estimate for c is used to calculate S_c, from which a new c is obtained. The iteration is repeated until convergence. Katul et al. (2006) described such a numeral solution for CO_2 and photosynthesis.

An analytical solution similar to the exponential wind-speed profile can be obtained for scalar concentrations (Finnigan 2004; Belcher et al. 2008, 2012; Harman and Finnigan 2008; Siqueira and Katul 2010). Unlike wind, the profile of temperature, water vapor, CO_2, and other scalars within a canopy is highly dependent on the vertical distribution of sources and sinks in the canopy. However, an exponential vertical profile is obtained from the equations for conservation of mass and momentum in combination with mixing length closure for the turbulent flux and a simple form of the scalar source flux. A simplified expression for the scalar source flux is

$$S_c(z) = g_{\ell c}(z)[c_\ell(z) - c(z)]a(z), \tag{16.50}$$

with the leaf conductance,

$$g_{\ell c}(z) = \rho_m \frac{c_d(z) r_c}{2} u(z). \tag{16.51}$$

The parameter r_c is the leaf Stanton number (or Nusselt number for heat) and quantifies the difference in transport of scalars and momentum across the leaf boundary layer. A typical value is $r_c = 0.1-0.2$. If leaf area density, drag coefficient, and leaf concentration are constant with height, the scalar conservation equation is

$$\frac{\partial}{\partial z}\left(l_c l_m \frac{\partial u}{\partial z} \frac{\partial c}{\partial z} \right) = \frac{c_d a r_c}{2} u(z)[c(z) - c_\ell]. \tag{16.52}$$

This is similar to that given previously by (16.7) for momentum. Expanding the $\partial / \partial z$ term:

$$l_c l_m \left(\frac{\partial u}{\partial z} \frac{\partial^2 c}{\partial z^2} + \frac{\partial c}{\partial z} \frac{\partial^2 u}{\partial z^2} \right) = \frac{c_d a r_c}{2} u(z)[c(z) - c_\ell]. \tag{16.53}$$

Wind speed is obtained from the previous derivation using (16.24). This replaces $u(z)$ and its spatial derivatives so that (16.53) can be rewritten as the second-order ordinary differential equation,

$$\frac{\beta l_m}{\text{Sc}} \frac{\partial^2 c}{\partial z^2} + \frac{\beta^2}{\text{Sc}} \frac{\partial c}{\partial z} = \frac{c_d a r_c}{2} [c(z) - c_\ell]. \tag{16.54}$$

The solution is the equation,

$$c(z) = c_\ell + [c(h_c) - c_\ell] \exp\left[\frac{f(z - h_c)}{l_m / \beta} \right], \tag{16.55}$$

with

$$f = \frac{1}{2}(1 + 4r_c \text{Sc})^{1/2} - \frac{1}{2}, \tag{16.56}$$

and $c(h_c)$ is the scalar concentration at height h_c.

Equation (16.55) is used along with (16.24) for wind speed in the Harman and Finnigan (2007, 2008) roughness sublayer parameterization described in Chapter 6. The solution differs from that for wind speed given by (16.24) in two respects. First, c approaches c_ℓ (rather than zero) with greater depth in the canopy from the top. The second difference is that the scalar concentration varies more slowly with depth than does wind speed. This is a manifestation of the leaf Stanton number r_c ($f = 0.09$ for $r_c = 0.1$ and Sc $= 1$) and arises from the difference in momentum and scalar fluxes across the leaf boundary layer. The same solution applies to temperature, but replacing the Schmidt number Sc with the Prandtl number Pr. This is similarly defined as the ratio of the diffusivities for momentum and heat so that Pr $=$ Sc.

The derivation of the scalar profile given by (16.55) requires several simplifying approximations to permit analytic tractability. The solution uses a first-order turbulence closure (K-theory) with constant mixing length for diffusivity. The scalar source S_c is a simplification of complex biophysical and physiological processes at the leaf scale. The leaf conductance for heat, for example, is a more complex function of wind speed than the linear form used in (16.51) and varies in relation to $u^{0.5}$ for laminar flow and $u^{0.8}$ for turbulent flow (Chapter 10). The conductance for water vapor additionally depends on stomatal conductance. The boundary condition is specified as a leaf concentration c_ℓ that is invariant with height. Despite these assumptions, (16.55) performs adequately for a variety of canopies (Harman and Finnigan 2007, 2008; Siqueira and Katul 2010).

$$\frac{\rho_m \Delta z_i}{\Delta t} c_p \left(\theta_i^{n+1} - \theta_i^n\right) - g_{a,i-1} c_p \theta_{i-1}^{n+1} + \left(g_{a,i-1} + g_{a,i}\right) c_p \theta_i^{n+1} - g_{a,i} c_p \theta_{i+1}^{n+1} = 2c_p \left(T_{\ell sun,i}^{n+1} - \theta_i^{n+1}\right) g_{bh,i} \Delta L_{sun,i}$$
$$+ 2c_p \left(T_{\ell sha,i}^{n+1} - \theta_i^{n+1}\right) g_{bh,i} \Delta L_{sha,i}$$

$$(16.57)$$

Question 16.11 Derive (16.54). Relate this equation to the scalar conservation equation given by (16.33) with K_c given by (16.41).

16.6 | An Implicit Flux–Profile Solution

For temperature and water vapor, the leaf energy balance provides an additional constraint and can be used with the scalar conservation equation in an implicit solution that avoids the need for iteration. The solution solves a system of linear equations for air temperature, water vapor, and leaf temperature simultaneously at time $n + 1$ from values at time n. Ryder et al. (2016) and Chen et al. (2016b)

2008) to calculate aerodynamic conductance. That implementation is described as follows.

With reference to Figure 16.6, the one-dimensional conservation equation for temperature in a volume of air extending from the ground to some height above the canopy is

The first term on the left-hand side of (16.57) is the storage of heat over the time interval Δt in a layer of air with thickness Δz. The next three terms describe the vertical flux divergence using conductance notation as in (16.37). Aerodynamic conductance is obtained from K_c with (16.42). Above the canopy, K_c is calculated from flux–profile relationships modified by the roughness sublayer (Harman and Finnigan 2007, 2008) as described in Chapter 6. Within the canopy, K_c is specified by (16.41) so that conductance is given by (16.43). The two terms on the right-hand side of (16.57) are the sensible heat flux for sunlit and shaded leaves (Chapter 10) with temperatures $T_{\ell sun}$ and $T_{\ell sha}$, respectively, and scaled by their respective leaf area index ΔL_{sun} and ΔL_{sha}. Equation (16.57) applies above and within the canopy, with $\Delta L = 0$ for layers without vegetation. The comparable equation for water vapor is

$$\frac{\rho_m \Delta z_i}{\Delta t} \left(q_i^{n+1} - q_i^n\right) - g_{a,i-1} q_{i-1}^{n+1} + \left(g_{a,i-1} + g_{a,i}\right) q_i^{n+1} - g_{a,i} q_{i+1}^{n+1} = \left[q_{sat}\left(T_{\ell sun,i}^n\right) + s_i^{sun}\left(T_{\ell sun,i}^{n+1} - T_{\ell sun,i}^n\right) - q_i^{n+1}\right] g_{\ell sun,i} \Delta L_{sun,i}$$
$$+ \left[q_{sat}\left(T_{\ell sha,i}^n\right) + s_i^{sha}\left(T_{\ell sha,i}^{n+1} - T_{\ell sha,i}^n\right) - q_i^{n+1}\right] g_{\ell sha,i} \Delta L_{sha,i},$$

$$(16.58)$$

implemented this solution in a multilayer version of ORCHIDEE. Bonan et al. (2018) revised the solution to separately resolve sunlit and shaded portions of the canopy and also used the roughness sublayer parameterization of Harman and Finnigan (2007,

with the right-hand side representing transpiration from sunlit and shaded leaves. The conductances $g_{\ell sun}$ and $g_{\ell sha}$ are the total leaf conductance to water vapor as in Chapter 10. An expression for $T_{\ell sun}$ is obtained from the leaf energy balance:

$$\frac{c_{L,i}}{\Delta t} \left(T_{\ell sun,i}^{n+1} - T_{\ell sun,i}^n\right) = R_{nsun,i} - 2c_p \left(T_{\ell sun,i}^{n+1} - \theta_i^{n+1}\right) g_{bh,i} - \lambda \left[q_{sat}\left(T_{\ell sun,i}^n\right) + s_i^{sun}\left(T_{\ell sun,i}^{n+1} - T_{\ell sun,i}^n\right) - q_i^{n+1}\right] g_{\ell sun,i}.$$

$$(16.59)$$

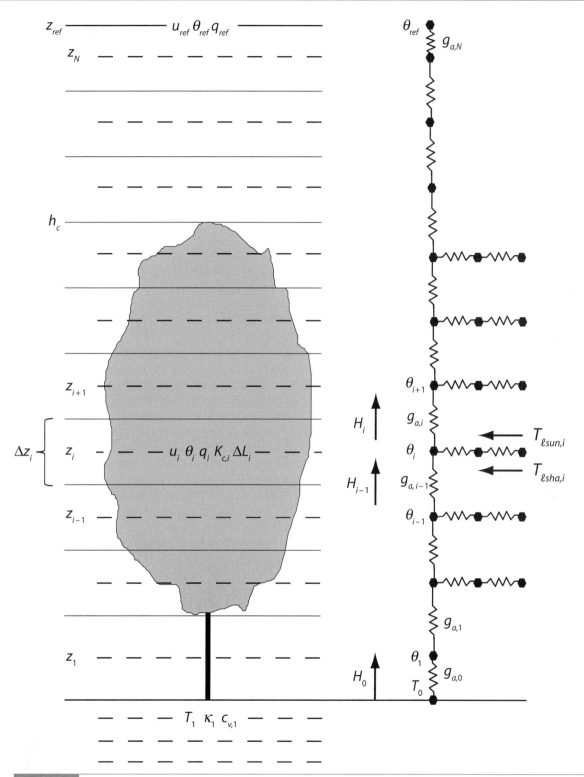

Figure 16.6 Numerical grid used to represent a multilayer canopy. The volume of air from z_{ref} to the ground consists of N layers, each with thickness Δz_i, leaf area index ΔL_i, and leaf area density $a_i = \Delta L_i / \Delta z_i$. Wind speed u_i, temperature θ_i, water vapor concentration q_i, and scalar diffusivity $K_{c,i}$ are centered in each layer at height z_i. The right-hand side of the figure depicts sensible heat fluxes as a network of conductances. The ground is an additional source flux with temperature T_0. Adapted from Bonan et al. (2018)

In this equation, the change in heat storage for a layer of leaves (the left-hand side of the equation) is the balance between net radiation absorbed by the sunlit leaves R_{nsun}, sensible heat flux, and latent heat flux as described in Chapter 10. A comparable equation pertains to shaded leaves with

$$\frac{c_{L,i}}{\Delta t}\left(T_{\ell sha,i}^{n+1} - T_{\ell sha,i}^{n}\right) = R_{nsha,i} - 2c_p\left(T_{\ell sha,i}^{n+1} - \theta_i^{n+1}\right)g_{bh,i} - \lambda\left[q_{sat}\left(T_{\ell sha,i}^{n}\right) + s_i^{sha}\left(T_{\ell sha,i}^{n+1} - T_{\ell sha,i}^{n}\right) - q_i^{n+1}\right]g_{\ell sha,i} .$$

(16.60)

In these equations, leaf transpiration uses the linear approximation,

$$q_{sat}\left(T_{\ell,i}^{n+1}\right) = q_{sat}\left(T_{\ell,i}^{n}\right) + s_i\left(T_{\ell,i}^{n+1} - T_{\ell,i}^{n}\right),$$

(16.61)

with $s_i = dq_{sat}/dT$ evaluate at $T_{\ell,i}^{n}$. In Chapter 10, q_{sat} was linearized about air temperature $(T_\ell - T_a)$ to obtain the Penman–Monteith equation; the equations given here linearize q_{sat} using $T_\ell^{n+1} - T_\ell^{n}$.

Equations (16.57) and (16.58) for scalar conservation and (16.59) and (16.60) for leaf temperatures represent a system of equations that can be solved for θ, q, $T_{\ell sun}$, and $T_{\ell sha}$ at each layer. At the lowest layer above the ground ($i = 1$), the ground fluxes H_0 and E_0 are additional source fluxes, and the ground surface energy balance must be solved to provide the lower boundary conditions with $\theta_{i-1}^{n+1} = T_0^{n+1}$ and $q_{i-1}^{n+1} = q_0^{n+1}$ the ground surface temperature and water vapor, respectively. T_0^{n+1} is calculated from the energy balance equation:

$$R_{n0} = c_p\left(T_0^{n+1} - \theta_1^{n+1}\right)g_{a,0} + \lambda\left\{h_{s0}\left[q_{sat}\left(T_0^{n}\right)\right.\right.$$
$$\left.\left. + s_0\left(T_0^{n+1} - T_0^{n}\right)\right] - q_1^{n+1}\right\}g_{s0} + \frac{\kappa_1}{\Delta z_{1/2}}\left(T_0^{n+1} - T_1^{n}\right).$$

(16.62)

The first term on the right-hand side is the ground surface sensible heat flux. The second term is the latent heat flux, similar to (7.26), with h_{s0} fractional humidity of the first soil layer, $s_0 = dq_{sat}/dT$ evaluated at T_0^{n}, and g_{s0} soil conductance to evaporation consisting of the aerodynamic conductance $g_{a,0}$ and soil surface conductance acting in series. The third term is the soil heat flux, which depends on the thermal conductivity κ_1, thickness $\Delta z_{1/2}$, and temperature T_1 of the first soil layer. Soil evaporation uses the linearization:

$$q_0^{n+1} = h_{s0}q_{sat}\left(T_0^{n+1}\right) = h_{s0}\left[q_{sat}\left(T_0^{n}\right) + s_0\left(T_0^{n+1} - T_0^{n}\right)\right].$$

(16.63)

The numerical solution involves rewriting the leaf energy balance given by (16.59) and (16.60) to obtain expressions for $T_{\ell sun,i}^{n+1}$ and $T_{\ell sha,i}^{n+1}$ and substituting these in the scalar conservation equations (16.57) and (16.58). Equations (16.62) and (16.63) provide T_0^{n+1} and q_0^{n+1} at $i = 1$. This results in a tridiagonal system of equations with the form:

$$a_{1,i}\theta_{i-1}^{n+1} + b_{11,i}\theta_i^{n+1} + b_{12,i}q_i^{n+1} + c_{1,i}\theta_{i+1}^{n+1} = d_{1,i}$$

(16.64)

$$a_{2,i}q_{i-1}^{n+1} + b_{21,i}\theta_i^{n+1} + b_{22,i}q_i^{n+1} + c_{2,i}q_{i+1}^{n+1} = d_{2,i}.$$

(16.65)

Table 16.1 gives the algebraic coefficients. Boundary conditions are θ_{ref}^{n+1} and q_{ref}^{n+1} at some height z_{ref} above the canopy. The system of equations can be solved for θ and q using the method of Richtmyer and Morton (1967, pp. 275–278) as described in Appendix A8.

This solution forms the multilayer canopy model of Bonan et al. (2018) discussed in Chapter 15 (Figures 15.11 and 15.12). The utility of the method is seen by comparing modeled fluxes with those obtained when turbulent transport is neglected. The well-mixed assumption is based on the notion that wind speed, temperature, and water vapor in the canopy airspace are the same as above the canopy (i.e., $u_i = u_{ref}$, $\theta_i = \theta_{ref}$, $q_i = q_{ref}$). This greatly simplifies a canopy model but can give large errors (Juang et al. 2008; Bonan et al. 2018). As shown in Figure 16.7, a multilayer canopy model that uses the well-mixed assumption has large biases in sensible and latent heat fluxes. Using a first-order turbulence closure (K-theory) to calculate aerodynamic conductances and solving for u, θ, and q greatly improves model performance. Additional modifications to leaf biophysics and turbulence further improve the model (CLM-ml; Figure 15.11a).

Question 16.12 Equations (16.64) and (16.65) give an implicit solution for θ, q, $T_{\ell sun}$, and $T_{\ell sha}$. What are some limitations that prevent this from being a fully implicit solution?

Table 16.1 Coefficients for (16.64) and (16.65)

$a_{1,i}$	$b_{11,i}$	$b_{12,i}$	$c_{1,i}$	$d_{1,i}$
$1 < i < N$				
$-g_{a,i-1}$	$\dfrac{\rho_m \Delta z_i}{\Delta t} + g_{a,i-1}$ $+g_{a,i}$ $+g_{H,i}^{sun}\left(1-\alpha_i^{sun}\right)$ $+g_{H,i}^{sha}\left(1-\alpha_i^{sha}\right)$	$-g_{H,i}^{sun}\beta_i^{sun} - g_{H,i}^{sha}\beta_i^{sha}$	$-g_{a,i}$	$\dfrac{\rho_m \Delta z_i}{\Delta t}\theta_i^n + g_{H,i}^{sun}\delta_i^{sun} + g_{H,i}^{sha}\delta_i^{sha}$
$i = N$				
same	same	same	0	same $+ g_{a,i}\theta_{ref}^{n+1}$
$i = 1$				
0	same $- g_{a,0}\alpha_0$	same $- g_{a,0}\beta_0$	same	same $+ g_{a,0}\delta_0$

$a_{2,i}$	$b_{21,i}$	$b_{22,i}$	$c_{2,i}$	$d_{2,i}$
$1 < i < N$				
$-g_{a,i-1}$	$-g_{E,i}^{sun}s_i^{sun}\alpha_i^{sun}$ $-g_{E,i}^{sha}s_i^{sha}\alpha_i^{sha}$	$\dfrac{\rho_m \Delta z_i}{\Delta t} + g_{a,i-1} + g_{a,i}$ $+g_{E,i}^{sun}\left(1-s_i^{sun}\beta_i^{sun}\right)$ $+g_{E,i}^{sha}\left(1-s_i^{sha}\beta_i^{sha}\right)$	$-g_{a,i}$	$\dfrac{\rho_m \Delta z_i}{\Delta t}q_i^n$ $+g_{E,i}^{sun}\left[q_{sat}\left(T_{\ell sun,i}^n\right) + s_i^{sun}\left(\delta_i^{sun} - T_{\ell sun,i}^n\right)\right]$ $+g_{E,i}^{sha}\left[q_{sat}\left(T_{\ell sha,i}^n\right) + s_i^{sha}\left(\delta_i^{sha} - T_{\ell sha,i}^n\right)\right]$
$i = N$				
same	same	same	0	same $+ g_{a,i}q_{ref}^{n+1}$
$i = 1$				
0	same $- h_{s0}s_0g_{s0}\alpha_0$	same $- h_{s0}s_0g_{s0}\beta_0$, and with $g_{a,i-1}=g_{s0}$	same	same $+ h_{s0}\left[q_{sat}\left(T_0^n\right) + s_0\left(\delta_0 - T_0^n\right)\right]g_{s0}$

Note: The chapter appendix provides the derivation. In this table, *same* refers to equations for layers $1 < i < N$ with additional terms added or subtracted as noted. $g_{H,i}^{sun} = 2g_{bh,i}\Delta L_{sun,i}$ and $g_{H,i}^{sha} = 2g_{bh,i}\Delta L_{sha,i}$ are sunlit and shaded leaf conductances for sensible heat scaled to the canopy layer. $g_{E,i}^{sun} = g_{\ell sun,i}\Delta L_{sun,i}$ and $g_{E,i}^{sha} = g_{\ell sha,i}\Delta L_{sha,i}$ are similar conductances for evapotranspiration.

16.7 | Localized Near-Field Theory

Lagrangian dispersion models provide an alternative to K-theory. These models compute scalar concentrations by tracking the vertical dispersion of a large number of fluid parcels that are transported with random velocities to represent turbulence. In Lagrangian models, scalar transport depends on the statistics of the turbulence; it does not depend on diffusion along the mean concentration gradient as in K-theory. As applied to plant canopies, individual leaves release a plume of scalar material that disperses throughout the canopy. Considering dispersion only in the vertical direction, Lagrangian theory describes the mean concentration at height z and time t statistically as the ensemble of these plumes whereby

$$c(z,t) = \frac{1}{\rho_m}\int\int P(z,t|z_0,t_0)\,S_c(z_0,t_0)\,dz_0\,dt_0. \quad (16.66)$$

In this equation, P is a probability density function such that $P\,dz$ is the probability that a source parcel released at z_0 and t_0 is observed at another time t and location z. The probability density function can be determined numerically by calculating the trajectories of an ensemble of particles.

In practice, (16.66) is solved numerically in a discrete form. Raupach (1988, 1989a) defined a Lagrangian dispersion matrix that relates scalar source fluxes to concentrations. The concentration at a given height is obtained from

$$c_i - c_{ref} = \sum_{j=1}^{N}\frac{S_{c,j}}{\rho_m}\Delta z_j D_{ij}, \quad (16.67)$$

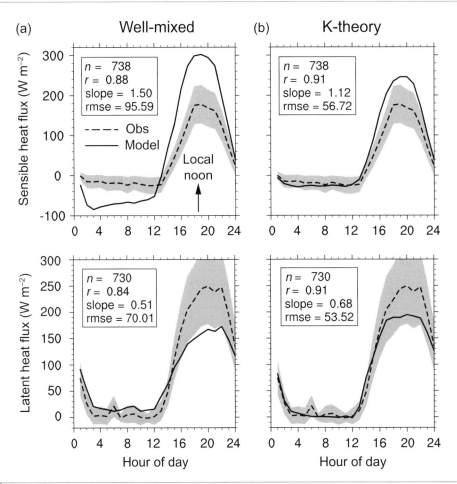

Figure 16.7 Simulations for the University of Michigan Biological Station (US-UMB) flux tower site (July 2006). These simulations are similar to those in Figure 15.11, but for (a) a multilayer canopy model with the well-mixed assumption for within-canopy wind speed, air temperature, and water vapor concentration and (b) the same model but with K-theory turbulence closure. Shown are the average diurnal cycle (GMT) of sensible heat flux and latent heat flux for the observations (dashed line) and model (solid line). Local noon is at approximately 1800. Shading denotes ± 1 standard deviation of the random flux error for the observations. Statistics show sample size (n), correlation coefficient (r), slope of the regression line, and root mean square error (rmse) between the model and observations. Unpublished data for the m0 and m1 simulations described by Bonan et al. (2018)

as given previously in (16.47), in which D_{ij} is the contribution of the source flux in layer j to the concentration of layer i (here with units s m^{-1}). If dispersion is entirely diffusive, D_{ij} relates to aerodynamic conductance as in (16.48). In Lagrangian theory, the dispersion matrix is a discrete form of the transition probability density function and consists of diffusive and non-diffusive components. The dispersion matrix is obtained numerically using a turbulence model that follows the dispersion of an ensemble of fluid parcels in a canopy. A unit source density $S_{c,j}$ is released in layer j with zero source in all other layers so that the simulated

concentration profile is the contribution of dispersion from layer j to the concentration at layer i, and

$$D_{ij} = \frac{c_i - c_{ref}}{S_{c,j}\Delta z_j/\rho_m}. \quad (16.68)$$

This process is repeated to fill in all elements of the matrix. Baldocchi (1992) used a Lagrangian random walk model to numerically simulate single-particle trajectories of several thousand fluid parcels with a stochastic differential equation for fluid parcel motion. The resulting dispersion matrix is used in CANOAK to calculate scalar concentrations

(Baldocchi and Harley 1995; Baldocchi and Meyers 1998; Baldocchi and Wilson 2001).

Use of higher-order turbulence models or numerical random walk simulations to calculate the dispersion matrix allows for incorporation of Lagrangian theory into canopy models but is computationally intensive. Raupach (1989a,b,c) applied Lagrangian concepts to develop the localized near-field theory, which provides an analytical expression for the concentration profile produced by a specified source density profile. The theory describes turbulent transport in relation to the standard deviation of vertical wind velocity σ_w and a Lagrangian time scale T_L. The Lagrangian time scale relates to the persistence of turbulence in plant canopies. With this theory, turbulent dispersion is decomposed into two different processes in the time limits when the travel time $t - t_0$ is smaller than T_L (termed near-field) and when $t - t_0$ is larger than T_L (termed far-field). In the near-field regime, dispersion is dominated by persistence of the turbulent motion (the tendency for air moving in a particular direction to continue moving with similar velocity) and cannot be described by diffusion. In the far-field regime, persistence has little effect; instead, dispersion is dominated by randomness as in diffusion. Each individual leaf in a canopy emits a scalar plume, which progresses though near-field dispersion close to the source and far-field dispersion at some distance from the source. The scalar concentration at any height is the sum of the many individual plumes arriving at that location. Each plume exists between the near-field and far-field limits depending on time of travel. The total scalar concentration is the sum of the far-field component c_f obtained from K-theory and the near-field component c_n that represents the deviation from diffusive behavior so that

$$c(z) = c_f(z) + c_n(z). \tag{16.69}$$

The far-field concentration follows K-theory in which

$$F_c(z) = -\rho_m K_f(z) \frac{\partial c_f}{\partial z}, \tag{16.70}$$

with the far-field diffusivity

$$K_f(z) = \sigma_w^2(z) T_L(z). \tag{16.71}$$

For steady state, the flux at height z relates to the source flux as

$$F_c(z) = F_{c,0} + \int_0^z S_c(z') dz', \tag{16.72}$$

and the far-field concentration is

$$c_f(z) = (c_{ref} - c_{n,ref}) + \frac{1}{\rho_m} \int_z^{z_{ref}} \frac{F_c(z')}{K_f(z')} dz'. \tag{16.73}$$

In practice, c_f is obtained from (16.67) with $c_i = c_{n,i} + c_{f,i}$ and with the dispersion matrix calculated from σ_w and T_L. The near-field concentration profile is more complex and is given by

$$c_n(z) = \frac{1}{\rho_m} \int_0^\infty \frac{S_c(z')}{\sigma_w(z')} \left\{ k_n \left[\frac{z - z'}{\sigma_w(z') T_L(z')} \right] + k_n \left[\frac{z + z'}{\sigma_w(z') T_L(z')} \right] \right\} dz', \tag{16.74}$$

where $k_n(x)$ is a near-field kernel function given by

$$k_n(\xi) = -0.39894 \ln \left(1 - e^{-|\xi|} \right) - 0.15623 e^{-|\xi|}. \tag{16.75}$$

In Lagrangian theory, σ_w and T_L must be parameterized. Raupach (1988, 1989a) used linear profiles (Figure 16.8) in which

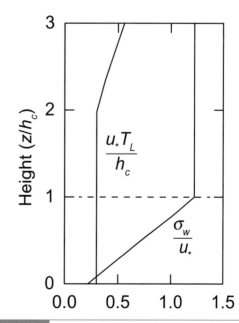

Figure 16.8 Idealized vertical profiles of σ_w and T_L. Adapted from Raupach (1988, 1989a)

$$\frac{\sigma_w(z)}{u_*} = \begin{cases} a_1 & z > h_c \\ a_0 + (a_1 - a_0)z/h_c & z \leq h_c \end{cases} \quad (16.76)$$

and

$$\frac{T_L(z)u_*}{h_c} = \max\left[c_0, \frac{k(z-d)}{a_1 h_c}\right], \quad (16.77)$$

with $a_0 = 0.25$, $a_1 = 1.25$, and $c_0 = 0.3$. In (16.77), k is the von Karman constant, and d is displacement height. The near-field regime critically influences the shape of the concentration profile but is neglected by K-theory. Figure 16.9 shows an example in which there is a large scalar source flux in the upper canopy and a smaller source flux in the understory. The far-field concentration has a smooth profile whereas the near-field concentration is shaped by the source flux and produces a large deviation from the far-field concentration in the upper canopy. The far-field concentration represents the large-scale background variation, and the near-field concentration provides local structure to the concentration profile. Large local gradients in the near-field concentration allow for counter-gradient fluxes.

Localized near-field theory has been used in an inverse sense to infer scalar source distributions from concentration profiles. The theory has also been implemented in plant canopy models in a forward sense to simulate scalar concentrations from source fluxes, but early work found no clear advantage compared with K-theory in simulating turbulent fluxes above canopies or scalar profiles within canopies (Dolman and Wallace 1991; McNaughton and van den Hurk 1995; van den Hurk and McNaughton 1995; Wu et al. 2001; Wilson et al. 2003). Localized near-field theory is used in a variety of models including CABLE with its dual-source canopy parameterization (Haverd et al. 2016) and multilayer models (Gu et al. 1999; Ogée et al. 2003). A difficulty with using the theory in models is that it does not provide a unified understanding for both momentum and scalar fluxes. Instead, the theory requires that vertical profiles of σ_w and T_L be specified. Raupach (1988, 1989a) proposed the linear functions (16.76) and (16.77), but the equations used in models can differ and are mostly heuristic (Raupach et al. 1997; Wilson et al. 2003; Haverd et al. 2009, 2016; Chen et al. 2016b; Makar et al. 2017). While the vertical profile of σ_w can be measured, T_L is more challenging to obtain. The Lagrangian time scale represents the time since emission at which a flux transitions from near-field to far-field dispersion and cannot be directly observed. The profiles of σ_w and T_L must also be adjusted for stability effects, resulting in complex expressions (Leuning 2000). An alternative is to obtain the profiles from higher-order turbulence models. Massman and Weil (1999), for example, extended the theory of Massman (1997) to obtain analytical expressions for σ_w/u_* and $T_L u_*/h_c$ from a second-order closure and this has been used in some canopy models (Ogée et al. 2003; Chen et al. 2016b; Ryder et al. 2016).

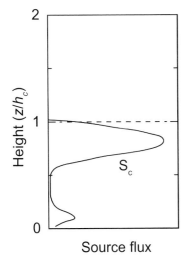
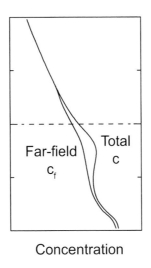

Figure 16.9 Illustration of concentration profiles in the localized near-field theory. Shown is (a) the source flux profile in which most of the emission is in the upper canopy and (b) the far-field concentration c_f and total concentration c. The near-field concentration c_n is the difference $c - c_f$. Adapted from Raupach (1989a).

Some models do not directly resolve near-field dispersion but, rather, represent the near-field effect by multiply the far-field diffusivity by a near-field correction factor. In these models, scalar profiles are calculated using (16.70) with $f(t/T_L)K_f$. Such a correction factor is used in several canopy–chemistry models (Makar et al. 1999; Stroud et al. 2005; Wolfe and Thornton 2011; Bryan et al. 2012; Saylor 2013) in which

$$f(x) = \frac{(1 - e^{-x})(x - 1)^{3/2}}{(x - 1 + e^{-x})^{3/2}}, \tag{16.78}$$

with $x = t/T_L$ the elapsed time since emission relative to the Lagrangian time scale. For $t \gg T_L$, $f \to 1$ and the effective diffusivity is that for the far-field. As t decreases to T_L (i.e., $t/T_L \to 1$), $f \to 0$ and the effective diffusivity decreases because of near-field effects. In practice, however, the correction factor is tuned by choosing an appropriate value for t/T_L (a representative value is four). The multilayer canopy version of ORCHIDEE calculates a similar effective scalar diffusivity, but several empirical parameters were introduced to better match model simulations with observations (Ryder et al. 2016; Chen et al. 2016b).

16.8 | Chapter Appendix – Terms in Table 16.1

Equations (16.64) and (16.65) are obtained by rewriting the leaf energy balance equations (16.59) and (16.60) to solve for $T_{\ell sun,i}^{n+1}$ and $T_{\ell sha,i}^{n+1}$ and substituting these in the scalar conservation equations (16.57) and (16.58). Equation (16.59) for the energy balance of a sunlit leaf can be rearranged in the form

$$T_{\ell sun,i}^{n+1} = \alpha_i^{sun}\theta_i^{n+1} + \beta_i^{sun}q_i^{n+1} + \delta_i^{sun}, \tag{16.79}$$

in which

$$\alpha_i^{sun} = \frac{2c_p g_{bh,i}}{\gamma_i^{sun}} \tag{16.80}$$

$$\beta_i^{sun} = \frac{\lambda g_{\ell sun,i}}{\gamma_i^{sun}} \tag{16.81}$$

$$\delta_i^{sun} = \frac{R_{n sun,i} - \lambda\left[q_{sat}(T_{\ell sun,i}^n) - s_i^{sun}T_{\ell sun,i}^n\right]g_{\ell sun,i} + c_{L,i}T_{\ell sun,i}^n/\Delta t}{\gamma_i^{sun}} \tag{16.82}$$

$$\gamma_i^{sun} = 2c_p g_{bh,i} + \lambda s_i^{sun} g_{\ell sun,i} + c_{L,i}/\Delta t. \tag{16.83}$$

Similar coefficients are found from (16.60) for the shaded leaf to give

$$T_{\ell sha,i}^{n+1} = \alpha_i^{sha}\theta_i^{n+1} + \beta_i^{sha}q_i^{n+1} + \delta_i^{sha}. \tag{16.84}$$

The ground surface energy balance (16.62) is similarly rewritten as

$$T_0^{n+1} = \alpha_0\theta_1^{n+1} + \beta_0 q_1^{n+1} + \delta_0, \tag{16.85}$$

in which

$$\alpha_0 = \frac{c_p g_{a,0}}{\gamma_0} \tag{16.86}$$

$$\beta_0 = \frac{\lambda g_{s0}}{\gamma_0} \tag{16.87}$$

$$\delta_0 = \frac{R_{n0} - \lambda h_{s0}\left[q_{sat}(T_0^n) - s_0 T_0^n\right]g_{s0} + T_1^n \kappa_1/\Delta z_{1/2}}{\gamma_0} \tag{16.88}$$

$$\gamma_0 = c_p g_{a,0} + \lambda h_{s0} s_0 g_{s0} + \kappa_1/\Delta z_{1/2}. \tag{16.89}$$

With these substitutions, the scalar conservation equations (16.57) and (16.58) are rewritten as (16.64) and (16.65) with the algebraic coefficients in Table 16.1.

16.9 | Supplemental Programs

16.1 Implicit Flux–Profile Solution: This program solves (16.64) and (16.65) in the implicit solution for θ, q, $T_{\ell sun}$, and $T_{\ell sha}$ as in Bonan et al. (2018). It uses the Harman and Finnigan (2007, 2008) roughness sublayer parameterization to obtain β. Leaf area density is specified from the two-parameter beta distribution given by (2.4). Wind speed is as in (16.24), and aerodynamic conductance is from (16.43). Leaf boundary layer conductances g_{bh} and g_{bw} are calculated from wind speed (Chapter 10). For convenience, the net radiation absorbed by leaves is specified using an exponential profile from the top of the canopy, and stomatal conductance is treated as a parameter. The complete model of Bonan et al. (2018) calculates solar radiation, longwave radiation, and stomatal conductance as part of the canopy solution.

16.10 | Modeling Projects

1 Write a program to calculate d and z_{0m} with the Massman (1997) model as in Figure 16.3. Compare the solution for variable leaf area density with that for uniform foliage distribution. How do the results compare with Choudhury and Monteith (1988) and Raupach (1994), which were introduced in Chapter 6 (see Question 6.9)? How do the results compare with the simple assumption that $d/h_c = 0.67$ and $z_{0m}/h_c = 0.13$?

2 Write a program that solves for air temperature given a specified profile of sensible heat flux from leaves and at the ground. Use a canopy that is 20 m tall with $\Delta z = 1$ m. Specify leaf fluxes in relation to height using the two-parameter beta distribution similar to that for leaf area density given by (2.4) but with a total canopy flux of 200 W m^{-2}. Calculate conductance using (16.43). (a) Compare results obtained using (16.45) and (16.47) to calculate temperature. Do they give the same answers? (b) Try the tridiagonal solution given by (16.37). (c) Repeat the calculations but now with conductance calculated from localized near-field theory using K_f (16.71) and again including the factor $f(t/T_L)$. How do the different methods for obtaining conductance affect the temperature profile?

3 Use Supplemental Program 16.1 to calculate θ, q, $T_{\ell sun}$, and $T_{\ell sha}$ and also total canopy fluxes of sensible and latent heat. Compare results using the three different conductances in the previous project.

Biogeochemical Models

Chapter Overview

Carbon gain from gross primary production is the single largest term in the terrestrial carbon budget, but the carbon balance is controlled not just by photosynthesis. Allocation of carbon to the growth of leaves, wood, and roots; loss of carbon during autotrophic respiration; and carbon turnover (comprising litterfall, background mortality, and disturbances) are critical determinants of carbon storage. Litter decomposition and resulting soil organic matter formation provide a long-term carbon store. Associated with the flows of carbon through an ecosystem is the parallel flow of nitrogen and other nutrients. This chapter develops the ecological foundation and mathematics to describe ecosystem carbon dynamics using biogeochemical models. The CASA-CNP model is used to illustrate the basic details of biogeochemical models.

17.1 | Introduction

Terrestrial ecosystems can be described in terms of a number of carbon pools and the transfers among these pools. Along a 154-year forest chronosequence in central Manitoba, for example, the forest ecosystem accumulates carbon in live biomass and the forest floor with greater age following fire (Figure 17.1). This is the classical biogeochemical view of an ecosystem. Ecosystems can also be described in terms of community composition, which considers the types of plants present, and population structure, which considers the age and size of individual plants. While broad assemblages of plants can be recognized at the biome scale (e.g., boreal forest, tropical rainforest, savanna, grassland, tundra), many types of species commonly co-occur across the landscape in a particular region. The plants in a particular community have a certain age or size structure. A forest, for example, may consist of trees with the same age that regenerated following fire; or an old-growth forest is likely to have trees of different sizes with a few large canopy dominants and many small trees in the understory. Size structure is an outcome of demographic processes such as regeneration, establishment, growth, competition for resources, and mortality. A class of models known as biogeochemical models ignores the complexities of species composition, size structure, and demography and instead abstracts an ecosystem as pools of carbon and the flows of carbon among these pools. This chapter develops the theory and mathematics of biogeochemical models, and Chapter 18 extends this framework specifically to soil biogeochemistry. Another class of models, known as individual plant or ecosystem demography models, retains the complexity of individual plants or cohorts of similar plants. In these models, ecosystem properties such as carbon storage are the outcome of demographic processes. Chapter 19 considers that class of models.

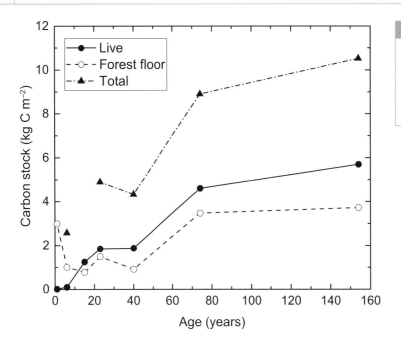

Figure 17.1 Post-disturbance successional trends in the carbon stocks of boreal forests in central Manitoba, Canada. Shown are live biomass, forest floor, and total carbon (the sum of live biomass, forest floor, and coarse woody debris). Adapted from Goulden et al. (2011).

17.2 | Model Structure

Biogeochemical models simulate processes of allocation of photosynthetic carbon gain to plant parts (e.g., foliage, fine root, wood), turnover of plant biomass as litterfall, transformation of litter to soil organic matter, and carbon loss during respiration. They are based on the foundational principles of the following: net carbon input is equal to gross primary production minus autotrophic respiration; carbon flows from donor to receiver pools at a rate that depends on the donor pool size and its chemical quality as modified by the environment; mass balance is maintained as carbon flows through the system of interconnected pools; and decay of litter and soil organic matter releases CO_2 as heterotrophic respiration. This conceptualization leads to a system of first-order, linear differential equations to represent multiple plant, litter, and soil pools (Luo et al. 2003, 2017; Manzoni and Porporato 2009; Sierra et al. 2012; Sierra and Müller 2015). Most models additionally simulate nitrogen dynamics to allow for coupled carbon–nitrogen cycling. These principles are common to all ecosystems, and biogeochemical models can be applied to any type of ecosystem such as grassland, savanna, forest, shrubland, and tundra. Biogeochemical models apply to cropland, too, though with unique biogeochemistry related to planting, harvesting, fertilizer application, and agricultural management. The generality of biogeochemical models has facilitated their use in global models of the terrestrial carbon cycle, and they are the dominant paradigm used in terrestrial biosphere models (Table 1.2).

Figure 17.2 shows the structure of a typical biogeochemical model, in this case the CASA-CNP model (Wang et al. 2010). Nine carbon pools – three plant, three litter, and three soil organic matter – describe the ecosystem. Gross primary production (GPP) provides the input of carbon to the system. Some fraction of this carbon is lost as autotrophic respiration R_A, which consists of separate terms of maintenance respiration R_m (the carbon cost of maintaining existing plant tissues) and growth respiration R_g (the carbon cost of growing new tissues). A simplifying assumption in some biogeochemical models is that autotrophic respiration is 50% of gross primary production. The remaining carbon is the net primary production (NPP; $GPP - R_A$) allocated to leaf, fine root, and wood plant biomass. A portion of the leaf and fine root mass

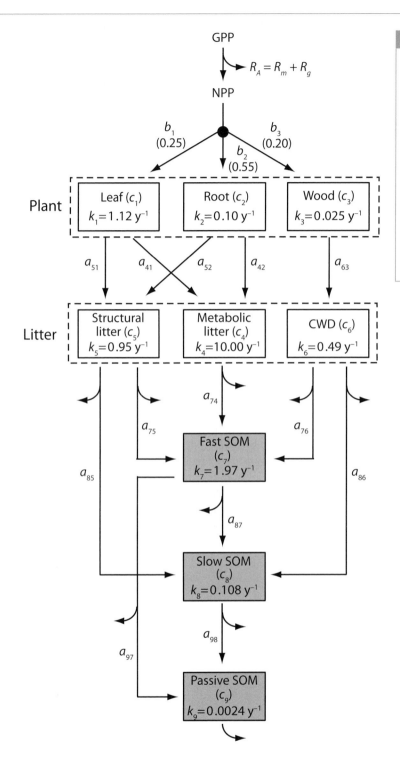

GPP

$R_A = R_m + R_g$

NPP

b_1 (0.25)

b_2 (0.55)

b_3 (0.20)

Plant

Leaf (c_1)
$k_1 = 1.12 \ y^{-1}$

Root (c_2)
$k_2 = 0.10 \ y^{-1}$

Wood (c_3)
$k_3 = 0.025 \ y^{-1}$

a_{51} a_{41} a_{52} a_{42} a_{63}

Litter

Structural litter (c_5)
$k_5 = 0.95 \ y^{-1}$

Metabolic litter (c_4)
$k_4 = 10.00 \ y^{-1}$

CWD (c_6)
$k_6 = 0.49 \ y^{-1}$

a_{74}

a_{75} a_{76}

a_{85}

Fast SOM (c_7)
$k_7 = 1.97 \ y^{-1}$

a_{86}

a_{87}

Slow SOM (c_8)
$k_8 = 0.108 \ y^{-1}$

a_{98}

a_{97}

Passive SOM (c_9)
$k_9 = 0.0024 \ y^{-1}$

Figure 17.2 Structure of the nine-pool CASA-CNP biogeochemical model (Wang et al. 2010). Plant carbon mass consists of leaf c_1, fine root c_2, and wood c_3. Net primary production is allocated to plant material in proportion to b_i. Plant residue becomes metabolic litter c_4, structural litter c_5, or coarse woody debris c_6. These pools decompose to fast c_7, slow c_8, and passive c_9 soil organic matter. Pools differ in base turnover rate k_i. Lines indicate carbon pathways, with a_{ij} the fraction of carbon turnover from pool j that enters pool i after heterotrophic respiration loss. Curved arrows denote heterotrophic respiration fluxes for each pathway. Shown are representative allocation and turnover rates for evergreen broadleaf forest.

turns over and becomes input to litter pools, in this example represented as separate metabolic and structural litter, as explained further in the next chapter. Twigs, branches, tree trunks, large roots, and other woody debris enter the coarse woody debris litter pool. Litter decomposes, during which microbes respire some carbon to the atmosphere as heterotrophic respiration. The remainder becomes soil organic matter, distinguished by fast, slow, and passive pools. The soil pools themselves decompose, with associated heterotrophic respiration and flows to other pools. The carbon residence time varies among pools, from several months for leaves and roots to centuries for soil organic matter.

Biogeochemical models use a system of first-order, linear differential equations (Chapter 4) to mathematically describe carbon pools and fluxes with a typical time step of one day. The system depicted in Figure 17.2 has nine equations. The equation

$$\frac{dc_1}{dt} = b_1 U - \xi_1 k_1 c_1 \qquad (17.1)$$

represents leaf carbon c_1 with units mass per area (e.g., kg C m^{-2}) as the difference between production and litterfall. Here, U is net primary production (mass per area per unit time; e.g., kg C m^{-2} d^{-1}), and b_1 is the fractional allocation of production to leaves. A portion of leaf mass is lost as litterfall, with k_1 the leaf turnover rate (per unit time; e.g., d^{-1}) and ξ_1 a scaling factor that modifies turnover for environmental conditions. Similar equations describe root c_2 and wood c_3 pools as

$$\frac{dc_2}{dt} = b_2 U - k_2 c_2 \qquad (17.2)$$

$$\frac{dc_3}{dt} = b_3 U - k_3 c_3. \qquad (17.3)$$

The partitioning of net primary production varies with light availability, soil temperature, soil moisture, and nutrients. Leaf allocation further varies with time of year depending on phenology. Turnover rates depend on the plant material and are fastest for leaves. Leaf turnover is increased for shedding in cold conditions and with drought stress using the factor ξ_1.

Litterfall is partitioned to three litter pools that vary in chemical quality and turnover rate. The carbon mass of each litter pool is the difference between carbon input from plant residue and carbon loss during decomposition. For metabolic litter x_4:

$$\frac{dc_4}{dt} = a_{41}\xi_1 k_1 c_1 + a_{42} k_2 c_2 - \xi_4 k_4 c_4. \qquad (17.4)$$

Here, $\xi_1 k_1 c_1$ and $k_2 c_2$ are the carbon loss from the leaf and root pools, respectively, and a_{41} and a_{42} are the fractional flow of that carbon to metabolic litter. The term $\xi_4 k_4 c_4$ is the decomposition loss of metabolic litter. Similar equations describe structural litter c_5 and coarse woody debris c_6 as

$$\frac{dc_5}{dt} = a_{51}\xi_1 k_1 c_1 + a_{52} k_2 c_2 - \xi_5 k_5 c_5 \qquad (17.5)$$

$$\frac{dc_6}{dt} = k_3 c_3 - \xi_6 k_6 c_6. \qquad (17.6)$$

The base turnover rates k_4–k_6 are modified for soil temperature and soil moisture through the environmental scaling factors ξ_4–ξ_6, which are the product of a temperature factor $f(T)$ and a moisture factor $f(\theta)$ as described further in Chapter 18. Structural litter is further modified for the lignin fraction of foliage (Table 17.1). Partitioning of leaves to the metabolic and structural pools depends on the lignin-to-nitrogen ratio of the foliage (Table 17.2). The lignin-to-nitrogen ratio is calculated as the product of the lignin fraction of leaf mass f_{lig} (g lignin g^{-1} C) and the C-to-N ratio of leaves (C/N; g C g^{-1} N). Root litter is similarly partitioned between the metabolic and structural pools based on its

Table 17.1	Baseline turnover rates in CASA-CNP
Pool	Rate
Leaf	k_1
Fine root	k_2
Wood	k_3
Metabolic litter	k_4
Structural litter	$k_5 \exp\left(-3 f_{\text{lig}}^{\text{leaf}}\right)$
Coarse woody debris	k_6
Fast soil organic matter	$k_7(1 - 0.75 \times \text{sand})$
Slow soil organic matter	k_8
Passive soil organic matter	k_9

Note: $f_{\text{lig}}^{\text{leaf}}$ is the lignin fraction (g lignin g^{-1} C) of leaves. The sand content of soil is expressed as a fraction.

Table 17.2	Transfer coefficients in CASA-CNP

Leaf → litter
$a_{41} = 0.85 - 0.018 \times (\text{lignin}/\text{N})$ | $a_{51} = 1 - a_{41}$

Root → litter
$a_{42} = 0.85 - 0.018 \times (\text{lignin}/\text{N})$ | $a_{52} = 1 - a_{42}$

Wood → CWD
$a_{63} = 1$

Metabolic litter → SOM
$a_{74} = 0.45$

Structural litter → SOM
$a_{75} = 0.45\left(1 - f_{lig}^{leaf}\right)$ | $a_{85} = 0.7 f_{lig}^{leaf}$

CWD → SOM
$a_{76} = 0.40\left(1 - f_{lig}^{wood}\right)$ | $a_{86} = 0.7 f_{lig}^{wood}$

Fast SOM → SOM
$a_{87} = (0.85 - 0.68 \times \text{sand})(0.997 - 0.032 \times \text{clay})$ | $a_{97} = (0.85 - 0.68 \times \text{sand})(0.003 + 0.032 \times \text{clay})$

Slow SOM → Passive SOM
$a_{98} = 0.45(0.003 + 0.009 \times \text{clay})$

Note: f_{lig}^{leaf} and f_{lig}^{wood} are the lignin fraction (g lignin g^{-1} C) of leaves and wood, respectively. The sand and clay contents of soil are expressed as a fraction.

chemical quality. Woody litter accumulates in the coarse woody debris pool.

The litter pools decompose to three types of soil organic matter that differ in chemical quality and turnover time. The carbon mass of each soil organic matter pool is the difference between carbon input from other pools and carbon loss during decomposition. Metabolic litter decomposes to fast soil organic matter. Structural litter and coarse woody debris decompose to fast and slow pools based on the lignin fraction of the material. Products from the fast and slow pools form the passive pool. The passive pool decomposes without transfers to other pools. For the fast pool c_7,

$$\frac{dc_7}{dt} = a_{74}\xi_4 k_4 c_4 + a_{75}\xi_5 k_5 c_5 + a_{76}\xi_6 k_6 c_6 - \xi_7 k_7 c_7,$$

(17.7)

where a_{ij} is the fractional flow from pool j to pool i. The first three terms on the right-hand side of (17.7) are the carbon products from litter pools c_4, c_5, and c_6 that flow to the fast pool after heterotrophic respiration loss, and the last term is the decomposition loss from the fast pool. Similar equations describe the slow c_8 and passive c_9 pools:

$$\frac{dc_8}{dt} = a_{85}\xi_5 k_5 c_5 + a_{86}\xi_6 k_6 c_6 + a_{87}\xi_7 k_7 c_7 - \xi_8 k_8 c_8$$

(17.8)

$$\frac{dc_9}{dt} = a_{97}\xi_7 k_7 c_7 + a_{98}\xi_8 k_8 c_8 - \xi_9 k_9 c_9.$$ (17.9)

In these equations, $1 - a_{ij}$ is the portion of the decomposition flow lost as heterotrophic respiration. All of the turnover from the passive pool additionally contributes to heterotrophic respiration. As will be described further in the next chapter, a_{ij} consists of two terms – the fraction f_{ij} of the carbon turnover from pool j that goes to pool i and the portion r_{ij} of this carbon that is lost as heterotrophic respiration such that $a_{ij} = (1 - r_{ij})f_{ij}$. The transfer coefficients in (17.7)–(17.9) depend on various factors (Table 17.2). The turnover rates are modified for soil temperature and soil moisture through the environmental scaling factors ξ_7–ξ_9. Turnover of the fast pool additionally depends on the sand fraction of the soil (Table 17.1).

Equations (17.1)–(17.9) are specific equations for each of the nine carbon pools. A generalized equation appropriate for any pool i is

$$\frac{dc_i}{dt} = b_i U + \sum_{\substack{j=1 \\ j \neq i}}^{9} a_{ij}\xi_j k_j c_j - \xi_i k_i c_i. \tag{17.10}$$

The first term on the right-hand side of (17.10) is the carbon gain from net primary production allocated to pool i, the second term is the carbon input to pool i from all other pools, and the third term is the carbon loss from pool i. The coefficient b_i defines the allocation of net primary production to each pool, and a_{ij} defines carbon transfer for each of the 9×9 possible pathways. Non-zero values for a_{ij} denote flows for particular pools. Total heterotrophic respiration is

$$R_H = \xi_9 k_9 c_9 + \sum_{j=4}^{8} \left(1 - \sum_{\substack{i=4 \\ i \neq j}}^{9} a_{ij}\right) \xi_j k_j c_j. \tag{17.11}$$

The total change in ecosystem carbon is the balance between litter inputs and respiration loss so that

$$\sum_{i=1}^{9} \frac{dc_i}{dt} = U - R_H = GPP - R_A - R_H. \tag{17.12}$$

Figure 17.3 shows the carbon dynamics of tropical broadleaf evergreen forest. Most of the carbon stored in the ecosystem accumulates in plant biomass. Equilibrium is attained after a few hundred years. Soil organic matter also accumulates significant carbon, but equilibrium requires several thousand years because of the low turnover rates. After 1000 years, soil carbon is only 87% of its equilibrium value; 3400 years are required to reach 99% of equilibrium. Considerably less carbon is stored as litter and coarse woody debris, both of which quickly attain equilibrium. This example illustrates the long time period needed for ecosystem carbon to reach steady state. Substantially more time to equilibrate is needed in cold boreal and tundra ecosystems with much lower turnover rates.

Question 17.1 A biogeochemical model uses gross primary production (GPP) as input rather than net primary production (NPP) and allocates GPP to leaf, root, and wood biomass with the factors b_1–b_3. Here, however, the allocation coefficients b_i sum to 0.5 rather than to 1 as when NPP is used. Why is this?

Question 17.2 Why do the transfer coefficients of leaves in CASA-CNP sum to 1 (i.e., $a_{41} + a_{51} = 1$), and similarly for roots ($a_{42} + a_{52} = 1$) and coarse woody debris ($a_{63} = 1$)? Why do the transfer coefficients of litter pools and soil organic matter pools not sum to 1?

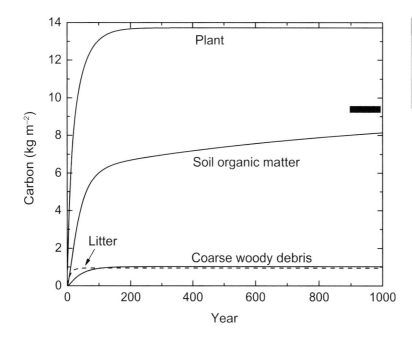

Figure 17.3 Carbon dynamics of evergreen broadleaf forest using parameters in Table 17.3 and Table 17.4 with net primary production equal to 1000 g C m^{-2} y^{-1}. The thick black line denotes soil organic matter at equilibrium.

Question 17.3 Why does the term $\xi_9 k_9 c_9$ appear separately in (17.11) for heterotrophic respiration?

17.3 | Generalized Matrix Form

Equation (17.10) can be written in matrix form (Appendix A6) as

$$\frac{d\mathbf{C}}{dt} = \mathbf{B}U + \mathbf{A}\xi\mathbf{K}\mathbf{C}, \qquad (17.13)$$

in which $\mathbf{C} = [c_1, c_2, \ldots, c_9]^T$ is a 9×1 column vector of carbon pools; $\mathbf{B} = [b_1, b_2, \ldots, b_9]^T$ is a 9×1 column vector describing allocation of net primary production U to each pool; $\mathbf{K}$ is a 9×9 diagonal matrix in which k_{ii} is the turnover rate for pool i and all other elements are zero ($k_{ij} = 0$ for $i \neq j$); ξ is a 9×9 diagonal matrix in which ξ_{ii} is the environmental scalar effect on turnover and all other elements are zero; and $\mathbf{A}$ is a 9×9 carbon transfer matrix in which a_{ij} is the fraction of carbon loss from pool j entering pool i. Equation (17.13) is easily solved for $d\mathbf{C}/dt$, and carbon pools are integrated forward in time from some initial condition. A common numerical technique is the Euler method (Appendix A3) in which carbon pools are stepped forward from time n to $n + 1$ over the time

and

$$\mathbf{K} = \begin{bmatrix} k_1 & 0 & 0 & 0 & 0 & 0 & 0 & 0 & 0 \\ 0 & k_2 & 0 & 0 & 0 & 0 & 0 & 0 & 0 \\ 0 & 0 & k_3 & 0 & 0 & 0 & 0 & 0 & 0 \\ 0 & 0 & 0 & k_4 & 0 & 0 & 0 & 0 & 0 \\ 0 & 0 & 0 & 0 & k_5 & 0 & 0 & 0 & 0 \\ 0 & 0 & 0 & 0 & 0 & k_6 & 0 & 0 & 0 \\ 0 & 0 & 0 & 0 & 0 & 0 & k_7 & 0 & 0 \\ 0 & 0 & 0 & 0 & 0 & 0 & 0 & k_8 & 0 \\ 0 & 0 & 0 & 0 & 0 & 0 & 0 & 0 & k_9 \end{bmatrix}.$$

$$(17.16)$$

The environmental scaling matrix ξ has a similar form as $\mathbf{K}$. The carbon transfer matrix is

$$\mathbf{A} = \begin{bmatrix} -1 & 0 & 0 & 0 & 0 & 0 & 0 & 0 & 0 \\ 0 & -1 & 0 & 0 & 0 & 0 & 0 & 0 & 0 \\ 0 & 0 & -1 & 0 & 0 & 0 & 0 & 0 & 0 \\ a_{41} & a_{42} & 0 & -1 & 0 & 0 & 0 & 0 & 0 \\ a_{51} & a_{52} & 0 & 0 & -1 & 0 & 0 & 0 & 0 \\ 0 & 0 & a_{63} & 0 & 0 & -1 & 0 & 0 & 0 \\ 0 & 0 & 0 & a_{74} & a_{75} & a_{76} & -1 & 0 & 0 \\ 0 & 0 & 0 & 0 & a_{85} & a_{86} & a_{87} & -1 & 0 \\ 0 & 0 & 0 & 0 & 0 & 0 & a_{97} & a_{98} & -1 \end{bmatrix}.$$

$$(17.17)$$

The diagonal elements $a_{ii} = -1$ describe carbon loss from pool i. The off-diagonal elements are non-zero for flows from pool j to pool i and otherwise are zero. The product $\mathbf{A}\xi\mathbf{K}$ is a 9×9 matrix:

$$\mathbf{A}\xi\mathbf{K} = \begin{bmatrix} -\xi_1 k_1 & 0 & 0 & 0 & 0 & 0 & 0 & 0 & 0 \\ 0 & -\xi_2 k_2 & 0 & 0 & 0 & 0 & 0 & 0 & 0 \\ 0 & 0 & -\xi_3 k_3 & 0 & 0 & 0 & 0 & 0 & 0 \\ a_{41}\xi_1 k_1 & a_{42}\xi_2 k_2 & 0 & -\xi_4 k_4 & 0 & 0 & 0 & 0 & 0 \\ a_{51}\xi_1 k_1 & a_{52}\xi_2 k_2 & 0 & 0 & -\xi_5 k_5 & 0 & 0 & 0 & 0 \\ 0 & 0 & a_{63}\xi_3 k_3 & 0 & 0 & -\xi_6 k_6 & 0 & 0 & 0 \\ 0 & 0 & 0 & a_{74}\xi_4 k_4 & a_{75}\xi_5 k_5 & a_{76}\xi_6 k_6 & -\xi_7 k_7 & 0 & 0 \\ 0 & 0 & 0 & 0 & a_{85}\xi_5 k_5 & a_{86}\xi_6 k_6 & a_{87}\xi_7 k_7 & -\xi_8 k_8 & 0 \\ 0 & 0 & 0 & 0 & 0 & 0 & a_{97}\xi_7 k_7 & a_{98}\xi_8 k_8 & -\xi_9 k_9 \end{bmatrix}. \qquad (17.18)$$

interval Δt as

$$\mathbf{C}^{n+1} = \mathbf{C}^n + \frac{d\mathbf{C}}{dt}\Delta t. \qquad (17.14)$$

In the example given in Figure 17.2,

$$\mathbf{B} = [b_1, b_2, b_3, 0, 0, 0, 0, 0, 0]^T \qquad (17.15)$$

Each column represents the loss from pool j, given by the diagonal element $\xi_{jj} k_{jj}$, and the transfer of that carbon to pool i, given by the off-diagonal elements $a_{ij}\xi_{jj}k_{jj}$. The product $(\mathbf{A}\xi\mathbf{K})\mathbf{C}$ is a 9×1 column vector, where each row represents the increment dc_i/dt resulting from internal carbon flows. $\mathbf{B}U$ is a similar 9×1 column vector, with each row representing the increment dc_i/dt resulting from net primary production.

An advantage of the matrix form given by (17.13) compared with the specific equations for each pool given by (17.1)–(17.9) is that it is a generalized form that represents any number of carbon pools and transfers. Figure 1.4, for example, shows a 12-pool biogeochemical model with more complex soil biogeochemistry. A further advantage is that the matrix approach lends itself to formal analysis of the mathematical properties of the system (Sierra and Müller 2015; Luo et al. 2017). Biogeochemical models can require several thousand model years to reach equilibrium, particularly in ecosystems with low turnover rates. Furthermore, net primary production and the parameters in (17.13) vary with environmental conditions (such a system is known in mathematics as a nonautonomous system). However, analysis of the steady-state solution provides key insight to carbon storage capacity. At steady state, $dC/dt = 0$ so that

$$-A\xi KC = BU, \qquad (17.19)$$

and the solution (see Appendix A7) is

$$C = -(A\xi K)^{-1}BU. \qquad (17.20)$$

Time-mean values $\bar{B}$, $\bar{\xi}$, $\bar{K}$, $\bar{A}$, and $\bar{U}$ can be used to approximate the steady-state solution, which provides a convenient means to spin up a model to equilibrium (Xia et al. 2012).

Carbon storage capacity can be decomposed into separate components of net primary production;

carbon residence time, which is separated into baseline turnover rates and environmental factors such as temperature and moisture; and environmental forcings (Xia et al. 2013; Rafique et al. 2016, 2017). This provides a formal mathematical framework known as traceability analysis to analyze model behavior and compare different models. Not all biogeochemical models are written in a form convenient for matrix algebra. Many models instead divide carbon flows into separate processes that are solved individually (and oftentimes sequentially) over a full time step rather than simultaneously. This mathematical technique, formally known as operator splitting, can simplify the numerics of solving the differential equations. However, such models can still be analyzed using a matrix emulator that provides terms for the traceability analysis. For example, the matrix approach can be applied to a 26-pool version of the Community Land Model (CLM4; Rafique et al. 2017) and also to a more detailed version of the model (CLM4.5) that uses vertically resolved soil carbon in 10 soil layers and has 70 carbon balance equations (Huang et al. 2018).

Using the matrix form makes the model states and their prognostic equations more evident and facilitates analysis of model structure, parameters, and forcings (Luo et al. 2017). In CASA-CNP, for example, the carbon allocation matrix **B** varies among biomes depending on phenology and other factors (Table 17.3). Carbon turnover depends on a

Table 17.3 CASA-CNP allocation, turnover, and environmental scaling parameters for evergreen broadleaf forest (EBF), evergreen needleleaf forest (ENF), and deciduous broadleaf forest (DBF)

Pool	EBF			ENF			DBF		
	b_i	k_i (y^{-1})	ξ_i	b_i	k_i (y^{-1})	ξ_i	b_i	k_i (y^{-1})	ξ_i
1-Leaf	0.25	1.12	1.01	0.20	0.59	1.10	0.14	1.09	1.16
2-Root	0.55	0.10	1.00	0.35	0.10	1.00	0.60	0.10	1.00
3-Wood	0.20	0.025	1.00	0.45	0.014	1.00	0.26	0.025	1.00
4-Metabolic litter	0	10.0	0.40	0	10.0	0.17	0	10.0	0.36
5-Structural litter	0	0.95	0.40	0	0.82	0.17	0	0.95	0.36
6-CWD	0	0.49	0.40	0	0.49	0.17	0	0.49	0.36
7-Fast SOM	0	1.97	0.40	0	1.41	0.17	0	2.15	0.36
8-Slow SOM	0	0.108	0.40	0	0.006	0.17	0	0.110	0.36
9-Passive SOM	0	0.0024	0.40	0	0.0015	0.17	0	0.0025	0.36

Source: From Xia et al. (2013).

Table 17.4	CASA-CNP carbon transfer parameters for evergreen broadleaf forest (EBF), evergreen needleleaf forest (ENF), and deciduous broadleaf forest (DBF)		
Parameter	EBF	ENF	DBF
a_{41}	0.67	0.40	0.60
a_{51}	0.33	0.60	0.40
a_{42}	0.58	0.46	0.57
a_{52}	0.42	0.54	0.43
a_{63}	1.00	1.00	1.00
a_{74}	0.45	0.45	0.45
a_{75}	0.36	0.34	0.36
a_{85}	0.14	0.18	0.14
a_{76}	0.24	0.24	0.24
a_{86}	0.28	0.28	0.28
a_{87}	0.39	0.47	0.42
a_{97}	0.006	0.005	0.006
a_{98}	0.003	0.002	0.003

Source: From Xia et al. (2013).

17.4 Allocation and Turnover

In biogeochemical models, partitioning of net primary production to plant components is achieved through allocation parameters (b_1–b_3 in CASA-CNP; Figure 17.2). These parameters determine the amount of carbon that accumulates in plants and so also affect structural characteristics such as leaf area, root system, and plant height. Leaf area and root biomass, in turn, determine photosynthetic capacity and resource acquisition (water, nutrients) to support photosynthesis. Leaves and fine roots have high nitrogen requirement, turn over rapidly, and produce litter of high quality that readily decomposes. Higher allocation to wood leads to lower turnover rates, more carbon accumulation in vegetation, less plant demand for nitrogen, and poor-quality litter that decomposes slowly. A key uncertainty is how allocation varies with changes in light availability (e.g., open versus closed canopy), soil moisture, and nutrients. Do plants allocate more carbon to roots to better acquire water and nutrients, or do they allocate carbon to foliage to increase their photosynthetic capacity? Current approaches to represent allocation in biogeochemical models include time-invariant partitioning coefficients that do not vary with environment (but which vary among biomes), empirical algorithms for variable allocation in response to environment, and optimization that balances multiple resource limitations. De Kauwe et al. (2014) reviewed carbon allocation algorithms in various models.

In CASA-CNP, the carbon allocation coefficients b_1–b_3 are specified by biome. Representative values are: evergreen broadleaf forest – 0.20, 0.45, and 0.35 for leaf, root, and wood; evergreen needleleaf forest – 0.25, 0.35, and 0.40; and deciduous broadleaf forest – 0.35, 0.40, and 0.25 (Xia et al. 2017). Allocation to foliage varies seasonally depending on leaf phenology (e.g., evergreen versus deciduous). Carbon allocation to foliage is 80% during a period of leaf onset; leaves have zero allocation during a senescence phase. Some models may also set a maximum leaf area index to avoid unrealistically high leaf area. To prevent leaf area from obtaining anomalously low or high values, CASA-CNP, for example, increases leaf allocation to 80% if leaf area index is below some minimum or decreases it to zero if leaf area index is above some maximum.

biome-specific base rate **K**, and the environmental matrix ξ varies the turnover rate for environmental conditions so that the effective turnover differs among biomes (Table 17.3). The carbon transfer matrix **A** varies with the lignin/N of leaves and roots (for transfer to litter pools), the lignin fraction of litter pools (for litter to soil pools), and soil texture (among soil pools) so that **A** also varies among biomes (Table 17.4). Much of the details and complexity of biogeochemical models lie in **B**, **K**, ξ, and **A**. Terrestrial ecosystems are thought to currently sequester a significant portion of anthropogenic CO_2 emissions. There is considerable uncertainty as to whether terrestrial ecosystems will continue to be a carbon sink in the future or whether changes in allocation or increases in autotrophic respiration, mortality, soil carbon turnover, and disturbance (e.g., wildfire, insects) will counteract the sink. This behavior is encapsulated in **B**, **K**, ξ, and **A**. In the remainder of this chapter, these matrices are described for plants. The next chapter provides further details on soil carbon.

Question 17.4 Discuss how allocation **B** and carbon transfer **A** affect carbon storage in CASA-CNP.

An alternative is to calculate carbon partitioning based on environmental conditions. Most such approaches increase allocation to roots when soil moisture or nutrient availability are suboptimal, increase allocation to leaves in open canopies where additional leaf area for photosynthesis is beneficial, and increase allocation to wood (stems) in closed canopies where light is limiting and additional leaves may not provide much photosynthetic gain because of low light availability. Such dynamic allocation is commonly parameterized using empirical relationships. An example is from Friedlingstein et al. (1999), who proposed dynamic allocation for the CASA biogeochemical model in which allocation to foliage, roots, and wood is calculated based on light availability, soil moisture, and nitrogen using the relationships:

$$b_1 = 1 - b_2 - b_3 \tag{17.21}$$

$$b_2 = 3r_0 \left[\frac{f_1}{f_1 + 2\min(f_2, f_3)} \right] \tag{17.22}$$

$$b_3 = 3s_0 \left[\frac{\min(f_2, f_3)}{2f_1 + \min(f_2, f_3)} \right]. \tag{17.23}$$

In these equations, $r_0 = s_0 = 0.3$ are carbon allocation parameters for root and wood, respectively, in non-limiting conditions. The terms in brackets adjust allocation for light f_1, soil moisture f_2, and nitrogen f_3 using relative scaling factors (0.1–1) to account for suboptimal conditions. The light factor uses canopy leaf area index L with $f_1 = \exp(-0.5L)$ so that wood allocation increases (and leaf allocation decreases) as the canopy closes. The soil moisture factor is $f_2 = S_e$, where S_e is soil moisture above wilting point relative to available soil moisture (field capacity minus wilting point). The temperature and moisture factors that affect soil organic matter decomposition are used as a surrogate for nitrogen availability so that $f_3 = f(T)f(\theta)$. Use of the minimum of the two belowground resources (water, nitrogen) assumes that only the most limiting of these resources controls allocation and that aboveground (light) and belowground resources (minimum of water or nitrogen) have equal weight. Comparison of fixed allocation and dynamic allocation in CASA-CNP showed a significant difference in the terrestrial carbon cycle between these two approaches (Xia et al. 2017). Arora and Boer (2005a) used a similar approach in the Canadian Terrestrial Ecosystem Model (CTEM) but neglected the nitrogen factor and with

$$b_1 = 1 - b_2 - b_3 \tag{17.24}$$

$$b_2 = \frac{r_1 + p_1(1 - f_2)}{1 + p_1(2 - f_1 - f_2)} \tag{17.25}$$

$$b_3 = \frac{s_1 + p_1(1 - f_1)}{1 + p_1(2 - f_1 - f_2)}. \tag{17.26}$$

The parameters r_1, s_1, and p_1 vary among biomes. Representative parameters for trees are $r_1 = 0.55$, $s_1 = 0.05$, and $p_1 = 0.8$ (Figure 17.4).

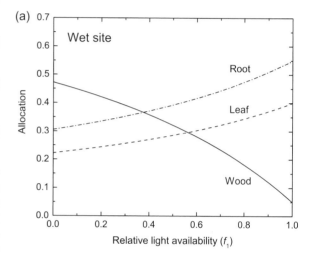

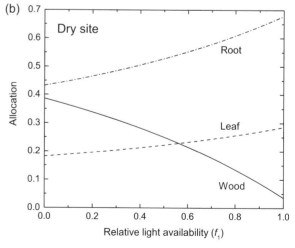

Figure 17.4 Biomass allocation to leaf, root, and wood in relation to relative light availability f_1 for (a) wet ($f_2 = 1$) and (b) dry ($f_2 = 0.5$) soil as in Arora and Boer (2005a).

Optimality theory can be used in biogeochemical models to determine allocation. Optimality models are based on the premise that plants optimally allocate resources to balance light acquisition (foliage), structural support and water transport (stems), and water and nutrient uptake (roots). Thomas and Williams (2014), for example, constructed a model in which carbon and nitrogen are allocated to plant tissues so as to optimize daily productivity based on the marginal change in net carbon or nitrogen uptake resulting from a change in carbon or nitrogen allocation and in which resources equally limit plant growth. In this way, allocation changes based on prevailing conditions such as light availability and soil resources.

Some models calculate allocation based on scaling relationships among plant components. These models maintain specified ratios of foliage, root, and wood biomass. Of particular importance is sapwood cross-sectional area to transport water to leaves and fine root mass to take up water and nutrients. Still other models separate wood into live stems (sapwood) and dead stems, roots into fine roots and coarse roots, and coarse roots into live pools and dead pools to account for the different physiological functioning of these biomass components (Figure 17.5). The Community Land Model (CLM4.5; Oleson et al. 2013) is one such model. Allocation of new growth is in relation to certain biomass ratios: a_1 is the ratio of new fine root to new leaf carbon; a_2 is the ratio of new coarse root to new stem; a_3 is the ratio of new stem to new leaf; a_4 is the portion of new wood that is live; and a_5 is the ratio of growth respiration carbon to new growth carbon. In this approach, the net carbon after maintenance respiration costs are subtracted from gross primary production ($GPP - R_m$) is the carbon allocated to growth. A fraction of the net carbon is lost as growth respiration during production of new biomass so that the net carbon increment is $\Delta C = GPP - R_m - R_g$. Representative values for trees are $a_1 = 1$, $a_2 = 0.3$, $a_3 = 2.2$, and $a_4 = 0.1$. Growth respiration is typically 30% of new growth so that $a_5 = 0.3$. Carbon allocation to foliage growth is

$$b_F = \{(1 + a_5)[1 + a_1 + a_3(1 + a_2)]\}^{-1}, \tag{17.27}$$

and to fine root growth is

$$b_{FR} = a_1 b_F. \tag{17.28}$$

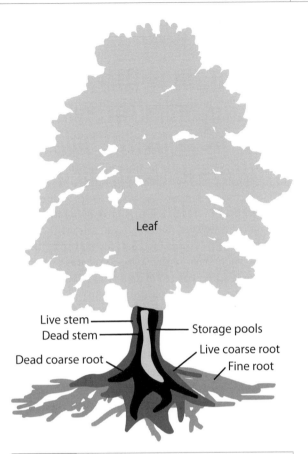

Figure 17.5 Representation of carbon pools among leaf, stem (live and dead), coarse roots (live and dead), fine roots, and storage pools.

Carbon allocation to live and dead stem is

$$b_{LS} = a_3 a_4 b_F \tag{17.29}$$

$$b_{DS} = a_3(1 - a_4)b_F, \tag{17.30}$$

and to live and dead coarse roots is

$$b_{LR} = a_2 a_3 a_4 b_F \tag{17.31}$$

$$b_{DR} = a_2 a_3(1 - a_4)b_F. \tag{17.32}$$

The fraction of carbon used for growth respiration is

$$b_{RG} = a_5(b_F + b_{FR} + b_{LS} + b_{DS} + b_{LR} + b_{DR}). \tag{17.33}$$

The total biomass increment is the carbon gain after autotrophic respiration is subtracted from gross primary production. A simplification is to specify autotrophic respiration as a proportion of gross primary production (e.g., 50%). In fact, autotrophic respiration is an outcome of two distinct physiological processes. Maintenance respiration is the carbon cost to support existing live biomass, and respiration loss increases with higher temperatures. Growth respiration is the CO_2 released during synthesis of new biomass. It is the carbon cost for the growth of new plant material and is independent of temperature. The value of 30% used for a_5 in the preceding allocation scheme is representative, though the precise value is uncertain. Some models, for example, take growth respiration as 25% of $GPP - R_m$; some models use a still lower value.

Maintenance respiration is proportional to the amount of biomass, with a base rate that varies among plant structures. In trees, for example, foliage and fine roots have higher base respiration rates for the same temperature than does wood. Many models relate maintenance respiration to nitrogen content rather than biomass. Ryan (1991) provided an early functional relationship between nitrogen and respiration. The two methods are related by the carbon-to-nitrogen ratio (C/N), with wood have considerably higher C/N (lower respiration) than leaves or fine roots. Some models now use empirical relationships developed from plant trait databases. One such equation for leaf respiration is from Atkin et al. (2015), as shown in (11.68), in which the base rate increases with greater leaf nitrogen (per unit leaf area) and also depends on growth temperature. Another critical difference among models is in the way temperature affects respiration. Respiration increases with warmer temperature. This is commonly represented using a Q_{10} or Arrhenius temperature function (Chapter 4), but the exact form can differ among models. CASA-CNP, for example, uses the relationship of Lloyd and Taylor (1994) in which

$$f(T) = \exp\left[308.56\left(\frac{1}{56.02} - \frac{1}{T - 227.13}\right)\right] \quad (17.34)$$

for temperature T in kelvin. Sitch et al. (2003) used the same function in LPJ (Chapter 18).

Many biogeochemical models include a storage pool of nonstructural carbohydrates, which accumulates the products of photosynthesis and from which respiration costs are subtracted (Figure 17.6). The storage pool provides a buffer to carbohydrate demand during periods of low or no photosynthesis. The carbon remaining in this pool after subtracting maintenance respiration loss is allocated to plant growth. The storage pool can also be used to provide

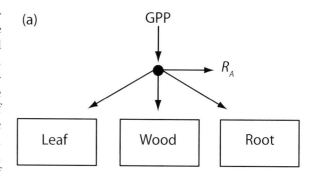

(a)

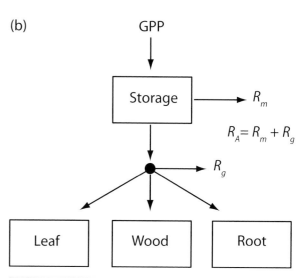

(b)

Figure 17.6 Two different representations of plant growth. (a) Autotrophic respiration R_A is subtracted from gross primary production (GPP) and the remaining carbon is allocated to the growth of leaves, wood, and roots. (b) GPP first enters a storage pool from which maintenance respiration R_m is subtracted. The remaining carbon is allocated to growth after accounting for growth respiration R_g.

the initial carbon for leaf onset in seasonally deciduous plants. In this way, leaf growth depends on carbon accumulation from the previous growing season.

Turnover rates (k_1–k_3 in CASA-CNP; Figure 17.2) vary depending on plant material and are specified as a fraction of biomass. Turnover rates are commonly estimated as the inverse of residence time or longevity (in years; converted to a daily rate as $1/(\text{age} \times 365)$). Wood is shed slowly, frequently as twigs and fine branches, with a residence time of 40–80 years so that $k_3 = 0.0125$–0.025 y^{-1}. Included in wood turnover is a background mortality rate. Fine roots turn over more rapidly with a residence time of 10–20 years for trees ($k_2 = 0.05–0.10$ y^{-1}) and less for grasses. Leaves have the shortest longevity (fastest turnover) – one year or less for deciduous trees (e.g., $k_1 = 1.25$ y^{-1}) and on the order of a few years for evergreen trees. Evergreen foliage has a constant background turnover throughout the year. Loss of deciduous leaves varies seasonally depending on phenology. In CASA-CNP, leaf loss increases in suboptimal conditions by adding a cold temperature rate and a drought rate to leaf turnover so that the realized rate is $k_1 + k_{cold} + k_{dry}$. The two additional terms vary between zero and some maximum rate as air temperature decreases below a cold tolerance and soil moisture stress increases, respectively, as developed by Arora and Boer (2005a) for CTEM (Figure 17.7). As used in (17.1), these terms are encapsulated in the environmental factor ξ_1.

Question 17.5 Calculate the allocation parameters b_1–b_3 for CASA-CNP using the dynamic allocation given by (17.21)–(17.23) and with optimal environmental conditions. How do these compare with the fixed allocation parameters? Compare results with the allocation used in CTEM given by (17.24)–(17.26).

Question 17.6 In the carbon allocation given by (17.27)–(17.33), what is the fraction of the available carbon GPP $- R_m$ that is used for new biomass growth? What is the fraction that is lost as growth respiration?

17.5 | Leaf Phenology

The seasonal emergence and senescence of leaves exerts a critical control on the terrestrial carbon cycle and must be represented in biogeochemical models. The fraction of net primary production allocated to leaf, fine root, or wood growth varies over the course of a year as determined by budburst and leaf senescence. Leaf turnover similarly depends on senescence or environmental stresses such as cold temperature and drought. Common phenology rules to govern leaf emergence and litter fall are

evergreen, in which plants maintain foliage throughout the year and have a constant background leaf turnover rate; summergreen, or seasonal deciduous, in which leaves are present during the warm season and drop in the cold season; and raingreen, or stress deciduous, in which foliage emerges during the rainy season and drops in the dry season.

A simple means to represent these types of phenology is to prescribe dates of leaf onset and offset. In CASA-CNP, for example, leaf phenology is specified using phenological dates obtained from satellite remote-sensing estimates of leaf area index. Leaf growth is divided into four phases: green-up – from budburst to the start of steady leaf growth (lasting 14 days); steady leaf area – from the start of steady leaf growth to the beginning of leaf senescence; senescence – a 14-day period of leaf drop; and from the end of leaf senescence to the onset of leaf growth. During green-up, 80% of production is allocated to leaves. During steady growth, allocation is based on rules that vary with light, water, temperature, and nutrients as described in the previous section. Between senescence and budburst, no carbon is allocated to foliage. For evergreen biomes, leaf phenology remains at steady leaf growth throughout the year. Leaves turnover at a background rate specified by phenology (e.g., lower for evergreen than for deciduous foliage) and which increases with onset of cold or moisture stress (Figure 17.7).

Other models utilize prognostic phenology in which leaf emergence and senescence depend on prevailing growing conditions. A complete leaf phenology model includes carbon allocation, leaf onset and growth, and litterfall as influenced by environmental conditions. A common predictor of springtime leaf onset in summergreen plants is a degree-day sum. This index is the temperature summation above some threshold (e.g., 5°C) based on the premise that plants break dormancy once sufficient spring warmth has accumulated. Accumulation of cold temperatures below a threshold (i.e., a chilling requirement) is an additional signal that the danger of late spring frost has passed. Day length is another determinant of leaf phenology. A day length requirement for autumn senescence, for example, ensures that plants prepare for cold temperatures prior to the onset of winter, while a spring photoperiod prevents release from dormancy arising from false warming signals. Another

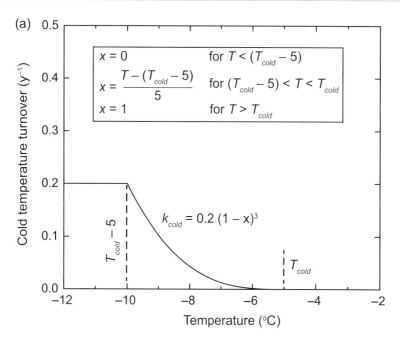

$$x = 0 \quad \text{for } T < (T_{cold} - 5)$$

$$x = \frac{T - (T_{cold} - 5)}{5} \quad \text{for } (T_{cold} - 5) < T < T_{cold}$$

$$x = 1 \quad \text{for } T > T_{cold}$$

$$k_{cold} = 0.2 (1 - x)^3$$

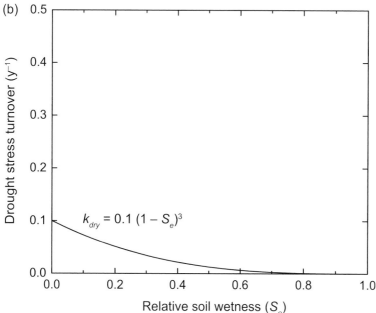

$$k_{dry} = 0.1 (1 - S_e)^3$$

Figure 17.7 Effect of (a) temperature and (b) soil moisture on leaf turnover rates as in Arora and Boer (2005a). S_e is soil moisture relative to available soil water above wilting point. Parameters are representative of evergreen needleleaf tree.

environmental driver of leaf emergence and senescence in raingreen plants is soil moisture. The precise way to model leaf phenology is highly uncertain. Satellite remote sensing of leaf area provides important observations to develop or improve prognostic phenology algorithms (Stöckli et al. 2008, 2011; MacBean et al. 2015). Nonetheless, biogeochemical models differ greatly in details of their phenology algorithms such as how heat or chilling requirements affect leaf emergence, whether daylength is included, and how soil moisture affects leaf onset and offset. Many of the equations used to represent environmental controls are designed to better fit the model to observations, require tunable parameters that are not necessarily observable and should be considered heuristic. Unsurprisingly, terrestrial biosphere models give vastly divergent predictions of leaf emergence, leaf senescence, and growing-season length (Richardson et al. 2012). Another process driven by phenology is that the photosynthetic capacity of leaves depends on their age. This is seen, for example, in seasonal changes in the photosynthetic parameter $V_{c\,max}$. The Community Land Model accounts for this by varying $V_{c\,max}$ in relation to daylength (Oleson et al. 2013). Other models such as ORCHIDEE account for separate age classes (MacBean et al. 2015).

17.6 | Nitrogen Cycle

The flow of nitrogen in an ecosystem parallels that of carbon so that each carbon pool and transfer represented in Figure 17.2 has an associated nitrogen pool and transfer. This cycling of nitrogen can be represented by a system of linear differential equations similar to that for carbon. In CASA-CNP, for example, the nitrogen balance for plant pools is

$$\frac{dn_i}{dt} = b_{n,i}F_{n,up} - \zeta_i k_i \left(1 - f_{n,i}\right)n_i. \tag{17.35}$$

The first term on the right-hand side of the equation describes allocation of plant nitrogen uptake $F_{n,up}$ to plant pools; the second term is the loss of nitrogen in litterfall of which a portion f_n is reabsorbed. The soil nitrogen cycle is more complex because nitrogen has a gaseous phase, occurring as diatomic nitrogen (N_2), ammonia (NH_3), nitrous oxide (N_2O), nitric oxide (NO), and nitrogen dioxide (NO_2), and because nitrogen used in plant growth is the inorganic ions of nitrate (NO_3^-) and ammonium (NH_4^+). The nitrogen cycle must additionally include inputs from biological nitrogen fixation, atmospheric deposition, and fertilizer, as well as gaseous and aqueous losses (Figure 17.8).

Nitrogen has a prominent cycling within an ecosystem. Plants take up NH_4^+ and NO_3^- from the soil

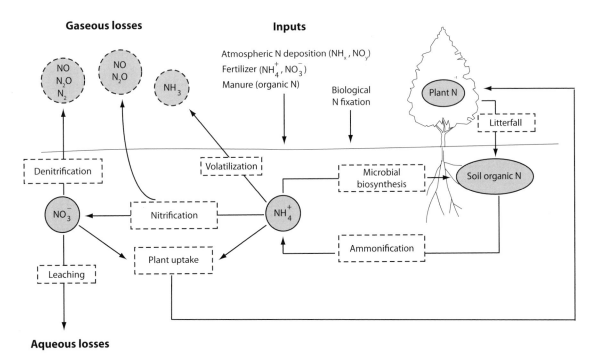

Figure 17.8 Depiction of the nitrogen cycle. Circles indicate various pools (solid lines) or gaseous losses (dashed lines). Boxes denote processes. Also shown are natural inputs from biological nitrogen fixation and anthropogenic inputs from nitrogen deposition, fertilizer, and manure. Reproduced from Bonan (2016)

solution during growth. Some of the nitrogen required for plant growth is internally recycled within plants in a process known as retranslocation, by which nutrients are withdrawn from senescing leaves and stored within the plant. Nitrogen in plant biomass is returned to the soil as litterfall. The organically bound nitrogen is released as inorganic nitrogen (NH_4^+) during decomposition in a process known as mineralization (also called ammonification). Some of this nitrogen is used by the microbes that decompose litter (immobilization). Mineralization and immobilization occur during decomposition of litter and soil organic matter as described in Chapter 18. Bacteria oxidize some of the NH_4^+ to NO_3^-, releasing NO and N_2O as by-products (nitrification). NO_3^- is lost to the atmosphere as N_2O and N_2 (denitrification) or is also leached from the soil solution. Additionally, ammonia volatilization is the loss of NH_3 to the atmosphere. The overall soil mineral nitrogen budget is

$$\frac{dN}{dt} = F_{n,dep} + F_{n,fix} + F_{n,fert} + F_{n,net} - F_{n,up} - F_{n,loss},$$

(17.36)

where the various terms are nitrogen deposition, fixation, fertilization, net mineralization, plant uptake, and gaseous and aqueous loss rates, respectively. Some models track separate NH_4^+ and NO_3^- pools.

The first biogeochemical models used to simulate the carbon cycle in Earth system models did not include the nitrogen cycle, despite its close coupling with and regulation of the carbon cycle (Friedlingstein et al. 2006, 2014). Low nitrogen availability can restrict plant growth, whereas additional nitrogen from atmospheric deposition or enhanced mineralization with warmer soils can stimulate plant growth. Failure to account for nitrogen, especially its role to limit plant productivity, is evident in the high simulated terrestrial carbon uptake over the twenty-first century in response to elevated atmospheric CO_2 concentration (Wieder et al. 2015c). Beginning in the late 2010s, the carbon cycle in biogeochemical models was expanded to include nitrogen (Sokolov et al. 2008; Jain et al. 2009; Thornton et al. 2007, 2009; Zaehle and Friend 2010), and most models now include such coupling. Some models additionally include phosphorus (Wang et al. 2010; Yang et al. 2014; Goll et al. 2017). However, the appropriate way to model carbon–nitrogen biogeochemistry, especially at the large spatial scales used in global models, is highly uncertain.

Different representations of the biogeochemistry result in divergent carbon cycle simulations. This is evident in comparison of the Community Land Model (CLM4; Thornton et al. 2007, 2009) and O-CN (Zaehle and Friend 2010) – two early implementations of the nitrogen cycle for global models. The models have very different responses to CO_2 fertilization, climate warming, and anthropogenic nitrogen deposition over the historical era (Bonan and Levis 2010; Zaehle et al. 2010) and differ also in the extent to which nitrogen availability limits plant productivity (Thomas et al. 2013b). In the Community Land Model, nitrogen control of plant productivity is strongly dependent on the particular assumptions used to represent plant uptake of nitrogen and the pathways of nitrogen loss (Thomas et al. 2013a). Comparison of multiple models in their simulated response to CO_2 enrichment also reveals the consequences of different biogeochemical representations (Zaehle et al. 2014). Key processes that differ among models include: how nitrogen availability restricts photosynthesis; plant nitrogen uptake; plant and soil carbon and nitrogen stoichiometry, especially whether C/N ratios for various pools are constant or vary in response to nitrogen availability; biological nitrogen fixation; nitrogen gas losses; and competition between plants and microbes for soil mineral nitrogen (Zaehle and Dalmonech 2011). The traceability framework to analyze the carbon cycle can be extended to include the nitrogen cycle so as to assess the importance of different model assumptions (Du et al. 2018). Rather than describing model differences in detail, the following text provides an overview of these processes.

All models simulate a decrease in plant growth when soil mineral nitrogen is insufficient to meet demand, but they differ in the manner in which this is implemented. One approach uses an instantaneous downregulation of photosynthesis. Potential productivity is calculated in the absence of nitrogen limitation; and the nitrogen demand required to support plant growth is assessed from potential gross primary production, allocation of carbon to the various plant components, the C/N ratio of plant biomass, and retranslocated nitrogen. If there is insufficient nitrogen, gross primary production is decreased such that plant nitrogen demand is equal to nitrogen supply. In this approach, the C/N stoichiometry of plant biomass is set through model parameters and does not change with environmental conditions. An alternative approach couples photosynthesis and nitrogen through foliar chemistry. Photosynthetic parameters such as $V_{c\max}$, $J_{\max}$, and R_d vary with leaf nitrogen as described in Chapter 11 so that gross primary production responds to nitrogen supply as it affects leaf chemistry. Leaf nitrogen is a prognostic variable that is calculated based on the balance among plant nitrogen uptake, its allocation during growth, and retranslocation. Leaf C/N stoichiometry is flexible and relates to prevailing conditions. The C/N ratio increases up to some maximum value if there is insufficient plant nitrogen uptake or decreases to some minimum value if there is excess nitrogen uptake. The excess photosynthetic carbon if the C/N ratio exceeds the maximum is added to autotrophic respiration loss, in contrast with the instantaneous downregulation in which gross primary productivity is reduced to satisfy nitrogen availability. In case of low C/N, respiration loss is decreased so as to maintain the necessary C/N ratio. Both methods reduce net primary production for low nitrogen availability but differ in how that downregulation is achieved, either by reducing GPP or increasing R_A, and so have differ consequences for carbon use efficiency (NPP/GPP). Comparisons of the two approaches in O-CN (Meyerholt and Zaehle 2015) and the Community Land Model (Ghimire et al. 2016) highlight the different carbon outcomes.

Yet another approach to model plant nitrogen and productivity is based on the requirement that plants pay for the carbon costs of nitrogen uptake. Nitrogen can be acquired by plants through passive and active uptake from the soil, retranslocation, and biological nitrogen fixation. An expenditure of energy (i.e., a carbon cost) occurs with each acquisition pathway. The Fixation and Uptake of Nitrogen (FUN) model calculates nitrogen uptake based on the carbon costs of nitrogen acquisition (Fisher et al. 2010a; Brzostek et al. 2014; Shi et al. 2016). Carbon is optimally allocated to maximize net primary production and nitrogen uptake while maintaining specified C/N ratios and minimizing the carbon cost.

An ecosystem gains nitrogen from atmospheric deposition and biological nitrogen fixation and loses nitrogen from gaseous emissions and leaching. Chapter 20 examines gaseous losses from nitrification, denitrification, and volatilization. Nitrogen loss from leaching is poorly represented in biogeochemical models. It is calculated assuming some portion of soil mineral nitrogen is dissolved inorganic nitrogen and is washed away in runoff. The addition of nitrogen from atmospheric deposition is specified as a model input dataset in which a specified amount of nitrogen is added to the soil mineral pool. Biological nitrogen fixation is an important source of nitrogen. It is commonly represented based on empirical relationships with evapotranspiration or net primary production or as a function of nitrogen demand. These relationships are highly uncertain but have a large impact on the simulated carbon cycle (Wieder et al. 2015b; Meyerholt et al. 2016).

Question 17.7 Derive an equation for plant nitrogen uptake for the carbon allocation given by (17.27)–(17.33) and in which the C/N ratio of plant pools is prescribed.

17.7 | Disturbance

The carbon balance given by (17.13) can include an additional carbon flux related to disturbance. Changes in land cover from anthropogenic land use such as forest clearing for agricultural, reforestation with farm abandonment, or plowing of grasslands for crops alter the carbon balance. Harvesting of wood or crops is an additional loss of carbon. Wildfires burn plant and litter biomass and emit carbon to the atmosphere. The details with which land use and land-cover change, wood harvest, and fire are represented in biogeochemical models are complex. The text that follows gives a brief introduction.

Temporal changes in land cover can be modeled by classifying a landscape into discrete types of vegetation and mathematically representing changes in land area in either continuous form as a system of linear differential equations or in discrete form based on the probability that vegetation of one type transitions to another type over some time interval. Shugart et al. (1973), for example, used differential equations to model forest succession, while Waggoner and Stephens (1970) and Horn (1975) are examples of transition probabilities. The latter approach can be used in terrestrial biosphere models to represent changes in land cover resulting from land use (Hurtt et al. 2006). The particular land-cover states represented are two types of natural vegetation (primary and secondary) and two types of agricultural land (crop and pasture). Primary vegetation is undisturbed by humans. Secondary vegetation has recovered from human disturbance. The area in a grid cell undergoing change from land use in each year is specified from a transition matrix that describes conversion from one vegetation to another. Land cover at time $n + 1$ for grid cell x is represented by

$$\mathbf{l}(x, n + 1) = \mathbf{T}(x, n)\mathbf{l}(x, n), \tag{17.37}$$

where $\mathbf{l}(x, n)$ is an $m \times 1$ vector that describes the area of land in each of the m land-use categories in grid cell x and year n, and $\mathbf{T}(x, n)$ is an $m \times m$ matrix of land-use transitions within grid cell x during year n. Each element t_{ij} in the transition matrix describes the conversion of land in category j to category i from time n to $n + 1$. Figure 17.9 illustrates the matrix. Primary forest, for example, can change to crop or pasture through deforestation; the latter two categories can convert to secondary forest through reforestation or afforestation. Logging can convert primary forest to secondary forest. Hurtt et al. (2011) used this framework to develop a global dataset of historical land use and land-cover change

over the past several hundred years and extended this through 2100 with projections of future land use. New land-use datasets provide more details to differentiate between forest and non-forest land, between pasture and rangeland, and among types of crop (Lawrence et al. 2016). A particular detail is whether the transitions represent gross or net change. Gross land-use changes among forest, crop, and pasture can exceed the net change in area of each land-cover type individually (e.g., with deforestation to cropland and reforestation of cropland in the same grid cell).

The land-use transitions produce a change in carbon (e.g., as trees are cleared), and the resulting change is a flux of carbon to the atmosphere that must be tracked. Various models differ in the details of how this is done (Arora and Boer 2010; Lawrence et al. 2012; Reick et al. 2013). The Community Land Model, for example, uses a mass balance booking approach (Lawrence et al. 2012). For vegetation that decreases in area, the carbon density (per unit area) is unchanged, but the total mass (density × area) has decreased. Some of the carbon is assumed to be removed as wood or fiber products, which slowly decay to the atmosphere at different rates (e.g., 1 year, 10 years, 100 years); some remains on-site as litter; and some is immediately lost to the atmosphere. If the area increases, the carbon density decreases in order to conserve mass. Harvest is a loss of carbon without a change in area. Wood harvest can be specified as an amount of carbon removed, with some carbon left on-site as litter. A similar approach is used for nitrogen.

Conversion of grassland to cropland typically reduces soil organic matter during the first few decades of cultivation. Reduced input of plant litter with harvesting and post-harvesting removal of plant residues contribute to decreased soil organic matter. Soil tillage also increases turnover rates. The DAYCENT biogeochemical model, for example, increases the turnover rate of litter and soil organic matter pools based on the occurrence of specific cultivation events (Figure 18.8). Terrestrial biosphere models have begun to account for the effects of tillage on soil organic matter (Levis et al. 2014; Pugh et al. 2015). Levis et al. (2014), for example, applied the DAYCENT methodology in the Community Land Model to increase turnover for 30 days after any of 11 different types of cultivation. Multiple cultivation events can occur throughout the year.

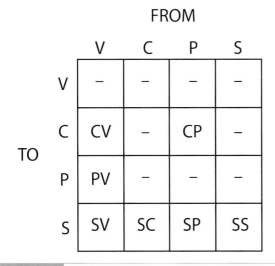

FROM

	V	C	P	S
V	–	–	–	–
C	CV	–	CP	–
P	PV	–	–	–
S	SV	SC	SP	SS

(TO)

Figure 17.9 Example transition matrix among primary forest (V), crop (C), pasture (P), and secondary forest (S). In this example, the transitions are: V → C and V → P; C → S and P → S; and P → C. Woody harvesting in primary forest is represented by V → S and on secondary forest by S → S.

Question 17.8 Waggoner and Stephens (1970) provided a 10-year forest transition matrix. Calculate the distribution of forests after 40 years if the initial land cover is 100% maple. What is the steady-state forest distribution?

To	From				
	Maple	Oak	Birch	Other	Minor
Maple	0.82	0.16	0.13	0.07	0.07
Oak	0.07	0.72	0.02	0.03	0.07
Birch	0.02	0.08	0.83	0.07	0.07
Other	0	0	0.02	0.69	0.07
Minor	0.09	0.04	0	0.14	0.72

Fires are another disturbance that removes carbon. The amount of carbon burned during fire is modeled in terms of fire occurrence, the area burned, and the severity of fire. A general approach for carbon emission during combustion is

$$F_c = A \cdot \sum c_i \cdot f_i, \tag{17.38}$$

where F_c is the total carbon emission, A is area burned, c_i is the particular carbon pool burned, and f_i represents the portion of the biomass burned and completeness of combustion. The latter term accounts for the fact that not all biomass is burned during fire; nor is the burnt biomass completely combusted.

Thonicke et al. (2001) developed one of the first global fire models, which was used with LPJ (Chapter 19) to calculate the annual area burned and subsequently incorporated into other models. Fire occurrence depends on fuel load and litter moisture. A minimum fuel load of 200 g C m^{-2} is needed for a fire to occur. Ignition is assumed to occur if conditions are favorable and the fuel is sufficiently dry. The daily probability of fire is

$$p(S_e) = \exp\left[-\pi(S_e/m_e)^2\right], \tag{17.39}$$

where S_e is relative soil water content of the top soil layer (used as a surrogate for fuel moisture) and m_e is an empirical parameter (0.3 for trees and 0.2 for grasses). The burnt area is assumed to relate to fire season length such that more area is burned as longer burning conditions prevail. The length of

the fire season is based on the annual sum of the daily probability of fire so that fire season length as a fraction of the year is

$$s = \frac{1}{365} \sum p(S_e). \tag{17.40}$$

The fractional area burned is

$$A(s) = s \cdot \exp\left[(s-1)/f(s)\right], \tag{17.41}$$

with

$$f(s) = 1.04 + 2.96(s-1) + 2.83(s-1)^2 + 0.45(s-1)^3. \tag{17.42}$$

The latter equation is obtained empirically to fit burnt area to fire season length. In this model, as litter moisture decreases, fire occurrence increases and more area is burned (Figure 17.10).

Newer global fire models now represent lightning and human ignition sources, human fire suppression, fire spread, and interactions between fuel structure, fire type, and vegetation adaptation to fire (Hantson et al. 2016). Many such models are variants of the fire model for CTEM (Arora and Boer 2005b). In this model, the probability of fire depends on biomass available for burning so that fires are more likely as fuel load increases, fuel moisture so that fires are more likely as fuel becomes drier, and the availability of an ignition source. The area burned is depicted by an ellipse with a point of ignition. Fire spreads in the upwind and downwind directions along the major axis of the ellipse in relation to wind speed and soil moisture, as well as in the perpendicular direction (Figure 17.11). The duration of fire depends on factors that limit fire spread. Kloster et al. (2010) expanded this approach to introduce anthropogenic sources of ignition and fire suppression based on population density and also to account for deforestation fires. In the Li et al. (2012a, 2013) variant, burnt area is determined as the product of the number of fires in a grid cell (rather than fire probability) and the average fire spread area; agricultural, deforestation, and peat fires are included, and gross domestic product is used with population density to account for socioeconomic influences on fires.

SPITFIRE (Thonicke et al. 2010) is a daily fire model that is a more complex and more physically based model in its representation of ignition, fire

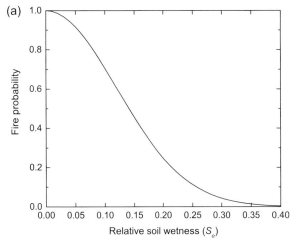

(a)

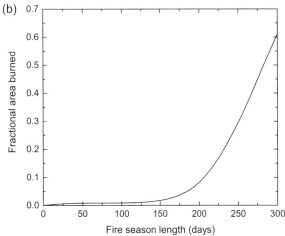

(b)

Figure 17.10 The global fire model of Thonicke et al. (2001) in which (a) the daily probability of at least one fire increases as litter moisture decreases and (b) the fractional area burned increases with greater fire season length.

spread and duration, fire intensity, biomass combustion, and fire mortality (Figure 17.12). It scales the number of potential fire ignitions from lightning and human activity by a fire danger index to obtain daily fire counts. The fire danger index describes the likelihood that an ignition will start a fire and depends on temperature, humidity, fuel moisture, and different classes of fuel (e.g., leaves, twigs, small branches). Mean fire size is based on elliptical fire spread. Fire impacts depend on fire intensity (e.g., surface or crown fire), scorch height, crown damage, bark thickness, and cambial heating. SPITFIRE was developed for LPJ, but has since been implemented in other models including

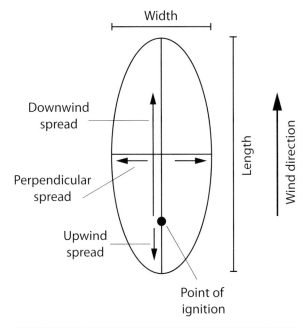

Figure 17.11 Fire spread as depicted by an ellipse. Adapted from Arora and Boer (2005b)

ORCHIDEE (Yue et al. 2014), JSBACH (Lasslop et al. 2014), and CLM(ED) (Fisher et al. 2015). However, the degree of complexity needed to model fires at the large spatial scales of global models is unclear (Hantson et al. 2016).

17.8 | Supplemental Programs

17.1 Biogeochemical Model: This program calculates the carbon cycle given by equations (17.1)–(17.9). **B**, **K**, ξ, and **A** are specified for evergreen broadleaf forest as in Table 17.3 and Table 17.4. The model time step is one day. Figure 17.3 shows model output.

17.9 | Modeling Projects

1 Compare the simulation for evergreen broadleaf forest in Figure 17.3 with evergreen needleleaf forest and deciduous broadleaf forest. How do the steady-state solutions differ? What is the time

Model input

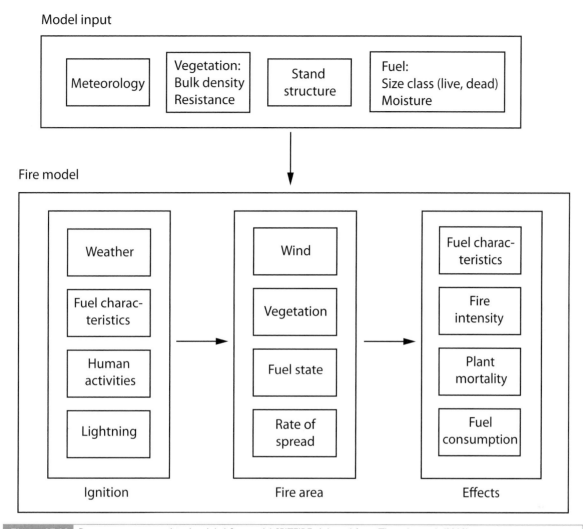

Figure 17.12 Processes represented in the global fire model SPITFIRE. Adapted from Thonicke et al. (2010)

scale over which steady state is achieved? What are the major differences among biomes in **B**, **K**, ξ, and **A**?

2 Examine how allocation of net primary production to foliage, root, and wood and allocation of litter between metabolic and structural pools affect carbon storage.

3 Examine the effect of disturbance on carbon dynamics. How does the timing (age), frequency,

and magnitude of carbon loss affect carbon accumulation?

4 The carbon cycle simulation shown in Figure 17.3 used Supplemental Program 17.1 with a time step $\Delta t = 1$ day. Repeat the calculations using a time step of one year. How do the transient dynamics and steady-state solutions compare? Try also the fourth-order Runge–Kutta method (Appendix A3) with a one-year time step.

Soil Biogeochemistry

Chapter Overview

Soils store vast quantities of carbon, more than in the atmosphere or in plant biomass. Decomposition loss of soil carbon is a large term in the global carbon budget and mineralizes nitrogen and other elements needed for plant growth. This chapter develops the biogeochemical foundation and mathematical theory to describe litter decomposition, soil organic matter formation, and nutrient mineralization. The DAYCENT model is used to illustrate the basic details of soil biogeochemical models. Advanced modeling concepts include vertically resolved soil carbon, microbial models, and competition among multiple nutrient consumers.

18.1 | Introduction

Decomposition of fresh litter begins when soil microorganisms (bacteria, fungi) consume organic matter to gain energy and nutrients for growth. The gain in mass by microbes is less than the mass of litter consumed because the microbes respire CO_2. The non-respired carbon is incorporated into microbial biomass, and this secondary material itself decomposes. Over long time periods, the litter decomposes to the point that it is no longer recognizable, at which point it is termed soil organic matter, or humus. Figure 18.1 illustrates the progressive loss of litter mass during decomposition.

The mass of slash pine litter decreases by about 50% over 10 years. The litter of drypetes, a tropical broadleaf tree, loses about 70% of its mass over 10 years. Decomposition proceeds more rapidly in a warm tropical site than in a cold arctic site.

Nitrogen contained in plant material recycles to the soil as litter decomposes and releases organically bound nutrients. However, decomposition does not immediately release nitrogen to the soil for plant use. Microbes require nitrogen to grow, and microbial growth creates demand for nitrogen. If there is not sufficient nitrogen in decomposing litter to meet microbial demand, microbes obtain the necessary nitrogen from the soil. This process, by which the net amount of nitrogen in decomposing litter increases as microbes take up nitrogen from soil and incorporate it into new microbial biomass, is known as immobilization. If more nitrogen is present in decomposing litter than is required for microbial growth, the excess nitrogen releases to the soil as mineral nitrogen. This process is known as mineralization. Typically, fresh litter immobilizes nitrogen while soil organic matter mineralizes nitrogen. Similar dynamics pertain to other chemical elements in litter and soil.

Figure 18.2 illustrates the dynamics of mass loss, nitrogen immobilization, and nitrogen mineralization for yellow birch leaf litter. The transition from immobilization to net mineralization is evident when the amount of nitrogen in the litter, expressed as a fraction of the initial nitrogen, is plotted in relation to the remaining litter mass. The amount of nitrogen in the material increases as the litter decreases in mass from 100% to 64%, after which further decay

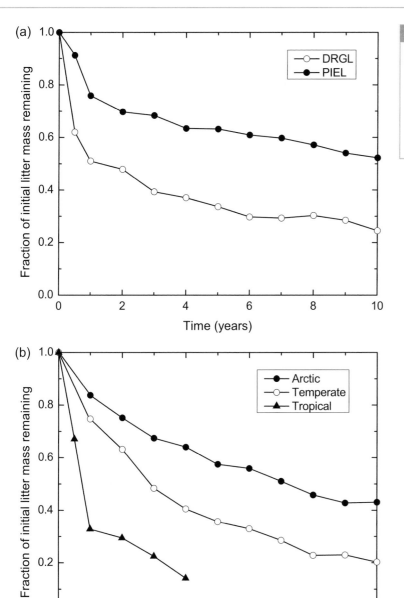

Figure 18.1 Mass loss in relation to litter quality and climate. (a) Mass loss for *Drypetes glauca* (DRGL) and *Pinus elliottii* (PIEL) leaf litter. Litter was decomposed at 27 sites that differed in climate. Shown is the mean across all sites. (b) Mass loss in arctic, temperate, and wet tropical sites. Shown is the mean across several litter types. Data from Adair et al. (2008)

produces a net release of nitrogen. As the mass of leaf litter decreases, the amount of nitrogen in the decaying material increases from an initial concentration of 0.85% of litter mass to 2.3% of litter mass. This linear relationship occurs in many tree species (Aber and Melillo 1980). In this example, net mineralization occurs when 36% of the original litter mass has been lost, at which point the nitrogen concentration is 1.75%. Litter decomposition studies across various ecosystems in a range of climates illustrate the generality of these concepts (Parton et al. 2007). The dynamics of leaf litter decomposition and nitrogen immobilization and mineralization can be predicted based on initial litter nitrogen concentration and the mass of original litter remaining.

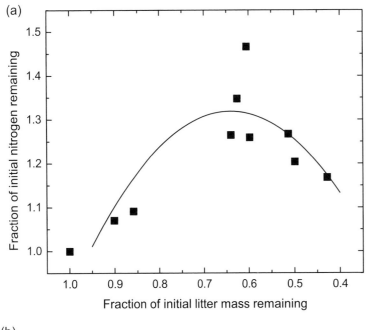

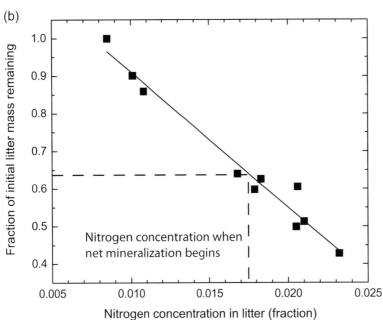

Figure 18.2 Decomposition and nitrogen dynamics of yellow birch (*Betula alleghaniensis*) leaves. (a) Relationship between nitrogen content in the remaining litter (relative to initial nitrogen) and remaining litter mass. (b) Relationship between litter mass and nitrogen concentration in the remaining litter. Data from Aber and Melillo (1980) and reproduced from Bonan (2016)

18.2 | Exponential Decay

Olson (1963) outlined the basic mathematical concepts of litter decomposition. Mass loss due to decomposition can be represented by a first-order,

linear differential equation that describes the instantaneous rate of change of organic mass m as

$$\frac{dm}{dt} = -k \cdot m. \tag{18.1}$$

This equation is similar to that presented in Chapter 17 for ecosystems in general, but here it is specific to litter decomposition with k an instantaneous fractional decay rate. Integration of (18.1) with respect to time gives the mass remaining at time t,

$$m(t) = m_0 e^{-kt},\qquad (18.2)$$

with m_0 the initial mass at $t = 0$. The units of k depend on the time interval (e.g., y^{-1} for annual decay; d^{-1} for daily). The loss rate varies with environment. Higher values of k mean more rapid decomposition and less litter accumulation on the ground. In tropical climates, annual decomposition is fast, and k can be greater than $1\ y^{-1}$. If $k = 1\ y^{-1}$, then 36.8% of the original litter mass remains after one year. In cold climates, k is much smaller; $k = 0.1$ y^{-1} means that 90.5% of the original litter mass remains after one year.

With constant litter input U, accumulation of organic matter is the balance between litterfall and decomposition and the governing equation is

$$\frac{dm}{dt} = U - k{\cdot}m.\qquad (18.3)$$

The mass at time t is, for an initial condition of no litter mass, given by

$$m(t) = \frac{U}{k}\left(1 - e^{-kt}\right).\qquad (18.4)$$

At steady state, $dm/dt = 0$ and

$$m = U/k.\qquad (18.5)$$

For example, 200 g m^{-2} accumulates at steady state with litter input equal to 200 g m^{-2} y^{-1} and $k = 1\ y^{-1}$. With $k = 0.1\ y^{-1}$, the soil accumulates 2000 g m^{-2}.

The rate k is the instantaneous, or continuous, decay rate. Oftentimes it is more convenient to represent mass loss over a discrete time interval Δt. Similar to (18.1), mass loss over the period Δt is represented as a fraction of the original material present in which

$$\Delta m = k'{\cdot}m_0,\qquad (18.6)$$

and $k' = \Delta m/m_0$ is the discrete loss rate over Δt. This loss can be related to the continuous rate k. From (18.2), the mass loss over the time interval Δt is

$$\Delta m = m_0\left(1 - e^{-k\Delta t}\right)\qquad (18.7)$$

and

$$k' = \frac{\Delta m}{m_0} = 1 - e^{-k\Delta t}.\qquad (18.8)$$

Over a unit of time ($\Delta t = 1$):

$$k = -\ln\left(1 - k'\right).\qquad (18.9)$$

For example, if a cohort of litter loses 63.2% of its mass over a year, the discrete decay rate is $k' = 0.632\ y^{-1}$ and the continuous decay rate is $k = 1\ y^{-1}$. The discrete rate k' is termed the turnover rate (the fractional loss per time interval) and the inverse $(1/k')$ is the turnover, or residence, time (the time the material resides in the system). Models commonly represent mass loss over a discrete time interval Δt, and hereafter k is used to represent the discrete decay rate unless otherwise noted.

Question 18.1 A cohort of litter placed on the ground loses 20% of its mass in the first year. (a) What is the discrete annual decay rate? What is the continuous decay rate? (b) A model solves for litter loss using (18.6) with a daily time step. What is the value of the decay rate that should be used? (c) What is the decay rate for a time step of 30 minutes?

18.3 | Rate Constants

The chemical quality of litter affects the rate of decomposition (Berg et al. 1982; Aber et al. 1990; Adair et al. 2008; Cornwell et al. 2008). Labile materials such as sugars and amino acids in leaves readily decompose. Cellulose and hemicellulose decompose more slowly. Lignin is recalcitrant structural material and decomposes the slowest. Nutrient-rich leaves decompose faster than nutrient-poor leaves because they have a high concentration of labile materials. One index of litter quality is the ratio of carbon to nitrogen (C/N) in litter. High litter nitrogen concentrations (low C/N) increase initial decomposition rates. Another index of litter quality is the ratio of lignin to nitrogen (lignin/N). In the example shown in Figure 18.1a, the drypetes litter (C/N = 24, lignin/N = 5) decomposes much faster than the pine litter (C/N = 164, lignin/N = 59). Leaf litter from deciduous broadleaf trees tends to have higher initial nitrogen concentration and lower lignin/N than foliage of evergreen needleleaf trees (Table 18.1). Meta-analyses of leaf trait data similarly show high

Table 18.1 | Litter chemistry for various broadleaf and needleleaf leaves

Species		Fraction				
		Labile	Cellulose	Lignin	%N	L/N
Broadleaf						
Yellow birch	*Betula lutea*	0.276	0.458	0.266	1.60	16.6
Black locust	*Robinia pseudoacacia*	0.419	0.405	0.177	2.45	7.2
Sugar maple	*Acer saccharum*	0.568	0.273	0.159	0.81	19.6
Beech	*Fagus grandifolia*	0.249	0.491	0.260	0.85	30.6
Tulip poplar	*Liriodendron tulipifera*	0.600	0.313	0.087	0.72	12.1
Aspen	*Populus tremuloides*	0.405	0.408	0.188	0.74	25.5
Chestnut oak	*Quercus prinus*	0.371	0.394	0.235	1.03	22.9
Needleleaf						
Slash pine	*Pinus elliottii*	0.372	0.413	0.214	0.36	59.5
Subalpine fir	*Abies lasiocarpa*	0.514	0.307	0.180	0.71	25.3
Red pine	*Pinus resinosa*	0.362	0.446	0.192	0.59	32.6
Eastern white pine	*Pinus strobus*	0.397	0.397	0.206	0.62	33.1
Western red cedar	*Thuja plicata*	0.374	0.359	0.267	0.62	42.9
Douglas fir	*Pseudotsuga menzesii*	0.354	0.373	0.274	0.82	33.4

Note: %N is percent nitrogen (by mass), and L/N denotes lignin/N.
Source: From Adair et al. (2008).

nitrogen concentration, low lignin, and high decomposability of broadleaf trees compared with needleleaf trees (Cornwell et al. 2008; Brovkin et al. 2012).

A single exponential equation captures the slowing of the absolute rate of mass loss over time. However, decay rates in later stages of decomposition slow beyond that represented by a single exponential equation. Decomposition proceeds in multiple phases, with rapid initial loss of labile compounds followed by slower loss of recalcitrant materials. Two- or three-pool exponential models are needed to capture these various phases (Adair et al. 2008; Harmon et al. 2009). Adair et al. (2008) analyzed decomposition over 10 years for leaf and root litter from 26 species across 27 sites in North and Central America representative of a wide range of ecosystems and climates. They found that a three-pool model captured most of the variation in the data. In a three-pool model, litter mass remaining at time t is

$$m(t) = m_1 e^{-k_1 t} + m_2 e^{-k_2 t} + m_3 e^{-k_3 t}, \qquad (18.10)$$

with m_1, m_2, and m_3 the initial litter mass of a rapidly decomposing labile pool, an intermediate decomposing cellulose pool, and a recalcitrant

lignin pool, respectively; and k_1, k_2, and k_3 are the corresponding decomposition rates for each pool. Litter chemistry determines the allocation of initial litter mass among the three pools. The initial size of the slowest pool m_3 is proportional to the lignin fraction; the size of pools $m_2 + m_3$ increases with higher lignin/N; and low initial lignin/N increases the size of m_1.

The rate of decomposition varies in relation to climate. Decomposition proceeds rapidly in a wet tropical forest, moderately in a temperate forest, and most slowly in arctic tundra (Figure 18.1b). Annual mean air temperature, annual precipitation, and annual evapotranspiration correlate with annual decay rates across wide geographic gradients (Meentemeyer 1978; Berg et al. 1993; Gholz et al. 2000; Trofymow et al. 2002; Adair et al. 2008; Currie et al. 2010). More precisely, soil temperature and soil moisture control microbial activity and decomposition rate. In general, organic material decomposes faster in warm soil than in cold soil because microbial activity increases with higher temperature. Dry soils do not have sufficient water to support microbial growth, but extremely wet soils do not provide enough oxygen for optimal microbial activity. Models that represent the combined effects of temperature and moisture provide a good predictor of observed decay rates across a range of climates from arctic to tropical and humid to arid (Gholz et al. 2000; Adair et al. 2008).

Biogeochemical models represent the effects of soil temperature and soil moisture on decomposition rates by multiplying a base rate k_{base} by the functions $f(T)$ and $f(\theta)$ whereby

$$k = k_{base}f(T)f(\theta). \tag{18.11}$$

A common approach uses the Q_{10} function given by (4.7) for $f(T)$, in which the rate increases by a factor equal to Q_{10} for every 10°C increase in temperature. Other models allow for Q_{10} to vary with temperature. The Lloyd and Taylor (1994) exponential function given by (17.34) used in some models (e.g., CASA-CNP, LPJ) is an example of a variable Q_{10} response. Del Grosso et al. (2005a) described the arctangent function used in CENTURY whereby

$$f(T) = 0.56 + \frac{1.46}{\pi}\tan^{-1}[0.031\pi(T - 288.85)]$$

$$\tag{18.12}$$

for temperature T in kelvin. Figure 18.3a compares the different temperature responses. In this

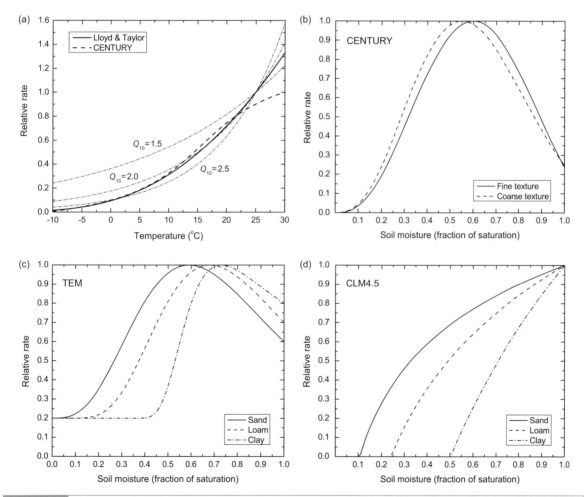

Figure 18.3 Effect of temperature and soil moisture on decomposition rate as represented in several biogeochemical models. (a) Comparison of the Q_{10} function (4.7) referenced to 25°C (with Q_{10} = 1.5, 2.0, 2.5), the Lloyd and Taylor (1994) exponential function (17.34), and the Del Grosso et al. (2005a) arctangent function (18.12). The Lloyd and Taylor (1994) function is scaled to one at 25°C. Soil moisture effects are shown for (b) CENTURY with fine and coarse texture soil (Kelly et al. 2000), (c) TEM with sand, loam, and clay soil (Raich et al. 1991), and (d) CLM4.5 (Oleson et al. 2013) with sand, loam, and clay soil using soil water potential as in Andrén and Paustian (1987)

example, the Q_{10} function is referenced to 25°C and is shown for three Q_{10} values (1.5, 2.0, 2.5). A prominent feature is the high values for $f(T)$ at low temperature with $Q_{10} = 1.5$. The various functions also differ at high temperatures, with lower values for the CENTURY arctangent function compared with the other functions. Equations (17.34) and (18.12) have large temperature sensitivity (i.e., derived Q_{10} values) at low temperature and reduced sensitivity at high temperature. Variable Q_{10} functions improve predicted decomposition rates across large climatic gradients compared with other functions (Adair et al. 2008).

Soil moisture also affects decomposition rates. Microbial activity is less in dry soils compared with wet soils, but also decreases when water fills the pore space. However, biogeochemical models differ greatly in how they represent the effect of soil moisture. Some models relate $f(\theta)$ to the fraction of pore space filled with water ($S_e = \theta/\theta_{sat}$ where θ is volumetric water content and θ_{sat} is porosity). For example, CENTURY uses a function from Kelly et al. (2000) in which

$$f(S_e) = \left(\frac{S_e - b}{a - b}\right)^e \left(\frac{S_e - c}{a - c}\right)^d, \quad (18.13)$$

with $e = d(b - a)/(a - c)$. For fine-texture soils, $a = 0.6$, $b = 1.27$, $c = 0.0012$, and $d = 2.84$. For coarse-texture soils, $a = 0.55$, $b = 1.7$, $c = -0.007$, and $d = 3.22$. Other models use matric potential rather than soil moisture (e.g., CLM4.5). Differences among models are considerable (Figure 18.3b–d). All models increase the rate constant as soils increase in moisture, but some models (e.g., CENTURY, TEM) have an optimum wetness above which excessively wet soil decreases decomposition.

Question 18.2 The effect of temperature on decomposition is represented by the function $f(T)$. Compare $f(T)$ between 0°C and 10°C and 10°C and 20°C using the Lloyd and Taylor (1994) and Del Grosso et al. (2005a) temperature functions shown in Figure 18.3. Calculate the corresponding Q_{10} values. How does Q_{10} differ with temperature?

18.4 | Litter Cohort Models

Aber and Mellilo (1982) used the relationship between mass remaining and nitrogen concentration (e.g., Figure 18.2b) to develop a model of mass loss, nitrogen immobilization, and nitrogen mineralization during litter decomposition. The governing equation for the model is the linear increase in litter nitrogen concentration as litter mass decreases, which is described by

$$\frac{m}{m_0} = a + b\frac{n}{m}, \quad (18.14)$$

where m_0 is the mass of the original organic residue at time $t = 0$, m is the mass of organic residue remaining at time t, and n is the mass of nitrogen in the residue at time t. The mass of nitrogen at time t can be obtained from (18.14) using

$$n = \left(\frac{m}{m_0}\right)m_0\left(\frac{n}{m}\right) = \left(a + b\frac{n}{m}\right)m_0\left(\frac{n}{m}\right), \quad (18.15)$$

and the fraction of original nitrogen n_0 remaining is

$$\frac{n}{n_0} = \left(a + b\frac{n}{m}\right)\left(\frac{n}{m}\right)\left(\frac{n_0}{m_0}\right)^{-1}. \quad (18.16)$$

The amount of nitrogen immobilized before net mineralization occurs is the maximum nitrogen. This is found by setting the derivative of (18.16) with respect to n equal to zero whereby

$$\frac{d}{dn}\left(\frac{n}{n_0}\right) = \left(\frac{a}{m} + \frac{2bn}{m^2}\right)\frac{m_0}{n_0} = 0. \quad (18.17)$$

The ratio of nitrogen to litter mass n/m that satisfies the equation is deemed the critical nitrogen concentration and is equal to

$$\frac{n_c}{m_c} = -\frac{a}{2b}. \quad (18.18)$$

The fraction of original mass remaining when $n = n_c$ is given by

$$\frac{m_c}{m_0} = a + b\frac{n_c}{m_c} = \frac{1}{2}a. \quad (18.19)$$

From (18.18) and (18.19), the amount of nitrogen immobilized at this point is

$$\Delta n = n_c - n_0 = -\frac{a^2}{4b}m_0 - n_0, \quad (18.20)$$

and the mass of nitrogen immobilized per unit mass loss is

$$\frac{\Delta n}{\Delta m} = \frac{\Delta n}{m_0 - m_c} = -\left(\frac{a^2}{4b} + \frac{n_0}{m_0}\right)\left(1 - \frac{a}{2}\right)^{-1}. \quad (18.21)$$

The fraction of initial nitrogen is

$$\frac{n_c}{n_0} = -\frac{a^2}{4b}\left(\frac{n_0}{m_0}\right)^{-1}. \quad (18.22)$$

In the example shown in Figure 18.2, $a = 1.28$, $b = -36.6$, and $n_0/m_0 = 0.0085$. Net mineralization occurs when $m_c = 0.64m_0$, at which point $n_c/m_c = 0.0175$, $n_c/n_0 = 1.32$, and $\Delta n/\Delta m = 7.5$ mg N immobilized for each gram of material decayed. This nitrogen concentration signifies the transition of decomposing material from litter, which immobilizes nitrogen, to humus (soil organic matter), which mineralizes nitrogen.

Aber et al. (1982) used these concepts in a forest ecosystem model to simulate decomposition and nitrogen dynamics by following annual litter cohorts. Pastor and Post (1986) extended this model to forests of eastern North America, and Bonan (1990a,b) adapted it for boreal forests (Chapter 19). Figure 18.4 illustrates the approach. Annual litterfall enters a litter pool as individual litter cohorts characterized by initial mass, nitrogen, and lignin. Distinct types of litter form separate cohorts. A cohort decomposes at a rate determined by its litter chemistry and site conditions, and immobilizes nitrogen following (18.21). A single cohort may be present for one or more years, so that several cohorts of the same litter type co-occur and represent different stages of decomposition. A cohort transfers to a humus pool upon reaching its critical nitrogen concentration (18.18). Fresh wood first passes through a well-decayed wood pool before becoming humus. Fresh wood does not immobilize nitrogen during decay until some fraction of its initial mass is lost, at which point it becomes well-

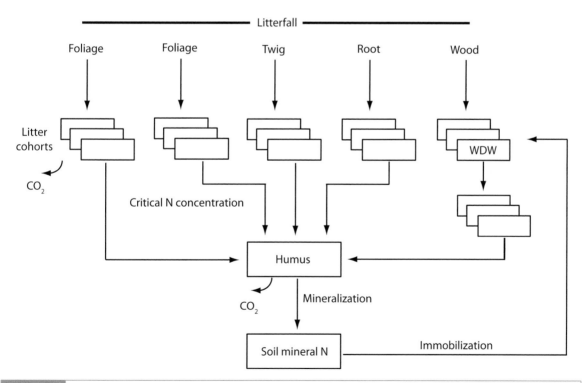

Figure 18.4 Decomposition as represented by a litter cohort model. Foliage, twig, root, and wood litter form individual cohorts with an initial carbon, nitrogen, and lignin mass. Each box represents an individual cohort for a particular year. Foliage litter can vary in initial quality, represented by multiple litter cohorts. The cohorts decompose over time, immobilize nitrogen, and transfer to humus upon reaching a critical nitrogen concentration. Fresh wood first passes through a well-decayed wood pool (WDW) before becoming humus. Humus decomposes and mineralizes nitrogen. Adapted from Bonan (1990a)

decayed wood where nitrogen immobilization occurs. Humus decomposes and mineralizes nitrogen.

In the model of Pastor and Post (1986) for forests of eastern North America, mass loss for litter cohort i with mass m_i and decay rate k_i is

$$\Delta m_i = k_i m_i \Delta t. \tag{18.23}$$

The annual decay rate relates to climate, specified in terms of annual evapotranspiration E (mm y^{-1}), and litter quality, specified by the lignin-to-nitrogen ratio l_i/n_i, in which

$$k_i = \frac{a}{100} - \frac{b}{100}\left(\frac{l_i}{n_i}\right), \tag{18.24}$$

with $a = 0.9804 + 0.09352E$ and $b = -0.4956 + 0.001927E$. Equation (18.24) has the form that the decay rate does not vary much with litter quality when evapotranspiration is low but does vary with litter quality as evapotranspiration increases (Figure 18.5). Canopy openings increase the decay rate. Wood decays are rates that depend on material: twigs, $k < 0.2$ y^{-1}; small trunks less than 10 cm diameter, $k = 0.1$ y^{-1}; large trunks greater than 10 cm diameter, $k = 0.03$ y^{-1}; and well-decayed wood, $k = 0.05$ y^{-1}.

The nitrogen and lignin content of a litter cohort vary with stage of decomposition. The mass of

nitrogen in a cohort increases because of immobilization as given by

$$\Delta n_i = \left(\frac{\Delta n}{\Delta m}\right)\Delta m_i, \tag{18.25}$$

with $\Delta n/\Delta m$ from (18.21). Lignin concentration increases with mass loss, represented by

$$\frac{l_i}{m_i} = c - d\frac{m_i}{m_0}. \tag{18.26}$$

A cohort transfers its organic matter and nitrogen to humus when its nitrogen concentration reaches n_c/m_c, as in (18.18). With m_i and n_i the mass of organic material and nitrogen at the beginning of the time step, this occurs when

$$\frac{n_i + \Delta n_i}{m_i - \Delta m_i} = \frac{n_i + (\Delta n/\Delta m)\Delta m_i}{m_i - \Delta m_i} = \frac{n_c}{m_c}. \tag{18.27}$$

Equation (18.27) solved for Δm_i gives the mass loss required for the current nitrogen concentration to reach the critical nitrogen concentration:

$$\Delta m_i = \frac{(n_c/m_c)m_i - n_i}{(\Delta n/\Delta m) + (n_c/m_c)}. \tag{18.28}$$

If mass loss Δm_i from (18.23) is less than that from (18.28), the critical nitrogen concentration has not been reached and mass loss proceeds following (18.23). If mass loss is greater, the critical nitrogen

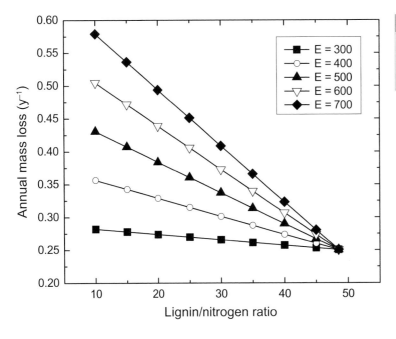

Figure 18.5 Annual fractional mass loss in relation to annual evapotranspiration (mm) and lignin-to-nitrogen ratio as represented in the model of Pastor and Post (1986).

concentration is attained and mass loss is specified by (18.28). Litter then transfers to humus – except for wood, which transfers to well-decayed wood and further decomposes before becoming humus. Humus decomposes at a specified rate and mineralizes nitrogen in relation to its C/N ratio.

Pastor and Post (1986) recognized multiple classes of litter that vary in initial nitrogen and lignin concentration, immobilization, and critical nitrogen concentration. Leaf litter varies among species; fine root, twig, and wood litter are independent of species. Fresh wood consists of small and large trunks, both with the same properties. Figure 18.6 illustrates organic matter mass and nitrogen dynamics over three years for a single cohort of maple and pine leaf litter. In this example, annual evapotranspiration is 700 mm, and humus decays at a rate of 0.5 y^{-1} for maple and 0.3 y^{-1} for

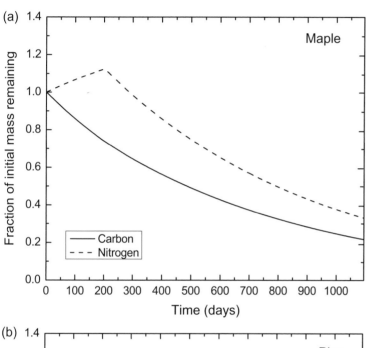

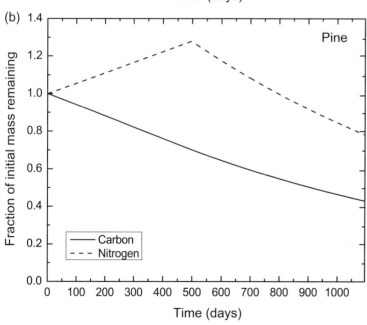

Figure 18.6 Fraction of initial carbon and nitrogen remaining for (a) maple (*Acer*) and (b) pine (*Pinus*) litter cohort simulations. Parameter values are from Pastor and Post (1986). The model has a daily time step in this example.

pine. Over the course of three years, maple litter loses 78% of its original mass. The litter initially immobilizes nitrogen, increasing its nitrogen mass by 12.5% relative to its initial nitrogen. At this point, the litter has lost 26% of its original mass, its nitrogen concentration is 1.6%, and mineralization begins. Pine litter has a higher initial lignin/N ratio. It decomposes more slowly, losing 57% of its original mass over three years. Nitrogen mineralization begins when the litter has lost 30% of its original mass, when the nitrogen concentration is 0.82% and nitrogen mass has increased by 28% relative to initial nitrogen.

Question 18.3 In the model of Pastor and Post (1986), aspen (*Populus*) leaf litter has $n_0/m_0 = 0.83\%$, immobilizes 9.5 mg N g^{-1}, and $n_c/m_c = 1.7\%$. Beech (*Fagus*) leaf litter has $n_0/m_0 = 0.90\%$, immobilizes 36.7 mg N g^{-1}, and $n_c/m_c = 4.8\%$. Which litter loses more mass at the point when mineralization begins?

18.5 | Discrete Pool Models

The litter cohort model tracks a large number of litter pools that change in chemical quality as litter decomposes, but the model represents only a single humus pool. Although the model accounts for the heterogeneous chemistry of litterfall, it does not account for heterogeneity of soil organic matter. Organic matter quality degrades during decomposition from the input of high-quality fresh litter towards more recalcitrant substances in soil organic matter. Decomposers readily consume carbon of high quality during the initial phases of decomposition and leave behind carbon of poorer quality that is harder to decompose. A homogenous substrate of uniform initial carbon quality degrades throughout decomposition into a heterogeneous substrate with a continuum of various qualities. A particular type of model represents this process in terms of a continuous distribution of carbon quality (Bosatta and Ågren 1985, 1991, 2003; Ågren and Bosatta 1987, 1996). In this model, litter with low initial quality loses less carbon over time than does better-quality litter. It also immobilizes more nitrogen and has

longer time to net mineralization. The continuous quality model allows analytical solutions. Although it has been applied in a forest ecosystem model (Bonan 1993), it has not been widely adopted.

The more common approach used in biogeochemical models to account for differences in substrate quality is to simulate a number of chemically distinct organic matter pools with varying decay rates. In these models, litter chemistry determines the initial allocation of plant residue among pools. The models follow the decomposition of litter pools into soil organic matter pools and the associated carbon released in heterotrophic respiration. This conceptualization of soil carbon results in a system of first-order, linear differential equations that describes pools and carbon flows among pools as in Chapter 17 and is the prevailing paradigm for biogeochemical models (Manzoni and Porporato 2009; Sierra et al. 2012; Sierra and Müller 2015).

One such model is the Rothamsted model, known as RothC (Jenkinson 1990; Jenkinson et al. 1991), which is used in JULES (Clark et al. 2011a). The model has four organic matter pools (Figure 18.7). Plant litter is classified as either decomposable plant material (DPM) or resistant plant material (RPM). These pools decompose to microbial biomass (BIO) and humus (HUM), releasing CO_2 in heterotrophic respiration. The latter two pools decompose to form more CO_2, microbial biomass, and humus. Annual base decay rates for these pools are 10, 0.3, 0.67, and 0.02 y^{-1}, corresponding to turnover times of 0.1, 3.3, 1.5, and 50 years. Soil temperature and soil moisture affect these base rates, and soil texture additionally influences decomposition processes.

CENTURY, and its variant DAYCENT, is a widely used biogeochemical model with multiple compartments for litter and soil organic matter (Parton et al. 1987, 1988, 1993, 1994). The model partitions plant litter into metabolic material that readily decomposes and structural material that is more resistant to decomposition. Soil organic matter is represented by three pools of active, slow, and passive carbon that represent material with increasingly longer turnover time. These pools have a surface and a belowground component, except the passive pool, which only occurs belowground. Leaf litter enters the surface pools, and root litter enters the belowground pools. The model additionally represents fine-branch and large-wood debris, which reside on

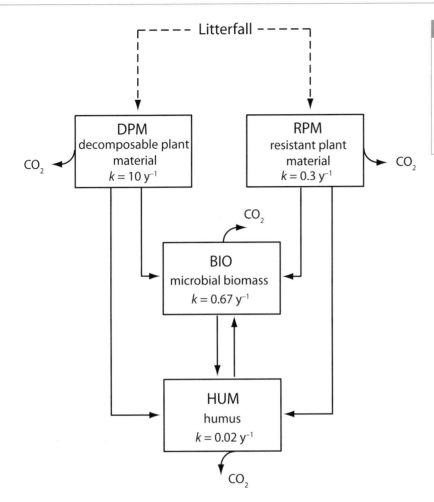

Figure 18.7 Decomposition of organic material in the RothC model. Shown is the annual base decomposition rate of each pool. The actual rate varies with temperature, soil moisture, and other factors. Solid lines indicate decomposition pathways. Curved arrows denote heterotrophic respiration loss for each pool.

the surface, and coarse root debris belowground. The model has a total of 5 litter fluxes and 12 pools to describe litter and soil organic matter. Each pool has a specific base decomposition rate, which is modified for abiotic factors. The description that follows, shown in Figure 18.8, is the version used by Bonan et al. (2013).

The total carbon flux from a donor pool equals the carbon mass of the pool c_i times a rate constant k_i such that the carbon loss is

$$\frac{dc_i}{dt} = k_i c_i. \qquad (18.29)$$

Soil temperature, soil moisture, pH, lignin, soil anaerobic conditions, soil texture, and cultivation influence the turnover of specific pools. The carbon turnover from a pool transfers to any number of receiver pools. Decomposition of organic material is assumed to be microbially mediated, and a fraction of the carbon flux is lost as microbial (or heterotrophic) respiration. The amount of carbon that flows from donor pool j to receiver pool i is

$$\frac{dc_{ij}}{dt} = \left(1 - r_{ij}\right) f_{ij} k_j c_j. \qquad (18.30)$$

The term f_{ij} is the fraction of the carbon turnover $k_j c_j$ that transfers from donor pool j to receiver pool i. The term $1 - r_{ij}$ is microbial growth efficiency, or the fraction of the carbon turnover that is assimilated into microbial biomass; the fraction r_{ij} of the flow to the receiver pool is lost as respiration. The carbon input to pool i is the m litter fluxes plus the flows to pool i from all p pools; the carbon loss is the turnover of pool i; and the carbon balance is

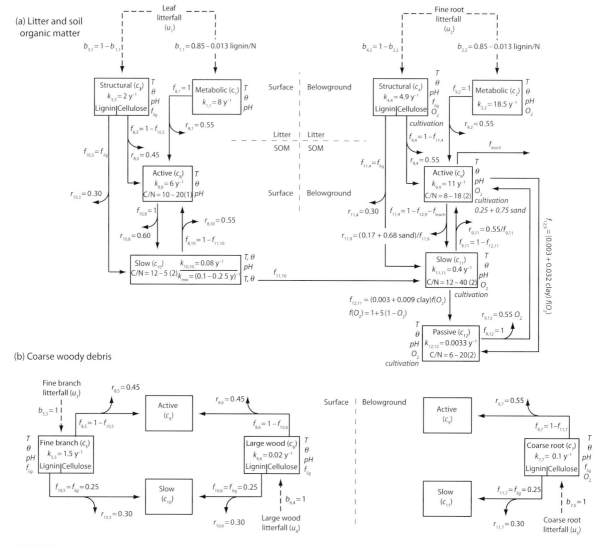

Figure 18.8 Litter, soil organic matter, and coarse woody debris pools and fluxes represented in DAYCENT, as used in Bonan et al. (2013). The model has leaf, fine root, and three coarse woody debris litter flux inputs (u_1–u_5) and twelve carbon pools (c_1–c_{12}). Shown are litter flux partitioning parameters b_{ij}, base decomposition rates k_{ii} (per year), fractional carbon transfer f_{ij}, and respiration fraction r_{ij}. Solid lines indicate decomposition pathways, with curved arrows denoting heterotrophic respiration fluxes for each pathway. DAYCENT allows for photodegradation from solar radiation, but that was not included in Bonan et al. (2013). Nor was leaching loss included. (a) Leaves decompose as surface material represented by two litter pools and two organic matter pools (shown on the left). Fine roots decompose as belowground material represented by two litter pools and three organic matter pools (shown on the right). The actual decomposition rate varies with soil temperature T, soil moisture θ, and pH. Belowground decomposition additionally varies with anaerobic conditions (O_2), cultivation, and soil texture. Structural litter decomposition also depends on lignin fraction f_{lig}. The total turnover of the surface slow pool depends on decomposition and mixing, with a fraction to the belowground slow pool and the remainder to the surface active pool. The C/N of organic matter differs among pools and varies with soil mineral nitrogen. Shown is the minimum and maximum value for each pool and (in parentheses) the soil mineral nitrogen (g N m^{-2}) for the minimum C/N. (b) Coarse woody debris decomposes to the active and slow pools. Fine-branch and large-wood debris flows to surface pools, and coarse root flows to belowground pools. Adapted from Bonan (2016)

$$\frac{dc_i}{dt} = \sum_{j=1}^{m} b_{ij}u_j + \sum_{\substack{j=1 \\ j \neq i}}^{p} (1 - r_{ij})f_{ij}k_jc_j - k_ic_i. \quad (18.31)$$

The first term on the right-hand side of the equation is the litter input summed for each of m types of litterfall (e.g., leaves, fine roots) with b_{ij} a factor for the partitioning of litter flux j to pool i. The next term is the total carbon input from other pools, and the last term is the carbon loss. Total heterotrophic respiration is

$$R_H = \sum_{i=1}^{p} \sum_{\substack{j=1 \\ j \neq i}}^{p} r_{ij}f_{ij}k_jc_j, \quad (18.32)$$

and the change in total carbon is balanced by litter inputs minus respiration loss whereby

$$\sum_{i=1}^{p} \frac{dc_i}{dt} = \sum_{j=1}^{m} u_j - R_H. \quad (18.33)$$

Equation (18.31) is easy to code, but further insight to soil carbon dynamics is gained by writing the equation in matrix form, as presented in Chapter 17. Equation (18.31) represents a system of p linear differential equations. The equivalent matrix equation (Appendix A6) is

$$\frac{d\mathbf{C}}{dt} = \mathbf{BU} + \mathbf{AKC}. \quad (18.34)$$

In this form, $\mathbf{C} = [c_1, c_2, \ldots, c_p]^T$ is a $p \times 1$ column vector of p carbon pools; $\mathbf{U} = [u_1, u_2, \ldots, u_m]^T$ is a $m \times 1$ column vector of litter fluxes for m types of litter; $\mathbf{B}$ is a $p \times m$ litter flux partitioning matrix in which b_{ij} is the partitioning of litter flux j to pool i; $\mathbf{A}$ is a $p \times p$ carbon transfer matrix in which $a_{ii} = -1$ for pool i and $a_{ij} = (1 - r_{ij})f_{ij}$ is the fraction of carbon loss from pool j entering pool i (for $j \neq i$); and $\mathbf{K}$ is a $p \times p$ diagonal matrix in which k_{ii} is the decay rate for pool i and all other elements are zero ($k_{ij} = 0$ for $j \neq i$). Equation (18.34) can represent many different model structures – e.g., RothC, as described by Sierra et al. (2012). At steady state, $d\mathbf{C}/dt = 0$ and

$$\mathbf{C} = -(\mathbf{AK})^{-1}\mathbf{BU}. \quad (18.35)$$

Equation (18.35) provides a convenient means to calculate equilibrium carbon pools for specified turnover rates, similar to that presented in Chapter 17.

Much of the detail of the model lies in the number of pools represented, their rate constants and the effects of abiotic factors on these rates, and the transfers among pools. The left portion of Figure 18.8a shows carbon flows for surface pools. Leaves become surface litter and enter either metabolic or structural litter pools. Metabolic litter is labile material that decomposes quickly. Structural material contains cellulose and lignin and decomposes more slowly. The partitioning between pools is a function of the lignin/N ratio of the residue. The fraction

$$f_{met} = 0.85 - 0.013\,(\text{lignin}/N) \quad (18.36)$$

is the portion to metabolic litter, and $1 - f_{met}$ goes to structural litter. With increases in the lignin/N ratio, more of the leaf debris is slowly decaying structural litter. The active soil organic matter pool represents live soil microbes and microbial products and decomposes rapidly. It consists of easily decomposable products from metabolic litter and the non-lignin fraction of structural litter, as well as products from the slow pool. The lignin fraction f_{lig} of structural litter decomposes to the slow pool. The slow pool includes decay-resistant plant material from the lignin portion of structural litter and soil-stabilized material from the active pool. A portion of the slow pool transfers belowground by physical mixing. Fine-branch and large-wood debris also contribute to the surface active and slow pools depending on their lignin fraction.

The base decomposition rates of surface pools are reduced by multiplicative functions of soil temperature, soil moisture, and pH in which

$$k_i = k_{base,i}f(T)f(\theta)f(\text{pH}), \quad (18.37)$$

where k_{base} is the base rate; $f(T)$ is the arctangent temperature function (18.12); $f(\theta)$ is the soil moisture function (18.13); and the effect of pH on decomposers is

$$f(\text{pH}) = b + \frac{c}{\pi}\tan^{-1}[d(\text{pH} - a)\pi]. \quad (18.38)$$

Parameters in (18.38) depend on the specific pool and the dominant decomposer (Figure 18.9). This function reduces the decay rate for pH less than 7 and becomes more pronounced with lower pH. The decomposition of structural litter and coarse woody debris additionally depends on the lignin fraction of the material so that

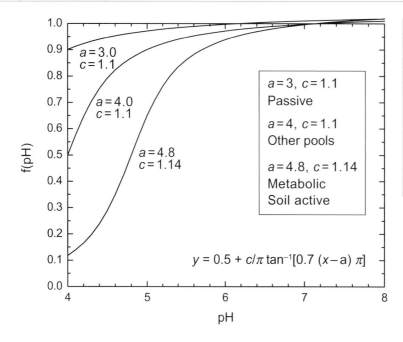

Figure 18.9 Effect of soil pH on decomposition rate. The passive pool has fungi-dominant decomposers and is least responsive to pH ($a = 3$, $b = 0.5$, $c = 1.1$, $d = 0.7$). The metabolic pools and the belowground active soil organic matter pool have bacteria-dominant decomposers and are strongly affected by pH ($a = 4.8$, $b = 0.5$, $c = 1.14$, $d = 0.7$). All other pools have a combination of decomposers and follow the intermediate curve ($a = 4$, $b = 0.5$, $c = 1.1$, $d = 0.7$).

$$k_i = k_{base, i} f(T) f(\theta) f(\text{pH}) \exp\left(-3 f_{lig}\right). \qquad (18.39)$$

Fine roots become belowground litter. Belowground pools and transfers are similar to surface pools but also include a passive soil organic matter pool. The passive pool is very resistant to decomposition, includes physically and chemically stabilized soil organic matter from the active and slow pools, and has the slowest decay rate. Some of the products from the active pool can be lost as leached organic carbon. The base decay rates of belowground pools are larger than the corresponding surface pools. Decomposition of belowground pools additionally includes an effect for anaerobic soils, represented by a rate factor denoted as O_2 that varies between zero and one and that decreases decomposition rates as soil becomes increasingly waterlogged. Cultivation also increases the decomposition rate of belowground structural and soil organic matter pools (e.g., by a factor 1.5 depending on the type of cultivation). Coarse root debris enters belowground pools.

Soil texture influences decomposition, and more soil carbon accumulates in clay than in sand. The model accounts for this through several means. The decomposition rate of the belowground active pool increases as the sand fraction increases (by a rate factor $0.25 + 0.75 \times$ sand), and the respiration fraction also increases with sand fraction (by $0.17 + 0.68 \times$ sand). The efficiency of stabilizing the active and slow pools into the passive pool varies with soil texture, with higher stabilization rates for clay soils. Consequently, the proportion of the decomposition products that enter the passive pool from the active pool increases with greater clay fraction and additionally increases with anaerobic conditions according to

$$f_{12,9} = (0.003 + 0.032 \text{ clay})[1 + 5(1 - O_2)],$$

$$(18.40)$$

with O_2 the relative effect (0–1) of soil anaerobic conditions on decomposition. The fractional flow of material to the passive pool is typically small ($f_{12,9} = 0.02$ for clay $= 0.5$ and $O_2 = 1$). For the slow pool, the fraction transferred to the passive pool is

$$f_{12,11} = (0.003 + 0.009 \text{ clay})[1 + 5(1 - O_2)],$$

$$(18.41)$$

which is similarly small (less than 0.01 for clay $= 0.5$).

Each carbon pool has an analogous organic nitrogen pool. The C/N ratio of the pool, a specified model parameter, determines the corresponding nitrogen for a given amount of carbon. The associated nitrogen flux for the carbon flow from donor pool j to receiver pool i is diagnosed from the carbon flow and the specified C/N ratios of the donor and receiver pools. The flux of nitrogen out of a particular donor pool equals the carbon flux divided by the C/N ratio of that pool. The flux of nitrogen required by the receiver pool is similarly determined from the carbon flux into the pool and the C/N ratio of the receiver pool. Nitrogen in excess of that required by the receiver pool is mineralized. Immobilization occurs when additional nitrogen is needed to satisfy the requirement of the receiver pool. Thus, mineralization of nitrogen occurs as carbon is lost in respiration and as carbon flows from pools with low C/N ratios to those with higher C/N ratios. Immobilization occurs when carbon flows from pools with high C/N ratios to those with lower C/N ratios.

Figure 18.10a–b illustrates carbon and nitrogen flows. A unit of carbon lost from donor pool j carries with it an amount of organic nitrogen related to its C/N ratio by $(C/N)_j^{-1}$. A portion r_{ij} of the carbon is lost as respiration as it flows to receiver pool i, and $r_{ij}(C/N)_j^{-1}$ is released as mineral nitrogen. The remaining carbon $1 - r_{ij}$ transfers to the receiver pool carrying with it an amount of organic nitrogen equal to $(1 - r_{ij})(C/N)_j^{-1}$. The receiver pool requires that incoming carbon have the C/N ratio specific for that pool, and the nitrogen required by receiver pool i as it gains a unit of carbon from donor pool j is $(1 - r_{ij})(C/N)_i^{-1}$. The difference between the nitrogen required by the receiver pool and the nitrogen input from the donor pool is the additional nitrogen (positive difference) or excess nitrogen (negative difference) required to satisfy the flow. The net mineral nitrogen flux from the soil associated with the flow from j to i is

$$\text{mineral } N_{j \to i} = -\frac{r_{ij}}{(C/N)_j} + (1 - r_{ij})\left[\frac{1}{(C/N)_i} - \frac{1}{(C/N)_j}\right].$$

(18.42)

The first term on the right-hand side of the equation is the nitrogen mineralization (a negative flux) during respiration, and the second term is the immobilization flux. A positive value denotes net gain of nitrogen (immobilization); a negative value denotes net loss of excess nitrogen (mineralization). Immobilization occurs when the C/N ratio of donor pool j exceeds that of receiver pool i; the flow $(1 - r_{ij})(C/N)_j^{-1}$ is less than the required flow $(1 - r_{ij})(C/N)_i^{-1}$ because of the lower C/N ratio of the receiver. The additional required nitrogen is obtained from soil mineral nitrogen. Mineralization occurs when the C/N ratio of donor pool j is less than the C/N ratio of receiver pool i; the flow

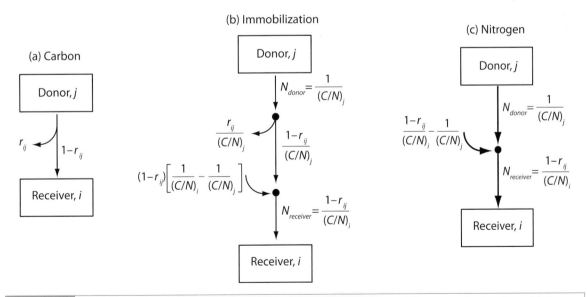

Figure 18.10 Carbon and nitrogen fluxes during transfer from a donor pool to a receiver pool. Shown are (a) carbon fluxes, (b) nitrogen fluxes specific for immobilization, and (c) generalized nitrogen fluxes. All fluxes are normalized per unit carbon loss ($f_{ij} k_j c_j$).

$(1 - r_{ij})(C/N)_j^{-1}$ exceeds the required flow $(1 - r_{ij})(C/N)_i^{-1}$ because of the higher C/N ratio of the receiver pool. The excess nitrogen is mineralized, and the second term in (18.42) is a negative flux.

Figure 18.10c shows a simplified form of these equations. The donor pool loses an amount of nitrogen equal to $(C/N)_j^{-1}$, and the receiver pools gains an amount of nitrogen equal to $(1 - r_{ij})(C/N)_i^{-1}$. The difference between these fluxes is the additional soil mineral nitrogen required to satisfy the flow whereby

$$\text{mineral } N_{j \to i} = \frac{1 - r_{ij}}{(C/N)_i} - \frac{1}{(C/N)_j}. \tag{18.43}$$

The net flux given by (18.42) is equivalent to (18.43). The total net immobilization or mineralization from all carbon flows is

$$F_{n,net} = \sum_{i=1}^{p} \sum_{\substack{j=1 \\ j \neq i}}^{p} \left[\frac{1 - r_{ij}}{(C/N)_i} - \frac{1}{(C/N)_j} \right] f_{ij} k_j c_j. \tag{18.44}$$

This is the nitrogen flux term that appears in the soil nitrogen balance given by (17.36).

Immobilization or mineralization occurs for each carbon flow among pools depending on the C/N ratios of the pools. The C/N ratio in structural pools is set to a high value (C/N = 200), while the ratio in metabolic pools and coarse woody debris varies with the nitrogen content of plant residues. Each soil organic matter pool has an allowable range of C/N ratios, specified by a linear function of soil mineral nitrogen (Paustian et al. 1992) as illustrated in Figure 18.11. Low soil nitrogen results in high C/N ratios in the various soil organic matter pools, so that less nitrogen is immobilized. High soil nitrogen results in low C/N ratios and more immobilization. Soil nitrogen limits litter decomposition if there is insufficient nitrogen to satisfy the immobilization demand.

Bonan et al. (2013) compared DAYCENT with results from litter decomposition experiments. Required litter flux inputs are the amount of litter, its carbon and nitrogen, the lignin/N ratio, and the lignin fraction f_{lig}. These were specified from observations for various types of leaves. The model simulates litter decomposition and nitrogen dynamics consistent with observations. Figure 18.12 shows observed leaf litter decomposition in conifer forest over 10 years, which the model closely replicates. Figure 18.13 shows the nitrogen dynamics. The nitrogen in leaf litter initially increases as the litter decomposes and immobilization occurs. Sugar maple litter with an initial nitrogen concentration of 0.81% (by mass) immobilizes less nitrogen than wheat litter with an initial nitrogen of 0.38%. At some critical concentration, net mineralization

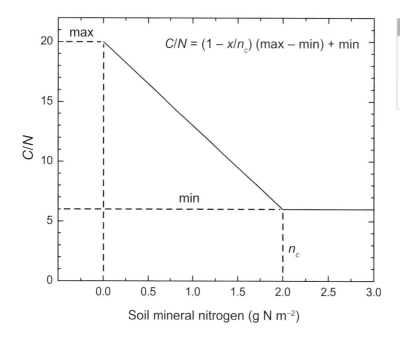

Figure 18.11 Soil organic matter C/N ratio in relation to soil mineral nitrogen. Figure 18.8 shows the minimum and maximum C/N for the five soil organic matter pools and the soil mineral nitrogen for the minimum C/N. This specific graphic is for the passive pool.

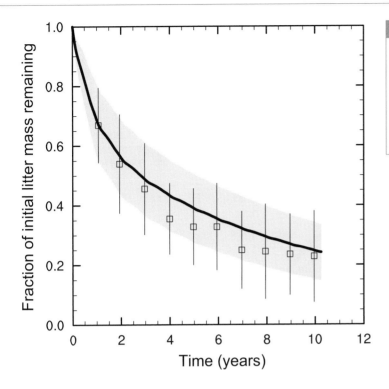

Figure 18.12 Fraction of initial carbon remaining for leaf litter placed in conifer forest. Data are averaged across six types of leaf litter decomposed at five different locations. Shown are the observations (symbols) with ± one standard deviation. Model output for DAYCENT is continuous in time, and the shading indicates ± one standard deviation. Adapted from Bonan et al. (2013)

occurs and nitrogen in the litter decreases. The model closely matches the observed maple litter, and results with low and high soil organic matter C/N ratios bracket the observations. The wheat litter simulation also corresponds to the observations, better for high C/N ratios than for low C/N ratios.

Question 18.4 Compare litter and soil carbon pools and flows in CASA-CNP (Figure 17.2) with DAYCENT (Figure 18.8).

Question 18.5 What are the site inputs needed for DAYCENT?

Question 18.6 Calculate turnover times for each of the carbon pools in DAYCENT, as shown in Figure 18.8.

Question 18.7 Discuss how lignin affects soil carbon in DAYCENT.

Question 18.8 Soil pH, texture, and cultivation are not commonly included in biogeochemical models. Discuss how these affect soil carbon in DAYCENT.

Question 18.9 With reference to Figure 18.8, which carbon flows in DAYCENT result in immobilization? Which flows produce net mineralization?

18.6 | Model Frontiers

Many terrestrial biosphere models use a discrete pool approach comparable to RothC or CENTURY to simulate soil biogeochemistry. This approach captures the overall dynamics of litter decomposition and soil organic matter formation, especially the effects of substrate quality and abiotic factors (e.g., temperature, moisture). However, it treats soil carbon as bulk pools without explicit vertical resolution. In fact, soil carbon has a pronounced profile in which substantial carbon is below 20 cm depth (Figure 2.12). Koven et al. (2013) developed a variant for the Community Land Model that applies the equations in a vertically resolved soil. The same equations apply at each soil layer, but litter inputs, pools, and decay rates vary with depth and vertical mixing between layers is possible (Figure 18.14). In this approach the carbon balance of pool i is

$$\frac{dc_i(z)}{dt} = u_i(z) + \sum_{\substack{j=1 \\ j \neq i}}^{p} \left(1 - r_{ij}\right) f_{ij} k_j(z) c_j(z) - k_i(z) c_i(z)$$
$$+ \frac{\partial}{\partial z} \left[D(z) \frac{\partial c_i}{\partial z} \right].$$

$$(18.45)$$

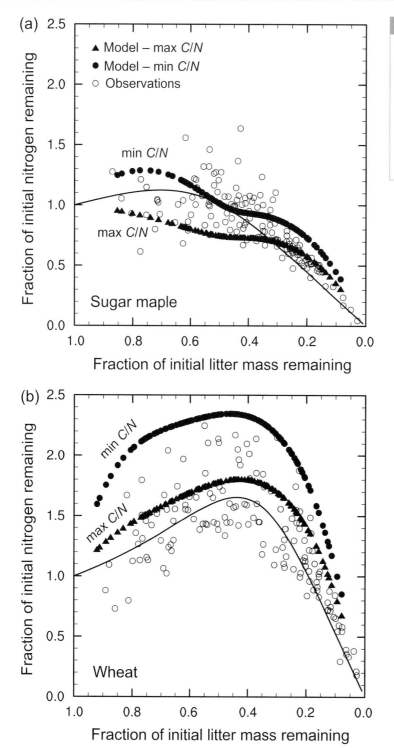

Figure 18.13 Fraction of initial nitrogen remaining in relation to carbon mass remaining for (a) sugar maple (*Acer saccharum*) leaf litter (initial nitrogen, 0.81%) and (b) wheat (*Triticum aestivum*) leaf litter (initial nitrogen, 0.38%). Shown are the observations (open symbols) and the best-fit equation for the observations (solid line). DAYCENT data show two different simulations with minimum (min) and maximum (max) C/N ratios. Adapted from Bonan et al. (2013)

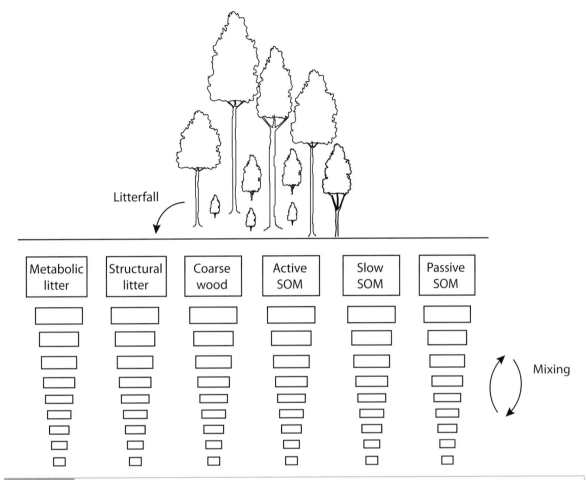

Figure 18.14 Depiction of soil carbon in a vertically resolved model. Carbon pools are defined for each soil layer. Carbon transfers from litter to soil organic matter and among soil organic matter pools within a layer. Carbon pools also mix vertically between layers. Adapted from Koven et al. (2013)

The first three terms on the right-hand side of the equation are as before, but pool size c_i (kg C m^{-3}) and decay rate k_i are defined for each soil layer. As used here, u_i is the vertically distributed litter input (kg C m^{-3} s^{-1}). The last term is vertical diffusion similar to that for soil temperature (Chapter 5) and soil moisture (Chapter 8) with D a diffusivity (m^2 s^{-1}) to account for vertical mixing. A similar approach can be used in other models – e.g., JULES with the RothC soil carbon (Burke et al. 2017). How the decay rate varies with depth is a critical control of soil carbon accumulation. As implemented in the Community Land Model, k_i decreases exponentially with depth. In addition, the diffusivity controls the rate of mixing between layers.

Question 18.10 Vertically resolved models allow carbon to move to deeper soil. Does DAYCENT allow for similar vertical mixing?

Biogeochemical models such as RothC and CENTURY follow a paradigm in which decomposition loss of carbon from a pool is controlled by the size of the pool and follows first-order decay as represented by (18.29). This results in a system of linear equations that describe litter decomposition and soil organic matter (18.34). However, the models do not explicitly represent soil microbes and their role in organic matter decomposition. The environmental rate modifiers that control decomposition are heuristic functions that adjust turnover rates for conditions in which microbial activity is enhanced or restricted. A newer class of microbial models authentically represents microbes, their physiology, and their role in decomposition. These models are consistent with theoretical and experimental evidence that soil microbes mediate the formation of soil organic matter through production of microbial products that form stabilized soil organic matter.

Microbial models are based on the premise that substrate availability for microbially mediated transformations limits decomposition rates. This is represented using Michaelis–Menten kinetics (Chapter 4) in which decomposition loss depends on both substrate availability (i.e., pool size c_i) and a catalyst (i.e., microbial biomass), but there are two contrasting formulations (Wieder et al 2015a). Some microbial models use the traditional Michaelis–Menten equation in which decomposition loss relates to microbial biomass by

$$\frac{dc_i}{dt} = c_{mic} \frac{v_{max}c_i}{K_m + c_i}, \qquad (18.46)$$

where c_{mic} is the mass of microbes, v_{max} is the maximum rate of substrate carbon assimilation per unit microbial biomass, and K_m is the half-saturation constant. In this equation, decomposition loss saturates as the substrate pool size increases. An alternative is to use a reverse Michaelis–Menten equation in which

$$\frac{dc_i}{dt} = c_i \frac{v_{max}c_{mic}}{K_m + c_{mic}}. \qquad (18.47)$$

In this form, decomposition loss depends nonlinearly on microbial biomass so that v_{max} and K_m have different definitions than in (18.46). The mathematical properties of microbial models are more complex than linear models and differ depending on use of Michaelis–Menten or reverse Michaelis–Menten kinetics (Wang et al. 2016). The models can have oscillatory response of carbon pools to a small increase in initial pool size, less so with reverse Michaelis–Menten kinetics.

One such microbial model is the Microbial-Mineralization Carbon Stabilization (MIMICS) model (Wieder et al. 2014, 2015d, 2018). The model represents metabolic and structural litter pools as in CENTURY but defines soil organic matter based on whether it is physically protected, chemically protected, or available for microbes (Figure 18.15). Microbes are represented by the biomass of two different functional groups (r-selected bacteria or K-selected fungi). Carbon transfers among these seven pools are described by a system of nonlinear equations. Decomposition of metabolic litter, structural

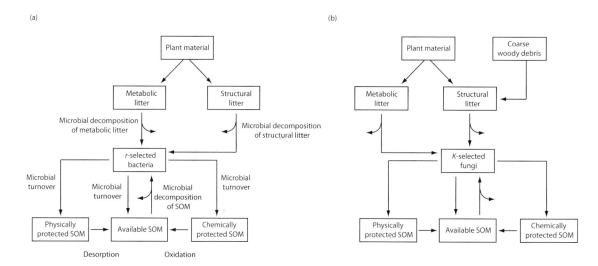

Figure 18.15 Depiction of carbon flows in MIMICS. (a) r-selected bacteria consume metabolic litter, structural litter, and available soil organic matter with CO_2 respiration loss. Microbial biomass turns over to soil organic matter (SOM). Physically and chemically protected soil organic matter can transfer to the available pool. (b) Similar flows involve K-selected fungi, shown separately for clarity. Adapted from supplementary material in Wieder et al. (2018)

litter, and available soil organic matter is controlled by reverse Michaelis–Menten kinetics in which consumption by either r-selected or K-selected live microbial pools is limited by microbial mass. Temperature determines the rate of decomposition through its effect on microbial physiology as represented by v_{max} and K_m. Heterotrophic respiration loss occurs during consumption of carbon by microbes. Microbial growth efficiency determines what portion of the consumed carbon is incorporated into microbial growth and what portion is lost as respiration. A low microbial growth efficiency means a greater portion of carbon is respired and less is incorporated into microbial biomass. Microbial turnover transfers carbon to soil organic matter. Desorption of physically protected soil organic matter and oxidation of chemically protected material transfers carbon to the available pool.

Whereas the prevailing theory of soil organic matter formation represented by RothC and CENTURY is based on chemical recalcitrance and a cascade from high- to progressively lower-quality material, MIMICS specifies decomposition in relation to microbial activity and access of microbes to soil organic matter. In addition, some of the by-products of microbial turnover are physically or chemically protected and are inaccessible to microbes. Microbial models can produce different responses of soil carbon to litter addition or soil warming compared with conventional models such as RothC and CENTURY. In particular, whereas conventional models simulate a decline in soil carbon with warming, microbial models simulate a decrease or increase in soil carbon depending on particular model assumptions. However, the specific details of representing microbes at large spatial scales are not well known, and particular assumptions about interactions between microbes and the soil environment produce divergent model results (Wieder et al. 2018).

The soil mineral nitrogen budget is a balance among nitrogen deposition, fixation, fertilization, net mineralization, plant uptake, and gaseous and aqueous losses, as given by (17.36). Many biogeochemical models solve this balance sequentially in which plant and microbes have first access to nitrogen, followed by nitrifiers and denitrifiers utilizing the remaining nitrogen and, finally, aqueous loss of any residual nitrogen. Competition between plants and microbes for nitrogen is especially problematic (Zhu et al. 2017). Some models give priority to immobilization so that mineralized nitrogen is first used by microbes, and the remainder is available for plants. Other models represent microbes and plants as being equally competitive for nitrogen. A newer approach attempts to represent these processes simultaneously with multiple consumers (plants, decomposing microbes, nitrifiers, denitrifiers, mineral surfaces) competing for NH_4^+ and NO_3^- (Zhu and Riley 2015; Zhu et al. 2016, 2017).

18.7 | Supplemental Programs

18.1 DAYCENT: This program simulates litter decomposition and soil organic matter dynamics using DAYCENT as described in Figure 18.8. The program follows carbon and nitrogen pools and fluxes associated with the addition of a single pulse of leaf litter on the surface.

18.8 | Modeling Projects

1 Use Supplemental Program 18.1 to simulate decomposition over ten years for various types of litter shown in Table 18.1. How does f_{met} and f_{lig} affect the carbon mass remaining?
2 Modify Supplemental Program 18.1 to calculate equilibrium carbon pools given a constant flux of litter. Contrast soil carbon accumulation for equivalent amounts of leaf litter, fine root litter, fine branches, large wood, and coarse roots.
3 How sensitive are the results of the preceding project to pH and soil texture?

Vegetation Demography

Chapter Overview

Terrestrial ecosystems undergo temporal dynamics in plant populations, community composition, and ecosystem structure. These changes in ecosystems are driven by demographic processes of recruitment, establishment, growth, and mortality and require models distinctly different from biogeochemical models. This chapter provides an overview of this class of models with three specific examples. Individual-based forest gap models track the birth, growth, and death of individual trees in an area of land. Dynamic global vegetation models simulate changes in the area occupied by discrete patches of plant functional types. Ecosystem demography models define patches based on age since disturbance and simulate the dynamics of cohorts of similar plant functional types rather than tracking every individual. Common to each model is the representation of vegetation demography, with age- and size-dependent growth and mortality and in which growth is constrained by allometric relationships of stem diameter with height, sapwood area, leaf area, and biomass. Rather than biogeochemical cycles as in Chapter 17, vegetation demography provides the dynamical core for the next generation of terrestrial biosphere models.

19.1 | Introduction

Biogeochemical models can be used to describe the cycling of carbon and nutrients within an ecosystem and their responses to changes in climate, atmospheric CO_2 concentration, or other perturbations. However, terrestrial ecosystems change with time in ways other than just biogeochemical cycles. For example, the height of trees, their size distribution, and species diversity affect not only the direction, rate, and magnitude of change in forest ecosystems but also determine habitat and forest products. This change occurs over time periods of decades to centuries in response to human disturbances from logging, forest clearing, or land abandonment and also from natural disturbances such as wildfires, windstorms, and insect infestations. This process of change is known as forest succession – a term broadly used by ecologists to characterize temporal changes in population, community composition, and ecosystem structure. A similar temporal dynamics occurs in grasslands and prairies. Woody encroachment in shrubland or savanna is another type of vegetation dynamics. Tree mortality in response to drought changes the composition and structure of forests. Warmer temperatures produce a greening of arctic tundra. Over long time periods of centuries to millennia, the biogeography of broad classes of biomes such as forest, shrubland, savanna, grassland, tundra, and desert rearranges in response to climate change. These changes in ecosystems are understood in terms of vegetation demography – the recruitment and establishment of plants, their growth, and their mortality – and require another class of models distinctly different from biogeochemical models to capture the temporal dynamics of vegetation.

The coupling of plant demography and vegetation dynamics with climate models to form Earth system models has been underway for several decades. The types of models proposed for coupling can be broadly grouped into three categories. Individual-based models track the birth, growth, and death of individual plants in a patch of land (Figure 19.1a). A landscape may consist of numerous such patches that vary in stage of development. Whereas biogeochemical models encode their calculations on an areal basis (e.g., kg C m^{-2}), individual plant models base their calculations on an individual (e.g., kg C per individual). Total plant biomass is not modeled per se as a fundamental property as in biogeochemical models, but rather is the sum of the biomass of all the individual plants growing in an area of land. Individual plant models avoid the conceptual fallacy of biogeochemical models: whereas growth, competition for resources, and mortality occur among individuals, there are no individual plants in biogeochemical models. This is most evident in the allocation rules used to distribute productivity to plant biomass, which are observed for individual plants but are applied in biogeochemical models at the scale of ecosystems. Allocation, however, varies with the size of trees in known allometric relationships (Chapter 2). Mortality, too, is age- or size-dependent, but biogeochemical models represent turnover of plant material as a proportion of biomass. Mortality from fire varies with the stem diameter and height of trees, and harvesting of wood products, too, is size-dependent. Dynamic global vegetation models were introduced in the 1990s and 2000s as a computationally efficient means to simulate vegetation dynamics without the burden of individual plant models. These models characterize vegetation as separate patches of plant functional types within a model grid cell and simulate changes in the area occupied by a plant functional type (Figure 19.1b). A third class of models, termed ecosystem demography models, simulates the dynamics of cohorts of similar-sized plant functional types rather than tracking every individual. Patches are defined by age since disturbance, and multiple cohorts can co-occur within a patch (Figure 19.1c). Ecosystem demography models are more similar to individual-based models than are area-based dynamic global vegetation models, allow vegetation composition to change with time, and are computationally efficient for global implementation.

This chapter provides an overview of these three classes of models. A particular type of individual-based model called forest gap models is discussed first (Shugart 1984; Botkin 1993; Bugmann 2001). These models are authentic in their representation of vegetation demography but are thought to be impractical for the global scale. The Lund–Potsdam–Jena (LPJ) model (Sitch et al. 2003) is an example of an area-based dynamic global vegetation model that has been used in Earth system models. A distinguishing feature of the model, in contrast with other such models, is that it incorporates demographic processes by separately simulating the average individual and population density within a patch. However, the model does not genuinely represent the arrangement of plants into communities and the vertical structure of plant canopies that drives the dynamics of forests and other ecosystems. The next-generation terrestrial biosphere model will use ecosystem demography concepts to represent cohorts of individuals of similar functional traits and life history in a vertically structured plant canopy (Fisher et al. 2018). Demography, not biogeochemical cycles per se, provides the dynamical core for the next generation of terrestrial biosphere models. Incorporation of vegetation demography at the global scale for use with Earth system models is in its infancy (Fisher et al. 2018). The prevailing paradigm uses cohorts of similarly sized individuals, though SEIB is a notable exception in is representation of individuals in a patch of vegetation (Sato et al. 2007).

19.2 | Forest Gap Models

Gap models simulate forest dynamics as a population process, following the annual establishment, growth, and death of individual trees in a small patch of land (Figure 1.5). Each tree is characterized by its species, stem diameter, height, and age. Within this patch, trees compete for light, soil moisture, and nutrients. The first gap model conceived of a patch as 10 m × 10 m (Botkin et al. 1972), but subsequent versions expanded the scale to 833 m^2 (1/12 ha), which is the size of an opening in the canopy created by the death of a large, dominant tree and is the scale at which neighboring trees compete for light and other resources (Shugart and West 1977; Shugart 1984).

(a) Gap model

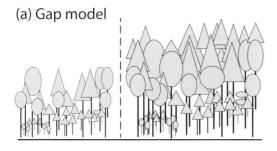

(b) Area-based DGVM

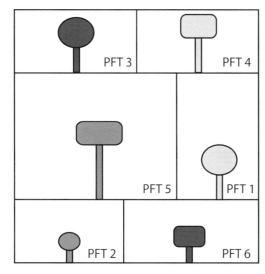

(c) Ecosystem demography

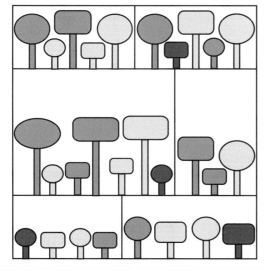

Figure 19.1 Representation of vegetation patches in models. (a) Forest gap models are individual-based models and represent a

The patch undergoes temporal changes in the density, size, and composition of trees with the formation of a gap in the canopy (Figure 19.2). Community composition, biomass, productivity, and biogeochemical cycles are emergent outcomes of individual trees interacting among themselves and with the environment to acquire the resources necessary for growth and survival. Some models additionally include disturbances such as wildfire. Gap models represent the landscape as a mosaic of hundreds of independent forest patches, each of which can differ in species composition and stage of development in response to disturbance that creates an opening in the canopy.

The first gap model, JABOWA (Botkin et al. 1972), was developed for northern hardwood forests in northeastern United States, followed by FORET (Shugart and West 1977) for forests of eastern Tennessee. The original formulation of JABOWA decreased tree growth as stand biomass approached some specified maximum so as to simulate competition for belowground resources. Pastor and Post (1986) added a nitrogen cycle to explicitly simulate the effect of nutrient availability on growth of trees in eastern North America (see the litter cohort model in Chapter 18) and also added a soil moisture factor (Figure 19.3). Bonan (1989, 1990a,b) expanded the model to boreal forests by including a soil temperature factor to account for permafrost and cold soils. Other developments improved the calculation of shading, formulated spatially explicit versions, devised new growth equations, and reconsidered environmental influences on growth (Bugmann 2001). Gap models have been implemented for numerous forests throughout the world (Shugart and Woodward 2011). While most studies have

landscape as patches that differ in stage of post-disturbance development. Shown are two patches, each comprised of multiple trees that differ in size and species. (b) Area-based dynamic global vegetation models (DGVMs) represent vegetation as discrete patches of plant functional types (PFTs) that differ in area. In this example, the model grid cell has six plant functional types (distinguished by canopy shape and shading) that differ in biomass (size of plant) and patch area (size of subgrid tile). (c) Ecosystem demography models define patches based on time elapsed since disturbance. Multiple plant functional types can exist within a patch and are represented as cohorts defined by plant type and size. Shown are six patches with different subgrid area and cohort assemblages.

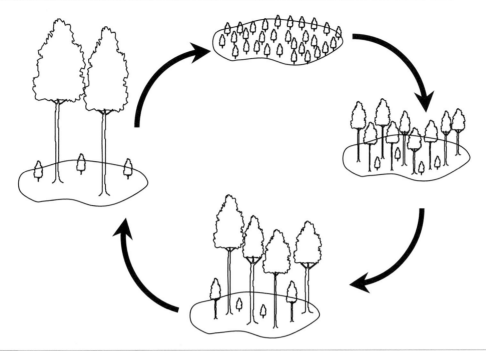

Figure 19.2 Cyclic growth and thinning of trees in a forest patch during gap dynamics. Reproduced from Bonan (2016)

been local in extent, recent studies have applied a model to forests across Russia (Shuman et al. 2013, 2014, 2015). Botkin et al. (1972) provided a thorough description of the equations used in JABOWA, and these were carried forth with subsequent models (Shugart 1984; Bugmann 2001).

In gap models, allometric relationships are a critical driver of individual tree growth. Height is particularly important for its effect on stem diameter increment, both directly through tree volume growth and indirectly through shading. The height h (m) of a tree is calculated from its stem diameter D (cm) using the equation

$$h = 1.37 + b_2 D - b_3 D^2. \tag{19.1}$$

As used by Botkin et al. (1972) and subsequent models (Shugart 1984), the parameters b_2 and b_3 vary among species and are estimated from the maximum diameter D_{max} and height h_{max} observed for the particular species (Figure 19.4a). If $dh/dD = 0$ and $h = h_{max}$ when $D = D_{max}$, then

$$b_2 = 2\left(\frac{h_{max} - 1.37}{D_{max}}\right) \tag{19.2}$$

and

$$b_3 = \frac{h_{max} - 1.37}{D_{max}^2}. \tag{19.3}$$

A second critical allometric relationship pertains to tree biomass. Biomass allocation is not modeled directly but rather using empirical equations that constrain foliage, stem, and root mass for a given size tree (e.g., Figure 2.11a). Of particular importance is the relationship between stem diameter and leaf area, which drives light extinction in the canopy (Figure 19.4b).

The annual growth of a tree is calculated from its diameter and height as modified by light, climate, and site conditions. An equation for diameter increment is obtained from tree volume, which relates to $D^2 h$. The annual volume increment is

$$\frac{d(D^2 h)}{dt} = rL\left(1 - \frac{Dh}{D_{max}h_{max}}\right), \tag{19.4}$$

where r is a growth rate and L is leaf area. In this equation, tree growth is directly proportional to its leaf area and decreases as the tree approaches its maximum size. If leaf area is proportional to stem diameter squared (i.e., $L = cD^2$), (19.4) can be replaced by

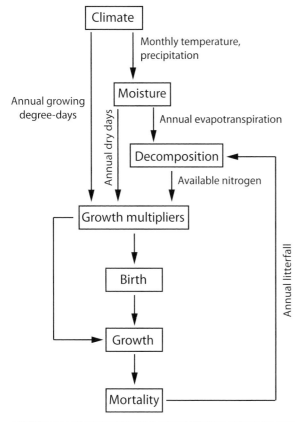

Figure 19.3 Structure of a forest gap model for eastern North America. Adapted from Pastor and Post (1986)

$$\frac{d\left(D^2h\right)}{dt} = G_0 D^2 \left(1 - \frac{Dh}{D_{\max}h_{\max}}\right), \qquad (19.5)$$

with $G_0 = rc$ a growth parameter. From the chain rule:

$$\frac{d\left(D^2h\right)}{dt} = \frac{d\left(D^2h\right)}{dD}\frac{dD}{dt}. \qquad (19.6)$$

Multiplying (19.1) by D^2 gives an expression for D^2h so that

$$\frac{d\left(D^2H\right)}{dD} = 2.74D + 3b_2 D^2 - 4b_3 D^3 \qquad (19.7)$$

and

$$\frac{dD}{dt} = \frac{G_0 D \left(1 - \dfrac{Dh}{D_{\max}h_{\max}}\right)}{2.74 + 3b_2 D - 4b_3 D^2}. \qquad (19.8)$$

The growth parameter G_0 can be estimated by assuming that a tree reaches two-thirds its

maximum diameter at one-half its maximum age (Botkin et al. 1972; Shugart 1984). Figure 19.5 illustrates the different growth strategies of two trees found in the forests of eastern United States. Blackjack oak is a small but long-lived deciduous tree growing up to 15 m tall, commonly found on poor, dry soils in southeastern United States. Basswood is a fast growing, larger tree that is found on mesic, nutrient-rich sites in more northern locations than blackjack oak.

Competition for light is a critical driver of forest dynamics and is represented through shading of smaller individuals by taller trees, which decreases the growth of suppressed trees in the understory. Gap models account for shading by creating a vertical profile of leaf area in the patch. There is no horizontal spatial structure within the patch, but there is vertical structure in which trees are arranged into canopy layers. In this representation of a canopy, all of a tree's leaf area L is located at its height h and is spread over the entire patch area (Figure 19.6a-b). The height of a tree determines its location in the cumulative leaf area profile, and in this way taller trees shade smaller trees. A particular tree with height h receives an amount of light determined by

$$I^{\downarrow}(h) = e^{-K \cdot L(h)},$$

with K a light extinction coefficient and $L(h)$ equal to the cumulative leaf area above h (Figure 19.4c). The diameter growth rate given by (19.8) is multiplied by the factor:

$$f\left[I^{\downarrow}(h)\right] = c_1\left\{1 - e^{-c2\left[I^{\downarrow}(h) - c_3\right]}\right\}.$$

Species that are intolerant of shade grow best with high light; other species can tolerate low light (Figure 19.4d).

Tree growth is adjusted for site conditions in addition to light availability (Figure 19.7). A temperature factor based on annual growing degree-day summation of daily temperature above some threshold was used in JABOWA (Botkin et al. 1972) and subsequent gap models. Tree growth is defined by some minimum and maximum number of degree-days, which are estimated from biogeographical range limits (e.g., northern and southern range limits for northern temperate and boreal trees). Growth response is assumed to be parabolic between these limits, increasing to a maximum at

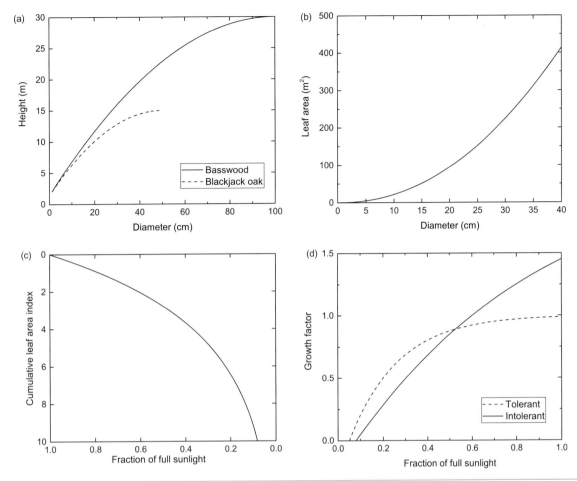

Figure 19.4 Tree height and light competition in a forest gap model. (a) Height in relation to stem diameter for two species of trees. (b) Leaf area in relation to stem diameter. (c) Light profile in relation to cumulative leaf area index. (d) Light growth factors for shade tolerant and intolerant species. Height is shown for basswood (*Tilia americana*) and blackjack oak (*Quercus marilandica*) with parameters from a model of forests in eastern North America (Pastor and Post 1985). Leaf area and light extinction are from FORET (Shugart 1984). See also Bugmann (2001).

midrange. Pastor and Post (1986) added a soil moisture factor in which growth decreases with greater drought and a nitrogen factor in which growth increases with greater nitrogen availability.

Mortality is represented by the death of individual trees and is treated as a stochastic process. Trees die with a constant probability each year. The probability that a tree will be dead by year n is

$$P_m = 1 - (1 - c)^n,$$

in which c is an annual probability of mortality. A value $c = 4.0/\text{age}_{max}$ means that approximately 2% of trees survive to reach their maximum age; $c = 4.605/\text{age}_{max}$ gives a 1% chance of survival.

The probability of mortality increases when tree growth is less than some minimum (e.g., 0.1 cm y^{-1}) for two consecutive years. The rationale is that suppressed trees in the understory (from low light) or slow-growing trees (e.g., from drought) have a greater likelihood of dying than do healthy trees. A suppressed tree has a 1% chance of surviving for 10 years with $c = 0.368$. Wildfire and insect outbreaks can be included in gap models (Bonan 1989). The occurrence of fire is treated stochastically with an annual probability of burning. An individual patch may, for example, have a 1% change of burning in any given year. Fuel load is used to classify fire severity as surface fires that kill small

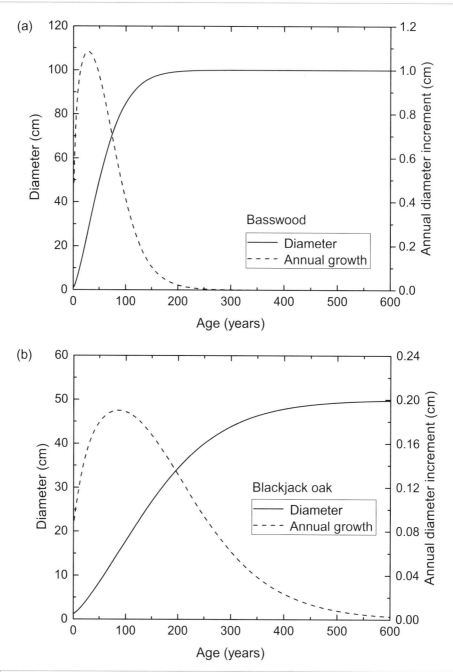

Figure 19.5 Stem diameter growth in relation to age as represented in gap models. Relationships are shown for basswood ($D_{max} = 100$ cm, $h_{max} = 30$ m, $G_0 = 188.7$ cm y^{-1}, maximum age 140 years) and blackjack oak ($D_{max} = 50$ cm, $h_{max} = 15$ m, $G_0 = 34.0$ cm y^{-1}, maximum age 400 years). Parameter values are from Pastor and Post (1985)

trees, intense fires that also kill larger trees, and lethal fires that kill all trees. Particular species of trees differ in their tolerance of fire. This method of determining fire occurrence treats individual patches separately. In fact, fires spread across the landscape, and spatially explicit methods are needed to account for this spread. Insect outbreaks can also be treated as a stochastic process.

(a) 3-D

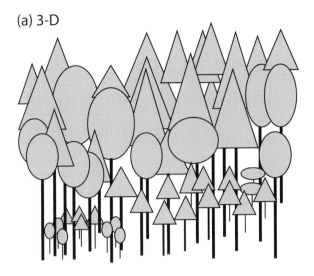

(b) Thin, flat crowns

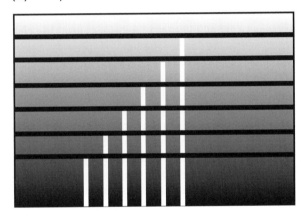

(c) PPA

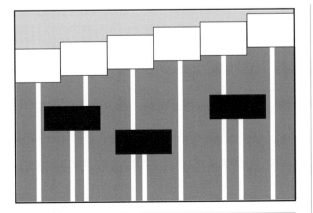

Figure 19.6 Representation of plant canopies in vegetation dynamics models. (a) Actual canopies have complex three-dimensional structure determined by the spatial location of trees and their crown geometry. (b) Gap models simplify the canopy so that the crown of an individual tree is a thin, flat layer of leaves at the top of the tree. Shown are six trees with different heights (white vertical lines). The leaf area of each tree (black horizontal lines) spreads over the area of the patch so that the tallest tree shades all others, and so forth, through the canopy. Horizontal positioning is for illustration only. The models do not represent spatial location, only the vertical dimension. Darker shading denotes progressively less light deeper in the canopy. The ecosystem demography model (ED) uses a similar concept, but applied to cohorts rather than individual trees. (c) The perfect plasticity approximation (PPA) organizes canopies into layers. All cohorts in the overstory (white boxes) receive identical light for that layer. The understory (black boxes) receives less light. Spatial location is for illustration only. See also Fisher et al. (2018).

Regeneration is based on sapling establishment from seed or by sprouting and is also treated stochastically. The seeds of all species are assumed to be present on-site, but available light at the forest floor, climate tolerances, and other site conditions determine which species become established. The number of trees that establish is treated as a random variable specified by a species-specific maximum number adjusted for site conditions. Species that can sprout do so once trees have reached a certain size. Initial stem diameter is treated as a random variable.

The strength of gap models is that they realistically simulate forest succession as resulting from demographic processes. Species are characterized by life history characteristics such as small or large stature, short-lived or long-lived, shade tolerant or intolerant, tolerant of drought or low soil nitrogen, fecundity, ability to establish in shade, tolerance of fire, and tolerance of warm or cold climate. Forest dynamics are an outcome of these life history characteristics operating in a gradient of environmental conditions. Figure 19.8, for example, shows simulated forest succession in interior Alaska following fire. The site is initially dominated by fast-growing birch trees (*Betula papyrifera*), but these trees are short-lived and are replaced by slower growing white spruce (*Picea glauca*). Gap models also give detailed information about stand structure such as basal area (Table 19.1). Gap models have been used to examine the response of forests to past climate change (Solomon et al. 1980, 1981; Bonan and

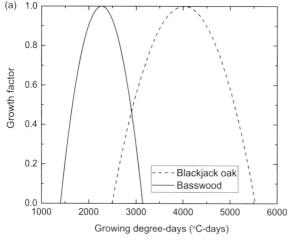

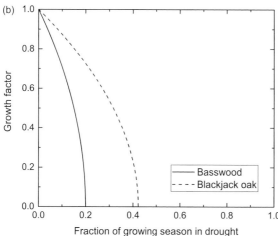

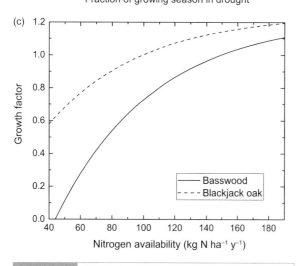

Figure 19.7 Growth factors for (a) temperature (growing degree-days), (b) soil moisture, and (c) soil nitrogen illustrated for basswood and blackjack oak (Pastor and Post 1985). Basswood grows at more northern locations than blackjack oak, on wetter soils, and is intolerant of low soil nitrogen.

Hayden 1990) and future climate change (Solomon 1986; Pastor and Post 1986, 1988).

Gap models are authentic in their representation of demography, life history characteristics, and forest dynamics, but they have not been used in Earth system models. Gap models are stochastic models of forest succession in which important demographic events (e.g.. mortality, regeneration) occur at random. Each individual patch may have a different age, community composition, or biomass as a result, but the landscape as a whole consists of an ensemble of patches and exhibits more stable properties in which multiple species co-occur. The richness of the model is obtained from this random-ness, but there is no deterministic or analytical solution. Instead, the dynamics of many individual forest patches are averaged to give the expected ensemble mean. Gap models do not simulate short-timescale leaf physiological processes. Biomass pro-duction is diagnosed from annual stem diameter increment using allometric relationships. This con-trasts with leaf-scale models of photosynthesis (Chapter 11) and stomatal conductance (Chapter 12) and big-leaf models that scale leaf processes to cal-culate gross primary production for a canopy of leaves (Chapter 15). Gap models can be reconciled with big-leaf canopy models so that tree growth is consistent with short-term leaf physiology (Friend et al. 1993, 1997), but the advent of dynamic global vegetation models led to interest in that new type of model for coupling to Earth system models.

Question 19.1 Derive the expressions for b_2 and b_3 given by (19.2) and (19.3).

Question 19.2 Gap models have been used to simulate the response of forests to climate change. Discuss how the use of species range maps to derive bioclimatic tolerances affects model response to a warmer climate.

Question 19.3 What aspects of gap models limit their application in Earth system models?

19.3 | Area-Based Dynamic Global Vegetation Models

Dynamic global vegetation models (DGVMs) were coupled with land surface models beginning in the late 1990s and early 2000s as a means to simulate vegetation dynamics at the large spatial scales used by global climate models. DGVMs simulate changes in vegetation composition, biomass, productivity, and biogeochemical cycling. The models take advantage of the subgrid tiling used in land surface models (Figure 9.15). Rather than describing vegetation composition by species as in gap models or by biomes as in biogeography models, DGVMs describe vegetation by distinct plant functional types. Plant functional types (PFTs) are distinguished by: growth form (woody or herbaceous); leaf longevity (deciduous or evergreen); leaf type (broadleaf or needleleaf); photosynthetic pathway (C_3 or C_4); and bioclimatic variants (e.g., arctic, boreal, temperate, tropical). Vegetation is characterized by homogenous patches of PFTs within a model grid cell so that, for example, savanna consists of trees and grasses, and forests can be comprised of deciduous broadleaf and evergreen needleleaf trees. DGVMs update the carbon balance in each subgrid tile similar to biogeochemical models (Chapter 17) and additionally use simplistic rules to change the area occupied by PFTs as determined by productivity, mortality, establishment, fire, and bioclimatic tolerances.

The Integrated Biosphere Simulator (IBIS) was the first DGVM developed for climate models (Foley et al. 1996; Kucharik et al. 2000) and was used to study vegetation feedbacks in past and future climate (Levis et al. 1999, 2000). The TRIFFID DGVM (Cox 2001) updates PFT area based on competition among PFTs using Lotka–Volterra equations for population dynamics with species competition and was coupled with the Hadley Center MOSES land

Table 19.1	Simulated and observed basal area and stand biomass for a 70-year-old aspen (*Populus tremuloides*) forest in the Boundary Waters Canoe Area, Minnesota

Basal area (m² ha⁻¹)	Simulated	Observed
Populus tremuloides	15.3	10.6
Betula papyrifera	1.0	2.7
Pinus banksiana	3.4	2.2
Abies balsamea	2.0	1.6
Picea glauca	0.7	0.8
Thuja occidentalis	0.1	0.1
Pinus strobus	0.1	1.4
Picea mariana	0.1	0.7
Pinus resinosa	0.0	0.4
Total basal area	22.4	20.5
Biomass (t ha⁻¹)	135.0	137

Source: From Bonan (1989).

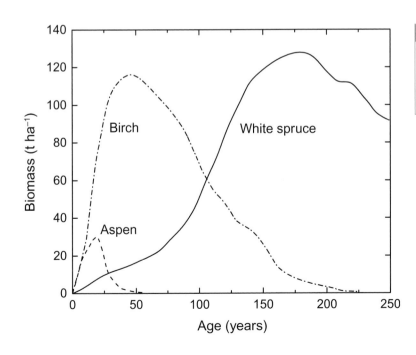

Figure 19.8 Simulated forest dynamics following fire on a well-drained, south-facing slope at Fairbanks, Alaska. Shown is plant biomass for three tree species ($1 \text{ t ha}^{-1} = 0.1 \text{ kg m}^{-2}$). Redrawn from Bonan (1989)

surface model in the first carbon cycle–climate simulation (Cox et al. 2000). The Lund–Potsdam–Jena (LPJ) model (Sitch et al. 2003) was unique among first-generation DGVMs to explicitly model demographic processes. The model represents tree PFTs by an average individual plant, the number of individuals, and the fractional cover in the grid cell. The model has been extended to included managed land (Schaphoff et al. 2018). A variant of the model (LPJ-GUESS) combines concepts of individual plants and gap dynamics using size-based cohorts with replicate patches (Smith et al. 2001, 2014).

Bonan et al. (2003) used concepts from LPJ to develop a dynamic global vegetation model that integrated the traditional focus on short-term (sub-hourly) biogeophysics and hydrometeorology found in land surface models with the longer term

biogeochemistry, plant community composition, and vegetation dynamics (Figure 19.9). Three time-scales (minutes, days, years) govern the model. Energy fluxes, the hydrologic cycle, and carbon assimilation occur quickly. Carbon is accumulated annually to update vegetation once per year. Changes in vegetation composition and structure occur annually in relation to productivity, allocation, mortality, establishment, and fire. Bioclimatic rules influence mortality and establishment. Leaf area index is updated daily based on prevailing conditions with a maximum value from the annual vegetation dynamics. Levis et al. (2004) described the annual vegetation dynamics portion of the model adapted from LPJ.

Trees grow in accordance with simple scaling relationships and are described by mass (leaf, fine

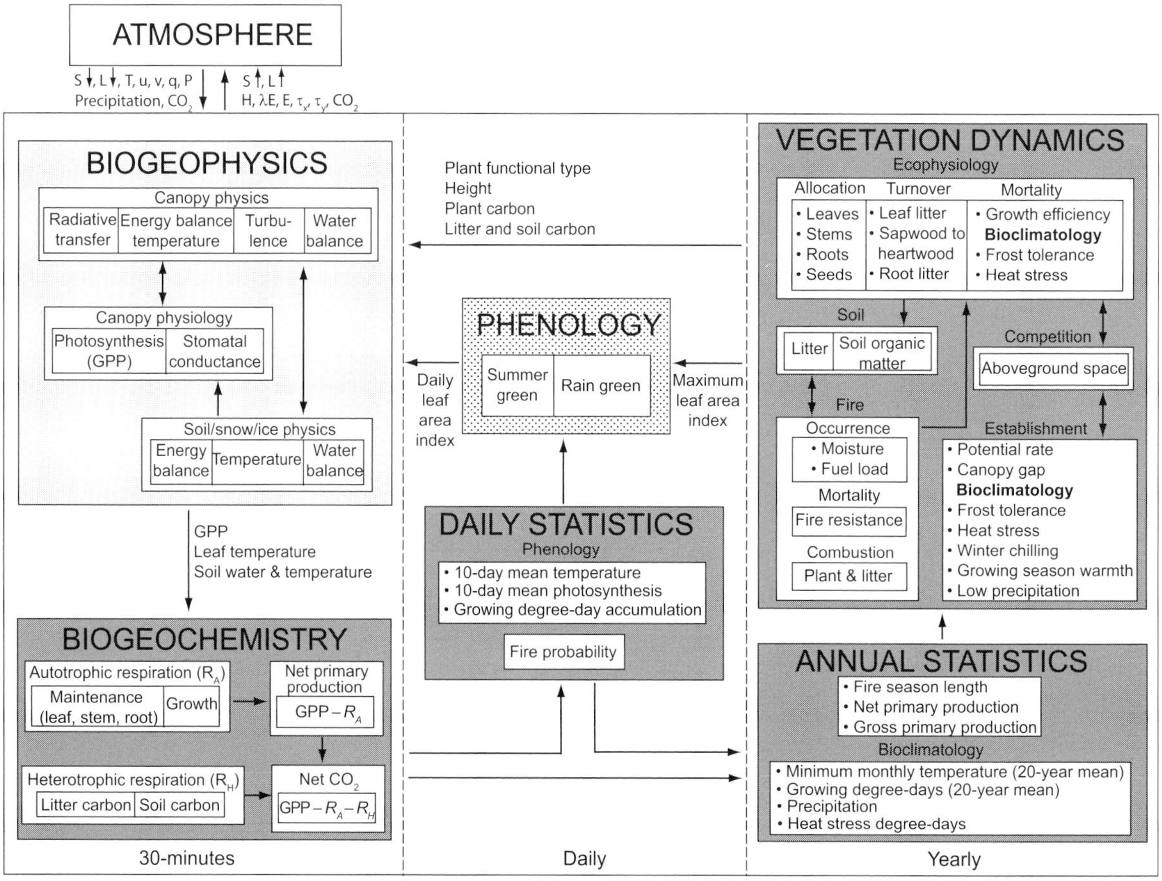

Figure 19.9 Coupling of a dynamic global vegetation model with a land surface model. Shown are linkages among the biogeophysics, biogeochemistry, and vegetation dynamics components of the model. The lightly shaded biogeophysical processes represent the traditional hydrometeorological scope of land surface models. The darker boxes represent the greening of land surface models with the introduction of dynamic vegetation and the carbon cycle. Reproduced from Bonan et al. (2003) as in Bonan (2016)

(a) Tree

Crown area

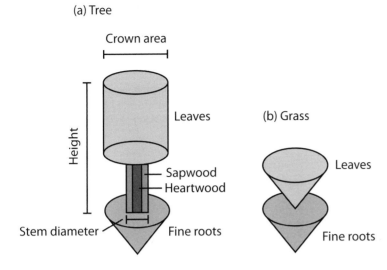

roots, sapwood, heartwood), stem diameter, height, leaf area, and crown area (Figure 19.10a). Several rules constrain the apportionment of biomass among leaves, fine roots, and sapwood for an individual. The pipe model theory holds that a unit of leaf area requires a certain amount of sapwood area through which water flows to leaves (Shinozaki et al. 1964; Lehnebach et al. 2018). This requirement is specified by a relationship in which the leaf area A_L (m^2) of an individual relates to sapwood cross-sectional area A_S (m^2) by

$$A_L = a_1 A_S,$$ (19.9)

in which a_1 is a proportionality constant. Plants must invest in fine roots to acquire water, and this investment increases in dry soils. This requirement is specified by a relationship between the leaf mass M_L and root mass M_R of an individual (kg) in which

$$M_L = a_2 M_R.$$ (19.10)

The proportionality constant a_2 can be adjusted for soil moisture stress so that allocation to leaves decreases in dry soils. Height h (m) relates to stem diameter with the relationship

$$h = a_3 D^{a_4},$$ (19.11)

where here diameter D is specified in meters. Another relationship is between mass and volume. Assuming a cylindrical stem, wood density ρ_w (kg m^{-3}) can be defined by

$$\rho_w = \frac{M_S}{A_S h},$$ (19.12)

which is the mass of sapwood divided by the volume of sapwood. Sapwood area relates to leaf area through (19.9), and leaf area relates to leaf mass through leaf mass per area m_a using

$$A_L = M_L / m_a$$ (19.13)

so that (19.12) can be rewritten as

$$h = \frac{a_1 m_a}{\rho_w} \frac{M_S}{M_L}$$ (19.14)

and

$$M_S = \frac{\rho_w}{a_1 m_a} M_L h.$$ (19.15)

Equation (19.15) provides the relationship between leaf mass and sapwood mass. Substituting (19.11) for h and rearranging terms gives an expression for stem diameter, whereby

$$D = \left(\frac{a_1 m_a}{a_3 \rho_w} \frac{M_S}{M_L} \right)^{1/a_4}.$$ (19.16)

A second mass–volume rule equivalently defines wood density in relation to bole mass (sapwood M_S plus heartwood M_H) and bole volume so that

$$\rho_w = \frac{M_S + M_H}{(\pi/4) D^2 h}.$$ (19.17)

Table 19.2	Representative allocation parameters in LPJ
Parameter	Value
a_1	$8000 \text{ m}^2 \text{ m}^{-2}$
a_2	1 kg kg^{-1}
a_3	40
a_4	0.5
a_5	100
a_6	1.6
ρ_w	200 kg m^{-3}

Source: From Sitch et al. (2003).

An additional scaling relationship defines crown area by

$$A_C = a_5 D^{a_6}, \tag{19.18}$$

with a maximum crown area of 15 m², and leaf area index is

$$L = \frac{M_L}{m_a A_C}.$$

Table 19.2 provides representative allocation parameters. Grasses are represented only by leaves and fine roots (Figure 19.10b).

Gross primary production (GPP) is calculated using leaf photosynthesis scaled to a big-leaf canopy. Maintenance respiration costs are subtracted, and 25% of the available carbon (GPP − R_m) is lost as growth respiration so that

$$\text{NPP} = (\text{GPP} - R_m)(1 - f_{RG}), \tag{19.19}$$

where $f_{RG} = 0.25$ is the fraction of carbon used for growth respiration. Maintenance respiration uses the Lloyd and Taylor (1994) function given by (17.34) with base rates specified by nitrogen rather than biomass. An additional 10% of the carbon remaining after respiration is deducted as a reproduction cost. Similar to biogeochemical models (Chapter 17), some portion of leaf, sapwood, and fine root carbon turns over each year. Leaf and fine root turnover becomes litterfall. Sapwood turnover becomes heartwood. Plant mass increment is

$$\Delta M = \Delta M_L + \Delta M_R + \Delta M_S, \tag{19.20}$$

where ΔM is allocated to satisfy the scaling relationships. The chapter appendix provides the necessary derivations. A minimum leaf production is required to maintain the current sapwood mass, and a similar minimum root production is needed to support leaf mass. This can result in special allocation under conditions of low growth. After incrementing biomass, height is obtained from (19.14), stem diameter from (19.16), and crown area from (19.18). The growth of grasses is distributed between leaves and roots according to (19.10).

The foliage projected cover of an individual is of particular importance and is defined based on its leaf area index using

$$F_{ind} = 1 - e^{-0.5L}. \tag{19.21}$$

The fractional cover of a PFT patch in a grid cell is calculated from the foliage projected cover and crown area of the individual and population density N (individuals per m²) whereby

$$F = A_C F_{ind} N. \tag{19.22}$$

Equation (19.22) governs vegetation dynamics by determining competitive dominance. If the total fractional cover of all PFTs exceeds 1, grasses are decreased to mimic the dominant position of trees in the canopy. The total fractional cover of woody PFTs is limited to 0.95; any excess tree cover is removed, favoring PFTs with higher annual increment in foliage projected cover (ΔF_{ind}).

Mortality occurs from several processes and manifests in a reduction of population density. Shading causes mortality and is represented as competition for space based on fractional cover in a grid cell as described in the preceding paragraph. A PFT also dies from negative net primary production, low growth efficiency, heat stress, or extreme cold. Some portion of trees die each year with a background mortality rate (1% per year). This rate increases when growth efficiency is low. Mortality of boreal trees increases with heat stress, represented by the annual degree-days above 23°C. A PFT has a cold-temperature tolerance and dies in years that are too cold. Grasses complete their life cycle in one year. Wildfires occur annually using the Thonicke et al. (2001) model described in Chapter 17. The fraction of the population killed is in proportion to the burnt area and depends on fire resistance. Some PFTs are more susceptible to fire mortality than are others.

The establishment rate of new trees depends on the total fractional cover of trees in the grid cell.

Each tree PFT that is capable of regenerating does so at a rate specified by

$$\Delta N = \frac{\Delta N_{\max}}{n_{\mathrm{PFT}}} \left[1 - e^{-5(1 - F_{\mathrm{tree}})} \right] (1 - F_{\mathrm{tree}}), \qquad (19.23)$$

where $\Delta N_{\max} = 0.24$ individuals per m^2 per year is the maximum number of trees that can establish in any year and n_{PFT} is the number of tree PFTs that can establish in the current year. The first two terms on the right-hand side of this equation are a maximum establishment rate that is reduced by shading as tree cover approaches 1. The total establishment is partitioned equally among regenerating tree PFTs. The last term scales establishment by the fraction of the grid cell not covered with trees, thereby limiting establishment to openings in the canopy. The biomass of the sapling is calculated from scaling rules using an initial leaf area index equal to 1.5 m^2 m^{-2} and a specified fraction of heartwood in relation to sapwood. Grasses establish in areas that are not vegetated.

Bioclimatic rules influence mortality and establishment as specified by climatic tolerances for each PFT. Each PFT has a frost tolerance and dies in a year in which the 20-year average temperature of the coldest month is below some minimum. Boreal trees also have a maximum temperature that cannot be exceeded without heat-stress mortality. Additional rules pertain for establishment. A winter chilling requirement is mimicked by precluding establishment in warm years, defined by requiring the 20-year mean of the coldest monthly temperature be below some maximum. Growing season warmth requirements are represented by a minimum annual growing degree-days for establishment, but there must also be zero heat-stress days with temperature above some maximum. Additionally, annual precipitation must exceed 100 mm.

DGVMs are tested on their ability to simulate global biogeography and transient dynamics. Figure 19.11, for example, illustrates forest succession in boreal forest. Grasses initially dominate the simulation, followed by a rapid decline in grasses and increase in deciduous trees. Deciduous trees attain peak abundance in less than 100 years and then decline in cover as evergreen trees gain dominance. The simulated succession is similar to observed succession in interior Alaska (Figure 19.8). However, recurring fires preclude very old forests and create a landscape that is a mosaic of forests of all ages. Area-based DGVMs, with their subgrid tiling based on PFTs, cannot replicate the age structure arising from disturbance.

Question 19.4   The allocation rules in the Community Land Model given by (17.27)–(17.32) are simple expressions based on proportionality parameters. The LPJ allocation to leaf, root, and sapwood growth are more complex equations (see chapter appendix). Discuss the differences. Why is the LPJ allocation more complex?

Question 19.5   Derive equations for the minimum leaf and root production needed to follow the LPJ allocation rules.

Question 19.6   DGVMs are tested on their ability to replicate global biogeography (e.g., Sitch et al. 2003; Bonan et al. 2003). Discuss how bioclimatic limits affect this model test.

Question 19.7   Why are 20-year means used to assess bioclimatic tolerances, in contrast with a particular year?

Question 19.8   Contrast the effects of fire in a gap model compared with an area-based DGVM such as LPJ.

19.4 Cohort-Based Ecosystem Demography Models

Earth system models simulate vegetation over large regions of hundreds of square kilometers. Heterogeneity is represented by discretizing the land surface into separate PFT tiles in which vegetation is homogenous within a tile. This characterization of vegetation does not represent fine-scale heterogeneity in canopy structure and disturbance. Nor does it allow for the competition for light that drives forest dynamics. The ecosystem demography (ED) model was formulated as a means to bridge the divide between fine-scale processes related to disturbance history and height-structured competition for light and the large spatial scales needed for global models. The model depicts age structure and height classes at fine scales as originally put forth by Moorcroft et al. (2001) and expanded upon

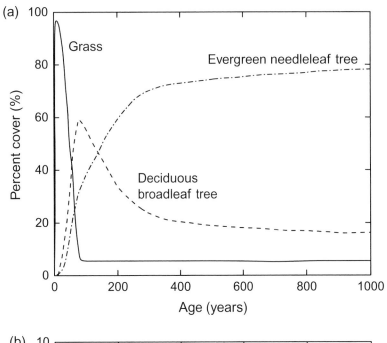

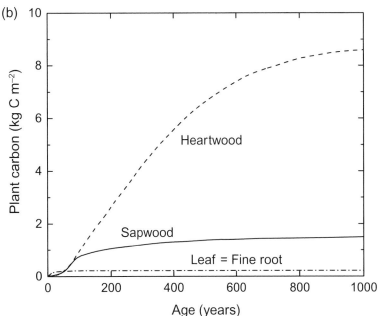

Figure 19.11 Boreal forest dynamics in terms of (a) percentage cover and (b) plant carbon pools as simulated by a dynamic global vegetation model. The simulation is from initially bare ground for a single model grid cell in the boreal forest over 1000 years in the absence of fire. Percentage cover is the annual extent of plant functional types in the grid cell. Adapted from Bonan et al. (2003)

by Medvigy et al. (2009) with ED2. Fisher et al. (2010b) coupled ED with a land surface model (JULES) for use with an Earth system model, and this has been extended to CLM(ED) (Fisher et al. 2015) and LM3-PPA (Weng et al. 2015). This coupling links short timescale canopy biophysics and physiology with daily changes in vegetation structure arising from growth, allocation, recruitment, mortality, and fire.

In the ecosystem demography framework, vegetation is depicted by patches and cohorts (Figure 19.1c). A patch is an area of land with a particular disturbance history arising from canopy gaps or wildfires. One patch may be recently disturbed and represents the early stages of post-disturbance succession; another patch may be in an intermediate stage of recover; and a third patch may be old-growth forest. Cohorts exist within a patch. A cohort is a population of plants with similar functional type, size, and stature in the canopy. A cohort is described by its PFT, stem diameter, height, biomass, and the number of individuals. Cohorts are assigned to the overstory or understory according to their height. In this way, overstory cohorts shade the understory. A PFT can have multiple cohorts in a patch based on position in the canopy. The subgrid tiling is based on disturbance history. Vegetation within a grid cell is represented by patches with different age and with cohorts of different PFTs and height classes within a patch. The creation of new patches by disturbance and the merging of existing patches that are similar causes the number of patches and their area to change over time. A diverse landscape with frequent disturbance may have many patches and numerous cohorts. A less complex landscape has fewer patches and cohorts.

Each cohort is described by an average individual, and allometric relationships govern biomass growth. A tree cohort has a stem diameter D (cm), height h (m), and biomass (kg C) for leaves M_L, live fine roots M_R, live sapwood M_S, and dead structural stem or heartwood M_H similar to LPJ (Figure 19.10a). The following description of allometry is from ED (Moorcroft et al. 2001), as used also by Fisher et al. (2015). The height–diameter allometry is

$$h = a_1 D^{a_2}. \tag{19.24}$$

The biomass of leaves for a tree with stem diameter D is

$$M_L = b_1 D^{b_2} \rho_w^{b_3}, \tag{19.25}$$

and structural stem biomass additionally depends on height using

$$M_H = c_1 D^{c_2} \rho_w^{c_3} h^{c_4}. \tag{19.26}$$

For these equations, ρ_w is wood density (g cm^{-3}) so that leaf and heartwood mass increase with more dense wood. The mass of roots is proportional to leaf mass so that

$$M_R = d_1 M_L, \tag{19.27}$$

where d_1 is the ratio of fine-root mass to leaf mass. Sapwood biomass is obtained assuming that leaf area is proportional to sapwood cross-sectional area as in (19.9) for LPJ. Upon converting sapwood area to sapwood mass,

$$M_S = d_2 M_L h, \tag{19.28}$$

with d_2 the ratio of sapwood mass to leaf mass per unit tree height (kg kg^{-1} m^{-1}), as in (19.15). The total living biomass is

$$M_A = M_L + M_R + M_S = M_L(1 + d_1 + d_2 h), \tag{19.29}$$

and the fraction of biomass in leaves, roots, and sapwood is

$$f_L = \frac{M_L}{M_A} = \frac{1}{1 + d_1 + d_2 h} \tag{19.30}$$

$$f_R = \frac{M_R}{M_A} = \frac{d_1}{1 + d_1 + d_2 h} \tag{19.31}$$

$$f_S = \frac{M_S}{M_A} = \frac{d_2 h}{1 + d_1 + d_2 h}. \tag{19.32}$$

Table 19.3 gives representative allocation parameters.

These relationships govern new growth, which is allocated such that a tree grows according to the specified allometry. The new production ΔM available for biomass is allocated to live pools

Table 19.3 | Representative allocation parameters in ecosystem demography models

Parameter	Value
a_1	2.34
a_2	0.64
b_1	0.0419
b_2	1.56
b_3	0.55
c_1	0.069
c_2	1.94
c_3	0.931
c_4	0.572
d_1	1 kg kg^{-1}
d_3	0.1 m cm^{-1}
ρ_w	0.7 g cm^{-3}

Source: From Moorcroft et al. (2001) and Fisher et al. (2015) with d_3 from Purves et al. (2008).

$(M_A = M_L + M_R + M_S)$ and the structural pool (M_H) with a fraction f_A available for live biomass growth and $1 - f_A$ for structural biomass growth. The prognostic growth equations are

$$\frac{\partial M_A}{\partial t} = f_A \Delta M \tag{19.33}$$

$$\frac{\partial M_H}{\partial t} = (1 - f_A)\Delta M. \tag{19.34}$$

Leaf, root, and sapwood biomass are diagnosed from M_A using (19.30)–(19.32). The fraction of growth allocated to M_A is obtained from

$$\frac{\partial M_A}{\partial M_H} = \frac{f_A}{1 - f_A} \tag{19.35}$$

so that

$$f_A = \frac{\partial M_A / \partial M_H}{1 + \partial M_A / \partial M_H}, \tag{19.36}$$

with

$$\frac{\partial M_A}{\partial M_H} = \frac{\partial M_L}{\partial M_H} + \frac{\partial M_R}{\partial M_H} + \frac{\partial M_S}{\partial M_H}. \tag{19.37}$$

The necessary partial derivatives are obtained from the allometric relationships (19.24)–(19.28), as shown in the chapter appendix. Equation (19.36) is a general expression so that allocation can accommodate different allometries.

The biomass of a cohort is the balance between carbon gain from gross primary production, carbon loss from autotrophic respiration, and turnover loss. As described by Fisher et al. (2015) for CLM (ED), the preceding equations govern the target allocation with optimal conditions. Additional allocation rules are needed during periods of negative carbon balance, when live biomass decreases but structural biomass remains constant. Further rules are required when productivity subsequently recovers to regrow live biomass and regain the target allometry. Net primary production is the carbon balance after subtracting maintenance and growth respiration, similar to (19.19). The carbon is first used to meet tissue maintenance, carbohydrate storage, seed production, and leaf phenology before allocation to biomass pools. Tissue-maintenance demand replenishes biomass lost in background turnover. Carbohydrate storage provides a carbon reserve. Allocation to seeds is specified as a fraction of available carbon.

An additional important allometric relationship is for crown area in relation to stem diameter. A simple expression is

$$A_C = \pi(d_3 D)^2. \tag{19.38}$$

In this equation, d_3 is the ratio of crown radius to stem diameter (m cm^{-1}) so that $d_3 D$ is the crown radius. A general value is $d_3 = 0.1$ m cm^{-1} (Purves et al. 2008). More complex expressions can be used instead of (19.38). Total canopy area (m^2) is the sum of each cohort's crown area multiplied by the number of individuals. When expressed using population density N (individuals per m^2),

$$f_C = \sum_i A_{C,i} N_i \tag{19.39}$$

is the fractional area of the patch covered by crowns. In ecosystem demography models, crown area provides a means to organize the canopy into layers so as to represent vertically structured light competition within a patch.

The original ED model represented crowns by thin, flat disks, similar to gap models in which each cohort's leaf area is spread over the entire patch at the cohort's height (Figure 19.6b). In this way, taller cohorts shade smaller cohorts and have a growth advantage through acquisition of light. Subsequent models (Fisher et al. 2015; Weng et al. 2015) organize canopies differently, based on the perfect plasticity approximation (Figure 19.6c). This theory holds that the crowns of trees fill the available space in the overstory so that the canopy is fully closed before an understory layer forms (Purves et al. 2008; Strigul et al. 2008). From the height and crown geometry of trees, a critical height z_* can be found such that trees taller than z_* are in the overstory and shorter trees are in the understory. This is described formally by the relationship

$$1 = \sum_i \int_{z_*}^{\infty} N_i(z) A_{C,i}(z_*, z) dz, \tag{19.40}$$

where $A_{C,i}$ is the crown area (m^2) above z_* at height z for PFT i and N_i is the population density (m^{-2}) for height z. As used here, A_C is defined as a continuous function of height to accommodate different crown shapes. In practice, tree crowns can be represented as a flat disk at the top of the tree, in which case A_C is a step function equal to (19.38) at $z = h$. For a flat-topped crown, (19.40) is implemented in practice with a simple algorithm, as shown in Figure 19.12:

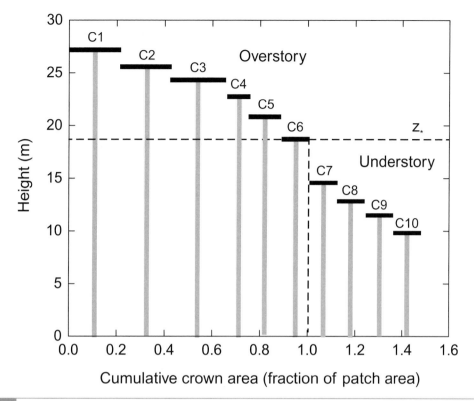

Figure 19.12 Canopy organization in the perfect plasticity approximation. Shown are 10 cohorts arranged from tallest to shortest. The cumulative crown area of the six tallest cohorts (C1–C6) sums to one, and these form the overstory. The remaining cohorts (C7–C10) form the understory. The height z_* separates the two canopy layers.

calculate the crown area $N_i A_{C,i}$ of the tallest cohort; do so for the next-tallest cohort and so forth down through the canopy; z_* is the height at which the cumulative crown area sums to 1. Trees below this height occur in the understory. Additional understory layers can be created, though, in practice, models represent only an overstory and one understory layer. Some model implementations allow for partial canopy closure so that the left-hand side of (19.40) is less than 1.

The perfect plasticity approximation provides a constraint on canopy organization into an overstory and understory based on crown space and facilitates depiction of vertically structured competition for light. In its simplest implementation, cohorts in the overstory experience full sunlight and shade understory cohorts. Some models additionally allow for multilayer radiative transfer within a canopy layer by having a profile of leaf area in each layer (Fisher et al. 2015). The canopy organization introduces a dynamics in which cohorts are continually demoted from or promoted to

the overstory based on growth (an increase in crown area) or mortality (a decrease in crown area) (Figure 19.13a,b). Models differ in the precise manner in which demotion or promotion occur (Fisher et al. 2018). For example, Fisher et al. (2010b, 2015) introduced a modification to allow less deterministic canopy layering. Rather than defining a critical height for canopy layer separation, that implementation uses a competitive exclusion coefficient to move some fraction of the excess crown area to the understory or new crown area to the overstory. The fraction of a cohort's crown area that is moved is based on its stem diameter weighted by the competitive exclusion factor. A higher weight favors large trees remaining in the overstory. A smaller weight means a greater likelihood that small cohorts are in the overstory.

The temporal dynamics of cohorts results in a change in population density, size, and canopy status (Table 19.4). Population density increases from recruitment and decreases due to mortality. Mortality can be represented by a number of processes including background mortality, carbon starvation, plant hydraulic stress, and wildfire (Moorcroft et al. 2001; Fisher et al. 2010b, 2015; Weng et al. 2015). Carbon starvation occurs from low carbon storage

Table 19.4	Processes governing cohort and patch dynamics
Cohort dynamics	
Establishment of new cohorts (recruitment)	
Cohorts grow vertically (height) and wider (crown area)	
Cohorts lose biomass and become smaller from mortality	
Promotion to or demotion from overstory based on crown space	
Merging of similar cohorts	
Elimination of small population cohorts	
Patch dynamics	
Creation of new patch from canopy opening	
Merging of similar patches	
Loss of small patches	

Source: From Fisher et al. (2015) for CLM(ED).

during periods of low productivity, such as with prolonged drought or suppression in the understory. Over time, the cohort grows, increasing its diameter, height, biomass, and crown area. Mortality reduces biomass and crown area. Promotion to the overstory may result to fill open crown space, and demotion to the understory occurs when the crown space is filled. As described by Fisher et al. (2015) for CLM(ED), two cohorts of the same PFT may merge together as they grow if they become similar (e.g., in height) so as to reduce the computational demand.

The number of patches and their area change over time in relation to patch age and disturbance (Table 19.4). New patches form from existing patches if there is a disturbance, and old patches may fuse into a single patch if they are sufficiently similar. Mortality of overstory cohorts creates space in the canopy. This space can be filled by promotion of understory cohorts to the overstory as described previously (Figure 19.13b), or the space can be used to create a new patch (Figure 19.13c). Both processes occur together so that mortality results in canopy reorganization and patch initiation. To prevent a larger number of patches from arising, similar patches can merge together. Fisher et al. (2015) based similarity on the distribution of aboveground biomass among PFTs and in stem diameter classes. When patches merge,

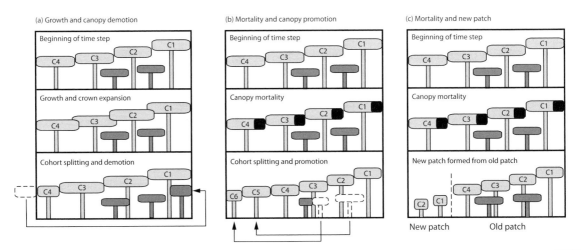

Figure 19.13 Depiction of cohort and patch dynamics. (a) Growth of cohorts leads to crown expansion, resulting in splitting of a cohort and demotion to the understory. Shown are four overstory cohorts (C1–C4). Part of the shortest cohort (C4) is demoted to the understory to form a new cohort. (b) Mortality leads to open canopy space that can be filled by promotion of understory cohorts to the overstory. In this example, the two understory cohorts form new cohorts (C5, C6) in the overstory. Patch area is unchanged. (c) The open canopy space upon mortality can also be used to create a new patch that is filled by new cohorts. The area of the old patch decreases, and a new patch is formed from the open canopy area. Adapted from Charlie Koven

state variables are averaged among patches to create a patch of the combined area. Patches cease to exist when their area is too small.

Ecosystem demography models simulate a richness in forest composition and structure similar to that of gap models. Figure 19.14, for example, shows results from LM3-PPA. The forest changes in composition from aspen to maple because aspen is short-lived (high background mortality) and cannot thrive in the understory. The forest after 80 years consists of many small trees and a few large trees. Ecosystem demography models allow for trait filtering in which predictions of community composition and ecosystem structure arise from plant traits, the environmental conditions in which plants grow, and selective filtering of traits that are successful for the prevailing environment (Fisher et al. 2018). For example, Fisher et al. (2015) examined the ability to simulate forest biogeography without use of the bioclimatic envelopes common to DGVMs and to instead determine biogeography in relation to leaf traits. A critical research frontier is to define the particular traits and to consider trade-offs among various traits. The original conceptualization of PFTs was based on physiological traits that link photosynthesis and carbon economics (e.g., leaf mass per area, leaf nitrogen, leaf longevity). With the addition of plant hydraulics, traits now include stem conductivity and xylem vulnerability. While PFTs capture broad differences in traits, the use of PFTs in global models is limited to a small number of functional types. A research frontier is to expand beyond the discrete representation of traits by PFTs to allow for continuous traits. For these reasons, CLM(ED) is better described as the Functionally Assembled Terrestrial Ecosystem Simulator (FATES; Fisher et al. 2018).

Question 19.9   A cohort is initialized with a height of 1.5 m. Calculate its initial biomass using the allometry given by (19.24)–(19.28).

Question 19.10   Compare height allometry among the equations given for gap models, LPJ, and ecosystem demography models.

Question 19.11   Both LPJ and ED use scaling relationships to allocate growth to biomass. Describe similarities and differences between these two models.

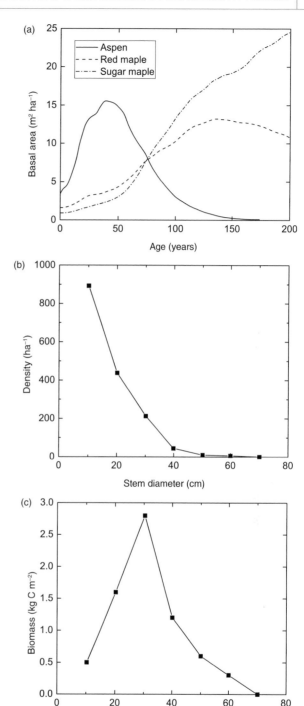

Figure 19.14 Forest dynamics simulated by LM3-PPA. Shown is (a) basal area of aspen (*Populus tremuloides*), red maple (*Acer rubrum*), and sugar maple (*Acer saccharum*) in relation to stand age. Also shown for year 80 are (b) stem diameter distribution and (c) biomass distribution. Adapted from Weng et al. (2015)

19.5 | Chapter Appendix: LPJ Allocation

The scaling relationships provide simple expressions for root and sapwood growth in relation to leaf growth. An increment in foliage mass ΔM_L and root mass ΔM_R must satisfy (19.10) in which

$$M_L + \Delta M_L = a_2(M_R + \Delta M_R) \tag{19.41}$$

and

$$\Delta M_R = \frac{M_L + \Delta M_L}{a_2} - M_R. \tag{19.42}$$

Substituting (19.42) into the mass balance of (19.20) gives an expression for sapwood growth whereby

$$\Delta M_S = \Delta M - \Delta M_L - \frac{M_L + \Delta M_L}{a_2} + M_R. \tag{19.43}$$

An expression for leaf growth is more complex. It is obtained from scaling relationships with height. According to (19.14), height upon an increment in leaf and sapwood mass must satisfy the relationship:

$$h = \frac{a_1 m_a}{\rho_w}\left(\frac{M_S + \Delta M_S}{M_L + \Delta M_L}\right). \tag{19.44}$$

Substituting (19.43) into (19.44):

$$h = \left(\frac{a_1 m_a}{\rho_w}\right)\frac{M_S + \Delta M - \Delta M_L - \frac{M_L + \Delta M_L}{a_2} + M_R}{M_L + \Delta M_L}. \tag{19.45}$$

This equation can be solved for ΔM_L provided plant height h is known. That is obtained using the relationship between height and stem diameter. With an increment in sapwood mass, diameter must satisfy the relationship given by (19.17) with

$$\rho_w = \frac{M_S + \Delta M_S + M_H}{(\pi/4)D^2 h}. \tag{19.46}$$

Substituting (19.43) for ΔM_S provides an expression for stem diameter,

$$D^2 = \frac{M_S + \Delta M - \Delta M_L - \frac{M_L + \Delta M_L}{a_2} + M_R + M_H}{(\pi/4)\rho_w h}, \tag{19.47}$$

which is then substituted into (19.11) to obtain an expression for height whereby

$$h = a_3^{2/(a_4+2)}\left[\frac{M_S + \Delta M - \Delta M_L - \frac{M_L + \Delta M_L}{a_2} + M_R + M_H}{(\pi/4)\rho_w}\right]^{a_4/(a_4+2)}. \tag{19.48}$$

Equating the two expressions for height given by (19.45) and (19.48) provides an equation that can be solved numerically for ΔM_L using root finding methods (Appendix A5).

19.6 | Chapter Appendix: ED Allocation

Equation (19.36) is the allocation of growth to live biomass. It requires the change in leaf, root, and sapwood biomass with respect to structural mass. These are obtained from the allometric relationships. Using (19.27) for roots:

$$\frac{\partial M_R}{\partial M_H} = d_1 \frac{\partial M_L}{\partial M_H}. \tag{19.49}$$

Using (19.28) for sapwood:

$$\frac{\partial M_S}{\partial M_H} = d_2\left(h\frac{\partial M_L}{\partial M_H} + M_L\frac{\partial h}{\partial M_H}\right). \tag{19.50}$$

An expression for leaves is obtained using

$$\frac{\partial M_L}{\partial M_H} = \frac{\partial M_L/\partial D}{\partial M_H/\partial D}. \tag{19.51}$$

The numerator in (19.51) is the change in leaf mass for a unit change in stem diameter from (19.25) with

$$\frac{\partial M_L}{\partial D} = b_2 b_1 D^{b_2-1}\rho_w^{b_3}. \tag{19.52}$$

The denominator is the comparable change in structural mass from (19.26) with

$$\frac{\partial M_H}{\partial D} = c_2 c_1 D^{c_2-1}\rho_w^{c_3}h^{c_4} + c_4 c_1 D^{c_2}\rho_w^{c_3}h^{c_4-1}\frac{\partial h}{\partial D}, \tag{19.53}$$

and from (19.24)

$$\frac{\partial h}{\partial D} = a_2 a_1 D^{a_2-1}. \tag{19.54}$$

The final expression needed is $\partial h/\partial M_H = 1/(\partial M_H/\partial h)$ and

$$\frac{\partial M_H}{\partial h} = c_4 c_1 D^{c_2}\rho_w^{c_3}h^{c_4-1} + c_2 c_1 D^{c_2-1}\rho_w^{c_3}h^{c_4}\frac{\partial D}{\partial h}. \tag{19.55}$$

Canopy Chemistry

Chapter Overview

Terrestrial ecosystems exchange many chemical gases and particles with the atmosphere. These emissions alter atmospheric composition and affect climate through radiative forcing. Some flux exchanges (CH_4, N_2O) alter the concentration of long-lived greenhouse gases. Others alter short-lived gases that affect atmospheric chemistry and air quality. Chemistry–climate interactions from these short-lived climate forcers (NO_x, biogenic volatile organic compounds, O_3, secondary organic aerosols) are significant and comparable in magnitude to other climate forcings. Stable isotopes are useful to diagnose biogeochemical and hydrologic cycles. A research frontier is to link the biogeophysical and carbon cycle influences of terrestrial ecosystems with a full depiction of biogeochemical feedbacks mediated through atmospheric chemistry.

20.1 | Dry Deposition

Transfer of gases from the atmosphere to the land surface is an important process in air quality modeling. Deposition of these gases as they settle out of the atmosphere or are absorbed by plant materials and the ground is called dry deposition and occurs from the same meteorological and physiological processes responsible for the fluxes of momentum, heat, and moisture (Wesely and Hicks 1977, 2000;

Hicks et al. 1987; Wesely 1989). The analogous dry deposition flux F_j (mol m^{-2} s^{-1}) of a chemical between the atmosphere and surface is expressed as the product of the concentration c_j (mol mol^{-1}) of the gas in the atmosphere and a deposition velocity v_d (m s^{-1}) whereby

$$F_j = -\rho_m v_d c_j. \tag{20.1}$$

This equation is similar to the surface flux equations for sensible heat and water vapor (Chapter 6), but the surface concentration for dry deposition is taken as zero. Equation (20.1) represents dry deposition using a big-leaf canopy with a series of conductances between the atmosphere, leaf boundary layer, and leaf surface, which are encapsulated in the deposition velocity (Figure 20.1). A value of 1 cm s^{-1} is considered moderately large and is typical for O_3 deposition (Wesely and Hicks 2000).

Deposition velocity consists of three conductances connected in series and is generally defined as

$$v_d = \left(g_{am}^{-1} + g_{bj}^{-1} + g_c^{-1} \right)^{-1}, \tag{20.2}$$

where the aerodynamic conductance g_{am} accounts for turbulent transfer to the surface and the boundary layer conductance g_{bj} accounts for diffusion through the quasi-laminar boundary layer around leaves. Dry deposition involves several additional pathways including deposition onto leaf cuticle surfaces, uptake through stomata, and deposition to the soil surface. Stomatal uptake is the most important pathway in vegetation because gases such as O_3, SO_2, NO_2, NO, NH_3, and other compounds diffuse

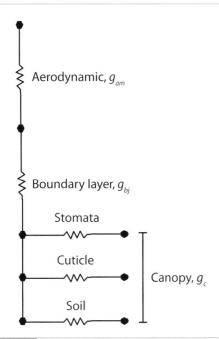

Figure 20.1 Conductance network used to represent dry deposition in plant canopies. Adapted from Baldocchi et al. (1987) and Hicks et al. (1987). Note the similarity with the big-leaf and dual-source canopies used to represent sensible heat flux and evapotranspiration in Figure 7.3.

through stomata when they are open. The surface or canopy conductance g_c represents these pathways as three parallel deposition fluxes so that

$$g_c = g_{\text{stomata}} + g_{\text{cuticle}} + g_{\text{soil}}. \tag{20.3}$$

More complex expressions for g_c can be formulated to account for the various pathways of deposition in the canopy (Wesely 1989; Wesely and Hicks 2000; Pleim and Ran 2011).

The conductance g_{bj} is referred to in atmospheric chemistry modeling literature as the quasi-laminar boundary layer conductance. The interpretation of g_{bj} is the same as given in Chapter 6 in the discussion of heat and water vapor transport between the surface and atmosphere, in contrast to momentum transfer. The same equations apply to any scalar – which, in the case of dry deposition, is the concentration of the gas (Wesely and Hicks 1977; Hicks et al. 1987). In molar units, the boundary layer conductance is

$$\frac{1}{g_{bj}} = \frac{1}{\rho_m k u_*} \ln\left(\frac{z_{0m}}{z_{0j}}\right), \tag{20.4}$$

as in (6.66) with z_{0j} the scalar roughness length specific to the particular gas. An alternative equation is

$$\frac{1}{g_{bj}} = \frac{2}{\rho_m k u_*} \left(\frac{\text{Sc}}{\text{Pr}}\right)^{2/3}, \tag{20.5}$$

with $\text{Pr} = v/D_h$ the Prandtl number and $\text{Sc} = v/D_j$ the Schmidt number (Chapter 10). In this latter formulation, a form of which is commonly used in atmospheric chemistry models, g_{bj} varies with the molecular diffusivity of the gas.

Dry deposition parameterizations based on the principles outlined earlier have been implemented in big-leaf canopy models used with air quality models (Pleim et al. 2001; Zhang et al. 2003) and in terrestrial biosphere models such as the Community Land Model (Val Martin et al. 2014). Other models extend the equations to include multilayer stomatal conductance to account for profiles of light (Baldocchi et al. 1987). Some models further account for wind profiles in the canopy and calculate stomatal and leaf boundary layer conductances for sunlit and shaded leaves at each layer in the canopy (Meyers et al. 1998; Wu et al. 2003; Saylor et al. 2014). Still other models additionally account for turbulent transfer within the canopy (Baldocchi 1988; Launiainen et al. 2013).

Question 20.1 Equations (20.2) and (20.3) are commonly used to represent dry deposition in plant canopies, as depicted in Figure 20.1. Discuss how the boundary layer conductance connects to the stomatal, cuticle, and soil conductances in this figure. How does this differ from the dual-source canopy used to represent sensible heat flux and evapotranspiration in Figure 7.3?

Question 20.2 Equation (20.4) is an expression for boundary layer conductance in which $kB^{-1} = \ln(z_0/z_{0j})$. Equation (20.5) is an alternative expression for g_{bj}. (a) Derive the corresponding equation for kB^{-1}. (b) The thermal diffusivity of air at $0°C$ is $D_h = 18.9 \text{ mm}^2 \text{ s}^{-1}$, and the molecular diffusivity of SO_2 is $D_j = 10.9 \text{ mm}^2 \text{ s}^{-1}$ (Massman 1998). Calculate kB^{-1} for SO_2.

20.2 | BVOC Emissions

Process models of biogenic volatile organic compound (BVOC) emissions (e.g., isoprene, monoterpenes, and other compounds) have been developed and adapted to terrestrial biosphere models. A common approach to modeling BVOC emissions is empirical and scales a specified leaf emission rate at standard conditions for the multiplicative effects of environmental factors such as canopy density, light, temperature, and other influences (Guenther et al. 1995, 2006, 2012). In this method, the flux F_j of chemical compound j is given as

$$F_j = \varepsilon_j f_j(I^{\downarrow}) f_j(T_{\ell}) f_j(\theta) f_j(c_a) f_j(\text{age}) m_a L. \qquad (20.6)$$

The term ε_j is a base leaf emission rate (per unit leaf mass) at standard conditions, which is adjusted using empirical relationships specific to the compound for the influence of light, temperature, soil moisture, CO_2 concentration, and leaf age. The leaf flux is scaled to the canopy by multiplying by leaf mass per area m_a and leaf area index L.

Several factors determine BVOC emissions. Chief among these is vegetation type; some plant species are high emitters of particular BVOCs, while others have low emissions. This is represented by the base leaf emission rate ε_j, which depends on vegetation type and chemical compound. The emission rate also depends on meteorological conditions. Isoprene emission, for example, increases with greater sunlight and warmer temperatures up to some temperature optimum, is inhibited at high atmospheric CO_2 concentration, and decreases with prolonged drought stress. These effects are represented by emission activity factors. The base emission rate is defined at a photosynthetically active radiation equal to 1000 μmol m^{-2} s^{-1} and leaf temperature equal to 30°C. The functions $f(I^{\downarrow})$ and $f(T_{\ell})$ are emission activity factors that use empirical relationships specific to a chemical compound to adjust the emission rate for nonstandard light and temperature, respectively (Guenther et al. 1995). The emission activity factors have been expanded to also include the effects of soil moisture, atmospheric CO_2 concentration, and leaf age on emissions (Guenther et al. 2012). This empirical model has been adapted to terrestrial biosphere models (Levis et al. 2003; Lathière et al. 2006; Heald et al. 2008). However, the simulated emissions are very sensitive to the canopy model that provides the light environment, temperature, and soil moisture used in (20.6), and the emissions rates have to be adjusted depending on canopy model (Heald et al. 2008; Guenther et al. 2012). The base emission rates are themselves subject to considerable uncertainty and are highly model dependent (Arneth et al. 2008, 2011).

An alternative approach models leaf isoprene emission directly from photosynthesis based on the electron requirement for isoprene synthesis (Niinemets et al. 1999b). This parameterization has been used in terrestrial biosphere models (Arneth et al. 2007, 2011; Pacifico et al. 2011; Unger 2013; Unger et al. 2013) and has since been further extended (Morfopoulos et al. 2013, 2014; Grote et al. 2014). In this parameterization, the leaf isoprene emission F_i (μmol m^{-2} s^{-1}) is

$$F_i = \varepsilon_i J \left[\frac{c_i - \Gamma_*}{6(4.67c_i + 9.33\Gamma_*)} \right] e^{0.1(T_{\ell} - 303.15)} \frac{c_{i*}}{c_i}. \qquad (20.7)$$

Here, J is the photosynthetic electron transport rate (μmol m^{-2} s^{-1}), and ε_i is a plant-specific factor that is the fraction of electrons available for isoprene production. Its use is similar to the leaf emission factors in (20.6). The term in brackets converts the electron flux to isoprene equivalents. The last two terms were introduced by Arneth et al. (2007) and adjust the flux for leaf temperature and CO_2. The temperature factor adjusts emission for warm temperatures, separate from the effect of temperature on electron transport. High atmosphere CO_2 concentration inhibits isoprene emission, with c_{i*} a reference value for c_i typically taken as the value of c_i at an atmospheric concentration of 370 μmol mol^{-1}.

Equation (20.7) is derived from the Farquhar et al. (1980) model that relates the light-limited photosynthesis rate A_j to electron transport rate J (Niinemets et al. 1999b; Morfopoulos et al. 2014). The light-limited photosynthesis rate is

$$A_j = \frac{J}{4} \left(\frac{c_i - \Gamma_*}{c_i + 2\Gamma_*} \right) \qquad (20.8)$$

so that the electron transport rate is

$$J = A_j \left(\frac{4c_i + 8\Gamma_*}{c_i - \Gamma_*} \right). \qquad (20.9)$$

Niinemets et al. (1999b) assumed that isoprene emission is limited by NADPH, in which isoprene synthesis requires 1.17 times more NADPH per CO_2 assimilated (2.33 NADPH per CO_2) compared with sugar synthesis (2 NADPH per CO_2) and six CO_2 molecules are assimilated for each isoprene molecule produced. With these requirements, the electron transport rate required to sustain the isoprene emission rate F_i is

$$J_i = (1.17 \times 6) F_i \left(\frac{4c_i + 8\Gamma_*}{c_i - \Gamma_*} \right). \qquad (20.10)$$

Defining the fraction of electrons available for isoprene production as $\varepsilon_i = J_i/J$, (20.10) can be rearranged to give the isoprene emission rate

$$F_i = \frac{\varepsilon J}{7.02} \left(\frac{c_i - \Gamma_*}{4c_i + 8\Gamma_*} \right), \qquad (20.11)$$

which is similar to (20.7).

Question 20.3 Equation (20.6) has been coupled to land surface models to calculate BVOC emissions, whereby solar radiation, leaf temperature, and soil moisture are provided by the host model. In presenting the BVOC model, Guenther et al. (2012) introduced a factor called the canopy environment coefficient. This has a value equal to 0.30 when coupled with the Community Land Model (CLM4) and 0.57 for the WRF-AQ model. This factor adjusts emissions so that they are similar for both models at standard conditions. Discuss the significance of this factor.

Question 20.4 One researcher implements a photosynthesis-based isoprene emission model using (20.7). Another researcher uses $F_i = \varepsilon_i A_j / 7.02$. Show that these equations are equivalent.

20.3 | Dust Emissions

Mobilization by wind provides a source of dust to the atmosphere. Mobilization depends on several factors, the most important of which are friction velocity, vegetation cover, and soil moisture. In general, the source regions of dust are dry, non-vegetated surfaces with strong winds. Zender et al. (2003) and Mahowald et al. (2006) developed a dust emission parameterization for the Community Land Model, which is representative of this class of model. Dust particles are transported within the atmosphere in four separate bins that vary in diameter (0.1–1.0, 1.0–2.5, 2.5–5.0, and 5.0–10.0 μm). The mass flux of dust F_j (kg m^{-2} s^{-1}) into transport bin j is

$$F_j = T \cdot A_m \cdot S \cdot \alpha \cdot Q_s \cdot \sum_i M_{ij}. \qquad (20.12)$$

The vertical dust flux into the atmosphere is equal to αQ_s. The horizontal mass flux of particles Q_s (kg m^{-1} s^{-1}) depends on friction velocity u_*, in which dust entrainment occurs above some critical threshold value. This threshold value depends on particle size and density and the roughness of the soil surface, and it increases with soil moisture to account for inhibition of mobilization with wet soils. α is a mass efficiency term (m^{-1}) that converts the horizontal mass flux Q_s to the vertical dust mass flux F_j. This factor depends on clay content and is larger as the clay fraction increases from sandy to sandy loam. The total dust flux consists of three separate particle size bins, and M_{ij} is the mass fraction of three different source modes i into the four transport bins j. Several factors adjust the calculated mass flux. A_m is the fraction of the surface that is exposed to bare soil from which dust mobilization can occur and depends on the snow-covered fraction of the surface (Chapter 9) and vegetation cover. Dust emissions primarily occur from surfaces with low vegetation cover, which is represented by inhibiting emissions when leaf area index exceeds some critical threshold (e.g., 0.1–0.3 m^2 m^{-2}). Emissions are further modified by a source erodibility factor S to account for preferential geographic sources. The dust flux depends on the horizontal and temporal resolution of the model because of the nonlinearity of Q_s with wind speed; T is a resolution-dependent global tuning factor used to match observations of dust emissions.

Question 20.5 The dust emission parameterization described by (20.12) has a global tuning factor T that depends on the spatial resolution of the model. Discuss why this factor is needed and what it represents.

Question 20.6 Discuss terms in the parameterization of dust emissions described by (20.12) that are dependent on the particular terrestrial biosphere model.

20.4 | Wildfire Emissions

Wildfires emit CO_2, CH_4, and other chemicals to the atmosphere. Fire emissions are commonly parameterized using an emission factor approach in which the emission of a chemical is the product of the amount of biomass burned and the emissions per amount of biomass burned. In this method, the emission F_j (kg) of chemical species j is

$$F_j = A \cdot B \cdot f \cdot \varepsilon_j, \qquad (20.13)$$

in which A is area burned (m^2), B is biomass or fuel load ($kg\,m^{-2}$), f is the fraction of biomass burned, and ε_j is an emission factor with the units mass of compound j emitted per mass of biomass burned (Hoelzemann et al. 2004; Schultz et al. 2008; van der Werf et al. 2010; Wiedinmyer et al. 2011). The emission factor depends on the type of biomass burning – such as agricultural, savanna, peatland, and tropical, temperate, or boreal forest (Andreae and Merlet 2001; Akagi et al. 2011). Equation (20.13) can be used to calculate chemical emissions for over 100 chemical species (Akagi et al. 2011). However, each of the terms in (20.13) has considerable uncertainty.

Question 20.7 Van der Werf et al. (2010) estimated that 1882 Tg C y^{-1} was emitted globally from wildfire during the period 2005–2009. Wiedinmyer et al. (2011) estimated emissions of 7188 Tg CO_2 y^{-1}. How similar are these two estimates of carbon (C) emission to the atmosphere? Discuss some of the uncertainties in wildfire chemical emissions as described by (20.13).

20.5 | Reactive Nitrogen

Several nitrogen (N) gases such as NO, N_2O, and N_2 are emitted from soils to the atmosphere as part of the carbon–nitrogen biogeochemical cycling (Figure 20.2). Nitrification is the oxidation of ammonium (NH_4^+) to nitrite (NO_2^-) and then to nitrate (NO_3^-), with production of NO and N_2O as byproducts. Nitrification is an aerobic process that requires high soil O_2 concentration. Nitrification dominates in moist, but not too wet, soil for which NO production is high. Denitrification is the reduction of NO_3^- to N_2, also with production of NO and N_2O. It is an anaerobic process that occurs in wet, O_2-depleted soils. Wet soil favors denitrification, with high N_2O emission. Much of the N_2O is further reduced to N_2 as soil becomes saturated and oxygen depleted. Terrestrial biosphere models differ in the complexity with which these gas losses are represented. Some models do not explicitly simulate denitrification and instead represent it as a bulk loss specified as a fraction of N mineralization or soil mineral N concentration. Other models use variants of DAYCENT and DNDC, which are two widely used

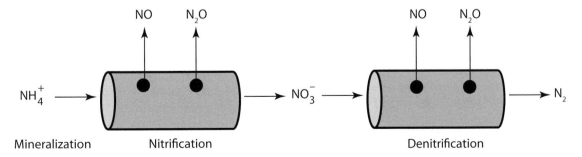

Figure 20.2 Conceptual representation of nitrogen gas losses in the hole-in-the-pipe model (Firestone and Davidson 1989; Davidson 1991). Gas losses during nitrification and denitrification depend on the rate at which nitrogen flows through the pipe, during which a percentage leaks out as NO and N_2O emissions (conceptualized as holes in the pipe). Adapted from Davidson et al. (2000) as used by Bonan (2016)

models that account for the site conditions that control nitrification and denitrification.

DAYCENT has an N gas flux submodel that simulates daily N_2O and N_2 emissions from nitrification and denitrification (Parton et al. 1996, 2001; Del Grosso et al. 2000). The model does not explicitly represent the biochemical and geochemical processes that control gas losses but, rather, accounts for the conditions under which nitrification and denitrification are expected to occur. Empirical functions modify a maximum potential rate for temperature, moisture, and other site conditions similar to those used for soil organic matter decomposition (Chapter 18). The rate of N loss in nitrification (g N m^{-2} d^{-1}) depends on the N mineralization rate, NH_4^+ concentration, soil temperature, soil moisture, and soil pH. This is represented by the equation:

$$F_N = \text{Net}_{\min}K_1 + K_{\max}\left[NH_4^+\right]f(T)f(\theta)f(\text{pH}). \quad (20.14)$$

The first term on the right-hand side is a fraction K_1 of the net N mineralization from soil organic matter decomposition (Chapter 18) that is assumed to be nitrified. $\left[NH_4^+\right]$ is the ammonium concentration (g N m^{-2}), and $K_{\max}$ is the maximum fraction of NH_4^+ that is nitrified (d^{-1}). The functions $f(T)$, $f(\theta)$, and $f(\text{pH})$ are rate modifiers that adjust the base rate for soil conditions. A prescribed fraction of the nitrification rate is assumed to be N_2O. The rate of denitrification is controlled by the concentration of NO_3^-, labile carbon, and O_2 in soil. Heterotrophic soil CO_2 respiration is used as a proxy for labile carbon, and soil moisture is a proxy for O_2 so that the N loss during denitrification is

$$F_D = \min\left[f\left(NO_3^-\right), f\left(CO_2\right)\right]f(\theta). \quad (20.15)$$

In this equation, a maximum potential rate determined as the minimum of NO_3^- and CO_2 functions is modified by soil moisture. Equation (20.15) describes the total N loss from denitrification ($N_2O + N_2$) but can be partitioned between N_2O and N_2. The ratio of N_2 to N_2O emitted from soil is modeled in relation to the ratio of soil NO_3^- concentration to soil CO_2 emission, in which

$$R|_{N_2:N_2O} = f\left(NO_3^-/CO_2\right)f(\theta). \quad (20.16)$$

As soils become more anoxic, a larger proportion of N_2O produced during denitrification is reduced to N_2. The various parameters and functions used in the model were determined empirically from laboratory or field studies. The N gas model relies on a soil decomposition model to mineralize NH_4^+ so that nitrification and denitrification are not independent of soil organic matter decomposition (Chapter 18). Variants of the DAYCENT N gas flux model are used in terrestrial biosphere models (e.g., the Community Land Model).

DAYCENT relies on empirical equations to calculate N gas losses. The Denitrification and Decomposition model (DNDC; Li et al. 1992, 2000) provides a more process-rich representation of nitrification and denitrification. The model represents soil biochemical and geochemical processes based on the kinetics of chemical reactions and production, consumption, and diffusion in soil. The model simulates decomposition of soil organic matter and production of NH_4^+ during N mineralization. Sorption onto clay surfaces removes some NH_4^+ from the soil solution. A portion of the NH_4^+ remaining in solution becomes ammonia (NH_3) and is volatilized to the atmosphere. NH_4^+ is also nitrified to NO_3^-, during which NO and N_2O are emitted in proportion to the nitrification rate. The rate of nitrification depends on the biomass and activity of nitrifying microbes. Denitrification similarly depends on the growth of denitrifying microbes, substrate consumption, and gas diffusion under anaerobic conditions. A series of sequential chemical reactions modeled using Michaelis–Menten kinetics govern the transformation from NO_3^- to NO, N_2O, and finally N_2. A feature of the model is that it allows aerobic and anaerobic microsites to co-occur within the soil so that nitrification and denitrification can occur simultaneously. The ratio between these microsites is controlled by the soil redox potential. The DNDC approach has been used in the O-CN terrestrial biosphere model (Zaehle and Friend 2010) and in a version of the Community Land Model (Saikawa et al. 2013).

Ammonia volatilization is the loss of NH_3 to the atmosphere. This gas loss is generally minor in natural soils because of low NH_4^+ concentration. It is much larger in agricultural systems with application of fertilizer and manure and is also large in livestock feedlots. Temperature, soil moisture, pH, and wind speed are factors that affect ammonia volatilization. The exchange of NH_3 between the surface and atmosphere can be parameterized in terms of emission factors, but meteorologically based flux parameterizations have also been developed (Sutton et al. 1995, 2013; Nemitz et al. 2001; Flechard et al. 2013). The flux of NH_3 to the atmosphere is calculated from the difference between the surface concentration c_s and the atmospheric concentration c_a multiplied by the total conductance so that

$$F_{NH_3} = \frac{c_s - c_a}{g_{am}^{-1} + g_b^{-1} + g_c^{-1}}, \qquad (20.17)$$

with g_{am} and g_b aerodynamic and boundary layer conductances similar to those used in dry deposition. The conductance g_c is a canopy conductance that represents loss of NH_3 from leaves through stomata. Ammonia is also deposited onto leaf surfaces and taken up by stomata so that the biosphere–atmosphere exchange is bidirectional and some of the emission is captured by the canopy. Net loss of NH_3 mainly occurs in agricultural sites, while net deposition usually occurs in natural vegetation. Riddick et al. (2016) described an implementation of NH_3 chemistry and emissions in the Community Land Model.

Question 20.8 $R|_{N_2:N_2O}$ is the ratio of N_2 to N_2O emitted from denitrification in DAYCENT. Give the equations by which this ratio is used to calculate the fluxes of N_2O and N_2.

Question 20.9   Denitrification is episodic, depends strongly on soil moisture, and occurs mostly from "hot spots" where conditions are favorable. Describe the challenges this presents to representing N gas losses in terrestrial biosphere models.

Question 20.10   Contrast (20.14) and (20.15) for nitrification and denitrification with (20.17) for NH_3 emission. What is a key difference between how these fluxes are formulated?

20.6 | Methane

Methane (CH_4) production occurs in wet, O_2-depleted soil, such as found in wetlands or waterlogged soil. Pathways by which CH_4 reaches the atmosphere include diffusion through soil and water; plant aerenchyma, by which CH_4 is transported through plants via vessel-like tubes; and ebullition, which is the release into the air of CH_4 bubbles that have been trapped in pockets in the soil. Methane fluxes can be modeled through emission factors, and this approach is commonly used for atmospheric chemistry models. Wania et al. (2010) developed one of the first process models for use with terrestrial biosphere models. Further information on CH_4 models is provided by Wania et al. (2013) and Melton et al. (2013) in a model intercomparison. The CH_4 model for the Community Land Model provides an example of the process detail (Riley et al. 2011), as shown in Figure 20.3. The overall governing equation is

$$\frac{\partial}{\partial t}[CH_4] = \frac{\partial F_D}{\partial z} + P(z) - E(z) - A(z) - O(z). \qquad (20.18)$$

This equation describes the mass balance of CH_4 (mol m^{-3}) in a layer of soil in relation to gas diffusion F_D, production P, ebullition E, aerenchyma transport A, and oxidation O. Methane production depends on the amount of substrate available for

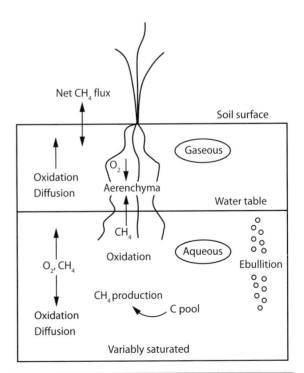

Figure 20.3 Processes represented in methane biogeochemistry for variably saturated soil. Adapted from Riley et al. (2011)

methanogens. Production is assumed to be proportional to heterotrophic respiration, with a base ratio of CH_4 to CO_2 adjusted for temperature, pH, and other factors. Bubbling occurs in relation to the partial pressure of CH_4 in the soil. Aerenchyma transport is represented as a diffusion process, with a concentration gradient between the soil and the atmosphere and with plant and aerodynamic conductances acting in series. Methane oxidation is modeled using Michaelis–Menten kinetics and depends on the gas concentrations of CH_4 and O_2 in soil. The diffusion flux between layers is specified in relation to the concentration gradient and a diffusion coefficient. Equation (20.18) is solved for each soil layer in a tridiagonal system of equations similar to soil temperature (Chapter 5) and soil moisture (Chapter 8). A critical component of CH_4 emissions is to calculate wetland dynamics and the inundated fraction of the surface.

Question 20.11 What are the units for P, E, A, and O in (20.18)? What are the units for F_D?

20.7 | Canopy–Chemistry Models

Meteorology, dry deposition, and chemical reactions in the plant canopy affect the concentrations and fluxes of BVOCs, NO_x, O_3, and other chemical compounds. In particular, the leaf-level emission of BVOCs may differ substantially from the above-canopy flux into the atmospheric boundary layer because of within-canopy meteorological and chemical processes. Similarly, the concentration of O_3 in the canopy may be different compared with that above the canopy. A class of models called canopy–chemistry models combines principles of boundary layer meteorology, plant canopies, leaf gas exchange, and atmospheric chemistry to simulate vertical profiles of temperature, water vapor, and gas concentrations in a column of air that extends from the ground through the forest canopy and the atmospheric boundary layer. These multilayer models solve prognostic equations for temperature and water vapor, similar to plant canopy models, and additionally include the vertical transport and transformation of chemical compounds above and within the plant canopy (Figure 20.4). The prognostic equation for a chemical compound c is

$$\frac{\partial c}{\partial t} = \frac{\partial}{\partial z}\left[K_c(z)\frac{\partial c}{\partial z}\right] + \frac{S_c(z)}{\rho_m} + R_c(z). \quad (20.19)$$

This is similar to that presented in Chapter 16 for scalars. The term S_c is the total sources and sinks from emission and deposition, and the additional term R_c represents the change in concentration

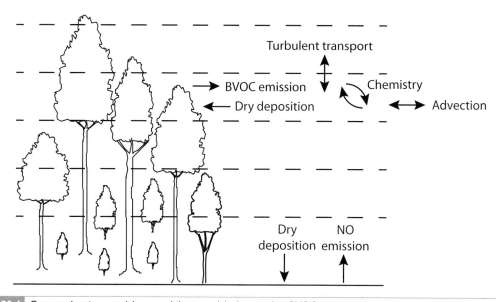

Figure 20.4 Canopy–chemistry models are multilayer models that simulate BVOC emission, dry deposition, chemical reactions, turbulent transport, and advection at each level in the canopy. Adapted from Gao et al. (1993) and Wolfe and Thornton (2011)

resulting from chemical production and loss at each level of the model. These chemical reactions include BVOCs, NO_x, and O_3 emitted or deposited to the foliage and soil. The critical elements of the models are the source and sink fluxes, the change in concentration of the chemical compounds as they react with one another, the vertical mixing and transport of these compounds within and above the canopy, and also advection of chemical compounds into the air column (included in some models in S_c). Example models include the following: Gao et al. (1993); Makar et al. (1999); Stroud et al. (2005); CACHE (Forkel et al. 2006; Bryan et al. 2012); SOSA (Boy et al. 2011); CAFE (Wolfe and Thornton 2011); ACCESS (Saylor 2013); and FORCAsT (Ashworth et al. 2015).

20.8 | Stable Isotopes

Isotopes are atoms of a particular element that differ in number of neutrons and therefore differ in mass. Stable isotopes do not undergo radioactive decay and are used in many hydrological and biogeochemical analyses as indicators of the water and carbon cycles. Hydrogen has two stable isotopes (^1H and ^2H; the latter is commonly called deuterium). Carbon has two naturally occurring stable isotopes (^{12}C and ^{13}C). Two stable isotopes of oxygen used in geochemical studies are ^{16}O and ^{18}O. Lighter isotopes are much more common than heavier isotopes (Table 20.1), but changes in the abundance of heavy isotopes can be used to study the hydrologic and carbon cycles.

Table 20.1 | Average abundance of stable isotopes

Element	Isotope	Abundance (%)
Hydrogen	^1H	99.985
	^2H	0.015
Carbon	^{12}C	98.90
	^{13}C	1.10
Oxygen	^{16}O	99.76
	^{17}O	0.04
	^{18}O	0.20

Source: From West et al. (2010).

The isotope composition of a product can differ from the source material. The process by which this occurs is termed isotope fractionation and consists of kinetic and equilibrium isotope effects. Lighter isotopes are favored in evaporation and photosynthesis, leaving the source material enriched in heavier isotopes relative to the product. During photosynthesis, for example, plants assimilate more of the lighter $^{12}CO_2$ relative to the heavier $^{13}CO_2$ so that the products of photosynthesis are depleted in ^{13}C relative to air. This fractionation occurs because diffusion through stomata to the chloroplast favors $^{12}CO_2$ and discriminates against $^{13}CO_2$ (this relates to the larger diffusivity in air of $^{12}CO_2$ compared with $^{13}CO_2$ and produces a small fractionation) and because the Rubisco enzyme that catalyzes carboxylation discriminates against $^{13}CO_2$ (which gives large fractionation). Both of these are examples of a kinetic isotope effect. Evaporation of water has both kinetic and equilibrium isotope effects. Kinetic isotope fractionation occurs because the diffusivity in air of the heavy isotopes (^{18}O, ^2H) is less than that of the corresponding lighter isotopes (^{16}O, ^1H). In addition, as water changes from liquid to vapor, the heavier isotopes remain in the liquid phase while the lighter isotopes favor the vapor phase because the vapor pressure of the heavy isotopes is less than that of the lighter $^1H_2^{16}O$. This is an equilibrium isotope effect.

The isotope ratio of a sample A is the mass of the heavy isotope in the sample relative to the light isotope. For ^{13}C, for example,

$$R_A = \frac{^{13}C_A}{^{12}C_A}.$$ (20.20)

Because isotope differences in the environment are typically small, isotope abundance is more commonly expressed using delta notation. The δ of the sample is given by its isotope ratio relative to that of a standard so that

$$\delta_A = \frac{R_A - R_{STD}}{R_{STD}} = \frac{R_A}{R_{STD}} - 1.$$ (20.21)

δ_A is commonly reported as parts per thousand (per mil, ‰), which is obtained by multiplying (20.21) by 1000. The standard is the isotope ratio at which $\delta_A = 0$. For ^{13}C, R_{STD} is the Vienna Pee Dee Belemnite standard ($R_{VPDB} = 1.12372 \cdot 10^{-2}$). A negative value indicates that the sample is depleted relative to the standard. $\delta^{13}C = -8‰$ is representative of atmospheric CO_2 so that $^{13}C/^{12}C = 1.1147 \cdot 10^{-2}$. A representative value for plant biomass is $\delta^{13}C = -26‰$ and $^{13}C/^{12}C = 1.0945 \cdot 10^{-2}$. For hydrogen and oxygen, isotope abundance is expressed relative to Vienna Standard Mean Ocean Water (VSMOW; $R_{VSMOW} = 2.0052 \cdot 10^{-3}$ for $^{18}O/^{16}O$ and $1.5576 \cdot 10^{-4}$ for $^{2}H/^{1}H$).

The difference between two samples is a measure of isotope fractionation, and the isotope partitioning between a source A and product B can be expressed in several ways. In Δ notation, the magnitude of fractionation is simply the difference in δ:

$$\Delta_{A-B} = \delta_A - \delta_B. \tag{20.22}$$

The fractionation factor α expresses fractionation as the ratio of the isotope ratios whereby

$$\alpha_{A-B} = \frac{R_A}{R_B} = \frac{\delta_A + 1}{\delta_B + 1}. \tag{20.23}$$

$\alpha_{A-B} > 1$ indicates that B is depleted in the heavy isotope relative to A, and $\alpha_{A-B} < 1$ indicates that B is enriched in the heavy isotope relative to A. The deviation of α_{A-B} from 1 expresses the magnitude of isotope fractionation as an enrichment factor (also called discrimination) in which

$$\varepsilon_{A-B} = \alpha_{A-B} - 1 = \frac{R_A}{R_B} - 1 \tag{20.24}$$

so that

$$\varepsilon_{A-B} = \frac{\delta_A - \delta_B}{1 + \delta_B}, \tag{20.25}$$

and because the denominator is approximately equal to 1, (20.25) can be simplified to $\varepsilon_{A-B} \approx \Delta_{A-B}$. A positive value shows that B is depleted in the heavy isotope relative to A; a negative value shows that B is enriched in the heavy isotope relative to A.

Photosynthetic fractionation for C_3 plants results from CO_2 diffusion to the chloroplast and CO_2 fixation by Rubisco (Figure 20.5a). This can be represented as a multistage process of the following: diffusion across the leaf boundary layer to the leaf surface; diffusion in the stomatal pore; dissolution in mesophyll water and aqueous transport to the chloroplast; and fixation by Rubisco (Farquhar and Richards 1984; Farquhar et al. 1989a). Each step results in isotope fractionation, and the total discrimination can be obtained by summing the individual discriminations at each stage weighted by the associated change in CO_2. The isotope composition of the products of leaf photosynthesis in contrast to air is

$$\alpha_p = \frac{R_a}{R_p} = 1 + \frac{2.9}{1000}\frac{(c_a - c_s)}{c_a} + \frac{4.4}{1000}\frac{(c_s - c_i)}{c_a}$$

$$+ \frac{(1.1 + 0.7)}{1000}\frac{(c_i - c_c)}{c_a} + \frac{27}{1000}\frac{c_c}{c_a}, \tag{20.26}$$

with c_a, c_s, c_i, and c_c the CO_2 mole fraction of ambient air, the leaf surface, intercellular space, and the chloroplast, respectively. The various terms in this equation describe fractionation during diffusion through the leaf boundary layer (2.9‰); diffusion through the stomatal pore (4.4‰); dissolution of CO_2 in mesophyll water (1.1‰) and aqueous transport to the chloroplast (0.7‰); and fixation by Rubisco (27‰). A simplified form of this equation is

$$\alpha_p = \frac{R_a}{R_p} = 1 + \frac{a + (b - a)c_i/c_a}{1000}, \tag{20.27}$$

with $a = 4.4‰$ and $b = 27‰$ the fractionations that occur during diffusion through the stomatal pore and fixation by Rubisco, respectively (Farquhar et al. 1982, 1989a; Farquhar and Richards 1984). C_4 photosynthesis also discriminates against ^{13}C, but much less than in C_3 plants. A simple approach is to consider only diffusion through stomata so that $\alpha_p = 1 + 4.4/1000$ (Suits et al. 2005). (Note that Farquhar et al. [1989a] and Farquhar and Richards [1984] described photosynthetic fractionation using the notation $\Delta = \alpha_p - 1$. In this chapter, this is denoted as ε, and Δ is used to represent the difference in δ between two substances, for consistency with geochemical isotopic usages. To avoid confusion between ε and Δ, the fractionation represented by [20.26] and [20.27] is given in terms of α_p.)

Equation (20.27) is derived by considering $^{13}CO_2$ and $^{12}CO_2$ photosynthesis fluxes, as described by Farquhar et al. (1989a). The isotope ratio of photosynthate is the same as the isotope ratio of the photosynthesis rates so that

$$R_p = \frac{^{13}A_n}{A_n}. \tag{20.28}$$

The $^{12}CO_2$ photosynthesis flux is

$$A_n = g_{\ell c}(c_a - c_i), \tag{20.29}$$

where here c_a and c_i are the $^{12}CO_2$ mole fraction of ambient air and intercellular space, respectively,

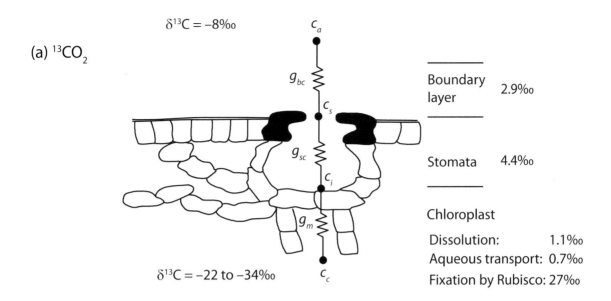

(a) $^{13}CO_2$

$\delta^{13}C = -8‰$

c_a

g_{bc}

c_s

g_{sc}

c_i

g_m

$\delta^{13}C = -22$ to $-34‰$

c_c

Boundary layer   2.9‰

Stomata   4.4‰

Chloroplast

Dissolution: 1.1‰
Aqueous transport: 0.7‰
Fixation by Rubisco: 27‰

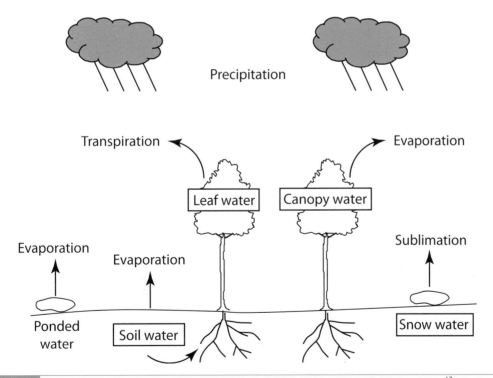

(b) $^{1}H_2{}^{18}O$, $^{1}H^{2}H^{16}O$

Precipitation

Transpiration

Evaporation

Leaf water

Canopy water

Evaporation

Evaporation

Sublimation

Ponded water

Soil water

Snow water

Figure 20.5 Processes involved in isotopic fractionation in the biosphere. (a) Photosynthetic fractionation of $^{13}CO_2$. Shown are representative $\delta^{13}C$ values for the atmosphere and plant biomass and the individual enrichment terms. (b) Fractionation of $^{1}H_2{}^{18}O$ and $^{1}H^{2}H^{16}O$ in the hydrologic cycle. Boxes show pools of water that are enriched in ^{18}O or ^{2}H, and arrows denote the process producing the fractionation.

and $g_{\ell c}$ is leaf conductance to $^{12}CO_2$, consisting of stomatal g_{sc} and boundary layer g_{bc} conductances in series. A similar equation pertains to the $^{13}CO_2$ flux. Using the isotope ratios $R_a = {}^{13}c_a/c_a$ and $R_i = {}^{13}c_i/c_i$ for air and intercellular space, respectively, the $^{13}CO_2$ photosynthesis flux is

$$^{13}A_n = {}^{13}g_{\ell c}(R_a c_a - R_i c_i). \tag{20.30}$$

Substituting (20.29) and (20.30) into (20.28) gives

$$R_p = \frac{{}^{13}A_n}{A_n} = \frac{{}^{13}g_{\ell c}(R_a c_a - R_i c_i)}{g_{\ell c}(c_a - c_i)}. \tag{20.31}$$

Rearranging terms,

$$\frac{g_{\ell c}}{{}^{13}g_{\ell c}}(c_a - c_i) = \frac{R_a}{R_p}c_a - \frac{R_i}{R_p}c_i \tag{20.32}$$

and

$$\alpha_p = \frac{R_a}{R_p} = \frac{g_{\ell c}}{{}^{13}g_{\ell c}}\left(1 - \frac{c_i}{c_a}\right) + \frac{R_i}{R_p}\frac{c_i}{c_a}. \tag{20.33}$$

Equation (20.33) represents two isotope fractionation effects. The ratio R_i/R_p is the isotope effect α_{rubisco} from carboxylation. The isotope effect for diffusion is $\alpha_k = g_{\ell c}/{}^{13}g_{\ell c}$, and this represents molecular diffusion in the stomatal pore and turbulent diffusion through the leaf boundary layer. Molecular diffusion results in kinetic fractionation because the lighter $^{12}CO_2$ (molecular mass 44 g mol^{-1}) has a larger diffusivity in air relative to the heavier $^{13}CO_2$ (molecular mass 45 g mol^{-1}). For a gas with isotopes x and y that have molecular mass M_x and M_y, the ratio of their diffusivities in air with molecular mass $M_a = 28.97$ g mol^{-1} is

$$\frac{D_x}{D_y} = \left(\frac{M_y}{M_x} \cdot \frac{M_x + M_a}{M_y + M_a}\right)^{1/2} \tag{20.34}$$

(Merlivat 1978; Farquhar and Lloyd 1993). For molecular diffusion through the stomatal pore, $\alpha_{ks} = D_x/D_y = 1.0044$ is the kinetic fractionation factor and is the ratio of the diffusivity of $^{12}CO_2$ to $^{13}CO_2$ in air. This factor needs to be modified, however, when applied to diffusion through the leaf boundary layer. If the boundary layer is treated as laminar, the diffusivity ratio is raised to the 2/3 power, as in (10.39) to calculate leaf boundary layer conductances, so that $\alpha_{kb} = (D_x/D_y)^{0.67} = 1.0029$. The total kinetic fractionation is

$$\alpha_k = \frac{\alpha_{kb}g_{bc}^{-1} + \alpha_{ks}g_{sc}^{-1}}{g_{bc}^{-1} + g_{sc}^{-1}}, \tag{20.35}$$

which can be simplified to $\alpha_k \approx 1.0044$ for $g_{bc} \gg g_{sc}$ (Farquhar et al. 1989b; Farquhar and Lloyd 1993). These fractionations can be more generally represented in the form $\alpha_k = 1 + a/1000$ and $\alpha_{\text{rubisco}} = 1 + b/1000$. With these substitutions, (20.33) simplifies to (20.27).

When c_i/c_a is small, there is less discrimination by Rubisco and α_p for C_3 plants approaches the diffusive fractionation term (1.0044). At high c_i/c_a, α_p approaches the carboxylation fractionation (1.027). With high vapor pressure deficit or drought stress, stomatal conductance decreases. In these conditions, c_i/c_a decreases and, therefore, α_p decreases. The ratio c_i/c_a relates to water-use efficiency (Chapter 12); α_p, therefore, is also an indicator of water-use efficiency and decreases with higher water-use efficiency (Farquhar and Richards 1984; Farquhar et al. 1989a).

For many applications, it is necessary to calculate whole-canopy photosynthetic discrimination. This is the difference in δ between air and the canopy so that

$$\Delta_c = \delta_a - \delta_c. \tag{20.36}$$

In a big-leaf canopy model, δ_c is calculated from gross primary production (Suits et al. 2005; Raczka et al. 2016). In a multilayer canopy model, the canopy-integrated photosynthetic discrimination is obtained by weighting leaf-scale values by the leaf photosynthetic rate and integrating over the canopy (Baldocchi and Bowling 2003; Flanagan et al. 2012). In these models,

$$\Delta_c = \frac{\int_0^{h_c} [\delta_a(z) - \delta_p(z)]A_n(z)dz}{\int_0^{h_c} A_n(z)dz}. \tag{20.37}$$

This calculation can be further refined using the weighted average of sunlit and shaded leaves at each canopy layer. A multilayer canopy model allows for study of how vertical gradients in photosynthetic capacity, light, and other factors affect isotope discrimination (Baldocchi and Bowling 2003). In particular, leaf photosynthetic fractionation Δ_p increases (δ_p decreases) from the top to the bottom of the canopy. This increase in isotope discrimination is related to the isotopic signal of respiration from leaves and soil but also arises from decreasing irradiance with greater depth from the top of the canopy (which leads to high c_i/c_a as photosynthesis decreases). The light dependence of photosynthetic fractionation also results in a diurnal cycle, with high canopy Δ_c in the early morning or late afternoon at low irradiance and less discrimination at high irradiance. Δ_c is also low in the midafternoon, when vapor pressure deficit is high and stomata close, reducing c_i/c_a. A challenge with multilayer models is to account for the turbulent transfer and mixing of ^{13}C within the canopy airspace (Baldocchi and Bowling 2003).

Some biogeochemical models separately track the flow of total carbon and ^{13}C into various plant, litter, and soil organic matter pools so that the isotope composition of biomass can be calculated (van der Velde et al. 2013; Raczka et al. 2016; Duarte et al. 2017; Keller et al. 2017). The ^{13}C photosynthesis flux is calculated, and the ^{13}C carbon input follows the same flows of total carbon, implemented in the model by treating it as a tracer in which the carbon cycle is duplicated for ^{13}C.

Question 20.12 Derive (20.23).
Question 20.13 Derive (20.25), which relates ε_{A-B} to Δ_{A-B}.
Question 20.14 The $\delta^{13}C$ of CO_2 in the atmosphere is approximately −8‰, and the $\delta^{13}C$ of plant biomass is approximately −26‰. Calculate the photosynthetic fractionation factor α_{A-B} and the discrimination ε_{A-B}.
Question 20.15 The fractional abundance of ^{13}C is equal to $^{13}C/(^{12}C+^{13}C)$. Some models calculate the ^{13}C photosynthetic flux as $^{13}A_n = A_n R_p/(1 + R_p)$. Others use $^{13}A_n = A_n R_a/\alpha_p$. Which method is correct?
Question 20.16 Show that (20.35) is the correct equation for kinetic fractionation during photosynthetic uptake of CO_2 through the leaf boundary layer and stomata connected in series.

Isotopes are also useful diagnostics of the hydrologic cycle. Three water isotopes are $^1H_2^{16}O$ (with abundance greater than 99%) and the heavier water

$^1H^2H^{16}O$ (commonly abbreviated HDO) and $^1H_2^{18}O$. The isotope composition of water is expressed as the isotope ratios $^2H/^1H$ and $^{18}O/^{16}O$, or as the deviation relative to Vienna Standard Mean Ocean Water (δ^2H, $\delta^{18}O$). Changes in the isotope composition of water arise from phase changes and mixing as water flows through the hydrologic cycle. Most importantly, isotope fractionation occurs naturally through evaporation and condensation. Lighter isotopes evaporate more easily than heavier isotopes and occur preferentially in the vapor phase. Thus, water vapor that evaporates is depleted in ^2H or ^{18}O (low δ^2H or $\delta^{18}O$). Heavier isotopes preferentially condense and occur with greater abundance in the liquid phase.

Over land, the isotope composition of evapotranspiration reflects precipitation inputs and surface processes that result in fractionation. Terrestrial biosphere models have been modified to simulate water isotopes (e.g., Riley et al. 2002; Yoshimura et al. 2006; Haese et al. 2013; Wong et al. 2017). Such models simulate the isotope composition of canopy-intercepted water, soil water, snow, and leaf water by following the mass balance of the isotopes $^1H_2^{18}O$ and $^1H^2H^{16}O$ (Figure 20.5b). They account for equilibrium and kinetic fractionation during evaporation of canopy-intercepted water, leaf transpiration, soil evaporation, and snow sublimation. In addition to their standard meteorological forcing data, isotope-enabled terrestrial biosphere models require the magnitude of each isotope in precipitation and atmospheric water vapor. In these models, isotope fractionation during evaporation is calculated using the conceptual framework developed by Craig and Gordon (1965). That approach separates evaporation from a liquid surface to the atmosphere into three distinct layers of the following: a saturated interface where equilibrium fractionation occurs during phase change from liquid to vapor; a laminar layer immediately above the saturated interface in which diffusion results in kinetic fractionation; and a third layer of turbulent atmospheric transport away from the surface and in which there is no fractionation.

Equilibrium fractionation during the liquid–vapor phase transition arises because of isotopic differences in vapor pressure. The equilibrium isotope effect is denoted by α_{l-v} and is defined by the ratio of the heavy to light isotope ($^{18}O/^{16}O$ or $^2H/^1H$) in liquid and vapor as

$$\alpha_{l-v} = \frac{R_l}{R_v}, \tag{20.38}$$

where the subscripts l and v denote liquid water and water vapor, respectively. The saturation vapor pressure of $^1H_2^{16}O$ is greater than that of $^1H_2^{18}O$ or $^1H^2H^{16}O$, and this difference causes fractionation in the liquid to vapor phase change. Equilibrium fractionation is equal to the ratio of saturation vapor pressure for $^1H_2^{16}O$ relative to that of the heavier isotope so that $\alpha_{l-v} > 1$, which means that the liquid phase is heavier than the vapor phase. Equilibrium fractionation is greatest at low temperatures so that α_{l-v} is close to one at high temperatures and increases with decreasing temperatures. From Majoube (1971), the relationship between α_{l-v} and temperature T (K) for $^{18}O/^{16}O$ is

$$\ln \alpha_{l-v} = \frac{1.137 \times 10^3}{T^2} - \frac{0.4156}{T} - 2.0667 \times 10^{-3}. \tag{20.39}$$

A representative value is α_{l-v} = 1.0102 at 15°C. Similar relationships apply for $^2H/^1H$ with

$$\ln \alpha_{l-v} = \frac{24.844 \times 10^3}{T^2} - \frac{76.248}{T} + 52.612 \times 10^{-3}, \tag{20.40}$$

which gives α_{l-v}= 1.0911 at 15°C.

Diffusion results in kinetic fractionation because the diffusivity of the lighter $^1H_2^{16}O$ in air is greater than the heavier $^1H_2^{18}O$ or $^1H^2H^{16}O$. Values for the ratio of the diffusivity of the light-to-heavy isotope are $^1H_2^{16}O/^1H_2^{18}O$ = 1.0285 and $^1H_2^{16}O/^1H^2H^{16}O$ = 1.025 (Merlivat 1978). These values pertain to molecular diffusion and need to be modified for turbulent conditions in the boundary layer immediately above the evaporating surface to give the kinetic fractionation factor α_k.

Craig and Gordon (1965) derived equations for isotope partitioning during evaporation from water bodies (see also Gat 1996), and these can be applied likewise to land evaporation. The $^1H_2^{16}O$ evaporative flux between the surface and atmosphere is

$$E = g_w(q_s - q_a), \tag{20.41}$$

with q_s and q_a the mole fraction of water vapor at the evaporating surface and in the ambient air, respectively, and g_w the evaporative conductance. A similar equation describes the evaporative flux for $^1H_2^{18}O$ or $^1H^2H^{16}O$. The isotope composition of

water vapor at the surface can be obtained from the isotope ratio R_v of water vapor at the evaporating surface so that $^{iso}q_s = R_v q_s$, and similarly for atmospheric water vapor with $^{iso}q_a = R_a q_a$. At the evaporating surface, R_v can be obtained from the isotope ratio R_l of liquid water at the site of evaporation as modified by equilibrium fractionation during the change from liquid to vapor so that $R_v = R_l/\alpha_{l-v}$. With these substitutions, the heavy isotope evaporative flux is

$$^{iso}E = {}^{iso}g_w \left(\frac{R_l}{\alpha_{l-v}} q_s - R_a q_a \right), \tag{20.42}$$

and with $\alpha_k = g_w/^{iso}g_w$ the kinetic isotope effect for diffusion, (20.42) can be written as

$$^{iso}E = \frac{g_w}{\alpha_k} \left(\frac{R_l}{\alpha_{l-v}} q_s - R_a q_a \right). \tag{20.43}$$

The isotope ratio R_E of the evaporative flux is then

$$R_E = \frac{^{iso}E}{E} = \frac{\frac{1}{\alpha_k} \left(\frac{R_l}{\alpha_{l-v}} q_s - R_a q_a \right)}{q_s - q_a}, \tag{20.44}$$

and q_a/q_s can be taken as the relative humidity h so that

$$R_E = \frac{^{iso}E}{E} = \frac{1}{\alpha_k} \left(\frac{R_l}{\alpha_{l-v}} \cdot \frac{1}{1-h} - R_a \cdot \frac{h}{1-h} \right). \tag{20.45}$$

This is the Craig and Gordon (1965) equation for R_E.

The same equations can be used to describe isotope fractionation during transpiration (Farquhar et al. 1989b, 2007; Flanagan et al. 1991; Flanagan 1993; Cernusak et al. 2016). During transpiration, leaf water becomes enriched in heavy isotopes relative to the source water from which it is derived (stem water or soil water). For transpiration,

$$^{iso}E = \frac{g_{\ell w}}{\alpha_k} \left(\frac{R_l}{\alpha_{l-v}} q_i - R_a q_a \right). \tag{20.46}$$

Here, q_i is the saturation water vapor mole fraction in the intercellular space, and $g_{\ell w}$ is the leaf conductance to water vapor, consisting of stomatal g_{sw} and boundary layer g_{bw} conductances in series. R_l is the isotope ratio of liquid water at the site of evaporation and is different from the isotope ratio of leaf water. Similar to that given previously for CO_2, α_k is the kinetic fractionation for molecular diffusion through the stomatal pore and turbulent diffusion through the leaf boundary layer. For molecular diffusion through stomata, the kinetic fractionation factor α_{ks} is the ratio of the diffusivity of the light-to-heavy isotope ($^1H_2^{16}O/^1H_2^{18}O$ = 1.0285 and $^1H_2^{16}O/^1H^2H^{16}O$ = 1.025; Merlivat 1978). For diffusion through a laminar leaf boundary layer, the diffusivity ratio is raised to the 2/3 power so that $\alpha_{kb} = \alpha_{ks}^{0.67}$. The total kinetic fractionation is

$$\alpha_k = \frac{\alpha_{kb}g_{bw}^{-1} + \alpha_{ks}g_{sw}^{-1}}{g_{bw}^{-1} + g_{sw}^{-1}} \tag{20.47}$$

(Farquhar et al. 1989b; Farquhar and Lloyd 1993; Cernusak et al. 2016). The isotope ratio of the transpiration flux is

$$R_E = \frac{^{iso}E}{E} = \frac{\frac{1}{\alpha_k}\left(\frac{R_l}{\alpha_{l-v}}q_i - R_a q_a\right)}{q_i - q_a}$$
$$= \frac{1}{\alpha_k}\left(\frac{R_l}{\alpha_{l-v}} \cdot \frac{1}{1-h} - R_a \cdot \frac{h}{1-h}\right), \tag{20.48}$$

and the isotope ratio at the site of evaporation relative to the transpired water is

$$\frac{R_l}{R_E} = \alpha_{l-v}\left[\alpha_k\left(\frac{q_i - q_a}{q_i}\right) + \frac{R_a q_a}{R_E q_i}\right] = \alpha_{l-v}\left[\alpha_k(1-h) + \frac{R_a}{R_E}h\right]. \tag{20.49}$$

An equivalent form of this equation, used in some models, is

$$\frac{R_l}{R_E} = \alpha_{l-v}\left[\alpha_{ks}\left(\frac{q_i - q_s}{q_i}\right) + \alpha_{kb}\left(\frac{q_s - q_a}{q_i}\right) + \frac{R_a q_a}{R_E q_i}\right], \tag{20.50}$$

which is similar in form to (20.26) for photosynthetic fractionation. At steady state, the isotope composition of transpiring water is the same as that of soil water so that (20.49) or (20.50) can be used to calculate the isotope composition of leaf water. In fact, however, the isotope composition of water is heterogeneous within a leaf, with maximum enrichment of heavy isotopes at the sites of evaporation. The observed heavy isotope composition of leaf water is less than that predicted by these equations (Flanagan et al. 1991; Farquhar et al. 2007; Cernusak et al. 2016). More complex models account for this using non–steady state conditions (Farquhar and Cernusak 2005) and by allowing for diffusion of water in the leaf itself (Farquhar and Lloyd 1993).

The isotope composition of water pools (intercepted canopy water, soil water, snow) can be calculated by treating the isotopes of water as tracers and tracking their movement. This means that the hydrologic cycle is replicated for each isotope ($^1H_2^{16}O$, $^1H_2^{18}O$, $^1H^2H^{16}O$). A simple means to do so is to expand every water pool and flux in the hydrologic cycle with a general n-tracer dimension (e.g., $\theta(z, n)$ for soil water). Surface soil water is enriched in ^{18}O and ^2H compared with deeper soil water because of evaporation and is replenished from precipitation. Vertical transport of water within the soil column does not produce fractionation, but the movement of water does mix the isotope composition of water. This is modeled by extending the Richards equation (Chapter 8) to include isotopes.

Question 20.17 Show that (20.49) and (20.50) are equivalent equations for the fractionation of leaf water relative to source water during steady-state transpiration.

Question 20.18 The isotope ratio of liquid water at the site of evaporation relative to the source water can be written as $\Delta_e = (1 + \varepsilon_{l-v})[(1 + \varepsilon_k)(1 - h) + h(1 + \Delta_v)] - 1$ (Farquhar and Lloyd 1993; Farquhar et al. 2007; Cernusak et al. 2016). Show that this equation is equivalent to (20.49).

Appendices

Table A.1 | Basic and derived scientific units

Quantity name	Unit name	Dimension symbol	Unit symbol	Base units
Length	meter	L	m	—
Mass	kilogram	M	kg	—
Time	second	T	s	—
	hour		h	
	year		y	
Temperature	kelvin	K	K	—
Amount	mole	—	mol	—
Area	square meter	L^2	m^2	m^2
Volume	cubic meter	L^3	m^3	m^3
Density	kilogram per cubic meter	$M\,L^{-3}$	$kg\,m^{-3}$	$kg\,m^{-3}$
Velocity	meter per second	$L\,T^{-1}$	$m\,s^{-1}$	$m\,s^{-1}$
Acceleration	meter per second per second	$L\,T^{-2}$	$m\,s^{-2}$	$m\,s^{-2}$
Force	newton	$M\,L\,T^{-2}$	N	$kg\,m\,s^{-2}$
Energy	joule	$M\,L^2\,T^{-2}$	J	$kg\,m^2\,s^{-2} = N\,m$
Power	watt	$M\,L^2\,T^{-3}$	W	$kg\,m^2\,s^{-3} = J\,s^{-1}$
Pressure	pascal	$M\,L^{-1}\,T^{-2}$	Pa	$kg\,m^{-1}\,s^{-2} = J\,m^{-3} = N\,m^{-2}$

Note: $°C = K - 273.15$ so that a change in temperature of $1°C = 1\,K$.

Table A.2 | Metric prefixes

Multiple	Prefix	Symbol	Multiple	Prefix	Symbol
10^{-1}	deci-	d	10^{1}	deca-	da
10^{-2}	centi-	c	10^{2}	hecto-	h
10^{-3}	milli-	m	10^{3}	kilo-	k
10^{-6}	micro-	μ	10^{6}	mega-	M
10^{-9}	nano-	n	10^{9}	giga-	G
10^{-12}	pico-	p	10^{12}	tera-	T
10^{-15}	femto-	f	10^{15}	peta-	P

Table A.3 | Diffusivity ($m^2\,s^{-1}$) at standard pressure (1013.25 hPa) for 0°C and 15°C

Variable	Temperature	
	0°C	15°C
Momentum, v	13.3×10^{-6}	14.7×10^{-6}
Heat, D_h	18.9×10^{-6}	20.8×10^{-6}
H_2O, D_w	21.8×10^{-6}	24.0×10^{-6}
CO_2, D_c	13.8×10^{-6}	15.2×10^{-6}

Note: The diffusivity D at temperature T and pressure P can be calculated from the diffusivity D_0 at temperature T_0 and pressure P_0 as:

$$D(T,P) = D_0 \left(\frac{P_0}{P} \right) \left(\frac{T}{T_0} \right)^{1.81}$$

with $T_0 = 273.15\,K$ and $P_0 = 101325\,Pa$. Reference values at 0°C (D_0) for momentum and heat are from Campbell and Norman (1998) and Monteith and Unsworth (2013). Reference values for H_2O and CO_2 are from Massman (1998).

Table A.4 Physical constants used in the Community Land Model

Symbol	Quantity	Value	Units
c_{ice}	Specific heat of ice	2117.27	$J\ kg^{-1}\ K^{-1}$
c_{wat}	Specific heat of water	4188	$J\ kg^{-1}\ K^{-1}$
c_{pd}	Specific heat of dry air	1004.64	$J\ kg^{-1}\ K^{-1}$
c_{pw}	Specific heat of water vapor	1810	$J\ kg^{-1}\ K^{-1}$
g	Gravitational acceleration	9.80616	$m\ s^{-2}$
k	von Karman constant	0.4	—
L_f	Latent heat of fusion	0.3337×10^6	$J\ kg^{-1}$
M_a	Molecular mass of dry air	28.966	$g\ mol^{-1}$
M_w	Molecular mass of water	18.016	$g\ mol^{-1}$
$\mathfrak{R}$	Gas constant	8.31447	$J\ K^{-1}\ mol^{-1}$
T_f	Freezing point of water	273.15	K
κ_{air}	Thermal conductivity of air	0.023	$W\ m^{-1}\ K^{-1}$
κ_{ice}	Thermal conductivity of ice	2.29	$W\ m^{-1}\ K^{-1}$
κ_{wat}	Thermal conductivity of water	0.57	$W\ m^{-1}\ K^{-1}$
λ	Latent heat of vaporization	2.501×10^6	$J\ kg^{-1}$
		45.06	$kJ\ mol^{-1}$
ρ_{ice}	Density of ice	917	$kg\ m^{-3}$
ρ_{wat}	Density of water	1000	$kg\ m^{-3}$
σ	Stefan–Boltzmann constant	5.67×10^{-8}	$W\ m^{-2}\ K^{-4}$

Source: From Oleson et al. (2013).

A. Numerical Methods

Many mathematical techniques and numerical methods are used in models. This appendix describes the techniques and methods discussed throughout this book.

A.1 | Linear Approximations

A function $f(x)$ can be expanded in a Taylor series about x as

$$f(x + \Delta x) = f(x) + \Delta x \frac{df}{dx} + \frac{\Delta x^2}{2} \frac{d^2f}{dx^2} + \frac{\Delta x^3}{6} \frac{d^3f}{dx^3}$$
$$+ \frac{\Delta x^4}{24} \frac{d^4f}{dx^4} + \cdots, \tag{A.1}$$

with the derivatives evaluated at x. Neglecting terms higher than Δx, this is approximated by

$$f(x + \Delta x) = f(x) + \frac{df}{dx} \Delta x. \tag{A.2}$$

Equation (A.2) is known as a first-order Taylor series expansion and provides a linear approximation for a nonlinear function. Consider, for example, leaf temperature raised to the fourth power, denoted T_ℓ^4. Leaf temperature is typically unknown and is solved as part of the leaf energy budget (Chapter 10). The term T_ℓ^4 can be linearized about a known air temperature T_a using the relationship $T_\ell = T_a + (T_\ell - T_a)$. Algebraically, it is easy to show that

$$T_\ell^4 = [T_a + (T_\ell - T_a)]^4 = T_a^4 + 4T_a^3(T_\ell - T_a)$$
$$+ 6T_a^2(T_\ell - T_a)^2 + 4T_a(T_\ell - T_a)^3 + (T_\ell - T_a)^4. \tag{A.3}$$

This is the Taylor series given by (A.1), and (A.2) is the linear approximation ignoring terms higher than $T_\ell - T_a$. Another common application in biometeorology is to linearize saturation vapor pressure with respect to temperature. Saturation vapor pressure at leaf temperature is approximated from that of air temperature using the relationship

$$e_{sat}(T_\ell) = e_{sat}(T_a) + s(T_\ell - T_a), \tag{A.4}$$

where $s = de_{sat}/dT$ is the derivative of saturation vapor pressure with respect to temperature evaluated at T_a. This linearization is used to solve for surface temperature (Chapter 7) and leaf temperature (Chapter 10) with the Penman–Monteith equation.

A.2 | Finite Difference Approximations for Derivatives

The Taylor series expansion of a function $f(x)$ is

$$f(x + \Delta x) = f(x) + \Delta x \frac{df}{dx} + \frac{\Delta x^2}{2} \frac{d^2f}{dx^2} + \frac{\Delta x^3}{6} \frac{d^3f}{dx^3} + \cdots, \tag{A.5}$$

and, similarly,

$$f(x - \Delta x) = f(x) - \Delta x \frac{df}{dx} + \frac{\Delta x^2}{2} \frac{d^2f}{dx^2} - \frac{\Delta x^3}{6} \frac{d^3f}{dx^3} + \cdots, \tag{A.6}$$

with the derivatives evaluated at x. These equations are used to numerically approximate the first derivative df/dx and the second derivative d^2f/dx^2. Neglecting terms higher than Δx, the forward difference approximation for the first derivative, obtained from (A.5), is

$$\frac{df}{dx} = \frac{f(x + \Delta x) - f(x)}{\Delta x} + O(\Delta x), \tag{A.7}$$

where $O(\Delta x)$ denotes that the error arising from truncating the Taylor series is on the order of Δx. The backward difference approximation for the first derivative, from (A.6), is

$$\frac{df}{dx} = \frac{f(x) - f(x - \Delta x)}{\Delta x} + O(\Delta x). \tag{A.8}$$

Subtracting (A.6) from (A.5) yields the central difference approximation, ignoring terms higher than Δx^2:

$$\frac{df(x)}{dx} = \frac{f(x + \Delta x) - f(x - \Delta x)}{2\Delta x} + O(\Delta x^2). \tag{A.9}$$

The truncation error of the central difference approximation is considerably less than that of the forward or backward difference approximations. The central difference approximation for the second derivative is found by adding (A.5) and (A.6) so that

$$\frac{d^2f}{dx^2} = \frac{f(x-\Delta x) - 2f(x) + f(x+\Delta x)}{\Delta x^2} + O(\Delta x^2).$$

(A.10)

A.3 | Numerical Integration of First-Order Differential Equations

One common use of (A.7) is to numerically integrate a differential equation forward in time. Consider, for example, a quantity y whose change in time is described by the function $f(t,y)$, which depends on the value of y at time t. This is the first-order differential equation

$$\frac{dy}{dt} = f(t,y),$$

(A.11)

with the initial condition $y = y_0$ at time $t = t_0$. This is known as an initial value problem, and analytical solutions are found for specific forms of $f(t,y)$. In many problems, however, there is not an analytical solution, and (A.11) must be stepped forward in time using numerical methods. Time is specified in discrete increments of Δt so that the time at the nth time step is $t_n = t_0 + n\Delta t$, and the value of y at time t_n is denoted y_n. Over the time interval Δt, y is stepped forward to y_{n+1} making use of (A.7) with

$$\frac{dy}{dt} = \frac{y_{n+1} - y_n}{\Delta t}.$$

(A.12)

Combining (A.11) and (A.12) gives an equation for y_{n+1} in which

$$y_{n+1} = y_n + f(t_n, y_n)\Delta t.$$

(A.13)

This is known as the Euler method. Starting with an initial condition $y(0) = y_0$, y is successively stepped forward in time using (A.13).

More commonly, however, advanced integration techniques are used instead of the Euler method. The fourth-order Runge–Kutta method evaluates the derivative dy/dt four times: at the beginning of the time interval (t_n); twice at the midpoint $(t_n + \Delta t/2)$; and again at the end of the interval $(t_n + \Delta t)$. Then,

$$y_{n+1} = y_n + \frac{\Delta t}{6}(k_1 + 2k_2 + 2k_3 + k_4),$$

(A.14)

with

$$\begin{aligned} k_1 &= f(t_n, y_n) \\ k_2 &= f(t_n + \Delta t/2, y_n + k_1\Delta t/2) \\ k_3 &= f(t_n + \Delta t/2, y_n + k_2\Delta t/2) \\ k_4 &= f(t_n + \Delta t, y_n + k_3\Delta t) \end{aligned}.$$

(A.15)

A specific example of (A.11) is the equation

$$\frac{dy}{dt} = a - by.$$

(A.16)

This is a first-order, linear differential equation and describes the temporal dynamics of biomass accumulation (4.23), leaf temperature (10.27), and leaf water potential (13.10). It can be solved numerically using (A.13) or (A.14) but is, in fact, easily integrated to give

$$y(t) = \frac{a}{b} + ce^{-bt}.$$

(A.17)

The term c is a constant of integration that is obtained with the initial condition $y(0) = y_0$. Then, $c = y_0 - a/b$ and

$$y(t) = \frac{a}{b} + \left(y_0 - \frac{a}{b}\right)e^{-bt}.$$

(A.18)

More complex differential equations are solved numerically using the above techniques.

A.4 | Finite Difference Methods to Solve the Diffusion Equation

Many processes in the biosphere–atmosphere system can be treated as a diffusion problem. Examples include soil temperature (Chapter 5), soil moisture (Chapter 8), and scalar canopy profiles (Chapter 16). In one dimension, the diffusion equation for a constituent c that varies in space and time, denoted as $c(z,t)$, is

$$\frac{\partial c}{\partial t} = \frac{\partial}{\partial z}\left(D_c \frac{\partial c}{\partial z}\right) = D_c \frac{\partial^2 c}{\partial z^2},$$

(A.19)

where D_c is a diffusion coefficient that is constant in space. A grid in space is defined for $z_1, z_2, \ldots, z_k$ with uniform spacing equal to Δz so that $z_i = z_1 - (i-1)\Delta z$. Time is specified in discrete increments of Δt so that the time at the nth time step is $t_n = t_0 + n\Delta t$. The concentrations at time n are denoted $c_1^n, c_2^n, \ldots, c_k^n$. The time derivative is given by a forward difference approximation at n using (A.7) so that

$$\frac{\partial c}{\partial t} = \frac{c_i^{n+1} - c_i^n}{\Delta t}, \tag{A.20}$$

and the second derivative is obtained using the central difference approximation (A.10), whereby

$$\frac{\partial^2 c}{\partial z^2} = \left(\frac{c_{i-1}^n - 2c_i^n + c_{i+1}^n}{\Delta z^2}\right). \tag{A.21}$$

Combining equations,

$$\frac{c_i^{n+1} - c_i^n}{\Delta t} = D_c \left(\frac{c_{i-1}^n - 2c_i^n + c_{i+1}^n}{\Delta z^2}\right), \tag{A.22}$$

which can be rearranged to give the expression,

$$c_i^{n+1} = c_i^n + Mc_{i-1}^n - 2Mc_i^n + Mc_{i+1}^n, \tag{A.23}$$

with $M = D_c \Delta t / \Delta z^2$. Equation (A.23) is referred to as an explicit solution. The concentration c_i^{n+1} is calculated using c_{i-1}^n, c_i^n, and c_{i+1}^n known from the previous time step, and the equation has only one unknown (c_i^{n+1}). The concentrations are stepped forward in time from the initial conditions $c_1^0, c_2^0, \ldots, c_k^0$. The equation is first-order accurate in time. It is numerically stable and converges to the exact solution for $M \leq 1/2$. This means that as the interval Δz is decreased to reduce truncation error the time step Δt must also decrease.

The implicit method evaluates the spatial derivative using the unknown concentrations at time $n+1$ and uses a backward difference approximation for the time derivative at $n+1$ so that

$$\frac{c_i^{n+1} - c_i^n}{\Delta t} = D_c \left(\frac{c_{i-1}^{n+1} - 2c_i^{n+1} + c_{i+1}^{n+1}}{\Delta z^2}\right) \tag{A.24}$$

and

$$c_i^{n+1} = c_i^n + Mc_{i-1}^{n+1} - 2Mc_i^{n+1} + Mc_{i+1}^{n+1}. \tag{A.25}$$

The latter equation can be rewritten:

$$-Mc_{i-1}^{n+1} + (1 + 2M)c_i^{n+1} - Mc_{i+1}^{n+1} = c_i^n. \tag{A.26}$$

Equation (A.26) is an implicit solution because calculation of c_i^{n+1} also requires c_{i-1}^{n+1} and c_{i+1}^{n+1}. The result is a system of k linear equations with k unknowns ($c_1^{n+1}, c_2^{n+1}, \ldots, c_k^{n+1}$). The tridiagonal form of the equations (with unknowns c_{i-1}^{n+1}, c_i^{n+1}, and c_{i+1}^{n+1}) means that the system of linear equations can be solved efficiently using common numerical methods (Appendix A8). The implicit method is first-order accurate in time, but its advantage over the explicit method is that it is numerically stable and convergent for $M > 0$ (i.e., for any choice of Δz and Δt).

A third method, the Crank–Nicolson method, increases the accuracy of the numerical solution by combining the explicit method with fluxes evaluated at time n and the implicit method with fluxes evaluated at $n+1$ (Crank and Nicolson 1947). The flux is taken as the average of these two times so that

$$\frac{c_i^{n+1} - c_i^n}{\Delta t} = D_c \left(\frac{c_{i-1}^n - 2c_i^n + c_{i+1}^n}{2\Delta z^2} + \frac{c_{i-1}^{n+1} - 2c_i^{n+1} + c_{i+1}^{n+1}}{2\Delta z^2}\right), \tag{A.27}$$

yielding a tridiagonal system of equations with the form:

$$-\frac{M}{2}c_{i-1}^{n+1} + (1 + M)c_i^{n+1} - \frac{M}{2}c_{i+1}^{n+1} = \frac{M}{2}c_{i-1}^n$$
$$+ (1 - M)c_i^n + \frac{M}{2}c_{i+1}^n. \tag{A.28}$$

The Crank-Nicolson method is second-order accurate in both space and time (the left-hand side and right-hand side are centered on time $n + 1/2$). It combines the numerical stability of the implicit method with the accuracy of a method that is second-order in space and time.

A.5 | Root Finding

A root-finding algorithm is a numerical method to find x such that $f(x) = 0$. Starting with an initial estimate, these algorithms use iteration to successively refine previous estimates of x until some convergence criteria is met. One simple numerical method is fixed-point iteration. This method solves the equation $f(x) = 0$ by rewriting it in the form $x = g(x)$. Starting with some initial value x_0, the recursive algorithm $x_{n+1} = g(x_n)$ for $n = 0, 1, 2, \ldots$ gives successive values for x_{n+1} at iteration $n + 1$ obtained from the prior estimate x_n. The iteration continues until $\Delta x = x_{n+1} - x_n$ is less than some convergence requirement. This value of x is called a fixed point of g because evaluating $g(x)$ gives the same value as x, and x is also the solution to $f(x) = 0$. As an example, consider the equation $x^5 - x - 1 = 0$, for which $x = 1.167304$. An expression for $g(x)$ is $g(x) = (x + 1)^{1/5}$. Starting with $x_0 = 0$, $x_{n+1} = g(x_n)$ gives the sequence of values 0, 1, 1.1487, 1.1653, 1.1671, 1.1673, ... that converges on the solution. This is a simple example of fixed-point iteration and converges, but not all functions can be evaluated with this method and convergence depends on the expression for $g(x)$ as well as the initial value x_0. The Picard iteration used to numerically solve the Richards equation (Chapter 8) is an example of fixed-point iteration to solve ordinary differential equations.

If the derivative of the function can be calculated, the Newton–Raphson method is another root-finding algorithm. This approach derives from the Taylor series expansion. For small values of Δx, the higher-order terms can be neglected, as in (A.2), so that $f(x + \Delta x) = 0$ for

$$\Delta x = -\frac{f(x)}{df/dx}. \qquad (A.29)$$

Given an initial value x_0, the calculation

$$x_{n+1} = x_n - \frac{f(x_n)}{df/dx} \qquad (A.30)$$

is iterated until $\Delta x = x_{n+1} - x_n$ is less than the convergence requirement. Newton–Raphson iteration is a particular case of fixed-point iteration in which $g(x) = x_n + \Delta x_n$. Using the previous example, $df/dx = 5x^4 - 1$, and starting with $x_0 = 1$, Newton–Raphson iteration yields the sequence 1, 1.25, 1.1785, 1.1675, 1.1673.

In many applications, the function is complex and its derivative cannot be calculated. Bracketing algorithms can find the root over the interval $[a, b]$ provided $f(a)$ and $f(b)$ have opposite signs. Then, the root is known to lie between a and b. The bisection method is a simple algorithm that refines the end points of the interval containing x. It repeatedly bisects the interval, choosing the correct subinterval that brackets x for use in the subsequent iteration. For a given iteration, the midpoint of the interval is $c = (a + b)/2$. The interval for the next iteration is $[a, c]$ or $[c, b]$ depending on whether $f(a)$ and $f(c)$

have opposite signs (and thereby bracketing the root by replacing b with c) or $f(c)$ and $f(b)$ have opposite signs (replacing a with c). Each successive step reduces the interval containing x by one-half. Bisection requires two initial estimates of the root (a and b for the first iteration), and these must bracket the root.

The secant method uses two prior values (x_{n-1} and x_n) to estimate x_{n+1}. The method approximates $f(x)$ with the line between two points $(x_{n-1}, f(x_{n-1}))$ and $(x_n, f(x_n))$ and uses linear interpolation to find x_{n+1} such that $f(x) = 0$. The linear approximation for $f(x)$ is

$$f(x) = \frac{f(x_n) - f(x_{n-1})}{x_n - x_{n-1}}(x - x_n) + f(x_n), \qquad (A.31)$$

and x that gives $f(x) = 0$ is the value x_{n+1} such that

$$x_{n+1} = x_n - f(x_n)\frac{x_n - x_{n-1}}{f(x_n) - f(x_{n-1})}. \qquad (A.32)$$

This is equivalent to the Newton–Raphson method, but with the derivative approximated as a finite difference – i.e.,

$$\frac{df}{dx} = \frac{f(x_n) - f(x_{n-1})}{x_n - x_{n-1}}. \qquad (A.33)$$

The secant method requires two initial values (x_0 and x_1).

Brent's method is an advanced algorithm that combines aspects of root bracketing, bisection, and the secant method. Bisection is robust but converges slowly; the secant method converges rapidly but does not always find the root. Brent's method replaces the linear interpolation of the secant method with inverse quadratic interpolation (using three prior root estimates to fit an inverse quadratic function) and reverts to bisection if that fails. It requires two initial values, which must bracket the root.

A.6 | Matrix Algebra

Matrices are commonly used to solve a system of linear equations. A matrix is an array of numbers arranged in rows and columns. Each individual entry in the matrix is called an element. A matrix with m rows and n columns is called a $m \times n$ matrix. For example,

$$\mathbf{A} = \begin{bmatrix} a_{11} & a_{12} & a_{13} \\ a_{21} & a_{22} & a_{23} \end{bmatrix} \qquad (A.34)$$

denotes $\mathbf{A}$ is a 2×3 matrix with the element of the ith row and jth column denoted by a_{ij}. A matrix with a single row $(1 \times n)$ is called a row vector, and a matrix with a single column $(m \times 1)$ is a column vector. A matrix with equal number of rows and columns $(n \times n)$ is called a square matrix. The transpose of a $m \times n$ matrix is the $n \times m$ matrix formed by making rows into columns and columns into rows. The transpose of the matrix $\mathbf{A}$ given by (A.34) is

$$\mathbf{A}^T = \begin{bmatrix} a_{11} & a_{21} \\ a_{12} & a_{22} \\ a_{13} & a_{23} \end{bmatrix}. \qquad (A.35)$$

Matrices can be added, subtracted, or multiplied. Two matrices must be of the same size to be added or subtracted. For example, the sum of the $m \times n$ matrices $\mathbf{A}$ and $\mathbf{B}$ is a $m \times n$ matrix found by adding the corresponding elements:

$$\mathbf{A} + \mathbf{B} = \begin{bmatrix} a_{11} & a_{12} & a_{13} \\ a_{21} & a_{22} & a_{23} \end{bmatrix} + \begin{bmatrix} b_{11} & b_{12} & b_{13} \\ b_{21} & b_{22} & b_{23} \end{bmatrix}$$

$$= \begin{bmatrix} a_{11} + b_{11} & a_{12} + b_{12} & a_{13} + b_{13} \\ a_{21} + b_{21} & a_{22} + b_{22} & a_{23} + b_{23} \end{bmatrix}. \qquad (A.36)$$

The scalar multiplication $c\mathbf{A}$ is found by multiplying each element a_{ij} by c:

$$c\mathbf{A} = \begin{bmatrix} ca_{11} & ca_{12} & ca_{13} \\ ca_{21} & ca_{22} & ca_{23} \end{bmatrix}. \qquad (A.37)$$

Two matrices can be multiplied if the number of columns of the left matrix equals the number of rows of the right matrix. If $\mathbf{A}$ is a $m \times n$ matrix and $\mathbf{B}$ is a $n \times p$ matrix, the product $\mathbf{AB}$ is a $m \times p$ matrix. The ijth element of the product matrix is computed by summing all products obtained by multiplying each element in the ith row of $\mathbf{A}$ by the corresponding element in the jth column of $\mathbf{B}$. Each column j of row i is given by

$$(ab)_{ij} = a_{i1}b_{1j} + a_{i2}b_{2j} + \ldots + a_{in}b_{nj} = \sum_{k=1}^{n} a_{ik}b_{kj}. \qquad (A.38)$$

Matrices cannot be divided. Instead, the inverse (reciprocal) of a matrix is used. The inverse of a square $(n \times n)$ matrix $\mathbf{A}$ is the $n \times n$ matrix $\mathbf{A}^{-1}$ given by

$$\mathbf{A}\mathbf{A}^{-1} = \mathbf{I}. \qquad (A.39)$$

The matrix $\mathbf{I}$ is the $n \times n$ identity matrix in which all the diagonal elements are equal to one and all other elements equal zero:

$$\mathbf{I} = \begin{bmatrix} 1 & 0 & \cdots & 0 \\ 0 & 1 & \cdots & 0 \\ \vdots & \vdots & \ddots & \vdots \\ 0 & 0 & \cdots & 1 \end{bmatrix}. \qquad (A.40)$$

Numerical methods are needed to calculate $\mathbf{A}^{-1}$.

A.7 | Solving a System of Linear Equations

Matrix algebra can be used to write a system of linear equations. If $\mathbf{A}$ is a $m \times n$ matrix, $\mathbf{X}$ is a $n \times 1$ column vector, and $\mathbf{B}$ is a $m \times 1$ column vector, then

$$\mathbf{AX} = \mathbf{B} \qquad (A.41)$$

describes the system of m equations with n unknown values x_i:

$$a_{11}x_1 + a_{12}x_2 + \ldots + a_{1n}x_n = b_1$$
$$\ldots \qquad (A.42)$$
$$a_{m1}x_1 + a_{m2}x_2 + \ldots + a_{mn}x_n = b_m.$$

One common example of this approach is a system of first-order, linear differential equations to represent biogeochemical cycles such that $d\mathbf{X}/dt = \mathbf{AX}$ (Chapter 17). The Euler method (A.13) can be used to advance the state variables such that $\mathbf{X}^{n+1} = \mathbf{X}^n + (d\mathbf{X}/dt)\Delta t$. If the timescale of the individual equations varies substantially, the equations are said to be numerically stiff, and they require sophisticated numerical methods to integrate forward in time. A second common use of (A.41) is to solve for $\mathbf{X}$. If $\mathbf{A}$ and $\mathbf{B}$ are known, the solution is

$$\mathbf{X} = \mathbf{A}^{-1}\mathbf{B}. \qquad (A.43)$$

A.8 | Tridiagonal Systems of Linear Equations

Some systems of n equations have the form

$$a_i u_{i-1} + b_i u_i + c_i u_{i+1} = d_i, \qquad (A.44)$$

with $a_1 = 0$ and $c_n = 0$. A common example is seen in one-dimensional diffusion problems, such as soil temperature (Chapter 5), soil moisture (Chapter 8), and scalar canopy profiles (Chapter 16). This is known as a tridiagonal system of equations because it has non-zero elements only on the diagonal plus or minus one column. In matrix notation, the equations are written as

$$
\begin{bmatrix}
b_1 & c_1 & & & & \\
a_2 & b_2 & c_2 & & & \\
& a_3 & b_3 & c_3 & & \\
& & \ddots & \ddots & \ddots & \\
& & & a_{n-1} & b_{n-1} & c_{n-1} \\
& & & & a_n & b_n
\end{bmatrix}
\times
\begin{bmatrix}
u_1 \\ u_2 \\ u_3 \\ \vdots \\ u_{n-1} \\ u_n
\end{bmatrix}
=
\begin{bmatrix}
d_1 \\ d_2 \\ d_3 \\ \vdots \\ d_{n-1} \\ d_n
\end{bmatrix}.
\tag{A.45}
$$

The solution is much simpler and more computationally efficient than for (A.42).

A simple numerical algorithm is obtained from linear algebra (Richtmyer and Morton (1967, p. 200). It is a two-phase algorithm that rewrites (A.44) in the form $u_i + e_i u_{i+1} = f_i$ (so that $u_i = f_i - e_i u_{i+1}$) and then solves for u_i. A forward elimination phase removes u_{i-1} from (A.44) and generates e_i and f_i. Then a backward substitution phase is used to sequentially solve for u_i given u_{i+1}. The coefficients e_i and f_i are found algebraically by substituting $u_{i-1} = f_{i-1} - e_{i-1} u_i$ for u_{i-1} in (A.44). The forward sweep produces the coefficients

$$
e_i = \frac{c_i}{b_i - a_i e_{i-1}}; \quad i = 1, \ldots, n-1
\tag{A.46}
$$

and

$$
f_i = \frac{d_i - a_i f_{i-1}}{b_i - a_i e_{i-1}}; \quad i = 1, \ldots, n,
\tag{A.47}
$$

with $e_0 = f_0 = 0$. The backward sweep provides the solution

$$
u_i = f_i - e_i u_{i+1}; \quad i = n-1, \ldots, 1,
\tag{A.48}
$$

profiles of temperature and water vapor in plant canopies (Chapter 16). Here, the equations are

$$
\begin{aligned}
a_{1,i} x_{i-1} + b_{11,i} x_i + b_{12,i} y_i + c_{1,i} x_{i+1} &= d_{1,i} \\
a_{2,i} y_{i-1} + b_{21,i} x_i + b_{22,i} y_i + c_{2,i} y_{i+1} &= d_{2,i}.
\end{aligned}
\tag{A.49}
$$

It is more convenient to write these equations in matrix notation, whereby the solution is the same as before (Richtmyer and Morton 1967, pp. 275–278). In matrix notation, (A.49) is

$$
\mathbf{A}_i \mathbf{u}_{i-1} + \mathbf{B}_i \mathbf{u}_i + \mathbf{C}_i \mathbf{u}_{i+1} = \mathbf{D}_i,
\tag{A.50}
$$

with $\mathbf{u}_i$ the vector

$$
\mathbf{u}_i = \begin{bmatrix} x_i \\ y_i \end{bmatrix}
\tag{A.51}
$$

and

$$
\mathbf{A}_i = \begin{bmatrix} a_{1,i} & 0 \\ 0 & a_{2,i} \end{bmatrix}, \quad
\mathbf{B}_i = \begin{bmatrix} b_{11,i} & b_{12,i} \\ b_{21,i} & b_{22,i} \end{bmatrix},
$$

$$
\mathbf{C}_i = \begin{bmatrix} c_{1,i} & 0 \\ 0 & c_{2,i} \end{bmatrix}, \quad
\mathbf{D}_i = \begin{bmatrix} d_{1,i} \\ d_{2,i} \end{bmatrix}.
\tag{A.52}
$$

The solution is the same as given previously with

$$
\mathbf{u}_i = \mathbf{F}_i - \mathbf{E}_i \mathbf{u}_{i+1}; \quad i = n-1, \ldots, 1,
\tag{A.53}
$$

and $\mathbf{u}_n = \mathbf{F}_n$. $\mathbf{E}_i$ is a 2×2 matrix and $\mathbf{F}_i$ is a 2×1 vector defined by

$$
\mathbf{E}_i = (\mathbf{B}_i - \mathbf{A}_i \mathbf{E}_{i-1})^{-1} \mathbf{C}_i; \quad i = 1, \ldots, n-1
\tag{A.54}
$$

$$
\mathbf{F}_i = (\mathbf{B}_i - \mathbf{A}_i \mathbf{E}_{i-1})^{-1} (\mathbf{D}_i - \mathbf{A}_i \mathbf{F}_{i-1}); \quad i = 1, \ldots, n
\tag{A.55}
$$

and $\mathbf{E}_0 = \mathbf{F}_0 = 0$. Carrying out the matrix mathematics gives

$$
\mathbf{E}_i = \frac{1}{\det} \begin{bmatrix} c_{1,i}(b_{22,i} - a_{2,i} e_{22,i-1}) & -c_{2,i}(b_{12,i} - a_{1,i} e_{12,i-1}) \\ -c_{1,i}(b_{21,i} - a_{2,i} e_{21,i-1}) & c_{2,i}(b_{11,i} - a_{1,i} e_{11,i-1}) \end{bmatrix}
\tag{A.56}
$$

$$
\mathbf{F}_i = \frac{1}{\det} \begin{bmatrix} (b_{22,i} - a_{2,i} e_{22,i-1})(d_{1,i} - a_{1,i} f_{1,i-1}) - (b_{12,i} - a_{1,i} e_{12,i-1})(d_{2,i} - a_{2,i} f_{2,i-1}) \\ -(b_{21,i} - a_{2,i} e_{21,i-1})(d_{1,i} - a_{1,i} f_{1,i-1}) + (b_{11,i} - a_{1,i} e_{11,i-1})(d_{2,i} - a_{2,i} f_{2,i-1}) \end{bmatrix},
\tag{A.57}
$$

with $u_n = f_n$. The algorithm is robust for $|b_i| > |a_i| + |c_i|$.

In some applications, the problem is written as a simultaneous system of two equations with two unknowns. Such a system of equations describes

with

$$
\begin{aligned}
\det = &(b_{11,i} - a_{1,i} e_{11,i-1})(b_{22,i} - a_{2,i} e_{22,i-1}) \\
&- (b_{12,i} - a_{1,i} e_{12,i-1})(b_{21,i} - a_{2,i} e_{21,i-1}).
\end{aligned}
\tag{A.58}
$$

A.9 | Newton–Raphson Iteration for Nonlinear Systems of Equations

Newton–Raphson iteration can be extended to a system of equations. Consider, for example, the system of m equations,

$$
\begin{aligned}
f_1(x_1, x_2, \ldots, x_m) &= 0 \\
f_2(x_1, x_2, \ldots, x_m) &= 0 \\
&\vdots \\
f_m(x_1, x_2, \ldots, x_m) &= 0,
\end{aligned}
\tag{A.59}
$$

which has m unknown values for x. Neglecting higher-order terms, the Taylor series expansion for f_i is

$$
f_i(x_1, x_2, \ldots, x_m) + \frac{\partial f_i}{\partial x_1}\Delta x_1 + \frac{\partial f_i}{\partial x_2}\Delta x_2 + \ldots + \frac{\partial f_i}{\partial x_m}\Delta x_m = 0.
\tag{A.60}
$$

This gives a linear system of m equations that can be solved for the m values of Δx. The matrix form of the equation is

$$
\begin{bmatrix}
\partial f_1/\partial x_1 & \partial f_1/\partial x_2 & \cdots & f_1/\partial x_m \\
\partial f_2/\partial x_1 & \partial f_2/\partial x_2 & \cdots & f_2/\partial x_m \\
\vdots & \vdots & \ddots & \vdots \\
\partial f_m/\partial x_1 & \partial f_m/\partial x_2 & \cdots & \partial f_m/\partial x_m
\end{bmatrix}
\times
\begin{bmatrix}
\Delta x_1 \\ \Delta x_2 \\ \vdots \\ \Delta x_m
\end{bmatrix}
= -
\begin{bmatrix}
f_1(x_1, x_2, \ldots, x_m) \\
f_2(x_1, x_2, \ldots, x_m) \\
\vdots \\
f_m(x_1, x_2, \ldots, x_m)
\end{bmatrix}
\tag{A.61}
$$

or, more generally,

$$
\mathbf{J}\Delta\mathbf{x} = -\mathbf{F} \quad \Rightarrow \quad \Delta\mathbf{x} = -\mathbf{J}^{-1}\mathbf{F},
\tag{A.62}
$$

where $\mathbf{J}$ is a $m \times m$ matrix of partial derivatives known as the Jacobian and $\Delta\mathbf{x}$ and $\mathbf{F}$ are each a $m \times 1$ vector. In Newton–Raphson iteration, (A.62) is used to give successive estimates for $\Delta\mathbf{x} = \mathbf{x}_{n+1} - \mathbf{x}_n$ whereby

$$
\mathbf{x}_{n+1} = \mathbf{x}_n - \mathbf{J}_n^{-1}\mathbf{F}_n
\tag{A.63}
$$

and the right-hand side terms are evaluated using the prior estimates at iteration n. Newton–Raphson iteration is a common method to solve a system of nonlinear equations such as the Richards equation for soil moisture (Chapter 8).

A.10 | Special Functions

The gamma function is

$$
\Gamma(x) = \int_0^\infty t^{x-1}e^{-t}dt
\tag{A.64}
$$

and is calculated using numerical methods. The beta function is

$$
B(x, y) = \int_0^1 t^{x-1}(1-t)^{y-1}dt.
\tag{A.65}
$$

It is calculated using the gamma function as:

$$
B(x, y) = \frac{\Gamma(x)\Gamma(y)}{\Gamma(x+y)}.
\tag{A.66}
$$

The error function is

$$
\mathrm{erf}(x) = \frac{2}{\sqrt{\pi}}\int_0^x e^{-t^2}dt.
\tag{A.67}
$$

A.11 | Probability Density Functions

A probability density function $f(x)$ describes the probability that a value x lies between two points a and b as

$$
\mathrm{Prob}(a \leq x \leq b) = \int_a^b f(x)dx.
\tag{A.68}
$$

The cumulative distribution function is the probability of a value less than or equal to x, obtained by

$$
F(x) = \int_{-\infty}^x f(u)du.
\tag{A.69}
$$

The cumulative distribution function is the area under the probability density function from $-\infty$ to x. Equation (A.68) can be equivalently written in the form

$$\text{Prob}(a \leq x \leq b) = F(b) - F(a). \tag{A.70}$$

The functions $f(x)$ and $F(x)$ are related by

$$f(x) = \frac{d}{dx}F(x). \tag{A.71}$$

As an example, consider the exponential probability density function

$$f(x) = u\exp(-ux) \tag{A.72}$$

for $x \geq 0$. The cumulative distribution function is

$$F(x) = 1 - \exp(-ux), \tag{A.73}$$

and the mean is

$$\bar{x} = \int_{0}^{\infty} xf(x)dx = \frac{1}{u}. \tag{A.74}$$

Probability density functions are used to characterize leaf angle distributions (Chapter 2) and spatial variability in precipitation and soil moisture for hydrologic scaling (Chapter 9).

References

Abbott, M. B., Bathurst, J. C., Cunge, J. A., O'Connell, P. E., and Rasmussen, J. (1986). An introduction to the European Hydrological System – Systeme Hydrologique Eur-opeen, "SHE", 2: Structure of a physically-based, distributed modelling system. *Journal of Hydrology*, 87, 61–77.

Aber, J. D., and Melillo, J. M. (1980). Litter decomposition: Measuring relative contributions of organic matter and nitrogen to forest soils. *Canadian Journal of Botany*, 58, 416–421.

Aber, J. D., and Melillo, J. M. (1982). Nitrogen immobilization in decaying hardwood leaf litter as a function of initial nitrogen and lignin content. *Canadian Journal of Botany*, 60, 2263–2269.

Aber, J. D., Melillo, J. M., and Federer, C. A. (1982). Predicting the effects of rotation length, harvest intensity, and fertilization on fiber yield from northern hardwood forests in New England. *Forest Science*, 28, 31–45.

Aber, J. D., Melillo, J. M., and McClaugherty, C. A. (1990). Predicting long-term patterns of mass loss, nitrogen dynamics, and soil organic matter formation from initial fine litter chemistry in temperate forest ecosystems. *Canadian Journal of Botany*, 68, 2201–2208.

Adair, E. C., Parton, W. J., Del Grosso, S. J., et al. (2008). Simple three-pool model accurately describes patterns of long-term litter decomposition in diverse climates. *Global Change Biology*, 14, 2636–2660.

Ågren, G. I., and Bosatta, E. (1987). Theoretical analysis of the long-term dynamics of carbon and nitrogen in soils. *Ecology*, 68, 1181–1189.

Ågren, G. I., and Bosatta, E. (1996). *Theoretical Ecosystem Ecology: Understanding Element Cycles*. Cambridge: Cambridge University Press.

Ainsworth, E. A., and Rogers, A. (2007). The response of photosynthesis and stomatal conductance to rising [CO_2]: Mechanisms and environmental interactions. *Plant, Cell and Environment*, 30, 258–270.

Ainsworth, E. A., Yendrek, C. R., Sitch, S., Collins, W. J., and Emberson, L. D. (2012). The effects of tropospheric ozone on net primary productivity and implications for climate change. *Annual Review of Plant Biology*, 63, 637–661.

Akagi, S. K., Yokelson, R. J., Wiedinmyer, C., et al. (2011). Emission factors for open and domestic biomass burning for use in atmospheric models. *Atmospheric Chemistry and Physics*, 11, 4039–4072.

Ali, A. A., Xu, C., Rogers, A., et al. (2016). A global scale mechanistic model of photosynthetic capacity (LUNA V1.0). *Geoscientific Model Development*, 9, 587–606.

Amenu, G. G., and Kumar, P. (2008). A model for hydraulic redistribution incorporating coupled soil–root moisture transport. *Hydrology and Earth System Sciences*, 12, 55–74.

Amthor, J. S., Goulden, M. L., Munger, J. W., and Wofsy, S. C. (1994). Testing a mechanistic model of forest-canopy mass and energy exchange using eddy correlation: Carbon dioxide and ozone uptake by a mixed oak–maple stand. *Australian Journal of Plant Physiology*, 21, 623–651.

Anderson, E. A. (1976). A Point Energy and Mass Balance Model of a Snow Cover, NOAA Technical Report NWS 19. Silver Spring, MD: Office of Hydrology, National Weather Service.

Anderson, J. L., Balaji, V., Broccoli, A. J., et al. (2004). The new GFDL global atmosphere and land model AM2–LM2: Evaluation with prescribed SST simulations. *Journal of Climate*, 17, 4641–4673.

Anderson, M. C., Kustas, W. P., Norman, J. M., et al. (2011). Mapping daily evapotranspiration at field to continental scales using geostationary and polar orbiting satellite imagery. *Hydrology and Earth System Sciences*, 15, 223–239.

Anderson, M. C., Norman, J. M., Mecikalski, J. R., Otkin, J. A., and Kustas, W. P. (2007). A climatological study of evapotranspiration and moisture stress across the continental United States based on thermal remote sensing: 1. Model formulation. *Journal of Geophysical Research*, 112, D10117, doi:10.1029/2006JD007506.

Andreae, M. O., and Merlet, P. (2001). Emission of trace gases and aerosols from biomass burning. *Global Biogeochemical Cycles*, 15, 955–966.

Andrén, O., and Paustian, K. (1987). Barley straw decomposition in the field: A comparison of models. *Ecology*, 68, 1190–1200.

Anten, N. P. R., Miyazawa, K., Hikosaka, K., Nagashima, H., and Hirose, T. (1998a). Leaf nitrogen distribution in relation to leaf age and photon flux density in dominant and subordinate plants in dense stands of a dicotyledonous herb. *Oecologia*, 113, 314–324.

Anten, N. P. R., Schieving, F., and Werger, M. J. A. (1995). Patterns of light and nitrogen distribution in relation to whole canopy carbon gain in C_3 and C_4 mono- and dicotyledonous species. *Oecologia*, 101, 504–513.

Anten, N. P. R., Werger, M. J. A., and Medina, E. (1998b). Nitrogen distribution and leaf area indices in relation to photosynthetic nitrogen use efficiency in savanna grasses. *Plant Ecology*, 138, 63–75.

Arneth, A., Lloyd, J., Šantrůčková, H., et al. (2002). Response of central Siberian Scots pine to soil water deficit and long-term trends in atmospheric CO_2 concentration. *Global Biogeochemical Cycles*, 16, 1005, doi:10.1029/2000GB001374.

Arneth, A., Mercado, L., Kattge, J., and Booth, B. B. B. (2012). Future challenges of representing land-processes in studies on land-atmosphere interactions. *Biogeosciences*, 9, 3587–3599.

Arneth, A., Monson, R. K., Schurgers, G., Niinemets, Ü., and Palmer, P. I. (2008). Why are estimates of global terrestrial isoprene emissions so similar (and why is this not so for monoterpenes)? *Atmospheric Chemistry and Physics*, 8, 4605–4620.

Arneth, A., Niinemets, Ü., Pressley, S., et al. (2007). Process-based estimates of terrestrial ecosystem isoprene emissions: Incorporating the effects of a direct CO_2-isoprene interaction. *Atmospheric Chemistry and Physics*, 7, 31–53.

Arneth, A., Schurgers, G., Lathiere, J., et al. (2011). Global terrestrial isoprene emission models: Sensitivity to variability in climate and vegetation. *Atmospheric Chemistry and Physics*, 11, 8037–8052.

Arora, V. K., and Boer, G. J. (2005a). A parameterization of leaf phenology for the terrestrial ecosystem component of climate models. *Global Change Biology*, 11, 39–59.

Arora, V. K., and Boer, G. J. (2005b). Fire as an interactive component of dynamic vegetation models. *Journal of Geophysical Research*, 110, G02008, doi:10.1029/2005JG000042.

Arora, V. K., and Boer, G. J. (2010). Uncertainties in the 20th century carbon budget associated with land use change. *Global Change Biology*, 16, 3327–3348.

Arora, V. K., Boer, G. J., Christian, J. R., et al. (2009). The effect of terrestrial photosynthesis down regulation on the twentieth-century carbon budget simulated with the CCCma Earth system model. *Journal of Climate*, 22, 6066–6088.

Ashworth, K., Chung, S. H., Griffin, R. J., et al. (2015). FORest Canopy Atmosphere Transfer (FORCAsT) 1.0: A 1-D model of biosphere–atmosphere chemical exchange. *Geoscientific Model Development*, 8, 3765–3784.

Asner, G. P., Scurlock, J. M. O., and Hicke, J. A. (2003). Global synthesis of leaf area index observations: Implications for ecological and remote sensing studies. *Global Ecology and Biogeography*, 12, 191–205.

Assouline, S. (2013). Infiltration into soils: Conceptual approaches and solutions. *Water Resources Research*, 49, 1755–1772, doi:10.1002/wrcr.20155.

Aston, A. R. (1985). Heat storage in a young eucalypt forest. *Agricultural and Forest Meteorology*, 35, 281–297.

Atkin, O. K. (2016). *Corrigendum. New Phytologist*, 211, 1142.

Atkin, O. K., Atkinson, L. J., Fisher, R. A., et al. (2008). Using temperature-dependent changes in leaf scaling relationships to quantitatively account for thermal acclimation of respiration in a coupled global climate–vegetation model. *Global Change Biology*, 14, 2709–2726.

Atkin, O. K., Bloomfield, K. J., Reich, P. B., et al. (2015). Global variability in leaf respiration in relation to climate, plant functional types and leaf traits. *New Phytologist*, 206, 614–636.

Aubinet, M., Grelle, A., Ibrom, A., et al. (2000). Estimates of the annual net carbon and water exchange of forests: The EUROFLUX methodology. *Advances in Ecological Research*, 30, 113–175.

Aubinet, M., Vesala, T., and Papale, D. (2012). *Eddy Covariance: A Practical Guide to Measurement and Data Analysis*. Dordrecht: Springer.

Bailey, R. L., and Dell, T. R. (1973). Quantifying diameter distributions with the Weibull function. *Forest Science*, 19, 97–104.

Baker, I. T., Prihodko, L., Denning, A. S., et al. (2008). Seasonal drought stress in the Amazon: Reconciling models and observations. *Journal of Geophysical Research*, 113, G00B01, doi:10.1029/2007JG000644.

Baker, I. T., Sellers, P. J., Denning, A. S., et al. (2017). Closing the scale gap between land surface parameterizations and GCMs with a new scheme, SiB3-Bins. *Journal of Advances in Modeling Earth Systems*, 9, 691–711, doi:10.1002/2016MS000764.

Baldocchi, D. (1988). A multi-layer model for estimating sulfur dioxide deposition to a deciduous oak forest canopy. *Atmospheric Environment*, 22, 869–884.

Baldocchi, D. (1989). Turbulent transfer in a deciduous forest. *Tree Physiology*, 5, 357–377.

Baldocchi, D. (1992). A Lagrangian random-walk model for simulating water vapor, CO_2 and sensible heat flux densities and scalar profiles over and within a soybean canopy. *Boundary-Layer Meteorology*, 61, 113–144.

Baldocchi, D. (1994). An analytical solution for coupled leaf photosynthesis and stomatal conductance models. *Tree Physiology*, 14, 1069–1079.

Baldocchi, D. (2003). Assessing the eddy covariance technique for evaluating carbon dioxide exchange rates of ecosystems: Past, present and future. *Global Change Biology*, 9, 479–492.

Baldocchi, D. D., and Bowling, D. R. (2003). Modelling the discrimination of $^{13}CO_2$ above and within a temperate broad-leaved forest canopy on hourly to seasonal time scales. *Plant, Cell and Environment*, 26, 231–244.

Baldocchi, D. D., and Harley, P. C. (1995). Scaling carbon dioxide and water vapour exchange from leaf to canopy in a deciduous forest. II. Model testing and application. *Plant, Cell and Environment*, 18, 1157–1173.

Baldocchi, D. D., and Meyers, T. P. (1988). A spectral and lag-correlation analysis of turbulence in a deciduous forest canopy. *Boundary-Layer Meteorology*, 45, 31–58.

Baldocchi, D. D., and Meyers, T. P. (1998). On using eco-physiological, micrometeorological and biogeochemical theory to evaluate carbon dioxide, water vapor and trace gas fluxes over vegetation: A perspective. *Agricultural and Forest Meteorology*, 90, 1–25.

Baldocchi, D. D., Hicks, B. B., and Camara, P. (1987). A canopy stomatal resistance model for gaseous deposition to vegetated surfaces. *Atmospheric Environment*, 21, 91–101.

Baldocchi, D. D., Hicks, B. B., and Meyers, T. P. (1988). Measuring biosphere–atmosphere exchanges of biologically related gases with micrometeorological methods. *Ecology*, 69, 1331–1340.

Baldocchi, D. D., Hutchison, B. A., Matt, D. R., and McMillen, R. T. (1985). Canopy radiative transfer models for spherical and known leaf inclination angle distributions: A test in an oak–hickory forest. *Journal of Applied Ecology*, 22, 539–555.

Baldocchi, D. D., and Wilson, K. B. (2001). Modeling CO_2 and water vapor exchange of a temperate broadleaved forest across hourly to decadal time scales. *Ecological Modelling*, 142, 155–184.

Baldocchi, D. D., Wilson, K. B., and Gu, L. (2002). How the environment, canopy structure and canopy physiological functioning influence carbon, water and energy fluxes of a temperate broad-leaved deciduous forest – an assessment with the biophysical model CANOAK. *Tree Physiology*, 22, 1065–1077.

Ball, J. T., Woodrow, I. E., and Berry, J. A. (1987). A model predicting stomatal conductance and its contribution to the control of photosynthesis under different environmental conditions. In *Progress in Photosynthesis Research*, vol. 4, ed. J. Biggins. Dordrecht: Martinus Nijhoff, pp. 221–224.

Ball, M. C., Cowan, I. R., and Farquhar, G. D. (1988). Maintenance of leaf temperature and the optimisation of carbon gain in relation to water loss in a tropical mangrove forest. *Australian Journal of Plant Physiology*, 15, 263–276.

Balsamo, G., Albergel, C., Beljaars, A., et al. (2015). ERA-Interim/Land: A global land surface reanalysis data set. *Hydrology and Earth System Sciences*, 19, 389–407.

Balsamo, G., Viterbo, P., Beljaars, A., et al. (2009). A revised hydrology for the ECMWF model: Verification from field site to terrestrial water storage and impact in the integrated forecast system. *Journal of Hydrometeorology*, 10, 623–643.

Band, L. E., Patterson, P., Nemani, R., and Running, S. W. (1993). Forest ecosystem processes at the watershed scale: Incorporating hillslope hydrology. *Agricultural and Forest Meteorology*, 63, 93–126.

Barman, R., and Jain, A. K. (2016). Comparison of effects of cold-region soil/snow processes and the uncertainties from model forcing data on permafrost physical characteristics. *Journal of Advances in Modeling Earth Systems*, 8, 453–466, doi:10.1002/2015MS000504.

Barnard, D. M., and Bauerle, W. L. (2013). The implications of minimum stomatal conductance on modeling water flux in forest canopies. *Journal of Geophysical Research Biogeosciences*, 118, 1322–1333, doi:10.1002/jgrg.20112.

Bauerle, W. L., Daniels, A. B., and Barnard, D. M. (2014). Carbon and water flux responses to physiology by environment interactions: A sensitivity analysis of variation in climate on photosynthetic and stomatal parameters. *Climate Dynamics*, 42, 2539–2554.

Bauerle, W. L., Oren, R., Way, D. A., et al. (2012). Photoperiodic regulation of the seasonal pattern of photosynthetic capacity and the implications for carbon cycling. *Proceedings of the National Academy of Sciences USA*, 109, 8612–8617.

Belcher, S. E., Finnigan, J. J., and Harman, I. N. (2008). Flows through forest canopies in complex terrain. *Ecological Applications*, 18, 1436–1453.

Belcher, S. E., Harman, I. N., and Finnigan, J. J. (2012). The wind in the willows: Flows in forest canopies in complex terrain. *Annual Review of Fluid Mechanics*, 44, 479–504.

Beljaars, A. C. M., and Holtslag, A. A. M. (1991). Flux parameterization over land surfaces for atmospheric models. *Journal of Applied Meteorology*, 30, 327–341.

Berg, B., Berg, M. P., Bottner, P., et al. (1993). Litter mass loss rates in pine forests of Europe and eastern United States: Some relationships with climate and litter quality. *Biogeochemistry*, 20, 127–159.

Berg, B., Hannus, K., Popoff, T., and Theander, O. (1982). Changes in organic chemical components of needle litter during decomposition. Long-term decomposition in a Scots pine forest. I. *Canadian Journal of Botany*, 60, 1310–1319.

Bernacchi, C. J., Bagley, J. E., Serbin, S. P., et al. (2013). Modelling C_3 photosynthesis from the chloroplast to the ecosystem. *Plant, Cell and Environment*, 36, 1641–1657.

Bernacchi, C. J., Pimentel, C., and Long, S. P. (2003). *In vivo* temperature response functions of parameters required to model RuBP-limited photosynthesis. *Plant, Cell and Environment*, 26, 1419–1430.

Bernacchi, C. J., Portis, A. R., Nakano, H., von Caemmerer, S., and Long, S. P. (2002). Temperature response of mesophyll conductance. Implications for the determination of rubisco enzyme kinetics and for limitations to photosynthesis in vivo. *Plant Physiology*, 130, 1992–1998.

Bernacchi, C. J., Singsaas, E. L., Pimentel, C., Portis, A. R. J., and Long, S. P. (2001). Improved temperature response functions for models of Rubisco-limited photosynthesis. *Plant, Cell and Environment*, 24, 253–259.

Best, M. J., Pryor, M. Clark, D. B., et al. (2011). The Joint UK Land Environment Simulator (JULES), model description – Part 1: Energy and water fluxes. *Geoscientific Model Development*, 4, 677–699.

Beven, K. (1993). Prophecy, reality and uncertainty in distributed hydrological modelling. *Advances in Water Resources*, 16, 41–51.

Beven, K. (2000). *Rainfall–Runoff Modelling: The Primer*. Chichester: Wiley.

Beven, K. (2002). Towards a coherent philosophy for modelling the environment. *Proceedings of the Royal Society London A*, 458, 2465–2484.

Beven, K. (2006). A manifesto for the equifinality thesis. *Journal of Hydrology*, 320, 18–36.

Beven, K., and Freer, J. (2001). Equifinality, data assimilation, and uncertainty estimation in mechanistic modelling of complex environmental systems using the GLUE methodology. *Journal of Hydrology*, 249, 11–29.

Beven, K. J., and Kirkby, M. J. (1979). A physically based, variable contributing area model of basin hydrology. *Hydrological Sciences Bulletin*, 24, 43–69.

Beven, K. J., Lamb, R., Quinn, P. F., Romanowicz, R., and Freer, J. (1995). TOPMODEL. In *Computer Models of Watershed Hydrology*, ed. V. P. Singh. Highlands Ranch, CO: Water Resources Publications, pp. 627–668.

Blackman, F. F. (1905). Optima and limiting factors. *Annals of Botany*, 19, 281–295.

Blanken, P. D., Black, T. A., Yang, P. C., et al. (1997). Energy balance and canopy conductance of a boreal aspen forest: Partitioning overstory and understory components. *Journal of Geophysical Research*, 102D, 28915–28927.

Bohrer, G., Mourad, H., Laursen, T. A., et al. (2005). Finite element tree crown hydrodynamics model (FETCH) using porous media flow within branching elements: A new representation of tree hydrodynamics. *Water Resources Research*, 41, W11404, doi:10.1029/2005WR004181.

Bolhàr-Nordenkampf, H. R., and Draxler, G. (1993). Functional leaf anatomy. In *Photosynthesis and Production in a Changing Environment: A Field and Laboratory Manual*, ed. D. O. Hall, J. M. O. Scurlock, H. R. Bolhàr-Nordenkampf, R. C. Leegood, and S. P. Long. New York, NY: Chapman and Hall, pp. 91–112.

Bonacina, C., Comini, G., Fasano, A., and Primicerio, M. (1973). Numerical solution of phase-change problems. *International Journal of Heat and Mass Transfer*, 16, 1825–1832.

Bonan, G. B. (1989). Environmental factors and ecological processes controlling vegetation patterns in boreal forests. *Landscape Ecology*, 3, 111–130.

Bonan, G. B. (1990a). Carbon and nitrogen cycling in North American boreal forests. I. Litter quality and soil thermal effects in interior Alaska. *Biogeochemistry*, 10, 1–28.

Bonan, G. B. (1990b). Carbon and nitrogen cycling in North American boreal forests. II. Biogeographic patterns. *Canadian Journal of Forest Research*, 20, 1077–1088

Bonan, G. B. (1991). A biophysical surface energy budget analysis of soil temperature in the boreal forests of interior Alaska. *Water Resources Research*, 27, 767–781.

Bonan, G. B. (1993). Physiological controls of the carbon balance of boreal forest ecosystems. *Canadian Journal of Forest Research*, 23, 1453–1471.

Bonan, G. B. (1995). Land-atmosphere CO_2 exchange simulated by a land surface process model coupled to an atmospheric general circulation model. *Journal of Geophysical Research*, 100D, 2817–2831.

Bonan, G. B. (1996). *A Land Surface Model (LSM Version 1.0) for Ecological, Hydrological, and Atmospheric Studies: Technical Description and User's Guide*, Technical Note NCAR/TN-417+STR. Boulder, CO: National Center for Atmospheric Research.

Bonan, G. B. (2008). Forests and climate change: Forcings, feedbacks, and the climate benefits of forests. *Science*, 320, 1444–1449.

Bonan, G. B. (2014). Connecting mathematical ecosystems, real-world ecosystems, and climate science. *New Phytologist*, 202, 731–733.

Bonan, G. B. (2016). *Ecological Climatology: Concepts and Applications*, 3rd edn. Cambridge: Cambridge University Press.

Bonan, G. B., Davis, K. J., Baldocchi, D., Fitzjarrald, D., and Neumann, H. (1997). Comparison of the NCAR LSM1 land surface model with BOREAS aspen and jack pine tower fluxes. *Journal of Geophysical Research*, 102D, 29065–29075.

Bonan, G. B., and Doney, S. C. (2018). Climate, ecosystems, and planetary futures: The challenge to predict life in Earth system models. *Science*, 359, eaam8328, doi:10.1126/science.aam8328.

Bonan, G. B., Hartman, M. D., Parton, W. J., and Wieder, W. R. (2013). Evaluating litter decomposition in earth system models with long-term litterbag experiments: An example using the Community Land Model version 4 (CLM4). *Global Change Biology*, 19, 957–974.

Bonan, G. B., and Hayden, B. P. (1990). Using a forest stand simulation model to examine the ecological and climatic significance of the late-Quaternary pine–spruce pollen zone in eastern Virginia, U.S.A. *Quaternary Research*, 33, 204–218.

Bonan, G. B., Lawrence, P. J., Oleson, K. W., et al. (2011). Improving canopy processes in the Community Land Model version 4 (CLM4) using global flux fields empirically inferred from FLUXNET data. *Journal of Geophysical Research*, 116, G02014, doi:10.1029/2010JG001593.

Bonan, G. B., and Levis, S. (2010). Quantifying carbon–nitrogen feedbacks in the Community Land Model (CLM4). *Geophysical Research Letters*, 37, L07401, doi:10.1029/2010GL042430.

Bonan, G. B., Levis, S., Kergoat, L., and Oleson, K. W. (2002). Landscapes as patches of plant functional types: An integrating concept for climate and ecosystem models. *Global Biogeochemical Cycles*, 16, 1021, doi:10.1029/2000GB001360.

Bonan, G. B., Levis, S., Sitch, S., Vertenstein, M., and Oleson, K. W. (2003). A dynamic global vegetation model for use with climate models: Concepts and description of simulated vegetation dynamics. *Global Change Biology*, 9, 1543–1566.

Bonan, G. B., Patton, E. G., Harman, I. N., et al. (2018). Modeling canopy-induced turbulence in the Earth system: A unified parameterization of turbulent exchange within plant canopies and the roughness sublayer (CLM-ml v0). *Geoscientific Model Development*, 11, 1467–1496.

Bonan, G. B., Williams, M., Fisher, R. A., and Oleson, K. W. (2014). Modeling stomatal conductance in the earth system: Linking leaf water-use efficiency and water transport along the soil–plant–atmosphere continuum. *Geoscientific Model Development*, 7, 2193–2222.

Boone, A., Samuelsson, P., Gollvik, S., et al. (2017). The interactions between soil–biosphere–atmosphere land surface model with a multi-energy balance (ISBA-MEB) option in SURFEXv8 – Part 1: Model description. *Geoscientific Model Development*, 10, 843–872.

Bosatta, E., and Ågren, G. I. (1985). Theoretical analysis of decomposition of heterogeneous substrates. *Soil Biology and Biochemistry*, 17, 601–610.

Bosatta, E., and Ågren, G. I. (1991). Dynamics of carbon and nitrogen in the organic matter of the soil: A generic theory. *American Naturalist*, 138, 227–245.

Bosatta, E., and Ågren, G. I. (2003). Exact solutions to the continuous-quality equation for soil organic matter turnover. *Journal of Theoretical Biology*, 224, 97–105.

Botkin, D. B. (1993). *Forest Dynamics: An Ecological Model*. Oxford: Oxford University Press.

Botkin, D. B., Janak, J. F., and Wallis, J. R. (1972). Some ecological consequences of a computer model of forest growth. *Journal of Ecology*, 60, 849–872.

Bouten, W., Schaap, M. G., Aerts, J., and Vermetten, A. W. M. (1996). Monitoring and modelling canopy water storage amounts in support of atmospheric deposition studies. *Journal of Hydrology*, 181, 305–321.

Boy, M., Sogachev, A., Lauros, J., et al. (2011). SOSA – a new model to simulate the concentrations of organic vapours and sulphuric acid inside the ABL – Part 1: Model description and initial evaluation. *Atmospheric Chemistry and Physics*, 11, 43–51.

Bras, R. L. (1990). *Hydrology: An Introduction to Hydrologic Science*. Reading, MA: Addison-Wesley.

Bresler, E., and Dagan, G. (1983). Unsaturated flow in spatially variable fields: 2. Application of water flow models to various fields. *Water Resources Research*, 19, 421–428.

Brooks, R. H., and Corey, A. T. (1964). *Hydraulic Properties of Porous Media*, Hydrology Papers No. 3. Fort Collins, CO: Colorado State University.

Brooks, R. H., and Corey, A. T. (1966). Properties of porous media affecting fluid flow. *Journal of the Irrigation and Drainage Division Proceedings of the American Society of Civil Engineers*, 92(IR2), 61–88.

Brovkin, V., van Bodegom, P. M., Kleinen, T., et al. (2012). Plant-driven variation in decomposition rates improves projections of global litter stock distribution. *Biogeosciences*, 9, 565–576.

Brutsaert, W. (1982). *Evaporation into the Atmosphere: Theory, History, and Applications*. Dordrecht: Kluwer.

Brutsaert, W. (2005). *Hydrology: An Introduction*. Cambridge: Cambridge University Press.

Bryan, A. M., Bertman, S. B., Carroll, M. A., et al. (2012). In-canopy gas-phase chemistry during CABINEX 2009: Sensitivity of a 1-D canopy model to vertical mixing and isoprene chemistry. *Atmospheric Chemistry and Physics*, 12, 8829–8849.

Brzostek, E. R., Fisher, J. B., and Phillips, R. P. (2014). Modeling the carbon cost of plant nitrogen acquisition: Mycorrhizal trade-offs and multipath resistance uptake improve predictions of retranslocation. *Journal of Geophysical Research: Biogeosciences*, 119, 1684–1697, doi:10.1002/2014JG002660.

Buckley, T. N. (2017). Modeling stomatal conductance. *Plant Physiology*, 174, 572–582.

Buckley, T. N., Cescatti, A., and Farquhar, G. D. (2013). What does optimization theory actually predict about crown profiles of photosynthetic capacity when models incorporate greater realism? *Plant, Cell and Environment*, 36, 1547–1563.

Buckley, T. N., Miller, J. M., and Farquhar, G. D. (2002). The mathematics of linked optimisation for water and nitrogen use in a canopy. *Silva Fennica*, 36(3), 639–669.

Buckley, T. N., Sack, L., and Farquhar, G. D. (2017). Optimal plant water economy. *Plant, Cell and Environment, 40*, 881–896.

Buckley, T. N., and Schymanski, S. J. (2014). Stomatal optimisation in relation to atmospheric CO_2. *New Phytologist*, 201, 372–377.

Bugmann, H. (2001). A review of forest gap models. *Climatic Change*, 51, 259–305.

Burke, E. J., Chadburn, S. E., and Ekici, A. (2017). A vertical representation of soil carbon in the JULES land surface scheme (vn4.3_permafrost) with a focus on permafrost regions. *Geoscientific Model Development*, 10, 959–975.

Calder, I. R. (1977). A model of transpiration and interception loss from a spruce forest in Plynlimon, central Wales. *Journal of Hydrology*, 33, 247–265.

Campbell, G. S. (1974). A simple method for determining unsaturated conductivity from moisture retention data. *Soil Science*, 117, 311–314.

Campbell, G. S. (1985). *Soil Physics with Basic: Transport Models for Soil–Plant Systems*. Amsterdam: Elsevier.

Campbell, G. S. (1986). Extinction coefficients for radiation in plant canopies calculated using an ellipsoidal inclination angle distribution. *Agricultural and Forest Meteorology*, 36, 317–321.

Campbell, G. S. (1990). Derivation of an angle density function for canopies with ellipsoidal leaf angle distributions. *Agricultural and Forest Meteorology*, 49, 173–176.

Canadell, J., Jackson, R. B., Ehleringer, J. R., et al. (1996). Maximum rooting depth of vegetation types at the global scale. *Oecologia*, 108, 583–595.

Campbell, G. S., and Norman, J. M. (1998). *An Introduction to Environmental Biophysics*, 2nd edn. New York, NY: Springer-Verlag.

Carlyle-Moses, D. E., and Gash, J. H. C. (2011). Rainfall interception loss by forest canopies. In *Forest Hydrology and Biogeochemistry: Synthesis of Past Research and Future Directions*, ed. D. F. Levia, D. Carlyle-Moses, and T. Tanaka. Dordrecht: Springer, pp. 407–423.

Carsel, R. F., and Parrish, R. S. (1988). Developing joint probability distributions of soil water retention characteristics. *Water Resources Research*, 24, 755–769.

Carswell, F. E., Meir, P., Wandelli, E. V., et al. (2000). Photosynthetic capacity in a central Amazonian rain forest. *Tree Physiology*, 20, 179–186.

Celia, M. A., Bouloutas, E. T., and Zarba, R. L. (1990). A general mass-conservative numerical solution for the unsaturated flow equation. *Water Resources Research*, 26, 1483–1496.

Cellier, P., and Brunet, Y. (1992). Flux–gradient relationships above tall plant canopies. *Agricultural and Forest Meteorology*, 58, 93–117.

Cernusak, L. A., Barbour, M. M., Arndt, S. K., et al. (2016). Stable isotopes in leaf water of terrestrial plants. *Plant, Cell and Environment*, 39, 1087–1102.

Cescatti, A. (1997). Modelling the radiative transfer in discontinuous canopies of asymmetric crowns. I. Model structure and algorithms. *Ecological Modelling*, 101, 263–274.

Cescatti, A., and Niinemets, Ü. (2004). Sunlight capture: Leaf to landscape. In *Photosynthetic Adaptation: Chloroplast to Landscape*, ed. W. K. Smith, T. C. Vogelmann, and C. Critchley. New York, NY: Springer, pp. 42–85.

Chamberlain, A. C. (1966). Transport of gases to and from grass and grass-like surfaces. *Proceedings of the Royal Society London A*, 290, 236–265.

Chen, B., Liu, J., Chen, J. M., et al. (2016a). Assessment of foliage clumping effects on evapotranspiration estimates in forested ecosystems. *Agricultural and Forest Meteorology*, 216, 82–92.

Chen, C. P., Zhu, X.-G., and Long, S. P. (2008). The effect of leaf-level spatial variability in photosynthetic capacity on biochemical parameter estimates using the Farquhar model: A theoretical analysis. *Plant Physiology*, 148, 1139–1147.

Chen, F., and Dudhia, J. (2001). Coupling an advanced land surface–hydrology model with the Penn State–NCAR MM5 modeling system. Part I: Model implementation and sensitivity. *Monthly Weather Review*, 129, 569–585.

Chen, J., and Kumar, P. (2001). Topographic influence on the seasonal and interannual variation of water and energy balance of basins in North America. *Journal of Climate*, 14, 1989–2014.

Chen, J.-L., Reynolds, J. F., Harley, P. C., and Tenhunen, J. D. (1993). Coordination theory of leaf nitrogen distribution in a canopy. *Oecologia*, 93, 63–69.

Chen, J. M., and Black, T. A. (1992). Defining leaf area index for non-flat leaves. *Plant, Cell and Environment*, 15, 421–429.

Chen, J. M., Menges, C. H., and Leblanc, S. G. (2005). Global mapping of foliage clumping index using multiangular satellite data. *Remote Sensing of Environment*, 97, 447–457.

Chen, J. M., Mo, G., Pisek, J., et al. (2012). Effects of foliage clumping on the estimation of global terrestrial gross primary productivity. *Global Biogeochemical Cycles*, 26, GB1019, doi:10.1029/2010GB003996.

Chen, Y., Ryder, J., Bastrikov, V., et al. (2016b). Evaluating the performance of land surface model ORCHIDEE-CAN v1.0 on water and energy flux estimation with a single- and multi- layer energy budget scheme. *Geoscientific Model Development*, 9, 2951–2972.

Chen, Z. Q., Govindaraju, R. S., and Kavvas, M. L. (1994). Spatial averaging of unsaturated flow equations under infiltration conditions over areally heterogeneous fields: 1. Development of models. *Water Resources Research*, 30, 523–533.

Cherkauer, K. A., and Lettenmaier, D. P. (1999). Hydrologic effects of frozen soils in the upper Mississippi River basin. *Journal of Geophysical Research*, 104D, 19599–19610.

Choat, B., Jansen, S., Brodribb, T. J., et al. (2012). Global convergence in the vulnerability of forests to drought. *Nature*, 491, 752–755.

Choudhury, B. J., and Monteith, J. L. (1988). A four-layer model for the heat budget of homogeneous land surfaces. *Quarterly Journal of the Royal Meteorological Society*, 114, 373–398.

Christoffersen, B. O., Gloor, M., Fauset, S., et al. (2016). Linking hydraulic traits to tropical forest function in a size-structured and trait-driven model (TFS v.1-Hydro). *Geoscientific Model Development*, 9, 4227–4255.

Cionco, R. M. (1965). A mathematical model for air flow in a vegetative canopy. *Journal of Applied Meteorology*, 4, 517–522.

Cionco, R. M. (1978). Analysis of canopy index values for various canopy densities. *Boundary-Layer Meteorology*, 15, 81–93.

Clapp, R. B., and Hornberger, G. M. (1978). Empirical equations for some soil hydraulic properties. *Water Resources Research*, 14, 601–604.

Clapp, R. B., Hornberger, G. M., and Cosby, B. J. (1983). Estimating spatial variability in soil moisture with a simplified dynamic model. *Water Resources Research*, 19, 739–745.

Clark, D. B., Mercado, L. M., Sitch, S., et al. (2011a). The Joint UK Land Environment Simulator (JULES), model description – Part 2: Carbon fluxes and vegetation dynamics. *Geoscientific Model Development*, 4, 701–722.

Clark, M. P., Hendrikx, J., Slater, A. G., et al. (2011b). Representing spatial variability of snow water equivalent in hydrologic and land-surface models: A review. *Water Resources Research*, 47, W07539, doi:10.1029/2011WR010745.

Clark, M. P., Fan, Y., Lawrence, D. M., et al. (2015a). Improving the representation of hydrologic processes in Earth System Models. *Water Resources Research*, 51, 5929–5956, doi:10.1002/2015WR017096.

Clark, M. P., Nijssen, B., Lundquist, J. D., et al. (2015b). A unified approach for process-based hydrologic modeling: 1. Modeling concept. *Water Resources Research*, 51, 2498–2514, doi:10.1002/2015WR017198.

Clark, M. P., Nijssen, B., Lundquist, J. D., (2015c). A unified approach for process-based hydrologic modeling: 2. Model implementation and case studies. *Water Resources Research*, 51, 2515–2542, doi:10.1002/2015WR017200.

Clarke, R. H., Dyer, A. J., Brook, R. R., Reid, D. G., and Troup, A. J. (1971). *The Wangara Experiment: Boundary Layer Data*, Division of Meteorological Physics Technical Paper Number 19. Melbourne: Commonwealth Scientific and Industrial Research Organization.

Collatz, G. J., Ball, J. T., Grivet, C., and Berry, J. A. (1991). Physiological and environmental regulation of stomatal conductance, photosynthesis and transpiration: A model that includes a laminar boundary layer. *Agricultural and Forest Meteorology*, 54, 107–136.

Collatz, G. J., Berry, J. A., Farquhar, G. D., and Pierce, J. (1990). The relationship between the Rubisco reaction mechanism and models of photosynthesis. *Plant, Cell and Environment*, 13, 219–225.

Collatz, G. J., Ribas-Carbo, M., and Berry, J. A. (1992). Coupled photosynthesis–stomatal conductance model for leaves of C_4 plants. *Australian Journal of Plant Physiology*, 19, 519–538.

Connor, D. J., Sadras, V. O., and Hall, A. J. (1995). Canopy nitrogen distribution and the photosynthetic performance of sunflower crops during grain filling – a quantitative analysis. *Oecologia*, 101, 274–281.

Cornwell, W. K, Cornelissen, J. H. C., Amatangelo, K., et al. (2008). Plant species traits are the predominant control on litter decomposition rates within biomes worldwide. *Ecology Letters*, 11, 1065–1071.

Cosby, B. J., Hornberger, G. M., Clapp, R. B., and Ginn, T. R. (1984). A statistical exploration of the relationships of soil moisture characteristics to the physical properties of soils. *Water Resources Research*, 20, 682–690.

Cowan, I. R. (1968). Mass, heat and momentum exchange between stands of plants and their atmospheric environment. *Quarterly Journal of the Royal Meteorological Society*, 94, 523–544.

Cowan, I. R. (1977). Stomatal behaviour and environment. *Advances in Botanical Research*, 4, 117–228.

Cowan, I. R. (1982). Regulation of water use in relation to carbon gain in higher plants. In *Encyclopedia of Plant Physiology, New Series*, vol. 12B. *Physiological Plant Ecology. II. Water Relations and Carbon Assimilation*, ed. O. L. Lange, P. S. Nobel, C. B. Osmond, and H. Ziegler. Berlin: Springer-Verlag, pp. 589–613.

Cowan, I. R., and Farquhar, G. D. (1977). Stomatal function in relation to leaf metabolism and environment. In *Integration of Activity in the Higher Plant*, ed. D. H. Jennings. Cambridge: Cambridge University Press, pp. 471–505.

Cox, P. M. (2001). *Description of the "TRIFFID" Dynamic Global Vegetation Model*, Technical Note 24. Bracknell: Met Office Hadley Centre.

Cox, P. M., Betts, R. A., Bunton, C. B., et al. (1999). The impact of new land surface physics on the GCM simulation of climate and climate sensitivity. *Climate Dynamics*, 15, 183–203.

Cox, P. M., Betts, R. A., Jones, C. D., Spall, S. A., and Totterdell, I. J. (2000). Acceleration of global warming due to carbon-cycle feedbacks in a coupled climate model. *Nature*, 408, 184–187.

Cox, P. M., Huntingford, C., and Harding, R. J. (1998). A canopy conductance and photosynthesis model for use in a GCM land surface scheme. *Journal of Hydrology*, 212/213, 79–94.

Craig, H., and Gordon, L. I. (1965). Deuterium and oxygen-18 variations in the ocean and the marine atmosphere. In *Stable Isotopes in Oceanographic Studies and Paleotemperatures*, ed. E. Tongiorgi. Pisa: Consiglio Nazionale delle Ricerche - Laboratorio di Geologia Nucleare, pp. 9–130.

Crank, J., and Nicolson, P. (1947). A practical method for numerical evaluation of solutions of partial differential equations of the heat-conduction type. *Mathematical Proceedings of the Cambridge Philosophical Society*, 43, 50–67.

Crawford, N. H., and Burges, S. J. (2004). History of the Stanford Watershed Model. *Water Resources IMPACT*, 6 (2), 3–5.

Crawford, N. H., and Linsley, R. K. (1966). *Digital Simulation in Hydrology: Stanford Watershed Model IV*, Technical Report No. 39, Department of Civil Engineering. Stanford, CA: Stanford University.

Currie, W. S., Harmon, M. E., Burke, I. C., et al. (2010). Cross-biome transplants of plant litter show decomposition models extend to a broader climatic range but lose predictability at the decadal time scale. *Global Change Biology*, 16, 1744–1761.

Dagan, G., and Bresler, E. (1983). Unsaturated flow in spatially variable fields: 1. Derivation of models of infiltration and redistribution. *Water Resources Research*, 19, 413–420.

Dai, Y., Dickinson, R. E., and Wang, Y.-P. (2004). A two-big-leaf model for canopy temperature, photosynthesis, and stomatal conductance. *Journal of Climate*, 17, 2281–2299.

Damour, G., Simonneau, T., Cochard, H., and Urban, L. (2010). An overview of models of stomatal conductance at the leaf level. *Plant, Cell and Environment*, 33, 1419–1438.

Dang, Q. L., Margolis, H. A., and Collatz, G. J. (1998). Parameterization and testing of a coupled photosynthesis–stomatal conductance model for boreal trees. *Tree Physiology*, 18, 141–153.

Dang, Q. L., Margolis, H. A., Coyea, M. R., Sy, M., and Collatz, G. J. (1997a). Regulation of branch-level gas exchange of boreal trees: Roles of shoot water potential and vapor pressure difference. *Tree Physiology*, 17, 521–535.

Dang, Q. L., Margolis, H. A., Sy, M., et al. (1997b). Profiles of photosynthetically active radiation, nitrogen and photosynthetic capacity in the boreal forest: Implications for scaling from leaf to canopy. *Journal of Geophysical Research*, 102D, 28845–28859.

Davidson, E. A. (1991). Fluxes of nitrous oxide and nitric oxide from terrestrial ecosystems. In *Microbial Production and Consumption of Greenhouse Gases: Methane, Nitrogen Oxides and Halomethanes*, ed. J. E. Rogers and W. B. Whitman. Washington, DC: American Society for Microbiology, pp. 219–235.

Davidson, E. A., Keller, M., Erickson, H. E., Verchot, L. V., and Veldkamp, E. (2000). Testing a conceptual model of soil emissions of nitrous and nitric oxides. *BioScience*, 50, 667–680.

Deardorff, J. W. (1978). Efficient prediction of ground surface temperature and moisture, with inclusion of a layer of vegetation. *Journal of Geophysical Research*, 83C, 1889–1903.

De Kauwe, M. G., Medlyn, B. E., Zaehle, S., et al. (2013). Forest water use and water use efficiency at elevated CO_2: A model-data intercomparison at two contrasting temperate forest FACE sites. *Global Change Biology*, 19, 1759–1779.

De Kauwe, M. G., Medlyn, B. E., Zaehle, S., (2014). Where does the carbon go? A model–data intercomparison of vegetation carbon allocation and turnover processes at two temperate forest free-air CO_2 enrichment sites. *New Phytologist*, 203, 883–899.

De Kauwe, M. G., Kala, J., Lin, Y.-S., et al. (2015). A test of an optimal stomatal conductance scheme within the CABLE land surface model. *Geoscientific Model Development*, 8, 431–452.

Del Grosso, S. J., Parton, W. J., Mosier, A. R., et al. (2000). General model for N_2O and N_2 gas emissions from soils due to denitrification. *Global Biogeochemical Cycles*, 14, 1045–1060.

Del Grosso, S. J., Parton, W. J., Mosier, A. R., (2005a). Modeling soil CO_2 emissions from ecosystems. *Biogeochemistry*, 73, 71–91.

Del Grosso, S. J., Mosier, A. R., Parton, W. J., and Ojima, D. S. (2005b). DAYCENT model analysis of past and contemporary soil N_2O and net greenhouse gas flux for major crops in the USA. *Soil & Tillage Research*, 83, 9–24.

Del Grosso, S. J., Ojima, D. S., Parton, W. J., et al. (2009). Global scale DAYCENT model analysis of greenhouse gas emissions and mitigation strategies for cropped soils. *Global and Planetary Change*, 67, 44–50.

Denmead, O. T., and Bradley, E. F. (1985). Flux–gradient relationships in a forest canopy. In *The Forest–Atmosphere Interaction*, ed. B. A. Hutchinson and B. B. Hicks. Dordrecht: Reidel, pp. 421–442.

de Pury, D. G. G., and Farquhar, G. D. (1997). Simple scaling of photosynthesis from leaves to canopies without the errors of big-leaf models. *Plant, Cell and Environment*, 20, 537–557.

De Ridder, K. (2010). Bulk transfer relations for the roughness sublayer. *Boundary-Layer Meteorology*, 134, 257–267.

de Vries, D. A. (1963). Thermal properties of soils. In *Physics of Plant Environment*, ed. W. R. van Wijk. Amsterdam: North-Holland, pp. 210–235.

Dewar, R., Mauranen, A., Mäkelä, A., et al. (2018). New insights into the covariation of stomatal, mesophyll and hydraulic conductances from optimization models incorporating nonstomatal limitations to photosynthesis. *New Phytologist*, 217, 571–585.

de Wit, C. T. (1965). *Photosynthesis of Leaf Canopies, Agricultural Research Reports Number 663*. Wageningen: Center for Agricultural Publications and Documentation.

Diaz-Espejo, A., Bernacchi, C. J., Collatz, G. J., and Sharkey, T. D. (2012). Models of photosynthesis. In *Terrestrial Photosynthesis in a Changing Environment: A Molecular, Physiological and Ecological Approach*, ed. J. Flexas, F. Loreto, and H. Medrano. Cambridge: Cambridge University Press, pp. 98–112.

Dickinson, R. E. (1983). Land surface processes and climate-surface albedos and energy balance. *Advances in Geophysics*, 25, 305–353.

Dickinson, R. E. (1988). The force-restore model for surface temperatures and its generalizations. *Journal of Climate*, 1, 1086–1097.

Dickinson, R. E., Jäger, J., Washington, W. M., and Wolski, R. (1981). *Boundary Subroutine for the NCAR Global Climate Model*, Technical Note NCAR/TN-173+IA. Boulder, CO: National Center for Atmospheric Research.

Dickinson, R. E., Henderson-Sellers, A., and Kennedy, P. J. (1993). *Biosphere–Atmosphere Transfer Scheme (BATS) Version 1e as Coupled to the NCAR Community Climate Model*, Technical Note NCAR/TN-387+STR. Boulder, CO: National Center for Atmospheric Research.

Dickinson, R. E., Henderson-Sellers, A., Kennedy, P. J., and Wilson, M. F. (1986). *Biosphere–Atmosphere Transfer Scheme (BATS) for the NCAR Community Climate Model*, Technical Note NCAR/TN-275+STR. Boulder, CO: National Center for Atmospheric Research.

Dickinson, R. E., Berry, J. A., Bonan, G. B., et al. (2002). Nitrogen controls on climate model evapotranspiration. *Journal of Climate*, 15, 278–295.

Dickinson, R. E., Shaikh, M., Bryant, R., and Graumlich, L. (1998). Interactive canopies for a climate model. *Journal of Climate*, 11, 2823–2836.

Dingman, S. L. (1994). *Physical Hydrology*. Upper Saddle River, NJ: Prentice Hall.

Dolman, A. J. (1993). A multiple-source land surface energy balance model for use in general circulation models. *Agricultural and Forest Meteorology*, 65, 21–45.

Dolman, A. J., and Gregory, D. (1992). The parametrization of rainfall interception in GCMs. *Quarterly Journal of the Royal Meteorological Society*, 118, 455–467.

Dolman, A. J., and Wallace, J. S. (1991). Lagrangian and K-theory approaches in modelling evaporation from sparse canopies. *Quarterly Journal of the Royal Meteorological Society*, 117, 1325–1340.

Domingues, T. F., Berry, J. A., Martinelli, L. A., Ometto, J. P. H. B., and Ehleringer, J. R. (2005). Parameterization of canopy structure and leaf-level gas exchange for an eastern Amazonian tropical rain forest (Tapajós National Forest, Pará, Brazil). *Earth Interactions*, 9(17), 1–23.

Donigian, A. S., Jr., and Imhoff, J. (2010). History and evolution of watershed modeling derived from the Stanford Watershed Model. In *Watershed Models*, ed. V. P. Singh and D. K. Frevert. Boca Raton, FL: CRC Press, pp. 21–45.

Dreccer, M. F., Schapendonk, A. H. C. M., van Oijen, M., Pot, C. S., and Rabbinge, R. (2000a). Radiation and nitrogen use at the leaf and canopy level by wheat and oilseed rape during the critical period for grain number definition. *Australian Journal of Plant Physiology*, 27, 899–910.

Dreccer, M. F., van Oijen, M., Schapendonk, A. H. C. M., Pot, C. S., and Rabbinge, R. (2000b). Dynamics of vertical leaf nitrogen distribution in a vegetative wheat canopy: Impact on canopy photosynthesis. *Annals of Botany*, 86, 821–831.

Drewry, D. T., and Albertson, J. D. (2006). Diagnosing model error in canopy-atmosphere exchange using empirical orthogonal function analysis. *Water Resources Research*, 42, W06421, doi:10.1029/2005WR004496.

Drewry, D. T., Kumar, P., Long, S., et al. (2010). Ecohydrological responses of dense canopies to environmental variability: 1. Interplay between vertical structure and photosynthetic pathway. *Journal of Geophysical Research*, 115, G04022, doi:10.1029/2010JG001340.

Drouet, J.-L., and Bonhomme, R. (1999). Do variations in local leaf irradiance explain changes to leaf nitrogen within row maize canopies? *Annals of Botany*, 84, 61–69.

Drouet, J.-L., and Bonhomme, R. (2004). Effect of 3D nitrogen, dry mass per area and local irradiance on canopy photosynthesis within leaves of contrasted heterogeneous maize crops. *Annals of Botany*, 93, 699–710.

Du, Z., Weng, E., Xia, J., et al. (2018). Carbon–nitrogen coupling under three schemes of model representation: Traceability analysis. *Geoscientific Model Development Discussions*, https://doi.org/10.5194/gmd-2018-41.

Duarte, H. F., Raczka, B. M., Ricciuto, D. M., et al. (2017). Evaluating the Community Land Model (CLM 4.5) at a coniferous forest site in northwestern United States using flux and carbon-isotope measurements. *Biogeosciences*, 14, 4315–4340.

Ducharne, A., Koster, R. D., Suarez, M. J., Stieglitz, M., and Kumar, P. (2000). A catchment-based approach to modeling land surface processes in a general circulation model: 2. Parameter estimation and model demonstration. *Journal of Geophysical Research*, 105D, 24823–24838.

Ducharne, A., Laval, K., and Polcher, J. (1998). Sensitivity of the hydrological cycle to the parameterization of soil hydrology in a GCM. *Climate Dynamics*, 14, 307–327.

Ducoudré, N. I., Laval, K., and Perrier, A. (1993). SECHIBA, a new set of parameterizations of the hydrologic exchanges at the land–atmosphere interface within the LMD atmospheric general circulation model. *Journal of Climate*, 6, 248–273.

Dümenil, L., and Todini, E. (1992). A rainfall-runoff scheme for use in the Hamburg climate model. In *Advances in Theoretical Hydrology: A Tribute to James Dooge*, ed. J. P. O'Kane. Amsterdam: Elsevier, pp. 129–157.

Duursma, R. A. (2015). Plantecophys – An R package for analysing and modelling leaf gas exchange data. *PLoS ONE*, 10(11), e0143346, doi:10.1371/journal.pone.0143346.

Duursma, R. A., Barton, C. V. M., Lin, Y.-S., et al. (2014). The peaked response of transpiration rate to vapour pressure deficit in field conditions can be explained by the temperature optimum of photosynthesis. *Agricultural and Forest Meteorology*, 189/190, 2–10.

Duursma, R. A., and Medlyn, B. E. (2012). MAESPA: A model to study interactions between water limitation, environmental drivers and vegetation function at tree and stand levels, with an example application to [CO_2] × drought interactions. *Geoscientific Model Development*, 5, 919–940.

Dyer, A. J. (1974). A review of flux–profile relationships. *Boundary-Layer Meteorology*, 7, 363–372.

Dyer, A. J., and Hicks, B. B. (1970). Flux–gradient relationships in the constant flux layer. *Quarterly Journal of the Royal Meteorological Society*, 96, 715–721.

Eagleson, P. S. (1978). Climate, soil, and vegetation 3. A simplified model of soil moisture movement in the liquid phase. *Water Resources Research*, 14, 722–730.

Egea, G., Verhoef, A., and Vidale, P. L. (2011). Towards an improved and more flexible representation of water stress in coupled photosynthesis–stomatal conductance models. *Agricultural and Forest Meteorology*, 151, 1370–1384.

Ek, M. B., Mitchell, K. E., Lin, Y., et al. (2003). Implementation of Noah land surface model advances in the National Centers for Environmental Prediction operational mesoscale Eta model. *Journal of Geophysical Research*, 108, 8851, doi:10.1029/2002JD003296.

Ellsworth, D. S., and Reich, P. B. (1993). Canopy structure and vertical patterns of photosynthesis and related leaf traits in a deciduous forest. *Oecologia*, 96, 169–178.

Eltahir, E. A. B., and Bras, R. L. (1993). A description of rainfall interception over large areas. *Journal of Climate*, 6, 1002–1008.

Entekhabi, D., and Eagleson, P. S. (1989). Land surface hydrology parameterization for atmospheric general circulation models including subgrid scale spatial variability. *Journal of Climate*, 2, 816–831.

Essery, R., Best, M., and Cox, P. (2001). *MOSES 2.2 Technical Documentation*, Technical Note 30. Bracknell: Met Office Hadley Centre.

Essery, R. L. H., Best, M. J., Betts, R. A., Cox, P. M., and Taylor, C. M. (2003). Explicit representation of subgrid heterogeneity in a GCM land surface scheme. *Journal of Hydrometeorology*, 4, 530–543.

Ethier, G. J., and Livingston, N. J. (2004). On the need to incorporate sensitivity to CO_2 transfer conductance into the Farquhar–von Caemmerer–Berry leaf photosynthesis model. *Plant, Cell and Environment*, 27, 137–153.

Evans, J. R. (1989). Photosynthesis and nitrogen relationships in leaves of C_3 plants. *Oecologia*, 78, 9–19.

Evans, J. R. (1993a). Photosynthetic acclimation and nitrogen partitioning within a lucerne canopy. I. Canopy characteristics. *Australian Journal of Plant Physiology*, 20, 55–67.

Evans, J. R. (1993b). Photosynthetic acclimation and nitrogen partitioning within a lucerne canopy. II. Stability through time and comparison with a theoretical optimum. *Australian Journal of Plant Physiology*, 20, 69–82.

Ewers, B. E., Gower, S. T., Bond-Lamberty, B., and Wang, C. K. (2005). Effects of stand age and tree species on canopy transpiration and average stomatal conductance of boreal forests. *Plant, Cell and Environment*, 28, 660–678.

Ewers, B. E., Oren, R., and Sperry, J. S. (2000). Influence of nutrient versus water supply on hydraulic architecture and water balance in *Pinus taeda*. *Plant, Cell and Environment*, 23, 1055–1066.

Famiglietti, J. S., and Wood, E. F. (1994). Multiscale modeling of spatially variable water and energy balance processes. *Water Resources Research*, 30, 3061–3078.

Farouki, O. T. (1981). *Thermal Properties of Soils*, CRREL Monograph 81–1. Hanover, NH: U.S Army Corps of Engineers, Cold Regions Research and Engineering Laboratory.

Farquhar, G. D., Buckley, T. N., and Miller, J. M. (2002). Optimal stomatal control in relation to leaf area and nitrogen content. *Silva Fennica*, 36, 625–637.

Farquhar, G. D., and Cernusak, L. A. (2005). On the isotopic composition of leaf water in the non-steady state. *Functional Plant Biology*, 32, 293–303.

Farquhar, G. D., Cernusak, L. A., and Barnes, B. (2007). Heavy water fractionation during transpiration. *Plant Physiology*, 143, 11–18.

Farquhar, G. D., Ehleringer, J. R., and Hubick, K. T. (1989a). Carbon isotope discrimination and photosynthesis. *Annual Review of Plant Physiology and Plant Molecular Biology*, 40, 503–537.

Farquhar, G. D., Hubick, K. T., Condon, A. G., and Richards, R. A. (1989b). Carbon isotope fractionation and plant water-use efficiency. In *Stable Isotopes in Ecological Research*, ed. P. W. Rundel, J. R. Ehleringer, and K. A. Nagy. New York, NY: Springer-Verlag, pp. 21–40.

Farquhar, G. D., and Lloyd, J. (1993). Carbon and oxygen isotope effects in the exchange of carbon dioxide between terrestrial plants and the atmosphere. In *Stable Isotopes and Plant Carbon–Water Relations*, ed. J. R. Ehleringer, A. E. Hall, and G. D. Farquhar. San Diego, CA: Academic Press, pp. 47–70.

Farquhar, G. D., O'Leary, M. H., and Berry, J. A. (1982). On the relationship between carbon isotope discrimination and the intercellular carbon dioxide concentration in leaves. *Australian Journal of Plant Physiology*, 9, 121–137.

Farquhar, G. D., and Richards, R. A. (1984). Isotopic composition of plant carbon correlates with water-use efficiency of wheat genotypes. *Australian Journal of Plant Physiology*, 11, 539–552.

Farquhar, G. D., and von Caemmerer, S. (1982). Modelling of photosynthetic response to environmental conditions. In *Encyclopedia of Plant Physiology, New Series, vol. 12B. Physiological Plant Ecology. II. Water Relations and Carbon Assimilation*, ed. O. L. Lange, P. S. Nobel, C. B. Osmond, and H. Ziegler. Berlin: Springer-Verlag, pp. 549–587.

Farquhar, G. D., von Caemmerer, S., and Berry, J. A. (1980). A biochemical model of photosynthetic CO_2 assimilation in leaves of C_3 species. *Planta*, 149, 78–90.

Farquhar, G. D., and Wong, S. C. (1984). An empirical model of stomatal conductance. *Australian Journal of Plant Physiology*, 11, 191–210.

Feddes, R. A., and Raats, P. A. C. (2004). Parameterizing the soil – water – plant root system. In *Unsaturated-zone Modeling: Progress, Challenges and Applications*, ed. R. A. Feddes, G. H. de Rooij, and J. C. van Dam. Dordrecht: Kluwer, pp. 95–141.

Federer, C. A. (1979). A soil–plant–atmosphere model for transpiration and availability of soil water. *Water Resources Research*, 15, 555–562.

Field, C. (1983). Allocating leaf nitrogen for the maximization of carbon gain: Leaf age as a control on the allocation program. *Oecologia*, 56, 341–347.

Field, C., and Mooney, H. A. (1986). The photosynthesis–nitrogen relationship in wild plants. In *On the Economy of Plant Form and Function*, ed. T. J. Givnish. Cambridge: Cambridge University Press, pp. 25–55.

Finnigan, J. (2004). Advection and modeling. In *Handbook of Micrometeorology: A Guide for Surface Flux Measurement and Analysis*, ed. X. Lee, W. Massman, and B. Law. Dordrecht: Kluwer, pp. 209–244.

Finnigan, J., Harman, I., Ross, A., and Belcher, S. (2015). First-order turbulence closure for modelling complex canopy flows. *Quarterly Journal of the Royal Meteorological Society*, 141, 2907–2916.

Finnigan, J. J., and Raupach, M. R. (1987). Transfer processes in plant canopies in relation to stomatal characteristics. In *Stomatal Function*, ed. E. Zeiger, G. D. Farquhar, and I. R. Cowan. Stanford, CA: Stanford University Press, pp. 385–429.

Finnigan, J. J., Shaw, R. H., and Patton, E. G. (2009). Turbulence structure above a vegetation canopy. *Journal of Fluid Mechanics*, 637, 387–424.

Firestone, M. K., and Davidson, E. A. (1989). Microbiological basis of NO and N_2O production and consumption in soil. In *Exchange of Trace Gases between Terrestrial Ecosystems and the Atmosphere*, ed. M. O. Andreae and D. S. Schimel. New York, NY: Wiley, pp. 7–21.

Fisher, J. B., Huntzinger, D. N., Schwalm, C. R., and Sitch, S. (2014). Modeling the terrestrial biosphere. *Annual Review of Environment and Resources*, 39, 91–123.

Fisher, J. B., Sitch, S., Malhi, Y., et al. (2010a). Carbon cost of plant nitrogen acquisition: A mechanistic, globally applicable model of plant nitrogen uptake, retranslocation, and fixation. *Global Biogeochemical Cycles*, 24, GB1014, doi:10.1029/2009GB003621.

Fisher, R., McDowell, N., Purves, D., et al. (2010b). Assessing uncertainties in a second-generation dynamic vegetation model caused by ecological scale limitations. *New Phytologist*, 187, 666–681.

Fisher, R. A., Koven, C. D., Anderegg, W. R. L., et al. (2018). Vegetation demographics in Earth System Models: A review of progress and priorities. *Global Change Biology*, 24, 35–54.

Fisher, R. A., Muszala, S., Vertenstein, M., et al. (2015). Taking off the training wheels: The properties of a dynamic vegetation model without climate envelopes, CLM4.5(ED). *Geoscientific Model Development*, 8, 3593–3619.

Fisher, R. A., Williams, M., Lola da Costa, A., et al. (2007). The response of an Eastern Amazonian rain forest to drought stress: Results and modelling analyses from a throughfall exclusion experiment. *Global Change Biology*, 13, 2361–2378.

Flanagan, L. B. (1993). Environmental and biological influences on the stable oxygen and hydrogen isotopic composition of leaf water. In *Stable Isotopes and Plant Carbon–Water Relations*, ed. J. R. Ehleringer, A. E. Hall, and G. D. Farquhar. San Diego, CA: Academic Press, pp. 71–90.

Flanagan, L. B., Cai, T., Black, T. A., et al. (2012). Measuring and modeling ecosystem photosynthesis and the carbon isotope composition of ecosystem-respired CO_2 in three boreal coniferous forests. *Agricultural and Forest Meteorology*, 153, 165–176.

Flanagan, L. B., Comstock, J. P., and Ehleringer, J. R. (1991). Comparison of modeled and observed environmental influences on the stable oxygen and hydrogen isotope composition of leaf water in *Phaseolus vulgaris* L. *Plant Physiology*, 96, 588–596.

Flanner, M. G., Zender, C. S., Randerson, J. T., and Rasch, P. J. (2007). Present-day climate forcing and response from black carbon in snow. *Journal of Geophysical Research*, 112, D11202, doi:10.1029/2006JD008003.

Flatau, P. J., Walko, R. L., and Cotton, W. R. (1992). Polynomial fits to saturation vapor pressure. *Journal of Applied Meteorology*, 31, 1507–1513.

Flechard, C. R., Massad, R.-S., Loubet, B., et al. (2013). Advances in understanding, models and parameterizations of biosphere–atmosphere ammonia exchange. *Biogeosciences*, 10, 5183–5225.

Flexas, J., Brugnoli, E., and Warren, C. R. (2012). Mesophyll conductance to CO_2. In *Terrestrial Photosynthesis in a Changing Environment: A Molecular, Physiological and Ecological Approach*, ed. J. Flexas, F. Loreto, and H. Medrano. Cambridge: Cambridge University Press, pp. 169–185.

Flexas, J., Ribas-Carbó, M., Diaz-Espejo, A., Galmés, J., and Medrano, H. (2008). Mesophyll conductance to CO_2: Current knowledge and future prospects. *Plant, Cell and Environment*, 31, 602–621.

Foken, T. (2006). 50 years of the Monin–Obukhov similarity theory. *Boundary-Layer Meteorology*, 119, 431–447.

Foken, T. (2008). The energy balance closure problem: An overview. *Ecological Applications*, 18, 1351–1367.

Foley, J. A., Prentice, I. C., Ramankutty, N., et al. (1996). An integrated biosphere model of land surface processes, terrestrial carbon balance, and vegetation dynamics. *Global Biogeochemical Cycles*, 10, 603–628.

Ford, E. D., and Newbould, P. J. (1971). The leaf canopy of a coppiced deciduous woodland: I. *Development and structure. Journal of Ecology*, 59, 843–862.

Forkel, R., Klemm, O., Graus, M., et al. (2006). Trace gas exchange and gas phase chemistry in a Norway spruce forest: A study with a coupled 1-dimensional canopy atmospheric chemistry emission model. *Atmospheric Environment*, 40, S28–S42.

Franks, P. J., Berry, J. A., Lombardozzi, D. L., and Bonan, G. B. (2017). Stomatal function across temporal and spatial scales: Deep-time trends, land–atmosphere coupling and global models. *Plant Physiology*, 174, 583–602.

Freeze, R. A., and Harlan, R. L. (1969). Blueprint for a physically-based, digitally-simulated hydrologic response model. *Journal of Hydrology*, 9, 237–258.

Friedlingstein, P., Cox, P., Betts, R., et al. (2006). Climate–carbon cycle feedback analysis: Results from the C^{4}MIP model intercomparison. *Journal of Climate*, 19, 3337–3353.

Friedlingstein, P., Joel, G., Field, C. B., and Fung, I. Y. (1999). Toward an allocation scheme for global terrestrial carbon models. *Global Change Biology*, 5, 755–770.

Friedlingstein, P., Meinshausen, M., Arora, V. K., et al. (2014). Uncertainties in CMIP5 climate projections due to carbon cycle feedbacks. *Journal of Climate*, 27, 511–526.

Friend, A. D. (1995). PGEN: An integrated model of leaf photosynthesis, transpiration, and conductance. *Ecological Modelling*, 77, 233–255.

Friend, A. D. (2001). Modelling canopy CO_2 fluxes: Are 'big-leaf' simplifications justified? *Global Ecology and Biogeography*, 10, 603–619.

Friend, A. D., and Kiang, N. Y. (2005). Land surface model development for the GISS GCM: Effects of improved canopy physiology on simulated climate. *Journal of Climate*, 18, 2883–2902.

Friend, A. D., Shugart, H. H., and Running, S. W. (1993). A physiology-based gap model of forest dynamics. *Ecology*, 74, 792–797.

Friend, A. D., Stevens, A. K., Knox, R. G., and Cannell, M. G. R. (1997). A process-based, terrestrial biosphere model of ecosystem dynamics (Hybrid v3.0). *Ecological Modelling*, 95, 249–287.

Fung, I. Y., Doney, S. C., Lindsay, K., and John, J. (2005). Evolution of carbon sinks in a changing climate. *Proceedings of the National Academy of Sciences USA*, 102, 11201–11206.

Gao, W., and Wesely, M. L. (1994). Numerical modeling of the turbulent fluxes of chemically reactive trace gases in the atmospheric boundary layer. *Journal of Applied Meteorology*, 33, 835–847.

Gao, W., Wesely, M. L., and Doskey, P. V. (1993). Numerical modeling of the turbulent diffusion and chemistry of NO_x, O_3, isoprene, and other reactive trace gases in and above a forest canopy. *Journal of Geophysical Research*, 98D, 18339–18353.

Gardner, W. R. (1960). Dynamic aspects of water availability to plants. *Soil Science*, 89, 63–73.

Garratt, J. R. (1978). Flux profile relations above tall vegetation. *Quarterly Journal of the Royal Meteorological Society*, 104, 199–211.

Garratt, J. R. (1980). Surface influence upon vertical profiles in the atmospheric near-surface layer. *Quarterly Journal of the Royal Meteorological Society*, 106, 803–819.

Garratt, J. R. (1983). Surface influence upon vertical profiles in the nocturnal boundary layer. *Boundary-Layer Meteorology*, 26, 69–80.

Garratt, J. R. (1992). *The Atmospheric Boundary Layer*. Cambridge: Cambridge University Press.

Garratt, J. R., and Hicks, B. B. (1973). Momentum, heat and water vapour transfer to and from natural and artificial surfaces. *Quarterly Journal of the Royal Meteorological Society*, 99, 680–687.

Gastal, F., and Lemaire, G. (2002). N uptake and distribution in crops: An agronomical and ecophysiological perspective. *Journal of Experimental Botany*, 53, 789–799.

Gastellu-Etchegorry, J. P., Demarez, V., Pinel, V., and Zagolski, F. (1996). Modeling radiative transfer in heterogeneous 3-D vegetation canopies. *Remote Sensing of Environment*, 58, 131–156.

Gat, J. R. (1996). Oxygen and hydrogen isotopes in the hydrologic cycle. *Annual Review of Earth and Planetary Sciences*, 24, 225–262.

Gates, D. M. (1962). *Energy Exchange in the Biosphere*. New York, NY: Harper and Row.

Gates, D. M. (1963). Leaf temperature and energy exchange. *Archiv für Meteorologie, Geophysik und Bioklimatologie*, 12B, 321–336.

Gates, D. M. (1965). Energy, plants, and ecology. *Ecology*, 46, 1–13.

Gates, D. M. (1966). Transpiration and energy exchange. *Quarterly Review of Biology*, 41, 353–364.

Gates, D. M. (1980). *Biophysical Ecology*. New York, NY: Springer-Verlag.

Gates, D. M., Alderfer, R., and Taylor, E. (1968). Leaf temperatures of desert plants. *Science*, 159, 994–995.

Gates, D. M., and Papian, L. E. (1971). *Atlas of Energy Budgets of Plant Leaves*. New York, NY: Academic Press.

Geiger, R. (1927). *Das Klima der bodennahen Luftschicht*. Braunschweig, Germany: Friedr. Vieweg & Sohn.

Gelhar, L. W. (1986). Stochastic subsurface hydrology from theory to applications. *Water Resources Research*, 22, 135S–145S.

Ghimire, B., Riley, W. J., Koven, C. D., Mu, M., and Randerson, J. T. (2016). Representing leaf and root physiological traits in CLM improves global carbon and nitrogen cycling predictions. *Journal of Advances in Modeling Earth Systems*, 8, 598–613, doi:10.1002/2015MS000538.

Gholz, H. L., Wedin, D. A., Smitherman, S. M., Harmon, M. E., and Parton, W. J. (2000). Long-term dynamics of pine and hardwood litter in contrasting environments: Toward a global model of decomposition. *Global Change Biology*, 6, 751–765.

Givnish, T. J., and Vermeij, G. J. (1976). Sizes and shapes of liane leaves. *American Naturalist*, 110, 743–778.

Goel, N. S., and Strebel, D. E. (1984). Simple beta distribution representation of leaf orientation in vegetation canopies. *Agronomy Journal*, 76, 800–802.

Goll, D. S., Vuichard, N., Maignan, F., et al. (2017). A representation of the phosphorus cycle for ORCHIDEE (revision 4520). *Geoscientific Model Development*, 10, 3745–3770.

Golley, F. B. (1993). *A History of the Ecosystem Concept in Ecology: More than the Sum of the Parts*. New Haven, CT: Yale University Press.

Goudriaan, J. (1977). *Crop Micrometeorology: A Simulation Study*. Wageningen: Center for Agricultural Publishing and Documentation.

Goudriaan, J. (1982). Potential production processes. In *Simulation of Plant Growth and Crop Production*, ed. F. W. T. Penning de Vries, and H. H. van Laar. Wageningen: Center for Agricultural Publishing and Documentation, pp. 98–113.

Goudriaan, J. (1988). The bare bones of leaf-angle distribution in radiation models for canopy photosynthesis and energy exchange. *Agricultural and Forest Meteorology*, 43, 155–169.

Goudriaan, J., and van Laar, H. H. (1994). *Modelling Potential Crop Growth Processes: Textbook with Exercises*. Dordrecht: Kluwer.

Goulden, M. L., McMillan, A. M. S., Winston, G. C., et al. (2011). Patterns of NPP, GPP, respiration, and NEP during boreal forest succession. *Global Change Biology*, 17, 855–871.

Grace, J. (1981). Some effects of wind on plants. In *Plants and their Atmospheric Environment*, ed. J. Grace, E. D. Ford, and P. G. Jarvis. Oxford: Blackwell, pp. 31–56.

Green, W. H., and Ampt, G. A. (1911). Studies on soil physics: Part I – The flow of air and water through soils. *Journal of Agricultural Science*, 4, 1–24.

Grindlay, D. J. C. (1997). Towards an explanation of crop nitrogen demand based on the optimization of leaf nitrogen per unit leaf area. *Journal of Agricultural Science*, 128, 377–396.

Grote, R., Morfopoulos, C., Niinemets, Ü., et al. (2014). A fully integrated isoprenoid emissions model coupling emissions to photosynthetic characteristics. *Plant, Cell and Environment*, 37, 1965–1980.

Gu, L., Shugart, H. H., Fuentes, J. D., Black, T. A., and Shewchuk, S. R. (1999). Micrometeorology, biophysical exchanges and NEE decomposition in a two-story boreal forest – development and test of an integrated model. *Agricultural and Forest Meteorology*, 94, 123–148.

Gu, L., Pallardy, S. G., Tu, K., Law, B. E., and Wullschleger, S. D. (2010). Reliable estimation of biochemical parameters from C_3 leaf photosynthesis–intercellular carbon dioxide response curves. *Plant, Cell and Environment*, 33, 1852–1874.

Guenther, A., Hewitt, C. N., Erickson, D., et al. (1995). A global model of natural volatile organic compound emissions. *Journal of Geophysical Research*, 100D, 8873–8892.

Guenther, A., Karl, T., Harley, P., et al. (2006). Estimates of global terrestrial isoprene emissions using MEGAN (Model of Emissions of Gases and Aerosols from Nature). *Atmospheric Chemistry and Physics*, 6, 3181–3210.

Guenther, A. B., Jiang, X., Heald, C. L., et al. (2012). The Model of Emissions of Gases and Aerosols from Nature version 2.1 (MEGAN2.1): An extended and updated framework for modeling biogenic emissions. *Geoscientific Model Development*, 5, 1471–1492.

Gutschick, V. P. (1991). Joining leaf photosynthesis models and canopy photon-transport models. In *Photon–Vegetation Interactions: Applications in Optical Remote Sensing and Plant Ecology*, ed. R. B. Myneni and J. Ross. Berlin: Springer-Verlag, pp 501–535.

Gutschick, V. P., and Wiegel, F. W. (1988). Optimizing the canopy photosynthetic rate by patterns of investment in specific leaf mass. *American Naturalist*, 132, 67–86.

Haese, B., Werner, M., and Lohmann, G. (2013). Stable water isotopes in the coupled atmosphere–land surface model ECHAM5-JSBACH. *Geoscientific Model Development*, 6, 1463–1480.

Hagemann, S., and Gates, L. D. (2003). Improving a subgrid runoff parameterization scheme for climate models by the use of high resolution data derived from satellite observations. *Climate Dynamics*, 21, 349–359.

Halldin, S. (1985). Leaf and bark area distribution in a pine forest. In *The Forest–Atmosphere Interaction*, ed. B. A. Hutchinson and B. B. Hicks. Dordrecht: Reidel, pp. 39–58.

Hantson, S., Arneth, A., Harrison, S. P., et al. (2016). The status and challenge of global fire modelling. *Biogeosciences*, 13, 3359–3375.

Hari, P., Mäkelä, A., Korpilahti, E., and Holmberg, M. (1986). Optimal control of gas exchange. *Tree Physiology*, 2, 169–175.

Harley, P. C., and Sharkey, T. D. (1991). An improved model of C_3 photosynthesis at high CO_2: Reversed O_2 sensitivity explained by lack of glycerate reentry into the chloroplast. *Photosynthesis Research*, 27, 169–178.

Harley, P. C., Thomas, R. B., Reynolds, J. F., and Strain, B. R. (1992). Modelling photosynthesis of cotton grown in elevated CO_2. *Plant, Cell and Environment*, 15, 271–282.

Harman, I. N. (2012). The role of roughness sublayer dynamics within surface exchange schemes. *Boundary-Layer Meteorology*, 142, 1–20.

Harman, I. N., and Finnigan, J. J. (2007). A simple unified theory for flow in the canopy and roughness sublayer. *Boundary-Layer Meteorology*, 123, 339–363.

Harman, I. N., and Finnigan, J. J. (2008). Scalar concentration profiles in the canopy and roughness sublayer. *Boundary-Layer Meteorology*, 129, 323–351.

Harmon, M. E., Silver, W. L., Fasth, B., et al. (2009). Long-term patterns of mass loss during the decomposition of leaf and fine root litter: An intersite comparison. *Global Change Biology*, 15, 1320–1338.

Hartman, M. D., Merchant, E. R., Parton, W. J., et al. (2011). Impact of historical land-use changes on greenhouse gas exchange in the U.S. Great Plains, 1883–2003. *Ecological Applications*, 21, 1105–1119.

Hartmann, D. L. (1994). *Global Physical Climatology*. San Diego, CA: Academic Press.

Haverd, V., Cuntz, M., Leuning, R., and Keith, H. (2007). Air and biomass heat storage fluxes in a forest canopy: Calculation within a soil vegetation atmosphere transfer model. *Agricultural and Forest Meteorology*, 147, 125–139.

Haverd, V., Cuntz, M., Nieradzik, L. P., and Harman, I. N. (2016). Improved representations of coupled soil–canopy processes in the CABLE land surface model (Subversion revision 3432). *Geoscientific Model Development*, 9, 3111–3122.

Haverd, V., Leuning, R., Griffith, D., van Gorsel, E., and Cuntz, M. (2009). The turbulent Lagrangian time scale in forest canopies constrained by fluxes, concentrations and source distributions. *Boundary-Layer Meteorology*, 130, 209–228.

Haverkamp, R., and Vauclin, M. (1979). A note on estimating finite difference interblock hydraulic conductivity values for transient unsaturated flow problems. *Water Resources Research*, 15, 181–187.

Haverkamp, R., Vauclin, M., Touma, J., Wierenga, P. J., and Vachaud, G. (1977). A comparison of numerical simulation models for one-dimensional infiltration. *Soil Science Society of America Journal*, 41, 285–294.

Heald, C. L., Henze, D. K., Horowitz, L. W., et al. (2008). Predicted change in global secondary organic aerosol concentrations in response to future climate, emissions, and land use change. *Journal of Geophysical Research*, 113, D05211, doi:10.1029/2007JD009092.

Hendricks Franssen, H. J., Stöckli, R., Lehner, I., Rotenberg, E., and Seneviratne, S. I. (2010). Energy balance closure of eddy-covariance data: A multisite analysis for European FLUXNET stations. *Agricultural and Forest Meteorology*, 150, 1553–1567.

Hetherington, A. M., and Woodward, F. I. (2003). The role of stomata in sensing and driving environmental change. *Nature*, 424, 901–908.

Hicks, B. B., Baldocchi, D. D., Meyers, T. P., Hosker, R. P., Jr., and Matt, D. R. (1987). A preliminary multiple resistance routine for deriving dry deposition velocities from measured quantities. *Water, Air, and Soil Pollution*, 36, 311–330.

Hill, T. C., Williams, M., Woodward, F. I., and Moncrieff, J. B. (2011). Constraining ecosystem processes from tower fluxes and atmospheric profiles. *Ecological Applications*, 21, 1474–1489.

Hills, R. G., Hudson, D. B., Porro, I., and Wierenga, P. J. (1989). Modeling one-dimensional infiltration into very dry soils: 2. Estimation of the soil water parameters and model predictions. *Water Resources Research*, 25, 1271–1282.

Hirose, T., and Werger, M. J. A. (1987). Maximizing daily canopy photosynthesis with respect to the leaf nitrogen allocation pattern in the canopy. *Oecologia*, 72, 520–526.

Hirose, T., and Werger, M. J. A. (1994). Photosynthetic capacity and nitrogen partitioning among species in the canopy of a herbaceous plant community. *Oecologia*, 100, 203–212.

Hirose, T., Werger, M. J. A., Pons, T. L., and van Rheenen, J. W. A. (1988). Canopy structure and leaf nitrogen distribution in a stand of *Lysimachia vulgaris* L. as influenced by stand density. *Oecologia*, 77, 145–150.

Hirose, T., Werger, M. J. A., and van Rheenen, J. W. A. (1989). Canopy development and leaf nitrogen distribution in a stand of *Carex acutiformis*. *Ecology*, 70, 1610–1618.

Hoelzemann, J. J., Schultz, M. G., Brasseur, G. P., Granier, C., and Simon, M. (2004). Global Wildland Fire Emission Model (GWEM): Evaluating the use of global area burnt satellite data. *Journal of Geophysical Research*, 109, D14S04, doi:10.1029/2003JD003666.

Hollinger, D. Y. (1989). Canopy organization and foliage photosynthetic capacity in a broad-leaved evergreen montane forest. *Functional Ecology*, 3, 5–62.

Hollinger, D. Y. (1996). Optimality and nitrogen allocation in a tree canopy. *Tree Physiology*, 16, 627–634.

Holtslag, A. A. M., and Beljaars, A. C. M. (1989). Surface flux parameterization schemes: Developments and experiences at KNMI. In *Parameterization of Fluxes over Land Surface: Proceedings of a Workshop Held at ECMWF 24–26 October 1988. European Centre for Medium-Range Weather Forecasts*, Reading, UK, pp. 121–147.

Holtslag, A. A. M., Svensson, G., Baas, P., et al. (2013). Stable atmospheric boundary layers and diurnal cycles: Challenges for weather and climate models. *Bulletin of the American Meteorological Society*, 94, 1691–1706.

Horn, H. S. (1971). *The Adaptive Geometry of Trees*. Princeton, NJ: Princeton University Press.

Horn, H. S. (1975). Forest succession. *Scientific American*, 232(5), 90–98.

Hornberger, G., and Wiberg, P. (2005). *Numerical Methods in the Hydrological Sciences*. Washington, DC: American Geophysical Union.

Hornberger, G. M., Raffensperger, J. P., Wiberg, P. L., and Eshleman, K. N. (1998). *Elements of Physical Hydrology*. Baltimore, MD: Johns Hopkins University Press.

Hrachowitz, M., and Clark, M. P. (2017). HESS Opinions: The complementary merits of competing modelling philosophies in hydrology. *Hydrology and Earth System Sciences*, 21, 3953–3973.

Huang, Y., Lu, X., Shi, Z., et al. (2018). Matrix approach to land carbon cycle modeling: A case study with the Community Land Model. *Global Change Biology*, 24, 1394–1404.

Hurtt, G. C., Chini, L. P., Frolking, S., et al. (2011). Harmonization of land–use scenarios for the period 1500–2100: 600 years of global gridded annual land–use transitions, wood harvest, and resulting secondary lands. *Climatic Change*, 109, 117–161.

Hurtt, G. C., Frolking, S., Fearon, M. G., et al. (2006). The underpinnings of land-use history: Three centuries of global gridded land-use transitions, wood-harvest activity, and resulting secondary lands. *Global Change Biology*, 12, 1208–1229.

Hurtt, G. C., Moorcroft, P. R., Pacala, S. W., and Levin, S. A. (1998). Terrestrial models and global change: Challenges for the future. *Global Change Biology*, 4, 581–590.

Hutchinson, B. A., Matt, D. R., McMillen, R. T., et al. (1986). The architecture of a deciduous forest canopy in eastern Tennessee, U.S.A. *Journal of Ecology*, 74, 635–646.

Innis, G. S. (1975). Role of total systems models in the grassland biome study. In *Systems Analysis and Simulation in Ecology*, vol. III. ed. B. C. Patten. New York, NY: Academic Press, pp. 13–47.

Innis, G. S. (1978). *Grassland Simulation Model*. New York, NY: Springer-Verlag.

Inoue, E. (1963). On the turbulent structure of airflow within crop canopies. *Journal of the Meteorological Society of Japan Ser. II*, 41, 317–326.

Jackson, R. B., Canadell, J., Ehleringer, J. R., et al. (1996). A global analysis of root distributions for terrestrial biomes. *Oecologia*, 108, 389–411.

Jackson, R. B., Mooney, H. A., and Schulze, E.-D. (1997). A global budget for fine root biomass, surface area, and nutrient contents. *Proceedings of the National Academy of Sciences USA*, 94, 7362–7366.

Jacobson, M. Z. (2005). *Fundamentals of Atmospheric Modeling*, 2nd edn. Cambridge: Cambridge University Press.

Jain, A., Yang, X., Kheshgi, H., et al. (2009). Nitrogen attenuation of terrestrial carbon cycle response to global environmental factors. *Global Biogeochemical Cycles*, 23, GB4028, doi:10.1029/2009GB003519.

Jarvis, P. G. (1976). The interpretation of the variations in leaf water potential and stomatal conductance found in canopies in the field. *Philosophical Transactions of the Royal Society London B*, 273, 593–610.

Jarvis, P. G., and McNaughton, K. G. (1986). Stomatal control of transpiration: Scaling up from leaf to region. *Advances in Ecological Research*, 15, 1–49.

Jenkinson, D. S. (1990). The turnover of organic carbon and nitrogen in soil. *Philosophical Transactions of the Royal Society London B*, 329, 361–368.

Jenkinson, D. S., Adams, D. E., and Wild, A. (1991). Model estimates of CO_2 emissions from soil in response to global warming. *Nature*, 351, 304–306.

Jetten, V. G. (1996). Interception of tropical rain forest: Performance of a canopy water balance model. *Hydrological Processes*, 10, 671–685.

Ji, J. (1995). A climate–vegetation interaction model: Simulating physical and biological processes at the surface. *Journal of Biogeography*, 22, 445–451.

Jobbágy, E. G., and Jackson, R. B. (2000). The vertical distribution of soil organic carbon and its relation to climate and vegetation. *Ecological Applications*, 10, 423–436.

Johnson, I. R., and Thornley, J. H. M. (1984). A model of instantaneous and daily canopy photosynthesis. *Journal of Theoretical Biology*, 107, 531–545.

Johnson, K. D., Entekhabi, D., and Eagleson, P. S. (1993). The implementation and validation of improved land-surface hydrology in an atmospheric general circulation model. *Journal of Climate*, 6, 1009–1026.

Jones, H. G. (2014). *Plants and Microclimate: A Quantitative Approach to Environmental Plant Physiology*, 3rd edn. Cambridge: Cambridge University Press.

Jordan, R. (1991). *A One-Dimensional Temperature Model for a Snow Cover: Technical Documentation for SNTHERM.89*, Special Report 91–16. Hanover, NH: U.S. Army Corps of Engineers Cold Regions Research and Engineering Laboratory.

Juang, J.-Y., Katul, G. G., Siqueira, M. B., Stoy, P. C., and McCarthy, H. R. (2008). Investigating a hierarchy of Eulerian closure models for scalar transfer inside forested canopies. *Boundary-Layer Meteorology*, 128, 1–32.

Juang, J.-Y., Katul, G., Siqueira, M., Stoy, P., and Novick, K. (2007). Separating the effects of albedo from eco-physiological changes on surface temperature along a successional chronosequence in the southeastern United States. *Geophysical Research Letters*, 34, L21408, doi:10.1029/2007GL031296.

Jumikis, A. R. (1966). *Thermal Soil Mechanics*. New Brunswick, NJ: Rutgers University Press.

Kala, J., De Kauwe, M. G., Pitman, A. J., et al. (2015). Implementation of an optimal stomatal conductance scheme in the Australian Community Climate Earth Systems Simulator (ACCESS1.3b). *Geoscientific Model Development*, 8, 3877–3889.

Kattge, J., and Knorr, W. (2007). Temperature acclimation in a biochemical model of photosynthesis: A reanalysis of data from 36 species. *Plant, Cell and Environment*, 30, 1176–1190.

Kattge, J., Knorr, W., Raddatz, T., and Wirth, C. (2009). Quantifying photosynthetic capacity and its relationship to leaf nitrogen content for global-scale terrestrial biosphere models. *Global Change Biology*, 15, 976–991.

Katul, G. G., Finnigan, J. J., Poggi, D., Leuning, R., and Belcher, S. E. (2006). The influence of hilly terrain on canopy-atmosphere carbon dioxide exchange. *Boundary-Layer Meteorology*, 118, 189–216.

Katul, G. G., Mahrt, L., Poggi, D., and Sanz, C. (2004). One- and two-equation models for canopy turbulence. *Boundary-Layer Meteorology*, 113, 81–109.

Katul, G., Manzoni, S., Palmroth, S., and Oren, R. (2010). A stomatal optimization theory to describe the effects of atmospheric CO_2 on leaf photosynthesis and transpiration. *Annals of Botany*, 105, 431–442.

Katul, G. G., Oren, R., Manzoni, S., Higgins, C., and Parlange, M. B. (2012). Evapotranspiration: A process driving mass transport and energy exchange in the soil-plant-atmosphere-climate system. *Reviews of Geophysics*, 50, RG3002, doi:10.1029/2011RG000366.

Katul, G. G., Palmroth, S., and Oren, R. (2009). Leaf stomatal responses to vapour pressure deficit under current and CO_2-enriched atmosphere explained by the economics of gas exchange. *Plant, Cell and Environment*, 32, 968–979.

Keenan, T., Sabate, S., and Gracia, C. (2010). Soil water stress and coupled photosynthesis–conductance models: Bridging the gap between conflicting reports on the relative roles of stomatal, mesophyll conductance and biochemical limitations to photosynthesis. *Agricultural and Forest Meteorology*, 150, 443–453.

Keenan, T. F., and Niinemets, Ü. (2016). Global leaf trait estimates biased due to plasticity in the shade. *Nature Plants*, 3, 16201, doi:10.1038/nplants.2016.201.

Keller, K. M., Lienert, S., Bozbiyik, A., et al. (2017). 20th century changes in carbon isotopes and water-use efficiency: Tree-ring-based evaluation of the CLM4.5 and LPX-Bern models. *Biogeosciences*, 14, 2641–2673.

Kelliher, F. M., Leuning, R., Raupach, M. R., and Schulze, E.-D. (1995). Maximum conductances for evaporation from global vegetation types. *Agricultural and Forest Meteorology*, 73, 1–16.

Kelly, R. H., Parton, W. J., Hartman, M. D., et al. (2000). Intra-annual and interannual variability of ecosystem processes in shortgrass steppe. *Journal of Geophysical Research*, 105D, 20093–20100.

Kloster, S., Mahowald, N. M., Randerson, J. T., et al. (2010). Fire dynamics during the 20th century simulated by the Community Land Model. *Biogeosciences*, 7, 1877–1902.

Knauer, J., Werner, C., and Zaehle, S. (2015). Evaluating stomatal models and their atmospheric drought response in a land surface scheme: A multibiome analysis. *Journal of Geophysical Research: Biogeosciences*, 120, 1894–1911, doi:10.1002/2015JG003114.

Knauer, J., Zaehle, S., Medlyn, B. E., et al. (2018). Towards physiologically meaningful water-use efficiency estimates from eddy covariance data. *Global Change Biology*, 24, 694–710.

Knorr, W., and Heimann, M. (2001). Uncertainties in global terrestrial biosphere modeling 1. A comprehensive sensitivity analysis with a new photosynthesis and energy balance scheme. *Global Biogeochemical Cycles*, 15, 207–225.

Knox, R. G., Longo, M., Swann, A. L. S., et al. (2015). Hydrometeorological effects of historical land-conversion in an ecosystem-atmosphere model of Northern South America. *Hydrology and Earth System Sciences*, 19, 241–273.

Kobayashi, H., Baldocchi, D. D., Ryu, Y., et al. (2012). Modeling energy and carbon fluxes in a heterogeneous oak woodland: A three-dimensional approach. *Agricultural and Forest Meteorology*, 152, 83–100.

Kondo, J., and Kawanaka, A. (1986). Numerical study on the bulk heat transfer coefficient for a variety of vegetation types and densities. *Boundary-Layer Meteorology*, 37, 285–296.

Konrad, W., Roth-Nebelsick, A., and Grein, M. (2008). Modelling of stomatal density response to atmospheric CO_2. *Journal of Theoretical Biology*, 253, 638–658.

Koster, R. D., and Milly, P. C. D. (1997). The interplay between transpiration and runoff formulations in land surface schemes used with atmospheric models. *Journal of Climate*, 10, 1578–1591.

Koster, R. D., and Suarez, M. J. (1992). A comparative analysis of two land surface heterogeneity representations. *Journal of Climate*, 5, 1379–1390.

Koven, C. D., Riley, W. J., Subin, Z. M., et al. (2013). The effect of vertically resolved soil biogeochemistry and alternate soil C and N models on C dynamics of CLM4. *Biogeosciences*, 10, 7109–7131.

Kowalczyk, E. A., Stevens, L., Law, R. M., et al. (2013). The land surface model component of ACCESS: Description and impact on the simulated surface climatology. *Australian Meteorological and Oceanographic Journal*, 63, 65–82.

Kowalczyk, E. A., Wang, Y. P., Law, R. M., et al. (2006). *The CSIRO Atmosphere Biosphere Land Exchange (CABLE) Model for Use in Climate Models and as an Offline Model*, Research Paper 13. Aspendale, Australia: CSIRO Marine and Atmospheric Research.

Krinner, G., Viovy, N., de Noblet-Ducoudré, N., et al. (2005). A dynamic global vegetation model for studies of the coupled atmosphere–biosphere system. *Global Biogeochemical Cycles*, 19, GB1015, doi:10.1029/2003GB002199.

Kucharik, C. J., Foley, J. A., Delire, C., et al. (2000). Testing the performance of a Dynamic Global Ecosystem Model: Water balance, carbon balance, and vegetation structure. *Global Biogeochemical Cycles*, 14, 795–825.

Kull, O., Broadmeadow, M., Kruijt, B., and Meir, P. (1999). Light distribution and foliage structure in an oak canopy. *Trees*, 14, 55–64.

Kull, O., and Kruijt, B. (1998). Leaf photosynthetic light response: A mechanistic model for scaling photosynthesis to leaves and canopies. *Functional Ecology*, 12, 767–777.

Kumagai, T. (2011). Transpiration in forest ecosystems. In *Forest Hydrology and Biogeochemistry: Synthesis of Past Research and Future Directions*, ed. D. F. Levia, D. Carlyle-Moses, and T. Tanaka. Dordrecht: Springer, pp. 389–406.

Kustas, W. P., Anderson, M. C., Norman, J. M, and Li, F. (2007). Utility of radiometric–aerodynamic temperature relations for heat flux estimation. *Boundary-Layer Meteorology*, 122, 167–187.

Kustas, W. P., and Norman, J. M. (1999). Evaluation of soil and vegetation heat flux predictions using a simple two-source model with radiometric temperatures for partial canopy cover. *Agricultural and Forest Meteorology*, 94, 13–29.

Kwa, C. (1993). Modeling the grasslands. *Historical Studies in the Physical and Biological Sciences*, 24, 125–155.

Kwa, C. (2005). Local ecologies and global science: Discourses and strategies of the International Geosphere–Biosphere Programme. *Social Studies of Science*, 35, 923–950.

Lasslop, G., Thonicke, K., and Kloster, S. (2014). SPITFIRE within the MPI Earth system model: Model development and evaluation. *Journal of Advances in Modeling Earth Systems*, 6, 740–755, doi:10.1002/2013MS000284.

Lathière, J., Hauglustaine, D. A., Friend, A. D., et al. (2006). Impact of climate variability and land use changes on global biogenic volatile organic compound emissions. *Atmospheric Chemistry and Physics*, 6, 2129–2146.

Launiainen, S., Katul, G. G., Grönholm, T., and Vesala, T. (2013). Partitioning ozone fluxes between canopy and forest floor by measurements and a multi-layer model. *Agricultural and Forest Meteorology*, 173, 85–99.

Launiainen, S., Katul, G. G., Kolari, P., Vesala, T., and Hari, P. (2011). Empirical and optimal stomatal controls on leaf and ecosystem level CO_2 and H_2O exchange rates. *Agricultural and Forest Meteorology*, 151, 1672–1689.

Launiainen, S., Katul, G. G., Lauren, A., and Kolari, P. (2015). Coupling boreal forest CO_2, H_2O and energy flows by a vertically structured forest canopy – Soil model with separate bryophyte layer. *Ecological Modelling*, 312, 385–405.

Lawrence, D., Fisher, R., Koven, C., et al. (2018). *Technical Description of Version 5.0 of the Community Land Model (CLM)*. https://escomp.github.io/ctsm-docs/.

Lawrence, D. M., Hurtt, G. C., Arneth, A., et al. (2016). The Land Use Model Intercomparison Project (LUMIP) contribution to CMIP6: Rationale and experimental design. *Geoscientific Model Development*, 9, 2973–2998.

Lawrence, P. J., Feddema, J. J., Bonan, G. B., et al. (2012). Simulating the biogeochemical and biogeophysical impacts of transient land cover change and wood harvest in the Community Climate System Model (CCSM4) from 1850 to 2100. *Journal of Climate*, 25, 3071–3095.

Lee, J.-E., Oliveira, R. S., Dawson, T. E., and Fung, I. (2005). Root functioning modifies seasonal climate. *Proceedings of the National Academy of Sciences USA*, 102, 17576–17581.

Lee, T. J., and Pielke, R. A. (1992). Estimating the soil surface specific humidity. *Journal of Applied Meteorology*, 31, 480–484.

Lehmann, F., and Ackerer, Ph. (1998). Comparison of iterative methods for improved solutions of the fluid flow equation in partially saturated porous media. *Transport in Porous Media*, 31, 275–292.

Lehnebach, R., Beyer, R., Letort, V., and Heuret, P. (2018). The pipe model theory half a century on: A review. *Annals of Botany*, 121, 773–795.

Leigh, A., Sevanto, S., Ball, M. C., et al. (2012). Do thick leaves avoid thermal damage in critically low wind speeds? *New Phytologist*, 194, 477–487.

Leij, F. J., Alves, W. J., van Genuchten, M. T., and Williams, J. R. (1996). *UNSODA – The UNSODA Unsaturated Soil Hydraulic Database User's Manual Version 1.0*, EPA/600/R-96/095. Cincinnati, OH: U.S. Environmental Protection Agency.

Lemaire, G., Onillon, B., Gosse, G., Chartier, M., and Allirand, J. M. (1991). Nitrogen distribution within a lucerne canopy during regrowth: Relation with light distribution. *Annals of Botany*, 68, 483–488.

Lemeur, R. (1973). A method for simulating the direct solar radiation regime in sunflower, Jerusalem artichoke, corn and soybean canopies using actual stand structure data. *Agricultural Meteorology*, 12, 229–247.

Lemeur, R. (1990). Modelling stomatal behaviour and photosynthesis of *Eucalyptus grandis*. *Australian Journal of Plant Physiology*, 17, 159–175.

Lemeur, R. (1995). A critical appraisal of a combined stomatal–photosynthesis model for C_3 plants. *Plant, Cell and Environment*, 18, 339–355.

Lemeur, R. (2000). Estimation of scalar source/sink distributions in plant canopies using Lagrangian dispersion analysis: Corrections for atmospheric stability and comparison with a multilayer canopy model. *Boundary-Layer Meteorology*, 96, 293–314.

Lemeur, R. (2002). Temperature dependence of two parameters in a photosynthesis model. *Plant, Cell and Environment*, 25, 1205–1210.

Leuning, R., Kelliher, F. M., de Pury, D. G. G., and Schulze, E.-D. (1995). Leaf nitrogen, photosynthesis, conductance and transpiration: Scaling from leaves to canopies. *Plant, Cell and Environment*, 18, 1183–1200.

Leuning, R., van Gorsel, E., Massman, W. J., and Issac, P. R. (2012). Reflections on the surface energy imbalance problem. *Agricultural and Forest Meteorology*, 156, 65–74.

Levis, S., Foley, J. A., and Pollard, D. (1999). CO_2, climate, and vegetation feedbacks at the Last Glacial Maximum. *Journal of Geophysical Research*, 104D, 31191–31198.

Levis, S., Foley, J. A., and Pollard, D. (2000). Large-scale vegetation feedbacks on a doubled CO_2 climate. *Journal of Climate*, 13, 1313–1325.

Levis, S., Bonan, G. B., Vertenstein, M., and Oleson, K. W. (2004). *The Community Land Model's Dynamic Global Vegetation Model (CLM-DGVM): Technical Description and User's Guide*, Technical Note NCAR/TN-459+IA. Boulder, CO: National Center for Atmospheric Research.

Levis, S., Hartman, M. D., and Bonan, G. B. (2014). The Community Land Model underestimates land-use CO_2 emissions by neglecting soil disturbance from cultivation. *Geoscientific Model Development*, 7, 613–620.

Levis, S., Wiedinmyer, C., Bonan, G. B., and Guenther, A. (2003). Simulating biogenic volatile organic compound emissions in the Community Climate System Model. *Journal of Geophysical Research*, 108, 4659, doi:10.1029/2002JD003203.

Li, C., Frolking, S., and Frolking, T. A. (1992). A model of nitrous oxide evolution from soil driven by rainfall events: 1. Model structure and sensitivity. *Journal of Geophysical Research*, 97D, 9759–9776.

Li, C., Aber, J., Stange, F., Butterbach-Bahl, K., and Papen, H. (2000). A process-oriented model of N_2O and NO emissions from forest soils: 1. Model development. *Journal of Geophysical Research*, 105D, 4369–4384.

Li, F., Levis, S., and Ward, D. S. (2013). Quantifying the role of fire in the Earth system – Part 1: Improved global fire modeling in the Community Earth System Model (CESM1). *Biogeosciences*, 10, 2293–2314.

Li, F., Zeng, X. D., and Levis, S. (2012a). A process-based fire parameterization of intermediate complexity in a

Dynamic Global Vegetation Model. *Biogeosciences*, 9, 2761–2780.

Li, H., Huang, M., Wigmosta, M. S., et al. (2011). Evaluating runoff simulations from the Community Land Model 4.0 using observations from flux towers and a mountainous watershed. *Journal of Geophysical Research*, 116, D24120, doi:10.1029/2011JD016276.

Li, L., Wang, Y.-P., Yu, Q., et al. (2012b). Improving the responses of the Australian community land surface model (CABLE) to seasonal drought. *Journal of Geophysical Research*, 117, G04002, doi:10.1029/2012JG002038.

Liang, X., Lettenmaier, D. P., and Wood, E. F. (1996). One-dimensional statistical dynamic representation of subgrid spatial variability of precipitation in the two-layer variable infiltration capacity model. *Journal of Geophysical Research*, 101D, 21403–21422.

Liang, X., Lettenmaier, D. P., Wood, E. F., and Burges, S. J. (1994). A simple hydrologically based model of land surface water and energy fluxes for general circulation models. *Journal of Geophysical Research*, 99D, 14415–14428.

Lin, Y.-S., Medlyn, B. E., Duursma, R. A., et al. (2015). Optimal stomatal behaviour around the world. *Nature Climate Change*, 5, 459–464.

Lindroth, A., Mölder, M., and Lagergren, F. (2010). Heat storage in forest biomass improves energy balance closure. *Biogeosciences*, 7, 301–313.

Liou, K. N. (2002). *An Introduction to Atmospheric Radiation*, 2nd edn. San Diego, CA: Academic Press.

Liston, G. E. (2004). Representing subgrid snow cover heterogeneities in regional and global models. *Journal of Climate*, 17, 1381–1397.

Liu, Y., and Gupta, H. V. (2007). Uncertainty in hydrologic modeling: Toward an integrated data assimilation framework. *Water Resources Research*, 43, W07401, doi:10.1029/2006WR005756.

Lloyd, C. R., Gash, J. H. C., and Shuttleworth, W. J. (1988). The measurement and modelling of rainfall interception by Amazonian rain forest. *Agricultural and Forest Meteorology*, 43, 277–294.

Lloyd, J. (1991). Modelling stomatal responses to environment in *Macadamia integrifolia*. *Australian Journal of Plant Physiology*, 18, 649–660.

Lloyd, J., and Farquhar, G. D. (2008). Effects of rising temperatures and [CO_2] on the physiology of tropical forest trees. *Philosophical Transactions of the Royal Society B*, 363, 1811–1817.

Lloyd, J., Patiño, S., Paiva, R. Q., et al. (2010). Optimisation of photosynthetic carbon gain and within-canopy gradients of associated foliar traits for Amazon forest trees. *Biogeosciences*, 7, 1833–1859.

Lloyd, J., and Taylor, J. A. (1994). On the temperature dependence of soil respiration. *Functional Ecology*, 8, 315–323.

Lombardozzi, D., Levis, S., Bonan, G., Hess, P. G., and Sparks, J. P. (2015b). The influence of chronic ozone exposure on global carbon and water cycles. *Journal of Climate*, 28, 292–305.

Lombardozzi, D., Sparks, J. P., and Bonan, G. (2013). Integrating O_3 influences on terrestrial processes: Photosynthetic and stomatal response data available for regional and global modeling. *Biogeosciences*, 10, 6815–6831.

Lombardozzi, D. L., Bonan, G. B., Smith, N. G., Dukes, J. S., and Fisher, R. A. (2015a). Temperature acclimation of photosynthesis and respiration: A key uncertainty in the carbon cycle-climate feedback. *Geophysical Research Letters*, 42, 8624–8631, doi:10.1002/2015GL065934.

Lombardozzi, D. L., Smith, N. G., Cheng, S. J., et al. (2018). Triose phosphate limitation in photosynthesis models reduces leaf photosynthesis and global terrestrial carbon storage. *Environmental Research Letters*, 13, 074025, doi:10.1088/1748-9326/aacf68.

Lombardozzi, D. L., Zeppel, M. J. B., Fisher, R. A., and Tawfik, A. (2017). Representing nighttime and minimum conductance in CLM4.5: Global hydrology and carbon sensitivity analysis using observational constraints. *Geoscientific Model Development*, 10, 321–331.

Long, S. P., Postl, W. F., and Bolhàr-Nordenkampf, H. R. (1993). Quantum yields for uptake of carbon dioxide in C_3 vascular plants of contrasting habitats and taxonomic groups. *Planta*, 189, 226–234.

Long, T. A. (2005). *The Forest and the Mainframe: The Dynamics of Modeling and Field Study in the Coniferous Forest Biome, 1969–1980*, M.S. Thesis. Corvallis, OR: Oregon State University.

Louis, J.-F. (1979). A parametric model of vertical eddy fluxes in the atmosphere. *Boundary Layer Meteorology*, 17, 187–202.

Louis, J. F., Tiedtke, M., and Geleyn, J. F. (1982). A short history of the operational PBL-parameterization at ECMWF. In *Workshop on Planetary Boundary Layer Parameterization 25–27 November 1981*. Reading, UK: European Centre for Medium Range Weather Forecasts, pp. 59–79.

Lu, S., Ren, T., Gong, Y., and Horton, R. (2007). An improved model for predicting soil thermal conductivity from water content at room temperature. *Soil Science Society of America Journal*, 71, 8–14.

Lunardini, V. J. (1981). *Heat Transfer in Cold Climates*. New York, NY: Van Nostrand Reinhold.

Luo, Y., Shi, Z., Lu, X., et al. (2017). Transient dynamics of terrestrial carbon storage: Mathematical foundation and its applications. *Biogeosciences*, 14, 145–161.

Luo, Y., White, L. W., Canadell, J. G., et al. (2003). Sustainability of terrestrial carbon sequestration: A case study in Duke Forest with inversion approach. *Global Biogeochemical Cycles*, 17, 1021, doi:10.1029/2002GB001923.

Luo, Y. Q., Randerson, J. T., Abramowitz, G., et al. (2012). A framework for benchmarking land models. *Biogeosciences*, 9, 3857–3874.

MacBean, N., Maignan, F., Peylin, P., et al. (2015). Using satellite data to improve the leaf phenology of a global terrestrial biosphere model. *Biogeosciences*, 12, 7185–7208.

Mahat, V., Tarboton, D. G., and Molotch, N. P. (2013). Testing above- and below-canopy representations of turbulent fluxes in an energy balance snowmelt model. *Water Resources Research*, 49, doi:10.1002/wrcr.20073.

Mahowald, N. M., Muhs, D. R., Levis, S., et al. (2006). Change in atmospheric mineral aerosols in response to climate: Last glacial period, preindustrial, modern, and doubled carbon dioxide climates. *Journal of Geophysical Research*, 111, D10202, doi:10.1029/2005JD006653.

Mahrt, L. (1987). Grid-averaged surface fluxes. *Monthly Weather Review*, 115, 1550–1560.

Majoube, M. (1971). Fractionnement en oxygène-18 et en deutérium entre l'eau et sa vapeur. *Journal de Chimie Physique*, 68, 1423–1436.

Makar, P. A., Fuentes, J. D., Wang, D., Staebler, R. M., and Wiebe, H. A. (1999). Chemical processing of biogenic hydrocarbons within and above a temperate deciduous forest. *Journal of Geophysical Research*, 104D, 3581–3603.

Makar, P. A., Staebler, R. M., Akingunola, A., et al. (2017). The effects of forest canopy shading and turbulence on boundary layer ozone. *Nature Communications*, 8, 15243, doi:10.1038/ncomms15243.

Manabe, S. (1969). Climate and the ocean circulation. I. The atmospheric circulation and the hydrology of the Earth's surface. *Monthly Weather Review*, 97, 739–774.

Manabe, S., Smagorinsky, J., and Strickler, R. F. (1965). Simulated climatology of a general circulation model with a hydrologic cycle. *Monthly Weather Review*, 93, 769–798.

Mantoglou, A., and Gelhar, L. W. (1987a). Stochastic modeling of large-scale transient unsaturated flow systems. *Water Resources Research*, 23, 37–46.

Mantoglou, A., and Gelhar, L. W. (1987b). Capillary tension head variance, mean soil moisture content, and effective specific soil moisture capacity of transient unsaturated flow in stratified soils. *Water Resources Research*, 23, 47–56.

Mantoglou, A., and Gelhar, L. W. (1987c). Effective hydraulic conductivities of transient unsaturated flow in stratified soils. *Water Resources Research*, 23, 57–67.

Manzoni, S., and Porporato, A. (2009). Soil carbon and nitrogen mineralization: Theory and models across scales. *Soil Biology and Biochemistry*, 41, 1355–1379.

Manzoni, S., Vico, G., Katul, G., et al. (2011). Optimizing stomatal conductance for maximum carbon gain under water stress: A meta-analysis across plant functional types and climates. *Functional Ecology*, 25, 456–467.

Manzoni, S., Vico, G., Katul, G., (2013). Hydraulic limits on maximum plant transpiration and the emergence of the safety–efficiency trade-off. *New Phytologist*, 198, 169–178.

Markesteijn, L., Poorter, L., Bongers, F., Paz, H., and Sack, L. (2011). Hydraulics and life history of tropical dry forest tree species: Coordination of species' drought and shade tolerance. *New Phytologist*, 191, 480–495.

Markkanen, T., Rannik, Ü., Marcolla, B., Cescatti, A., and Vesala, T. (2003). Footprints and fetches for fluxes over forest canopies with varying structure and density. *Boundary-Layer Meteorology*, 106, 437–459.

Massman, W. J. (1982). Foliage distribution in old-growth coniferous tree canopies. *Canadian Journal of Forest Research*, 12, 10–17.

Massman, W. J. (1997). An analytical one-dimensional model of momentum transfer by vegetation of arbitrary structure. *Boundary-Layer Meteorology*, 83, 407–421.

Massman, W. J. (1998). A review of the molecular diffusivities of H_2O, CO_2, CH_4, CO, O_3, SO_2, NH_3, N_2O, NO, and NO_2 in air, O_2 and N_2 near STP. *Atmospheric Environment*, 32, 1111–1127.

Massman, W. J. (1999). A model study of kB_H^{-1} for vegetated surfaces using 'localized near-field' Lagrangian theory. *Journal of Hydrology*, 223, 27–43.

Massman, W. J., and Weil, J. C. (1999). An analytical one-dimensional second-order closure model of turbulence statistics and the Lagrangian time scale within and above plant canopies of arbitrary structure. *Boundary-Layer Meteorology*, 91, 81–107.

Matheny, A. M., Fiorella, R. P., Bohrer, G., et al. (2017). Contrasting strategies of hydraulic control in two co-dominant temperate tree species. *Ecohydrology*, 10, e1815, doi:10.1002/eco.1815.

Maurer, E. P., Wood, A. W., Adam, J. C., Lettenmaier, D. P., and Nijssen, B. (2002). A long-term hydrologically based dataset of land surface fluxes and states for the conterminous United States. *Journal of Climate*, 15, 3237–3251.

McCaughey, J. H. (1985). Energy balance storage terms in a mature mixed forest at Petawawa, Ontario – a case study. *Boundary-Layer Meteorology*, 31, 89–101.

McCaughey, J. H., and Saxton, W. L. (1988). Energy balance storage terms in a mixed forest. *Agricultural and Forest Meteorology*, 44, 1–18.

McGuire, A. D., Melillo, J. M., Joyce, L. A., et al. (1992). Interactions between carbon and nitrogen dynamics in estimating net primary productivity for potential vegetation in North America. *Global Biogeochemical Cycles*, 6, 101–124.

McNaughton, K. G., and van den Hurk, B. J. J. M. (1995). A 'Lagrangian' revision of the resistors in the two-layer model for calculating the energy budget of a plant canopy. *Boundary-Layer Meteorology*, 74, 261–288.

Meador, W. E., and Weaver, W. R. (1980). Two-stream approximations to radiative transfer in planetary atmospheres: A unified description of existing methods and a new improvement. *Journal of the Atmospheric Sciences*, 37, 630–643.

Medlyn, B. E. (2004). A MAESTRO retrospective. In *Forests at the Land–Atmosphere Interface*, ed. M. Mencuccini, J. Grace, J. Moncrieff, and K. G. McNaughton. Wallingford: CAB International, pp. 105–121.

Medlyn, B. E., Barton, C. V. M., Broadmeadow, M. S. J., et al. (2001). Stomatal conductance of forest species after long-term exposure to elevated CO_2 concentration: A synthesis. *New Phytologist*, 149, 247–264.

Medlyn, B. E., Dreyer, E., Ellsworth, D., et al. (2002). Temperature response of parameters of a biochemically based model of photosynthesis. II. A review of experimental data. *Plant, Cell and Environment*, 25, 1167–1179.

Medlyn, B. E., Duursma, R. A., De Kauwe, M. G., and Prentice, I. C. (2013). The optimal stomatal response to atmospheric CO_2 concentration: Alternative solutions, alternative interpretations. *Agricultural and Forest Meteorology*, 182/183, 200–203.

Medlyn, B. E., Duursma, R. A., Eamus, D., et al. (2011). Reconciling the optimal and empirical approaches to modelling stomatal conductance. *Global Change Biology*, 17, 2134–2144.

Medlyn, B. E., McMurtrie, R. E., Dewar, R. C., and Jeffreys, M. P. (2000). Soil processes dominate the long-term response of forest net primary productivity to increased temperature and atmospheric CO_2 concentration. *Canadian Journal of Forest Research*, 30, 873–888.

Medvigy, D., Wofsy, S. C., Munger, J. W., Hollinger, D. Y., and Moorcroft, P. R. (2009). Mechanistic scaling of ecosystem function and dynamics in space and time: Ecosystem Demography model version 2. *Journal of Geophysical Research*, 114, G01002, doi:10.1029/2008JG000812.

Meentemeyer, V. (1978). Macroclimate and lignin control of litter decomposition rates. *Ecology*, 59, 465–472.

Meinzer, F. C., Goldstein, G., Jackson, P., et al. (1995). Environmental and physiological regulation of transpiration in tropical forest gap species: The influence of boundary layer and hydraulic properties. *Oecologia*, 101, 514–522.

Meir, P., Kruijt, B., Broadmeadow, M., et al. (2002). Acclimation of photosynthetic capacity to irradiance in tree canopies in relation to leaf nitrogen concentration and leaf mass per unit area. *Plant, Cell and Environment*, 25, 343–357.

Melillo, J. M., McGuire, A. D., Kicklighter, D. W., et al. (1993). Global climate change and terrestrial net primary production. *Nature*, 363, 234–240.

Melton, J. R., and Arora, V. K. (2014). Sub-grid scale representation of vegetation in global land surface schemes: Implications for estimation of the terrestrial carbon sink. *Biogeosciences*, 11, 1021–1036.

Melton, J. R., Wania, R., Hodson, E. L., et al. (2013). Present state of global wetland extent and wetland methane modelling: Conclusions from a model inter-comparison project (WETCHIMP). *Biogeosciences*, 10, 753–788.

Merlivat, L. (1978). Molecular diffusivities of $H_2{}^{16}O$, $HD^{16}O$, and $H_2{}^{18}O$ in gases. *Journal of Chemical Physics*, 69, 2864–2871.

Meyerholt, J., and Zaehle, S. (2015). The role of stoichiometric flexibility in modelling forest ecosystem responses to nitrogen fertilization. *New Phytologist*, 208, 1042–1055.

Meyerholt, J., Zaehle, S., and Smith, M. J. (2016). Variability of projected terrestrial biosphere responses to elevated levels of atmospheric CO_2 due to uncertainty in biological nitrogen fixation. *Biogeosciences*, 13, 1491–1518.

Meyers, T., and Paw U, K. T. (1986). Testing of a higher-order closure model for modeling airflow within and above plant canopies. *Boundary-Layer Meteorology*, 37, 297–311.

Meyers, T. P., Finkelstein, P., Clarke, J., Ellestad, T. G., and Sims, P. F. (1998). A multilayer model for inferring dry deposition using standard meteorological measurements. *Journal of Geophysical Research*, 103D, 22645–22661.

Meyers, T. P., and Hollinger, S. E. (2004). An assessment of storage terms in the surface energy balance of maize and soybean. *Agricultural and Forest Meteorology*, 125, 105–115.

Michaletz, S. T., Weiser, M. D., McDowell, N. G., et al. (2016). The energetic and carbon economic origins of leaf thermoregulation. *Nature Plants*, 2, 16129, doi:10.1038/nplants.2016.129.

Michaletz, S. T., Weiser, M. D., Zhou, J., et al. (2015). Plant thermoregulation: Energetics, trait–environment interactions, and carbon economics. *Trends in Ecology and Evolution*, 30, 714–724.

Michiles, A. A. S., and Gielow, R. (2008). Above-ground thermal energy storage rates, trunk heat fluxes and surface energy balance in a central Amazonian rainforest. *Agricultural and Forest Meteorology*, 148, 917–930.

Miller, S. D., Goulden, M. L., Hutyra, L. R., et al. (2011). Reduced impact logging minimally alters tropical rainforest carbon and energy exchange. *Proceedings of the National Academy of Sciences USA*, 108, 19431–19435.

Millington, R. J. (1959). Gas diffusion in porous media. *Science*, 130, 100–102.

Milly, P. C. D. (1985). A mass-conservative procedure for time-stepping in models of unsaturated flow. *Advances in Water Resources*, 8, 32–36.

Milly, P. C. D. (1988). Advances in modeling of water in the unsaturated zone. *Transport in Porous Media*, 3, 491–514.

Milly, P. C. D., Malyshev, S. L., Shevliakova, E., et al. (2014). An enhanced model of land water and energy for global hydrologic and Earth-system studies. *Journal of Hydrometeorology*, 15, 1739–1761.

Milly, P. C. D., and Shmakin, A. B. (2002). Global modeling of land water and energy balances. Part I: The land dynamics (LaD) model. *Journal of Hydrometeorology*, 3, 283–299.

Miner, G. L., Bauerle, W. L., and Baldocchi, D. D. (2017). Estimating the sensitivity of stomatal conductance to photosynthesis: A review. *Plant, Cell and Environment*, 40, 1214–1238.

Mintz, Y., and Walker, G. K. (1993). Global fields of soil moisture and land surface evapotranspiration derived from observed precipitation and surface air temperature. *Journal of Applied Meteorology*, 32, 1305–1334.

Miralles, D. G., Gash, J. H., Holmes, T. R. H., de Jeu, R. A. M., and Dolman, A. J. (2010). Global canopy interception from satellite observations. *Journal of Geophysical Research*, 115, D16122, doi:10.1029/2009JD013530.

Mirfenderesgi, G., Bohrer, G., Matheny, A. M., et al. (2016). Tree level hydrodynamic approach for resolving aboveground water storage and stomatal conductance and modeling the effects of tree hydraulic strategy. *Journal of Geophysical Research: Biogeosciences*, 121, 1792–1813, doi:10.1002/2016JG003467.

Mölder, M., Grelle, A., Lindroth, A., and Halldin, S. (1999). Flux-profile relationships over a boreal forest – roughness sublayer corrections. *Agricultural and Forest Meteorology*, 98/99, 645–658.

Monsi, M., and Saeki, T. (1953). Über den Lichtfaktor in den Pflanzengesellschaften und seine Bedeutung für die Stoffproduktion. *Japanese Journal of Botany*, 14, 22–52.

Monsi, M., and Saeki, T. (2005). On the factor light in plant communities and its importance for matter production. *Annals of Botany*, 95, 549–567.

Monteith, J. L. (1965). Evaporation and environment. In *The State and Movement of Water in Living Organisms (19th Symposia of the Society for Experimental Biology)*, ed. G. E. Fogg. New York: Academic Press, pp. 205–234.

Monteith, J. L. (1981a). Evaporation and surface temperature. *Quarterly Journal of the Royal Meteorological Society*, 107, 1–27.

Monteith, J. L. (1981b). Coupling of plants to the atmosphere. In *Plants and their Atmospheric Environment*, ed. J. Grace, E. D. Ford, and P. G. Jarvis. Oxford: Blackwell, pp. 1–29.

Monteith, J. L., and Unsworth, M. H. (2013). *Principles of Environmental Physics*, 4th edn. Amsterdam: Elsevier.

Moorcroft, P. R., Hurtt, G. C., and Pacala, S. W. (2001). A method for scaling vegetation dynamics: The ecosystem demography model (ED). *Ecological Monographs*, 71, 557–586.

Moore, C. J., and Fisch, G. (1986). Estimating heat storage in Amazonian tropical forest. *Agricultural and Forest Meteorology*, 38, 147–168.

Moore, R. J. (2007). The PDM rainfall-runoff model. *Hydrology and Earth System Sciences*, 11, 483–499.

Morfopoulos, C., Prentice, I. C., Keenan, T. F., et al. (2013). A unifying conceptual model for the environmental responses of isoprene emissions from plants. *Annals of Botany*, 112, 1223–1238.

Morfopoulos, C., Sperlich, D., Peñuelas, J., et al. (2014). A model of plant isoprene emission based on available reducing power captures responses to atmospheric CO_2. *New Phytologist*, 203, 125–139.

Morison, J. I. L. (1987). Intercellular CO_2 concentration and stomatal response to CO_2. In *Stomatal Function*, ed. E. Zeiger, G. D. Farquhar, and I. R. Cowan. Stanford, CA: Stanford University Press, pp. 229–251.

Morison, J. I. L., and Jarvis, P. G. (1983). Direct and indirect effects of light on stomata. II. In *Commelina communis* L. *Plant, Cell and Environment*, 6, 103–109.

Muzylo, A., Llorens, P., Valente, F., et al. (2009). A review of rainfall interception modelling. *Journal of Hydrology*, 370, 191–206.

Nemitz, E., Milford, C., and Sutton, M. A. (2001). A two-layer canopy compensation point model for describing bi-directional biosphere–atmosphere exchange of ammonia. *Quarterly Journal of the Royal Meteorological Society*, 127, 815–833.

Niinemets, Ü. (1998). Adjustment of foliage structure and function to a canopy light gradient in two co-existing deciduous trees. Variability in leaf inclination angles in relation to petiole morphology. *Trees*, 12, 446–451.

Niinemets, Ü. (1999). Components of leaf dry mass per area – thickness and density – alter leaf photosynthetic capacity in reverse directions in woody plants. *New Phytologist*, 144, 35–47.

Niinemets, Ü. (2007). Photosynthesis and resource distribution through plant canopies. *Plant, Cell and Environment*, 30, 1052–1071.

Niinemets, Ü. (2012). Optimization of foliage photosynthetic capacity in tree canopies: Towards identifying missing constraints. *Tree Physiology*, 32, 505–509.

Niinemets, Ü., and Anten, N. P. R. (2009). Packing the photosynthetic machinery: From leaf to canopy. In *Photosynthesis in silico: Understanding Complexity from Molecules to Ecosystems*, ed. A. Laisk, L. Nedbal, and Govindjee. Dordrecht: Springer, pp. 363–399.

Niinemets, Ü., Bilger, W., Kull, O., and Tenhunen, J. D. (1999a). Responses of foliar photosynthetic electron transport, pigment stoichiometry, and stomatal conductance to interacting environmental factors in a mixed species forest canopy. *Tree Physiology*, 19, 839–852.

Niinemets, Ü., Diaz-Espejo, A., Flexas, J., Galmés, J., and Warren, C. R. (2009). Importance of mesophyll diffusion conductance in estimation of plant photosynthesis in the field. *Journal of Experimental Botany*, 60, 2271–2282.

Niinemets, Ü., Keenan, T. F., and Hallik, L. (2015). A worldwide analysis of within-canopy variations in leaf structural, chemical and physiological traits across plant functional types. *New Phytologist*, 205, 973–993.

Niinemets, Ü., and Kull, O. (1995). Effects of light availability and tree size on the architecture of assimilative surface in the canopy of *Picea abies*: Variation in needle morphology. *Tree Physiology*, 15, 307–315.

Niinemets, Ü., Kull, O., and Tenhunen, J. D. (1998). An analysis of light effects on foliar morphology, physiology, and light interception in temperate deciduous woody species of contrasting shade tolerance. *Tree Physiology*, 18, 681–696.

Niinemets, Ü., Kull, O., and Tenhunen, J. D. (2004). Within-canopy variation in the rate of development of photosynthetic capacity is proportional to integrated quantum flux density in temperate deciduous trees. *Plant, Cell and Environment*, 27, 293–313.

Niinemets, Ü., and Tenhunen, J. D. (1997). A model separating leaf structural and physiological effects on carbon gain along light gradients for the shade-tolerant species *Acer saccharum*. *Plant, Cell and Environment*, 20, 845–866.

Niinemets, Ü., Tenhunen, J. D., Harley, P. C., and Steinbrecher, R. (1999b). A model of isoprene emission based on energetic requirements for isoprene synthesis and leaf photosynthetic properties for *Liquidambar* and *Quercus*. *Plant, Cell and Environment*, 22, 1319–1335.

Nijssen, B., O'Donnell, G. M., Lettenmaier, D. P., Lohmann, D., and Wood, E. F. (2001). Predicting the discharge of global rivers. *Journal of Climate*, 14, 3307–3323.

Nilson, T. (1971). A theoretical analysis of the frequency of gaps in plant stands. *Agricultural Meteorology*, 8, 25–38.

Niu, G.-Y., and Yang, Z.-L. (2004). Effects of vegetation canopy processes on snow surface energy and mass balances. *Journal of Geophysical Research*, 109, D23111, doi:10.1029/2004JD004884.

Niu, G.-Y., and Yang, Z.-L. (2006). Effects of frozen soil on snowmelt runoff and soil water storage at a continental scale. *Journal of Hydrometeorology*, 7, 937–952.

Niu, G.-Y., and Yang, Z.-L. (2007). An observation-based formulation of snow cover fraction and its evaluation over large North American river basins. *Journal of Geophysical Research*, 112, D21101, doi:10.1029/2007JD008674.

Niu, G.-Y., Yang, Z.-L., Dickinson, R. E., and Gulden, L. E. (2005). A simple TOPMODEL-based runoff parameterization (SIMTOP) for use in global climate models. *Journal of Geophysical Research*, 10, D21106, doi:10.1029/2005JD006111.

Niu, G.-Y., Yang, Z.-L., Mitchell, K. E., et al. (2011). The community Noah land surface model with multiparameterization options (Noah-MP): 1. Model description and evaluation with local-scale measurements. *Journal of Geophysical Research*, 116, D12109, doi:10.1029/2010JD015139.

Noilhan, J., and Planton, S. (1989). A simple parameterization of land surface processes for meteorological models. *Monthly Weather Review*, 117, 536–549.

Norman, J. M. (1979). Modeling the complete crop canopy. In *Modification of the Aerial Environment of Plants*, ed. B. J. Barfield and J. F. Gerber. St. Joseph, MI: American Society of Agricultural Engineers, pp. 249–277.

Norman, J. M. (1982). Simulation of microclimates. In *Biometeorology in Integrated Pest Management*, ed. J. L. Hatfield and I. J. Thomason. New York, NY: Academic Press, pp. 65–99.

Norman, J. M. (1989). Synthesis of canopy processes. In *Plant Canopies: Their Growth, Form and Function*, ed. G. Russell, B. Marshall, and P. G. Jarvis. Cambridge: Cambridge University Press, pp. 161–175.

Norman, J. M. (1993). Scaling processes between leaf and canopy levels. In *Scaling Physiological Processes: Leaf to Globe*, ed. J. R. Ehleringer and C. B. Field. New York, NY: Academic Press, pp. 41–76.

Norman, J. M., and Campbell, G. (1983). Application of a plant–environment model to problems in irrigation. In *Advances in Irrigation*, vol. 2, ed. D. Hillel. New York, NY: Academic Press, pp. 155–188.

Norman, J. M., Kustas, W. P., and Humes, K. S. (1995). Source approach for estimating soil and vegetation energy fluxes in observations of directional radiometric surface temperature. *Agricultural and Forest Meteorology*, 77, 263–293.

O'Connell, P. E. (1991). A historical perspective. In *Recent Advances in the Modeling of Hydrologic Systems*, ed. D. S. Bowles and P. E. O'Connell. Dordrecht: Kluwer, pp. 3–30.

Odum, H. T. (1957). Trophic structure and productivity of Silver Springs, Florida. *Ecological Monographs*, 27, 55–112.

Odum, H. T. (1960). Ecological potential and analogue circuits for the ecosystem. *American Scientist*, 48, 1–8.

Ogée, J., Brunet, Y., Loustau, D., Berbigier, P., and Delzon, S. (2003). MuSICA, a CO_2, water and energy multilayer, multileaf pine forest model: Evaluation from hourly to yearly time scales and sensitivity analysis. *Global Change Biology*, 9, 697–717.

Oleson, K. W., Bonan, G. B., Feddema, J. J., Vertenstein, M., and Kluzek, E. (2010a). *Technical Description of an Urban Parameterization for the Community Land Model (CLMU)*, Technical Note NCAR/TN-480+STR. Boulder, CO: National Center for Atmospheric Research.

Oleson, K. W., Dai, Y., Bonan, G., et al. (2004). *Technical Description of the Community Land Model (CLM)*, Technical Note NCAR/TN-461+STR. Boulder, CO: National Center for Atmospheric Research.

Oleson, K. W., Lawrence, D. M., Bonan, G. B., et al. (2010b). *Technical Description of Version 4.0 of the Community Land Model (CLM)*, Technical Note NCAR/TN-478+STR. Boulder, CO: National Center for Atmospheric Research.

Oleson, K. W., Lawrence, D. M., Bonan, G. B., et al. (2013). *Technical Description of Version 4.5 of the Community Land Model (CLM)*, Technical Note NCAR/TN-503+STR. Boulder, CO: National Center for Atmospheric Research.

Oliphant, A. J., Grimmond, C. S. B., Zutter, H. N., et al. (2004). Heat storage and energy balance fluxes for a temperate deciduous forest. *Agricultural and Forest Meteorology*, 126, 185–201.

Olson, J. S. (1963). Energy storage and the balance of producers and decomposers in ecological systems. *Ecology*, 44, 322–331.

Olson, J. S. (1965). Equations for cesium transfer in a *Liriodendron* forest. *Health Physics*, 11, 1385–1392.

Oren, R., Sperry, J. S., Katul, G. G., et al. (1999). Survey and synthesis of intra- and interspecific variation in stomatal sensitivity to vapour pressure deficit. *Plant, Cell and Environment*, 22, 1515–1526.

Pacala, S. W., Canham, C. D., Saponara, J., et al. (1996). Forest models defined by field measurements: Estimation, error analysis and dynamics. *Ecological Monographs*, 66, 1–43.

Pacifico, F., Harrison, S. P., Jones, C. D., et al. (2011). Evaluation of a photosynthesis-based biogenic isoprene emission scheme in JULES and simulation of isoprene emissions under present-day climate conditions. *Atmospheric Chemistry and Physics*, 11, 4371–4389.

Paniconi, C., Aldama, A. A., and Wood, E. F. (1991). Numerical evaluation of iterative and noniterative methods for the solution of the nonlinear Richards equation. *Water Resources Research*, 27, 1147–1163.

Paniconi, C., and Putti, M. (1994). A comparison of Picard and Newton iteration in the numerical solution of multidimensional variably saturated flow problems. *Water Resources Research*, 30, 3357–3374.

Panofsky, H. A. (1963). Determination of stress from wind and temperature measurements. *Quarterly Journal of the Royal Meteorological Society*, 89, 85–94.

Parkhurst, D. F., and Loucks, O. L. (1972). Optimal leaf size in relation to environment. *Journal of Ecology*, 60, 505–537.

Parton, W. J., Hartman, M., Ojima, D., and Schimel, D. (1998). DAYCENT and its land surface submodel: Description and testing. *Global and Planetary Change*, 19, 35–48.

Parton, W. J., Holland, E. A., Del Grosso, S. J., et al. (2001). Generalized model for NO_x and N_2O emissions from soils. *Journal of Geophysical Research*, 106D, 17403–17419.

Parton, W. J., Mosier, A. R., Ojima, D. S., et al. (1996). Generalized model for N_2 and N_2O production from nitrification and denitrification. *Global Biogeochemical Cycles*, 10, 401–412.

Parton, W. J., Ojima, D. S., Cole, C. V., and Schimel, D. S. (1994). A general model for soil organic matter dynamics: Sensitivity to litter chemistry, texture and management. In *Quantitative Modeling of Soil Forming Processes*, ed. R. B. Bryant and R. W. Arnold. Madison, WI: Soil Science Society of America, pp. 147–167.

Parton, W. J., Schimel, D. S., Cole, C. V., and Ojima, D. S. (1987). Analysis of factors controlling soil organic matter levels in Great Plains grasslands. *Soil Science Society of America Journal*, 51, 1173–1179.

Parton, W. J., Scurlock, J. M. O., Ojima, D. S., et al. (1993). Observations and modeling of biomass and soil organic matter dynamics for the grassland biome worldwide. *Global Biogeochemical Cycles*, 7, 785–809.

Parton, W., Silver, W. L., Burke, I. C., et al. (2007). Global-scale similarities in nitrogen release patterns during long-term decomposition. *Science*, 315, 361–364.

Parton, W. J., Stewart, J. W. B., and Cole, C. V. (1988). Dynamics of C, N, P and S in grassland soils: A model. *Biogeochemistry*, 5, 109–131.

Pastor, J., and Post, W. M. (1985). *Development of a Linked Forest Productivity–Soil Process Model*, ORNL/TM-9519. Oak Ridge, TN: Oak Ridge National Laboratory.

Pastor, J., and Post, W. M. (1986). Influence of climate, soil moisture, and succession on forest carbon and nitrogen cycles. *Biogeochemistry*, 2, 3–27.

Pastor, J., and Post, W. M. (1988). Response of northern forests to CO_2-induced climate change. *Nature*, 334, 55–58.

Pastor, J., and Post, W. M. (1996). LINKAGES – An individual-based forest ecosystem model. *Climatic Change*, 34, 253–261.

Patten, B. C. (1975). *Systems Analysis and Simulation in Ecology*, vol. III. New York, NY: Academic Press.

Patton, E. G., Horst, T. W., Sullivan, P. P., et al. (2011). The Canopy Horizontal Array Turbulence Study. *Bulletin of the American Meteorological Society*, 92, 593–611.

Patton, E. G., Sullivan, P. P., Shaw, R. H., Finnigan, J. J., and Weil, J. C. (2016). Atmospheric stability influences on coupled boundary layer and canopy turbulence. *Journal of the Atmospheric Sciences*, 73, 1621–1647.

Paulson, C. A. (1970). The mathematical representation of wind speed and temperature profiles in the unstable atmospheric surface layer. *Journal of Applied Meteorology*, 9, 857–861.

Paustian, K., Parton, W. J., and Persson, J. (1992). Modeling soil organic matter in organic-amended and nitrogen-fertilized long-term plots. *Soil Science Society of America Journal*, 56, 476–488.

Paw U, K. T. (1987). Mathematical analysis of the operative temperature and energy budget. *Journal of Thermal Biology*, 12, 227–233.

Peltoniemi, M. S., Duursma, R. A., and Medlyn, B. E. (2012). Co-optimal distribution of leaf nitrogen and hydraulic conductance in plant canopies. *Tree Physiology*, 32, 510–519.

Penman, H. L. (1940). Gas and vapour movements in the soil: I. The diffusion of vapours through porous solids. *Journal of Agricultural Science*, 30, 437–462.

Penman, H. L. (1948). Natural evaporation from open water, bare soil and grass. *Proceedings of the Royal Society London A*, 193, 120–145.

Pereira, F. L., Gash, J. H. C., David, J. S., and Valente, F. (2009). Evaporation of intercepted rainfall from isolated evergreen oak trees: Do the crowns behave as wet bulbs? *Agricultural and Forest Meteorology*, 149, 667–679.

Peters-Lidard, C. D., Blackburn, E., Liang, X., and Wood, E. F. (1998). The effect of soil thermal conductivity parameterization on surface energy fluxes and temperatures. *Journal of the Atmospheric Sciences*, 55, 1209–1224.

Philip, J. R. (1957a). Evaporation, and moisture and heat fields in the soil. *Journal of Meteorology*, 14, 354–366.

Philip, J. R. (1957b). The theory of infiltration: 4. Sorptivity and algebraic infiltration equations. *Soil Science*, 84, 257–264.

Physick, W. L., and Garratt, J. R. (1995). Incorporation of a high-roughness lower boundary into a mesoscale model for studies of dry deposition over complex terrain. *Boundary-Layer Meteorology*, 74, 55–71.

Pisek, J., Sonnentag, O., Richardson, A. D., and Mõttus, M. (2013). Is the spherical leaf inclination angle distribution a valid assumption for temperate and boreal broadleaf tree species? *Agricultural and Forest Meteorology*, 169, 186–194.

Pitman, A. J., Henderson-Sellers, A., and Yang, Z.-L. (1990). Sensitivity of regional climates to localized precipitation in global models. *Nature*, 346, 734–737.

Pleim, J., and Ran, L. (2011). Surface flux modeling for air quality applications. *Atmosphere*, 2, 271–302.

Pleim, J. E., Xiu, A., Finkelstein, P. L., and Otte, T. L. (2001). A coupled land-surface and dry deposition model and comparison to field measurements of surface heat, moisture, and ozone fluxes. *Water, Air, and Soil Pollution: Focus*, 1, 243–252.

Polcher, J., McAvaney, B., Viterbo, P., et al. (1998). A proposal for a general interface between land surface schemes and general circulation models. *Global and Planetary Change*, 19, 261–276.

Pollard, D., and Thompson, S. L. (1995). Use of a land-surface-transfer scheme (LSX) in a global climate model: The response to doubling stomatal resistance. *Global and Planetary Change*, 10, 129–161.

Poorter, H., Niinemets, Ü, Poorter, L., Wright, I. J., and Villar, R. (2009). Causes and consequences of variation in leaf mass per area (LMA): A meta-analysis. *New Phytologist*, 182, 565–588.

Potter, C. S., Randerson, J. T., Field, C. B., et al. (1993). Terrestrial ecosystem production: A process model based on global satellite and surface data. *Global Biogeochemical Cycles*, 7, 811–841.

Prentice, I. C., Bondeau, A., Cramer, W., et al. (2007). Dynamic Global Vegetation Modeling: Quantifying terrestrial ecosystem responses to large-scale environmental change. In *Terrestrial Ecosystems in a Changing World*, ed. J. G. Canadell, D. E. Pataki, and L. F. Pitelka. Berlin: Springer, pp. 175–192.

Prentice, I. C., Dong, N., Gleason, S. M., Maire, V., and Wright, I. J. (2014). Balancing the costs of carbon gain and water transport: Testing a new theoretical framework for plant functional ecology. *Ecology Letters*, 17, 82–91.

Prentice, I. C., Liang, X., Medlyn, B. E., and Wang, Y.-P. (2015). Reliable, robust and realistic: The three R's of next-generation land-surface modelling. *Atmospheric Chemistry and Physics*, 15, 5987–6005.

Prince, S. D., and Goward, S. N. (1995). Global primary production: A remote sensing approach. *Journal of Biogeography*, 22, 815–835.

Pugh, T. A. M., Arneth, A., Olin, S., et al. (2015). Simulated carbon emissions from land-use change are substantially enhanced by accounting for agricultural management. *Environmental Research Letters*, 10, 124008, doi:10.1088/1748-9326/10/12/124008.

Purves, D. W., Lichstein, J. W., Strigul, N., and Pacala, S. W. (2008). Predicting and understanding forest dynamics using a simple tractable model. *Proceedings of the National Academy of Sciences USA*, 105, 17018–17022.

Pyles, R. D., Weare, B. C., and Paw U, K. T. (2000). The UCD Advanced Canopy–Atmosphere–Soil Algorithm:

Comparisons with observations from different climate and vegetation regimes. *Quarterly Journal of the Royal Meteorological Society*, 126, 2951–2980.

Qualls, R. J., and Brutsaert, W. (1996). Effect of vegetation density on the parameterization of scalar roughness to estimate spatially distributed sensible heat fluxes. *Water Resources Research*, 32, 645–652.

Raats, P. A. C. (2007). Uptake of water from soils by plant roots. *Transport in Porous Media*, 68, 5–28.

Raczka, B., Duarte, H. F., Koven, C. D., et al. (2016). An observational constraint on stomatal function in forests: Evaluating coupled carbon and water vapor exchange with carbon isotopes in the Community Land Model (CLM4.5). *Biogeosciences*, 13, 5183–5204.

Raddatz, T. J., Reick, C. H., Knorr, W., et al. (2007). Will the tropical land biosphere dominate the climate–carbon cycle feedback during the twenty-first century? *Climate Dynamics*, 29, 565–574.

Rafique, R., Xia, J., Hararuk, O., et al. (2016). Divergent predictions of carbon storage between two global land models: Attribution of the causes through traceability analysis. *Earth System Dynamics*, 7, 649–658.

Rafique, R., Xia, J., Hararuk, O., et al. (2017). Comparing the performance of three land models in global C cycle simulations: A detailed structural analysis. *Land Degradation and Development*, 28, 524–533.

Raich, J. W., Rastetter, E. B., Melillo, J. M., et al. (1991). Potential net primary productivity in South America: Application of a global model. *Ecological Applications*, 1, 399–429.

Rajaram, H., Bahr, J. M., Blöschl, G., et al. (2015). A reflection on the first 50 years of Water Resources Research. *Water Resources Research*, 51, 7829–7837, doi:10.1002/2015WR018089.

Ran, L., Pleim, J., Song, C., et al. (2017). A photosynthesis-based two-leaf canopy stomatal conductance model for meteorology and air quality modeling with WRF/CMAQ PX LSM. *Journal of Geophysical Research: Atmospheres*, 122, 1930–1952, doi:10.1002/2016JD025583.

Randerson, J. T., Hoffman, F. M., Thornton, P. E., et al. (2009). Systematic assessment of terrestrial biogeochemistry in coupled climate–carbon models. *Global Change Biology*, 15, 2462–2484.

Randerson, J. T., Thompson, M. V., Malmstrom, C. M., Field, C. B., and Fung, I. Y. (1996). Substrate limitations for heterotrophs: Implications for models that estimate the seasonal cycle of atmospheric CO_2. *Global Biogeochemical Cycles*, 10, 585–602.

Raschke, K. (1960). Heat transfer between the plant and the environment. *Annual Review of Plant Physiology*, 11, 111–126.

Raupach, M. R. (1979). Anomalies in flux-gradient relationships over forest. *Boundary Layer Meteorology*, 16, 467–486.

Raupach, M. R. (1988). Canopy transport processes. In *Flow and Transport in the Natural Environment: Advances and Applications*, ed. W. L. Steffen and O. T. Denmead. Berlin: Springer-Verlag, pp. 95–127.

Raupach, M. R. (1989a). Applying Lagrangian fluid mechanics to infer scalar source distributions from concentration profiles in plant canopies. *Agricultural and Forest Meteorology*, 47, 85–108.

Raupach, M. R. (1989b). A practical Lagrangian method for relating scalar concentrations to source distributions in vegetation canopies. *Quarterly Journal of the Royal Meteorological Society*, 115, 609–632.

Raupach, M. R. (1989c). Stand overstorey processes. *Philosophical Transactions of the Royal Society London B*, 324, 175–190.

Raupach, M. R. (1994). Simplified expressions for vegetation roughness length and zero-plane displacement as functions of canopy height and area index. *Boundary-Layer Meteorology*, 71, 211–216.

Raupach, M. R. (1995). Corrigenda. *Boundary-Layer Meteorology*, 76, 303–304.

Raupach, M. R., Finkele, K., and Zhang, L. (1997). *SCAM (Soil-Canopy-Atmosphere Model): Description and Comparisons with Field Data*, Technical Report No. 132. Canberra: CSIRO Centre for Environmental Mechanics.

Raupach, M. R., and Finnigan, J. J. (1988). 'Single-layer models of evaporation from plant canopies are incorrect but useful, whereas multilayer models are correct but useless': Discuss. *Australian Journal of Plant Physiology*, 15, 705–716.

Raupach, M. R., Finnigan, J. J., and Brunet, Y. (1996). Coherent eddies and turbulence in vegetation canopies: The mixing-layer analogy. *Boundary-Layer Meteorology*, 78, 351–382.

Rawls, W. J., Ahuja, L. R., Brakensiek, D. L., and Shirmohammadi, A. (1993). Infiltration and soil water movement. In *Handbook of Hydrology*, ed. D. R. Maidment. New York, NY: McGraw-Hill, pp. 5.1–5.51.

Rawls, W. J., and Brakensiek, D. L. (1985). Prediction of soil water properties for hydrologic modeling. In *Watershed Management in the Eighties*, ed. E. B. Jones and T. J. Ward. New York, NY: American Society of Civil Engineers, pp. 293–299.

Rawls, W. J., Brakensiek, D. L., and Saxton, K. E. (1982). Estimation of soil water properties. *Transactions of the ASAE*, 25, 1316–1320.

Reich, P. B., Ellsworth, D. S., and Walters, M. B. (1998). Leaf structure (specific leaf area) modulates photosynthesis–nitrogen relations: Evidence from within and across species and functional groups. *Functional Ecology*, 12, 948–958.

Reick, C. H., Raddatz, T., Brovkin, V., and Gayler, V. (2013). Representation of natural and anthropogenic land cover change in MPI-ESM. *Journal of Advances in Modeling Earth Systems*, 5, 459–482, doi:10.1002/jame.20022.

Richards, L. A. (1931). Capillary conduction of liquids through porous mediums. *Physics*, 1, 318–333.

Richardson, A. D., Anderson, R. S., Arain, M. A., et al. (2012). Terrestrial biosphere models need better representation of vegetation phenology: Results from the North American Carbon Program Site Synthesis. *Global Change Biology*, 18, 566–584.

Richtmyer, R. D., and Morton, K. W. (1967). *Difference Methods for Initial-Value Problems*, 2nd edn. New York, NY: Wiley.

Riddick, S., Ward, D., Hess, P., et al. (2016). Estimate of changes in agricultural terrestrial nitrogen pathways and ammonia emissions from 1850 to present in the Community Earth System Model. *Biogeosciences*, 13, 3397–3426.

Riley, W. J., Still, C. J., Torn, M. S., and Berry, J. A. (2002). A mechanistic model of $H_2^{18}O$ and $C^{18}OO$ fluxes between ecosystems and the atmosphere: Model description and sensitivity analyses. *Global Biogeochemical Cycles*, 16, 1095, doi:10.1029/2002GB001878.

Riley, W. J., Subin, Z. M., Lawrence, D. M., et al. (2011). Barriers to predicting changes in global terrestrial methane fluxes: Analyses using CLM4Me, a methane biogeochemistry model integrated in CESM. *Biogeosciences*, 8, 1925–1953.

Rogers, A. (2014). The use and misuse of $V_{c,max}$ in Earth System Models. *Photosynthesis Research*, 119, 15–29.

Rogers, A., Medlyn, B. E., Dukes, J. S., et al. (2017). A roadmap for improving the representation of photosynthesis in Earth system models. *New Phytologist*, 213, 22–42.

Rosenzweig, C., and Abramopoulos, F. (1997). Land-surface model development for the GISS GCM. *Journal of Climate*, 10, 2040–2054.

Ross, J. (1975). Radiative transfer in plant communities. In *Vegetation and the Atmosphere*: vol. 1. *Principles*, ed. J. L. Monteith. New York, NY: Academic Press, pp. 13–55.

Ross, J. (1981). *The Radiation Regime and Architecture of Plant Stands*. The Hague: Dr. W. Junk.

Running, S. W., and Coughlan, J. C. (1988). A general model of forest ecosystem processes for regional applications. I. Hydrological balance, canopy gas exchange and primary production processes. *Ecological Modelling*, 42, 125–154.

Running, S. W., and Gower, S. T. (1991). FOREST-BGC, a general model of forest ecosystem processes for regional applications. II. Dynamic carbon allocation and nitrogen budgets. *Tree Physiology*, 9, 147–160.

Running, S. W., and Hunt, E. R., Jr. (1993). Generalization of a forest ecosystem process model for other biomes, BIOME-BGC, and an application for global-scale models. In *Scaling Physiological Processes: Leaf to Globe*, ed. J. R. Ehleringer and C. B. Field. New York, NY: Academic Press, pp. 141–158.

Running, S. W., Nemani, R. R., Heinsch, F. A., et al. (2004). A continuous satellite-derived measure of global terrestrial primary production. *BioScience*, 54, 547–560.

Running, S. W., Thornton, P. E., Nemani, R., and Glassy, J. M. (2000). Global terrestrial gross and net primary productivity from the Earth Observing System. In *Methods in Ecosystem Science*, ed. O. E. Sala. New York, NY: Springer-Verlag, pp. 44–57.

Running, S. W., Waring, R. H., and Rydell, R. A. (1975). Physiological control of water flux in conifers: A computer simulation model. *Oecologia*, 18, 1–16.

Rutter, A. J. (1967). An analysis of evaporation from a stand of Scots pine. In *Forest Hydrology: Proceedings of a National Science Foundation Advanced Science Seminar held at The Pennsylvania State University, University Park, Pennsylvania, Aug 29–Sept 10, 1965*, ed. W. E. Sopper and H. W. Lull. Oxford: Pergamon Press, pp. 403–417.

Rutter, A. J., Kershaw, K. A., Robins, P. C., and Morton, A. J. (1971). A predictive model of rainfall interception in forests, 1. Derivation of the model from observations in a plantation of Corsican pine. *Agricultural Meteorology*, 9, 367–384.

Rutter, A. J., Morton, A. J., and Robins, P. C. (1975). A predictive model of rainfall interception in forests. II. Generalization of the model and comparison with observations in some coniferous and hardwood stands. *Journal of Applied Ecology*, 12, 367–380.

Ryan, M. G. (1991). Effects of climate change on plant respiration. *Ecological Applications*, 1, 157–167.

Ryder, J., Polcher, J., Peylin, P., et al. (2016). A multi-layer land surface energy budget model for implicit coupling with global atmospheric simulations. *Geoscientific Model Development*, 9, 223–245.

Ryel, R. J., Caldwell, M. M., Yoder, C. K., Or, D., Leffler, A. J. (2002). Hydraulic redistribution in a stand of *Artemisia tridentata*: Evaluation of benefits to transpiration assessed with a simulation model. *Oecologia*, 130, 173–184.

Ryu, Y., Baldocchi, D. D., Kobayashi, H., et al. (2011). Integration of MODIS land and atmosphere products with a coupled-process model to estimate gross primary productivity and evapotranspiration from 1 km to global scales. *Global Biogeochemical Cycles*, 25, GB4017, doi:10.1029/2011GB004053.

Saikawa, E., Schlosser, C. A., and Prinn, R. G. (2013). Global modeling of soil nitrous oxide emissions from natural processes. *Global Biogeochemical Cycles*, 27, 972–989, doi:10.1002/gbc.20087.

Sakaguchi, K., and Zeng, X. (2009). Effects of soil wetness, plant litter, and under-canopy atmospheric stability on ground evaporation in the Community Land Model (CLM3.5). *Journal of Geophysical Research*, 114, D01107, doi:10.1029/2008JD010834.

Sands, P. J. (1995). Modelling canopy production. II. From single-leaf photosynthetic parameters to daily canopy photosynthesis. *Australian Journal of Plant Physiology*, 22, 603–614.

Sato, H., Itoh, A., and Kohyama, T. (2007). SEIB–DGVM: A new Dynamic Global Vegetation Model using a spatially explicit individual-based approach. *Ecological Modelling*, 200, 279–307.

Saylor, R. D. (2013). The Atmospheric Chemistry and Canopy Exchange Simulation System (ACCESS): Model description and application to a temperate deciduous forest canopy. *Atmospheric Chemistry and Physics*, 13, 693–715.

Saylor, R. D., Wolfe, G. M., Meyers, T. P., and Hicks, B. B. (2014). A corrected formulation of the Multilayer Model (MLM) for inferring gaseous dry deposition to vegetated surfaces. *Atmospheric Environment*, 92, 141–145.

Schaap, M. G., Leij, F. J., and van Genuchten, M. T. (1998). Neural network analysis for hierarchical prediction of soil hydraulic properties. *Soil Science Society of America Journal*, 62, 847–855.

Schaap, M. G., Leij, F. J., and van Genuchten, M. T. (2001). ROSETTA: A computer program for estimating soil hydraulic parameters with hierarchical pedotransfer functions. *Journal of Hydrology*, 251, 163–176.

Schaphoff, S., von Bloh, W., Rammig, A., et al. (2018). LPJmL4 – a dynamic global vegetation model with managed land – Part 1: Model description. *Geoscientific Model Development*, 11, 1343–1375.

Schieving, F., Werger, M. J. A., and Hirose, T. (1992). Canopy structure, nitrogen distribution and whole canopy photosynthetic carbon gain in growing and flowering stands of tall herbs. *Vegetatio*, 102, 173–181.

Schimel, D. S., Kittel, T. G. F., Knapp, A. K., et al. (1991). Physiological interactions along resource gradients in a tallgrass prairie. *Ecology*, 72, 672–684.

Schmidt, G. A., Kelley, M., Nazarenko, L., et al. (2014). Configuration and assessment of the GISS ModelE2 contributions to the CMIP5 archive. *Journal of Advances in Modeling Earth Systems*, 6, 141–184, doi:10.1002/2013MS000265.

Schuepp, P. H. (1993). Leaf boundary layers. *New Phytologist*, 125, 477–507.

Schultz, M. G., Heil, A., Hoelzemann, J. J., et al. (2008). Global wildland fire emissions from 1960 to 2000. *Global Biogeochemical Cycles*, 22, GB2002, doi:10.1029/2007GB003031.

Schultz, N. M., Lee, X., Lawrence, P. J., Lawrence, D. M., and Zhao, L. (2016). Assessing the use of subgrid land model output to study impacts of land cover change. *Journal of Geophysical Research: Atmospheres*, 121, 6133–6147, doi:10.1002/2016JD025094.

Schwarz, P. A., Law, B. E., Williams, M., et al. (2004). Climatic versus biotic constraints on carbon and water fluxes in seasonally drought-affected ponderosa pine ecosystems. *Global Biogeochemical Cycles*, 18, GB4007, doi:10.1029/2004GB002234.

Schymanski, S. J., Or, D., and Zwieniecki, M. (2013). Stomatal control and leaf thermal and hydraulic capacitances under rapid environmental fluctuations. *PLoS ONE*, 8(1), e54231, doi:10.1371/journal.pone.0054231.

Séférian, R., Delire, C., Decharme, B., et al. (2016). Development and evaluation of CNRM Earth system model – CNRM-ESM1. *Geoscientific Model Development*, 9, 1423–1453.

Sellers, P. J. (1985). Canopy reflectance, photosynthesis and transpiration. *International Journal of Remote Sensing*, 6, 1335–1372.

Sellers, P. J., Berry, J. A., Collatz, G. J., Field, C. B., and Hall, F. G. (1992). Canopy reflectance, photosynthesis, and transpiration. III. A reanalysis using improved leaf models and a new canopy integration scheme. *Remote Sensing of Environment*, 42, 187–216.

Sellers, P. J., Fennessy, M. J., and Dickinson, R. E. (2007). A numerical approach to calculating soil wetness and evapotranspiration over large grid areas. *Journal of Geophysical Research*, 112, D18106, doi:10.1029/2007JD008781.

Sellers, P. J., and Lockwood, J. G. (1981). A computer simulation of the effects of differing crop types on the water balance of small catchments over long time periods. *Quarterly Journal of the Royal Meteorological Society*, 107, 395–414.

Sellers, P. J., Los, S. O., Tucker, C. J., et al. (1996b). A revised land surface parameterization (SiB2) for atmospheric GCMs. Part II: The generation of global fields of terrestrial biophysical parameters from satellite data. *Journal of Climate*, 9, 706–737.

Sellers, P. J., Mintz, Y., Sud, Y. C., and Dalcher, A. (1986). A simple biosphere model (SiB) for use within general circulation models. *Journal of the Atmospheric Sciences*, 43, 505–531.

Sellers, P. J., Randall, D. A., Collatz, G. J., et al. (1996a). A revised land surface parameterization (SiB2) for atmospheric GCMs. *Part I: Model formulation. Journal of Climate*, 9, 676–705.

Seth, A., Giorgi, F., and Dickinson, R. E. (1994). Simulating fluxes from heterogeneous land surfaces: Explicit sub-grid method employing the biosphere–atmosphere transfer scheme (BATS). *Journal of Geophysical Research*, 99D, 18651–18667.

Sharkey, T. D., Bernacchi, C. J., Farquhar, G. D., and Singsaas, E. L. (2007). Fitting photosynthetic carbon dioxide response curves for C_3 leaves. *Plant, Cell and Environment*, 30, 1035–1040.

Shaw, R. H., and Pereira, A. R. (1982). Aerodynamic roughness of a plant canopy: A numerical experiment. *Agricultural Meteorology*, 26, 51–65.

Shevliakova, E., Pacala, S. W., Malyshev, S., et al. (2009). Carbon cycling under 300 years of land use change: Importance of the secondary vegetation sink. *Global Biogeochemical Cycles*, 23, GB2022, doi:10.1029/2007GB003176.

Shi, M., Fisher, J. B., Brzostek, E. R., and Phillips, R. P. (2016). Carbon cost of plant nitrogen acquisition: Global carbon cycle impact from an improved plant nitrogen cycle in the Community Land Model. *Global Change Biology*, 22, 1299–1314.

Shinozaki, K., Yoda, K., Hozumi, K., and Kira, T. (1964). A quantitative analysis of plant form – the pipe model theory. I. Basic analyses. *Japanese Journal of Ecology*, 14, 97–105.

Shiraiwa, T., and Sinclair, T. R. (1993). Distribution of nitrogen among leaves in soybean canopies. *Crop Science*, 33, 804–808.

Shugart, H. H. (1984). *A Theory of Forest Dynamics: The Ecological Implications of Forest Succession Models*. New York, NY: Springer-Verlag.

Shugart, H. H., Crow, T. R., and Jett, J. M. (1973). Forest succession models: A rationale and methodology for modeling forest succession over large regions. *Forest Science*, 19, 203–212.

Shugart, H. H., and West, D. C. (1977). Development of an Appalachian deciduous forest succession model and its application to assessment of the impact of the chestnut blight. *Journal of Environmental Management*, 5, 161–179.

Shugart, H. H., and Woodward, F. I. (2011). *Global Change and the Terrestrial Biosphere: Achievements and Challenges*. Chichester: Wiley-Blackwell.

Shuman, J. K., Shugart, H. H., and Krankina, O. N. (2013). Assessment of carbon stores in tree biomass for two management scenarios in Russia. *Environmental Research Letters*, 8, 045019, doi:10.1088/1748-9326/8/4/045019.

Shuman, J. K., Shugart, H. H., and Krankina, O. N. (2014). Testing individual-based models of forest dynamics: Issues and an example from the boreal forests of Russia. *Ecological Modelling*, 293, 102–110.

Shuman, J. K., Tchebakova, N. M., Parfenova, E. I., et al. (2015). Forest forecasting with vegetation models across Russia. *Canadian Journal of Forest Research*, 45, 175–184.

Shuttleworth, W. J. (1976). A one-dimensional theoretical description of the vegetation-atmosphere interaction. *Boundary-Layer Meteorology*, 10, 273–302.

Shuttleworth, W. J. (1978). A simplified one-dimensional theoretical description of the vegetation-atmosphere interaction. *Boundary-Layer Meteorology*, 14, 3–27.

Shuttleworth, W. J. (1988). Macrohydrology – the new challenge for process hydrology. *Journal of Hydrology*, 100, 31–56.

Shuttleworth, W. J. (2012). *Terrestrial Hydrometeorology*. Chichester: Wiley-Blackwell.

Shuttleworth, W. J., and Gurney, R. J. (1990). The theoretical relationship between foliage temperature and canopy resistance in sparse canopies. *Quarterly Journal of the Royal Meteorological Society*, 116, 497–519.

Shuttleworth, W. J., and Wallace, J. S. (1985). Evaporation from sparse crops – an energy combination theory. *Quarterly Journal of the Royal Meteorological Society*, 111, 839–855.

Sierra, C. A., and Müller, M. (2015). A general mathematical framework for representing soil organic matter dynamics. *Ecological Monographs*, 85, 505–524.

Sierra, C. A., Müller, M., and Trumbore, S. E. (2012). Models of soil organic matter decomposition: the SOILR package, version 1.0. *Geoscientific Model Development*, 5, 1045–1060.

Sinclair, T. R., Murphy, C. E., and Knoerr, K. R. (1976). Development and evaluation of simplified models for simulating canopy photosynthesis and transpiration. *Journal of Applied Ecology*, 13, 813–829.

Siqueira, M. B., and Katul, G. G. (2010). An analytical model for the distribution of CO_2 sources and sinks, fluxes, and mean concentration within the roughness sub-layer. *Boundary-Layer Meteorology*, 135, 31–50.

Sitch, S., Smith, B., Prentice, I. C., et al. (2003). Evaluation of ecosystem dynamics, plant geography and terrestrial carbon cycling in the LPJ dynamic global vegetation model. *Global Change Biology*, 9, 161–185.

Sitch, S., Cox, P. M., Collins, W. J., and Huntingford, C. (2007). Indirect radiative forcing of climate change through ozone effects on the land-carbon sink. *Nature*, 448, 791–794.

Sivapalan, M., Beven, K., and Wood, E. F. (1987). On hydrologic similarity: 2. A scaled model of storm runoff production. *Water Resources Research*, 23, 2266–2278.

Skillman, J. B. (2008). Quantum yield variation across the three pathways of photosynthesis: Not yet out of the dark. *Journal of Experimental Botany*, 59, 1647–1661.

Smith, B., Prentice, I. C., and Sykes, M. T. (2001). Representation of vegetation dynamics in the modelling of terrestrial ecosystems: Comparing two contrasting approaches within European climate space. *Global Ecology and Biogeography*, 10, 621–637.

Smith, B., Wårlind, D., Arneth, A., et al. (2014). Implications of incorporating N cycling and N limitations on primary production in an individual-based dynamic vegetation model. *Biogeosciences*, 11, 2027–2054.

Smith, N. G., and Dukes, J. S. (2013). Plant respiration and photosynthesis in global-scale models: Incorporating acclimation to temperature and CO_2. *Global Change Biology*, 19, 45–63.

Smith, N. G., Malyshev, S. L., Shevliakova, E., et al. (2016). Foliar temperature acclimation reduces simulated carbon sensitivity to climate. *Nature Climate Change*, 6, 407–411.

Smits, K. M., Ngo, V. V., Cihan, A., Sakaki, T., and Illangasekare, T. H. (2012). An evaluation of models of bare soil evaporation formulated with different land surface boundary conditions and assumptions. *Water Resources Research*, 48, W12526, doi:10.1029/2012WR012113.

Sokolov, A. P., Kicklighter, D. W., Melillo, J. M., et al. (2008). Consequences of considering carbon–nitrogen interactions on the feedbacks between climate and the terrestrial carbon cycle. *Journal of Climate*, 21, 3776–3796.

Sollins, P., Brown, A. T., and Swartzman G. L. (1979). *CONIFER: A Model of Carbon and Water Flow through a Coniferous Forest – Revised Documentation*, Coniferous Forest Biome Bulletin No. 15. Seattle, WA: University of Washington.

Solomon, A. M. (1986). Transient response of forests to CO_2-induced climate change: Simulation modeling experiments in eastern North America. *Oecologia*, 68, 567–579.

Solomon, A. M., Delcourt, H. R., West, D. C., and Blasing, T. J. (1980). Testing a simulation model for reconstruction of prehistoric forest-stand dynamics. *Quaternary Research*, 14, 275–293.

Solomon, A. M., West, D. C., and Solomon, J. A. (1981). Simulating the role of climate change and species immigration in forest succession. In *Forest Succession: Concepts and Application*, ed. D. C. West, H. H. Shugart, and D. B. Botkin. New York, NY: Springer-Verlag, pp. 154–177.

Sperry, J. S., Adler, F. R., Campbell, G. S., and Comstock, J. P. (1998). Limitation of plant water use by rhizosphere and xylem conductance: Results from a model. *Plant, Cell and Environment*, 21, 347–359.

Sperry, J. S., Venturas, M. D., Anderegg, W. R. L., et al. (2017). Predicting stomatal responses to the environment from the optimization of photosynthetic gain and hydraulic cost. *Plant, Cell and Environment*, 40, 816–830.

Spitters, C. J. T. (1986). Separating the diffuse and direct component of global radiation and its implications for modeling canopy photosynthesis. Part II. Calculation of canopy photosynthesis. *Agricultural and Forest Meteorology*, 38, 231–242.

Stamm, J. F., Wood, E. F., and Lettenmaier, D. P. (1994). Sensitivity of a GCM simulation of global climate to the representation of land-surface hydrology. *Journal of Climate*, 7, 1218–1239.

Staudt, K., Falge, E., Pyles, R. D., Paw U, K. T., and Foken, T. (2010). Sensitivity and predictive uncertainty of the ACASA model at a spruce forest site. *Biogeosciences*, 7, 3685–3705.

Staudt, K., Serafimovich, A., Siebicke, L., Pyles, R. D., and Falge, E. (2011). Vertical structure of evapotranspiration at a forest site (a case study). *Agricultural and Forest Meteorology*, 151, 709–729.

Steele, M. J., Coutts, M. P., and Yeoman, M. M. (1989). Developmental changes in Sitka spruce as indices of physiological age I. Changes in needle morphology. *New Phytologist*, 113, 367–375.

Stewart, J. B. (1977). Evaporation from the wet canopy of a pine forest. *Water Resources Research*, 13, 915–921.

Stieglitz, M., Rind, D., Famiglietti, J., and Rosenzweig, C. (1997). An efficient approach to modeling the topographic control of surface hydrology for regional and global climate modeling. *Journal of Climate*, 10, 118–137.

Stöckli, R., Rutishauser, T., Dragoni, D., et al. (2008). Remote sensing data assimilation for a prognostic phenology model. *Journal of Geophysical Research*, 113, G04021, doi:10.1029/2008JG000781.

Stöckli, R., Rutishauser, T., Baker, I., Liniger, M. A., and Denning, A. S. (2011). A global reanalysis of vegetation phenology. *Journal of Geophysical Research*, 116, G03020, doi:10.1029/2010JG001545.

Stoy, P. C., Katul, G. G., Siqueira, M. B., et al. (2006). Separating the effects of climate and vegetation on evapotranspiration along a successional chronosequence in the southeastern US. *Global Change Biology*, 12, 2115–2135.

Stoy, P. C., Mauder, M., Foken, T., et al. (2013). A data-driven analysis of energy balance closure across FLUXNET research sites: The role of landscape scale heterogeneity. *Agricultural and Forest Meteorology*, 171/172, 137–152.

Strigul, N., Pristinski, D., Purves, D., Dushoff, J., and Pacala, S. (2008). Scaling from trees to forests: Tractable macroscopic equations for forest dynamics. *Ecological Monographs*, 78, 523–545.

Stroud, C., Makar, P., Karl, T., et al. (2005). Role of canopy-scale photochemistry in modifying biogenic-atmosphere exchange of reactive terpene species: Results from the CELTIC field study. *Journal of Geophysical Research*, 110, D17303, doi:10.1029/2005JD005775.

Su, H.-B., Paw U, K. T., and Shaw, R. H. (1996). Development of a coupled leaf and canopy model for the simulation of plant–atmosphere interaction. *Journal of Applied Meteorology*, 35, 733–748.

Suits, N. S., Denning, A. S., Berry, J. A., et al. (2005). Simulation of carbon isotope discrimination of the terrestrial biosphere. *Global Biogeochemical Cycles*, 19, GB1017, doi:10.1029/2003GB002141.

Sun, Y., Gu, L., Dickinson, R. E., et al. (2014a). Asymmetrical effects of mesophyll conductance on fundamental photosynthetic parameters and their relationships estimated from leaf gas exchange measurements. *Plant, Cell and Environment*, 37, 978–994.

Sun, Y., Gu, L., Dickinson, R. E., (2014b). Impact of mesophyll diffusion on estimated global land CO_2 fertilization. *Proceedings of the National Academy of Sciences USA*, 111, 15774–15779.

Sutton, M. A., Reis, S., Riddick, S. N., et al. (2013).Towards a climate-dependent paradigm of ammonia emission and deposition. *Philosophical Transactions of the Royal Society B*, 368, 20130166, doi:10.1098/rstb.2013.0166.

Sutton, M. A., Schjørring, J. K., and Wyers, G. P. (1995). Plant–atmosphere exchange of ammonia. *Philosophical Transactions of the Royal Society London A*, 351, 261–278.

Swenson, S. C., and Lawrence, D. M. (2012). A new fractional snow-covered area parameterization for the Community Land Model and its effect on the surface energy balance. *Journal of Geophysical Research*, 117, D21107, doi:10.1029/2012JD018178.

Swenson, S. C., and Lawrence, D. M. (2014). Assessing a dry surface layer-based soil resistance parameterization for the Community Land Model using GRACE and FLUXNET-MTE data. *Journal of Geophysical Research: Atmospheres*, 119, 10299–10312, doi:10.1002/2014JD022314.

Tague, C. L., and Band, L. E. (2004). RHESSys: Regional Hydro-Ecologic Simulation System – an object-oriented approach to spatially distributed modeling of carbon, water, and nutrient cycling. *Earth Interactions*, 8(19), 1–42.

Takata, K., Emori, S., and Watanabe, T. (2003). Development of the minimal advanced treatments of surface interaction and runoff. *Global and Planetary Change*, 38, 209–222.

Tang, J. Y., and Riley, W. J. (2013). A new top boundary condition for modeling surface diffusive exchange of a generic volatile tracer: Theoretical analysis and application to soil evaporation. *Hydrology and Earth System Sciences*, 17, 873–893.

Ter-Mikaelian, M. T., and Korzukhin, M. D. (1997). Biomass equations for sixty-five North American tree species. *Forest Ecology and Management*, 97, 1–24.

Thom, A. S. (1971). Momentum absorption by vegetation. *Quarterly Journal of the Royal Meteorological Society*, 97, 414–428.

Thom, A. S. (1972). Momentum, mass and heat exchange of vegetation. *Quarterly Journal of the Royal Meteorological Society*, 98, 124–134.

Thom, A. S. (1975). Momentum, mass and heat exchange of plant communities. In *Vegetation and the Atmosphere: vol. 1. Principles*, ed. J. L. Monteith. New York, NY: Academic Press, pp. 57–109.

Thomas, R. Q., Bonan, G. B., and Goodale, C. L. (2013a). Insights into mechanisms governing forest carbon response to nitrogen deposition: A model–data comparison using observed responses to nitrogen addition. *Biogeosciences*, 10, 3869–3887.

Thomas, R. Q., and Williams, M. (2014). A model using marginal efficiency of investment to analyze carbon and nitrogen interactions in terrestrial ecosystems (ACONITE Version 1). *Geoscientific Model Development*, 7, 2015–2037.

Thomas, R. Q., Zaehle, S., Templer, P. H., and Goodale, C. L. (2013b). Global patterns of nitrogen limitation: Confronting two global biogeochemical models with observations. *Global Change Biology*, 19, 2986–2998.

Thomas, S. C., and Winner, W. E. (2000). A rotated ellipsoidal angle density function improves estimation of foliage inclination distributions in forest canopies. *Agricultural and Forest Meteorology*, 100, 19–24.

Thomsen, J. E., Bohrer, G., Matheny, A. M., et al. (2013). Contrasting hydraulic strategies during dry soil conditions in *Quercus rubra* and *Acer rubrum* in a sandy site in Michigan. *Forests*, 4, 1106–1120.

Thonicke, K., Spessa, A., Prentice, I. C., et al. (2010). The influence of vegetation, fire spread and fire behaviour on biomass burning and trace gas emissions: Results from a process-based model. *Biogeosciences*, 7, 1991–2011.

Thonicke, K., Venevsky, S., Sitch, S., and Cramer, W. (2001). The role of fire disturbance for global vegetation dynamics: Coupling fire into a Dynamic Global Vegetation Model. *Global Ecology and Biogeography*, 10, 661–677.

Thornley, J. H. M. (2002). Instantaneous canopy photosynthesis: Analytical expressions for sun and shade leaves based on exponential light decay down the canopy and an acclimated non-rectangular hyperbola for leaf photosynthesis. *Annals of Botany*, 89, 451–458.

Thornton, P. E., Doney, S. C., Lindsay, K., et al. (2009). Carbon–nitrogen interactions regulate climate–carbon cycle feedbacks: Results from an atmosphere–ocean general circulation model. *Biogeosciences*, 6, 2099–2120.

Thornton, P. E., Lamarque, J.-F., Rosenbloom, N. A., and Mahowald, N. M. (2007). Influence of carbon–nitrogen cycle coupling on land model response to CO_2 fertilization and climate variability. *Global Biogeochemical Cycles*, 21, GB4018, doi:10.1029/2006GB002868.

Thornton, P. E., Law, B. E., Gholz, H. L., et al. (2002). Modeling and measuring the effects of disturbance history and climate on carbon and water budgets in evergreen needleleaf forests. *Agricultural and Forest Meteorology*, 113, 185–222.

Tjoelker, M. G., Oleksyn, J., and Reich, P. B. (2001). Modelling respiration of vegetation: Evidence for a general temperature-dependent Q_{10}. *Global Change Biology*, 7, 223–230.

Todini, E. (2007). Hydrological catchment modelling: Past, present and future. *Hydrology and Earth System Sciences*, 11, 468–482.

Toon, O. B., McKay, C. P., Ackerman, T. P., and Santhanam, K. (1989). Rapid calculation of radiative heating rates and photodissociation rates in inhomogeneous multiple scattering atmospheres. *Journal of Geophysical Research*, 94D, 16287–16301.

Trofymow, J. A., Moore, T. R., Titus, B., et al. (2002). Rates of litter decomposition over 6 years in Canadian forests: Influence of litter quality and climate. *Canadian Journal of Forest Research*, 32, 789–804.

Tuzet, A., Perrier, A., and Leuning, R. (2003). A coupled model of stomatal conductance, photosynthesis and transpiration. *Plant, Cell and Environment*, 26, 1097–1116.

Twine, T. E., Kustas, W. P., Norman, J. M., et al. (2000). Correcting eddy-covariance flux underestimates over a grassland. *Agricultural and Forest Meteorology*, 103, 279–300.

Tyree, M. T. (1988). A dynamic model for water flow in a single tree: Evidence that models must account for hydraulic architecture. *Tree Physiology*, 4, 195–217.

Tyree, M. T., and Ewers, F. W. (1991). The hydraulic architecture of trees and other woody plants. *New Phytologist*, 119, 345–360.

Unger, N. (2013). Isoprene emission variability through the twentieth century. *Journal of Geophysical Research: Atmospheres*, 118, 13606–13613, doi:10.1002/2013JD020978.

Unger, N., Harper, K., Zheng, Y., et al. (2013). Photosynthesis-dependent isoprene emission from leaf to planet in a global carbon–chemistry–climate model. *Atmospheric Chemistry and Physics*, 13, 10243–10269.

Urbanski, S., Barford, C., Wofsy, S., et al. (2007). Factors controlling CO_2 exchange on timescales from hourly to decadal at Harvard Forest. *Journal of Geophysical Research*, 112, G02020, doi:10.1029/2006JG000293.

Valente, F., David, J. S., and Gash, J. H. C. (1997). Modelling interception loss for two sparse eucalypt and pine forests in central Portugal using reformulated Rutter and Gash analytical models. *Journal of Hydrology*, 190, 141–162.

Val Martin, M., Heald, C. L., and Arnold, S. R. (2014). Coupling dry deposition to vegetation phenology in the Community Earth System Model: Implications for the simulation of surface O_3. *Geophysical Research Letters*, 41, 2988–2996, doi:10.1002/2014GL059651.

van den Honert, T. H. (1948). Water transport in plants as a catenary process. *Discussions of the Faraday Society*, 3, 146–153.

van den Hurk, B. J. J. M., and McNaughton, K. G. (1995). Implementation of near-field dispersion in a simple two-layer surface resistance model. *Journal of Hydrology*, 166, 293–311.

van den Hurk, B. J. J. M., Viterbo, P., Beljaars, A. C. M., and Betts, A. K. (2000). *Offline Validation of the ERA40 Surface Scheme*, Technical Memorandum Number 295. Reading: European Centre for Medium-Range Weather Forecasts.

van der Velde, I. R., Miller, J. B., Schaefer, K., et al. (2013). Biosphere model simulations of interannual variability in terrestrial $^{13}C/^{12}C$ exchange. *Global Biogeochemical Cycles*, 27, 637–649, doi:10.1002/gbc.20048.

van der Werf, G. R., Randerson, J. T., Giglio, L., et al. (2010). Global fire emissions and the contribution of deforestation, savanna, forest, agricultural, and peat fires (1997–2009). *Atmospheric Chemistry and Physics*, 10, 11707–11735.

van Dijk, A. I. J. M., Gash, J. H. van Gorsel, E., et al. (2015). Rainfall interception and the coupled surface water and energy balance. *Agricultural and Forest Meteorology*, 214/215, 402–415.

van Genuchten, M. T. (1980). A closed-form equation for predicting the hydraulic conductivity of unsaturated soils. *Soil Science Society of America Journal*, 44, 892–898.

Verhoef, A., De Bruin, H. A. R., and van den Hurk, B. J. J. M. (1997). Some practical notes on the parameter kB^{-1} for sparse vegetation. *Journal of Applied Meteorology*, 36, 560–572.

Verhoef, A., and Egea, G. (2014). Modeling plant transpiration under limited soil water: Comparison of different plant and soil hydraulic parameterizations and preliminary implications for their use in land surface models. *Agricultural and Forest Meteorology*, 191, 22–32.

Verseghy, D. L. (1991). CLASS – A Canadian land surface scheme for GCMs. I. Soil model. *International Journal of Climatology*, 11, 111–133.

Verseghy, D. L., McFarlane, N. A., and Lazare, M. (1993). CLASS – A Canadian land surface scheme for GCMs, II.

Vegetation model and coupled runs. *International Journal of Climatology*, 13, 347–370.

Vico, G., Manzoni, S., Palmroth, S., Weih, M., and Katul, G. (2013). A perspective on optimal leaf stomatal conductance under CO_2 and light co-limitations. *Agricultural and Forest Meteorology*, 182/183, 191–199.

Vidale, P. L., and Stöckli, R. (2005). Prognostic canopy air space solutions for land surface exchanges. *Theoretical and Applied Climatology*, 80, 245–257.

Vogel, S. (2009). Leaves in the lowest and highest winds: Temperature, force and shape. *New Phytologist*, 183, 13–26.

von Caemmerer, S. (2000). *Biochemical Models of Leaf Photosynthesis.* Collingwood, Victoria: CSIRO Publishing.

von Caemmerer, S. (2013). Steady-state models of photosynthesis. *Plant, Cell and Environment*, 36, 1617–1630.

von Caemmerer, S., and Evans. J. R. (2015). Temperature responses of mesophyll conductance differ greatly between species. *Plant, Cell and Environment*, 38, 629–637.

von Caemmerer, S., Evans, J. R., Hudson, G. S., and Andrews, T. J. (1994). The kinetics of ribulose-1,5-bisphosphate carboxylase/oxygenase in vivo inferred from measurements of photosynthesis in leaves of transgenic tobacco. *Planta*, 195, 88–97.

von Caemmerer, S., Farquhar, G., and Berry, J. (2009). Biochemical model of C_3 photosynthesis. In *Photosynthesis in silico: Understanding Complexity from Molecules to Ecosystems*, ed. A. Laisk, L. Nedbal and Govindjee. Dordrecht: Springer, pp. 209–230.

Vrettas, M. D., and Fung, I. Y. (2015). Toward a new parameterization of hydraulic conductivity in climate models: Simulation of rapid groundwater fluctuations in Northern California. *Journal of Advances in Modeling Earth Systems*, 7, doi:10.1002/2015MS000516.

Waggoner, P. E. (1975). Micrometeorological models. In *Vegetation and the Atmosphere: vol. 1. Principles*, ed. J. L. Monteith. New York, NY: Academic Press, pp. 205–228.

Waggoner, P. E., Furnival, G. M., and Reifsnyder, W. E. (1969). Simulation of the microclimate in a forest. *Forest Science*, 15, 37–45.

Waggoner, P. E., and Reifsnyder, W. E. (1968). Simulation of the temperature, humidity and evaporation profiles in a leaf canopy. *Journal of Applied Meteorology*, 7, 400–409.

Waggoner, P. E., and Stephens, G. R. (1970). Transition probabilities for a forest. *Nature*, 225, 1160–1161.

Walcroft, A. S., Brown, K. J., Schuster, W. S. F., et al. (2005). Radiative transfer and carbon assimilation in relation to canopy architecture, foliage area distribution and clumping in a mature temperate rainforest canopy in New Zealand. *Agricultural and Forest Meteorology*, 135, 326–339.

Walker, A. P., Quaife, T., van Bodegom, P. M., et al. (2017). The impact of alternative trait-scaling hypotheses for the maximum photosynthetic carboxylation rate (V_{cmax}) on global gross primary production. *New Phytologist*, 215, 1370–1386.

Wang, A., Li, K. Y., and Lettenmaier, D. P. (2008). Integration of the variable infiltration capacity model soil hydrology scheme into the community land model. *Journal of Geophysical Research*, 113, D09111, doi:10.1029/2007JD009246.

Wang, G. (2011). Assessing the potential hydrological impacts of hydraulic redistribution in Amazonia using a numerical modeling approach. *Water Resources Research*, 47, W02528, doi:10.1029/2010WR009601.

Wang, H., Prentice, I. C., and Davis, T. W. (2014). Biophsyical constraints on gross primary production by the terrestrial biosphere. *Biogeosciences*, 11, 5987–6001.

Wang, H., Prentice, I. C., Keenan, T. F., et al. (2017). Towards a universal model for carbon dioxide uptake by plants. *Nature Plants*, 3, 734–741.

Wang, Y.-P. (2000). A refinement to the two-leaf model for calculating canopy photosynthesis. *Agricultural and Forest Meteorology*, 101, 143–150.

Wang, Y.-P. (2003). A comparison of three different canopy radiation models commonly used in plant modelling. *Functional Plant Biology*, 30, 143–152.

Wang, Y. P., and Jarvis, P. G. (1990). Description and validation of an array model – MAESTRO. *Agricultural and Forest Meteorology*, 51, 257–280.

Wang, Y. P., Jarvis, P. G., and Benson, M. L. (1990). Two-dimensional needle-area density distribution within the crowns of *Pinus radiata. Forest Ecology and Management*, 32, 217–237.

Wang, Y. P., Jiang, J., Chen-Charpentier, B., et al. (2016). Responses of two nonlinear microbial models to warming and increased carbon input. *Biogeosciences*, 13, 887–902.

Wang, Y. P., Kowalczyk, E., Leuning, R., et al. (2011). Diagnosing errors in a land surface model (CABLE) in the time and frequency domains. *Journal of Geophysical Research*, 116, G01034, doi:10.1029/2010JG001385.

Wang, Y. P., Law, R. M., and Pak, B. (2010). A global model of carbon, nitrogen and phosphorus cycles for the terrestrial biosphere. *Biogeosciences*, 7, 2261–2282.

Wang, Y.-P., and Leuning, R. (1998). A two-leaf model for canopy conductance, photosynthesis and partitioning of available energy. I: Model description and comparison with a multi-layered model. *Agricultural and Forest Meteorology*, 91, 89–111.

Wania, R., Ross, I., and Prentice, I. C. (2010). Implementation and evaluation of a new methane model within a dynamic global vegetation model: LPJ-WHyMe v1.3.1. *Geoscientific Model Development*, 3, 565–584.

Wania, R., Melton, J. R., Hodson, E. L., et al. (2013). Present state of global wetland extent and wetland methane modelling: Methodology of a model inter-comparison project (WETCHIMP). *Geoscientific Model Development*, 6, 617–641.

Waring, R. H., and Running, S. W. (1976). Water uptake, storage and transpiration by conifers: A physiological model. In *Water and Plant Life: Problems and Modern Approaches*, ed. O. L. Lange, L. Kappen, and E.-D. Schulze. Berlin: Springer-Verlag, pp. 189–202.

Waring, R. H., and Running, S. W. (2007). *Forest Ecosystems: Analysis at Multiple Scales*, 3rd edn. Amsterdam: Elsevier.

Warrick, A. W. (1991). Numerical approximations of darcian flow through unsaturated soil. *Water Resources Research*, 27, 1215–1222.

Warrick, A. W. (2003). *Soil Water Dynamics*. Oxford: Oxford University Press.

Webb, E. K., Pearman, G. I., and Leuning, R. (1980). Correction of flux measurements for density effects due to heat and water vapour transfer. *Quarterly Journal of the Royal Meteorological Society*, 106, 85–100.

Weiler, M., and Beven, K. (2015). Do we need a Community Hydrological Model? *Water Resources Research*, 51, 7777–7784, doi:10.1002/2014WR016731.

Weng, E. S., Malyshev, S., Lichstein, J. W., et al. (2015). Scaling from individual trees to forests in an Earth system modeling framework using a mathematically tractable model of height-structured competition. *Biogeosciences*, 12, 2655–2694.

Werger, M. J. A., and Hirose, T. (1991). Leaf nitrogen distribution and whole canopy photosynthetic carbon gain in herbaceous stands. *Vegetatio*, 97, 11–20.

Wesely, M. L. (1989). Parameterization of surface resistances to gaseous dry deposition in regional-scale numerical models. *Atmospheric Environment*, 23, 1293–1304.

Wesely, M. L., and Hicks, B. B. (1977). Some factors that affect the deposition rates of sulfur dioxide and similar gases on vegetation. *Journal of the Air Pollution Control Association*, 27, 1110–1116.

Wesely, M. L., and Hicks, B. B. (2000). A review of the current status of knowledge on dry deposition. *Atmospheric Environment*, 34, 2261–2282.

West, J. B., Bowen, G. J., Dawson, T. E., and Tu, K. P. (2010). *Isoscapes: Understanding Movement, Pattern, and Process on Earth through Isotope Mapping*. Dordrecht: Springer.

Whitehead, D. (1998). Regulation of stomatal conductance and transpiration in forest canopies. *Tree Physiology*, 18, 633–644.

Whittaker, R. H., Bormann, F. H., Likens, G. E., and Siccama, T. G. (1974). The Hubbard Brook Ecosystem Study: Forest biomass and production. *Ecological Monographs*, 44, 233–252.

Wieder, W. R., Allison, S. D., Davidson, E. A., et al. (2015a). Explicitly representing soil microbial processes in Earth system models. *Global Biogeochemical Cycles*, 29, 1782–1800, doi:10.1002/2015GB005188.

Wieder, W. R., Cleveland, C. C., Lawrence, D. M., and Bonan, G. B. (2015b). Effects of model structural uncertainty on carbon cycle projections: Biological nitrogen fixation as a case study. *Environmental Research Letters*, 10, 044016, doi:10.1088/1748-9326/10/4/044016.

Wieder, W. R., Cleveland, C. C., Smith, W. K., and Todd-Brown, K. (2015c). Future productivity and carbon storage limited by terrestrial nutrient availability. *Nature Geoscience*, 8, 441–444.

Wieder, W. R., Grandy, A. S., Kallenbach, C. M., and Bonan, G. B. (2014). Integrating microbial physiology and physio-chemical principles in soils with the MIcrobial-MIneral Carbon Stabilization (MIMICS) model. *Biogeosciences*, 11, 3899–3917.

Wieder, W. R., Grandy, A. S., Kallenbach, C. M., Taylor, P. G., and Bonan, G. B. (2015d). Representing life in the Earth system with soil microbial functional traits in the MIMICS model. *Geoscientific Model Development*, 8, 1789–1808.

Wieder, W. R., Hartman, M. D., Sulman, B. N., et al. (2018). Carbon cycle confidence and uncertainty: Exploring variation among soil biogeochemical models. *Global Change Biology*, 24, 1563–1579.

Wiedinmyer, C., Akagi, S. K., Yokelson, R. J., et al. (2011). The Fire INventory from NCAR (FINN): A high resolution global model to estimate the emissions from open burning. *Geoscientific Model Development*, 4, 625–641.

Williams, M., Bond, B. J., and Ryan, M. G. (2001a). Evaluating different soil and plant hydraulic constraints on tree function using a model and sap flow data from ponderosa pine. *Plant, Cell and Environment*, 24, 679–690.

Williams, M., Eugster, W., Rastetter, E. B., McFadden, J. P., and Chapin, F. S. III (2000). The controls on net ecosystem productivity along an Arctic transect: A model comparison with flux measurements. *Global Change Biology*, 6 (S1), 116–126.

Williams, M., Malhi, Y., Nobre, A. D., et al. (1998). Seasonal variation in net carbon exchange and evapotranspiration in a Brazilian rain forest: A modelling analysis. *Plant, Cell and Environment*, 21, 953–968.

Williams, M., Law, B. E., Anthoni, P. M., and Unsworth, M. H. (2001b). Use of a simulation model and ecosystem flux data to examine carbon–water interactions in ponderosa pine. *Tree Physiology*, 21, 287–298.

Williams, M., Rastetter, E. B., Fernandes, D. N., et al. (1996). Modelling the soil–plant–atmosphere continuum in a

Quercus–Acer stand at Harvard Forest: The regulation of stomatal conductance by light, nitrogen and soil/plant hydraulic properties. *Plant, Cell and Environment*, 19, 911–927.

Williams, M., Richardson, A. D., Reichstein, M., et al. (2009). Improving land surface models with FLUXNET data. *Biogeosciences*, 6, 1341–1359.

Williamson, D. L., Kiehl, J. T., Ramanathan, V., Dickinson, R. E., and Hack, J. J. (1987). *Description of NCAR Community Climate Model (CCM1)*, Technical Note NCAR/TN-285 +STR. Boulder, CO: National Center for Atmospheric Research.

Wilson, K. B., and Baldocchi, D. D. (2000). Seasonal and interannual variability of energy fluxes over a broad-leaved temperate deciduous forest in North America. *Agricultural and Forest Meteorology*, 100, 1–18.

Wilson, K. B., Baldocchi, D. D., and Hanson, P. J. (2000). Spatial and seasonal variability of photosynthetic parameters and their relationship to leaf nitrogen in a deciduous forest. *Tree Physiology*, 20, 565–578.

Wilson, T. B., Norman, J. M., Bland, W. L., and Kucharik, C. J. (2003). Evaluation of the importance of Lagrangian canopy turbulence formulations in a soil–plant–atmosphere model. *Agricultural and Forest Meteorology*, 115, 51–69.

Witkowski, E. T. F., and Lamont, B. B. (1991). Leaf specific mass confounds leaf density and thickness. *Oecologia*, 88, 486–493.

Wittig, V. E., Ainsworth, E. A., and Long, S. P. (2007). To what extent do current and projected increases in surface ozone affect photosynthesis and stomatal conductance of trees? A meta-analytic review of the last 3 decades of experiments. *Plant, Cell and Environment*, 30, 1150–1162.

Wohlfahrt, G., Bahn, M., Tappeiner, U., and Cernusca, A. (2001). A multi-component, multi-species model of vegetation–atmosphere CO_2 and energy exchange for mountain grasslands. *Agricultural and Forest Meteorology*, 106, 261–287.

Wohlfahrt, G., and Cernusca, A. (2002). Momentum transfer by a mountain meadow canopy: A simulation analysis based on Massman's (1997) model. *Boundary-Layer Meteorology*, 103, 391–407.

Wolf, A., Anderegg, W. R. L., and Pacala, S. W. (2016). Optimal stomatal behavior with competition for water and risk of hydraulic impairment. *Proceedings of the National Academy of Sciences USA*, 113, E7222–E7230.

Wolfe, G. M., and Thornton, J. A. (2011). The Chemistry of Atmosphere–Forest Exchange (CAFE) model – Part 1: Model description and characterization. *Atmospheric Chemistry and Physics*, 11, 77–101.

Wolock, D. M. (1993). *Simulating the Variable-Source-Area Concept of Streamflow Generation with the Watershed Model TOPMODEL*, Water-Resources Investigations Report 93–4124. Lawrence, KS: U.S. Geological Survey.

Wolz, K. J., Wertin, T. M., Abordo, M., Wang, D., and Leakey, A. D. B. (2017). Diversity in stomatal function is integral to modelling plant carbon and water fluxes. *Nature Ecology and Evolution*, 1, 1292–1298.

Wong, S. C., Cowan, I. R., and Farquhar, G. D. (1978). Leaf conductance in relation to assimilation in *Eucalyptus pauciflora* Sieb. ex Spreng: Influence of irradiance and partial pressure of carbon dioxide. *Plant Physiology*, 62, 670–674.

Wong, S. C., Cowan, I. R., and Farquhar, G. D. (1979). Stomatal conductance correlates with photosynthetic capacity. *Nature*, 282, 424–426.

Wong, T. E., Nusbaumer, J., and Noone, D. C., (2017). Evaluation of modeled land-atmosphere exchanges with a comprehensive water isotope fractionation scheme in version 4 of the Community Land Model. *Journal of Advances in Modeling Earth Systems*, 9, 978–1001, doi:10.1002/2016MS000842.

Wood, E. F., Lettenmaier, D. P., and Zartarian, V. G. (1992). A land-surface hydrology parameterization with sub-grid variability for general circulation models. *Journal of Geophysical Research*, 97D, 2717–2728.

Woodward, F. I., and Lomas, M. R. (2004). Vegetation dynamics – simulating responses to climatic change. *Biological Reviews*, 79, 643–670.

Woodward, F. I., Smith, T. M., and Emanuel, W. R. (1995). A global land primary productivity and phytogeography model. *Global Biogeochemical Cycles*, 9, 471–490.

Wright, I. J., Reich, P. B., Westoby, M., et al. (2004). The worldwide leaf economics spectrum. *Nature*, 428, 821–827.

Wu, A., Black, A., Verseghy, D. L., and Bailey, W. G. (2001). Comparison of two-layer and single-layer canopy models with Lagrangian and K-theory approaches in modelling evaporation from forests. *International Journal of Climatology*, 21, 1821–1839.

Wu, T., Li, W., Ji, J., et al. (2013). Global carbon budgets simulated by the Beijing Climate Center Climate System Model for the last century. *Journal of Geophysical Research: Atmospheres*, 118, 4326–4347, doi:10.1002/jgrd.50320.

Wu, Y., Brashers, B., Finkelstein, P. L., and Pleim, J. E. (2003). A multilayer biochemical dry deposition model. 1. Model formulation. *Journal of Geophysical Research*, 108D, 4013, doi:10.1029/2002JD002293.

Wullschleger, S. D. (1993). Biochemical limitations to carbon assimilation in C_3 plants - A retrospective analysis of the A/C_i curves from 109 species. *Journal of Experimental Botany*, 44, 907–920.

Xu, L., and Baldocchi, D. D. (2003). Seasonal trends in photosynthetic parameters and stomatal conductance of blue oak (*Quercus douglasii*) under prolonged summer drought and high temperature. *Tree Physiology*, 23, 865–877.

Xia, J., Luo, Y., Wang, Y.-P., and Hararuk, O. (2013). Traceable components of terrestrial carbon storage capacity in biogeochemical models. *Global Change Biology*, 19, 2104–2116.

Xia, J. Y., Luo, Y. Q., Wang, Y.-P., Weng, E. S., and Hararuk, O. (2012). A semi-analytical solution to accelerate spin-up of a coupled carbon and nitrogen land model to steady state. *Geoscientific Model Development*, 5, 1259–1271.

Xia, J., Yuan, W., Wang, Y.-P., and Zhang, Q. (2017). Adaptive carbon allocation by plants enhances the terrestrial carbon sink. *Scientific Reports*, 7, 3341, doi:10.1038/s41598-017-03574-3.

Xu, C., Fisher, R., Wullschleger, S. D., et al. (2012). Toward a mechanistic modeling of nitrogen limitation on vegetation dynamics. *PLoS ONE*, 7(5), e37914, doi:10.1371/journal.pone.0037914.

Xu, L., Pyles, R. D., Paw U, K. T., Chen, S. H., and Monier, E. (2014). Coupling the high-complexity land surface model ACASA to the mesoscale model WRF. *Geoscientific Model Development*, 7, 2917–2932.

Xu, X., Medvigy, D., Powers, J. S., Becknell, J. M., and Guan, K. (2016). Diversity in plant hydraulic traits explains seasonal and inter-annual variations of vegetation dynamics in seasonally dry tropical forests. *New Phytologist*, 212, 80–95.

Yan, B., and Dickinson, R. E. (2014). Modeling hydraulic redistribution and ecosystem response to droughts over the Amazon basin using Community Land Model 4.0 (CLM4). *Journal of Geophysical Research: Biogeosciences*, 119, 2130–2143, doi:10.1002/2014JG002694.

Yan, X., and Shugart, H. H. (2005). FAREAST: A forest gap model to simulate dynamics and patterns of eastern Eurasian forests. *Journal of Biogeography*, 32, 1641–1658.

Yang, X., Thornton, P. E., Ricciuto, D. M., and Post, W. M. (2014). The role of phosphorus dynamics in tropical forests – a modeling study using CLM-CNP. *Biogeosciences*, 11, 1667–1681.

Yeh, T.-C. J., Gelhar, L. W., and Gutjahr, A. L. (1985a). Stochastic analysis of unsaturated flow in heterogeneous soils: 1. Statistically isotropic media. *Water Resources Research*, 21, 447–456.

Yeh, T.-C. J., Gelhar, L. W., and Gutjahr, A. L. (1985b). Stochastic analysis of unsaturated flow in heterogeneous soils: 2. Statistically anisotropic media with variable α. *Water Resources Research*, 21, 457–464.

Yeh, T.-C. J., Gelhar, L. W., and Gutjahr, A. L. (1985c). Stochastic analysis of unsaturated flow in heterogeneous soils: 3. Observations and applications. *Water Resources Research*, 21, 465–471.

Yoshimura, K., Miyazaki, S., Kanae, S., and Oki, T. (2006). Iso-MATSIRO, a land surface model that incorporates stable water isotopes. *Global and Planetary Change*, 51, 90–107.

Yuan, H., Dickinson, R. E., Dai, Y., et al. (2014). A 3D canopy radiative transfer model for global climate modeling: Description, validation, and application. *Journal of Climate*, 27, 1168–1192.

Yuan, W., Liu, S., Zhou, G., et al. (2007). Deriving a light use efficiency model from eddy covariance flux data for predicting daily gross primary production across biomes. *Agricultural and Forest Meteorology*, 143, 189–207.

Yue, C., Ciais, P., Cadule, P., et al. (2014). Modelling the role of fires in the terrestrial carbon balance by incorporating SPITFIRE into the global vegetation model ORCHIDEE – Part 1: Simulating historical global burned area and fire regimes. *Geoscientific Model Development*, 7, 2747–2767.

Zaehle, S., and Dalmonech, D. (2011). Carbon–nitrogen interactions on land at global scales: Current understanding in modelling climate biosphere feedbacks. *Current Opinion in Environmental Sustainability*, 3, 311–320.

Zaehle, S., and Friend, A. D. (2010). Carbon and nitrogen cycle dynamics in the O-CN land surface model: 1. Model description, site-scale evaluation, and sensitivity to parameter estimates. *Global Biogeochemical Cycles*, 24, GB1005, doi:10.1029/2009GB003521.

Zaehle, S., Friend, A. D., Friedlingstein, P., et al. (2010). Carbon and nitrogen cycle dynamics in the O-CN land surface model: 2. Role of the nitrogen cycle in the historical terrestrial carbon balance. *Global Biogeochemical Cycles*, 24, GB1006, doi:10.1029/2009GB003522.

Zaehle, S., Medlyn, B. E., De Kauwe, M. G., et al. (2014). Evaluation of 11 terrestrial carbon–nitrogen cycle models against observations from two temperate Free-Air CO_2 Enrichment studies. *New Phytologist*, 202, 803–822.

Zender, C. S., Bian, H., and Newman, D. (2003). Mineral Dust Entrainment and Deposition (DEAD) model: Description and 1990s dust climatology. *Journal of Geophysical Research*, 108, 4416, doi:10.1029/2002JD002775.

Zeppel, M., Macinnis-Ng, C., Palmer, A., et al. (2008). An analysis of the sensitivity of sap flux to soil and plant variables assessed for an Australian woodland using a soil–plant–atmosphere model. *Functional Plant Biology*, 35, 509–520.

Zhang, L., Brook, J. R., and Vet, R. (2003). A revised parameterization for gaseous dry deposition in air-quality models. *Atmospheric Chemistry and Physics*, 3, 2067–2082.

Zhao, M., Heinsch, F. A., Nemani, R. R., and Running, S. W. (2005). Improvements of the MODIS terrestrial gross and net primary production global data set. *Remote Sensing of Environment*, 95, 164–176.

Zhao, R.-J. (1992). The Xinanjiang model applied in China. *Journal of Hydrology*, 135, 371–381.

Zhao, R.-J., and Liu, X.-R. (1995). The Xinanjiang model. In *Computer Models of Watershed Hydrology*, ed. V. P. Singh. Highlands Ranch, CO: Water Resources Publications, pp. 215–232.

Zhao, R.-J., Zuang, Y.-L., Fang, L. R., Liu, X.-R., and Zhang, Q.-S. (1980). The Xinanjiang model. In *Hydrological Forecasting: Proceedings of the Oxford Symposium, 15–18 April 1980*, Publication No. 129. Wallingford: International Association of Hydrological Sciences, pp. 351–356.

Zheng, Z., and Wang, G. (2007). Modeling the dynamic root water uptake and its hydrological impact at the Reserva Jaru site in Amazonia. *Journal of Geophysical Research*, 112, G04012, doi:10.1029/2007JG000413.

Zhou, S., Duursma, R. A., Medlyn, B. E., Kelly, J. W. G., and Prentice, I. C. (2013). How should we model plant responses to drought? An analysis of stomatal and non-stomatal responses to water stress. *Agricultural and Forest Meteorology*, 182/183, 204–214.

Zhu, Q., and Riley, W. J. (2015). Improved modelling of soil nitrogen losses. *Nature Climate Change*, 5, 705–706.

Zhu, Q., Riley, W. J., and Tang, J. (2017). A new theory of plant–microbe nutrient competition resolves inconsistencies between observations and model predictions. *Ecological Applications*, 27, 875–886.

Zhu, Q., Riley, W. J., Tang, J., and Koven, C. D. (2016). Multiple soil nutrient competition between plants, microbes, and mineral surfaces: Model development, parameterization, and example applications in several tropical forests. *Biogeosciences*, 13, 341–363.

Index